Plant Growth Regulators to Manage Biotic and Abiotic Stress in Agroecosystems

Editors

Kamel A. Abd-Elsalam

Plant Pathology Research Institute
Agricultural Research Center
Giza, Egypt

Heba I. Mohamed

Department of Biological and Geological Sciences
Faculty of Education, Ain Shams University
Cairo, Egypt

CRC Press

Taylor & Francis Group
Boca Raton London New York

CRC Press is an imprint of the
Taylor & Francis Group, an **informa** business

A SCIENCE PUBLISHERS BOOK

First edition published 2024
by CRC Press
2385 NW Executive Center Drive, Suite 320, Boca Raton FL 33431

and by CRC Press
4 Park Square, Milton Park, Abingdon, Oxon, OX14 4RN

© 2024 Kamel A. Abd-Elsalam and Heba I. Mohamed

CRC Press is an imprint of Taylor & Francis Group, LLC

Library of Congress Cataloging-in-Publication Data (applied for)

ISBN: 978-1-032-48530-0 (hbk)
ISBN: 978-1-032-48531-7 (pbk)
ISBN: 978-1-003-38950-7 (ebk)

DOI: 10.1201/9781003389507

Typeset in Times New Roman
by Prime Publishing Services

Preface

Agriculture is confronted with several biotic and abiotic issues in today's quickly changing planet. Population growth, climate change, and decreasing natural resources have all put enormous strain on the world's agroecosystems. As a result, farmers and researchers are always looking for new ways to alleviate the detrimental effects of these stressors while also ensuring sustainable and efficient crop production.

The use of plant growth regulators (PGRs) is one such solution that has received a lot of attention. Plant growth regulators (PGRs) are natural or synthetic compounds that can control plant development processes such as cell division, elongation, and differentiation. These regulators have shown considerable promise in controlling both biotic and abiotic stresses in agroecosystems, making them a potentially useful tool for improving crop resilience and productivity.

The application of growth regulators to modulate plant growth is a method for achieving this aim. Plant growth regulators (PGRs) are compounds that, when applied to plants or seeds, can promote, hinder, or affect physiological processes, production, and/or stress responses. PGRs have been increasingly used in conventional agriculture over the last 20 years as farmers recognize their benefits. Phytohormones are chemical messengers that regulate the normal development of plants and their responses to environmental stimuli.

The biosynthesis of phytohormones by plants provides them with a mechanism for adapting to changing conditions. Previous research efforts have discovered the important roles of phytohormones and their interactions in the regulation of plant adaptation to various stresses. Molecular studies have revealed numerous plant hormonal pathways, each of which consists of several signaling components that relate a particular hormone perception to its regulation. The signal transduction pathways of auxin, abscisic acid, cytokinin, gibberellins, and ethylene have been thoroughly investigated. Recently, interesting signaling pathways for comprehending their various roles in plant physiological processes have been shown for brassinosteroids, jasmonates, salicylic acid, tocotrienol, melatonin, phenolic compounds, and strigolactones.

Biotic stressors, such as diseases, pests, and weeds, have long been a source of concern for farmers. Traditional pest control and disease management strategies frequently rely on the use of chemical pesticides, which can be harmful to the environment and human health. PGRs, on the other hand, provide a more environmental friendly and sustainable method. They have the potential to improve plant defense mechanisms and induce resistance to pests and diseases. Furthermore,

PGRs have the potential to lessen weed competition, allowing crops to thrive without relying too heavily on herbicides.

Abiotic stressors such as drought, salt, high temperatures, and nutrient deficiencies all pose substantial hazards to crop yields. These stressors can result in lower yields and poor crop quality, severely affecting farmers' livelihoods. PGRs have been reported to reduce the negative impacts of these stresses by increasing plant water efficiency, boosting nutrient uptake and utilization, and encouraging stress tolerance. Furthermore, PGRs can operate as signaling molecules that cause specialized responses in plants, allowing them to adapt and survive in harsh environments.

Scientists have made tremendous progress in understanding the mechanisms of action of PGRs and their potential applications in agroecosystems. Numerous studies have shown that PGRs improve crop growth, yield, and stress tolerance. However, there is still more to discover and investigate in this sector.

This book explores the biosynthesis of phytohormones and their role in reducing abiotic and biotic stress and increasing plant growth and productivity. It provides a summary of recent research in plant growth regulators, molecular and physiological interactions, and field monitoring of abiotic stress in plants. This will be the first comprehensive book to provide a summary of the latest research on plant growth regulators assisting in biotic and abiotic stress management in crops, with a focus on sustainability in agriculture. The current volume is of an interdisciplinary nature and will be very useful for students, teachers, researchers, plant physiologists, environmental scientists, and agrochemical companies. It will also be of immense value to graduate students in agriculture, biotechnology, and in environmental university programs. Professionals or practitioners in research institutes involved in plant science, material science, and crop production research will also benefit from this book.

Kamel A. Abd-Elsalam
Agricultural Research Center, Egypt

Heba I. Mohamed
Ain Shams University, Egypt

Contents

1

Overview of Plant Growth Regulators and their Synthesis

Humberto Aguirre-Becerra,[1] Diana Sáenz-de-la-O,[3]
Noelia Isabel Ferrusquía-Jiménez,[3] Joel Ernesto Martínez-Camacho,[3]
Ma. Cristiina Vázquez-Hernández,[2] Valeria Caltzontzin-Rabell[3] and
Ana Angélica Feregrino-Pérez[1],*

Introduction

Plant physiological processes are regulated by compounds known as plant growth regulators (PGRs). The compounds are present endogenously or can be obtained by laboratory synthesis or from the metabolism of microorganisms (Bisht et al. 2018). Phytohormones (referred to in this chapter as PGRs) can act specifically in fruit filling, germination, photosynthesis, the antioxidant enzyme system, senescence, and during stressful situations. They can be classified according to their molecular structure, activity, and inhibitory or stimulating effects. Among the major PGRs are abscisic acid (ABA), auxin, brassinosteroid (BR), ethylene (ET), cytokinin (CK), salicylic acid (SA), gibberellin (GA), strigolactone (SL), and jasmonate (JA) (Zahid et al. 2023). Other groups influencing plant growth, such as amino acids, macro-

---

[1] Cuerpo Académico de Bioingeniería Básica y Aplicada. Facultad de Ingeniería. Campus Amazcala. Universidad Autónoma de Querétaro. Chichimequillas-Amazcala Road Km 1 S/N, Amazcala, C.P. 76265, El Marqués, Querétaro, México.

[2] Cuerpo Académico de Innovación en Bioprocesos Sustentables. Depto. Ingenierías. Tecnológico Nacional de México en Roque. Carretera Celaya-Juventino Rosas Km 8. Roque-Celaya, Guanajuato, México. C.P. 38110.

[3] Doctorado en Ingeniería de Biosistemas. Facultad de Ingeniería. Campus Amazcala. Universidad Autónoma de Querétaro. Chichimequillas-Amazcala Road Km 1 S/N, Amazcala, C.P. 76265, El Marqués, Querétaro, México.

* Corresponding author: feregrino.angge@hotmail.com

and micro-nutrients, carbohydrates, and microorganisms have also been described (Zhang et al. 2022).

The biosynthesis of PGRs is notably related to the phenological stages of plants and fruits (Babu et al. 2022). Their interactions are relevant, as they can present synergetic or antagonistic effects (Li et al. 2021). The understanding of these interactions can be aimed at ensuring desired responses, such as higher yield and quality, using PGRs as agricultural tools. In addition, ambient conditions, such as heat, cold, radiation, salinity, and drought can affect the production of these compounds and ultimately affect plant development (Hossain et al. 2022). In this regard, plant responses to growth regulators under stress conditions aim to mitigate the detrimental effects of inadequate environments and yield-limiting factors (Zahid et al. 2023). The exogenous application (i.e., foliar spraying) is the most common method to apply PGRs, however, other modes of application are drenching, pre-plant sowing, capillary string, and seed priming, the method is selected based on the specific crop's needs (Ajmi et al. 2020).

Overall, the use of PGRs has been a matter of interest and become a relevant topic in agricultural practices due to the vast physiological processes regulated by them. The study and understanding of how they are synthesized and their interactions within plant physiology have become relevant in improving crop productivity. The current chapter encompasses updated knowledge about PGR, their biosynthesis pathways, their impact on the phenological stage, their effect against stress, and their interaction with the microbiome.

PGRs Biosynthesis

Biosynthetic pathways meticulously control the pool sizes of PGRs. The major biosynthetic pathways of these compounds have been elucidated thanks to the many studies using chemical inhibitors, exogenous hormone applications, and hormone-deficient and overproductive mutants. Despite being structurally diverse, it is not unusual that these hormones share standard features in their biosynthetic and catabolic pathways, and their interaction is essential in coordinating plant processes (Kaur et al. 2021). This section reviews the big picture of the biochemical pathways of the nine classical phytohormone groups and briefly discusses their interaction.

Abscisic Acid Biosynthesis

ABA, discovered in the early 1960s, is a category of metabolites called isoprenoids (or terpenoids). Many biochemical studies have elucidated the ABA biosynthetic pathway in higher plants in great detail (Fig. 1) (Xiong and Zhu 2003). ABA-deficient mutants have been crucial in understanding the functions and synthesis of this PGR (Schwartz et al. 2003). This PGR is synthesized via a mevalonic-acid-independent (MEP–plastidial 2-C-methyl-D-erythritol-4-phosphate) pathway (Nambara and Marion-Poll 2005). ABA biosynthesis begins in plastids by converting Pyruvate into isopentenyl diphosphate (IPP) via glyceraldehyde 3-phosphate as an intermediate (Trivedi et al. 2016). IPP is converted into the C40 carotenoid Zeaxanthin via β-carotene and subsequently into trans-violaxanthin by the ZEP enzyme (zeaxanthin

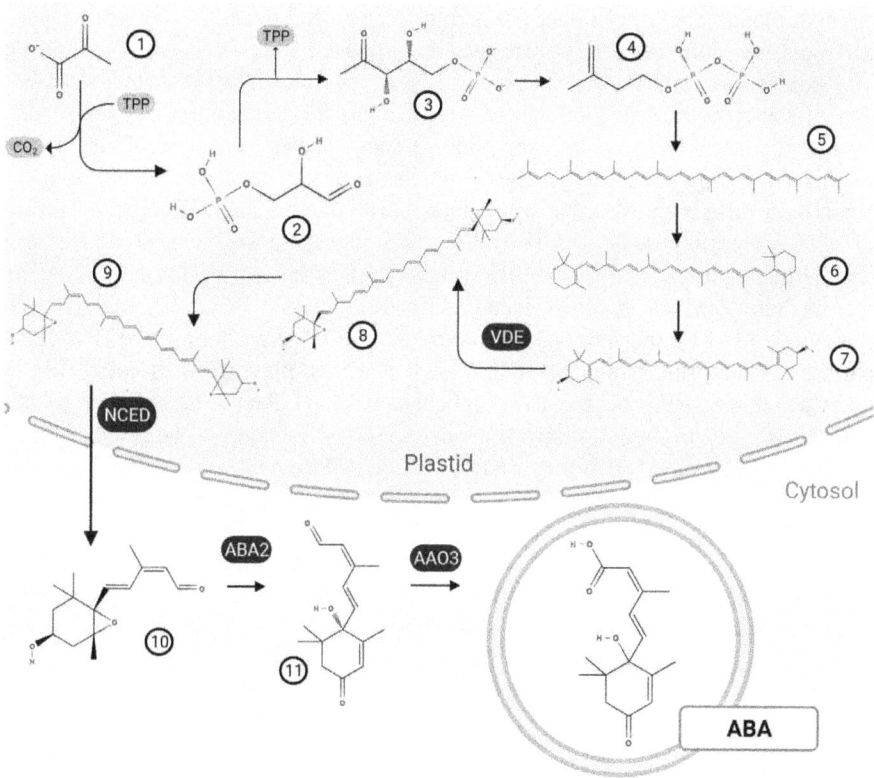

Fig. 1. The C15 backbone of ABA is the final product, in the cytosol, of the C40 carotenoid zeaxanthin cleavage in MEP, in the plastid. NCED: 9-cis-epoxycarotenoid dioxygenases, ABA2: short-chain alcohol dehydrogenase/reductase, VDE: Violaxanthin de-epoxidase, AAO3: ABA aldehyde oxidase. (1) Piruvate, (2) Glyceraldehyde 3-Phosphate, (3) 1-Deoxyxylulose-5-Phosphate, (4) Iso pentenyl diphosphate (IPP), (5) Lycopene, (6) ß-carotene, (7) Zeaxanthin, (8) trans-Violaxanthin, (9) cis-Violaxanthin, (10) Xanthoxin, (11) ABA aldehyde. Image source: Own elaboration created in BioRender.com

epoxidase) (Marin et al. 1996). In the presence of light, a reverse reaction occurs in chloroplasts, facilitated by the enzyme violaxanthin de-epoxidase (VDE) (Trivedi et al. 2016). Trans-violaxanthin is converted to cis-violaxanthin through an isomerization process. The oxidative cleavage of cis-violaxanthin, catalyzed by a family of 9-cis-epoxy carotenoid dioxygenases (NCEDs), produces xanthoxin, a C15 intermediate, which is translocated into the cytosol and converted to ABA through a reaction consisting of two steps (Schwartz et al. 1997). First, a short-chain alcohol dehydrogenase/reductase (SDR) yields ABA aldehyde, and next, abscisic aldehyde oxidase (AAO) catalyzes the final step (Rook et al. 2001, Schwartz et al. 1997, Seo et al. 2000).

Auxin Biosynthesis

In the early 1940s, Indole-3-acetic acid (IAA) was characterized (Haagen-Smit et al. 1941). IAA has been considered the predominant and most potent naturally occurring auxin in plants. These hormones are mainly synthesized in actively growing leaves and roots, and from there, they are transported to other plant parts (Ljung et al. 2001).

Auxin biosynthesis in plants is highly complex. While plants possess evolutionarily conserved fundamental mechanisms for this process, they may also employ distinct strategies and adaptations among species to enhance the synthesis of this hormone (Zhao 2010). Early biochemical and genetic analysis of plant auxin biosynthesis showed that tryptophan (Trp) is an amino acid precursor for de novo IAA biosynthesis (Zhao 2010). Four interconnected Trp-dependent pathways for the biosynthesis of this compound have been proposed, each named after its intermediate product: indole-3-acetamide (IAM), indole-3-pyruvic acid (IPyA), tryptamine (TAM), and indole-3-acetaldoxime (IAOx) (Kasahara 2016). The IPyA pathway is the primary source of free IAA in plants, playing an essential role in many developmental processes (Mashiguchi et al. 2011, Zhao 2012). Studies in both angiosperms and bryophytes have shown a conserved function of the IPyA pathway in IAA biosynthesis (Morffy and Strader 2020). The IPyA pathway involves two steps (Fig. 2). First, Trp is converted to IPyA via Tryptophan Aminotransferase (Stepanova et al. 2008). Subsequently, the YUCCA (YUC) family of flavin monooxygenases converts IPyA to IAA (Stepanova et al. 2011). On the other hand, pathways involving the IAM, TAM, and IAOx intermediates are still largely unknown. Whether plants use them as critical precursors for auxin biosynthesis is still inconclusive (Kasahara 2016, Morffy and Strader 2020). Although there is a lack of information regarding enzymes and proteins involved, some steps of these pathways are partially known for *Arabidopsis* (Table 1).

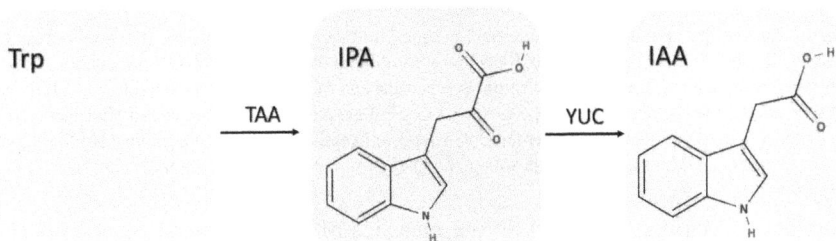

Fig. 2. The IPyA pathway is a tryptophan-dependent pathway for auxin (IAA) synthesis. Image source: Own elaboration.

Table 1. Current information know of the IAM, TAM, and IAOx biosynthesis pathways in *Arabidopsis*.

Pathways	Enzymes	Substrate → Product		Reference
Indole-3-acetamide (IAM)	IAM hydrolases/ Amidases	IAM	IAA	Gao et al. 2020, Pollmann et al. 2003
Tryptamine (TAM)	Aldehyde oxidase	IAAld	IAA	Seo et al. 1998
Indole-3-acetaldoxime (IAOx)	Cytochrome P450 monooxygenase	Trp	IAOx	Kasahara 2016
	Nitrilases	IAN	IAA	Bartel and Fink 1994

L-Tryptophan (Trp), Indole-3-acetamide (IAM), Indole-3-acetic acid (IAA), Tryptamine (TAM), N-Hydroxytryptamine (NHT), Indole-3-acetaldoxime (IAOx), Indole-3-acetaldehyde (IAAld), Indole-3-acetonitrile (IAN).

Trp-independent pathways also contribute significantly to auxin biosynthesis in plants. When analyzing mutants unable to synthesize Trp in maize and Arabidopsis, no disparities in the levels of free IAA were observed (Wright et al. 1991, Normanly et al. 1993, Wang et al. 2015). However, Trp-independent pathways need to be better described and studied. Moreover, the plant rhizosphere microbiota can produce IAA, a signaling method for inter-and intra-species communication (Tariq and Ahmed 2022). A considerable effort has been made to elucidate the complete structure of the IAA biosynthesis, but the complete picture remains to be understood.

Brassinosteroids Biosynthesis

BRs are natural plant steroids first isolated in the 1970s from *Brassica napus* pollen grains (Grove et al. 1979). Research on BRs rapidly expanded, and by the 1990s decade, significant advances in elucidating BR biosynthesis were made (Fujioka and Sakurai 1997, Sasse 1997). BRs dwarf mutants of *Arabidopsis* have contributed significantly to expanding biochemical knowledge of BRs (Kwon and Choe 2005). More than 70 BRs have been identified in Algae, Bryophyta, Angiosperms, and Pteridophyta, and in all plant organs (Bajguz 2019, Bajguz and Tretyn 2003, Yokota et al. 2017, Zullo and Bajguz 2019).

The BRs structure present at least one oxygen moiety at the C-3 position and additional ones, at one or more, in the C-2, C-6, C-22, and C-23 carbon atoms (Bishop and Yokota 2001). Their presence can occur as free molecules or combined with fatty acids and glucose (Bajguz and Tretyn 2003). The free BRs are grouped into C_{27}, C_{28}, and C_{29} based on the total carbons in the structure of their C_{24} alkyl groups (Fujioka and Yokota 2003). The basic configuration of C27-BRs is a C28-BRs: 5a-ergostane, 5a-cholestane skeleton, and C29-BRs: 5a-stigmastane. Their structural variations occur in different types and orientations of oxygenated functions in the A- and B-ring and the side chain of the molecules. These alterations occur during reduction and oxidation reactions.

The first significant step during BRs biosynthesis is expected for each type (C27-, C28-, or C29-type) and may result via the mevalonate (MVA) or non-MVA pathway. Then, subsequent stages diversify the BR biosynthesis pathways (Fujioka et al. 2002). The most abundant BRs in plants are the C28 BRs such as Castasterone (CS) and BL; these biosynthetic pathways have been elucidated primarily vis-à-vis *Arabidopsis* (Fig. 3) (Kim et al. 2018). Indeed, when *Arabidopsis* requires an increment of BR activity for growth and development regulation, CS undergoes conversion into the biologically more active BR BL (Noguchi et al. 2000). The initial conversion of campesterol results in the formation of campestanol, which can further undergo either early C-6 oxidation or late C-6 oxidation to produce CS. Eventually, CS is transformed into BL. Additionally, C27 BRs are synthesized via cholesterol, which finally transforms into 28-norBL. The direct substrate of C29 BRs is sitosterol which leads to 28-homoBL. Later, both C27- and C29- pathways may go through a process similar to C28 Brs, however, identifying only a subset of the indirect compounds participating in these two pathways has been achieved (Fujioka and Yokota 2003).

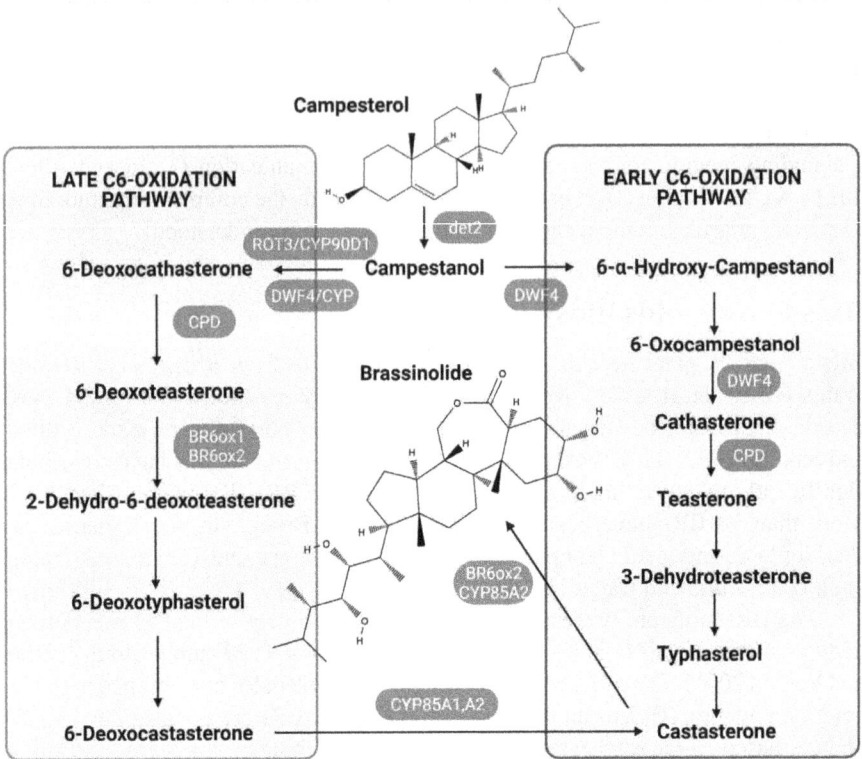

Fig. 3. Diagram of the BR biosynthetic pathway showing the early C-6- and late C-6- oxidation pathway. Multiple BR-biosynthesis genes and enzymes are shown by Kour et al. (2021). Image source: Own elaboration created in BioRender.com

Cytokinins Biosynthesis

In the early 1960s, the cytokinin zeatin was first isolated from immature maize endosperm (Letham 1963). CKs are phytohormones, structurally defined as adenine derivatives with an isoprene-derived side chain attached to the N6 amino group (Takei et al. 2001). The typical active forms of isoprenoid-side-chain CKs in plants comprise trans-zeatin (tZ), isopentenyl adenine (iP, N6-(Δ2-isopentenyl) adenine), dihydrozeatin (DZ), and cis-zeatin (cZ) (Ashihara et al. 2020). CKs with an aromatic side chain, such as benzyl or hydroxy benzyl group at N6, also occur but are rare, and their biosynthetic pathway is poorly characterized (Strnad 1997). CKs are synthesized as ribonucleotides in plant cells since they are present as nucleosides, nucleotides, and glycosidic conjugates (Sakakibara 2006). Research indicates that CKs biosynthesis starts in various plant organs and tissues (e.g., phloem of roots and shoots, immature seeds, and lateral root primordia) (Buchanan and Gruissen 2015).

The side chains of iP- and tZ-type CKs derive from the MEP pathway, and a significant part of the side chain of cZ-type CKs derive from the MVA pathway (Kasahara et al. 2004). The isoprenoid CKs biosynthesis pathway starts with the substrates 5-Phosphate adenosine (AMP, ATP, and ADP) and dimethylallyl diphosphate (DMAPP) or hydroxy-methylbutenyl diphosphate (Krall et al. 2002). The latter are catalyzed by adenosine phosphate-isopentenyl transferases (IPTs) to create N6-(2-isopentenyl Δ) adenine riboside 5'-triphosphate (iPRTP) or N6-(2-isopentenyl Δ) adenine riboside 5'-diphosphate (iPRDP) (Kakimoto 2001). IP and tZ nucleotides are converted to a nucleobase to become biologically active. First, dephosphorylation of iPRDP and iPRTP nucleotides to N6-(2-isopentenyl Δ) adenine riboside 5'-monophosphate (iPRMP) is catalyzed by 5'-ribonucleotide phosphohydrolase (5'-nucleotidase), and then deribosylation step is catalyzed by adenosine nucleosidase. Subsequently, iPRMP or iPRDP is hydroxylated by CYP450 (cytochrome P450) monooxygenases CYP735A2 and CYP735A1 to form the prenyl side chain in tZ-type CKs biosynthesis (Fig. 4). It has been found that the tZ formation can be regulated by the hormones auxin or ABA by repressing the expression of CYP735A (Buchanan and Gruissen 2015). A second pathway to synthesize CKs with a lower metabolic rate is by degradation of prenylated adenine in a subgroup of tRNA species in plants. Degradation of tRNA is catalyzed by tRNA-specific adenylate isopentenyltransferase, further resulting in several CK species such as iP riboside, cis-zeatin riboside (cZR), tZR, and their 2-methylthio derivates (Buchanan and Gruissen 2015, Li et al. 2021). In addition, a more effective synthetic pathway was discovered in *Arabidopsis*, where iPRMP is directly produced from AMP and DMAPP. This iPRMP is then converted into zeatin riboside-5'-monophosphate with the assistance of an internal hydroxylase (Åstot et al. 2000).

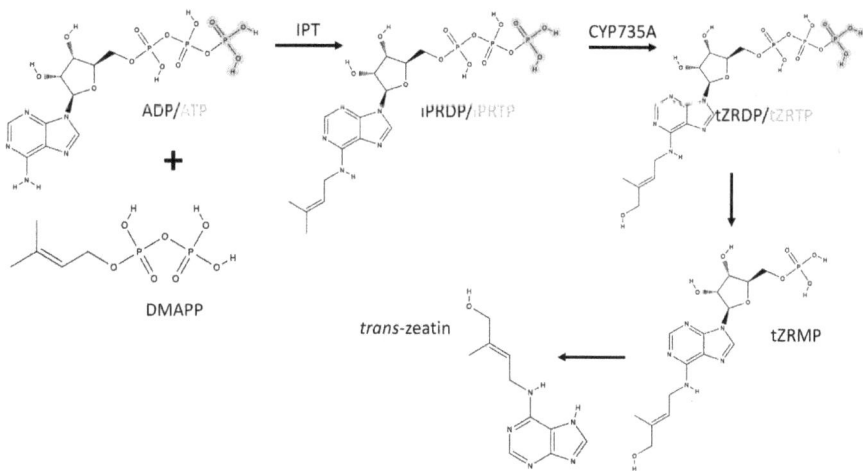

Fig. 4. Simplified biosynthetic pathway for the CK isoprenoid side chain of trans-Zeatin (tZ). Image source: Own elaboration.

Ethylene Biosynthesis

ET is a gas-based plant hormone generated in low quantities by all plants and characterized by a simple chemical structure comprising two carbon atoms (Xu and Zhang 2015). In addition to its fruit-ripening function, this compound also acts as an endogenous PGR (Buchanan and Gruissen 2015). ET biosynthesis in higher plants follows a simple pathway that was majorly elucidated from the mid-1960s to the 1980s by identifying important pathway precursors and intermediates (Xu and Zhang 2015). Methionine (MET) is a precursor of ET involved in other crucial physiological processes (e.g., sulfation, protein and nucleic acid methylation, and protein synthesis) (Lieberman et al. 1966). ET intermediate S-adenosylmethionine (SAM) is an activated form of MET involved in a broad spectrum of methylation reactions and a common precursor to the biosynthetic pathway of polyamines, and 1-aminocyclopropane-1-carboxylic acid (ACC) precedes during the synthesis of ET (Adams and Yang 1979, 1977). During the two-step biosynthetic pathway, first, SAM is converted to ACC by ACC synthase (ACS) and then oxidized by ACC oxidase (ACO) to form ET (Fig. 5) (Kende 1993, Yang and Hoffman 1984).

Other plant hormones and ET itself influence its biosynthesis rate. Auxins can promote ET production since they enhance the conversion of SAM to ACC by expressing ACS 6, 8, and 11 genes (Paponov et al. 2008, Yoshii and Imaseki 1982). An increased ACS protein stability and BRs- and CKs-induction of ET production through managing ACS and ACO has also been reported (Hansen et al. 2009).

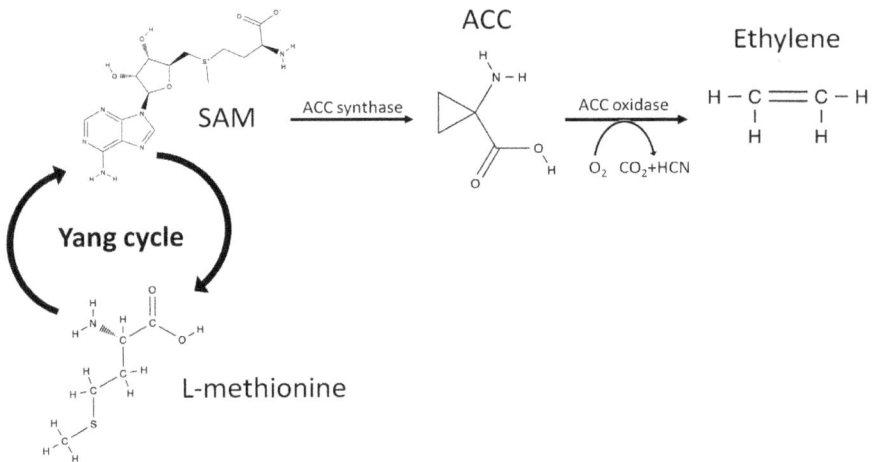

Fig. 5. The ethylene biosynthetic pathway. L-methionine is synthesized via the Yang cycle. Image source: Own elaboration.

Salicylic Acid Biosynthesis

SA (2-hydroxy benzoic acid) was acknowledged as the sixth plant hormone in 1992 and is usually related to defense activity (Brunswick 1992). Recently, the role of SA during plant morphogenesis was confirmed. SA is a phenolic compound with an aromatic ring bearing a hydroxyl group or its functional derivative. Through two studies in 2019, the biosynthesis of SA in plants was understood entirely (Fig. 6) (Rekhter et al. 2019, Torrens-Spence et al. 2019). The isochorismate synthase (ICS) pathway and the phenylalanine ammonia-lyase (PAL) pathway contribute to the biosynthesis of SA, both originating from chorismate, which is the end product of

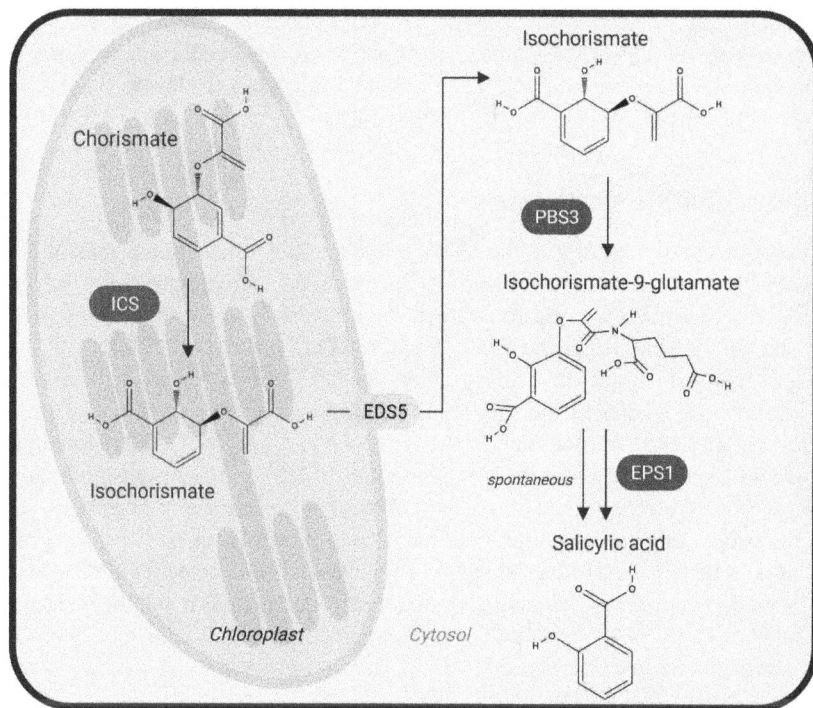

Fig. 6. Biosynthesis route for SA in plants. The ENHANCED DISEASESUSCEPTIBILITY 5 (EDS5) protein is a MATE transporter. Image source: Own elaboration created in BioRender.com

the shikimate pathway (Mishra and Baek 2021). Chorismate is a central metabolic precursor of primary and secondary metabolism, serving as the principal supplier for the biosynthesis of phenylalanine, tyrosine, and tryptophan, and an extensive array of secondary aromatic metabolites (Hubrich et al. 2021, Tzin and Galili 2010). ICS facilitates the beginning of the ICS pathway by transforming chorismate into its isomer, isochorismate, this step is synthesized in plastids (Yokoo et al. 2018). Isochorismate moves to the cytosol through the plastidial MATE (multidrug and toxic compound extrusion) transporter enzyme (Yamasaki et al. 2013). ICS is combined with L-glutamate and transformed into isochorismate-9-glutamate (ICS-Glu) through the cytosolic amidotransferase enzyme encoded by avrPphB Susceptible 3 (PBS3). Finally, ICS-Glu is either spontaneously decomposed into SA and 2-hydroxy-acryloyl-N-Glutamate, or transformed into SA by an acyltransferase encoded by Enhanced Pseudomonas Susceptibility 1 (EPS1) (Ding and Ding 2020, Rekhter et al. 2019). A minor SA fraction (~10%) is synthesized through the PAL pathway (Mishra and Baek 2021). Here, there are two ways of prephenate conversion into phenylalanine. The first occurs in plastids; the transition from prephenate to arogenate is catalyzed by prephenate aminotransferases, followed by the conversion of arogenate to phenylalanine through the action of arogenate dehydratase. The second occurs in the cytosol. The conversion of prephenate to phenylalanine occurs via phenylpyruvate, facilitated by prephenate dehydratase and phenylpyruvate aminotransferase (Ding and Ding 2020). Then, the PAL enzyme converts phenylalanine to trans-cinnamic acid (t-CA) in the cytosol. Finally, the formation of SA occurs via two possible intermediates, either benzoic acid or ortho-coumaric acid (Lefevere et al. 2020, Mishra and Baek 2021).

Gibberellins Biosynthesis

The first GA was discovered in the 1930s while studying the active principle of *G. fujikuroi*. However, it wasn't until the 1950s that GAs were suggested as intrinsic PGRs when the gibberellic acid structure, now known as GA_3, was elucidated (Curtis and Cross 1954, MacMillan and Suter 1958). GAs are present in bacteria and fungi that associate with plants and all vascular plants. Their synthesis is believed to occur in juvenile tissues of the shoot and developing seeds (McAdam et al. 2018, Miyazaki et al. 2018, Tanaka et al. 2014, Wiemann et al. 2013). GAs are tetracyclic diterpenoid carboxylic acids with ent-gibberellane (C20) or 20-nor-ent-gibberellane (C19) carbon skeletons. One hundred and thirty-six GA plant hormones have been described, but only a limited number, called bioactive GAs, work as growth regulators in higher plants (MacMillan and Takahashi 1968). Specific non-bioactive GAs precede synthesizing bioactive forms or are inactivated metabolites (Hedden 2012). GA1, GA3, GA4, and GA7 are the most commonly found bioactive GAs which contain a hydroxyl group on C-3β, a carboxyl group on C-6, and a γ-lactone between C-10 and C-4 (Yamaguchi 2008).

In plants, GAs are synthesized from acetyl CoA via the MEP pathway in plastids. The cytosolic mevalonate pathway also leads to GAs biosynthesis with a minor contribution (Kasahara et al. 2002). The biosynthesis of this compound starts from trans-geranylgeranyl diphosphate (GGPP). This molecule that precedes various terpenoids that act as the source of all carbon atoms in GA (Buchanan and Gruissen 2015). Three types of enzymes are necessary for GA biosynthesis from GGDP in plants: terpene synthases (cyclases), 2-oxoglutarate- dependent dioxygenases, and Cytochrome P450 monooxygenases (Buchanan and Gruissen 2015, Sun and Kamiya 1994). GGPP is converted from ent-copalyl diphosphate to ent-kaurene, which is catalyzed by the ent-kaurene synthase (KSp) and the copalyl diphosphate synthase (CPSp) (Tudzynski et al. 2001). Posterior oxidations produce kaurenol (alcohol form), kaurenal (aldehyde form), and kaurenoic acid, respectively (Castellaro et al. 1990, Sherwin and Coates 1982). Kaurenoic acid is converted to the aldehyde form of GA12 via GA12-aldehyde and 7β-hydroxy-ent-kaurenoic acid (Helliwell et al. 2001). GA12 is the initial GA ring system with 20 carbons (Fig.7).

Jasmonates Biosynthesis

JAs are recognized as endogenous growth-regulating substances initially classified as hormones associated with stress in higher plants (Campos et al. 2014). These compounds derivate from a class of fatty acids and encompass jasmonic acid (JA) and its precursors' isoleucine conjugate (JA-Ile) and methyl ester (MeJA). JA biosynthesis has been conducted during the last decades, mainly in tomatoes and *Arabidopsis thaliana* (Ruan et al. 2019). In the latter, three pathways for the synthesis

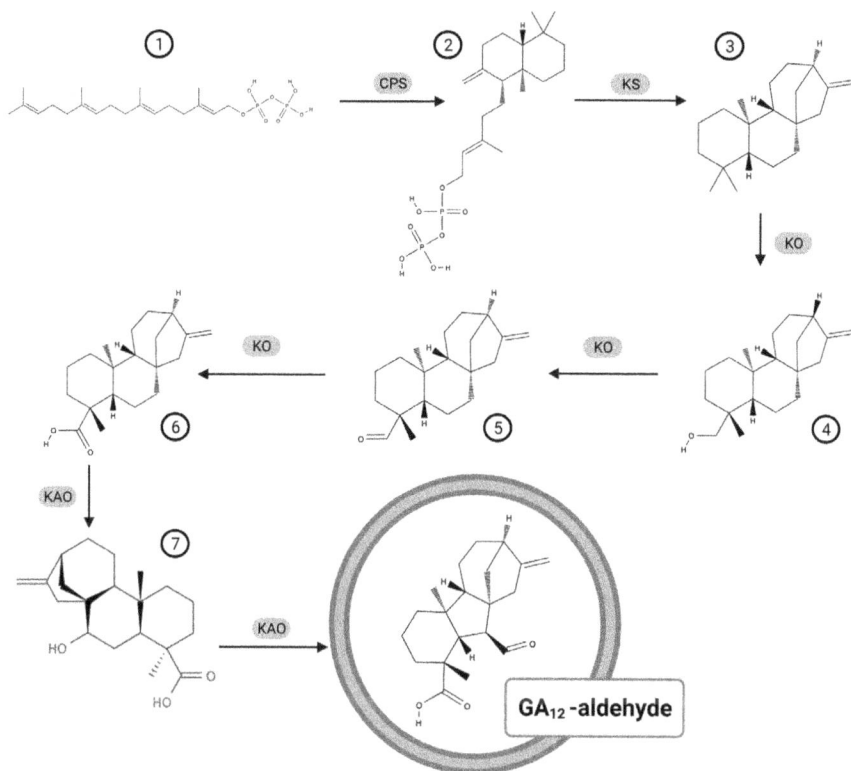

Fig. 7. The gibberellin (GA) biosynthesis pathway. (1) geranylgeranyl diphosphate (GGDP), (2) ent-copalyl diphosphate (CPP), (3) ent-kaurene, (4) ent-kaurenol, (5) ent-kaurenal, (6) ent-kaurenoic acid, (7) ent-7 α-hydroxy-kaurenoic acid. CPS, ent-copalyl diphosphate synthase, KO, ent-kaurene oxidase, KS, ent-kaurene synthase, KAO, ent-kaurenoic acid oxidase. Image source: Own elaboration.

of these molecules have been identified, including the octadecane pathway, where linolenic acid is transformed into 12-oxo-phytodienoic acid (12-OPDA), and the hexadecane pathway starting from hexadecatrienoic acid to produce deoxymethylated vegetable dienic acid (dn-OPDA) (Chini et al. 2018). The synthesis of JAs takes place at three reaction sites, no matter which of the proposed pathways. It starts in the chloroplast, converting unsaturated fatty acid into 12-OPDA or dn-OPDA. OPDA is later transported out by the chloroplastic inner envelope (IE) and through the channel protein JASSY on the chloroplastic outer envelope (OE) membranes (Guan et al. 2019, Simm et al. 2013). Import of OPDA into the peroxisome occurs by the ATP-dependent Comatose (CTS), an ABC transporter of the peroxisomal membrane, or in minor proportion, by an ion trapping mechanism due to a pH of ~8.2 in peroxisomes and ~7.2 in the cytosol (Theodoulou et al. 2005, Wasternack and Hause 2019). Then, OPDA is first transformed into 3-oxo-2-(cis-2′-pentenyl) cyclopentane-1-octanoic acid (OPC-8:0) by OPDA reductase3 (OPR3) inside the peroxisome. Later, OPC-8:0 is activated by a carboxyl-CoA ligase and shortened in the carboxylic acid side chain to produce JA by β-oxidation enzymes (Fig. 8)

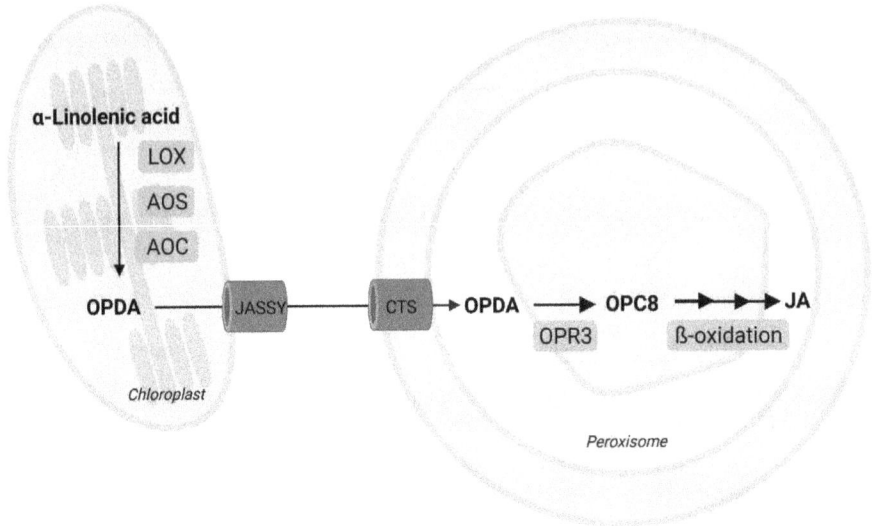

Fig. 8. Representaion of the JAs biosynthesis pathway in *Arabidopsis thaliana*. JA, jasmonic acid, AOS, allene oxide synthase, LOX, lipoxygenase, OPDA, 12-oxo-phytodienoic acid, AOC allene oxide cyclase, JASSY, chanel protein on the chloroplastic outer envelope membranes, CTS, ATP-dependent COMATOSE, OPR3, OPDA reductase, OPC8, 8-(3-oxo-2-(pent-2-enyl)cyclopentyl) octanoic acid. Image source: Own elaboration created in BioRender.com.

(Breithaupt et al. 2006, Kombrink 2012). The JA-amino acid synthetase catalyzes the final step to create JA's biologically active conjugated form (Staswick and Tiryaki 2004). Finally, in the cytoplasm, JA undergoes various chemical reactions that result in its metabolism leading to the creation of different structures, such as JA-Ile, MeJA, 12-hydroxyjasmonic acid, and cis-jasmone (Ruan et al. 2019).

Strigolactones (SLs) Biosynthesis

SLs are terpenoid lactones derived from carotenoids originally classified as germination stimulants for parasitic root plants (Cook et al. 1966). In 2008, two groups of independent researchers discovered the hormonal function of these compounds as suppressors of plant branching (Gomez-Roldan et al. 2008, Umehara et al. 2008). Investigations with mutants of *S. lycopersicum* demonstrated that carotenoids are the direct precursors of SLs (López-Ráez et al. 2008). During the initial phase, all-trans-β-carotene is converted by the carotene isomerase DWARF27 into 9-cis-β-carotene (Alder et al. 2012, Lin et al. 2009). The latter is cleaved and modified by two carotenoid cleavage dioxygenases. First, CCD7 produces the volatile compound b-ionone and 9-cis-b-apo-100-carotenal and then is modified by CCD8 to carlactone, the precursor for all known SLs (Alder et al. 2012, Bruno et al. 2014). CL can be converted into carlactonoic acid (CLA) or other derivatives by cytochrome P450 enzymes of the MAX1 family (Abe et al. 2014, Zhang et al. 2014). The conversion of CL to CLA in vascular plants is catalyzed by the CYP711A subfamily and the diversity of non-canonical and canonical SLs (Fig. 9) (Mashiguchi et al. 2021,

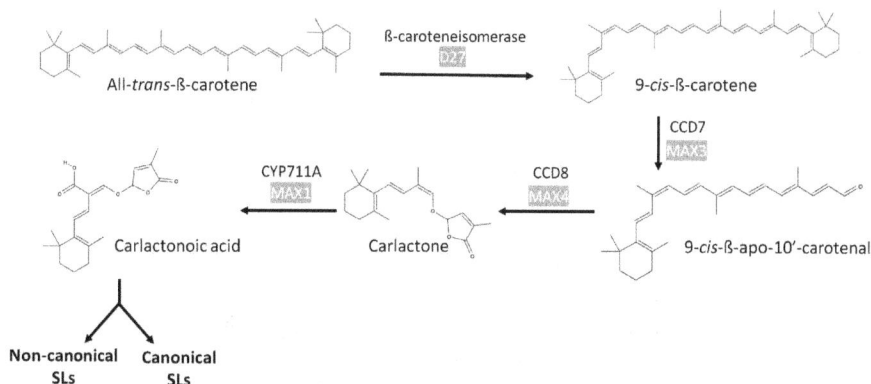

Fig. 9. SL biosynthetic pathway from all-trans-ß-carotene to carlactonoic acid, the precursor to the biosynthesis of canonical and non-canonical SLs. Image source: Own elaboration

Yoneyama et al. 2018). Conversion of CLA to methyl-CLA (MeCLA) in *Arabidopsis* is catalyzed by the late-acting enzyme Lateral Branching Oxido-reductase (LBO), but other enzymes might exist (Brewer et al. 2016). Many vital questions still await answers for us to fully comprehend the complete process of SL biosynthesis, and identifying additional enzymes will be essential.

Interaction between Biosynthetic Pathways

Phytohormones do not act alone. Functional and metabolic communication between them impacts plant growth and development. Enormous efforts have been put into elucidating biosynthetic pathways for individual hormones. In the path, it has been found that they are interconnected in a complex network. Metabolic cross-talk examples include common enzymes, precursors, biosynthetic pathways, or regulatory mechanisms among one or more phytohormones. The family of the CYP450 monooxygenases contains members that participate in the biosynthesis or deactivation of several hormones, including GAs, ABA, CKs, IAA, BRs, JA, and SLs (Helliwell et al. 2001, Fujioka and Yokota 2003, Nambara and Marion-Poll 2005, Zhao 2012,Campos et al. 2014, Mashiguchi et al. 2021).

Additionally, the biosynthetic origins of GAs, CKs, BRs, and ABA belong to the terpenoid biosynthesis pathway (Åstot et al. 2000, Hedden 2012, Kwon and Choe 2005, Nambara and Marion-Poll 2005). An example of a common regulatory mechanism is the methylation of the carboxyl group by SAM-dependent methyltransferases, which participate in the homeostasis of multiple phytohormones (Buchanan and Gruissen 2015). A metabolic cross-talk occurs between IAAs and SA through their common precursor chorismate (Mishra and Baek 2021, Tariq and Ahmed 2022). Indeed, IAAs and SA also share roles regulating fruit ripening and accumulation of bioactive compounds (Pérez-Llorca et al. 2019). Functional cross-talk also includes the interaction between CK and IAA that ultimately can regulate each other's levels by mutually inhibiting their biosynthesis via a metabolic feedback loop (Jones and Ljung 2011). The advances in hormone biosynthesis confirm that these small molecules govern all aspects of plant biology.

Impact of Growth Regulators on the Phenological Stage of Plants

Phenology refers to the study of natural phenomena in living organisms related to their growth and development stages and their relationship with the environment, aiming to describe the causes of stage variations, relating the beginning, ending, and duration between them. With the study of plant phenology, their behavior in different environments could be predicted, helping in the planning and design of cultivation sites and the selection of the best variants according to the environmental conditions (e.g., climate, soil, and water quality and availability) (Keller 2020). Therefore, understanding plant phenology allows adequate planning for irrigation, fertilization, harvesting, and growth modeling (Zhang et al. 2022).

The phenological stages of plants have been changing due to climate alteration, a consequence of human activities. Climate change promotes natural selection, modifying landscapes, and plant populations, potentially endangering certain organisms and putting them at risk of extinction (Mohan et al. 2019). Stages differ depending on the studied crop and have been described using scales with letters or numbers, but generally, they are classified as germination, flowering, budburst, fruit development, and senescence (Paradinas et al. 2022, Sosa-Zuniga et al. 2017). In this context, plant regulators govern plant phenological responses (Sayed Shourbalal et al. 2019). A description of the effect of phytohormones on plants' phenological stages is briefly described.

Abscisic Acid. This compound is synthesized in the early stages of the plant from farnesyl pyrophosphate, which is also present in some phytopathogenic fungi. This phytohormone is a plant-growth inhibitor, given its ability to regulate seed dormancy, stop germination, and promote senescence (Colebrook et al. 2014). It regulates the metabolic processes of amino acids and lipids, promotes secondary metabolite production, such as flavonoids, and modulates enzymatic activity and photosynthesis (Huan et al. 2020).

Auxins. The first discovered PGRs and most studied are auxins. Its predominant form is IAA, from which indole-butyric acid (IBA) can be obtained. They are endogenous compounds of plants, although IBA is a more efficient promoter in forming lateral roots than AIA. 2,4-dichlorophenoxyacetic acid and o-naphthalene acetic acid are synthetic PGRs with auxin action, capable of manipulating the processes of cell elongation, division, and differentiation. In agriculture, they stop the outbreak of buds in potatoes, prevent premature fruit fall and abortion of flowers, produce seedless fruits, and stimulate root growth in cuttings. Microorganisms such as bacteria and fungi that metabolize L-tryptophan to AIA can synthesize this PGR (Garay-Arroyo et al. 2014). Auxins have a synergistic effect with cytokinins, improving and accelerating plant growth.

Brassinosteroids. Brassinolide-based lactone polyhydroxysteroids, or BR, share similarities with animal steroid hormones. This phytohormone can be obtained from algae and some ancient plants. They regulate elongation, cell division, root growth, germination, and reproduction (Kanwar et al. 2017, Tang et al. 2016).

Cytokinins. CKs, compounds derived from adenine, are involved in senescence and induce cell division (Salazar-Cerezo et al. 2018, Zhang et al. 2021). They participate in development regulation, initiating buds and flowering, promoting cell expansion, and transportation of nutrients (Giannakoula et al. 2012, Yadav et al. 2021).

Ethylene. Gaseous compounds such as ET are also related to plant metabolic processes. ET, a volatile compound synthesized in diverse species and plant organs from methionine, has been associated with maturation and senescence, the beginning of flowering and fruit formation. ET controls the latter's color, texture, and aroma, therefore, its endogenous and exogenous action is particularly interesting for post-harvest handling. This compound induces a synergistic effect with auxin, ABA, and CKs, improving leaf maturation and development processes and exerts an inhibitory effect on GBs and JA (Dubois et al. 2018, Iqbal et al. 2017).

Salicylic Acid. This regulator helps tolerate extreme climates and oxidative, salt, and heavy metal stress. SA increases floral longevity and inhibits ethylene biosynthesis. This compound is linked to auxins and gibberellins, exerting a favorable effect for germination, even at low temperatures, and resistance to drought due to its control over the photosynthetic activity and the stomatal conductivity (Nazar et al. 2015).

Gibberellins. GAs are obtained from the metabolism of plants, fungi, and bacteria capable of metabolizing isoprene to terpenoids by symbiotic or parasitic interactions. They are associated with the constant growth of tissues such as leaves, roots, germination, and flowering and are vital in the fertility processes of male and female plants (Gupta and Chakrabarty 2013). They are responsible for stem lengthening, leaf development, beginning and development of reproductive organs, flower number, and fruit set (Giannakoula et al. 2012, Sayed Shourbalal et al. 2019).

Jasmonates. JA, the most widely used PGR in the plant kingdom, is produced by plants, algae, molds, pteridophytes, gymnosperms, and some fungi. This phytohormone can act both as an inhibitor and as a stimulant. Its function ranges from aspects of cell development to immune activity, including germination and tuber and root formation. It also generates synergy or antagonism with other PGRs, inducing organ promotion and inhibition or stimulation of root elongation (Huang et al. 2017, Hussain et al. 2020, Schuman et al. 2018)

Strigolactones. SLs are biomolecules with a terpenoid lactone structure. This phytohormone, derived from carotenoids, increases primary root growth, inhibits the development of lateral roots, helps in the adaptive process during nutrient deficiency, and promotes symbiosis with mycorrhizae (Smith 2014).

Other molecules. The ascorbic acid hormone is essential in various phenological stages, such as growth, development, and flowering. It modifies phytohormone signaling from the vegetative to the reproductive stage, which initiates blooming. Its application improved growth and fruit nutrient content (Zahid et al. 2021). Thiourea generates tolerance to stress, can terminate dormancy, and promotes growth and development (Waqas et al. 2017, Zahid et al. 2021).

Other PGRs have been recently studied. Melatonin is a universal molecule in living organisms with recently discovered biological functions and exhibiting evolutionary conservation (Debnath et al. 2019). Studies indicate that exogenously applying this compound enhances plant growth, photosynthesis, antioxidant activity, and drought resistance (Hu et al. 2020, Qiao et al. 2020).

Low molecular weight aliphatic polycations or polyamines (PAs) are other PGRs generally derived from arginine and found in all living beings. Plants' most abundant and studied PAs are spermidine, putrescine, and spermine. They are related to various biochemical and physiological processes, including root development, seed germination and elongation, flowering, photosynthesis, sugar accumulation, and molecular reactions such as protein translation, cell death, DNA synthesis, ion transport, and enzyme activation (Chen et al. 2019, Doneva et al. 2021, Igarashi and Kashiwagi 2019).

Amino acids such as proline, arginine, and gamma and beta aminobutyric acids, which are widely distributed in higher plants, are essential in the osmotic adjustment, improve photosynthesis, and enhance the production of antioxidant compounds (Abd El-Gawad et al. 2021, Hussein et al. 2019). Macronutrients (i.e., N, P, and K) and carbohydrates such as trehalose and chitosan are also considered PGRs. They have been associated with improving the photosynthetic processes, water use, sugar accumulation, and antioxidant enzymatic activity related to plant adaptation to harsh conditions (Gou et al. 2017, Veroneze-Júnior et al. 2019). Furthermore, rhizobacteria are plant growth promoters as they positively affect the host plant's development, nutrition, and stress adaptation (Vurukonda et al. 2016). Noticeably, bacteria in the rhizosphere do not need physical contact with the plant to induce the production of volatile metabolites that regulates plant growth, in addition to acting as positive and negative regulators of other PGRs production such as ABA, GA, IAA, and ethylene (Hashem et al. 2019, Lakshmanan et al. 2017).

PGRs Impact against Biotic and Abiotic Stress

PGRs significantly mediate plant tolerance against different stressors through signal transduction pathways organization (Bürger and Chory 2019). In this regard, PGRs act synergistically or antagonistically to modulate plant responses. This intimate crosstalk allows plants to generate a sophisticated and efficient response depending on the severity and type of stress (Khan et al. 2020). Among the significant plant PGRs, SA, JA, ET, and ABA perform vital functions in facilitating the defense response against stress (Bari and Jones 2009, Bürger and Chory 2019, Verma et al. 2016). There is also evidence for the crosstalk of those hormones with GAs, auxins, BRs, and CKs in regulating plant defense response stress (Khan et al. 2020, Verma et al. 2016). The final result of increasing or suppressing PGRs is altering the plant growth pattern to withstand stress factors.

Biotic Stress Factors

Among the factors of biotic stress are those living organisms, such as bacteria, fungi, viruses, oomycetes, herbivores, and nematodes, or their derivatives (extracts, proteins, DNA) that influence or modify plant metabolism, growth, and development (Vázquez-Hernández et al. 2019). These factors activate a series of responses that leads to the induction and interaction of several PGRs. In addition, the signaling mediated by biotic factors mediates abiotic stress tolerance (Khan et al. 2020). Some examples of the biotic stress effect on the regulation of phytohormones are summarized in Table 2.

Table 2. Participation of plant growth regulators in the response to biotic stress.

Biotic stress factor	Plant species	Response	Reference
Pseudomonas syringae (Bacteria)	*Arabidopsis thaliana*	Activation of JA-dependent defense response and suppression of SA signaling defense	Cui et al. 2005
Xanthomonas oryzae (Bacteria)	Oryza sativa	Suppression of IAA accumulation	Ding et al. 2008
Erysiphe cichoracearum (Fungi)	*Arabidopsis thaliana*	Induction of SA signaling	Li et al. 2006
Alternaria brassicola (Fungi)	*Arabidopsis thaliana*	Suppression of ABA accumulation	Flors et al. 2007
TYLCV (Virus)	*Solanum lycopersicum*	Expression of SA responsive PR genes (PR1, PR2 and PR5) and ROS scavenging enzymes (SOD, POD, CAT)	Li et al. 2019
PVY (Virus)	*Solanum tuberosum*	Induction of SA signaling pathway	Baebler et al. 2014
Nilaparavata lugens (Hervibore)	Oryza sativa	JA signaling positively regulates resistance against *Nilaparavata lugens* attack	Xu et al. 2021
Spodoptera frugiperda (Hervibore)	Z. mays	JA signaling and expression of defense-related genes.	Chuang et al. 2014
Meloidogyne spp., *Heterodera and Globodera* spp. (Nematodes)	*Arabidopsis thaliana,* *Solanum lycopersicum and Solanum tuberosum*	Induction of SA and JA signaling pathways	Manosalva et al. 2015

Microbes: Bacteria and Fungi

Infections by pathogenic bacteria and fungi promote a significant increase in SA, JA, and ET concentration, therefore, they are considered essential hormones for plants' response to biotic stress (Bari and Jones 2009) (Fig. 10). SA stimulates the defense response against biotrophic and hemibiotrophic pathogens, whereas ET and JA activate an immune response against necrotrophic pathogens (Bürger and Chory 2019, Vlot et al. 2021). SA is synthesized and activates the systemic acquired resistance (SAR) during pathogen infection, triggering pathogenesis-related (PR) genes to induce defense responses against a broad spectrum of pathogens (Vlot et al. 2021). Additionally, JA causes induced systemic resistance (ISR), which activates mutualistic microbes (Bürger and Chory 2019, Vlot et al. 2021). SA and JA pathways intersect at different points because these two pathways antagonistically regulate factors related to biotic stress (Bürger and Chory 2019, Verma et al. 2016). Nevertheless, synergistic interactions have been observed in the simultaneous

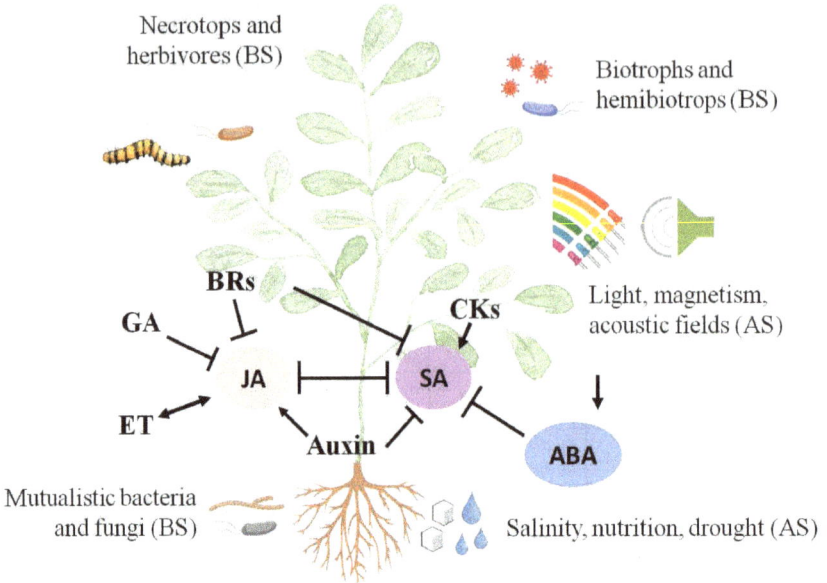

Fig. 10. Overview of abiotic and biotic stress factors and their effects on PGRs. JA, SA, ET, and ABA are key players in stress response. SA, JA, and ET are dominant in controlling biotic responses, while ABA mainly regulates abiotic stress. (AS): Abiotic Stress, (BS): Biotic Stress. Image source: Own elaboration.

induction of both pathways and at low SA-JA concentrations (Li et al. 2004, Vlot et al. 2021). Generally, ET and JA regulate plant immune response after microbial infection, for example, through necrotroph resistance and root hair development (Zhu et al. 2011). Other PGRs, such as GAs, ABA, and auxins, regulate the antagonistic relationship between SA and JA (Bürger and Chory 2019).

Herbivorous Insects

JA is produced during the immune response against necrotroph pathogens but also against herbivorous insects and is usually an antagonist of the SA signaling pathway (Fig. 10) (Bürger and Chory 2019). When herbivores feed on plants, they produce lesions and introduce substances such as oral secretions that trigger immune responses where the JA pathway is mainly involved, and also ET seems to operate synergistically to module the defense response (Bari and Jones 2009, Duran-Flores and Heil 2016, Xu et al. 2021). Herbivores that induce the JA signaling pathway include spider mites, caterpillars, beetles, mirid bugs, and thrips (Wasternack and Hause 2013). Research has shown that JA is fundamental for plant resistance against herbivores (Chuang et al. 2014, Huffaker et al. 2013, Verma et al. 2016, Xu et al. 2021). Moreover, SA, cytokinins, and strigolactones synergize with JA signaling by participating in the immune feedback to herbivore attack (Li et al. 2020, Ma et al. 2020, Schäfer et al. 2015).

Virus

Several PGRs participate during viral infections to achieve an adequate immune response in plants (Alazem and Lin 2015, Gupta et al. 2021, Paudel and Sanfaçon 2018). SA signaling constitutes a crucial pathway against viruses. After viral infection, the SA pathway is activated, inducing SAR and small interfering RNA (siRNA) machinery at distal sites (Paudel and Sanfaçon 2018, Vlot et al. 2009). Both cytokinins and brassinosteroids have been demonstrated to positively affect the defense against viruses. The main difference is that cytokinins improve plant defenses in an SA-mediated manner, whereas brassinosteroids act independently (Alazem and Lin 2015, Argueso et al. 2012, Li et al. 2019).

Auxins and ET have antagonistic effects on plant defenses to viruses. Auxins counteract the SA pathway and activate the auxin response crucial for viral replication (Padmanabhan et al. 2008). Some studies report that ET antagonizes the SA signaling pathway and participates in symptom development, systemic viral dispersion, and necrotic lesions induced by virus infections (Chen et al. 2013, Geri et al. 2004). ABA and JA have shown synergistic and antagonistic activities in defense against viruses. During the initial stages of infection, ABA can exert a positive regulatory effect on plant defense through callose deposition on plasmodesmata and the suppression of virus movement from cell to cell. Nevertheless, in later stages of infection, ABA suppresses SA and JA signaling and ROS production, avoiding immune responses regulated by these pathways (Alazem and Lin 2015). Similar effects have been described in the case of induced or applied JA as it supports plant defense in the early stages, but it decreases resistance to viruses in later stages (Alazem and Lin 2015, García-Marcos et al. 2013). Several cross-communication levels among SA, JA, and ABA signaling pathways highlight a complex immune response process against viruses (Paudel and Sanfaçon 2018).

Abiotic Stress Factors

Abiotic stress (e.g., temperature, drought, salinity, waterlogging, and nutritional imbalances) has been widely studied in plants (Hasanuzzaman et al. 2019). Abiotic stress hinders plants from achieving their maximum potential, adversely affecting plant growth and development (Upreti and Sharma 2016). Phytohormones act as signaling molecules by directing or redirecting critical plant processes, such as growth, reproduction, and fruit ripening (Bagher et al. 2021). These phytohormone responses depend on the stress severity and present a significant role in plant defense (Nadeem et al. 2016). Even though phytohormones have been reported to interact, they play different roles within different stresses to prevent or remediate adverse effects (Tuteja et al. 2010). Examples of the abiotic stress effect on the regulation of phytohormones are summarized in Table 3.

Table 3. Participation of plant growth regulators in the response to abiotic stress.

Abiotic stress factor	Plant species	Response	Reference
High light exposure	*Arabidopsis thaliana*	Upregulation of JA and ABA biosynthetic genes, SA accumulation, decrease of auxin and CK	Müller and Munné-Bosch 2021
Low light exposure	*Magnolia sinostellata*	Downregulation of JA, activation of ET mediated signaling pathway	Lu et al. 2021
Drought	Various	Upregulation of ABA signaling and biosynthesis pathways, auxin biosynthesis inhibition	Aslam et al. 2022
Drought	*Vigna unguiculata*	Lower ABA and higher IAA content in leaves	Tankari et al. 2021
Low nitrogen fertilization	*Zea mays*	Increase in content and gene expression for BRs biosynthesis	Xing et al. 2022
Boron deficiency	*Brassica napus*	Decrease in active CK forms, increase in ABA and IAA precursors	Eggert and von Wirén 2017
Heavy metals (Pb, Cr, Cu, Ni, and Al)	*Glycine max*	Higher production of GAs and IAA	Bilal et al. 2019
Heavy metals (Al)	*Citrus limonia* L.	Increase of ABA concentration in leaves	Gavassi et al. 2021
Single-walled nanotubes (SWNTs)	*Solanum lycopersicum* L.	Significant increase in SA content	Jordan et al. 2020
Silver nanoparticles	*Arabidopsis thaliana*	Significant increase in JA, ABA and JA levels.	Angelini et al. 2022
Traffic noise	*Tagetes patula* L. and *Salvia splendens*	Reduced content of SA, IAA and GA. Increase of ABA and JA content	Kafash et al. 2022
Magnetic fields	*Pisum sativum*	Increase of IAA and GA content	Podleśny et al. 2021

Temperature

Temperature affects plant development by modifying key processes like photosynthesis and pollination, negatively affecting crop yield and production (Hatfield and Prueger 2015, Moore et al. 2021). Responses to heat and cold can include changes in enzyme activities and overall adjustments in the metabolic profiles and phytohormone levels (Shao et al. 2021). High temperatures can activate phytohormone responses of auxins, ABA, BRs, and ET, so plants can adapt to increasing temperatures and relieve heat stress (Jha et al. 2021). Similarly, low temperatures can also affect plant germination and development, ABA, IAA, BRs, and ET have been reported to increase dehydration toleration and preserve plant cells' integrity (Aslam et al. 2022).

Light

Many physiological responses are mediated by plant photoreceptors (Kami et al. 2010). Light quality (i.e., light spectrum in nm), quantity (i.e., μmol m^{-2} s^{-1}), and photoperiod (i.e., light: dark hours) can be modulated to regulate plant processes (Aguirre-Becerra et al. 2021, 2020). However, not optimal light conditions can cause adverse effects on the overall capacity of plants to modulate specific and vital processes such as photosynthesis, tolerance to hot and cold temperatures, and biological responses to pathogens and insects (Roeber et al. 2021). The regulation of phytohormones like ET, CKs, GBs, auxins, and BRs has been closely related to light responses (Folta and Childers 2008). Responses to light excess regulated by phytohormones mainly focus on avoiding damage to the photosynthetic machinery and photomorphogenesis for light adaptation (Li et al. 2019, Wang et al. 2019).

Drought

Drought is a constraining factor for crop productivity, adversely impacting plant photosynthesis and growth (Fahad et al. 2017). This stress influences various phytohormone pathways that affect the production of auxins, CKs, GBs, BRs, SA, and ET (Iqbal et al. 2022). In general, water deficiency triggers multiple responses related to their gene expression, overall signaling pathways, and production to increase drought tolerance by improving the efficiency of water uptake and use for the biological processes of plants (Ilyas et al. 2021).

Nutrition

Nutrients are essential for regulating critical biomass, yield, and survival processes under stress conditions. Therefore, a nutrient limitation can result in physiological alterations preventing plants from reaching their total growth and development potential (Pandey et al. 2021). Most of the physiological responses in plants to this stress, such as the translocation and mobilization of scarce nutrients and changes in roots to facilitate their acquisition, are mediated by PGRs (Romera et al. 2021). In this sense, the root architecture of plants is intricately connected to their ability to uptake water and nutrients, especially during nutrient deficiency. Phytohormones like auxins, GAs, and CKs act as signaling mechanisms to enhance nutrient efficiency and mitigate stress by altering the root system structure (Divte et al. 2021, Xing et al. 2022).

Other Abiotic Factors

Heavy metal accumulation in the soil has negatively affected the agricultural systems (Anjum et al. 2015, Sytar et al. 2019). Their excessive concentration inside plant cells leads to physiological disorders affecting critical processes like photosynthesis and respiration (Angulo-Bejarano et al. 2021, Morkunas et al. 2018). PGRs are crucial for plant primary metabolism under metal toxicity. For example, auxin helps detoxify plants, preventing oxidative damage by reducing the accumulation of excessive ROS due to metalloid toxicity (Singh et al. 2021). ABA regulates plant responses during this stress as it accumulates because of the upregulation of genes involved in ABA biosynthesis (Bücker-Neto et al. 2017). A more detailed description of the role of each phytohormone during metal toxicity is described by Khan et al. (2021)

The implementation of nanoparticles in agriculture has shown potential benefits by promoting certain activities in plant metabolism, such as increased nutrient uptake and enzymatic systems activation (Landa 2021, Vera-Reyes et al. 2018). Nevertheless, there remains uncertainty surrounding their potentially detrimental effects due to their persistence and hazardous environmental interactions (Ranjan et al. 2021). Research suggests that these compounds affect antioxidant systems in plants, but data on their effects on phytohormones is limited (Etesami et al. 2021). Even so, using nanoparticles on stressed plants has been shown to affect CKs, BRs, ABA, and GBs levels (Hao et al. 2018, Jordan et al. 2020, Manickavasagam et al. 2019).

Plants have evolved mechanisms to interact with their environment and detect ambient signals, such as electric and magnetic fields that naturally exist on Earth (Belyavskaya 2004). With the increase in electricity and communication transmission systems, the potential effect of these signals on plants has become a topic of interest, as how plants perceive these stimuli has yet to be entirely understood. It has been reported that continuous magnetic fields and electric stimulation can significantly change the content of PGRs such as JA, ABA, and ET (Liu et al. 2021, Saletnik et al. 2022). For example, sunflower seeds reduced the ABBA content by over 50% when exposed to electromagnetic field radiation (EFR) (Mildažienė et al. 2019). In this regard, EFR interferes with dipolar molecules and charged ions and alters the plant's electric properties, as they could serve as living antennae (Bonnet et al. 2006, Senavirathna and Asaeda 2018). The electric polarity of plants would be easily affected by EFR by changing the ion movement inside plants, inducing systemic metabolic and morphological abnormalities (S. Kaur et al. 2021, Volkov and Ranatunga 2006) that phytohormones regulate the most.

Interaction of Microbiome and PGRs

The microbiome is an ecosystem where various microorganisms coexist with plants and soil. Many studies have examined the effective and regulated communication of microbiome components that generates synergy by producing compounds that serve as PGRs (Cheng et al. 2019, Levy et al. 2018). The microbiome composition depends on geolocation, flora, soil, fertilization, and crop management. The rhizosphere (i.e., the layer of soil or substrate where plant roots and microbiome interact) contains carbon-rich photosynthesis products that serve as an energy source for the survival and maintenance of the microorganisms in this region (Rodriguez et al. 2019). Depending on the plant-microbiome balance, many biochemical reactions trigger the production of plant metabolites that stimulate the development or deterioration of certain microorganisms (Qu et al. 2020).

The soil is among the most diverse habitats on Earth, with a great reservoir of living organisms, organic matter, and inorganic compounds. This component is of utmost importance for food production and provides a readily available source of nutrients and microorganisms that can support healthy and high-yielding crops (Expósito et al. 2017). Currently, microorganisms are used for the biological control of pests and for improving crop nutrient absorption (Rodriguez et al. 2019). Soil microbes that promote plant growth are of high interest in agriculture as they play critical roles in nutrient cycles (e.g., nitrogen and phosphorus), crop protection against stress, and biomass production (Qu et al. 2020, Yadav et al. 2018). Land management significantly impacts soil and the structure and stability of the root microbial community and, consequently, microbiome-associated functions (Rodriguez et al. 2019). Arbuscular mycorrhizal fungi, biotrophic organisms that need a host plant to complete their life cycle, comprise about 80% of the fungi associated symbiotically with plants (Levy et al. 2018, Yadav et al. 2018). Additionally, beneficial microbes generate metabolites that promote plant growth by modifying their hormonal balance (Hartman and Tringe 2019).

Microbes can enhance plant resilience to stress by modulating the production and accumulation of PGRs such as ABA, IAA, CKs, GAs, ET, JA, methyl jasmonate, SA, auxins, ACC, some polysaccharides, inorganic compounds, and other signaling molecules (Pascale et al. 2020, Petrov and Van Breusegem 2012, Spence and Bais 2015). They act by activating the defense system of plants, both SAR and ISR (Eichmann et al. 2021, H. Zhang et al. 2022). Bacteria and fungi that promote plant growth have been described as organisms that activate the ISR (Pascale et al. 2020, Rodriguez et al. 2019). IAA and Auxins induce tryptophan biosynthesis in bacteria from Rhizobia (Koo et al. 2020). Some bacteria can degrade IAA for a carbon and nitrogen source and auxins when high levels of IAA are present, affecting root growth and exudate production. Moreover, an excess of locally accumulated auxins in the roots can alter plant auxin transport and host root morphology (Fitzpatrick et al. 2020, Pascale et al. 2020).

ET is critical in the colonization of plant tissue by bacteria. ET can restrict fungal root colonization affecting hyphal growth and spore germination (Yadav 2017). JA and ET induce the ISR through interaction with rhizobacteria (Uroz et al. 2019), and are also generated in response to necrotrophic pathogens. SA assists in constructing root microbial communities (Trivedi et al. 2022). Alterations in the microbial community systems can restrict or enhance the production of these growth factors, causing biochemical and morphological changes in crops (Pascale et al. 2020). Several studies indicate that SA may interact via canonical signaling pathways with other PGRs, such as JA and ET (Eichmann et al. 2021, Koo et al. 2020, Rodriguez et al. 2019). Exogenous application of ET increased rhizosphere alpha diversity and reduced acidobacteria (Eichmann et al. 2021, T. Kaur et al. 2021). AX and CK are the main growth regulators of primary roots, stimulating cell division and differentiation. In addition, CKs, GAs, and BRs promote interactions between beneficial and pathogenic microbes in microbial consortia. CKs participate in nodule formation in legumes and induce pseudo-nodule development when applied exogenously (T. Kaur et al. 2021, Rodriguez et al. 2019). Additionally, they help crops in nitrogen fixation and better absorption of nutrients, promoting plant growth, possibly due to the exudates generated in the root in response to the interaction of microbes with other PGRs (Eichmann et al. 2021).

Conclusion

PGRs can enhance plant growth, development, defense, and productivity. They influence physiological and biochemical pathways that result in plant phenological changes, improving survival and adaptation to harsh environmental stimuli. Considerable progress has been achieved in comprehending the biochemical pathways involved in their biosynthesis, however, the full effects and cross-communication still need to be fully elucidated.

Considering the current agricultural problems (e.g., the changing and unpredictable climatic conditions, soil quality, and water scarcity), future research should develop a combination-based technology for multi-stress conditions to achieve growth and productivity goals. Future perspectives should consider that the elicitation results depend on the synergy of phytohormones, the application conditions, the concentration, and the type of cultivation. Additionally, new technologies for plant elicitation generate new possibilities for sustainable crop improvement. For example, abiotic factors (e.g., acoustic emissions, electric and magnetic fields, volatile organic compounds, and nanotechnology) can enhance phytohormone production and trigger plant secondary metabolism during crop production or postharvest. In the case of biotic elicitation, implementing specific miRNAs is an emerging technique currently being researched and debated within the scientific community. New tools such as bioinformatics enable the prediction of miRNA molecule modulation through algorithms, contributing to understanding their involvement in an intricate regulatory network encompassing PGRs and secondary metabolites.

On the other hand, inadequate agricultural practices (e.g., short crop rotation cycles and excessive use of agrochemicals) harms soil microbiota, disrupting the structure and stability of soil and root microbial communities and negatively impacting microbiome synergy with crops (Rodriguez et al. 2019). The complexity of the interactions between the microbiome and the chemical signals that promote the growth of crops causes a controversy about how communication between these intricate systems is carried out and how to take advantage of the knowledge generated to be able to use it in the production of healthy crops with a high yield. Conservation agriculture and the rational use of resources, including the exogenous application of PGRs and microorganisms, is a resource that can help farmers prevent soil deterioration and the proliferation of pathogenic microorganisms. The combination of several areas related to genetics (oomic sciences) can help provide information to know the different microorganisms found in the microbiome and to be able to evaluate the relationship between each of them with the various crops and how they interact by regulating the production of metabolites involved in the maintenance and defense of plants.

References

Abd El-Gawad, H.G., Mukherjee, S., Farag, R., Abd Elbar, O.H., Hikal, M., Abou El-Yazied, A. et al. 2021. Exogenous γ-aminobutyric acid (GABA)-induced signaling events and field performance associated with mitigation of drought stress in Phaseolus vulgaris L. Plant Signal. Behav., 16: 1853384.

Adams, D.O. and Yang, S. 1979. Ethylene biosynthesis: identification of 1-aminocyclopropane-1-carboxylic acid as an intermediate in the conversion of methionine to ethylene. Proc. Natl. Acad. Sci., 76: 170–174.

Adams, D.O. and Yang, S.F. 1977. Methionine metabolism in apple tissue: implication of S-adenosylmethionine as an intermediate in the conversion of methionine to ethylene. Plant Physiol., 60: 892–896.

Aguirre-Becerra, H., García-Trejo, J.F., Vázquez-Hernández, C., Alvarado, A.M., Feregrino-Pérez, A.A., Contreras-Medina, L.M. et al. 2020. Effect of Extended Photoperiod with a Fixed Mixture of Light Wavelengths on Tomato Seedlings. HortScience, 1: 1–8.

Aguirre-Becerra, H., Vazquez-Hernandez, M.C., Alvarado-Mariana, A., Guevara-Gonzalez, R.G., Garcia-Trejo, J.F. and Feregrino-Perez, A.A. 2021. Role of Stress and Defense in Plant Secondary Metabolites Production. In: Bioactive Natural Products for Pharmaceutical Applications. Springer, pp. 151–195.

Ajmi, A., Larbi, A., Morales, M., Fenollosa, E., Chaari, A. and Munné-Bosch, S. 2020. Foliar paclobutrazol application suppresses olive tree growth while promoting fruit set. J. Plant Growth Regul., 39: 1638–1646.

Alazem, M. and Lin, N.S. 2015. Roles of plant hormones in the regulation of host-virus interactions. Mol. Plant Pathol.

Alder, A., Jamil, M., Marzorati, M., Bruno, M., Vermathen, M., Bigler, P. et al. 2012. The path from β-carotene to carlactone, a strigolactone-like plant hormone. Science (80-.). 335: 1348–1351.

Angelini, J., Klassen, R., Široká, J., Novák, O., Záruba, K., Siegel, J. et al. 2022. Silver Nanoparticles Alter Microtubule Arrangement, Dynamics and Stress Phytohormone Levels. Plants, 11: 313.

Angulo-Bejarano, P.I., Puente-Rivera, J. and Cruz-Ortega, R. 2021. Metal and Metalloid Toxicity in Plants: An Overview on Molecular Aspects. Plants, 10: 635.

Anjum, N.A., Singh, H.P., Khan, M.I.R., Masood, A., Per, T.S., Negi, A. et al. 2015. Too much is bad—an appraisal of phytotoxicity of elevated plant-beneficial heavy metal ions. Environ. Sci. Pollut. Res., 22: 3361–3382.

Argueso, C.T., Ferreira, F.J., Epple, P., To, J.P.C., Hutchison, C.E., Schaller, G.E. et al. 2012. Two-component elements mediate interactions between cytokinin and salicylic acid in plant immunity. PLoS Genet., 8.

Ashihara, H., Ludwig, I.A. and Crozier, A. 2020. Plant Nucleotide Metabolism - Biosynthesis, Degradation, and Alkaloid Formation, 1st Editio. ed. Wiley.

Aslam, M., Fakher, B., Ashraf, M.A., Cheng, Y., Wang, B. and Qin, Y. 2022. Plant Low-Temperature Stress: Signaling and Response. Agronomy, 12: 702.

Aslam, M.M., Waseem, M., Jakada, B.H., Okal, E.J., Lei, Z., Saqib, H.S.A. et al., 2022. Mechanisms of Abscisic Acid-Mediated Drought Stress Responses in Plants. Int. J. Mol. Sci., 23: 1084.

Åstot, C., Dolezal, K., Nordström, A., Wang, Q., Kunkel, T., Moritz, T. et al. 2000. An alternative cytokinin biosynthesis pathway. Proc. Natl. Acad. Sci. U.S.A., 97: 14778–14783.

Babu, R.S.H., Srilatha, V. and Joshi, V. 2022. Plant Growth Regulators in Mango (Mangifera indica L.). In: Plant Growth Regulators in Tropical and Sub-Tropical Fruit Crops. CRC Press, pp. 315–468.

Baebler, Š., Witek, K., Petek, M., Stare, K., Tušek-Žnidarič, M., Pompe-Novak, M. et al. 2014. Salicylic acid is an indispensable component of the Ny-1 resistance-gene-mediated response against Potato virus y infection in potato. J. Exp. Bot., 65: 1095–1109.

Bagher, M., Bajguz, A., Saud, S., Khan, F.A., Sabagh, A.E.L., Sabagh, A.E. et al. 2021. Potential Role of Plant Growth Regulators in Administering Crucial Processes Against Abiotic Stresses. Front. Agron. 3.

Bajguz, A. 2019. Brassinosteroids in microalgae: application for growth improvement and protection against abiotic stresses. In: Brassinosteroids: Plant Growth and Development. Springer, pp. 45–58.

Bajguz, A. and Tretyn, A. 2003. The chemical characteristic and distribution of brassinosteroids in plants. Phytochemistry, 62: 1027–1046.

Bari, R. and Jones, J.D.G. 2009. Role of plant hormones in plant defence responses. Plant Mol. Biol.

Bartel, B. and Fink, G.R. 1994. Differential regulation of an auxin-producing nitrilase gene family in Arabidopsis thaliana. Proc. Natl. Acad. Sci., 91: 6649–6653.

Belyavskaya, N.A. 2004. Biological effects due to weak magnetic field on plants. Adv. Sp. Res., 34: 1566–1574.

Bilal, S., Shahzad, R., Khan, A.L., Al-Harrasi, A., Kim, C.K. and Lee, I.-J. 2019. Phytohormones enabled endophytic Penicillium funiculosum LHL06 protects Glycine max L. from synergistic toxicity of heavy metals by hormonal and stress-responsive proteins modulation. J. Hazard. Mater., 379: 120824.

Bishop, G.J. and Yokota, T. 2001. Plants Steroid Hormones, Brassinosteroids: Current Highlights of Molecular Aspects on their Synthesis/Metabolism, Transport, Perception and Response. Plant Cell Physiol., 42: 114–120.

Bisht, T.S., Rawat, L., Chakraborty, B. and Yadav, V. 2018. A Recent Advances in Use of Plant Growth Regulators (PGRs) in Fruit Crops—A Review.

Bonnet, P., Vian, A. and Beaubois, É. 2006. Plants as Living Antennas? In: The European Conference on Antennas and Propagation: EuCAP 2006, p. 653.

Breithaupt, C., Kurzbauer, R., Lilie, H., Schaller, A., Strassner, J., Huber, R. et al. 2006. Crystal structure of 12-oxophytodienoate reductase 3 from tomato: self-inhibition by dimerization. Proc. Natl. Acad. Sci., 103: 14337–14342.

Brewer, P.B., Yoneyama, K., Filardo, F., Meyers, E., Scaffidi, A., Frickey, T. et al. 2016. Lateral Branching Oxidoreductase acts in the final stages of strigolactone biosynthesis in Arabidopsis. Proc. Natl. Acad. Sci., 113: 6301–6306.

Bruno, M., Hofmann, M., Vermathen, M., Alder, A., Beyer, P. and Al-Babili, S. 2014. On the substrate- and stereospecificity of the plant carotenoid cleavage dioxygenase 7. FEBS Lett., 588: 1802–1807.

Brunswick, N. 1992. Salicylate, A New Plant Hormone1. Plant Physiol., 99: 799–803.

Buchanan, B.B. and Gruissen W, J.R.L. 2015. Biochemistry and Molecular Biology of Plants (second edition), Wiley, Blackwell (Chichester, UK).

Bücker-Neto, L., Paiva, A.L.S., Machado, R.D., Arenhart, R.A. and Margis-Pinheiro, M. 2017. Interactions between plant hormones and heavy metals responses. Genet. Mol. Biol., 40: 373–386.

Bürger, M. and Chory, J. 2019. Stressed Out About Hormones: How Plants Orchestrate Immunity. Cell Host Microbe.

Campos, M.L., Kang, J.H. and Howe, G.A. 2014. Jasmonate-Triggered Plant Immunity. J. Chem. Ecol., 40: 657–675.

Castellaro, S.J., Dolan, S.C., Hedden, P., Gaskin, P. and MacMillan, J. 1990. Stereochemistry of the m3 sx n etabolic steps from kaurenoic acids to kaurenolides and gibberellins. Phytochemistry, 29: 1833–1839.

Chen, D., Shao, Q., Yin, L., Younis, A. and Zheng, B. 2019. Polyamine function in plants: metabolism, regulation on development, and roles in abiotic stress responses. Front. Plant Sci., 1945.

Chen, L., Zhang, L., Li, D., Wang, F. and Yu, D. 2013. WRKY8 transcription factor functions in the TMV-cg defense response by mediating both abscisic acid and ethylene signaling in Arabidopsis. Proc. Natl. Acad. Sci. USA, 110.

Cheng, Y.T., Zhang, L. and He, S.Y. 2019. Plant-Microbe Interactions Facing Environmental Challenge. Cell Host Microbe.

Chini, A., Monte, I., Zamarreño, A.M., Hamberg, M., Lassueur, S., Reymond, P. et al. 2018. An OPR3-independent pathway uses 4,5-didehydrojasmonate for jasmonate synthesis. Nat. Chem. Biol., 14: 171–178.

Chuang, W.P., Ray, S., Acevedo, F.E., Peiffer, M., Felton, G.W. and Luthe, D.S. 2014. Herbivore cues from the fall armyworm (spodoptera frugiperda) larvae trigger direct defenses in maize. Mol. Plant-Microbe Interact., 27: 461–470.

Clarke, S.M., Mur, L.A.J., Wood, J.E. and Scott, I.M. 2004. Salicylic acid dependent signaling promotes basal thermotolerance but is not essential for acquired thermotolerance in Arabidopsis thaliana. Plant J., 38: 432–447.

Colebrook, E.H., Thomas, S.G., Phillips, A.L. and Hedden, P. 2014. The role of gibberellin signalling in plant responses to abiotic stress. J. Exp. Biol., 217: 67–75.

Cook, C.E., Whichard, L.P., Turner, B., Wall, M.E. and Egley, G.H. 1966. Germination of witchweed (Striga lutea Lour.): isolation and properties of a potent stimulant. Science, (80-.). 154: 1189–1190.

Cui, J., Bahrami, A.K., Pringle, E.G., Hernandez-Guzman, G., Bender, C.L., Pierce, N.E. et al. 2005. *Pseudomonas syringae* manipulates systemic plant defenses against pathogens and herbivores. Proc. Natl. Acad. Sci., 102: 1791–1796.

Curtis, P.J. and Cross, B.E. 1954. Gibberellic Acid-a new Metabolite From the Culture Filtrates of Gibberella-Fujikuroi. Chem. Ind.

Debnath, B., Islam, W., Li, M., Sun, Y., Lu, X., Mitra, S. et al. 2019. Melatonin mediates enhancement of stress tolerance in plants. Int. J. Mol. Sci., 20: 1040.

Ding, P. and Ding, Y. 2020. Stories of salicylic acid: a plant defense hormone. Trends Plant Sci., 25: 549–565.

Ding, X., Cao, Y., Huang, L., Zhao, J., Xu, C., Li, X. et al. 2008. Activation of the Indole-3-Acetic Acid–Amido Synthetase GH3-8 Suppresses Expansin Expression and Promotes Salicylate- and Jasmonate-Independent Basal Immunity in Rice. Plant Cell, 20: 228–240.

Divte, P.R., Yadav, P., Pawar, A.B., Sharma, V., Anand, A., Pandey, R. et al. 2021. Crop response to iron deficiency is guided by cross-talk between phytohormones and their regulation of the root system architecture. Agric. Res., 10: 347–360.

Doneva, D., Pál, M., Brankova, L., Szalai, G., Tajti, J., Khalil, R. et al. 2021. The effects of putrescine pre-treatment on osmotic stress responses in drought-tolerant and drought-sensitive wheat seedlings. Physiol. Plant., 171: 200–216.

Dubois, M., Van den Broeck, L. and Inzé, D. 2018. The pivotal role of ethylene in plant growth. Trends Plant Sci., 23: 311–323.

Duran-Flores, D. and Heil, M. 2016. Sources of specificity in plant damaged-self recognition. Curr. Opin. Plant Biol.

Eggert, K. and von Wirén, N. 2017. Response of the plant hormone network to boron deficiency. New Phytol., 216: 868–881.

Eichmann, R., Richards, L. and Schäfer, P 2021. Hormones as go-betweens in plant microbiome assembly. Plant J., 105: 518–541.

Etesami, H., Fatemi, H. and Rizwan, M. 2021. Interactions of nanoparticles and salinity stress at physiological, biochemical and molecular levels in plants: A review. Ecotoxicol. Environ. Saf., 225: 112769.

Expósito, R.G., de Bruijn, I., Postma, J. and Raaijmakers, J.M. 2017. Current insights into the role of Rhizosphere bacteria in disease suppressive soils. Front. Microbiol.

Fahad, S., Bajwa, A.A., Nazir, U., Anjum, S.A., Farooq, A., Zohaib, A. et al. 2017. Crop production under drought and heat stress: plant responses and management options. Front. Plant Sci., 1147.

Fitzpatrick, C.R., Salas-González, I., Conway, J.M., Finkel, O.M., Gilbert, S., Russ, D. et al. 2020. The Plant Microbiome: From Ecology to Reductionism and Beyond.

Flors, V., Ton, J., Doorn, R. Van, Jakab, G., García-Agustín, P. and Mauch-Mani, B. 2007. Interplay between JA, SA and ABA signalling during basal and induced resistance against Pseudomonas syringae and Alternaria brassicicola. Plant J., 54: 81–92.

Folta, K.M. and Childers, K.S. 2008. Light as a growth regulator: controlling plant biology with narrow-bandwidth solid-state lighting systems. HortScience, 43: 1957–1964.

Fujioka, S. and Sakurai, A. 1997. Biosynthesis and metabolism of brassinosteroids. Physiol. Plant. 100, 710–715.

Fujioka, S., Takatsuto, S. and Yoshida, S. 2002. An early C-22 oxidation branch in the brassinosteroid biosynthetic pathway. Plant Physiol., 130: 930–939.

Fujioka, S. and Yokota, T. 2003. Biosynthesis and Metabolism of Brassinosteroids. Annu. Rev. Plant Biol., 54: 137–164.

Gao, Y., Dai, X., Aoi, Y., Takebayashi, Y., Yang, L., Guo, X. et al. 2020. Two homologous INDOLE-3-ACETAMIDE (IAM) HYDROLASE genes are required for the auxin effects of IAM in Arabidopsis. J. Genet. Genomics, 47: 157–165.

Garay-Arroyo, A., de la Paz Sánchez, M., García-Ponce, B., Álvarez-Buylla, E.R. and Gutiérrez, C. 2014. La homeostasis de las auxinas y su importancia en el desarrollo de Arabidopsis thaliana. Rev. Educ. bioquímica, 33: 13–22.

García-Marcos, A., Pacheco, R., Manzano, A., Aguilar, E. and Tenllado, F. 2013. Oxylipin Biosynthesis Genes Positively Regulate Programmed Cell Death during Compatible Infections with the Synergistic Pair *Potato Virus X-Potato Virus Y* and Tomato Spotted Wilt Virus. J. Virol., 87: 5769–5783.

Gavassi, M.A., Silva, G.S., de Marchi Santiago da Silva, C., Thompson, A.J., Macleod, K., Oliveira, P.M.R. et al. 2021. NCED expression is related to increased ABA biosynthesis and stomatal closure under aluminum stress. Environ. Exp. Bot., 185: 104404.

Geri, C., Love, A.J., Cecchini, E., Barrett, S.J., Laird, J., Covey, S.N. et al. 2004. Arabidopsis mutants that suppress the phenotype induced by transgene-mediated expression of cauliflower mosaic virus (CaMV) gene VI are less susceptible to CaMV-infection and show reduced ethylene sensitivity. Plant Mol. Biol., 56: 111–124.

Giannakoula, A.E., Ilias, I.F., Dragišić Maksimović, J.J., Maksimović, V.M. and Živanović, B.D. 2012. The effects of plant growth regulators on growth, yield, and phenolic profile of lentil plants. J. Food Compos. Anal., 28: 46–53.

Gomez-Roldan, V., Fermas, S., Brewer, P.B., Puech-Pagès, V., Dun, E.A., Pillot, J.P. et al. 2008. Strigolactone inhibition of shoot branching. Nature, 455: 189–194.

Gou, W., Zheng, P., Tian, L., Gao, M., Zhang, L., Akram, N.A. et al. 2017. Exogenous application of urea and a urease inhibitor improves drought stress tolerance in maize (Zea mays L.). J. Plant Res., 130: 599–609.

Grove, M.D., Spencer, G.F., Rohwedder, W.K., Mandava, N., Worley, J.F., Warthen, J.D. et al. 1979. Brassinolide, a plant growth-promoting steroid isolated from Brassica napus pollen. Nature 281, 216–217.

Guan, L., Denkert, N., Eisa, A., Lehmann, M., Sjuts, I., Weiberg, A. et al. 2019. JASSY, a chloroplast outer membrane protein required for jasmonate biosynthesis. Proc. Natl. Acad. Sci., 116: 10568–10575.

Gupta, N., Reddy, K., Bhattacharyya, D. and Chakraborty, S. 2021. Plant responses to geminivirus infection: guardians of the plant immunity. Virol. J.

Gupta, R. and Chakrabarty, S.K. 2013. Gibberellic acid in plant: still a mystery unresolved. Plant Signal. Behav., 8: e25504.

Haagen-Smit, A.J., Leech, W.D. and Bergen, W.R. 1941. Estimation, isolation and identification of auxins in plant material. Science, (80-.). 93: 624–625.

Hansen, M., Chae, H.S. and Kieber, J.J. 2009. Regulation of ACS protein stability by cytokinin and brassinosteroid. Plant J., 57: 606–614.

Hao, Y., Yuan, W., Ma, C., White, J.C., Zhang, Z., Adeel, M. et al. 2018. Engineered nanomaterials suppress Turnip mosaic virus infection in tobacco (Nicotiana benthamiana). Environ. Sci. Nano, 5: 1685–1693.

Hartman, K. and Tringe, S.G. 2019. Interactions between plants and soil shaping the root microbiome under abiotic stress. Biochem. J. 476: 2705–2724.

Hasanuzzaman, M., Hakeem, K.R., Nahar, K.F. and Alharby, H. 2019. Plant Abiotic Stress Tolerance. Springer.

Hashem, A., Tabassum, B. and Abd_Allah, E.F. 2019. Bacillus subtilis: A plant-growth promoting rhizobacterium that also impacts biotic stress. Saudi J. Biol. Sci., 26: 1291–1297.

Hatfield, J.L. and Prueger, J.H. 2015. Temperature extremes: Effect on plant growth and development. Weather Clim. Extrem. 10: 4–10.

Hedden, P. 2012. Gibberellin Biosynthesis. eLS.

Helliwell, C.A., Chandler, P.M., Poole, A., Dennis, E.S. and Peacock, W.J. 2001. The CYP88A cytochrome P450, ent-kaurenoic acid oxidase, catalyzes three steps of the gibberellin biosynthesis pathway. Proc. Natl. Acad. Sci., 98: 2065–2070.

Hossain, A., Pamanick, B., Venugopalan, V.K., Ibrahimova, U., Rahman, M.A., Siyal, A.L. et al. 2022. Emerging roles of plant growth regulators for plants adaptation to abiotic stress–induced oxidative stress. pp. 1–72. In: Emerging Plant Growth Regulators in Agriculture. Elsevier.

Hu, W., Cao, Y., Loka, D.A., Harris-Shultz, K.R., Reiter, R.J., Ali, S. et al. 2020. Exogenous melatonin improves cotton (Gossypium hirsutum L.) pollen fertility under drought by regulating carbohydrate metabolism in male tissues. Plant Physiol. Biochem., 151: 579–588.

Huan, L., Jin-Qiang, W. and Qing, L. 2020. Photosynthesis product allocation and yield in sweet potato with spraying exogenous hormones under drought stress. J. Plant Physiol., 253: 153265.

Huang, H., Liu, B., Liu, L. and Song, S. 2017. Jasmonate action in plant growth and development. J. Exp. Bot., 68: 1349–1359.

Hubrich, F., Müller, M. and Andexer, J.N. 2021. Chorismate- And isochorismate converting enzymes: versatile catalysts acting on an important metabolic node. Chem. Commun., 57: 2441–2463.

Huffaker, A., Pearce, G., Veyrat, N., Erb, M., Turlings, T.C.J., Sartor, R. et al. 2013. Plant elicitor peptides are conserved signals regulating direct and indirect antiherbivore defense. Proc. Natl. Acad. Sci. USA, 110: 5707–5712.

Hussain, I., Rasheed, R., Ashraf, M.A., Mohsin, M., Shah, S.M.A., Rashid, D.A. et al. 2020. Foliar applied acetylsalicylic acid induced growth and key-biochemical changes in chickpea (Cicer arietinum L.) under drought stress. Dose-Response 18: 1559325820956801.

Hussein, H.-A.A., Mekki, B.B., Abd El-Sadek, M.E. and El Lateef, E.E. 2019. Effect of L-Ornithine application on improving drought tolerance in sugar beet plants. Heliyon, 5: e02631.

Igarashi, K. and Kashiwagi, K. 2019. The functional role of polyamines in eukaryotic cells. Int. J. Biochem. Cell Biol., 107: 104–115.

Ilyas, M., Nisar, M., Khan, N., Hazrat, A., Khan, A.H., Hayat, K. et al. 2021. Drought tolerance strategies in plants: a mechanistic approach. J. Plant Growth Regul., 40: 926–944.

Iqbal, N., Khan, N.A., Ferrante, A., Trivellini, A., Francini, A. and Khan, M.I.R. 2017. Ethylene role in plant growth, development and senescence: interaction with other phytohormones. Front. Plant Sci., 8: 475.

Iqbal, S., Wang, X., Mubeen, I., Kamran, M., Kanwal, I., D\'\iaz, G.A. et al. 2022. Phytohormones Trigger Drought Tolerance in Crop Plants: Outlook and Future Perspectives. Front. Plant Sci., 3378.

Jha, U.C., Nayyar, H. and Siddique, K.H.M. 2021. Role of Phytohormones in Regulating Heat Stress Acclimation in Agricultural Crops. J. Plant Growth Regul.

Jones, B. and Ljung, K. 2011. Auxin and cytokinin regulate each other's levels via a metabolic feedback loop. Plant Signal. Behav., 6: 901–904.

Jordan, J.T., Oates, R.P., Subbiah, S., Payton, P.R., Singh, K.P., Shah, S.A. et al. 2020. Carbon nanotubes affect early growth, flowering time and phytohormones in tomato. Chemosphere, 256: 127042.

Kafash, Z.H., Khoramnejadian, S., Ghotbi-Ravandi, A.A. and Dehghan, S.F. 2022. Traffic noise induces oxidative stress and phytohormone imbalance in two urban plant species. Basic Appl. Ecol., 60: 1–12.

Kakimoto, T. 2001. Identification of plant cytokinin biosynthetic enzymes as dimethylallyl diphosphate: ATP/ADP isopentenyltransferases. Plant Cell Physiol., 42: 677–685.

Kami, C., Lorrain, S., Hornitschek, P. and Fankhauser, C. 2010. Light-regulated plant growth and development. Curr. Top. Dev. Biol., 91: 29–66.

Kanwar, M.K., Bajguz, A., Zhou, J. and Bhardwaj, R. 2017. Analysis of brassinosteroids in plants. J. Plant Growth Regul., 36: 1002–1030.

Kasahara, H. 2016. Current aspects of auxin biosynthesis in plants. Biosci. Biotechnol. Biochem. 80, 34–42.

Kasahara, H., Hanada, A., Kuzuyama, T., Takagi, M., Kamiya, Y. and Yamaguchi, S. 2002. Contribution of the mevalonate and methylerythritol phosphate pathways to the biosynthesis of gibberellins inArabidopsis. J. Biol. Chem., 277: 45188–45194.

Kasahara, H., Takei, K., Ueda, N., Hishiyama, S., Yamaya, T., Kamiya, Y. et al. 2004. Distinct Isoprenoid Origins of cis- and trans-Zeatin Biosyntheses in Arabidopsis. J. Biol. Chem., 279: 14049–14054.

Kaur, H., Ozga, J.A. and Reinecke, D.M. 2021. Balancing of hormonal biosynthesis and catabolism pathways, a strategy to ameliorate the negative effects of heat stress on reproductive growth. Plant. Cell Environ., 44: 1486–1503.

Kaur, S., Vian, A., Chandel, S., Singh, H.P., Batish, D.R. and Kohli, R.K. 2021. Sensitivity of plants to high frequency electromagnetic radiation: cellular mechanisms and morphological changes. Rev. Environ. Sci. Bio/Technology, 20: 55–74.

Kaur, T., Devi, R., Kour, D., Yadav, A. and Yadav, A.N. 2021. Plant growth promoting soil microbiomes and their potential implications for agricultural and environmental sustainability.

Keller, M. 2020. Phenology and growth cycle, The Science of Grapevines.

Kende, H. 1993. Ethylene biosynthesis. Annu. Rev. Plant Physiol. Plant Mol. Biol., 44: 283–307.

Khan, M.I.R., Chopra, P., Chhillar, H., Ahanger, M.A., Hussain, S.J. and Maheshwari, C. 2021. Regulatory hubs and strategies for improving heavy metal tolerance in plants: Chemical messengers, omics and genetic engineering. Plant Physiol. Biochem., 164: 260–278.

Khan, N., Bano, A., Ali, S. and Babar, M.A. 2020. Crosstalk amongst phytohormones from planta and PGPR under biotic and abiotic stresses. Plant Growth Regul.

Kim, S., Moon, J., Roh, J. and Kim, S.K. 2018. Castasterone Can be Biosynthesized from 28-homodolichosterone in Arabidopsis thaliana. J. Plant Biol., 61: 330–335.

Kombrink, E. 2012. Chemical and genetic exploration of jasmonate biosynthesis and signaling paths. Planta, 236: 1351–1366.

Koo, Y.M., Heo, A.Y. and Choi, H.W. 2020. Salicylic acid as a safe plant protector and growth regulator. Plant Pathol. J.

Kour, J., Kohli, S.K., Khanna, K., Bakshi, P., Sharma, P., Singh, A.D. et al. 2021. Brassinosteroid Signaling, Crosstalk and, Physiological Functions in Plants Under Heavy Metal Stress. Front. Plant Sci., 12.

Krall, L., Raschke, M., Zenk, M.H. and Baron, C. 2002. The Tzs protein from Agrobacterium tumefaciens C58 produces zeatin riboside 5′-phosphate from 4-hydroxy-3-methyl-2-(E)-butenyl diphosphate and AMP. FEBS Lett., 527: 315–318.

Kwon, M. and Choe, S. 2005. Brassinosteroid biosynthesis and dwarf mutants. J. Plant Biol., 48: 1–15.

Lakshmanan, V., Ray, P. and Craven, K.D. 2017. Toward a resilient, functional microbiome: drought tolerance-alleviating microbes for sustainable agriculture. pp. 69–84. In: Plant Stress Tolerance. Springer.

Landa, P. 2021. Positive effects of metallic nanoparticles on plants: Overview of involved mechanisms. Plant Physiol. Biochem., 161: 12–24.

Lefevere, H., Bauters, L. and Gheysen, G. 2020. Salicylic Acid Biosynthesis in Plants. Front. Plant Sci., 11: 1–7.

Letham, D.S. 1963. Zeatin, a factor inducing cell division isolated from Zea mays. Life Sci., 2: 569–573.

Levy, A., Conway, J.M., Dangl, J.L. and Woyke, T. 2018. Elucidating Bacterial Gene Functions in the Plant Microbiome. Cell Host Microbe.

Li, J., Brader, G., Kariola, T. and Palva, E.T. 2006. WRKY70 modulates the selection of signaling pathways in plant defense. Plant J., 46: 477–491.

Li, J., Brader, G. and Palva, E.T. 2004. The WRKY70 Transcription Factor: A Node of Convergence for Jasmonate-Mediated and Salicylate-Mediated Signals in Plant Defense. Plant Cell, 16: 319–331.

Li, Q., Xu, F., Chen, Z., Teng, Z., Sun, K., Li, X. et al. 2021. Synergistic interplay of ABA and BR signal in regulating plant growth and adaptation. Nat. Plants, 7: 1108–1118.

Li, S., Joo, Y., Cao, D., Li, R., Lee, G., Halitschke, R. et al. 2020. Strigolactone signaling regulates specialized metabolism in tobacco stems and interactions with stem-feeding herbivores. PLOS Biol., 18: e3000830.

Li, S.M., Zheng, H.X., Zhang, X.S. and Sui, N. 2021. Cytokinins as central regulators during plant growth and stress response. Plant Cell Rep., 40: 271–282.

Li, T., Huang, Y., Xu, Z.S., Wang, F. and Xiong, A.S. 2019. Salicylic acid-induced differential resistance to the Tomato yellow leaf curl virus among resistant and susceptible tomato cultivars. BMC Plant Biol., 19.

Lieberman, M., Kunishi, A., Mapson, L.W. and Wardale, D.A. 1966. Stimulation of ethylene production in apple tissue slices by methionine. Plant Physiol., 41: 376–382.

Lin, H., Wang, R., Qian, Q., Yan, M., Meng, X., Fu, Z. et al. 2009. DWARF27, an iron-containing protein required for the biosynthesis of strigolactones, regulates rice tiller bud outgrowth. Plant Cell, 21: 1512–1525.

Liu, X., Wan, B., Hua, H. and Li, X. 2021. Electric field generated by high-voltage transmission system is beneficial to cotton growth. Acta Ecol. Sin., 41: 552–559.

Ljung, K., Bhalerao, R.P. and Sandberg, G. 2001. Sites and homeostatic control of auxin biosynthesis in Arabidopsis during vegetative growth. Plant J., 28: 465–474.

López-Ráez, J.A., Charnikhova, T., Gómez-Roldán, V., Matusova, R., Kohlen, W., De Vos, R. et al. 2008. Tomato strigolactones are derived from carotenoids and their biosynthesis is promoted by phosphate starvation. New Phytol., 178: 863–874.

Lu, D., Liu, B., Ren, M., Wu, C., Ma, J. and Shen, Y. 2021. Light Deficiency Inhibits Growth by Affecting Photosynthesis Efficiency as well as JA and Ethylene Signaling in Endangered Plant Magnolia sinostellata. Plants, 10: 2261.

Ma, F., Yang, X., Shi, Z. and Miao, X. 2020. Novel crosstalk between ethylene- and jasmonic acid-pathway responses to a piercing–sucking insect in rice. New Phytol., 225: 474–487.

MacMillan, J. and Suter, P.J. 1958. The occurrence of gibberellin A1 in higher plants: isolation from the seed of runner bean (Phaseolus multiflorus). Naturwissenschaften, 45: 46.

MacMillan, J. and Takahashi, N. 1968. Proposed procedure for the allocation of trivial names to the gibberellins. Nature, 217: 170–171.

Manickavasagam, M., Pavan, G. and Vasudevan, V. 2019. A comprehensive study of the hormetic influence of biosynthesized AgNPs on regenerating rice calli of indica cv. IR64. Sci. Rep., 9: 1–11.

Manosalva, P., Manohar, M., Reuss, S.H. Von, Chen, S., Koch, A., Kaplan, F. et al. 2015. Conserved nematode signalling molecules elicit plant defenses and pathogen resistance. Nat. Commun., 6.

Marin, E., Nussaume, L., Quesada, A., Gonneau, M., Sotta, B., Hugueney, P. et al. 1996. Molecular identification of zeaxanthin epoxidase of Nicotiana plumbaginifolia, a gene involved in abscisic acid biosynthesis and corresponding to the ABA locus of Arabidopsis thaliana. EMBO J., 15: 2331–2342.

Mashiguchi, K., Seto, Y. and Yamaguchi, S. 2021. Strigolactone biosynthesis, transport and perception. Plant J., 105: 335–350.

Mashiguchi, K., Tanaka, K., Sakai, T., Sugawara, S., Kawaide, H., Natsume, M. et al. 2011. The main auxin biosynthesis pathway in *Arabidopsis*. Proc. Natl. Acad. Sci., 108: 18512–18517.

McAdam, E.L., Reid, J.B. and Foo, E. 2018. Gibberellins promote nodule organogenesis but inhibit the infection stages of nodulation. J. Exp. Bot., 69: 2117–2130.

Mildažienė, V., Aleknavičiūtė, V., Žūkienė, R., Paužaitė, G., Naučienė, Z., Filatova, I. et al. 2019. Treatment of common sunflower (Helianthus annus L.) seeds with radio-frequency electromagnetic field and cold plasma induces changes in seed phytohormone balance, seedling development and leaf protein expression. Sci. Rep., 9: 1–12.

Mishra, A.K. and Baek, K.H. 2021. Salicylic acid biosynthesis and metabolism: A divergent pathway for plants and bacteria. Biomolecules 11.

Miyazaki, S., Hara, M., Ito, S., Tanaka, K., Asami, T., Hayashi, K. et al. 2018. An ancestral gibberellin in a moss Physcomitrella patens. Mol. Plant, 11: 1097–1100.

Mohan, J.E., Wadgymar, S.M., Winkler, D.E., Anderson, J.T., Frankson, P.T., Hannifin, R. et al. 2019. Plant reproductive fitness and phenology responses to climate warming: Results from native populations, communities, and ecosystems, Ecosystem Consequences of Soil Warming: Microbes, Vegetation, Fauna and Soil Biogeochemistry. Elsevier Inc.

Moore, C.E., Meacham-Hensold, K., Lemonnier, P., Slattery, R.A., Benjamin, C., Bernacchi, C.J. et al. 2021. The effect of increasing temperature on crop photosynthesis: From enzymes to ecosystems. J. Exp. Bot., 72: 2822–2844.

Morffy, N. and Strader, L.C. 2020. Old Town Roads: routes of auxin biosynthesis across kingdoms. Curr. Opin. Plant Biol., 55: 21–27.

Morkunas, I., Woźniak, A., Mai, V.C., Rucińska-Sobkowiak, R., Jeandet, P. et al. 2018. The role of heavy metals in plant response to biotic stress. Molecules, 23: 2320.

Müller, M. and Munné-Bosch, S. 2021. Hormonal impact on photosynthesis and photoprotection in plants. Plant Physiol., 185: 1500–1522.

Nadeem, S.M., Ahmad, M., Zahir, Z.A. and Kharal, M.A. 2016. Role of phytohormones in stress tolerance of plants. Plant, Soil Microbes.

Nambara, E. and Marion-Poll, A. 2005. Abscisic acid biosynthesis and catabolism. Annu. Rev. Plant Biol., 56: 165–185.

Nazar, R., Umar, S., Khan, N.A. and Sareer, O. 2015. Salicylic acid supplementation improves photosynthesis and growth in mustard through changes in proline accumulation and ethylene formation under drought stress. South African J. Bot., 98: 84–94.

Noguchi, T., Fujioka, S., Choe, S., Takatsuto, S., Tax, F.E., Yoshida, S. et al. 2000. Biosynthetic pathways of brassinolide in Arabidopsis. Plant Physiol., 124: 201–210.

Normanly, J., Cohen, J.D. and Fink, G.R. 1993. Arabidopsis thaliana auxotrophs reveal a tryptophan-independent biosynthetic pathway for indole-3-acetic acid. Proc. Natl. Acad. Sci., 90: 10355–10359.

Padmanabhan, M.S., Kramer, S.R., Wang, X. and Culver, J.N. 2008. Tobacco Mosaic Virus Replicase-Auxin/Indole Acetic Acid Protein Interactions: Reprogramming the Auxin Response Pathway To Enhance Virus Infection. J. Virol., 82: 2477–2485.

Pandey, R., Vengavasi, K. and Hawkesford, M.J. 2021. Plant adaptation to nutrient stress. Plant Physiol. Reports.

Paponov, I.A., Paponov, M., Teale, W., Menges, M., Chakrabortee, S., Murray, J.A.H. et al. 2008. Comprehensive transcriptome analysis of auxin responses in Arabidopsis. Mol. Plant, 1: 321–337.

Paradinas, A., Ramade, L., Mulot-Greffeuille, C., Hamidi, R., Thomas, M. and Toillon, J. 2022. Phenological growth stages of 'Barcelona' hazelnut (Corylus avellana L.) described using an extended BBCH scale. Sci. Hortic. (Amsterdam). 296: 110902.

Pascale, A., Proietti, S., Pantelides, I.S. and Stringlis, I.A. 2020. Modulation of the Root Microbiome by Plant Molecules: The Basis for Targeted Disease Suppression and Plant Growth Promotion. Front. Plant Sci.

Paudel, D.B. and Sanfaçon, H., 2018. Exploring the diversity of mechanisms associated with plant tolerance to virus infection. Front. Plant Sci.

Pérez-Llorca, M., Muñoz, P., Müller, M. and Munné-Bosch, S. 2019. Biosynthesis, metabolism and function of auxin, salicylic acid and melatonin in climacteric and non-climacteric fruits. Front. Plant Sci., 10: 1–10.

Petrov, V.D. and Van Breusegem, F. 2012. Hydrogen peroxide-a central hub for information flow in plant cells. AoB Plants, 12: 1–13.

Podleśny, J., Podleśna, A., Gładyszewska, B. and Bojarszczuk, J. 2021. Effect of pre-sowing magnetic field treatment on enzymes and phytohormones in pea (Pisum sativum L.) seeds and seedlings. Agronomy 11: 494.

Pollmann, S., Neu, D. and Weiler, E.W. 2003. Molecular cloning and characterization of an amidase from Arabidopsis thaliana capable of converting indole-3-acetamide into the plant growth hormone, indole-3-acetic acid. Phytochemistry, 62: 293–300.

Qiao, Y., Ren, J., Yin, L., Liu, Y., Deng, X., Liu, P. and Wang, S. 2020. Exogenous melatonin alleviates PEG-induced short-term water deficiency in maize by increasing hydraulic conductance. BMC Plant Biol., 20: 1–14.

Qu, Q., Zhang, Z., Peijnenburg, W.J.G.M., Liu, W., Lu, T., Hu, B. et al. 2020. Rhizosphere Microbiome Assembly and Its Impact on Plant Growth. J. Agric. Food Chem.

Ranjan, A., Rajput, V.D., Minkina, T., Bauer, T., Chauhan, A. and Jindal, T. 2021. Nanoparticles induced stress and toxicity in plants. Environ. Nanotechnology, Monit. Manag., 15: 100457.

Rekhter, D., Lüdke, D., Ding, Y., Feussner, K., Zienkiewicz, K., Lipka, V. et al. 2019. Isochorismate-derived biosynthesis of the plant stress hormone salicylic acid. Science (80-.). 365: 498–502.

Rodriguez, P.A., Rothballer, M., Chowdhury, S.P., Nussbaumer, T., Gutjahr, C. and Falter-Braun, P. 2019. Systems Biology of Plant-Microbiome Interactions. Mol. Plant.

Roeber, V.M., Bajaj, I., Rohde, M., Schmülling, T. and Cortleven, A. 2021. Light acts as a stressor and influences abiotic and biotic stress responses in plants. Plant. Cell Environ., 44: 645–664.

Romera, F.J., Lucena, C., Garc\'\ia, M.J., Alcántara, E., Angulo, M., Aparicio, M.Á. et al. 2021. Plant hormones and nutrient deficiency responses. Horm. Plant Response.

Rook, F., Corke, F., Card, R., Munz, G., Smith, C. and Bevan, M.W. 2001. Impaired sucrose-induction mutants reveal the modulation of sugar-induced starch biosynthetic gene expression by abscisic acid signalling. Plant J., 26: 421–433.

Ruan, J., Zhou, Y., Zhou, M., Yan, J., Khurshid, M., Weng, W. et al. 2019. Jasmonic acid signaling pathway in plants. Int. J. Mol. Sci., 20.

Sakakibara, H. 2006. Cytokinins: Activity, biosynthesis, and translocation. Annu. Rev. Plant Biol., 57: 431–449.

Salazar-Cerezo, S., Martínez-Montiel, N., García-Sánchez, J., Pérez-y-Terrón, R. and Martínez-Contreras, R.D., 2018. Gibberellin biosynthesis and metabolism: A convergent route for plants, fungi and bacteria. Microbiol. Res., 208: 85–98.

Saletnik, B., Zaguła, G., Saletnik, A., Bajcar, M., Słysz, E. and Puchalski, C. 2022. Effect of Magnetic and Electrical Fields on Yield, Shelf Life and Quality of Fruits. Appl. Sci., 12: 3183.

Sasse, J.M. 1997. Recent progress in brassinosteroid research. Physiol. Plant., 100: 696–701.

Sayed Shourbalal, S.K., Soleymani, A. and Javanmard, H.R. 2019. Shortening vernalization in winter wheat (Triticum aestivum L.) using plant growth regulators and cold stratification. J. Clean. Prod., 219: 443–450.

Schäfer, M., Meza-Canales, I.D., Brütting, C., Baldwin, I.T. and Meldau, S. 2015. Cytokinin concentrations and CHASE-DOMAIN CONTAINING HIS KINASE 2 (NaCHK2)-and <scp>N</scp> a <scp>CHK</scp> 3-mediated perception modulate herbivory-induced defense signaling and defenses in *<scp>N</scp> icotiana attenuata* . New Phytol., 207: 645–658.

Schuman, M.C., Meldau, S., Gaquerel, E., Diezel, C., McGale, E., Greenfield, S. et al. 2018. The active jasmonate JA-Ile regulates a specific subset of plant jasmonate-mediated resistance to herbivores in nature. Front. Plant Sci., 9: 787.

Schwartz, S.H., Qin, X. and Zeevaart, J.A.D. 2003. Elucidation of the indirect pathway of abscisic acid biosynthesis by mutants, genes, and enzymes. Plant Physiol., 131: 1591–1601.

Schwartz, S.H., Tan, B.C., Gage, D.A., Zeevaart, J.A.D. and McCarty, D.R. 1997. Specific oxidative cleavage of carotenoids by VP14 of maize. Science (80-.). 276: 1872–1874.

Senavirathna, M. and Asaeda, T. 2018. Microwave radiation alters burn injury-evoked electric potential in Nicotiana benthamiana. Plant Signal. Behav., 13: e1486145.

Seo, M., Akaba, S., Oritani, T., Delarue, M., Bellini, C., Caboche, M. et al. 1998. Higher Activity of an Aldehyde Oxidase in the Auxin-Overproducing superroot1 Mutant of Arabidopsis thaliana1. Plant Physiol., 116: 687–693.

Seo, M., Peeters, A.J.M., Koiwai, H., Oritani, T., Marion-Poll, A., Zeevaart, J.A.D. et al. 2000. The Arabidopsis aldehyde oxidase 3 (AAO3) gene product catalyzes the final step in abscisic acid biosynthesis in leaves. Proc. Natl. Acad. Sci., 97: 12908–12913.

Shao, C., Shen, L., Qiu, C., Wang, Y., Qian, Y., Chen, J. et al. 2021. Characterizing the impact of high temperature during grain filling on phytohormone levels, enzyme activity and metabolic profiles of an early indica rice variety. Plant Biol., 23: 806–818.

Sherwin, P.F. and Coates, R.M. 1982. Stereospecificity of the oxidation of ent-kauren-19-ol to ent-kaurenal by a microsomal enzyme preparation from Marah macrocarpus. J. Chem. Soc. Chem. Commun., 1013–1014.

Simm, S., Papasotiriou, D., Ibrahim, M., Leisegang, M., Müller, B., Schorge, T. et al. 2013. Defining the Core Proteome of the Chloroplast Envelope Membranes. Front. Plant Sci., 4.

Singh, H., Bhat, J.A., Singh, V.P., Corpas, F.J. and Yadav, S.R. 2021. Auxin metabolic network regulates the plant response to metalloids stress. J. Hazard. Mater., 405: 124250.

Smith, S.M. 2014. Q&A: What are strigolactones and why are they important to plants and soil microbes? BMC Biol., 12: 1–7.

Sosa-Zuniga, V., Brito, V., Fuentes, F. and Steinfort, U. 2017. Phenological growth stages of quinoa (Chenopodium quinoa) based on the BBCH scale. Ann. Appl. Biol., 171: 117–124.

Souza, A.C., Olivares, F.L., Peres, L.E.P., Piccolo, A. and Canellas, L.P. 2022. Plant hormone crosstalk mediated by humic acids. Chem. Biol. Technol. Agric., 9: 1–25.

Spence, C. and Bais, H. 2015. Role of plant growth regulators as chemical signals in plant-microbe interactions: A double edged sword. Curr. Opin. Plant Biol.

Staswick, P.E. and Tiryaki, I. 2004. The oxylipin signal jasmonic acid is activated by an enzyme that conjugate it to isoleucine in Arabidopsis W inside box sign. Plant Cell, 16: 2117–2127.

Stepanova, A.N., Robertson-Hoyt, J., Yun, J., Benavente, L.M., Xie, D.-Y., Doležal, K. et al. 2008. TAA1-Mediated Auxin Biosynthesis Is Essential for Hormone Crosstalk and Plant Development. Cell, 133: 177–191.

Stepanova, A.N., Yun, J., Robles, L.M., Novak, O., He, W., Guo, H. et al. 2011. The Arabidopsis YUCCA1 Flavin Monooxygenase Functions in the Indole-3-Pyruvic Acid Branch of Auxin Biosynthesis. Plant Cell, 23: 3961–3973.

Strnad, M. 1997. The aromatic cytokinins. Physiol. Plant., 101: 674–688.

Sun, T.P. and Kamiya, Y. 1994. The Arabidopsis GA1 locus encodes the cyclase ent-kaurene synthetase A of gibberellin biosynthesis. Plant Cell, 6: 1509–1518.

Sytar, O., Kumari, P., Yadav, S., Brestic, M. and Rastogi, A. 2019. Phytohormone priming: regulator for heavy metal stress in plants. J. Plant Growth Regul., 38: 739–752.

Takei, K., Sakakibara, H. and Sugiyama, T. 2001. Identification of genes encoding adenylate isopentenyltransferase, a cytokinin biosynthesis enzyme, in Arabidopsis thaliana. J. Biol. Chem., 276: 26405–26410.

Tanaka, J., Yano, K., Aya, K., Hirano, K., Takehara, S., Koketsu, E. et al. 2014. Antheridiogen determines sex in ferns via a spatiotemporally split gibberellin synthesis pathway. Science (80-.). 346: 469–473.

Tang, J., Han, Z. and Chai, J. 2016. Q&A: what are brassinosteroids and how do they act in plants? BMC Biol., 14: 1–5.

Tankari, M., Wang, C., Ma, H., Li, X., Li, L., Soothar, R.K. et al. 2021. Drought priming improved water status, photosynthesis and water productivity of cowpea during post-anthesis drought stress. Agric. Water Manag., 245: 106565.

Tariq, A. and Ahmed, A. 2022. Auxins-Interkingdom Signaling Molecules. Plant Horm. - Recent Adv. New Perspect. Appl. [Working Title].

Theodoulou, F.L., Job, K., Slocombe, S.P., Footitt, S., Holdsworth, M., Baker, A. et al. 2005. Jasmonic acid levels are reduced in COMATOSE ATP-binding cassette transporter mutants. Implications for transport of jasmonate precursors into peroxisomes. Plant Physiol., 137: 835–840.

Torrens-Spence, M.P., Bobokalonova, A., Carballo, V., Glinkerman, C.M., Pluskal, T., Shen, A. et al. 2019. PBS3 and EPS1 Complete Salicylic Acid Biosynthesis from Isochorismate in Arabidopsis. Mol. Plant, 12: 1577–1586.

Trivedi, D.K., Gill, S.S. and Tuteja, N. 2016. Abscisic Acid (ABA): Biosynthesis, Regulation, and Role in Abiotic Stress Tolerance. Abiotic Stress Response Plants, 315–326.

Trivedi, P., Batista, B.D., Bazany, K.E. and Singh, B.K. 2022. Plant–microbiome interactions under a changing world: responses, consequences and perspectives. New Phytol.

Tudzynski, B., Hedden, P., Carrera, E. and Gaskin, P. 2001. The P450-4 gene of Gibberella fujikuroi encodes ent-kaurene oxidase in the gibberellin biosynthesis pathway. Appl. Environ. Microbiol., 67: 3514–3522.

Tuteja, N., Gill, S.S., Trivedi, P.K., Asif, M.H. and Nath, P. 2010. Plant growth regulators and their role in stress tolerance. Plant Nutr. Abiotic Stress Toler. I Plant Stress, 4: 1–18.

Tzin, V. and Galili, G. 2010. New Insights into the shikimate and aromatic amino acids biosynthesis pathways in plants. Mol. Plant, 3: 956–972.

Umehara, M., Hanada, A., Yoshida, S., Akiyama, K., Arite, T., Takeda-Kamiya, N. et al. 2008. Inhibition of shoot branching by new terpenoid plant hormones. Nature, 455: 195–200.

Upreti, K.K. and Sharma, M. 2016. Role of plant growth regulators in abiotic stress tolerance. Abiotic Stress Physiol. Hortic. Crop.

Uroz, S., Courty, P.E. and Oger, P. 2019. Plant Symbionts Are Engineers of the Plant-Associated Microbiome. Trends Plant Sci.

Vázquez-Hernández, M.C., Parola-Contreras, I., Montoya-Gómez, L.M., Torres-Pacheco, I., Schwarz, D. and Guevara-González, R.G. 2019. Eustressors: Chemical and physical stress factors used to enhance vegetables production. Sci. Hortic. (Amsterdam)., 250: 223–229.

Vera-Reyes, I., Vázquez-Núñez, E., Lira-Saldivar, R.H. and Méndez-Argüello, B. 2018. Effects of nanoparticles on germination, growth, and plant crop development. Agric. Nanobiotechnology.

Verma, V., Ravindran, P. and Kumar, P.P. 2016. Plant hormone-mediated regulation of stress responses. BMC Plant Biol., 16.

Veroneze-Júnior, V., Martins, M., Mc Leod, L., Souza, K.R.D., Santos-Filho, P.R., Magalhães, P.C. et al. 2019. Leaf application of chitosan and physiological evaluation of maize hybrids contrasting for drought tolerance under water restriction. Brazilian J. Biol., 80: 631–640.

Vlot, A.C., Dempsey, D.A. and Klessig, D.F. 2009. Salicylic Acid, a Multifaceted Hormone to Combat Disease. Annu. Rev. Phytopathol., 47: 177–206.

Vlot, A.C., Sales, J.H., Lenk, M., Bauer, K., Brambilla, A., Sommer, A. et al. 2021. Systemic propagation of immunity in plants. New Phytol.

Volkov, A.G. and Ranatunga, D.R.A. 2006. Plants as environmental biosensors. Plant Signal. Behav., 1: 105–115.

Vurukonda, S.S.K.P., Vardharajula, S., Shrivastava, M. and SkZ, A. 2016. Enhancement of drought stress tolerance in crops by plant growth promoting rhizobacteria. Microbiol. Res., 184: 13–24.

Wang, B., Chu, J., Yu, T., Xu, Q., Sun, X., Yuan, J. et al. 2015. Tryptophan-independent auxin biosynthesis contributes to early embryogenesis in Arabidopsis. Proc. Natl. Acad. Sci. USA, 112: 4821–4826.

Wang, W., Chen, Q., Botella, J.R. and Guo, S. 2019. Beyond light: insights into the role of constitutively photomorphogenic1 in plant hormonal signaling. Front. Plant Sci., 10: 557.

Waqas, M.A., Khan, I., Akhter, M.J., Noor, M.A. and Ashraf, U. 2017. Exogenous Application of Plant Growth Regulators (PGRs) Induces Chilling Tolerance in Short-Duration Hybrid Maize, 11459–11471.

Wasternack, C. and Hause, B. 2013. Jasmonates: Biosynthesis, perception, signal transduction and action in plant stress response, growth and development. An update to the 2007 review in Annals of Botany. Ann. Bot.

Wasternack, C. and Hause, B. 2019. The missing link in jasmonic acid biosynthesis. Nat. Plants, 5: 776–777.

Wiemann, P., Sieber, C.M.K., Von Bargen, K.W., Studt, L., Niehaus, E.-M., Espino, J.J. et al. 2013. Deciphering the cryptic genome: genome-wide analyses of the rice pathogen Fusarium fujikuroi reveal complex regulation of secondary metabolism and novel metabolites. PLoS Pathog., 9: e1003475–e1003475.

Wright, A.D., Sampson, M.B., Neuffer, M.G., Michalczuk, L., Slovin, J.P. and Cohen, J.D. 1991. Indole-3-acetic acid biosynthesis in the mutant maize orange pericarp, a tryptophan auxotroph. Science (80-.). 254: 998–1000.

Xing, J., Wang, Y., Yao, Q., Zhang, Y., Zhang, M. and Li, Z. 2022. Brassinosteroids modulate nitrogen physiological response and promote nitrogen uptake in maize (Zea mays L.). Crop J., 10: 166–176.

Xiong, L. and Zhu, J.K. 2003. Regulation of abscisic acid biosynthesis. Plant Physiol., 133: 29–36.

Xu, J., Wang, X., Zu, H., Zeng, X., Baldwin, I.T., Lou, Y. et al. 2021. Molecular dissection of rice phytohormone signaling involved in resistance to a piercing-sucking herbivore. New Phytol., 230: 1639–1652.

Xu, J. and Zhang, S. 2015. Ethylene Biosynthesis and Regulation in Plants. In: Ethylene in Plants. Springer Netherlands, Dordrecht, pp. 1–25.

Yadav, A.N. 2017. Plant Microbiomes and Its Beneficial Multifunctional Plant Growth Promoting Attributes. Int. J. Environ. Sci. Nat. Resour., 3.

Yadav, A.N., Kumar, V., Dhaliwal, H.S., Prasad, R. and Saxena, A.K. 2018. Microbiome in Crops: Diversity, Distribution, and Potential Role in Crop Improvement. pp. 305–332. In: New and Future Developments in Microbial Biotechnology and Bioengineering: Crop Improvement through Microbial. Biotechnology. Elsevier.

Yadav, V., Kushwaha, S., Yadav, V.K., Tripathi, S.K., Srivastava, S.K. and Pratap, M. 2021. Influence of growth regulators on growth, phenological traits and yield attributes of wheat [Triticum aestivum L.] under restricted irrigation condition. Int. J. Chem. Stud., 9: 727–731.

Yamaguchi, S. 2008. Gibberellin metabolism and its regulation. Annu. Rev. Plant Biol., 59: 225–251.

Yamasaki, K., Motomura, Y., Yagi, Y., Nomura, H., Kikuchi, S., Nakai, M. et al. 2013. Chloroplast envelope localization of EDS5, an essential factor for salicylic acid biosynthesis in Arabidopsis thaliana. Plant Signal. \& Behav., 8: e23603–e23603.

Yang, S.F. and Hoffman, N.E. 1984. Ethylene Biosynthesis and its Regulation in Higher Plants. Annu. Rev. Plant Physiol., 35: 155–189.

Yokoo, S., Inoue, S., Suzuki, N., Amakawa, N., Matsui, H., Nakagami, H. et al. 2018. Comparative analysis of plant isochorismate synthases reveals structural mechanisms underlying their distinct biochemical properties. Biosci. Rep., 38: 1–13.

Yokota, T., Ohnishi, T., Shibata, K., Asahina, M., Nomura, T., Fujita, T. et al. 2017. Occurrence of brassinosteroids in non-flowering land plants, liverwort, moss, lycophyte and fern. Phytochemistry, 136: 46–55.

Yoneyama, K., Mori, N., Sato, T., Yoda, A., Xie, X., Okamoto, M. et al. 2018. Conversion of carlactone to carlactonoic acid is a conserved function of MAX 1 homologs in strigolactone biosynthesis. New Phytol., 218: 1522–1533.

Yoshii, H. and Imaseki, H. 1982. Regulation of auxin-induced ethylene biosynthesis. Repression of inductive formation of 1-aminocyclopropane-1-carboxylate synthase by ethylene. Plant Cell Physiol., 23: 639–649.

Zahid, A., Yike, G., Kubik, S., Fozia, Ramzan, M., Sardar, H., Akram, M.T. et al. 2021. Plant growth regulators modulate the growth, physiology, and flower quality in rose (Rosa hybirda). J. King Saud. Univ. - Sci., 33: 101526.

Zahid, G., Iftikhar, S., Shimira, F., Ahmad, H.M. and Kaçar, Y.A. 2023. An overview and recent progress of plant growth regulators (PGRs) in the mitigation of abiotic stresses in fruits: A review. Sci. Hortic. (Amsterdam)., 309: 111621.

Zhang, C., Xie, Z., shang, J., Liu, J., Dong, T., Tang, M. et al. 2022. Detecting winter canola (Brassica napus) phenological stages using an improved shape-model method based on time-series UAV spectral data. Crop J.

Zhang, H., Sun, X. and Dai, M. 2022. Improving crop drought resistance with plant growth regulators and rhizobacteria: Mechanisms, applications, and perspectives. Plant Commun.

Zhang, H., Zhang, L., Wu, S., Chen, Y., Yu, D. and Chen, L. 2021. AtWRKY75 positively regulates age-triggered leaf senescence through gibberellin pathway. Plant Divers., 43: 331–340.

Zhao, Y. 2010. Auxin biosynthesis and its role in plant development. Annu. Rev. Plant Biol., 61: 49–64.

Zhao, Y. 2012. Auxin biosynthesis: A simple two-step pathway converts tryptophan to indole-3-Acetic acid in plants. Mol. Plant, 5: 334–338.

Zhu, Z., An, F., Feng, Y., Li, P., Xue, L., A, M., Jiang, Z. et al. 2011. Derepression of ethylene-stabilized transcription factors (EIN3/EIL1) mediates jasmonate and ethylene signaling synergy in Arabidopsis. Proc. Natl. Acad. Sci. USA, 108: 12539–12544.

Zullo, M.A.T. and Bajguz, A. 2019. The Brassinosteroids Family—Structural Diversity of Natural Compounds and Their Precursors. pp. 1–44. In: Hayat, S., Yusuf, M., Bhardwaj, R. and Bajguz, A. (Eds.). Brassinosteroids: Plant Growth and Development. Springer Singapore, Singapore.

2

Plant Growth Regulators and Senescence

Nihal Gören Sağlam,[1,]* *Orkun Yaycılı*[1] *and Celal Doruk Arıcı*[2]

Introduction

Senescence is the final phase in plant growth resulting in death from the cellular level to the whole organism that occurs with age (Gregersen et al. 2013, Gully et al. 2015, Schippers et al. 2015, Wang et al. 2023). It is an important transportation and recycling procedure for nutrients such as sulfur, nitrogen, potassium and phosphorus. They are carried from senescent leaves to young growing tissues, thereby promoting the plant growth (Gregersen et al. 2013, Gully et al. 2015, Woo et al. 2019, Zhang et al. 2021, Zhu et al. 2022, Mandal et al. 2023). Researchers have reported that senescence occurs genetically in plants that grow under optimal growth conditions, as well as due to unfavourable environmental conditions (i.e., drought, temperature, nitrogen deficiency, insufficient light, disease and pathogen attacks) (He and Gan 2002, Christiansen and Gregersen 2014, Bresson et al. 2018, Miryeganeh 2022, Narayanan and Ma 2023). Senescence is generally controlled via developmental age, although it is also affected by various environmental stimulators such as water status, nutritional signals, temperature, and light change (Lim et al. 2007, Guo et al. 2021).

Plant growth regulators are internal factors that play an important role in the effects of developmental and environmental factors to regulate senescence as well as affecting all physiological processes in plants (Asif et al. 2022, Zahid et al. 2023). Plant hormones are divided into two groups: promoters and suppressors. Ethylene, salicylic acid (SA), abscisic acid (ABA), and jasmonic acid (JA) act as senescence promoters, while cytokinins and gibberellins act as suppressors (Li et al. 2023, Chen et al. 2023). Indole-3-acetic acid (IAA) is the most extensively studied, abundant

---
[1] Istanbul University, Faculty of Science, Department of Biology, Istanbul, Turkey.
[2] Istanbul University, Institute of Science, Istanbul, Turkey.
* Corresponding author: gorenn@istanbul.edu.tr

and physiologically effective auxin derivative that modulates plant growth and development (Sampedro-Guerrero et al. 2023). Researchers showed that exogenous IAA application might retard or accelerate senescence in leaves and/or cotyledons (Goren and Çağ 2007). The link between hormone signaling pathways during senescence is not yet fully understood. Many stress-related pathways, including stress-related hormones (such as ethylene, ABA, SA, and JA) are induced as a protection mechanism during developmental senescence, and thus disruption of this balance by treatment with auxin may result in the promotion of senescence (Zhu et al. 2015, Guo et al. 2021). Polyamines (PAs) are low molecular weight amines found in every living organism. Spermin (Spm) and Spermidine (Spd) are the most common and effective polyamines in plants (Gonzalez et al. 2021). Exogenous applications of Spm and Spd have been shown to delay leaf senescence (Gören-Sağlam and Çekiç 2019). Brassinosteroids are plant hormones with multiple effects as they influence various developmental processes in plants. They are also effective in the resistance of plants to abiotic stresses (Baghel et al. 2019, Sharma et al. 2022).

The effect of plant growth regulators on senescence may differ according to the plant, organ, tissue, and cell, as well as with the applied concentrations. This chapter focuses on interactions between plant growth regulators and plant senescence.

What is Plant Senescence?

Senescence is a programmed developmental stage that causes the death of a cell, tissue, organ, or whole plant (Lim and Nam 2007). It is a process in which not only does a part of the plant die, but various macromolecules are also recycled to developing parts of the plant (Sarwat et al. 2013, Shippers et al. 2015). Senescence is vital in the development of a plant in terms of transporting and recycling important nutrients such as phosphorus, nitrogen, potassium, and sulphur. These nutrients are transported from senescent leaves to actively growing tissues, thus promoting plant growth and reproduction (Balazadeh et al. 2008, Zhang and Zhou 2013). Some researchers have reported that senescence occurs not only as a result of unfavorable environmental conditions (drought, temperature, deficiency of nitrogen, light/dark conditions, disease, and pathogen attacks) but also genetically in healthy plants grown under optimal growing conditions (He and Gan 2002, Woo et al. 2019).

Aging is a process that includes catabolic reactions and ends in death, and was associated with senescence for a long time because of these features (Takahashi et al. 2000). It has been determined that aging is actually a passive and long process that occurs without a certain genetic program and includes senescence of some tissues (Noodén et al. 1997, Ok Lim et al. 2003). Conversely, it is known that senescence occurs due to a genetic program and can be prevented by external factors. However, it is not possible to prevent aging by external factors (Munne-Bosch 2007). Nowadays, senescence and aging are examined as different physiological events.

Programmed cell death (PCD) is a form of cellular suicide that serves to selectively eliminate damaged or harmful cells that are no longer needed (Locato and De Gara 2018, Romero-Bueno et al. 2019). In plants as in animals, programmed cell death or apoptosis is the final stage of genetically controlled cell differentiation. In plants, PCD can occur during the normal development process of a plant from embryogenesis to death, or it can occur when the plant is exposed to various biotic and abiotic stress factors (Conway and McCabe 2018, Romero-Bueno et al. 2019). In some cases, the cells acquire special functions when they die (such as vascular tissues, fibers) or, on the contrary, the cells die after completing their task. This type of programmed cell death is called developmental cell death and is controlled by an internal program. Stress-induced programmed cell death, on the other hand, occurs as a result of different environmental signals or pathogen attacks, including biotic and abiotic stimuli, and they change the original cell program (Wu et al. 2012).

Senescence is often associated with apoptosis. Programmed cell death in plants occurs in response to specialized conditions involving the hypersensitivity response to pathogens and the development of tracheal elements. Senescence may differ from programmed cell death as it can be delayed and reversible (Del Duca et al. 2014). There are two successive phases in senescence. While the first stage is reversible and the cells continue to live, the second stage is cell death (Sachdev et al. 2023). For example, cytokinin application can promote re-greening in a yellowed leaf and re-regulate protein synthesis and photosynthetic activity (Popov et al. 2022). Programmed cell death is an irreversible event. The target of senescence is not the death of the leaf, but after the end of senescence, death occurs (Jalil et al. 2022). This indicates that programmed cell death is a non-essential part of the senescence process.

Leopold (1961) classically categorized the types of senescence as follows:

- Seasonal leaf senescence, which usually occurs on trees that shed their leaves in autumn.
- Monocarpic senescence, which occurs in once-blooming (monocarpic). plants and causes the death of the entire plant except the seed.
- Stem senescence, which is usually seen in herbaceous plants and causes only the death of the stem.
- Sequential leaf senescence causing sequential death of leaves starting with the oldest.

It has been stated that sequential senescence occurs as a result of competition between young leaves and mature leaves, while seasonal senescence occurs as a result of decreasing temperature and sunlight (Smart 1994, Zentgraf 2022). Seasonal senescence plays a fundamental role in the life of a plant. In winter, leaves cannot maximize their production, so they become a burden to the plant and fall off. It transports plant nutrients to other parts of the plant. Leaf senescence, as is clearly known, becomes visible due to chlorophyll destruction and consequent yellowing of the leaf. In addition, it is generally believed that senescence is initiated before visible symptoms appear (Smart 1994).

Leaf Senescence

Leaf senescence is an extremely well-organized, complex process that includes more than leaf death, and constitutes the last stage of leaf development. Senescence causes the destruction of macromolecules and the transport of the formed products to developing organs of the plant in a coordinated manner (Galan et al. 2023). These events result in leaf yellowing and death. Leaf senescence is one of the important stages in plant leaf development and includes the expression of senescence-related genes (SAG= Senescence-Associated Genes). During senescence, leaf cells undergo regular changes in gene expression, metabolism, and structure (Lim et al. 2007).

The process of transporting nitrogen, carbon, and minerals from the mature leaf to other parts of the plant consists of a series of well-organized processes. Figure 1 shows some of the senescence-related processes, including the arrest of photosynthesis, disruption of chloroplast structure, loss of chlorophyll, degradation of leaf proteins, and transport of amino acids (Buchanan-Wollaston 1997).

Studies have shown that leaf cells undergo highly regulated changes in their metabolism, structure and gene expression during senescence. The most important and first change in cell structure is chloroplast disruption (Zhao et al. 2023, Li et al. 2023). Accordingly, the first symptom is a decrease in the rate of photosynthesis (Ghosh et al. 2001). Chloroplasts are rich in protein, Rubisco, protein-binding chlorophyll a/b, and membrane lipids. Metabolically, carbon assimilation (photosynthesis) is replaced by the breakdown of macromolecules such as chlorophyll and protein, membrane lipids and RNA (Pic et al. 2002, Dubey 2018). The decrease in photosynthetic activity during leaf senescence is characterized by the activity and amount of Rubisco, the rate of CO_2 exchange, and the level of mRNA expression of photosynthesis-related genes (Jiang et al. 1993, Humbeck et al. 1996). When photosynthesis decreases during senescence, chloroplasts produce less ATP. But energy is required for catabolic processes and increased respiration in leaves has been observed during senescence (Launay et al. 2019). This increase is very important for the supply of ATP required for catabolic processes (Launay et al. 2019). During leaf senescence, some changes occur at the cellular level as well. These cellular changes are summarized in the Table 1 (Sillanpaa 2003).

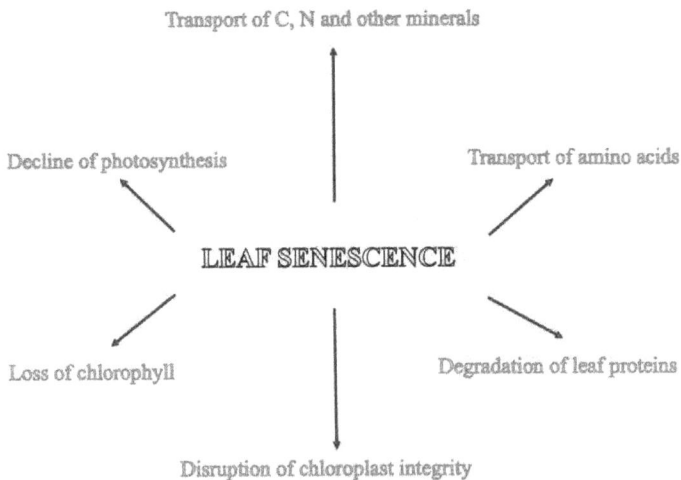

Fig. 1. Schematic representation of events during leaf senescence.

Table 1. Changes in organelles during leaf senescence.

Organelle	Changes
Nucleus	• Condensation of chromatin • Breakage of DNA
Chloroplasts	• Degradation of DNA and RNA • Destruction of thylakoid membranes • Degradation of pigments • Degradation of proteins • Increase in the number and size of plastoglobuli
Mitochondria	• Increase in the mithocondria • Increase of the respiration • Increase in the production of ROS
Peroxisome	• Increase in the number of the peroxisomes • Conversion of some peroxisomes to glyoxisomes • Increase in the ROS production • Lipid peroxidation • Degradation of proteins

Plant Growth Regulators

The first information about the existence of plant growth regulators dates back to the beginning of the twentieth century. With the studies carried out after this period, the roles of growth regulators in plant growth and development have been revealed and important natural and synthetic substances for fruits/vegetables have been discovered (Halloran and Kasım 2002). The importance of plant growth regulators (PGR) was first understood in the 1930s, and since then, their function in agricultural products has been investigated. The most thoroughly studied and best-known plant growth regulators are auxins, cytokinins, gibberellins, abscisic acid, and ethylene.

In recent years, detailed studies have been carried out on other plant growth agents such as salicylic acid, jasmonic acid, brassinosteroids, polyamines, karrikines, and strigolactones. Studies on plant physiology have revealed the roles of PGR in plant growth and development, and over time, it has been understood that not only substances that promote growth, but also substances that inhibit growth are synthesized in the plant (Gaba 2005, Kumar et al. 2016). In general, plant growth regulators are divided into two groups as plant growth promoters and inhibitors. Their roles in plants are summarazied in Table 2.

Plant Growth Regulators and Senescence

The onset of senescence is controlled by internal and external factors. Some environmental and biological stresses can promote senescence, such as heat, drought, nutrient deficiency, insufficient light, shade, darkness, and pathogen infection. Internal factors affecting senes` cence include age, plant hormones, and many developmental processes (Gan 2003, Yu et al. 2022, Sasi et al. 2022). Many factors are involved in the initiation, control, and regulation of developmental and stress-induced senescence. These include internal processes controlled by the plant, initiated by age and reproductive development, and external factors created

Table 2. A summary of the effects of plant growth regulators in plants (Westfall et al. 2013).

Plant Growth Regulators	Effects
Auxin	Lateral roots; cell elongation; apical dominance; branching; tropisms
Cytokinin	Releases lateral buds from apical dominance; cell division; delays senescence; root growth
Gibberellin	Seed germination; stem elongation; flowering; root growth; fruit growth; floral development; stresses
Abscisic Acid	Root/shoot growth; stomatal closure; germination; seed maturation; storage; drought tolerance; leaf senescence
Ethylene	Stress response; seed germination; flowering/fruit ripening; leaf senescence
Salicylic Acid	Biotrophic pathogen responses; Systemic acquired resistance to pathogens
Jasmonic Acid	Plant defense; root growth inhibition; plant immune system
Polyamines	Stem elongation and flowering; somatic embryogenesis; tuber development; fruit ripening; root growth; abiotic stresses, leaf senescence
Brassinosteroids	Cell division and elongation; reproductive development; stress responses; photomorphogenesis; leaf senescence;
Strigolactones	Leaf senescence; plant-microbe interactions; root development
Karrikins	Germination; inhibition of hypocotic elongation; inhibition of cotyledon expansion

by environmental factors such as injury, shading, sunlight, and temperature. A better understanding of senescence can be achieved by knowing more about the physical and chemical processes related to senescence. Some of the factors associated with senescence are known as signaling compounds and these include ethylene, ROS, salicylic acid, and jasmonic acid (Jalil et al. 2022, Sasi et al. 2022, Zhang et al. 2023).

Plant growth regulators (PGRs) are very important endogenous factors in the regulation of senescence. PGRs play a role in the promotion and inhibition of senescence (Fig. 2). At the same time, the endogenous levels of these substances affect senescence, and their exogenous application can promote or inhibit senescence.

Ethylene

Ethylene is a gaseous plant hormone involved in many aspects of plant growth and development, including senescence. Ethylene has been shown to have a role in abscission, seedling development, fruit ripening, and flower and leaf senescence (Huang et al. 2022). Ethylene has been reported to stimulate many of the physiological processes associated with senescence (Wang et al. 2001, Jing et al. 2005). Ethylene production in plant systems is controlled by two enzymes, ACC oxidase and ACC synthase. The conversion of S-adenosyl-L-methionine to 1-aminocyclopropane-1-carboxylic acid (ACC) and then to ethylene is catalyzed by ACC synthase and ACC oxidase, respectively (Tsuchisaka et al. 2009). Leaf senescence is delayed in leaves of *etr1*, the ethylene-insensitive mutant of *Arabidopsis*, and in mutants lacking signal transport (Grbic and Bleecker 1995, Oh et al. 1997). Inhibition of ACC oxidase

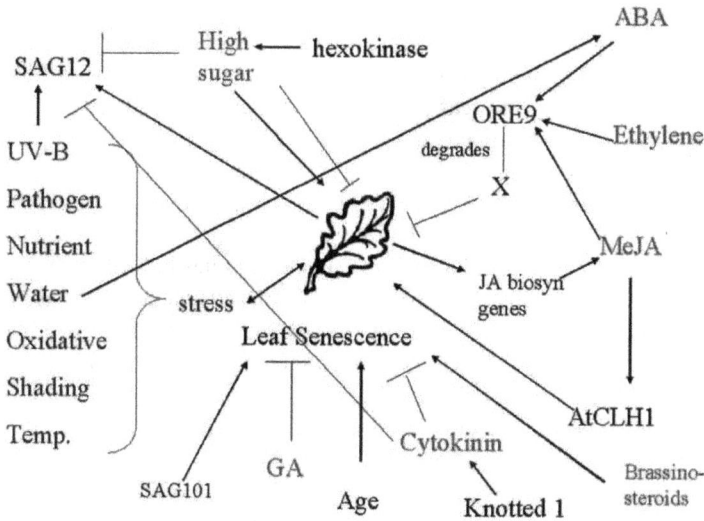

Fig. 2. Some pathways involved in leaf senescence.

expression and hence, ethylene synthesis via antisense technology resulted in delayed leaf senescence in tomatoes (John et al. 1995).

In transgenic clove plants containing the antisense ACC oxidase gene, its flowers showed low climacteric ethylene production and delayed petal senescence (Savin et al. 1995). Applying ethylene to leaves from the outside causes epinasticity. Ethylene also produces senescence-related changes in leaves and flower petals. The role of ethylene in these events has been demonstrated using ethylene inhibitors such as silver thiosulfate or 1-methylcyclopropene. Senescence was inhibited in cut flowers and leaves using these inhibitors (Ha and In 2022). In addition to all these, the leaves of transgenic *Arabidopsis* and tomato plants that produce excess ethylene do not show early senescence, suggesting that ethylene alone is not sufficient to initiate leaf senescence. Age-related factors are hypothesized to be necessary for leaf senescence regulated by ethylene (Gan 2003).

Antisense and mutant studies focused on delayed leaf senescence have shown that ethylene changes senescence time (Buchanan-Wollaston 1997). Increasing ethylene level programs the rate of senescence and the onset of senescence in leaves. It appears that ethylene does not directly activate senescence-related genes, but alters the activation of genes, possibly through other signaling pathways. Ethylene may have a role in suppressing the expression of genes involved in photosynthesis (Grbic and Bleecker 1995).

Jasmonic Acid

Jasmonic Acid (JA), first shown to induce senescence in cut oat leaves, is a class of oxylipin phytohormones (Li et al. 2021). In recent years, it has been reported that these phytohormones play a common role in different stages of plant development (Gan 2003). The ability of jasmonates to cause chlorosis has led to the idea that these structures play a role in plant senescence (Creelman and Mullet 1997). The application of JA and Me-Jasmonate (MeJA) promotes changes in plant gene expression. JA can affect several aspects of plant growth and development. Application of this hormone can promote senescence and leaf abscission and inhibit germination. In plant cell culture and cut leaves, JA induces senescence at concentrations greater than 50 μM. The senescence response includes the loss of chlorophyll, loss of chloroplast proteins such as ribulose biphosphate carboxylase, and accumulation of new proteins. JA specifically inhibits the conversion of ribulose biphosphate carboxylase (Creelman and Mullet 1997, Stoynova-Bakalova et al. 2008). Studies indicate that the external application of JA causes early senescence in cut and uncut leaves of wild type *Arabidopsis*. However, the JA-insensitive coil mutant also fails to induce early senescence, suggesting that the JA signaling pathway needs JA to induce senescence (He et al. 2002, Castillo and Leon 2008).

During leaf senescence, the expression of genes involved in JA biosynthesis gradually increased, and the concentration of JA in senescent leaves was four times higher than in non-senescence leaves (He et al. 2002). Leaf yellowing induced by jasmonic acid is indeed an active senescence process. The findings show that many genes whose expression is increased during leaf senescence, called SAG (senescence-associated genes), are also expressed in yellowing leaves (Gan 2003). In addition, many *Arabidopsis* mutants lacking JA production and signal transport do not show a delayed leaf senescence phenotype (He et al. 2002), rejecting the idea that JA plays a role in senescence. Possibly, other factors may promote leaf senescence in the absence of JA (Gan 2003).

Salicylic Acid

Salicylic acid (SA) is a phenolic substance produced by almost all plant species. Its concentration varies from species to species and from tissue to tissue. The leaves and bark of the willow tree contain very high amounts of acetylsalicylic acid (aspirin), a derivative of SA, used by Native Americans and ancient Greeks to treat pain and fever. In plants, SA is a signaling molecule involved in plant defense against pathogens and also plays a role in senescence. SA application promotes early leaf senescence, although a tablet aspirin dissolved in water has been observed to prolong the life of cut flowers. The intrinsic SA concentration is four times higher in senescent leaves than in non-senescent leaves (Morris et al. 2000, Zhang et al. 2017). As a result of increasing the expression of a SA-degrading enzyme, *Arabidopsis* mutants were produced where the amount of SA was significantly reduced compared to the wild type. These plants show delayed leaf senescence, as do *Arabidopsis* mutants with SA signaling pathway defects (Morris et al. 2000, Abreu and Munne-Bosch 2009). These observations indicate that SA promotes leaf senescence (Gan 2003).

Brassinosteroids

This class of plant steroid hormones plays an important role in a variety of developmental programs, including leaf senescence (Clouse 2002, Peres et al. 2019). The data obtained as a result of the studies revealed the role of brassinosteroids (BR) in promoting leaf senescence. Exogenous application of 2,4-epi-brassinolide (eBR) induced senescence in bean leaves (He et al. 1996), cucumber cotyledons (Zhao et al. 1990), and tomato fruits (Vardhini and Rao 2002) show *Arabidopsis* mutants deficient in BR biosynthesis (e.g. *det2*). The BR signaling pathway (e.g., *bri1*) shows a delayed leaf senescence phenotype (Clouse and Sasse 1998). A mutation suppressing the *bri1* phenotype has been identified. Plants containing this mutation show enhanced leaf senescence (Yin et al. 2002). The increase in the concentration of malondialdehyde and the inhibition of the activities of the antioxidant enzymes such as Superoxide Dismutase (SOD) and Catalase (CAT) after BR application in cucumber may mediate senescence promoted by BRs (He et al. 1996, Ding and Zhao 1995).

Abscisic Acid

Abscisic Acid (ABA) is a plant hormone composed of three isoprene rings. It is so named because it promotes abscission in leaves, flowers, and fruits. Initial studies have shown that the external application of ABA promotes senescence in cut leaves, but this effect is less in uncut leaves (Gan 2003). Environmental stress conditions such as drought, high salt concentration, and low temperature often promote leaf senescence, and ABA concentration in leaves increases when plants are exposed to these conditions. Genetic studies show that altered leaf senescence phenotypes are elicited in various *Arabidopsis* mutants lacking ABA biosynthesis or signaling pathways (Gan 2003).

Cytokinins

Plant hormones in this class include kinetin, isopentenyl adenine, and adenine derivatives such as zeatins. Kinetin was first discovered as a cell division promoting factor in plant cells and tissue culture (Miller et al. 1956). Interestingly, kinetin also delays mitotic senescence in cultured human cells and prolongs the lifespan of fruit flies (Sharma et al. 1997). Cytokinins regulate many plant developmental processes such as seed development, photomorphogenesis, root-stem differentiation, apical dominance, cell division, and senescence (Eckardt 2003, Perilli et al. 2010, Wu et al. 2021).

The role of cytokinins in delaying senescence was first studied by applying them externally to plant tissues, and it was observed that cut leaves delay senescence (Richmond and Lang 1957). Later, it was shown that the endogenous production of cytokinin regulates the senescence program of the whole plant (Gan and Amasino 1995). The amount of cytokinin decreases in senescent leaves (Singh et al. 1992). However, after senescence begins, the normal process continues until cytokinin application. In addition to the senescence-delaying effect of cytokinins, studies have shown that Benzyl Aminopurine (BAP) applied at high concentrations promotes senescence (Carimi et al. 2003, 2004).

Senescence-like phenotypes were observed with a decreased level of endogenous cytokinin production due to the increased expression of FPS15 in transgenic *Arabidopsis* plants created by an increasing expression of the famesil diphosphate synthase (FPS15, a mevalonic acid pathway gene) gene (Masferrer et al. 2002). In addition, the increased expression of genes belonging to the signal transduction pathway delays leaf senescence in *Arabidopsis* (Hwang and Sheen 2001). Kinetin application to barley leaf segments has been shown to attenuate the production of senescence-specific chloroplast polypeptides (Guéra and Sabater 1992). Gören and Çağ (2007) showed that the external application of Benzyl Adenine, a cytokinin derivative, delayed senescence in sunflower cotyledons. External application of cytokinin also reduces lipoxygenase activity in *Pisum sativum* L. leaves and delays leaf senescence (Liu and Huang 2002). These results confirm the inhibitory role of cytokinins in leaf senescence.

Polyamines

Polyamines (PA), including putresin, spermidine, and spermine, are cellular elements that play an important role in cell proliferation, cell growth, and the synthesis of proteins and nucleic acids in various organisms including yeast, plants, and animals (Jeevanandam and Peterson 2001). In plants, the PA concentration is high in actively dividing cells and low in non-growing and non-dividing cells (Kaur Shawney et al. 2003), suggesting that PAs may have a role in inhibiting mitotic senescence in plants (Walden et al. 1997).

It has been shown in many plant species and experimental systems that polyamines play a role in inhibiting post-mitotic senescence in plants. This is supported by the fact that exogenously applied PAs can prevent leaf senescence through chlorophyll loss and the inhibition of both membrane peroxidation and ribonuclease and protease activities (Evans and Malmberg 1989, Gan 2003).

S-adenosyl methionine is a common substance used for the biosynthesis of polyamines and ethylene in plants. When ethylene production in transgenic tobacco plants is blocked by antisense technology, S-adenosyl methionine is diverted to the PA biosynthesis pathway, greatly increasing the PA concentration and delaying stress-induced leaf senescence (Wi and Park 2002). In addition, ethylene is known to induce senescence, so blocking ethylene biosynthesis may contribute to the delayed leaf senescence phenotype in these transgenic tobacco plants (Gan 2003). In detached wheat leaves, Spermin and spermidin delayed dark-induced senescence (Goren-Saglam and Çekiç 2019).

Auxins

Auxin, a phytohormone, is essential for plant growth and regulates many developmental processes. Auxin participates in the regulation of various processes such as tropical responses to light and gravity, root and stem remodeling, organ regulation, vascular development, and growth (Woodward and Bartel 2005, Vanneste and Firml 2009).

Indole-3-acetic acid (IAA) is the most extensively studied, most abundant and physiologically valid auxin (Woodward and Bartel 2005). IAA biosynthesis is associated with rapidly dividing and fast-growing tissues, especially in shoots. Although all plant tissues produce low levels of IAA, it is primarily produced in stem apical meristems, young leaves, and developing fruit and seeds (Ljung et al. 2002). *Arabidopsis* seedlings can synthesize IAA in leaves, cotyledons, and roots; young leaves have the highest biosynthetic capacity (Ljung et al. 2001). IAA regulates various physiological and developmental processes, including apical dominance, tropic responses, lateral root formation, vascular differentiation, embryo formation, and stem elongation (Okushima et al. 2005).

Synthetic and natural auxins can delay senescence in most tissues (Biswas and Choudhuri 1980, Chamarro et al. 2001). Auxin affects senescence-related events such as loss of chlorophyll, RNA degradation, RNA synthesis, protein degradation, protein synthesis, membrane disruption, and many enzyme levels, thus delaying senescence. External application of auxin may delay senescence by altering the pattern of senescence (Osborne 1959). However, there are many cases where auxin does not delay senescence (Richmond and Lang 1957, Even-Chen et al. 1978, Goren-Saglam et al. 2020). Auxin has been shown to induce senescence in tissues such as flower petals, xylem differentiation, and some leaves (Mishra and Gaur 1980, Singh et al. 1992, Gören and Çağ 2007). A recent study showed that IAA treatment accelerated the progression of senescence-related changes (Goren-Saglam et al. 2020).

Gibberellins

Gibberellins (GAs) are a class of tetracyclic diterpenoid. They regulate many aspects of plant growth and development, such as stem elongation, leaf expansion, seed dormancy and germination, plant flowering, and the response to abiotic and biotic stresses (Hedden et al. 2012, Niharika et al. 2021).

Gibberellins (GAs) are considered as leaf senescence inhibitors and are able to avoid or delay leaf yellowing. GAs are commonly used as postharvest treatments in several cut flowers to prevent leaf yellowing (Ferrante et al. 2009). The reduction of functional gibberellins content or the conjugation of them with glucose (inactivation)-induced leaf yellowing in several sensitive species. The content of endogenous GAs leads to a decline in leaves' age, and the exogenous application of GA3 retards the degredation of chlorophyll (Li et al. 2010). The exogenous applications are able to delay senescence and reduce ethylene biosynthesis (Iqbal et al. 2022). Additionally, the decrease in GA level is associated with the increase in ABA content, the exogenous application of GA suppresses the increase in ABA during leaf senescence, and delays senescence and decreases ethylene biosynthesis, suggesting that GA indirectly regulates leaf senescence through other hormones (Huang et al. 2022, Iqbal et al. 2017).

Conclusions

The effect of plant growth regulators on the growth and development of plants has been investigated for many years. Senescence is an important stage of plant development and is a dynamic process influenced by internal and external factors. Plant growth regulators are one of the internal factors affecting this important process. While examining the effect of a plant growth regulator on the plant, there are many studies on senescence, which is one of the parameters. Today, thanks to the effective use of molecular biology techniques, the effect of PGR on senescence has begun to be investigated at the genomic level. The obtained data provide a better understanding of the interaction of PGRs with each other, as well as revealing the effect of PGRs on senescence.

Leaf senescence is very important for crop yield in agriculture. Delaying leaf senescence provides a very important economic contribution, especially in green leafy crops. For this reason, revealing the positive and negative effects of plant growth regulators on leaf senescence is very important in terms of increasing crop yield.

References

Abreu, M.E. and Munne-Bosch, S. 2009. Salicylic acid deficiency in NahG transgenic lines and sid2 mutants increases seed yield in the annual plant *Arabidopsis thaliana*. J. Exp. Bot., 60(4): 1261–1271.

Aeong Oh, S., Park, J.H., In Lee, G., Hee Paek, K., Ki Park, S. and Gil Nam, H. 1997. Identification of three genetic loci controlling leaf senescence in Arabidopsis thaliana. The Plant Journal, 12(3): 527–535.

Andrade Galan, A.G., Doll, J., Saile, S.C., Wünsch, M., Roepenack-Lahaye, E.V., Pauwels, L. et al. 2023. The Non-JAZ TIFY Protein TIFY8 of Arabidopsis thaliana interacts with the HD-ZIP III transcription factor REVOLUTA and Regulates Leaf Senescence. International Journal of Molecular Sciences, 24(4): 3079.

Asif, R., Yasmin, R., Mustafa, M., Ambreen, A., Mazhar, M., Rehman, A. et al. 2022. Phytohormones as plant growth regulators and safe protectors against biotic and abiotic stress. Plant Hormones: Recent Advances, New Perspectives and Applications, 115.

Baghel, M., Nagaraja, A., Srivastav, M., Meena, N.K., Senthil Kumar, M., Kumar, A. et al. 2019. Pleiotropic influences of brassinosteroids on fruit crops: a review. Plant Growth Regulation. 87: 375–388.

Balazadeh, S., Parlitz, S., Mueller-Roeber, B. and Meyer, R.C. 2008. Natural developmental variations in leaf and plant senescence in Arabidopsis thaliana. Plant Biology, 10: 136–147.

Biswas, A.K. and Choudhuri, M.A. 1980. Mechanism of monocarpic senescence in rice. Plant Physiology. 65(2): 340–345.

Bresson, J., Bieker, S., Riester, L., Doll, J. and Zentgraf, U. 2018. A guideline for leaf senescence analyses: from quantification to physiological and molecular investigations. Journal of Experimental Botany, 69(4): 769–786.

Buchanan-Wollaston, V. 1997. The molecular biology of leaf senescence. Journal of Experimental Botany, 48(2): 181–199.

Carimi, F., Terzi, M., De Michele, R., Zottini, M. and Schiavo, F.L. 2004. High levels of the cytokinin BAP induce PCD by accelerating senescence. Plant Science, 166(4): 963–969.

Carimi, F., Zottini, M., Formentin, E., Terzi, M. and Lo Schiavo, F. 2003. Cytokinins: new apoptotic inducers in plants. Planta, 216: 413–421.

Castillo, M.C. and Leon, J. 2008. Expression of the β-oxidation gene 3-ketoacyl-CoA thiolase 2 (KAT2) is required for the timely onset of natural and dark-induced leaf senescence in Arabidopsis. Journal of Experimental Botany, 59(8): 2171–2179.

Chamarro, J., Östin, A. and Sandberg, G. 2001. Metabolism of indole-3-acetic acid by orange (*Citrus sinensis*) flavedo tissue during fruit development. Phytochemistry, 57(2): 179–187.

Chen, L., Liu, C., Hao, J., Fan, S. and Han, Y. 2023. GA signaling protein LsRGL1 interacts with the abscisic acid signaling-related gene LsWRKY70 to affect the bolting of leaf lettuce. Horticulture Research, 10(05): uhad054.

Christiansen, M.W. and Gregersen, P.L. 2014. Members of the barley NAC transcription factor gene family show differential co-regulation with senescence-associated genes during senescence of flag leaves. Journal of Experimental Botany, 65(14): 4009–4022.

Clouse, S.D. 2002. Brassinosteroid signal transduction: clarifying the pathway from ligand perception to gene expression. Molecular Cell, 10(5): 973–982.

Clouse, S.D. and Sasse, J.M. 1998. Brassinosteroids: essential regulators of plant growth and development. Annual Review of Plant Biology, 49(1): 427–451.

Conway, T.J. and McCabe, P.F. 2018. Plant programmed cell death. Encyclopedia of Life Science, 1–11.

Creelman, R.A. and Mullet, J.E. 1997. Biosynthesis and action of jasmonates in plants. Annual Review of Plant Biology, 48(1): 355–381.

Del Duca, S., Serafini-Fracassini, D. and Cai, G. 2014. Senescence and programmed cell death in plants: polyamine action mediated by transglutaminase. Frontiers in Plant Science, 5: 120.

Ding, W.M. and Zhao, Y.J. 1995. Effect of Epi-BR on activity of peroxidase and soluble protein content of cucumber cotyledon. Acta Phytophysiologica Sinica, 21(3): 259–264.

Dubey, R.S. 2018. Photosynthesis in plants under stressful conditions. pp. 629–649. In Handbook of photosynthesis. CRC Press.

Eckardt, N.A. 2003. A new classic of cytokinin research: Cytokinin-deficient arabidopsis plants provide new insights into cytokinin biology. The Plant Cell. 15(11): pp. 2489–2492.

Evans, P.T. and Malmberg, R.L. 1989. Do polyamines have roles in plant development? Annual Review of Plant Biology, 40(1): 235–269.

Ferrante, A., Mensuali-Sodi, A. and Serra, G. 2009. Effect of thidiazuron and gibberellic acid on leaf yellowing of cut stock flowers. Central European Journal of Biology, 4: 461–468.

Gaba, V.P. 2005. Plant growth regulators in plant tissue culture and development. pp. 87–99. In Plant Development and Biotechnology. Boca Raton, FL: CRC Press.

Gan, S. 2003. Mitotic and postmitotic senescence in plants. Science of Aging Knowledge Environment. 38: re7–re7.

Gan, S. and Amasino, R.M. 1995. Inhibition of leaf senescence by autoregulated production of cytokinin. Science. 270(5244): 1986–1988.

Ghosh, S., Mahoney, S.R., Penterman, J.N., Peirson, D. and Dumbroff, E.B. 2001. Ultrastructural and biochemical changes in chloroplasts during *Brassica napus* senescence. Plant Physiology and Biochemistry, 39(9): 777–784.

Gonzalez, M.E., Jasso-Robles, F.I., Flores-Hernández, E., Rodríguez-Kessler, M. and Pieckenstain, F.L. 2021. Current status and perspectives on the role of polyamines in plant immunity. Annals of Applied Biology, 178(2): 244–255.

Goren, N. and Çağ, S. 2007. The effect of indole-3-acetic acid and benzyladenine on sequential leaf senescence on *Helianthus annuus L.* seedlings. Biotechnology & Biotechnological Equipment, 21(3): 322–327.

Goren-Saglam, N. and Çekiç, F.Ö. 2019. Roles of some plant growth regulators on natural and dark-induced senescence in wheat leaf segments. Notulae Botanicae Horti Agrobotanici Cluj-Napoca, 47(4): 1230–1237.

Goren-Saglam, N., Duygun, K., Kaya, G. and Vardar, F. 2019. Karrikinolide Promotes Seed Germination but Has no Effect on Leaf Segment Senescence in *Triticum aestivum L.* European Journal of Biology, 78(2): 69–74.

Gören-Sağlam, N., Harrison, E., Breeze, E., Öz, G. and Buchanan-Wollaston, V. 2020. Analysis of the impact of indole-3-acetic acid (IAA) on gene expression during leaf senescence in *Arabidopsis thaliana*. Physiology and Molecular Biology of Plants, 26: 733–745.

Grbić, V. and Bleecker, A.B. 1995. Ethylene regulates the timing of leaf senescence in Arabidopsis. The Plant Journal, 8(4): 595–602.

Gregersen, P. L., Culetic, A., Boschian, L. and Krupinska, K. 2013. Plant senescence and crop productivity. Plant Molecular Biology, 82: 603–622.

Guéra, A. and Sabater, B. 1992. Synthesis of chloroplast proteins by barley leaf segments: effects of senescence induction and kinetin treatment. Acta Botanica Neerlandica, 41(1): 43–49.

Gully, K., Hander, T., Boller, T. and Bartels, S. 2015. Perception of Arabidopsis At Pep peptides, but not bacterial elicitors, accelerates starvation-induced senescence. Frontiers in Plant Science, 6: 14.

Guo, Y., Ren, G., Zhang, K., Li, Z., Miao, Y. and Guo, H. 2021. Leaf senescence: progression, regulation, and application. Molecular Horticulture, 1(1): 1–25.

Ha, S.T.T. and In, B.C. 2022. Combined Nano Silver, α-Aminoisobutyric Acid, and 1-Methylcyclopropene Treatment Delays the Senescence of Cut Roses with Different Ethylene Sensitivities. Horticulturae, 8(6): 482.

Halloran, N. and Kasım, M.U. 2002. Meyve ve sebzelerde büyüme düzenleyici madde kullanımı ve kalıntı düzeyleri. Gıda, 27(5).

Harris, K.R. 2010. Genetic analysis of the Sorghum bicolor stay-green drought tolerance trait (Doctoral dissertation, Texas A & M University).

He, Y. and Gan, S. 2002. A gene encoding an acyl hydrolase is involved in leaf senescence in Arabidopsis. The Plant Cell., 14(4): 805–815.

He, Y. and Gan, S. 2002. A gene encoding an acyl hydrolase is involved in leaf senescence in Arabidopsis. The Plant Cell., 14(4): 805–815.

He, Y., Fukushige, H., Hildebrand, D. F. and Gan, S. 2002. Evidence supporting a role of jasmonic acid in Arabidopsis leaf senescence. Plant Physiology, 128(3): 876–884.

Hedden, P. and Thomas, S.G. 2012. Gibberellin biosynthesis and its regulation. Biochemical Journal, 444(1): 11–25.

Huang, P., Li, Z. and Guo, H. 2022. New advances in the regulation of leaf senescence by classical and peptide hormones. Frontiers in Plant Science, 13.

Hwang, I. and Sheen, J. 2001. Two-component circuitry in Arabidopsis cytokinin signal transduction. Nature, 413(6854): 383–389.

Iqbal, M.S., Zahoor, M., Akbar, M., Ahmad, K., Hussain, S., Munir, S. et al. 2022. Alleviating the deleterious effects of salt stress on wheat (*Triticum aestivum L.*) By foliar application of gibberellic acid and salicylic acid. Applied Ecology and Environmental Research, 20: 119–134.

Iqbal, N., Khan, N.A., Ferrante, A., Trivellini, A., Francini, A. and Khan, M.I.R. 2017. Ethylene role in plant growth, development and senescence: interaction with other phytohormones. Frontiers in Plant Science, 8: 475.

Jalil, S.U., Ansari, S.A. and Ansari, M.I. 2022. Role of environment stress leaf senescence and crop productivity. pp. 13–31. In: Augmenting Crop Productivity in Stress Environment. Singapore: Springer Nature Singapore.

Jeevanandam, M. and Petersen, S.R. 2001. Clinical role of polyamine analysis: problem and promise. Current Opinion in Clinical Nutrition & Metabolic Care, 4(5): 385–390.

Jing, H.C., Schippers, J.H., Hille, J. and Dijkwel, P.P. 2005. Ethylene-induced leaf senescence depends on age-related changes and OLD genes in Arabidopsis. Journal of Experimental Botany, 56(421): 2915–2923.

John, I., Drake, R., Farrell, A., Cooper, W., Lee, P., Horton, P. et al. 1995. Delayed leaf senescence in ethylene-deficient ACC-oxidase antisense tomato plants: molecular and physiological analysis. The Plant Journal, 7(3): 483–490.

Kaur-Sawhney, R., Tiburcio, A.F., Altabella, T. and Galston, A.W. 2003. Polyamines in plants: an overview. Journal of Cell Molecular Biology, 2: 1–12.

Kumar, S., Singh, R., Kalia, S., Sharma, S. and Kalia, R.K. 2016. Recent advances in understanding the role of growth regulators in plant growth and development in vitro-I. conventional growth regulators. Indian Forester, 142(5): 459–470.

Launay, A., Cabassa-Hourton, C., Eubel, H., Maldiney, R., Guivarc'h, A., Crilat, E. et al 2019. Proline oxidation fuels mitochondrial respiration during dark-induced leaf senescence in *Arabidopsis thaliana*. Journal of Experimental Botany, 70(21): 6203–6214.

Leopold, A.C. 1961. Senescence in Plant Development: The death of plants or plant parts may be of positive ecological or physiological value. Science, 134(3492): 1727–1732.

Li, J.R., Yu, K., Wei, J.R., Ma, Q., Wang, B.Q. and Yu, D. 2010. Gibberellin retards chlorophyll degradation during senescence of Paris polyphylla. Biologia Plantarum, 54: 395–399.

Li, M., Yu, G., Cao, C. and Liu, P. 2021. Metabolism, signaling, and transport of jasmonates. Plant Communications, 2(5): 100231.

Li, Z., Zhao, T., Liu, J., Li, H. and Liu, B. 2023. Shade-Induced Leaf Senescence in Plants. Plants, 12(7): 1550.

Lim, P.O. and Nam, H.G. 2007. Aging and senescence of the leaf organ. Journal of Plant Biology, 50: 291–300.

Lim, P.O., Kim, H.J. and Gil Nam, H. 2007. Leaf senescence. Annual Review of Plant Biology, 58: 115–136.

Liu, X. and Huang, B. 2002. Cytokinin effects on creeping bentgrass response to heat stress: II. Leaf senescence and antioxidant metabolism. Crop Science, 42(2): 466–472.

Ljung, K., Bhalerao, R.P. and Sandberg, G. 2001. Sites and homeostatic control of auxin biosynthesis in Arabidopsis during vegetative growth. The Plant Journal, 28(4): 465–474.

Ljung, K., Hull, A.K., Kowalczyk, M., Marchant, A., Celenza, J., Cohen, J.D. et al. 2002. Biosynthesis, conjugation, catabolism and homeostasis of indole-3-acetic acid in Arabidopsis thaliana. Plant Molecular Biology, 49: 249–272.

Locato, V. and De Gara, L. 2018. Programmed cell death in plants: an overview. Plant Programmed Cell Death: Methods and Protocols, 1–8.

Mandal, S., Gupta, S.K., Ghorai, M., Patil, M.T., Biswas, P., Kumar, M. et al. 2023. Plant nutrient dynamics: a growing appreciation for the roles of micronutrients. Plant Growth Regulation, 1–18.

Masferrer, A., Arró, M., Manzano, D., Schaller, H., Fernández-Busquets, X., Moncaleán, P. et al. 2002. Overexpression of *Arabidopsis thaliana* farnesyl diphosphate synthase (FPS1S) in transgenic Arabidopsis induces a cell death/senescence-like response and reduced cytokinin levels. The Plant Journal, 30(2): 123–132.

Miller, C.O., Skoog, F., Okumura, F.S., Von Saltza, M.H. and Strong, F.M. 1956. Isolation, structure and synthesis of kinetin, a substance promoting cell division1, 2. Journal of the American Chemical Society, 78(7): 1375–1380.

Miryeganeh, M. 2022. Epigenetic mechanisms of senescence in plants. Cells, 11(2): 251.

Mishra, S.D. and Gaur, B.K. 1980. Growth regulator control of senescence in discs of betel (*Piper betle L.*) leaf: effect of rate & degree of senescence. Indian Journal of Experimental Biology, 18(3): 297–298.

Morris, K., -Mackerness, S.A.H., Page, T., John, C.F., Murphy, A.M., Carr, J.P. et al. 2000. Salicylic acid has a role in regulating gene expression during leaf senescence. The Plant Journal, 23(5): 677–685.

Munné-Bosch, S. 2007. Aging in perennials. Critical Reviews in Plant Sciences, 26(3): 123–138.

Narayanan, M. and Ma, Y. 2023. Metal tolerance mechanisms in plants and microbe-mediated bioremediation. Environmental Research, 222: 115413.

Niharika, Singh, N.B., Singh, A., Khare, S., Yadav, V., Bano, C. and Yadav, R.K. 2021. Mitigating strategies of gibberellins in various environmental cues and their crosstalk with other hormonal pathways in plants: a review. Plant Molecular Biology Reporter, 39: 34–49.

Noodén, L.D., Guiamét, J.J. and John, I. 1997. Senescence mechanisms. Physiologia Plantarum, 101(4): 746–753.

Okushima, Y., Mitina, I., Quach, H.L. and Theologis, A. 2005. Auxin Response Factor 2 (ARF2): a pleiotropic developmental regulator. The Plant Journal. 43(1): 29–46.

Peres, A.L.G., Soares, J.S., Tavares, R.G., Righetto, G., Zullo, M.A., Mandava, N.B. et al. 2019. Brassinosteroids, the sixth class of phytohormones: a molecular view from the discovery to hormonal interactions in plant development and stress adaptation. International Journal of Molecular Sciences, 20(2): 331.

Perilli, S., Moubayidin, L. and Sabatini, S. 2010. The molecular basis of cytokinin function. Current Opinion in Plant Biology, 13(1): 21–26.

Pic, E., de La Serve, B.T., Tardieu, F. and Turc, O. 2002. Leaf senescence induced by mild water deficit follows the same sequence of macroscopic, biochemical, and molecular events as monocarpic senescence in pea. Plant Physiology, 128(1): 236–246.

Popov, V.N., Syromyatnikov, M.Y., Franceschi, C., Moskalev, A.A. and Krutovsky, K.V. 2022. Genetic mechanisms of aging in plants: What can we learn from them?. Ageing Research Reviews, 77: 101601.

Richmond, A.E. and Lang, A. 1957. Effect of kinetin on protein content and survival of detached Xanthium leaves. Science, 125(3249): 650–651.

Romero-Bueno, R., Ruiz, P.D.L.C., Artal-Sanz, M., Askjaer, P. and Dobrzynska, A. 2019. Nuclear organization in stress and aging. Cells, 8(7): 664.

Sachdev, S., Ansari, S.A. and Ansari, M.I. 2023. Senescence and apoptosis: ROS contribution to stress tolerance or cellular impairment. In Reactive Oxygen Species in Plants: The Right Balance, pp. 61–74. Singapore: Springer Nature Singapore.

Sampedro-Guerrero, J., Vives-Peris, V., Gomez-Cadenas, A. and Clausell-Terol, C. 2023. Efficient strategies for controlled release of nanoencapsulated phytohormones to improve plant stress tolerance. Plant Methods, 19(1): 1–20.

Sarwat, M., Naqvi, A.R., Ahmad, P., Ashraf, M. and Akram, N.A. 2013. Phytohormones and microRNAs as sensors and regulators of leaf senescence: assigning macro roles to small molecules. Biotechnology Advances, 31(8): 1153–1171.

Sasi, J.M., Gupta, S., Singh, A., Kujur, A., Agarwal, M. and Katiyar-Agarwal, S. 2022. Know when and how to die: gaining insights into the molecular regulation of leaf senescence. Physiology and Molecular Biology of Plants, 28(8): 1515–1534.

Savin, K.W., Baudinette, S.C., Graham, M.W., Michael, M.Z., Nugent, G.D., Lu, C.Y. et al. 1995. Antisense ACC oxidase RNA delays carnation petal senescence. HortScience, 30(5): 970–972.

Schippers, J.H., Schmidt, R., Wagstaff, C. and Jing, H.C. 2015. Living to die and dying to live: the survival strategy behind leaf senescence. Plant Physiology, 169(2): 914–930.

Sharma, R.R., Baghel, M.M. and Nagaraja, A. 2022. Brassinosteroids and their Use in Fruit Crops. In Plant Growth Regulators in Tropical and Sub-tropical Fruit Crops, pp. 14–34. CRC Press.

Sharma, S.P., Kaur, J. and Rattan, S.I. 1997. Increased longevity of kinetin-fed Zaprionus fruitflies is accompanied by their reduced fecundity and enhanced catalase activity. IUBMB Life, 41(5): 869–875.

Sillanpää, M. 2003. Leaf senescence in silver birch (Betula pendula Roth). Dissertationes Biocentri Viikki Universitatis Helsingiensis. 8/2003.

Singh, S., Letham, D.S. and Palni, L.M.S. 1992. Cytokinin biochemistry in relation to leaf senescence. VII. Endogenous cytokinin levels and exogenous applications of cytokinins in relation to sequential leaf senescence of tobacco. Physiologia Plantarum, 86(3): 388–397.

Smart, C.M. 1994. Gene expression during leaf senescence. New Phytologist, 126(3): 419–448.

Stoynova-Bakalova, E., Petrov, P.I., Gigova, L. and Baskin, T.I. 2008. Differential effects of methyl jasmonate on growth and division of etiolated zucchini cotyledons. Plant Biology, 10(4): 476–484.

Takahashi, Y., Kuro-o, M. and Ishikawa, F. 2000. Aging mechanisms. Proceedings of the National Academy of Sciences, 97(23): 12407–12408.

Tsuchisaka, A., Yu, G., Jin, H., Alonso, J.M., Ecker, J.R., Zhang, X. et al. 2009. A combinatorial interplay among the 1-aminocyclopropane-1-carboxylate isoforms regulates ethylene biosynthesis in *Arabidopsis thaliana*. Genetics, 183(3): 979–1003.

Vanneste, S. and Friml, J. 2009. Auxin: a trigger for change in plant development. Cell, 136(6): 1005–1016.

Vardhini, B.V. and Rao, S.S.R. 2002. Acceleration of ripening of tomato pericarp discs by brassinosteroids. Phytochemistry, 61(7): 843–847.

Walden, R., Cordeiro, A. and Tiburcio, A.F. 1997. Polyamines: small molecules triggering pathways in plant growth and development. Plant Physiology, 113(4): 1009.

Wang, Y., Shirakawa, M. and Ito, T. 2023. Arrest, senescence and death of shoot apical stem cells in *Arabidopsis thaliana*. Plant and Cell Physiology, 64(3): 284–290.

Wang, N.N., Yang, S.F. and Charng, Y.Y. 2001. Differential expression of 1-aminocyclopropane-1-carboxylate synthase genes during orchid flower senescence induced by the protein phosphatase inhibitor okadaic acid. Plant Physiology, 126(1): 253–260.

Westfall, C.S., Muehler, A.M. and Jez, J.M. 2013. Enzyme action in the regulation of plant hormone responses. Journal of Biological Chemistry, 288(27): 19304–19311.

Wi, S.J. and Park, K.Y. 2002. Antisense expression of carnation cDNA encoding ACC synthase or ACC oxidase enhances polyamine content and abiotic stress tolerance in transgenic tobacco plants. Molecules and Cells, 13(2): 209–220.

Woo, H.R., Kim, H.J., Lim, P.O. and Nam, H.G. 2019. Leaf senescence: systems and dynamics aspects. Annual Review of Plant Biology, 70: 347–376.

Woo, H. R., Kim, H. J., Lim, P. O. and Nam, H.G. 2019. Leaf senescence: systems and dynamics aspects. Annual Review of Plant Biology, 70: 347–376.

Woo, H.R., Kim, H.J., Nam, H.G. and Lim, P.O. 2013. Plant leaf senescence and death–regulation by multiple layers of control and implications for aging in general. Journal of Cell Science, 126(21): 4823–4833.

Woodward, A.W. and Bartel, B. 2005. Auxin: regulation, action, and interaction. Annals of Botany, 95(5): 707–735.

Wu, W., Du, K., Kang, X. and Wei, H. 2021. The diverse roles of cytokinins in regulating leaf development. Horticulture Research. 8.

Wu, X.Y., Kuai, B.K., Jia, J.Z. and Jing, H.C. 2012. Regulation of leaf senescence and crop genetic improvement F. Journal of Integrative Plant Biology, 54(12): 936–952.

Yin, Y., Wang, Z.Y., Mora-Garcia, S., Li, J., Yoshida, S., Asami, T. et al. 2002. BES1 accumulates in the nucleus in response to brassinosteroids to regulate gene expression and promote stem elongation. Cell, 109(2): 181–191.

Yu, H., Zhou, G., Lv, X., He, Q. and Zhou, M. 2022. Environmental factors rather than productivity drive autumn leaf senescence: Evidence from a grassland *in situ* simulation experiment. Agricultural and Forest Meteorology, 327: 109221.

Yujiong, H., Rujuan, X. and Yuju, Z. 1996. Enhancement of senescence by epibrassinolide in leaves of mung bean seedling. Acta Phytophysiologica Sinica, 22(1): 58–62.

Zahid, G., Iftikhar, S., Shimira, F., Ahmad, H.M. and Kaçar, Y.A. 2023. An overview and recent progress of plant growth regulators (PGRs) in the mitigation of abiotic stresses in fruits: A review. Scientia Horticulturae, 309: 111621.

Zentgraf, U., Andrade-Galan, A.G. and Bieker, S. 2022. Specificity of H2O2 signaling in leaf senescence: is the ratio of H2O2 contents in different cellular compartments sensed in Arabidopsis plants?. Cellular and Molecular Biology Letters, 27(1): 1–19.

Zhang, H. and Zhou, C. 2013. Signal transduction in leaf senescence. Plant Molecular Biology, 82: 539–545.

Zhang, Y., Wu, Z., Feng, M., Chen, J., Qin, M., Wang, W. et al. 2021. The circadian-controlled PIF8–BBX28 module regulates petal senescence in rose flowers by governing mitochondrial ROS homeostasis at night. The Plant Cell, 33(8): 2716–2735.

Zhang, Y., Zhao, L., Zhao, J., Li, Y., Wang, J., Guo, R. et al. 2017. S5H/DMR6 encodes a salicylic acid 5-hydroxylase that fine-tunes salicylic acid homeostasis. Plant Physiology, 175(3): 1082–1093.

Zhang, Y., Berman, A. and Shani, E. 2023. Plant Hormone Transport and Localization: Signaling Molecules on the Move. Annual Review of Plant Biology, 74.

Zhao, C., Yue, Y., Wu, J., Scullion, J., Guo, Q. et al. 2023. Panicle removal delays plant senescence and enhances vegetative growth improving biomass production in switchgrass. Biomass and Bioenergy, 174: 106809.

Zhao, Y.J., Xu, R.J. and Luo, W.H. 1990. Inhibitory effects of abscissic acid on epibrassinolide induced senescence of detached cotyledons in cucumber seedlings. Chinese Science Bulletin, 35: 928–931.

Zhu, F., Alseekh, S., Koper, K., Tong, H., Nikoloski, Z., Naake, T. et al. 2022. Genome-wide association of the metabolic shifts underpinning dark-induced senescence in Arabidopsis. The Plant Cell, 34(1): 557–578.

3

Fungal Phytohormones
Plant Growth Regulators and Crop Productivity

C. J. Mendoza-Meneses, Betsie Martínez-Cano,
María Isabel Nieto-Ramírez, Amanda Kim Rico-Chávez,
Karen Esquivel-Escalante and Ana A. Feregrino-Pérez*

Introduction

Phytohormones are molecules involved in the biological processes of plants and their response to environmental stimuli (Imtiaz et al. 2023). This interaction occurs in the metabolism, development, signaling, response to stress, and death of the specimen (Salvi et al. 2021, Šimura et al. 2018). In the mechanism of action of these compounds, a phytohormone can act in several processes. However, different phytohormones can be a part of a single process (Sytar et al. 2019). Although its production is in low concentrations (fmol to pmol g^{-1} fresh weight), its importance lies in solving problems associated with biology and in the conceptualization of modern agriculture (Šimura et al. 2018, Waadt 2020, Wani et al. 2016).

The hormones synthesized by plants considered as stimulators are auxins, gibberellic acids, cytokinins, strigolactone, brassinosteroids and melatonin. Meanwhile, the hormones with inhibitory potential are abscisic acid, ethylene, salicylic acid, and jasmonic acid (Li et al. 2021, Salvi et al. 2021, Sytar et al. 2019, Kaya et al. 2023). However, together these compounds modify the life cycle, production, and quality of plants. Therefore, the understanding and application of phytohormones is crucial for improvement in the agricultural sector (Li et al. 2021).

Plants are the main source of phytohormones; however, these compounds can also be synthesized from bacteria and fungi (Tudzynski and Sharon 2002).

Graduate and Research Division, Engineering Faculty, Universidad Autónoma de Querétaro, Cerro de las Campanas, C.P. 76010, Santiago de Querétaro Qro, México.
* Corresponding author: feregrino.angge@hotmail.com

Compounds that regulate growth, physiology, and immunity in plants are similar to the hormones produced directly by plants, and produce health benefits in the face of negative stimuli (Anand et al. 2022). Phytohormones produced by fungi have different applications (Chanclud and Morel 2016, Liao et al. 2018, Ozimek and Hanaka 2021, Vincent et al. 2020, Zhang et al. 2018). However, the fungal-plant symbiotic or pathogenic relationship changes the growth and development of plants with increases in productivity or physiological destruction, respectively.

In this sense, symbiotic fungal-plant interactions can generate benefits in crop production and quality, plant health, and increased plant resistance to disease. Therefore, this chapter compiles information about phytohormones synthesized from fungi with potential application as plant growth regulators and to increase crop productivity.

Importance of Fungi and Phytohormones

The relationship between plants and fungi has been recorded for many years, however, recent studies focus on the benefits of fungi on the different stages of physiological development of plants (Baron and Rigobelo 2022). Fungi are microorganisms capable of benefiting plant growth through nutrient solubilization mechanisms, production of plant growth regulators, and even the creation of antagonistic substances (Devi et al. 2020).

Fungi have an important function for agro-environmental sustainability due to the distribution of the microorganism and the metabolic functions of three main phyla—*Mucoromycota, Basidiomycota* and *Ascomycota*—which are divided into seven main classes (Yadav et al. 2022). Recent research focuses on the application of environmentally friendly techniques to maintain health in plants. In this sense, phytohormones are the regulators that mark the association between microorganisms and host plants (Sethi et al. 2023).

Mycorrhizal fungi are widely used in agriculture due to their benefits as enhancers of plant health (Devi et al. 2020). Specifically, jasmonic acid is related to the response of the immune system of plants to act quickly and efficiently against stressful stimuli through mycorrhizal networks (Olson et al. 2022). Even strigolactones are hormones that improve plant resistance to biotic and abiotic stress factors (Xiubing et al. 2023). Phytohormones such as cytokinin, gibberellin, and auxin form complex relationships with other phytohormones such as salicylic acid, jasmonate, and ethylene to form disease resistance and regulate other physiological functions in the plant (Gupta et al. 2023).

Phytohormone Classification and the Relevance of Fungal Phytohormones in Plant Systems

Phytohormones are chemical compounds synthesized by plant organisms for regulating developmental processes. The six major classes of phytohormones are auxins, cytokinins, gibberellins, brassinosteroids, ethylene, and abscisic acid (Dilworth et al. 2017). These hormones are typically classified as promoters or inhibitors according to their primary effect on plant growth, as shown in Fig. 1 (Barrington 2020).

Moreover, recent additions to that classification include stress response regulators, such as salicylic acid, jasmonates, and strigolactones (Brewer et al. 2013, Bürger and Chory 2019). Nevertheless, phytohormone function is more complex, and research on plant stress shows that all phytohormones are crucial in plant adaptability and act as mediators of plant defense before abiotic and biotic stress incidence (Khan et al. 2020, Souza et al. 2017).

Stress is a determinant factor of plant biochemistry. Plants defend themselves from biotic challenges through a two-layered immune system, which is activated by pathogen molecular patterns or effectors (Jones and Dangl 2006). After microbial perception, two different hormone-dependent systemic responses can be triggered to protect the plant, the systemic acquired resistance (SAR) and the induced systemic resistance (ISR), which are primarily regulated by salicylic acid and jasmonic acid, respectively (Maithani et al. 2020). These two hormone pathways are antagonistic but not independent from one another (Robert-Seilaniantz et al. 2011).

The plant immune response consists of an intricate crosstalk that includes other hormones linked to developmental plant functions, such as auxins, brassinosteroids, ethylene, abscisic acid, and gibberellins (Ahanger et al. 2018, Verma et al. 2016). The manipulation of plant hormones is a door for controlling the whole plant system and, thus, an objective of the evolutionary strategies of microorganisms (Bürger and Chory 2019). In particular, plant-fungi associations are among the most ancient and significant ecological interactions of land plants, although they are also frequently overlooked (Zanne et al. 2020).

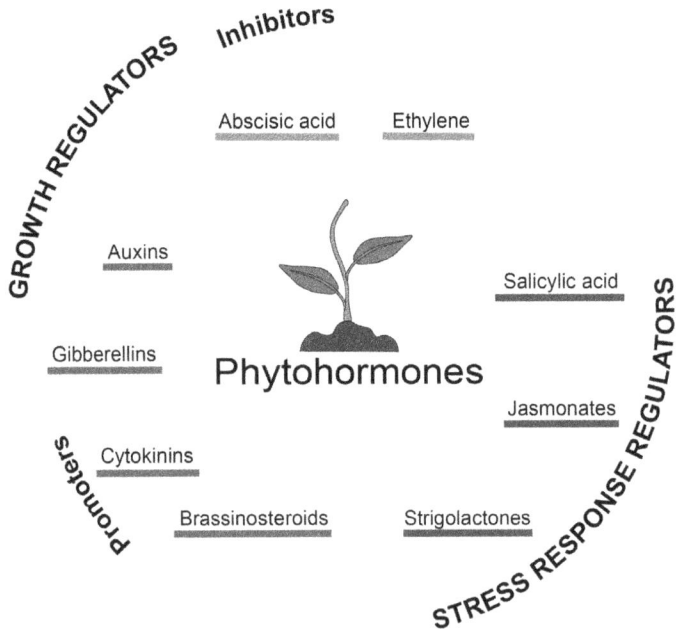

Fig. 1. Phytohormone typical classification according to their primary function on plant development.

Fungi synthesize phytohormones, proteins, and signaling molecules that play a role in hormone biosynthesis in plants (Tudzynski and Sharon 2002). This way, pathogenic and symbiotic fungi alter plant systems to dodge defensive responses and increase colonization success (Chanclud and Morel 2016). Fungal cultures and products interact with plant systems, increasing aerial and root growth, nutrient intake, stress tolerance, and competing with pathogens through the stimulation of primary and specialized metabolism and plant immunity (Pusztahelyi et al. 2015, Quesada Moraga 2020, Wei et al. 2016). As a result, fungal products and formulations represent an exciting alternative to conventional pesticides, fertilizers, and biostimulants for agricultural purposes (Ahmad et al. 2018).

Plant Growth Regulators

A plant growth regulator (PGR) was defined as a synthetic compound that can act as a natural hormone in plants (Davies 1995). Nowadays, these compounds are defined as natural or synthetic compounds that modify a plant's processes such as metabolic or developmental activities (Rademacher 2015). In natural conditions, PGR can be found in low concentration with high effect (George et al. 2008).

PGR biosynthesis can be in plants but also in microorganism such as bacteria, yeast, or fungi (Nutaratat et al. 2016). Fungi PGR production has been reported in mycorrhizal with different effects such as enhancing root production, improve plant nutrition, and improve soil structure and texture to enhance plant health (Begum et al. 2019). However, other fungi have been reported as a plant growth regulator, *Trichoderma* spp. (Tariq Javeed et al. 2021). Figure 2 shows the physiological responses in the plant due to the action of fungal phytohormones.

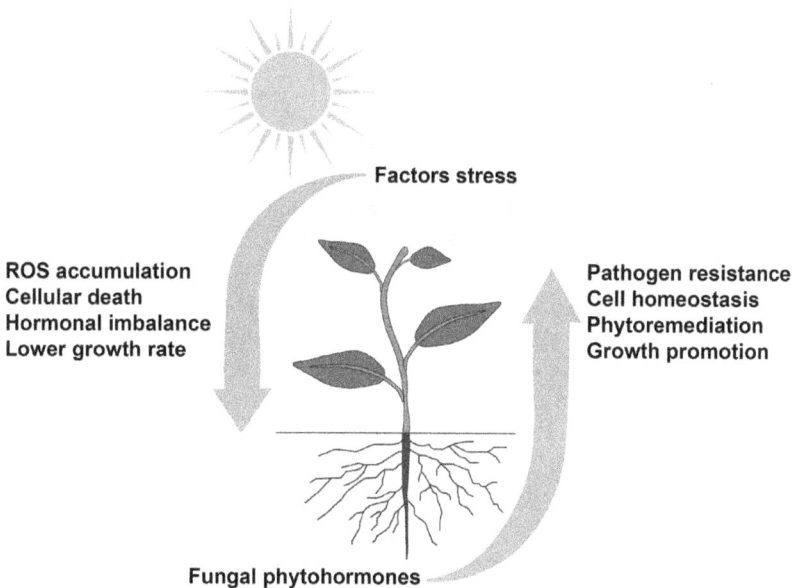

Factors stress

ROS accumulation
Cellular death
Hormonal imbalance
Lower growth rate

Pathogen resistance
Cell homeostasis
Phytoremediation
Growth promotion

Fungal phytohormones

Fig. 2. Physiological responses of the plant in the presence of fungal phytohormones.

Difference between Phytohormones and Synthesized Products

As already mentioned, phytohormones are small molecules, ubiquitous in vascular plants, responsible for regulating aspects of plant growth and development, besides coordinating adaptive responses to environmental signals and stress (Checker et al. 2018, Stirk and van Staden 2020). Phytohormone biosynthesis can occur at different morphological and cytological sites, usually in meristems, leaves, and developing fruits, and their actions can be observed at their production sites or in distant target tissues (Bhatla 2018).

Phytohormones act individually but also influence other signaling pathways of other plant hormones. Therefore, the hormonal signaling cascades in plants involve feedback regulation of their biosynthesis, additive, synergistic, or antagonistic interactions with other phytohormones (Depuydt and Hardtke 2011). Such interactions increase the complexity of hormonal actions in plants.

Auxin is involved in many processes for plant development that include interactions with other hormones, such as gibberellins, brassinosteroids, cytokinins, and ethylene. Some reports show that auxin and gibberellin overlap their regulation in multiple aspects of plant development and participate in a positive crosstalk. Likewise, gene expression in response to auxin signals is regulated at the transcriptional level in a similar form to that gibberellin (Yuan et al. 2019). Brassinosteroid and auxin positively regulate cell elongation and vascular development, while gibberellin contributes to the directional expansion of organs and plant development (Best et al. 2016, Kim et al. 2014).

On the other hand, Achard et al. (2003) showed that DELLA proteins are responsible for ethylene, auxin, and gibberellin responses and can be considered coordinators of multiple phytohormonal signals that regulate plant growth. Besides, a hormone may elicit different responses in different tissues or at different development stages in the same tissue (Khodanitska et al. 2021).

It is already well known that the physiological activity of plants is not determined by the content of specific groups of phytohormones but by the balance of biologically active substances. Phytohormones have a multifunctional action; they regulate many biochemical processes in plants (Kuryata and Khodanitska 2018). The regulation of physiological and biochemical processes in plants is given by the affinity of hormones towards receptor proteins present in cells that form a complex, thus causing gene expression that affects cell metabolism. In short, the plant genome controls the biosynthesis of the phytohormones themselves (Khodanitska et al. 2021). The above data suggest that plant development occurs through a complex hormonal interaction, which complicates determining the net response of the application of exogenous substances.

Therefore, the search for synthetic plant hormones and hormone mimics with higher stability and activity has been essential for the agrochemical industry. Currently, there are already chemically synthesized hormonal derivatives and analogs, such as 1-naphthaleneacetic acid (1-NAA), 2,4-dichlorophenoxyacetic acid (2,4-D), and benzothiadiazole (BTH) widely used as promoters or herbicides (Grossmann 2010).

The exogenous treatment of analogs compounds of phytohormones significantly affects the metabolic processes, which improve the crops' productivity. Growth regulators are substances of synthetic or natural origin that have biological activity and can cause changes in the morphophysiological processes of plants, such as cell division and differentiation and increase in plant size and productivity (Vijay et al. 2020).

Synthetic analogs of phytohormones, unlike natural biologically active substances, are more stable in the plant organism and are characterized by a prolonged action over time. Furthermore, chemical products can overcome functional redundancy by inhibiting multiple members of a family of redundant proteins (Fonseca et al. 2014).

In several studies, it has been observed that the application of synthetic growth regulator compounds increases the simultaneity and the germination energy of the seeds accelerates the growth of shoots and roots, which allows better absorption of nutrients and resistance to environmental factors (Calvo et al. 2014, Khodanitska et al. 2021). It is important to note that the combination of phytohormones and analogous substances is widely used to improve different crops, stimulate biomass yield, increase the production of commercially valuable compounds, and modulate plant responses to stress factors (Han et al. 2018).

However, despite the contribution of synthetic analogs of phytohormones, these substances have relevant limitations: obtaining them and their success are based on the chance of identifying a structurally compatible product from a small number of natural sources and large collections. Hormonal derivatives lack chirality, and their structural diversity is limited to variations in common backbone bonding. In addition to that, there are no synthetic products that affect the development of plants in all their phenological stages (Fonseca et al. 2014, Khodanitska et al. 2021). Finally, it is fundamental that the analogs phytohormone acts as a typical plant hormone that modulates all plant responses and provides immunity against different types of stress too (Arif et al. 2020).

Impact on Crop Productivity and Quality

Phytohormones regulate plant development and stress response. A significant number of phytohormones have been used in exogenous applications for modifying crop development, growth, resistance, and morphology in various phenological stages of production (Chhaya et al. 2021, Kosakivska et al. 2022, Xiang et al. 2021). The effects of such compounds on desirable agronomic traits, such as crop productivity and quality, are relevant to improving the sustainability and economic profit of food production systems (Vázquez-Hernández et al. 2019). Table 1 summarizes the most recent findings of agricultural applications of phytohormones in horticulture.

Agronomic Applications of Growth Regulators

Plant growth regulation has agricultural applications to obtain specific benefits such as major vegetative production, higher secondary metabolites content, and decrease stress susceptibility (Rademacher 2015). The production of these compounds is in plants and in other organisms such as yeast, bacterium, or fungi. Fungal plant growth

Table 1. Effects of exogenous application of phytohormones on crop productivity and quality.

Phytohormone class	Compound	Effect	Application	References
Abscisic acid		Regulates plant responses to abiotic stress	Improvement of plant stress tolerance	Yu et al. 2021
Abscisic acid		Regulates flowering locus T expression.	Floral induction under stress conditions.	Endo et al. 2018
Auxins	Indole-3-acetic acid (IAA)	Promotes apical root growth and cell division, and cell differentiation.	Rooting in cell cultures and stem cuttings.	Keswani et al. 2020
Auxins	Indole-3-acetic acid (IAA)	Regulates ethylene and abscisic acid signaling and biosynthesis.	Inhibition of fruit abscission.	Xie et al. 2018
Brassinosteroids	24-Epibrassinolide	Improves antioxidant enzyme activity and increases antioxidative metabolites.	Improves plant growth under stress conditions.	Zhang and Liao, 2021
Ethylene		Promotes fruit ripening by increasing sugar accumulation and organic acid degradation.	Improvement of fruit quality and acceleration of the postharvest ripening process.	Fan et al. 2018
Gibberellins	GA_3	Promotes abscisic acid degradation.	Improvement of seed germination rate.	Song et al. 2019
Gibberellins	GA_3	Increases antioxidant enzyme activity.	Improvement of salt stress tolerance.	Shahzad et al. 2021
Gibberellins	GA_3	Impairs ethylene biosynthesis and the expression of its receptors.	Increase of post-harvest longevity.	Khatami et al. 2020
Jasmonates	Methyl jasmonate	Enhances antioxidant enzyme activities and antioxidative metabolites content.	Alleviates salt stress.	Lang et al. 2020
Jasmonates	Methyl jasmonate	Improves resistance to pathogens.	Increases post-harvest longevity.	Pan et al. 2020
Salicylic acid and jasmonates	Salicylic acid and methyl jasmonate	Activate plant immune responses.	Increases specialized metabolites biosynthesis.	Rodríguez-Sánchez et al. 2020
Salycilic acid		Increases phenolics content and phenylalanine ammonium lyase activity.	Improves post-harvest longevity.	Martínez-Camacho et al. 2022
Strigolactone	GR24	Regulates minerals intake from the soil.	Decreases Cd toxicity.	Qiu et al. 2021
Strigolactones	GR24 analog	Improves antioxidant enzyme activity and increases photosynthetic pigments.	Alleviation of drought stress.	Sattar et al. 2022

IAA, Indole-3-acetic acid, **GA**, Gibberellin

regulators like auxin, cytokinin, gibberellin, and abscisic acid have been found in many fungi species (Tudzynski and Sharon 2002).

Different fungi species have been applied in agricultural production and have shown a positive effect on biomass production, tolerance to stress, and control of pests as shown in Table 2 (Al-Ani et al. 2021). Nowadays, fungal specific auxin such as indole-3-acetic acid have been applied to control *Cladosporium cladosporiodies* in *Piper nigrum* leaves (Munasinghe et al. 2017), to promote growth in *Zea mays* (Mehmood et al. 2019), to control wilt disease in *Solanum lycopersicum* L. (Bader et al. 2020).

Additionally, gibberellins are produce in *Aspergillus*, *Fusarium*, *Penicillium*, and *Rhizopus* to stimulate the growth and development of plants in salinity stress (Solyman et al. 2019, Waqas et al. 2012). Gibberellin application stimulates rooting and propagation in temperate fruit crops (Ghosh and Halder 2018), promotes host-plant growth during stress (Waqas et al. 2012), reduces brunch incidence in Vignoles grapes (Hed and Centinari 2021), and enhancse vegetative production (Tyśkiewicz et al. 2022). Cytokinin produced in fungi species have also been utilized in agriculture, to reduce the water stress in wheat plants (Illescas et al. 2021), enhance nitrogen absorption in rice (Li et al. 2018), and stimulate growth in *Arabidopsis thaliana* (Bean et al. 2021).

Table 2. Effect of fungal plant growth regulators on agricultural applications.

Fungi specie	PGR	Effect	
Colletotrichum siamense	IAA	High antifungal activity against *Cladosporium cladosporiodies*.	Munasinghe et al. 2017
Aspergillus awamori	IAA	Improve growth parameters.	Mehmood et al. 2019
Trichoderma harzianum	IAA	Stimulate plant growth by increasing leaf area, the uptake of phosphorous, and positive effect against *F. oxysporum*.	Bader et al. 2020
Aspergillus niger	IAA, GA	Tolerance of salt stress.	Solyman et al. 2019
Phoma glomerate	GA1, GA3, GA4, GA7, and IAA	Increase plant height, chlorophyll, and reduce the stress-responsive endogenous ABA content.	Waqas et al. 2012
Penicillium sp.	GA1, GA3, and IAA	Enhance leaf area and height under salinity stress.	Waqas et al. 2012
Trichoderma	GA	Enhance vegetative production and control of fungal phytopathogens.	Tyśkiewicz et al. 2022
Trichoderma	CK, GA, IAA	Increase weight under water stress.	Illescas et al. 2021
Phomopsis liquidambari	Auxin, CK	Improve nitrogen absorption.	X. Li et al. 2018
Trichoderma	CK	Plant growth stimulation and resistance against pathogens.	Bean et al. 2021

PGR, Plan Growth Regulator, **IAA**, Indole-3-acetic acid, **GA**, Gibberellin, **CK**, Cytokinin

Perspectives on the Use of Phytohormones to Increase Productivity

Today, agricultural production is challenged by the growing demand for food from a growing population, overwhelmed natural resources, and uncertainty because of climate change. In addition to these factors, abiotic stress represents a limitation to increasing crop productivity. Therefore, there is great interest in producing crops resistant to different types of stress.

Phytohormonal regulation of plant growth and development is a leading perspective in the development of agricultural technologies (Ullah et al. 2018). Specifically, phytohormone engineering could be a promising tool for improving plant productivity and stress tolerance (Ciura and Kruk 2018, Khodanitska et al. 2021). As well as the application of exogenous phytohormones to crops of commercial interest is a potential technology to reduce production costs.

Recent research studies on the production of phytohormones by fungal communities and analyzes their influence to manipulate the homeostasis of phytohormones and take advantage of their benefits on the growth and productivity of different crops (Goyal and Kalia 2020). As well as the genetic engineering of these substances to improve tolerance to abiotic stress in plants (Tiwari et al. 2020).

Some phytohormones improve, in different proportions, various physiological functions such as seed germination, photosynthesis, plant growth, and development. They also increase antioxidant activity and response to stress factors such as salinity, heavy metals, drought, cold, heat, and UV stress (Arif et al. 2020). Some studies suggest that the application of exogenous phytohormones, such as auxins, abscisic acid, cytokinins, gibberellins, jasmonate, salicylic acid, ethylene, brassinosteroids, and strigolactones, mitigate the effects of drought stress, so their use plays a relevant role in crop development in an environment under drought stress (Jogawat et al. 2021).

However, the type and dose of phytohormone besides the crosstalk with endogenous hormones are variables that must be studied in detail. As well as the influence of environmental conditions on the response of plants to treatment, it has been observed that when plants are subjected to different types of abiotic stress and phytohormones are added, it is possible to improve the yield and accumulation of high-value products in crops.

Currently, the objective is to study the effect of the combination and manipulation of phytohormones, abiotic stress, optimized cultivation processes, and genetic modification techniques to improve the productivity of different crops economically and efficiently (Zhao et al. 2019). As well as understanding the role of phytohormones in plants and in the rhizosphere's communities will help improve their agricultural applications (Checker et al. 2018, Ciura and Kruk 2018, Khodanitska et al. 2021).

Therefore, it is necessary to elucidate the interaction mechanisms of the different plant hormones and solve the plant immune network, to fundamentally understand how plants are part of the orchestrated function of the immune system. Then carry out large-scale tests with variable environmental conditions to assess whether the application of phytohormones is an economically viable strategy (Checker et al.

2018, Rhaman et al. 2021, Stirk and van Staden 2020). Once there is a better understanding of both endogenous and exogenous hormonal interactions, more appropriate strategies can be proposed for the application of hormones from other sources on different crops and obtaining high yields at a lower cost.

Conclusion

The importance of fungal phytohormones is linked to the possibility of solving problems associated with different areas related to living beings. The various phytohormones produced by fungi have the ability to influence the growth and development of hosts, the effects that can be differentiated between promoters and inhibitors allow directing future research to have optimization points in the production and performance of plants and crops.

Changes in crop productivity and quality are associated with the ability of phytohormones to modify the physiological development of plants, as well as crop resistance to pathogens. The main result of improving plant health in crops is productivity and quality; however, the economic benefit is usually the main reason for improving the overall health of an agricultural system.

The implications of the use of fungal phytohormones within the field of agriculture to mitigate the negative effects of stress and on pest control are an important topic to be addressed. This provides the opportunity to carry out new investigations with the central axis in the interactions of the compounds produced by the fungi as phytohormones and regulators on the productivity, yield and quality of crops in specific species, even considering the repercussions for a prolonged exposure in its application.

Acknowledgments

The authors thank the Faculty of Engineering for their support in carrying out this research, and the Consejo Nacional de Ciencia y Tecnología (CONACyT) for the scholarships awarded for postgraduate studies (B. M-C 804133, C.J. M-M 796218, M.I. N-R 620684, A.K. R-CH 636395).

References

Achard, P., Vriezen, W.H., Van Der Straeten, D. and Harberd, N.P. 2003. Ethylene Regulates Arabidopsis Development via the Modulation of DELLA Protein Growth Repressor Function. Plant Cell, 15(12): 2816–2825. https://doi.org/10.1105/tpc.015685
Ahanger, M.A., Ashraf, M., Bajguz, A. and Ahmad, P. 2018. Brassinosteroids Regulate Growth in Plants Under Stressful Environments and Crosstalk with Other Potential Phytohormones. Journal of Plant Growth Regulation, 37(4): 1007–1024. https://doi.org/10.1007/s00344-018-9855-2
Ahmad, M., Pataczek, L., Hilger, T.H., Zahir, Z.A., Hussain, A., Rasche, F. et al. 2018. Perspectives of Microbial Inoculation for Sustainable Development and Environmental Management. Frontiers in Microbiology, 9. https://www.frontiersin.org/article/10.3389/fmicb.2018.02992
Al-Ani, L.K.T., Surono, Aguilar-Marcelino, L., Salazar-Vidal, V.E., Becerra, A.G. and Raza, W. 2021. Role of Useful Fungi in Agriculture Sustainability. pp. 1–44. In: A.N. Yadav (Ed.). Recent Trends in Mycological Research: Volume 1: Agricultural and Medical Perspective. Springer International Publishing. https://doi.org/10.1007/978-3-030-60659-6_1

Anand, G., Gupta, R., Marash, I., Leibman-Markus, M. and Bar, M. 2022. Cytokinin production and sensing in fungi. Microbiological Research, 262: 127103. https://doi.org/10.1016/j.micres.2022.127103

Arif, Y., Sami, F., Siddiqui, H., Bajguz, A. and Hayat, S. 2020. Salicylic acid in relation to other phytohormones in plant: A study towards physiology and signal transduction under challenging environment. Environmental and Experimental Botany, 175 (November 2019), 104040. https://doi.org/10.1016/j.envexpbot.2020.104040

Bader, A.N., Salerno, G.L., Covacevich, F. and Consolo, V.F. 2020. Native Trichoderma harzianum strains from Argentina produce indole-3 acetic acid and phosphorus solubilization, promote growth and control wilt disease on tomato (Solanum lycopersicum L.). Journal of King Saud University - Science, 32(1): 867–873. https://doi.org/10.1016/j.jksus.2019.04.002

Baron, N.C. and Rigobelo, E.C. 2022. Endophytic fungi: a tool for plant growth promotion and sustainable agriculture. Mycology, 13: 39–55. https://doi.org/10.1080/21501203.2021.1945699

Barrington, E.J.W. 2020. Hormone. Encyclopedia Britannica. https://www.britannica.com/science/hormone

Bean, K.M., Kisiala, A.B., Morrison, E.N. and Emery, R.J.N. 2021. Trichoderma Synthesizes Cytokinins and Alters Cytokinin Dynamics of Inoculated Arabidopsis Seedlings. Journal of Plant Growth Regulation. https://doi.org/10.1007/s00344-021-10466-4

Begum, N., Qin, C., Ahanger, M.A., Raza, S., Khan, M.I., Ashraf, M. et al. 2019. Role of Arbuscular Mycorrhizal Fungi in Plant Growth Regulation: Implications in Abiotic Stress Tolerance. Frontiers in Plant Science, 10. https://www.frontiersin.org/article/10.3389/fpls.2019.01068

Best, N.B., Hartwig, T., Budka, J., Fujioka, S., Johal, G., Schulz, B. et al. 2016. Nana plant2 encodes a maize ortholog of the arabidopsis brassinosteroid biosynthesis gene DWARF1, identifying developmental interactions between brassinosteroids and gibberellins. Plant Physiology, 171(4): 2633–2647. https://doi.org/10.1104/pp.16.00399

Bhatla, S.C. 2018. Plant Growth Regulators: An Overview. Plant Physiology, Development and Metabolism, 559–568. https://doi.org/10.1007/978-981-13-2023-1_14

Brewer, P.B., Koltai, H. and Beveridge, C.A. 2013. Diverse Roles of Strigolactones in Plant Development. Molecular Plant, 6(1): 18–28. https://doi.org/10.1093/mp/sss130

Bürger, M. and Chory, J. 2019. Stressed Out About Hormones: How Plants Orchestrate Immunity. Cell Host & Microbe, 26(2): 163–172. https://doi.org/10.1016/j.chom.2019.07.006

Calvo, P., Nelson, L. and Kloepper, J.W. 2014. Agricultural uses of plant biostimulants. Plant and Soil, 383(1–2): 3–41. https://doi.org/10.1007/s11104-014-2131-8

Chanclud, E. and Morel, J.-B. 2016. Plant hormones: A fungal point of view. Molecular Plant Pathology, 17(8): 1289–1297. https://doi.org/10.1111/mpp.12393

Checker, V.G., Kushwaha, H.R., Kumari, P. and Yadav, S. 2018. Role of Phytohormones in Plant Defense: Signaling and Cross Talk BT - Molecular Aspects of Plant-Pathogen Interaction. Molecular Aspects of Plant-Pathogen Interaction, 159–184.

Chhaya, Yadav, B., Jogawat, A., Gnanasekaran, P., Kumari, P., Lakra, N., Lal, S.K. et al. 2021. An overview of recent advancement in phytohormones-mediated stress management and drought tolerance in crop plants. Plant Gene, 25: 100264. https://doi.org/10.1016/j.plgene.2020.100264

Ciura, J. and Kruk, J. 2018. Phytohormones as targets for improving plant productivity and stress tolerance. Journal of Plant Physiology, 229(February): 32–40. https://doi.org/10.1016/j.jplph.2018.06.013

Davies, P.J. 1995. *Plant Hormones*. Springer. https://link.springer.com/book/10.1007/978-94-011-0473-9

Depuydt, S. and Hardtke, C.S. 2011. Hormone signalling crosstalk in plant growth regulation. Current Biology, 21(9): R365–R373. https://doi.org/10.1016/j.cub.2011.03.013

Devi, R., Kaur, T., Kour, D., Rana, K.L., Yadav, A. and Yadav, A.N. 2020. Beneficial fungal communities from different habitats and their roles in plant growth promotion and soil health. International Scientific Journal of Microbial Biology, 5(1): 21–47. 10.21608/MB.2020.32802.1016

Dilworth, L.L., Riley, C.K. and Stennett, D.K. 2017. Plant Constituents: Carbohydrates, Oils, Resins, Balsams, and Plant Hormones. pp. 61–80. In: S. Badal and R. Delgoda (Eds.). Pharmacognosy. Academic Press. https://doi.org/10.1016/B978-0-12-802104-0.00005-6

Endo, T., Shimada, T., Nakata, Y., Fujii, H., Matsumoto, H., Nakajima, N. et al. 2018. Abscisic acid affects expression of citrus FT homologs upon floral induction by low temperature in Satsuma mandarin (Citrus unshiu Marc.). Tree Physiology, 38(5): 755–771. https://doi.org/10.1093/treephys/tpx145

Fan, X., Shu, C., Zhao, K., Wang, X., Cao, J. and Jiang, W. 2018. Regulation of apricot ripening and softening process during shelf life by post-storage treatments of exogenous ethylene and 1-methylcyclopropene. Scientia Horticulturae, 232: 63–70. https://doi.org/10.1016/j.scienta.2017.12.061

Fonseca, S., Rosado, A., Vaughan-Hirsch, J., Bishopp, A. and Chini, A. 2014. Molecular locks and keys: The role of small molecules in phytohormone research. Frontiers in Plant Science, 5(DEC): 1–16. https://doi.org/10.3389/fpls.2014.00709

George, E.F., Hall, M.A. and Klerk, G.-J.D. 2008. Plant Growth Regulators I: Introduction, Auxins, their Analogues and Inhibitors. pp. 175–204. In: E.F. George, M.A. Hall and G.-J.D. Klerk (Eds.). Plant Propagation by Tissue Culture: Volume 1. The Background. Springer Netherlands. https://doi.org/10.1007/978-1-4020-5005-3_5

Ghosh, S. and Halder, S. 2018. Effect of different kinds of gibberellin on temperate fruit crops: A review. 315–319.

Goyal, A. and Kalia, A. 2020. Fungal Phytohormones: Plant Growth-Regulating Substances and Their Applications in Crop Productivity. Agriculturally Important Fungi for Sustainable Agriculture. (pp. 143–169). Springer. https://doi.org/10.1007/978-3-030-45971-0_7

Grossmann, K. 2010. Auxin herbicides: Current status of mechanism and mode of action. Pest Management Science, 66(2): 113–120. https://doi.org/10.1002/ps.1860

Gupta, R., Anand, G. and Bar, M. 2023. Developmental Phytohormones: Key Players in Host-Microbe Interactions. J Plant Growth Regul. https://doi.org/10.1007/s00344-023-11030-y

Han, X., Zeng, H., Bartocci, P., Fantozzi, F. and Yan, Y. 2018. Phytohormones and effects on growth and metabolites of microalgae: A review. Fermentation, 4(2): 1–15. https://doi.org/10.3390/fermentation4020025

Hed, B. and Centinari, M. 2021. Gibberellin Application Improved Bunch Rot Control of Vignoles Grape, but Response to Mechanical Defoliation Varied Between Training Systems. Plant Disease, 105(2): 339–345. https://doi.org/10.1094/PDIS-06-20-1184-RE

Illescas, M., Pedrero-Méndez, A., Pitorini-Bovolini, M., Hermosa, R. and Monte, E. 2021. Phytohormone Production Profiles in Trichoderma Species and Their Relationship to Wheat Plant Responses to Water Stress. Pathogens, 10(8): 991. https://doi.org/10.3390/pathogens10080991

Imtiaz, H., Arif, Y., Alam, P. and Hayat, S. 2023. Apocarotenoids biosynthesis, signaling regulation, crosstalk with phytohormone, and its role in stress tolerance. Environmental and Experimental Botany, 210: 105337.

Jogawat, A., Yadav, B., Chhaya, Lakra, N., Singh, A.K. and Narayan, O.P. 2021. Crosstalk between phytohormones and secondary metabolites in the drought stress tolerance of crop plants: A review. Physiologia Plantarum, 172(2): 1106–1132. https://doi.org/10.1111/ppl.13328

Jones, J.D.G. and Dangl, J.L. 2006. The plant immune system. Nature, 444(7117): 323–329. https://doi.org/10.1038/nature05286

Kaya, C., Ugurlar, F., Ashraf, M. and Ahmad, P. 2023. Salicylic acid interacts with other plant growth regulators and signal molecules in response to stressful environments in plants. Plant Physiology and Biochemistry, 196: 431–443.

Keswani, C., Singh, S.P., Cueto, L., García-Estrada, C., Mezaache-Aichour, S., Glare, T.R. et al. 2020. Auxins of microbial origin and their use in agriculture. Applied Microbiology and Biotechnology, 104(20): 8549–8565. https://doi.org/10.1007/s00253-020-10890-8

Khan, N., Bano, A., Ali, S. and Babar, Md. A. 2020. Crosstalk amongst phytohormones from planta and PGPR under biotic and abiotic stresses. Plant Growth Regulation, 90(2): 189–203. https://doi.org/10.1007/s10725-020-00571-x

Khatami, F., Najafi, F., Yari, F. and Khavari-Nejad, R.A. 2020. Expression of etr1-1 gene in transgenic Rosa hybrida L. increased postharvest longevity through reduced ethylene biosynthesis and perception. Scientia Horticulturae, 263, 109103. https://doi.org/10.1016/j.scienta.2019.109103

Khodanitska, O., Shevchuk, O., Tkachuk, O. and Matviichuk, O. 2021. Physiological Activity of Plant Growth Stimulators. The Scientific Heritage, 58(58): 6.

Kim, B., Kwon, M., Jeon, J., Schulz, B., Corvalán, C., Jeong, Y.J. et al. 2014. The arabidopsis gulliver2/phyB mutant exhibits reduced sensitivity to brassinazole. Journal of Plant Biology, 57(1): 20–27. https://doi.org/10.1007/s12374-013-0380-3

Kosakivska, I.V., Vedenicheva, N.P., Babenko, L.M., Voytenko, L.V., Romanenko, K.O. and Vasyuk, V.A. 2022. Exogenous phytohormones in the regulation of growth and development of cereals under abiotic stresses. Molecular Biology Reports, 49(1): 617–628. https://doi.org/10.1007/s11033-021-06802-2

Kuryata, V.G. and Khodanitska, O.O. 2018. Features of Anatomical Structure, Formation and Functioning of 8(1): 918–926. https://doi.org/10.15421/2018

Lang, D., Yu, X., Jia, X., Li, Z. and Zhang, X. 2020. Methyl jasmonate improves metabolism and growth of NaCl-stressed Glycyrrhiza uralensis seedlings. Scientia Horticulturae, 266: 109287. https://doi.org/10.1016/j.scienta.2020.109287

Li, X., Zhou, J., Xu, R.-S., Meng, M.-Y., Yu, X. and Dai, C.-C. 2018. Auxin, Cytokinin, and Ethylene Involved in Rice N Availability Improvement Caused by Endophyte Phomopsis liquidambari. Journal of Plant Growth Regulation, 37(1): 128–143. https://doi.org/10.1007/s00344-017-9712-8

Li, Z.-G., Xiang, R.-H. and Wang, J.-Q. 2021. Hydrogen Sulfide–Phytohormone Interaction in Plants Under Physiological and Stress Conditions. Journal of Plant Growth Regulation, 40(6): 2476–2484. https://doi.org/10.1007/s00344-021-10350-1

Liao, D., Wang, S., Cui, M., Liu, J., Chen, A. and Xu, G. 2018. Phytohormones Regulate the Development of Arbuscular Mycorrhizal Symbiosis. International Journal of Molecular Sciences, 19(10): 3146. https://doi.org/10.3390/ijms19103146

Maithani, D., Singh, H. and Sharma, A. 2020. Stress Alleviation in Plants Using SAR and ISR: Current Views on Stress Signaling Network. https://doi.org/10.1007/978-981-15-7094-0_2

Martínez-Camacho, J.E., Guevara-González, R.G., Rico-García, E., Tovar-Pérez, E.G. and Torres-Pacheco, I. 2022. Delayed Senescence and Marketability Index Preservation of Blackberry Fruit by Preharvest Application of Chitosan and Salicylic Acid. Frontiers in Plant Science, 13. https://www.frontiersin.org/article/10.3389/fpls.2022.796393

Mehmood, A., Hussain, A., Irshad, M., Hamayun, M., Iqbal, A. and Khan, N. 2019. *In vitro* production of IAA by endophytic fungus Aspergillus awamori and its growth promoting activities in Zea mays. Symbiosis, 77(3): 225–235. https://doi.org/10.1007/s13199-018-0583-y

Munasinghe, M.V.K., Kumar, N.S., Jayasinghe, L. and Fujimoto, Y. 2017. Indole-3-Acetic Acid Production by Colletotrichum siamense, An Endophytic Fungus from Piper nigrum Leaves. Journal of Biologically Active Products from Nature, 7(6): 475–479. https://doi.org/10.1080/22311866.2017.1408429

Nutaratat, P., Srisuk, N., Arunrattiyakorn, P. and Limtong, S. 2016. Indole-3-acetic acid biosynthetic pathways in the basidiomycetous yeast Rhodosporidium paludigenum. Archives of Microbiology, 198(5): 429–437. https://doi.org/10.1007/s00203-016-1202-z

Olson, D., Berry, H.M., Riggs, J.D., Argueso, C.T. and Gomez, S.K. 2022. Phytohormone Profile of Medicago in Response to Mycorrhizal Fungi, Aphids, and Gibberellic Acid. Plants, 11: 720. https://doi.org/10.3390/plants11060720

Ozimek, E. and Hanaka, A. 2021. Mortierella Species as the Plant Growth-Promoting Fungi Present in the Agricultural Soils. Agriculture, 11(1): 7. https://doi.org/10.3390/agriculture11010007

Pan, L., Zhao, X., Chen, M., Fu, Y., Xiang, M. and Chen, J. 2020. Effect of exogenous methyl jasmonate treatment on disease resistance of postharvest kiwifruit. Food Chemistry, 305: 125483. https://doi.org/10.1016/j.foodchem.2019.125483

Pusztahelyi, T., Holb, I. and Pócsi, I. 2015. Secondary metabolites in fungus-plant interactions. Frontiers in Plant Science, 6. https://www.frontiersin.org/article/10.3389/fpls.2015.00573

Qiu, C.-W., Zhang, C., Wang, N.-H., Mao, W. and Wu, F. 2021. Strigolactone GR24 improves cadmium tolerance by regulating cadmium uptake, nitric oxide signaling and antioxidant metabolism in barley (Hordeum vulgare L.). Environmental Pollution, 273: 116486. https://doi.org/10.1016/j.envpol.2021.116486

Quesada Moraga, E. 2020. Entomopathogenic fungi as endophytes: Their broader contribution to IPM and crop production. Biocontrol Science and Technology, 30(9): 864–877. https://doi.org/10.1080/09583157.2020.1771279

Rademacher, W. 2015. Plant Growth Regulators: Backgrounds and Uses in Plant Production. Journal of Plant Growth Regulation, 34(4): 845–872. https://doi.org/10.1007/s00344-015-9541-6

Rhaman, M.S., Imran, S., Rauf, F., Khatun, M., Baskin, C.C., Murata, Y. and Hasanuzzaman, M. 2021. Seed Priming with Phytohormones: An Effective Approach for the Mitigation of Abiotic Stress. Plants, 10(37): 1–17.

Robert-Seilaniantz, A., Grant, M. and Jones, J.D.G. 2011. Hormone crosstalk in plant disease and defense: More than just jasmonate-salicylate antagonism. Annual Review of Phytopathology, 49: 317–343. https://doi.org/10.1146/annurev-phyto-073009-114447

Rodríguez-Sánchez, L.K., Pérez-Bernal, J.E., Santamaría-Torres, M.A., Marquínez-Casas, X., Cuca-Suárez, L.E., Prieto-Rodríguez, J.A. et al. 2020. Effect of methyl jasmonate and salicylic acid on the production of metabolites in cell suspensions cultures of Piper cumanense (Piperaceae). Biotechnology Reports, 28: e00559. https://doi.org/10.1016/j.btre.2020.e00559

Salvi, P., Manna, M., Kaur, H., Thakur, T., Gandass, N., Bhatt, D. et al. 2021. Phytohormone signaling and crosstalk in regulating drought stress response in plants. Plant Cell Reports, 40(8): 1305–1329. https://doi.org/10.1007/s00299-021-02683-8

Sattar, A., Ul-Allah, S., Ijaz, M., Sher, A., Butt, M., Abbas, T. et al. 2022. Exogenous application of strigolactone alleviates drought stress in maize seedlings by regulating the physiological and antioxidants defense mechanisms. Cereal Research Communications, 50(2): 263–272. https://doi.org/10.1007/s42976-021-00171-z

Sethi, M., Kaur, C., Hagroo, R.P. and Singh, M.P. 2023. Endophyte mediated plant health via phytohormones and biomolecules. pp. 151–166. In: M.K. Solanki, M.K. Yadav, B.P. Singh and V.K. Gupta (Eds.). Microbial Endophytes and Plant Growth. Academic Press. https://doi.org/10.1016/B978-0-323-90620-3.00017-9

Shahzad, K., Hussain, S., Arfan, M., Hussain, S., Waraich, E.A., Zamir, S. et al. 2021. Exogenously Applied Gibberellic Acid Enhances Growth and Salinity Stress Tolerance of Maize through Modulating the Morpho-Physiological, Biochemical and Molecular Attributes. Biomolecules, 11(7): 1005. https://doi.org/10.3390/biom11071005

Šimura, J., Antoniadi, I., Široká, J., Tarkowská, D., Strnad, M., Ljung, K. et al. 2018. Plant Hormonomics: Multiple Phytohormone Profiling by Targeted Metabolomics. Plant Physiology, 177(2): 476–489. https://doi.org/10.1104/pp.18.00293

Solyman, S.N.E.-D., Abdel-Monem, M.O., Abou-Taleb, K.A., Osman, H.S. and El-Sharkawy, R.M. 2019. Production of Plant Growth Regulators by Some Fungi Isolated under Salt Stress. South Asian Journal of Research in Microbiology, 1–10. https://doi.org/10.9734/sajrm/2019/v3i130076

Song, Q., Cheng, S., Chen, Z., Nie, G., Xu, F., Zhang, J. et al. 2019. Comparative transcriptome analysis revealing the potential mechanism of seed germination stimulated by exogenous gibberellin in Fraxinus hupehensis. BMC Plant Biology, 19(1): 199. https://doi.org/10.1186/s12870-019-1801-3

Souza, L.A., Monteiro, C.C., Carvalho, R.F., Gratão, P.L. and Azevedo, R.A. 2017. Dealing with abiotic stresses: An integrative view of how phytohormones control abiotic stress-induced oxidative stress. Theoretical and Experimental Plant Physiology, 29(3): 109–127. https://doi.org/10.1007/s40626-017-0088-8

Stirk, W.A. and van Staden, J. 2020. Potential of phytohormones as a strategy to improve microalgae productivity for biotechnological applications. Biotechnology Advances, 44(August): 107612. https://doi.org/10.1016/j.biotechadv.2020.107612

Sytar, O., Kumari, P., Yadav, S., Brestic, M. and Rastogi, A. 2019. Phytohormone Priming: Regulator for Heavy Metal Stress in Plants. Journal of Plant Growth Regulation, 38(2): 739–752. https://doi.org/10.1007/s00344-018-9886-8

TariqJaveed, M., Farooq, T., Al-Hazmi, A.S., Hussain, M.D. and Rehman, A.U. 2021. Role of Trichoderma as a biocontrol agent (BCA) of phytoparasitic nematodes and plant growth inducer. Journal of Invertebrate Pathology, 183: 107626. https://doi.org/10.1016/j.jip.2021.107626

Tiwari, P., Bajpai, M., Singh, L.K., Mishra, S. and Yadav, A.N. 2020. Phytohormones Producing Fungal Communities: Metabolic Engineering for Abiotic Stress Tolerance in Crops. Agriculturally Important Fungi for Sustainable Agriculture. (pp. 171–197). Springer. https://doi.org/10.1007/978-3-030-45971-0_8

Tudzynski, B. and Sharon, A. 2002. Biosynthesis, Biological Role and Application of Fungal Phytohormones. pp. 183–211. In: H.D. Osiewacz (Ed.). Industrial Applications. Springer. https://doi.org/10.1007/978-3-662-10378-4_9

Tyśkiewicz, R., Nowak, A., Ozimek, E. and Jaroszuk-Ściseł, J. 2022. Trichoderma: The Current Status of Its Application in Agriculture for the Biocontrol of Fungal Phytopathogens and Stimulation of Plant Growth. International Journal of Molecular Sciences, 23(4): 2329. https://doi.org/10.3390/ijms23042329

Ullah, A., Manghwar, H., Shaban, M., Khan, A.H., Akbar, A., Ali, U. et al. 2018. Phytohormones enhanced drought tolerance in plants: A coping strategy. Environmental Science and Pollution Research, 25(33): 33103–33118. https://doi.org/10.1007/s11356-018-3364-5

Vázquez-Hernández, M.C., Parola-Contreras, I., Montoya-Gómez, L.M., Torres-Pacheco, I., Schwarz, D. and Guevara-González, R.G. 2019. Eustressors: Chemical and physical stress factors used to enhance vegetables production. Scientia Horticulturae, 250: 223–229. https://doi.org/10.1016/j.scienta.2019.02.053

Verma, V., Ravindran, P. and Kumar, P.P. 2016. Plant hormone-mediated regulation of stress responses. BMC Plant Biology, 16(1): 86. https://doi.org/10.1186/s12870-016-0771-y

Vijay, A.K., Prabha, S., Thomas, J., Kurian, J.S. and George, B. 2020. Effect of auxin and its synthetic analogues on the biomass production and biochemical composition of freshwater microalga Ankistrodesmus falcatus CMSACR1001. Journal of Applied Phycology, 32(6): 3787–3797. https://doi.org/10.1007/s10811-020-02247-5

Vincent, D., Rafiqi, M. and Job, D. 2020. The Multiple Facets of Plant–Fungal Interactions Revealed Through Plant and Fungal Secretomics. Frontiers in Plant Science, 10. https://www.frontiersin.org/article/10.3389/fpls.2019.01626

Waadt, R. 2020. Phytohormone signaling mechanisms and genetic methods for their modulation and detection. Current Opinion in Plant Biology, 57: 31–40. https://doi.org/10.1016/j.pbi.2020.05.011

Wani, S.H., Kumar, V., Shriram, V. and Sah, S.K. 2016. Phytohormones and their metabolic engineering for abiotic stress tolerance in crop plants. The Crop Journal, 4(3): 162–176. https://doi.org/10.1016/j.cj.2016.01.010

Waqas, M., Khan, A.L., Kamran, M., Hamayun, M., Kang, S.-M., Kim, Y.-H. et al. 2012. Endophytic Fungi Produce Gibberellins and Indoleacetic Acid and Promotes Host-Plant Growth during Stress. Molecules, 17(9): 10754–10773. https://doi.org/10.3390/molecules170910754

Wei, X., Chen, J., Zhang, C. and Pan, D. 2016. A New Oidiodendron maius Strain Isolated from Rhododendron fortunei and its Effects on Nitrogen Uptake and Plant Growth. Frontiers in Microbiology, 7. https://www.frontiersin.org/article/10.3389/fmicb.2016.01327

Xiang, W., Wang, H.-W. and Sun, D.-W. 2021. Phytohormones in postharvest storage of fruit and vegetables: Mechanisms and applications. Critical Reviews in Food Science and Nutrition, 61(18): 2969–2983. https://doi.org/10.1080/10408398.2020.1864280

Xie, R., Ge, T., Zhang, J., Pan, X., Ma, Y., Yi, S. et al. 2018. The molecular events of IAA inhibiting citrus fruitlet abscission revealed by digital gene expression profiling. Plant Physiology and Biochemistry, 130: 192–204. https://doi.org/10.1016/j.plaphy.2018.07.006

Xiubing, G., Yan, L., Chunyan, L., Can, G., Zhang, Y., Ma, C. et al. 2023. Individual and combined effects of arbuscular mycorrhizal fungi and phytohormones on the growth and physiobiochemical characteristics of tea cutting seedlings. Front Plant Sci. 14: 1140267. https://doi.org/10.3389/fpls.2023.1140267

Yadav, A.N., Kour, D., Kaur, T., Devi, R. and Yadav, A. 2022. Endophytic fungal communities and their biotechnological implications for agro-environmental sustainability. Folia Microbiol., 67: 203–232. https://doi.org/10.1007/s12223-021-00939-0

Yu, T., Liu, Y., Fu, J., Ma, J., Fang, Z., Chen, J. et al. 2021. The NF-Y-PYR module integrates the abscisic acid signal pathway to regulate plant stress tolerance. Plant Biotechnology Journal, 19(12): 2589–2605. https://doi.org/10.1111/pbi.13684

Yuan, H., Zhao, L., Guo, W., Yu, Y., Tao, L., Zhang, L. et al. 2019. Exogenous application of phytohormones promotes growth and regulates expression of wood formation-related genes in Populus simonii × P. nigra. International Journal of Molecular Sciences, 20(3). https://doi.org/10.3390/ijms20030792

Zanne, A.E., Abarenkov, K., Afkhami, M.E., Aguilar-Trigueros, C.A., Bates, S., Bhatnagar, J.M. et al. 2020. Fungal functional ecology: Bringing a trait-based approach to plant-associated fungi. Biological Reviews, 95(2): 409–433. https://doi.org/10.1111/brv.12570

Zhang, S., Deng, Y.Z. and Zhang, L.-H. 2018. Phytohormones: The chemical language in Magnaporthe oryzae-rice pathosystem. Mycology, 9(3): 233–237. https://doi.org/10.1080/21501203.2018.14834 41

Zhang, Y. and Liao, H. 2021. Epibrassinolide improves the growth performance of Sedum lineare upon Zn stress through boosting antioxidative capacities. PLOS ONE, 16(9): e0257172. https://doi.org/10.1371/journal.pone.0257172

Zhao, Y., Wang, H.P., Han, B. and Yu, X. 2019. Coupling of abiotic stresses and phytohormones for the production of lipids and high-value by-products by microalgae: A review. Bioresource Technology, 274(October): 549–556. https://doi.org/10.1016/j.biortech.2018.12.030

4

Implication of Ethylene as a Regulator of Disease Resistance in Plants

Sumi Sarkar,[1] Nor Anis Nadhirah Md Nasir,[2] Irnis Azura Zakarya[2] and A.K.M. Aminul Islam[1],*

Introduction

Plants face numerous biotic and abiotic stresses in their lifetime. Plants stimulate effective defense responses by producing phytohormones which help in activating defense gene expression. Previously, hormones were known to contribute to the growth and expansion of plants but later the function of these hormones in defense activity was revealed through analysis of the genetic constituent of the model plant *Arabidopsis thaliana*. In both natural ecosystems and man-induced ecosystems, plants are endlessly subjected to several microbial pathogens. Particular metabolic pathways of plants help them to give rise to distinct resources of defense compounds allowing plants to resist pathogen attacks (Handrick et al. 2016). Plants under biotic stress, most specifically under pathogen attack, activate defense signaling networks through hormonal regulation. Phytohormones, comprising ethylene (ET), auxin (AUX), cytokinin (CK), gibberellic acid (GA), abscisic acid (ABA), salicylic acid (SA), jasmonic acid (JA), brassinosteroid (BR), and the recently recognized strigolactones (SLs) activate different genes to express effectual defense responses under various kind of stresses (Ku et al. 2018).

[1] Department of Genetics and Plant Breeding, Faculty of Agriculture, Bangabandhu Sheikh Mujibur Rahman Agricultural University, Gazipur 1706, Bangladesh.
[2] School of Environmental Engineering, Universiti Malaysia Perlis (UNIMAP), Kompleks Pusat Pengajian Jejawi 3, 02600, Arau, Perlis, Malaysia.
* Corresponding author: aminulgpb@bsmrau.edu.bd

Among all of these beneficial hormones, ethylene has been significantly found to prompt a number of pathogenesis-associated proteins that help plants defend against several pathogen attacks. The influence of ethylene on this resistance network depends on the multi-gene families that are responsible for ET biosynthesis and signal transduction that contributes to producing plant responses against pathogens (Khan et al. 2017). A complex signaling network is activated by plants under pathogen attack where ethylene (ET) takes part in a crucial part of the plant defenses activation through coordinating with jasmonic acid (JA) and salicylic acid (SA). ET along with JA activates the defenses against necrotrophic pathogens whereas ET alone has been found to be involved in defense activation in opposition to some biotrophic and hemibiotrophic pathogens as well (Berens et al. 2019). Therefore, keeping in mind this network of hormones, it is fundamental to have a perception of the significance of ET to know how the coordination of cells takes place in response to pathogen attacks. There are some transcription factors that control a number of genes under stress conditions. Under pathogen attack, up-regulation or down-regulation of these genes' expression is controlled by the ethylene signaling pathway, thus providing plants either resistance or susceptibility based on the pathogenic behavior as well ethylene-induced genes expression capability against the disease-causing pathogen (Wang et al. 2022). Proper understanding of the regulation factors of ethylene signaling pathways can help us to understand the concept of how and when ethylene production by the plants, as well as exogenous application of ethylene, can be helpful in changing the genetic makeup effectively to engender resistance in plants against disease-causing organisms. So, the aim of this study is to focus on the physiological and genetic aspects of plant response in terms of ethylene production under biotic stresses and reveal how plants develop resistance against major disease-causing organisms through the factors of ethylene.

Plant Hormones

Phytochemicals are naturally produced by plants and control the physiological developments of plants. Plant hormones are responsible for regulating different cellular processes and to fulfill these processes, they are either produced in that particular cell where they work or they move to another organ to perform a specific function (Asami and Nakagawa 2018). Several research and investigations preceded the detection of the major hormones of plants such as auxin, cytokinin (CK), ethylene (ET), abscisic acid (ABA), gibberellins (GA), brassinosteroids (BR), jasmonic acid (JA), salicylic acid (SA), and strigolactones (Jiang and Asami 2018). All the biological processes of a plant are affected by phytohormones either in a direct or indirect way. Each hormone has identical nature of occurrence, movement, and impacting cells (Table 1). Hormones are involved in the growth and development of plants specially called plant growth regulators and they may be applied synthetically as well as for promoting the growth of the plant, for example, ethylene (Zahid et al. 2023). They also enable plants to adapt to variable environments by interceding growth, expansion, and distribution of nutrients.

Table 1. Important phytohormones and thenature of their impact on plants.

Name of hormones	Base	Character	Biosynthesis in plant	Movement	Role	References
Ethylene (ET)	Methionine	Studied as the stress hormone and fruit ripening hormonal properties produced by many plant tissues under stress	Most of the tissues under stress as well as tissues, that are enduring senescence or ripening	Transfers through diffusion and has an effect at a distance from where it has been produced	• Prompt defense responses to injury or disease • Dormancy release • Growth and differentiation of shoot and root • Abscission of leaves and fruits • Induction of femaleness and flower, leaf senescence, and fruits ripening	Zahid et al. 2023, Asami and Nakagawa 2018, Campos-Rivero et al. 2017
Auxin	Indole Acetic Acid (IAA)	Primarily found in plants and derived from tryptophan metabolism	• Leaf primordial • Young leaves • Developing seeds	• Vascular cambium • Epidermal cells • Procambial strands • Root	• Cell division and enlargement • Growth of flower parts • differentiation of vascular tissue • Delayed ripening	Zahid et al. 2023, Campos-Rivero et al. 2017, Blázquez et al. 2020
Cytokinin (CK)	Zeatin	Originated from modification of adenine derivatives	Mostly in root tips as well as developing seeds	From roots to shoots through the xylem	• Cell division and shoot formation from callus • Morphogenesis—in tissue culture like the formation of adventitious bud in culture and cutting • Development of chloroplast as well as nutrient signaling • Delayed senescence	Zahid et al. 2023 Fahad et al. 2015 Uniyal et al. 2022

Gibberellins (GA)	Ent-gibberellane	Diterpenoid compounds formed both in plants and fungi and bacteria as well	Starts in the chloroplast. Besides, in new tissues of the shoot and emerging seed.	Xylem and phloem tissues	• Seed germination through producing a number of enzymes • Differentiation of pollen in angiosperms and induction of maleness in dioecious flowers • Stem elongation	Zahid et al. 2023, Fahad et al. 2015 Orozco-Mosqueda et al. 2023
Abscisic acid (ABA)	Abscisin	Isolated first in cotton from carotenoids association with abscission and dormancy	In mature, leaves, seeds, and roots. Most specifically under water stress	From leaves to roots through phloem and it may circulate from roots to leaves through xylem.	• Regulation of stomatal closure • Response to environmental stresses • Cell division and elongation • Control of organ size • Seed and bud dormancy	Zahid et al. 2023, Rubio et al. 2009, Muhammad Aslam et al. 2022
Brassinosteroids (BR)	Brassinolide	This group contains around 60 steroidal compounds characterized by brassinolide	First extracted from the pollen of *Brassica napus*	All parts of the plant	• Growth promoter steroid • Seedling photomorphogenesis • Male fertility • Cell elongation • Promotes gene transcription for encoding xyloglucanases that promotes wall relaxing thus leading to stem elongation	Zhabinskii et al. 2015, Rubio et al. 2009
Salicylic acid (SA)	Amino acid phenylalanine	Potential regulatory phenolic compounds produced by plants	First isolated from Salix (willow) bark extract and biosynthesized from the amino acid phenylalanine	Different plant parts	• Prompts the 'pathogenesis-related proteins' production thus enhancing resistance to pathogens • Augments longevity of flower • Obstructs germination of seed • Impedes response to wound and inverts the effects of ABA. • Functions as a growth regulator	Asami and Nakagawa 2018, Campos-Rivero et al. 2017, Janda et al. 2020

Table 1. contd....

... *Table 1. contd.*

Name of hormones	Base	Character	Biosynthesis in plant	Movement	Role	References
Jasmonic acid (JA)	Jasmonates	Represented by its methyl ester and termed by the name of jasmine plant. In this plant, methyl ester is a component of aroma.	The main sites of biosynthesis are Cellular organelles such as chloroplasts and peroxisomes	Leaves and roots	• Plant defense • Promote senescence and abscission • Pigment formation, tuber formation, and fruit ripening • Responses to water deficit	Asami and Nakagawa 2018, Campos-Rivero et al. 2017, Blázquez et al. 2020
Strigolactones	Strigol	Carotenoid derivative	First isolated from the root of cotton	Recognition of the plant by symbiotic fungi and help those fungi to supply phosphate and other soil nutrients to plants	• Stimulant for the root parasite • Regulate axillary shoot growth and leaf senescence • Inhibit shoot branching in plants • Lateral root formation and root hair elongation • Play significant roles in plant responses to biotic and abiotic stresses	Banerjee and Bhadra 2020, Zwanenburg and Pospíšil 201, Blázquez et al. 2020

Hormones Involved in Disease Resistance

Phytohormones largely control plant growth and reaction of the plant to diverse biotic and abiotic components. The plant hormones, most specifically ethylene (ET), abscisic acid (ABA), jasmonic acid (JA), and salicylic acid (SA), are crucial for creating immune responses in plants (Asif et al. 2022). Signaling pathways that are regulated by ethylene/jasmonic acid and salicylic acid play significant roles in resistance to plant pathogens (Robert-Seilaniantz et al. 2011). The defense mechanism of plants in opposition to biotrophic pathogens (pathogens that need living tissue to fulfill their life cycle) is regulated in a positive way by SA signaling, whereas ET/JA pathways regulate the upward resistance against necrotrophic pathogens (pathogens that decompose plant tissue during infection) (Derksen et al. 2013). Nonspecific disease resistance by ethylene and jasmonic acid through signaling pathways are imperative for plants and this process is found apparent from the typical systemic resistance regulated by salicylic acid (Fig. 1) (Xu et al. 2022).

Resistance against biotrophic and hemibiotrophic pathogens like *Hyaloperonospora arabidopsidis, P. syringae*, is regulated by SA (Asif et al. 2022). Both JA and ethylene were found to be participating in plant disease resistance to the necrotrophic fungus *Alternaria brassicicola* when the infection of *Arabidopsis* plants was inspected. An augmented level of JA and ethylene due to this pathogen attack helped in the accumulation of the plant defense as antifungal activity against *A. brassicicola* (Xu et al. 2022). Augmented production of ethylene is an active and early reaction of plants under pathogen attack which is allied with the initiation of defense responses (Coatsworth et al. 2023). SA was found to be synthesized in

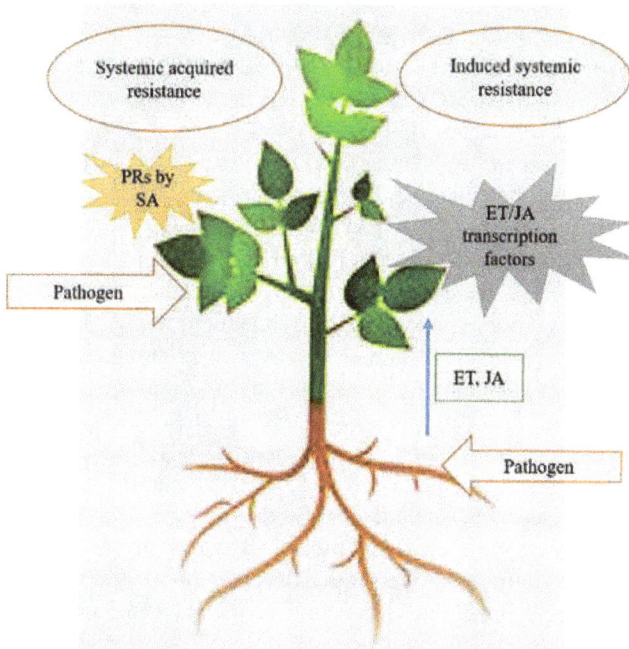

Fig. 1. Different resistance strategies by ET, JA, and SA.

Arabidopsis from chorismate which is a precursor of tryptophan as well as auxins and this would happen either via phenylalanine or through isochorismate (Peng et al. 2021). Besides, as a central immunity system for inducing proper resistance responses and reducing associated fitness costs, SA acts together with ET and JA signaling pathways (Zhao et al. 2021). Mediation of resistance (R) gene and basal resistance by SA evolves a positive link between SA-mediated plant defense and the antiviral machinery of small interfering RNA (siRNA) (Baebler et al. 2014). The stimulation of systemic acquired resistance (SAR) in distal tissues of plant is operated by SA that helps in reducing the impacts of secondary attacks by the pathogen (Alazem and Lin 2015).

Systemic acquired resistance (SAR) is naturally triggered in healthy systemic plant tissues of an infected plant. For the inception of SAR, salicylic acid (SA) contributes as a vital signaling molecule because genes that carry antimicrobial properties and encode for pathogenesis-related proteins (PRs) require SA for their activation (Sofy et al. 2021). On the other hand, ethylene and jasmonic acid-mediated signaling pathways control the induced systemic resistance (ISR) which is not directly associated with PR genes activation because, in this kind of resistance method, a long-distance signal moves across the vascular system to the distal tissues upon pathogen infection and initiate systemic immunity in the plant parts above the soil through colonization beneficial microorganisms at plant roots (Hillmer et al. 2017). Both SAR and ISR can effectively act against wide-ranging virulent plant pathogens.

Ethylene Production by Plants

Ethylene production in plants is basically controlled by internal signals in response to different environmental factors such as biotic stress. For example, a pathogen attack as well as abiotic stresses like salt, hypoxia, chilling, wounding, ozone, etc. To know how ethylene plays a vital role in the functioning of plants, it is important to have a clear concept of synthesizing ad signal-transducing mechanism of ethylene. In a model chemical coordination comprising ascorbic acid and Cu^{2+}, the amino acid was readily converted into ethylene, and based on this observation, one amino acid—methionine—was first suggested as a possible biological precursor of ethylene (Binder 2020). There are major three enzymes that are involved in ethylene production and regulation namely, S-adenosylmethionine (SAM) synthetase which helps methionine to convert into SAM, 1-aminocyclopropane-1-carboxylic acid (ACC) synthase that is needed to transform the SAM into ACC, and finally, ACC oxidase that alters ACC to ethylene (Fig. 2) (Bakshi et al. 2015).

Ethylene production level in the plant is generally found low but the level proliferates both under developmental and environmental signals. In most cases, ethylene biosynthesis and accumulation levels increase dramatically during the ripening of fruits of ethylene and under biotic and abiotic stresses because of what is also called the stress hormone. ACC synthase is the major regulation site for ethylene biosynthesis. In *Arabidopsis*, one of the ethylene mutant's eto1 was identified which has a constitutive phenotype to ethylene response and it has been shown in consequent studies that two other mutants—eto2 and eto3—with similar traits have higher

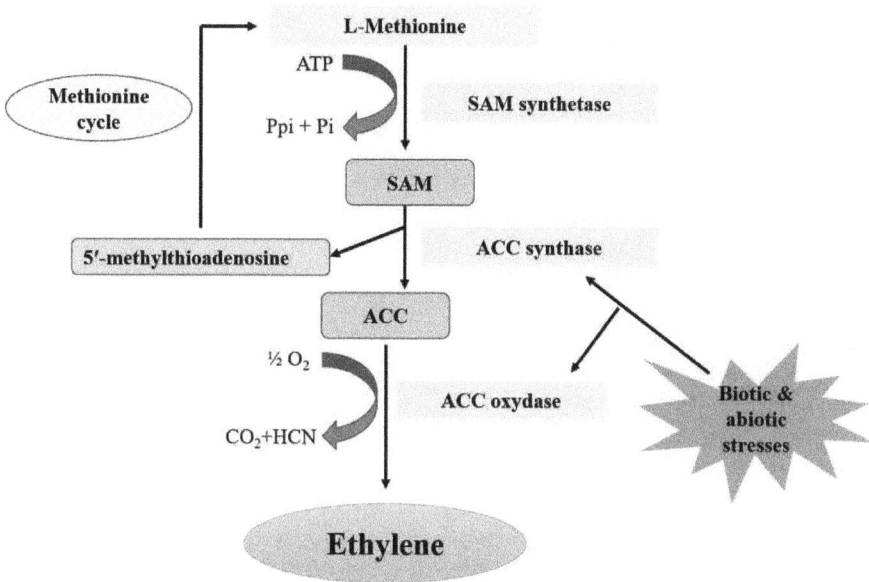

Fig. 2. Synthesis of ethylene under different stress in plant. Source: (Wang et al. 2002, Bakshi et al. 2015, Fatma et al. 2022).

activities for ACC synthase (Fatma et al. 2022). Ethylene disperses throughout the plant after biosynthesis and after that binds to the receptors of ethylene to accelerate ethylene responses (Binder 2020).

Regulatory Factors of Ethylene Production

Various developmental and environmental factors regulate ethylene production. It has been recognized in diverse studies that ethylene plays an essential function in the climacteric fruits ripening (Kou et al. 2021). Different kinds of biotic as well as abiotic stresses, including temperature, drought, chemicals, waterlogging, radiation, damage by insects, mechanical wounding, or disease by pathogen attack increase ethylene production (Das et al. 2016). Ethylene that is derived from stress produces adaptive responses triggering signals as well as influences signaling pathways of other hormones (Pieterse et al. 2009).

Abiotic Factors

The concentration of ethylene in plants is morpho-physiologically and genetically regulated by various abiotic stresses. Genes that are involved in ethylene regulation are changed in expression by chilling stress (Dong et al. 2022). The cold signaling pathway regulates the expression of various hormones including ethylene, salicylic acid, and abscisic acid (Yadav 2010). In rice, the level of ethylene was found to be altered under cold stress whereas under flooding and submergence, an acclimation is found at the ethylene level. Under submergence, the ethylene concentration rapidly rises in rice. Due to the slower rate of ethylene diffusion in water than in air and due to such mechanism conditions. mesocotyl, leaf, coleoptile, internode, and petiole elongation are promoted (Fukao et al. 2009, Abiri et al. 2017). Plants that are found in flood-prone areas activated imperative adaptive responses through flooding-induced ethylene accumulation (Nasrullah et al. 2022).

Ethylene biosynthesis is promoted under salinity stress, which can trigger the downstream network and alter the expression of the genes' salt tolerance in plants is regulated by ethylene either in a positive or negative way (Peng et al. 2014). Both at the transcriptional and post-transcriptional levels, ethylene biosynthesis regulation was operated by a rate-limiting enzyme, Acyl-CoA synthetase (ACS) under salinity stress and in different monocotyledonous and dicotyledonous plants were facing salt stress at a different level, these genes found to be up-regulated (Tao et al. 2015). Under salinity stress, ethylene precursor ACC level increased in rice roots, thus promoting inhibition of shoot elongation. In water-deficit conditions, biosynthesis of ethylene found to increase in rice. The restriction in growth and shoot elongation of rice occurs mostly before the water potential declines in the aerial parts of plant which is a kind of defense mechanism against drought and it may occur in response to ABA and ethylene-induced signals in the root (Kazan 2015). Under hypoxia, hypocotyls, and roots of *Arabidopsis* can form lysigenous aerenchyma through H_2O_2 and ethylene signaling that helps in the gaseous exchange of the plant (Feifei et al. 2017). Reactive oxygen species (ROS) are produced by ethylene under many abiotic stresses and AP2/ERF, which is an ethylene-responsive transcription factor, regulates the primary synthesis and signaling of ROS (Fig. 3). Under abiotic stresses, ethylene and ROS signaling are linked by this AP2/ERF gene family (Iqbal et al. 2017).

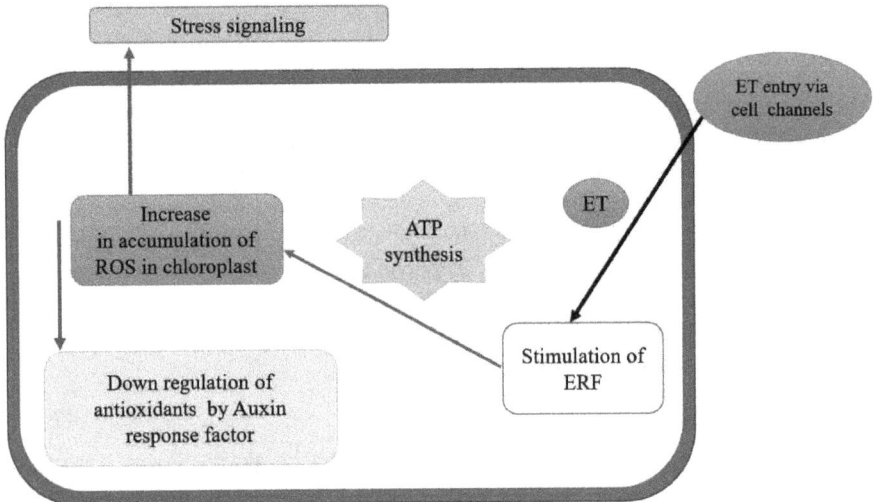

Fig. 3. Response of ET signaling pathway under abiotic stress. Source: (Husai et al. 2020).

Biotic Factors

Under a number of biotic stresses, most specifically, insects and microbial infections, ethylene was found to be regulating plant biological mechanisms (Müller and Munné-

Bosch 2015). Itacts either positively or negatively depending on the conditions. Under biotic stress, the biosynthesis of ethylene participates in mitigating adverse effects. One of the most significant plant immune responses to pathogen attack is the biosynthesis of ethylene that induces the plant's defense mechanism (Gamalero and Glick 2015).

As a biotic factor, microbes have the potential to affect the steps of the ethylene regulation pathway. Plant-associated several microbes are capable of inducing the level of ethylene either through direct synthesis of ethylene or enhancing the activity of the ACS enzyme. In the pathogen *Ralstonia solanacearum*, it was first reported that ethylene is also produced by microbes (de Pedro-Jové et al. 2022). Ethylene regulated normal fruit ripening become complicated due to response of pathogens (Seymour et al. 2013). Ethylene responsive pathway was regulated in tomato fruit infected by *Colletotrichum* as well as both tomato and grapes with *Botrytis* (Agudelo-Romero et al. 2015). The resistant unripe fruit of tomato responded to such biotrophic as well as nectrotrophic stage of pathogen mainly through the activation of SA, JA, ET, and ABA (Fig. 4) (Alkan et al. 2015).

Resistance would be reduced or stimulated depends on the timing of plant exposure to ethylene (Erofeeva 2022). The ethylene regulatory factor AP2/ERF plays a vital role in pathogenesis-related (PR) gene expression (Huh 2022). Ethylene signaling components in rice can distinguish distinct receptor molecules; mespecially—EIN3 (ETHYLENE INSENSITIVE 3) orthologue, EIN2, EIN5, RTE1, EBF1/2, and CTR1 homologues (CONSTITUTIVE TRIPLE RESPONSE 1) (Abiri et al. 2017). Under pathogen attack, ET binds to its receptors in endoplasmic reticulum that ultimately results into increased activation of EIN2 and reduced CTR1 activation as well. Thus, EIN3/EIL1 expression is stimulated by the cleavage and migration of the EIN2-C terminal domain to nucleus that later helps to build up

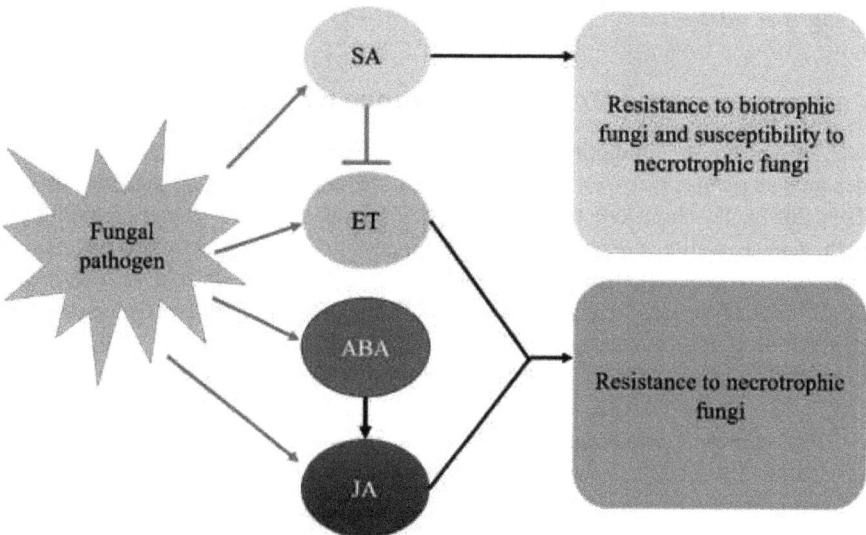

Fig. 4. Basic response of major plant hormones under fungal attack. Source (Alkan and Fortes 2015).

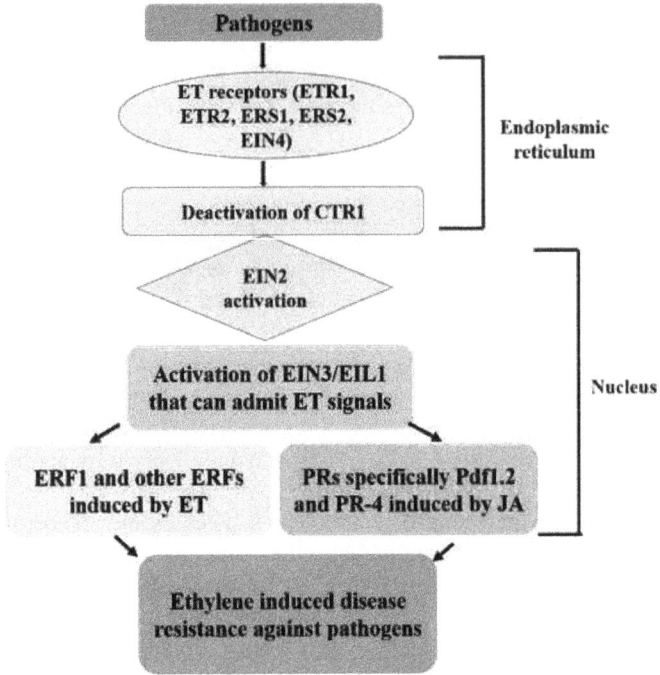

Fig. 5. Ethylene-induced signaling pathway in response to pathogen attack, Source: (Abiri et al. 2017).

resistance along with JA signaling factors (Fig. 5). In rice, ERF1 is induced by both jasmonic acid and ethylene as well as exploits downstream of EIN3/EIL1 (Aerts et al. 2021). As a reaction to infections by bacteria and fungus, ERF1 triggers the genes those are related to defense such as Pdf1.2 and PR-4 and JA-inducible genes as well (Yang et al. 2015).

Pathogenesis and Disease Resistance by Ethylene

Ethylene promotes immunity but whether its effect would be positive or negative depends on plant species, environmental situation, and pathogen type (Gorshkov and Tsers 2022). This hormone can provide as a virulence factor for bacterial and fungal pathogens as well as contributes in disease resistance through acting as a signaling compound (Chagué et al. 2006). Ethylene show induced resistance against pathogen of different lifestyles. Several evidences have been found of the resistance capacity of ethylene against fungal, bacterial, and viral pathogens in various crops. It has long been comprehended that the plant growth-promoting fungi (PGPF) contributes to improving plant expansion and suppressing diseases in plant (Chamkhi et al. 2022). A PGPF namely *Penicillium viridicatum* GP15-1 found to create induced systemic resistance (ISR) against *Pseudomonas syringae* in *Arabidopsis thaliana*. In such growth promotion of plant and regulation of ISR by the PGPF in *Arabidopsis*, the direct involvement of the ethylene signaling pathway has been found that provides evidence of the substantial role of ethylene in the plant's growth and resistance to plant diseases (Hossain et al. 2017).

Fungal Diseases

Most of the plant diseases are caused by various fungi. Ethylene plays a significant role in the plant resistance against a number of plant fungi (Jasrotia and Jasrotia 2022). For example, due to ethylene insensitivity, disease severity increased in soybean by the fungus *Rhizoctonia solani*; on the other hand, resistance against *R. solani* in rice was promoted by the enhanced expression of ACS2 (Helliwell et al. 2013). *Botrytis cinerea*, a fungus that is accountable for causing gray mold, is also competent to cause infection in an inclusive variety of vegetable crops and ornamental and fruit plants. But, ethylene was found to be involved in resistance to this fungus in case of carrot (Coatsworth et al. 2023). In angiosperms, ET seems to act positively in the resistance against necrotrophic pathogens as reported in *A. thaliana* (Shekhawat et al. 2023). In soybean, *A. thaliana*, tobacco, and rice, ethylene was found to contribut to the resistance against pathogens that were biotrophic and hemibiotrophic (Nico et al. 2013). A relative abundance of resistance-related metabolites of *F. graminearum* is regulated by ethylene signaling as well as in the roots of maize seedling, regulation of resistance to *F. graminearum* (the most familiar pathogens that cause harm to maize seedling) take place by ethylene (Munkvold and White 2016). To develop resistance in maize against this pathogen, acetylated feruloylsucrose accumulation through biosynthesis of ethylene is obligatory (Fig. 6).

Two major metabolites of maize—3,6-diferuloyl-3′,6′-diacetylsucrose (Smilaside A) and 3,6-diferuloyl-2′,3′,6′-triacetylsucrose (smiglaside C)—were found to provide resistance against *Fusarium graminearum* as well as many other fungal pathogens through establishing a co-segregated association with the ethylene signaling gene ZmEIN2 (ETHYLENE INSENSITIVE 2) (Zhou et al. 2019). Plant

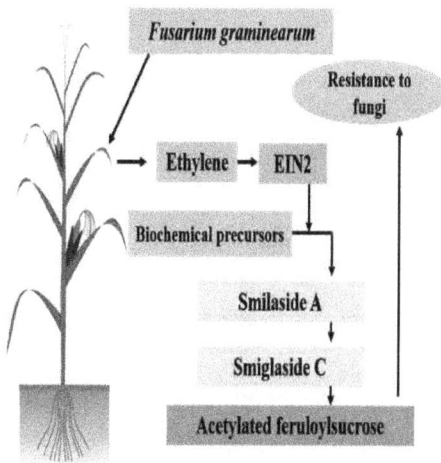

Fig. 6. Role of ethylene to induce resistance in maize against *Fusarium graminearum* fungus. Source: (Pattyn et al. 2021).

resistance to *Magnaporthe oryzae* invasion is positively regulated by endogenous accumulation of ethylene. In different rice varieties, higher ethylene levels may contribute as one of the key factors to disease resistance against *M. oryzae* (Yang et al. 2017). The plant's immune system can be activated through overexpression of several ERF proteins. From different plant species, many ethylene regulatory factor (ERF) genes have been derived such us *Arabidopsis*, *Vitis vinifera*, *Glycine max*, and *Oryza sativa*. A greater part of these genes are associated with biotic and abiotic stresses (Khoudi 2022). An ERF, TaERF3 participates in resistance to *R. cerealis* and *F. graminearum* through ET-JA signaling pathways (Zhang et al. 2007).

Bacterial Diseases

Ethylene provides a dominant function in regulating colonization of different bacteria such as endophytic, phyllospheric, and rhizospheric by the inflection of plant immune as well as symbiotic system. Ethylene participates in plant growth as well as resistance and symbiotic systems prompted by bacteria through influencing bacterial assembly in plant (Guinel 2015). Besides ethylene, involvement of precursor of ET like 1-aminocyclopropane-1-carboxylate (ACC) in plant protection responses has been found in several reports (Vanderstraeten and Straeten 2017). As a consequence of the major role of ET and ACC, many bacteria that are closely associated with plant reveals complicated mechanism in order to spot and regulate the ET and ACC levels within plant tissues and in the rhizosphere (Nascimento et al. 2018). *Azospirillum* sp. B510 is one well-characterized beneficial strain that is acquired from the stems of *Oryza sativa* cv. Nipponbare (Bao et al. 2013). When this bacterial strain comes under endophytic-colonization with plants, both the growth and yield is augmented and resistance is stimulated against *Piricularia oryzae* (syn. *Magnaporthe oryzae*) and *Xanthomonas oryzae* pv. *Oryzae*; those are the causal organisms of bacterial blight and rice blast disease of rice, respectively (Yasuda et al. 2009). This same strain was also obtained to stimulate resistance to bacterial speck in dicotyledonous tomato caused by *Pseudomonas syringae* pv. Tomato (Fujita et al. 2017). ET signaling is required for having a proper perception about the molecular mechanism of resistance to diseases prompted by *Azospirillum* endophytic-colonization (Poorni et al. 2023).

Many destructive plant diseases are caused by the major hemibiotrophic bacterial pathogen *Pseudomonas syringae*. *Arabidopsis thaliana* initiated resistance to Pst DC3000 (*Pseudomonas syringae* pv. tomato DC3000) and this process was found to be stimulated by the BT4 (BTB AND TAZ DOMAIN PROTEIN 4) protein function. The defense response to Pst DC3000 in bt4 mutants induced by salicylic acid (SA) was found to be weakened due to the disruption of BT4. But, BT4 transcription was controlled by both the signaling pathways of SA and ET under infection by Pst (Zheng et al. 2019). In the promoter of BT4, the particular binding modules of ERF (ETHYLENE RESPONSE FACTOR) proteins along with GCC box and dehydration responsive/C-repeat elements were found. Different transcription factors (TFs) play pivotal roles in diverse hormone signaling pathways regulation and signal transduction in order to mediate manifestation of genes responsible for defense (Campos et al. 2022). A number of hormone stimuli (such ET, SA, JA, and different signaling pathways of defense) are related to resistance and assimilated by most of the ERF genes during downstream of the ET signaling pathway (Zander et al. 2014). One of the key steps in ET/SA- mediated resistance of plant to Pst DC3000 is the transcriptional activation of BT4 by ERF11 (Catinot et al. 2015).

Viral Diseases

Ethylene influences most of the phases of plant development but does not activate defense-related genes or induce resistance normally. At infection site of plant, high levels of ethylene produced and surrounding cells sensed these gradients which helped in activating a proper defense mechanism as a response to the locally concentrated ethylene (Coatsworth et al. 2023). Accumulation of ethylene may happen during viral infection in plants. Numerous strategies are utilized by plant viruses that are favorable for the replication and spread of virus inside the cell of plant (Islam et al. 2017). Phytohormones may act against viruses to provide resistance to plant against viral infection. On the other hand, accumulation of phytohormone and their signaling pathways may be disrupted by virus infections either directly or indirectly. For example, AGD2-LIKE DEFENCE RESPONSE PROTEIN 1 (ALD1), was found in wild *Arabidopsis*, where it was found to trigger a defense against pathogens by salicylic acid (SA) pathway regulation. Later, it was found that the regulation of ALD1-mediated resistance in *Nicotiana benthamiana* (NbALD1) against viruses was linked with ethylene signaling pathways (Cecchini et al. 2015). When analyzing the role of ethylene pathway against turnip mosaic virus (TuMV), ethylene precursor content like ACC as well as ERF3 expression level were increased when NbALD1 was silenced whereas, expression of ERF3 and accumulation of ACC were hindered through the overexpression of NbALD1 (Wang et al. 2013, Yang et al. 2017). So, here we can find an adverse control of ethylene signaling mechanism by NbALD1, which indicated the enhancement of *N. benthamiana* resistance to TuMV through the suppression of ethylene pathway (Casteel et al. 2015). In plants infected by the cauliflower mosaic virus (CaMV), expression of the main symptom determinant protein CaMV-P6 was found to be dependent on the connections between ethylene-associated components and P6 (Love et al. 2007). Besides, reduced susceptibility of *Arabidopsis* mutants to CaMV found when ethylene signaling showed defects (Wang et al. 2019). When the tobacco mosaic virus (TMV) was analyzed, more resistance found in ethylene-insensitive plants. Ethylene insensitive Tetr tobacco plants were unable to express ET and JA inducible genes- PR-1g and PR-5c respectively, although ethylene production was greatly stimulated in these plants in response to TMV infection (Coatsworth et al. 2023). So, here the plants that were ethylene insensitive showed defensive expression to the virus.

Additionally, ethylene was obtained to take part in a major role through systemic resistance in plants against the chili veinal mottle virus (ChiVMV) because susceptibility to this virus was intensely augmented when both the genes for ethylene biosynthesis and signaling were silenced (Zhu et al. 2014). In order to induce resistance against two unrelated plant viruses such as potyvirus and camivirus, suppression of RNA silencing of these viral proteins (potyvirus HC-Pro and carmovirus P38 respectively) is needed, thus requiring RAV2 which is an rthylene-inducible transcription factor (Agaoua et al. 2022). So, it can be say that under viral infection, ethylene shows a complicated and variable role depending on the association of viral as well as ethylene-induced transcription factors.

Ethylene as a Defense Regulator against Plant Disease and Defense-signaling Pathways

Ethylene signaling pathways activation

In case of defense signaling, different proteins as well as small molecules contribute to resistance against plant pathogens and ethylene is one of them (Mignolet-Spruyt et al. 2016). It regulates specific areas in defense signaling where several signaling pathways interact to execute appropriate defense responses. ET contributes wound- or pathogen-induced defenses. One of the highest urgencies of plants after getting injured by a pathogen is to shut the site, thus constraining ingress of devious pathogen (Bruxelles and Roberts 2001). Ethylene is such an enzyme that helps in this process. Ethylene was found to be a stimulating gel in the leaves of castor bean which blocks the entrance of vascular pathogen like *Fusarium oxysporum* when the plant became affected by this pathogen (Dong et al. 2020). Blocking of the xylem vessels to guard the plant from contamination, ethylene plays a vital role in *Agrobacterium tumefaciens* to enhance the development of gall. Wild-type tomato created a reaction to the ethylene stimulated by *Agrobacterium* through restricting the diameter of vessel over the gall thus producing a rough and unorganized surface of callus (Narayan et al. 2022). Such alterations might take place to protect the callus from outer infection (Bruce Adie 2007).

Ethylene signal transduction pathways are common in *Arabidopsis thaliana* where ethylene molecules combine with receptors of ethylene such as ERS1, ETR1, ERS2, ETR2, and EIN4 (Riyazuddin et al. 2022). This combination incapacitate the constitutive triple response 1 (CTR1)-receptors complex that can't further phosphorylate the downstream signal component Ethylene-insensitive2 (EIN2) (Zhao and Guo 2011). However, the breakdown happens to the non-degraded EIN2 protein carboxy terminus (EIN2 CEND) that released later into the nucleus causing an inhibition to protein-mediated transcription factor Ethylene-insensitive3 (EIN3)/ Ethylene-insensitive-like1 (EIL1). This degradation continues to stimulate the accretion of EIN3/EIL1 in nucleus, thus producing an ethylene response through activating the expression of downstream ERF target gene (Ji and Guo 2013). In the defense response regulation to the fungus *phytophthora capsici*, the involvement of the ethylene-responsive factor CaPTI1 gene found in pepper (*Capsicum annuum* L.) (Jin et al. 2016). Ethylene-responsive factors (ERFs) contain several transcription factors as well as reveal imperative parts in response to stress (Bhattacharjee and Hallan 2022). AcERF2, which is a unique AP2/ERF transcription factor, was characterized from a halophyte *Atriplex canescens* (Sun et al. 2018). In *Arabidopsis*, overexpression of AcERF2 prompted the accumulation of transcript for genes related to plant defense (ERF1, ERF3, PR1, PR2, and PR5) and enhanced the resistance of *Arabidopsis* to *Botrytis cinerea* and *Pseudomonas syringae* pv. tomato DC3000, which are a necrotrophic fungal pathogen and bacterial pathogen, respectively (Wang et al. 2020).

JA and Ethylene-Dependent Signaling Pathway

It is believed from several studies that ET cooperate with JA in activating signal transduction against wounds caused by pathogen and working antagonistically againsta pathogen (Yang et al. 2022). Many observations has connected ET and JA signaling pathways together in their resistance to disease in plants and it has been shown that these two signaling pathways modulate each other during working against disease. The coi1 and jar1 JA-signaling patterns found parallel to the ein2 ET-signaling mutant that altered pathogen responses in *A. thaliana* (Li et al. 2019). Many defense-related genes are JA-dependent such as THI2.1, PDF1.2, CHIB, and HEL, which require EIN2 for their expression. A form of systemic resistance triggered by the root-colonizing bacterium *P. fluorescens* that is a kind of induced systemic resistance (ISR) under the contribution of both JA and ET signaling pathways (Zhu et al. 2011).

The expression of a number of genes related to defense against pathogen was found to be operated by the synergistic action of ethylene and JA (Yan et al. 2023). These genes include PR3 (chitinases), PR1b, PR4 (hevein-like proteins), PDF1.2, and PR5 (osmotin). Among them, PDF1.2 is one of the most significantly reviewed fungicidal peptide genes dependent on ET and JA signaling simultaneously (Lorenzo et al. 2003). Under biotic stress like pathogen attack, ET- and JA-dependent signaling and defense responses activation acts as a classical marker and this was evidenced when a fungal infection by *Alternaria brassicicola* studied (Birkenbihl et al. 2017). The gene, PDF1.2 is operated through ERF1 along with the signaling of these two hormones. This kind of transcription factor (TF) is stimulated by the coordination of both ET and JA. Whereas, if any of these signaling pathways are blocked by mutation, induction of the transcription factor ERF1 along with other anti-pathogenic genes are prevented. During an analysis of transcriptome, it has been found that a high percentage of defense related responses those are dependent on ET/JA, are regulated by ERF1 (Amorim et al. 2017). The activation of defense relevant genes like PDF1.2 and PR3 are triggered by the overexpression of ERF1 that augments resistance against many pathogens those are necrotrophic in nature (Ellis and Turner 2001). Signaling pathways of both ET and JA found to be activated by The *Arabidopsis* mutant cev1 when defense-related various genes like b-CHI, Thi2.1, PDF1.2, VSP1 and VSP2 are expression constitutively as well. Exhibition of resistance to powdery mildew diseases was also found by the study of this mutant (Vidhyasekaran 2015).

Interplay among SA, JA and Ethylene Signaling

Sometimes, an effective defense of plants against a number of pathogens happens through the combination of ethylene (ET), jasmonic acid (JA), and salicylic acid (SA) signaling pathways rather than ET solely (Cao et al. 2006). Synthesis of these hormones is prompted by biotic stress such as herbivory, infection by pathogen and damaging. The irregular stimulation as well as interaction of these pathways allows modified defense responses against a specific threat factor to plants (Table 2). When disease symptoms developed in tomato by *Xanthomonas campestris* pv. *vesicatoria*, both SA and ET were required for defense signaling whereas in infected plants, SA

Table 2. Hormonal interactions and their role in disease resistance., Source: (Shigenaga et al. 2017).

Hormonal interaction	Found Initially in Plants	Result	References
Jasmonic acid/ Ethylene	*Arabidopsis*	Resistance to nectrotrophic pathogen increased	Lorenzo et al. 2004 Wang et al. 2022
Abscisic acid/Ethylene	Rice	Resistance to nectrotrophic pathogen increased	Vleesschauwer et al. 2010
Salycylic acid–Jasmonic acid/ Ethylene	Different monocot and dicot plant species	Resistance to pathogens- both biotrophic and nectrotrophic	Zhao et al. 2021
Abscisic acid–Jasmonic acid/ Ethylene	*Arabidopsis*	Resistance to necrotrophic pathogens reduced	Anderson et al. 2004 Li et al. 2022

accumulation found dependent on synthesis of ET (Gupta et al. 2021). In *A. thaliana*, several genes were defense-associated, stimulated by the coordination of SA and ET (Mishra et al. 2017).

Some ERF1 help to unite ET and JA signaling pathways aiming at repressing wounds and responses of biotrophic pathogen as well as inducing defenses against necrotrophic pathogens (Marassatto 2022). On the other hand, repression of defenses against necrotrophic pathogens and defenses against biotrophic pathogens as well is induced by salicylic acid through WRKY70, which is a transcription factor of salicylic acid. Additionally, jasmonic acid operates the defense response against wounding through positively regulating genes of AtMYC2 (Lox, Thi2.1, and VSP) as well as genes like PDF1.2, HEL, and b-CHIare negatively regulated to pathogen response (Fig. 7).

Genetic Consequences of Disease Resistance by Ethylene

The role of ethylene in disease resistance was first revealed when ERF genes were classified basically in *Arabidopsis* and then in other plants (Zhang et al. 2022). There are some genetic consequences of plants that are dependent on ethylene responsive factors whereas there are some genes as well which are responsible for controlling the production of ethylene. Both types of genes may contribute both negatively and positively in plant defense against various pathogens those cause disease in plants. Several studies with various types of plants have been conducted to identify the genes or transcription factors that work as the regulators behind ethylene production and disease resistance. Although the basic study is conducted with the model plant *Arabidopsis*, the behavior and consequences of these genes may alter and vary depending on the plant type, pathogen type, and other factors as well. Even the mode of regulatory functions of ethylene against disease may change depending on these factors (Lahlali et al. 2022). So it is very important to first understand the specific genetic behavior that is controlled by ethylene and/or that controls ethylene for developing resistance against a particular disease caused by a particular pathogen. The collaboration among ethylene responsive factors as well as plants' own genetic makeup helps ethylene to work positively and can be utilized to stabilize resistance significantly against disease causing pathogens.

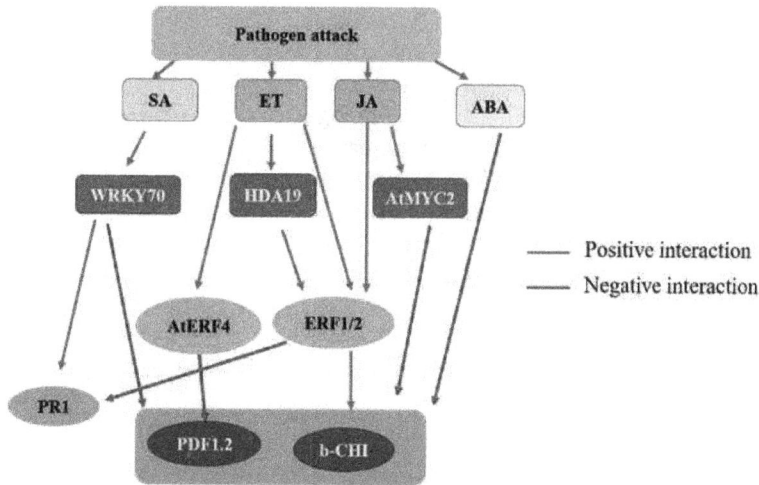

Fig. 7. Network of plant defense response through the interplay among ET, JA, SA, and ABA signaling pathways. Source: (Adie et al. 2007, Aerts et al. 2021).

Genes Regulating the Activity of Ethylene

Three types of proteins may lead the defense against a pathogen attack—the first types aid in building physical obstacles of plants against pathogen; the second group of proteins contribute to the biosynthesis of secondary metabolites having antimicrobial activities; and the last kinds may be called pathogenesis-related (PR) proteins which create major quantitative changes in defense responses. When the role of ET signaling proteins in developing structural barriers was revealed, it was found that the proteins were hydroxyproline-rich and gathered in plants when treated with ethylene for strengthening the cell wall of plants (Liu et al. 2023). But later this action was found less effective against pathogen because in most of the plants, rather than ethylene, abscisic acid was found to be regulating the callose deposition in the sites of cell wall where pathogens attempted penetration. On the other hand, pathogenesis-related (PR) proteins found to be the most active molecules in regulating ET-induced defense (Elkobrosy et al. 2022). ET and JA pathways were cooperatively found to control these PR-genes induction through the element of GCC-box in the regions of their promoter comprising distinct PR gene classes like PR-12 and PDFs (plant defensins), PR-4 (acidic hevein-like proteins), PR-3 (vacuolar basic-chitinases), and PR-2 (vacuolar β-1,3-glucanases). Whereas SA-dependent pathway control the PR-1 proteins and the extracellular chitinases and β-1,3-glucanases, particularly in tobacco and *Arabidopsis* (Li et al. 2019). So the ET stimulated PR genes are generated in the infested region to generate resistance against the pathogen.

ERFs (Ethylene Response Factors) are also called Ethylene Responsive Element Binding Proteins (EREBPs) and located in several plant species (Ogata et al. 2022). These proteins bind to the GCC-box. This box is an 11-base pair conserved sequence (TAAGAGCCGCC) of cis-acting ET response elements and consists of ET-responsive effector genes that are fundamental for ET regulation in different plant species. Besides, this box encodes proteins that are pathogenesis-related (PR) through being a part of ET-inducible gene promoters. ERFs bind to GCC-box through APETALA2 (AP2), which is a floral homeotic protein (Feng et al. 2020). Diverse ERF transcription factors take part in triggering or suppressing certain genes in response to defense, presenting a level of gene modification under different kind of biotic stress, including disease-causing pathogen attacks. This kind of defense response adjustment by ERF transcription factors can lead plants to avoid taking unnecessary action against pathogen attacks. Through the ERF transcription factors, ethylene can trigger the appearance of numerous defense-oriented plant genes against particular pathogens that may be term as pathogen-induced defense responses by ethylene (Dubois et al. 2018). In EIN3-ERF1–mediated ET signaling, one of the histone deacetylases of *Arabidopsis*- HDA19 is crucial for the regulation of eukaryotic gene expression. The ERF proteins mostly perform as transcriptional activators such as GCC-box containing genes are transcriptionally activated by AtERF1, AtERF2, and AtERF5. ERF proteins such as tobacco ERF3, AtERF10–12, AtERF7, AtERF4, AtERF3, and act as transcriptional repressors thus repressing the expression of a reporter gene present in GCC-box (Srivastava and Kumar 2019). The ET-insensitive phenotype is conferred by AtERF4 that leads in repressing genes present in GCC-box are responsible for encoding basic β-glucanase and chitinase. AtERF4 and AtERF2 work oppositely in phenotypes of disease-resistance against the contamination caused by *Fusarium oxysporum* like necrotrophic pathogen (Broekgaarden et al. 2015).

Gene Induction by Ethylene

Several genes have been derived from different plant species and are basically expressed through the induction of transcription factor of ethylene AP2/ERF (Table 3). These genes act against various pathogens and disease to develop resistance in plants. Sometimes, these genes may become active either in mutant or in wild type or in transgenic plant. Even, a single type of ERF-dependent gene that is derived from one plant may be utilized to develop ethylene- based resistance against the same pathogen or disease in another plant depending on the plant and pathogen type. Besides ET-based genes, there are many other genes those expressed through the combined implication of ET and JA transcription factors.

Factors Limiting Ethylene Action against Disease

The productions of ethylene in plant may alter based on type of plant, developmental stage, plant organ (e.g., leaf, flower root, etc.), biotic as well as abiotic factors, and plant pathogen interaction (Cristescu et al. 2013). The PR genes regulated by ethylene response factors (ERFs) may only be applicable against limited pathogens, thus indicating their shortcoming in activating the entire defense response as well as degrading the fitness of plants through overexpression (Zhang et al. 2023). With the evolution of modern approaches of plant improvements, several characterized mutants or transgenic lines have been developed in diverse plant species along with indigenous defensive mechanisms as well as augmented resistance to certain pathogens. But, such characterization in plant may lead to the misregulation of

Table 3. Major ethylene mediated genes found in various plants induce resistance against specific pathogens.

Genes induced by Ethylene/Genes Inhibit Ethylene Synthesis	Gene	Found in Plant	Resistance to Pathogen	References
AP2/ERF	HaAP2/ERF	Sunflower	Abiotic stresses	Najafi et al. 2018
	AtERF1	*Arabidopsis*	*Botrytis cinerea, F. oxysporum* f.sp. *lycopersici, Sclerotinia sclerotiorum, Erysiphe orontii* and *Fusarium oxysporum* f.sp. *conglutinans*	Gutterson and Reuber 2004, Amorim et al. 2017
	Tobacco stress-induced1 (Tsi1)	*Arabidopsis*	*Phytophthora capsici, Xanthomonas campestris*, cucumber mosaic virus (CMV), pepper mild mottle virus (PMMV),	Gutterson and Reuber 2004
ET/JA mediated PR gene	SlPti5	*Arabidopsis*, Tomato	*Botrytis cinerea*	Tang et al. 2022
Ethylene-insensitive mutant gene in *Arabidopsis*	ein2-1	*Arabidopsis*	*X. campestris* pv. *campestris*	Coatsworth et al. 2023
ACC Inhibitor	Aminoethoxyvinylglycine (AVG)	Potato, tomato, eggplant, pipper	Vector-borne diseases: tomato spotted wilt virus, potato leafroll virus, beet curly top virus,	Bak et al. 2021
AP2/ERF	OsERF83, OsEIN2	Rice	Rice blast fungus- *Magnaporthe oryzae*	Tezuka et al. 2019
	Atriplex canescens ERF (AcERF2)	*Arabidopsis*	Bacterial: *Pseudomonas syringae* pv. tomato DC3000 Fungal: pathogen *Botrytis cinerea*	Sun et al. 2018
	MeRAV1 and MeRAV2	Cassava (*Manihot esculenta*)	Bacterial blight of *Xanthomonas axonopodis* pv. *manihotis* (Xam).	Wei et al. 2017
	Octadecanoid-responsive *Arabidopsis* 59 (ORA59)	*Arabidopsis*	Necrotrophic pathogen *Pectobacterium carotovorum*	Kim et al. 2018

Table 3. contd. ...

... Table 3. contd.

Genes induced by Ethylene/Genes Inhibit Ethylene Synthesis	Gene	Found in Plant	Resistance to Pathogen	References
AP2/ERF	MdERF11	Apple	Fungal pathogen *Botryosphaeria dothidea*	Wang et al. 2020
	GmERF113, GmERF5	Soybean	Oomycete pathogen *Phytophthora sojae* induced *Phytophthora* root rot	Zhao et al. 2017
AP2/ERF from *Haynaldia villosa*	ERF1-V	Wheat	Powdery mildew triggered by obligate biotrophic fungus *Blumeria graminis* f. sp. *tritici* (Bgt)	Xing et al. 2017
AP2/ERF	CsERF004	Cucumber	*Pseudoperonospora cubensi*-induced downy mildew of cucumber, *Corynespora cassiicola* induced target spot in cucumber	Liu et al. 2017
	VvERF1	Grape	*B. cinerea* infection downstream of ethylene	Dong et al. 2020
	VaERF20	*Vitis amurensis* (Amur grape)	*Pseudomonas syringae* and *Botrytis cinerea*	Baillo et al. 2019
	CaERF5	Pepper	Protecting transgenic tobacco plants against *R. solanacearum*	Javed et al. 2020
	MYB102	*Arabidopsis*	Aphids in cruciferous plants such as *Brevicoryne brassicae* (cabbage aphid) and *Lipaphis erysimi* (Mustard aphid). Also for GPA, *Myzus persicae* (green peach aphid)	Zhu et al. 2018
	GhERF6	Cotton	*Verticillium* wilt caused by *Verticillium dahliae*	Yang et al. 2015, Xiong et al. 2020
	OsEBP2	Rice	Blast fungus	Lin et al. 2007 Hong et al. 2022
	SlERF	Tomato	Resistance to tomato wilt disease	Kusajima et al. 2018

signaling pathways of a particular hormone. Overexpression of a definite hormone-reliant pathway (e.g., ET, JA, SA, ET/JA, etc.) in *Arabidopsis* mutants was found to exhibit enhanced resistance to an instance type of pathogens (Denancé et al. 2013). However, such increased resistance affected plant fitness adversely thus leading to the alteration in phenotypes like enhanced senescence, dwarfism, belated flowering, sterility, or poor seed production (Thaler et al. 2012). Because of an intricate regulatory network among plant hormones, modification of a specific signaling pathway can misregulate other signaling pathways as well. Thus, enhanced resistance in mutants that is regulated by a particular hormone like ethylene against a distinct pathogen may later lead to intensified susceptibility to a different one.

Conclusion

After analyzing the factors, genes and other regulators are responsible for controlling ethylene biosynthesis as well as being controlled by ethylene signaling under pathogen attack, the remaining perspective of sustainable disease resistance may be balancing all these factors in plant system as perfectly as possible. And this may only be possible through the molecular breeding aspects. As multiple phenotypic alterations take place due to the pleiotropic impacts of ethylene on plants, controlling ethylene levels precisely in plants may help to prevent altering other signaling pathways linked to ethylene, thus activating resistance against pathogen along with maintaining major phenotypes of plant effectively.

References

Abiri, R., Shaharuddin, N. A., Maziah, M., Yusof, Z. N. B., Atabaki, N., Sahebi, M. et al. 2017. Role of ethylene and the APETALA 2/ethylene response factor superfamily in rice under various abiotic and biotic stress conditions. Environmental and Experimental Botany, 134: 33–44.

Adie, B., Chico, J. M., Rubio-Somoza, I. and Solano, R. 2007. Modulation of plant defenses by ethylene. Journal of Plant Growth Regulation, 26: 160–177.

Aerts, N., Pereira Mendes, M. and Van Wees, S.C. 2021. Multiple levels of crosstalk in hormone networks regulating plant defense. The Plant Journal, 105(2): 489–504.

Aerts, N., Pereira Mendes, M. and Van Wees, S.C. 2021. Multiple levels of crosstalk in hormone networks regulating plant defense. The Plant Journal, 105(2): 489–504.

Agaoua, A., Rittener, V., Troadec, C., Desbiez, C., Bendahmane, A., Moquet, F. et al. 2022. A single substitution in Vacuolar protein sorting 4 is responsible for resistance to Watermelon mosaic virus in melon. Journal of Experimental Botany, 73(12): 4008–4021.

Agudelo-Romero, P., Erban, A., Rego, C. and Pablo Carbonell-Bejerano, T.N.L.S.J.M.M.-Z.J.K.A.M.F. 2015. Transcriptome and metabolome reprogramming in Vitis vinifera cv. Trincadeira berries upon infection with Botrytis cinerea. Journal of Experimental Botany, 66(7): 1769–1785.

Alazem, M. and Lin, N.-S. 2015. Roles of plant hormones in the regulation of host–virus interactions. Molecular Plant Pathology, 16(5): 529–540.

Alkan, N. and Fortes, A.M. 2015. Insights into molecular and metabolic events associated with fruit response to post-harvest fungal pathogens. Frontiers in Plant Science, p. 889.

Alkan, N., Friedlander, G., Ment, D., Prusky, D. and Fluhr, R. 2015. Simultaneous transcriptome analysis of C olletotrichum gloeosporioides and tomato fruit pathosystem reveals novel fungal pathogenicity and fruit defense strategies. New Phytologist, 205(2): 801–815.

Amorim, L.L.B., da Fonseca Dos Santos, R., Neto, J.P.B., Guida-Santos, M., Crovella, S. and Benko-Iseppon, A.M. 2017. Transcription Factors Involved in Plant Resistance to Pathogens. Current Protein and Peptide Science, 18(4): 335–51.

Anderson, J.P., Badruzsaufari, E., Schenk, P.M., Manners, J.M., Desmond, O.J., Ehlert, C. et al. 2004. Antagonistic Interaction between Abscisic Acid and Jasmonate-Ethylene Signaling Pathways Modulates Defense Gene Expression and Disease Resistance in Arabidopsis. The Plant Cell, 16(12): 3460–3479.

Asami, T. and Nakagawa, Y. 2018. Preface to the Special Issue: Brief review of plant hormones and their utilization in agriculture. Journal of Pesticide Science, 43(3): 154–158.

Asif, R., Yasmin, R., Mustafa, M., Ambreen, A., Mazhar, M., Rehman, A. et al. 2022. Phytohormones as plant growth regulators and safe protectors against biotic and abiotic stress. *In*: Plant Hormones: Recent Advances, New Perspectives and New Perspectives and Applications. s.l.:s.n., p. 115.

Baebler, Š., Witek, K., Petek, M., Stare, K., Tušek-Žnidarič, M., Pompe-Novak, M. et al. 2014. Salicylic acid is an indispensable component of the Ny-1 resistance-gene-mediated response against Potato virus Y infection in potato. Journal of Experimental Botany, 65(4): 1095–1109.

Baillo, E.H., Kimotho, R.N., Zhang, Z. and Xu, P. 2019. Transcription Factors Associated with Abiotic and Biotic Stress Tolerance and Their Potential for Crops Improvement. Genes, 10(10): 771.

Bak, A., Nihranz, C.T., Patton, M.F., Aegerter, B.J. and Casteel, C.L. 2021. Evaluation of aminoethoxyvinylglycine (AVG) for control of vector-borne diseases in solanaceous crops. Crop Protection, 145: 105640.

Bakshi, A., Shemansky, J.M., Chang, C. and Binder, B.M. 2015. History of Research on the Plant Hormone Ethylene. Journal of Plant Growth Regulation, 34(4): 809–827.

Banerjee, P. and Bhadra, P. 2020. Mini-Review on Strigolactones: Newly Discovered Plant Hormones. Bioresource Biotechnology Research Communications, p. 13(3).

Bao, Z., Sasaki, K., Okubo, T., Ikeda, S., Anda, M., Hanzawa, E. et al. 2013. Impact of *Azospirillum* sp. B510 Inoculation on Rice-Associated Bacterial Communities in a Paddy Field. Microbes and Environments, 28: 487–490.

Berens, M.L. et al. 2019. Balancing trade-offs between biotic and abiotic stress responses through leaf age-dependent variation in stress hormone cross-talk. Proceedings of the National Academy of Sciences, 116(6): 2364–2373.

Bhattacharjee, B. and Hallan, V. 2022. Role of Plant Transcription Factors in Virus Stress. In: Transcription Factors for Biotic Stress Tolerance in Plants. s.l.: Cham: Springer International Publishing, pp. 79–102.

Binder, B.M. 2020. Ethylene signaling in plants. Journal of Biological Chemistry, 295(22): 7710–7725.

Birkenbihl, R.P., Liu, S. and Somssich, I.E. 2017. Transcriptional events defining plant immune responses. Current Opinion in Plant Biology, 38: 1–9.

Blázquez, M.A., Nelson, D.C. and Weijers, D. 2020 . Evolution of Plant Hormone Response Pathways. Annual Review of Plant Biology, 71: 4.1–4.27.

Broekgaarden, C., Caarls, L., Vos, I.A., Pieterse, C.M. and Van Wees, S.C. 2015. Ethylene: Traffic Controller on Hormonal Crossroads to Defense. Plant Physiology, 169(4): 2371–2379.

Bruce Adie, J.M. C. I. R.-S. a. R.S. 2007. Modulation of Plant Defenses by Ethylene. Jornal of Plant Growth Regulation, 26: 160–177.

Bruxelles, G.L.D. and Roberts, M.R. 2001. Signals Regulating Multiple Responses to Wounding and Herbivores. Critical Reviews in Plant Sciences, 20(5): 487–521.

Campos, M.D., Félix, M.D.R., Patanita, M., Materatski, P., Albuquerque, A., Ribeiro, J.A. et al. 2022. Defense strategies: The role of transcription factors in tomato–pathogen interaction. Biology, 11(2): 235.

Campos-Rivero, G., Osorio-Montalvo, P., Sánchez-Borges, R., Us-Camas, R., Duarte-Aké, F. and De-la-Peña, C. 2017. Plant hormone signaling in flowering: an epigenetic point of view. Journal of Plant Physiology, 214: 16–27.

Cao, Y., Wu, Y., Zheng, Z. and Song, F. 2006. Overexpression of the rice EREBP-like gene OsBIERF3 enhances disease resistance and salt tolerance in transgenic tobacco. Physiological and Molecular Plant Pathology, 67(3–5): 202–211.

Casteel, C.L., De Alwis, M., Bak, A., Dong, H., Whitham, S.A. and Jander, G. 2015. Disruption of Ethylene Responses by Turnip mosaic virus Mediates Suppression of Plant Defense against the Green Peach Aphid Vector. Plant Physiology, 169(1): 209–218.

Catinot, J., Huang, J.B., Huang, P.Y., Tseng, M.Y., Chen, Y.L., Gu, S.Y. et al. 2015. Ethylene Response Factor 96 positively regulates Arabidopsis resistance to necrotrophic pathogens by direct binding to GCC elements of jasmonate—and ethylene-responsive defence genes. Plant, Cell & Environment, 38(12): 2721–2734.

Cecchini, N.M., Jung, H.W., Engle, N.L., Tschaplinski, T.J., Greenberg, J.T. et al. 2015. ALD1 Regulates basal immune components and early inducible defense responses in Arabidopsis. Molecular Plant-Microbe Interactions, 28(4): 455–466.

Chagué, V., Danit, L.V., Siewers, V., Schulze-Gronover, C., Tudzynski, P., Tudzynski, B. et al. 2006. Ethylene sensing and gene activation in Botrytis cinerea: a missing link in ethylene regulation of fungus–plantinteractions? Molecular Plant-Microbe Interactions, 19(1): 33–42.

Chamkhi, I., El Omari, N., Balahbib, A., El Menyiy, N., Benali, T., Ghoulam, C. et al. 2022. Is—the rhizosphere a source of applicable multi-beneficial microorganisms for plant enhancement? Saudi Journal of Biological Sciences, 29(2): 1246–1259.

Coatsworth, P., Collins, A.S., Bozkurt, T. and Güder, F. 2023. Continuous monitoring of chemical signals in plants under stress. Nature Reviews Chemistry, 7(1): 7–25.

Cristescu, S.M., Mandon, J., Arslanov, D., De Pessemier, J., Hermans, C. and Harren, F.J. 2013. Current methods for detecting ethylene in plants. Annals of Botany, 111(3): 347–360.

Das, S.K., Patra, J.K. and Thatoi, H. 2016. Antioxidative response to abiotic and biotic stresses in mangrove plants: A review. International Review of Hydrobiology, 101 3–19.

de Pedro-Jové, R., Puigvert, M., Sebastià, P., Macho, A.P., Monteiro, J.S. and Coll, N.S. 2022. Dynamic expression of *Ralstonia solanacearum* virulence factors and metabolism-controlling genes during plant infection. BMC genomics, Sep. 28, 22(1): 1–18.

Denancé, N., Sánchez-Vallet, A., Goffner, D. and Molina, A. 2013. Disease resistance or growth: the role of plant hormones in balancing immune responses and fitness costs. Frontiers in Plant Science, 4: 155.

Derksen, H., Rampitsch, C. and Daayf, F. 2013. Signaling cross-talk in plant disease resistance. Plant Science, 207: 79–87.

Dong, J., Wang, Y., Xian, Q., Chen, X. and Xu, J. 2020. Transcriptome analysis reveals ethylene-mediated defense responses to *Fusarium oxysporum* f. sp. cucumerinum infection in *Cucumis sativus* L. BMC Plant Biol., 2020 Jul 16; 20(1): 334.

Dong, T., Zheng, T., Fu, W., Guan, L., Jia, H. and Fang, J. 2020. The effect of ethylene on the color change and resistance to Botrytis cinerea infection in 'Kyoho' grape fruits. Foods, 9(7): 892.

Dong, Y., Tang, M., Huang, Z., Song, J., Xu, J., Ahammed, G.J. et al. 2022. The miR164a-NAM3 module confers cold tolerance by inducing ethylene production in tomato. The Plant Journal, 111(2): 440–56.

Dubois, M., Broeck, L.V.d. and Inzé, D. 2018. The Pivotal Role of Ethylene in Plant Growth. Trends in Plant Science, 23(4): 311–323.

Elkobrosy, D.H., Aseel, D.G., Hafez, E.E., El-Saedy, M.A., Al-Huqail, A.A., Ali, H.M. et al. 2022. Quantitative detection of induced systemic resistance genes of potato roots upon ethylene treatment and cyst nematode, Globodera rostochiensis, infection during plant–nematode interactions. Saudi Journal of Biological Sciences, 29(5): 3617–3625.

Ellis, C. and Turner, J.G. 2001. The Arabidopsis Mutant cev1 Has Constitutively Active Jasmonate and Ethylene Signal Pathways and Enhanced Resistance to Pathogens. The Plant Cell, 13(5): 1025–1033.

Erofeeva, E.A. 2022. Hormesis in plants: its common occurrence across stresses. Current Opinion in Toxicology.

Fahad, S., Nie, L., Chen, Y., Wu, C., Xiong, D., Saud, S. et al. 2015. Crop plant hormones and environmental stress. Sustainable Agriculture Reviews, pp. 371–400.

Fatma, M., Asgher, M., Iqbal, N., Rasheed, F., Sehar, Z., Sofo, A. et al. 2022. Ethylene signaling under stressful environments: Analyzing collaborative knowledge. Plants, 11(17): 2211.

Feifei, W., Zhong-Hua, C. and Sergey, S. 2017. Hypoxia sensing in plants: on a quest for ion channels as putative oxygen sensors. Plant and Cell Physiology, 58: 1126–1142.

Feng, K., Hou, X.L., Xing, G.M., Liu, J.X., Duan, A.Q., Xu, Z.S. et al. 2020. Advances in AP2/ERF super-family transcription factors in plant. Critical Reviews in Biotechnology, 40(6): 750–776.

Fujita, M., Kusajima, M., Okumura, Y., Nakajima, M., Minamisawa, K., Nakashita, H. et al. 2017. Effects of colonization of a bacterial endophyte, *Azospirillum* sp. B510, on disease resistance in tomato. Bioscience, Biotechnology, and Biochemistry, 81(8): 1657–1662.

Fukao, T., Harris, T. and Bailey-Serres, J. 2009. Evolutionary analysis of the Sub1 gene cluster that confers submergence tolerance to domesticated rice. Annals of Botany, 103(2): 143–150.

Gamalero, E. and Glick, B. 2015. Bacterial modulation of plant ethylene levels. Plant Physiology, 169(1): 13–22.

Gorshkov, V. and Tsers, I. 2022. Plant susceptible responses: The underestimated side of plant–pathogen interactions. Biological Reviews, 97(1): 45–66.

Guinel, F.C. 2015. Ethylene, a Hormone at the Center-Stage of Nodulation. Frontiers in Plant Science, 6: 1121.

Gupta, R., Leibman-Markus, M., Pizarro, L. and Bar, M. 2021. Cytokinin induces bacterial pathogen resistance in tomato. Plant Pathology, 70(2): 318–325.

Gutterson, N. and Reuber, T.L. 2004. Regulation of disease resistance pathways by AP2/ERF transcription factors. Current Opinion in Plant Biology, 7(4): 465–471.

Handrick, V., Robert, C.A., Ahern, K.R., Zhou, S., Machado, R.A., Maag, D. et al. 2016. Biosynthesis of 8-O-Methylated Benzoxazinoid Defense Compounds in Maize. The Plant Cell, 28(7): 1682–1700.

Helliwell, E.E., Wang, Q. and Yang, Y. 2013. Transgenic rice with inducible ethylene production exhibits broad-spectrum disease resistance to the fungal pathogens M agnaporthe oryzae and R hizoctonia solani. Plant Biotechnology Journal, 11(1): 33–42.

Hillmer, R.A., Tsuda, K., Rallapalli, G., Asai, S., Truman, W., Papke, M.D., Sakakibara, H. et al. 2017. The highly buffered Arabidopsis immune signaling network conceals the functions of its components. Plos Genetics, 13(5): e1006639.

Hong, Y. et al. 2022. ERF transcription factor OsBIERF3 positively contributes to immunity against fungal and bacterial diseases but negatively regulates cold tolerance in rice. International Journal of Molecular Sciences, 23(2): 606.

Hossain, M.M., Sultana, F. and Hyakumachi, M. 2017. Role of ethylene signalling in growth and systemic resistance induction by the plant growth-promoting fungus Penicillium viridicatum in Arabidopsis. Journal of Phytopathology, 165(7–8): 432–441.

Huh, S.U. 2022. Functional analysis of hot pepper ethylene responsive factor 1A in plant defense. Plant Signaling & Behavior, 17(1): 2027137.

Husain, T., Fatima, A., Suhel, M., Singh, S., Sharma, A., Prasad, S.M. et al. 2020. A brief appraisal of ethylene signaling under abiotic stress in plants. Plant Signaling & Behavior, 15(9): 1782051.

Iqbal, N., Khan, N.A., Ferrante, A., Trivellini, A., Francini, A. and R. Khan, M. I. (2017). Ethylene Role in Plant Growth, Development and Senescence: Interaction with Other Phytohormones. Frontiers in Plant Science, 8: 475.

Islam, W., Zaynab, M., Qasim, M. and Wu, Z. 2017. Plant-Virus Interactions: Disease Resistance in Focus. Hosts Virus, 4: 5–20.

Janda, T., Szalai, G. and Pál, M. 2020. Salicylic Acid Signalling in Plants. International Journal of Molecular Sciences, 21(7): 2655.

Jasrotia, S. and Jasrotia, R. 2022. Role of ethylene in combating biotic stress. Ethylene in Plant Biology, pp. 388–397.

Javed, T., Shabbir, R., Ali, A., Afzal, I., Zaheer, U. and Gao, S.J. 2020. Transcription Factors in Plant Stress Responses: Challenges and Potential for Sugarcane Improvement. Plants, 9(4): 491.

Jiang, K. and Asami, T. 2018. Chemical regulators of plant hormones and their applications in basic research. Bioscience, Biotechnology, and Biochemistry, 82(8): 1265–1300.

Jin, J.H., Zhang, H.X., Tan, J.Y., Yan, M.J., Li, D.W., Khan, A. et al. 2016. A New Ethylene-Responsive Factor CaPTI1 Gene of Pepper (*Capsicum annuum* L.) Involved in the Regulation of Defense Response to Phytophthora capsici. Frontiers in Plant Science 6: 1217.

Ji, Y. and Guo, H. 2013. From Endoplasmic Reticulum (ER) to Nucleus: EIN2 Bridges the Gap in Ethylene Signaling. Molecular Plant, 6(1): 11–14.

Kazan, K. 2015. Diverse roles of jasmonates and ethylene in abiotic stress tolerance. Trends in Plant Science, 20(4): 219–229.

Khan, N.A., Khan, M.I.R., Ferrante, A. and Poor, P. 2017. Editorial: Ethylene: A Key Regulatory Molecule in Plants. Frontiers in Plant Science, 8: 1782.

Khoudi, H. 2022. SHINE clade of ERF transcription factors: A significant player in abiotic and biotic stress tolerance in plants. Plant Physiology and Biochemistry.

Kim, N.Y., Jang, Y.J. and Park, O.K. 2018. AP2/ERF Family Transcription Factors ORA59 and RAP2.3 Interact in the Nucleus and Function Together in Ethylene Responses. Frontiers in Plant Science, 1675.

Kou, X., Feng, Y., Yuan, S., Zhao, X., Wu, C., Wang, C. et al. 2021. Different regulatory mechanisms of plant hormones in the ripening of climacteric and non-climacteric fruits: a review. Plant Molecular Biology, pp. 1–21.

Kusajima, M., Shima, S., Fujita, M., Minamisawa, K., Che, F.S., Yamakawa, H. et al. 2018. Involvement of ethylene signaling in *Azospirillum* sp. B510-induced disease resistance in rice. Bioscience, Biotechnology, and Biochemistry, 82(9): 1522–1526.

Ku, Y.-S., Sintaha, M., Cheung, M.-Y. and Lam, H.-M. 2018. Plant Hormone Signaling Crosstalks between Biotic and Abiotic Stress Responses. International Journal of Molecular Sciences, 19(10): 3206.

Lahlali, R., Ezrari, S., Radouane, N., Kenfaoui, J., Esmaeel, Q., El Hamss, H. et al. 2022. Biological control of plant pathogens: A global perspective. Microorganisms, 10(3): 596.

Li, J., Chen, L., Ding, X., Fan, W. and Liu, J. 2022. Transcriptome analysis reveals crosstalk between the abscisic acid and jasmonic acid signaling pathways in rice-mediated defense against Nilaparvata lugens. International Journal of Molecular Sciences, 23(11): 6319.

Li, N., Han, X., Feng, D., Yuan, D. and Huang, L.J. 2019. Signaling Crosstalk between Salicylic Acid and Ethylene/Jasmonate in Plant Defense: Do We Understand What They Are Whispering? International Journal of Molecular Sciences, 20(3): 671.

Lin, R., Zhao, W., Meng, X. and Peng, Y.-L. 2007. Molecular cloning and characterization of a rice gene encoding AP2/EREBP-type transcription factor and its expression in response to infection with blast fungus and abiotic stresses. Physiological and Molecular Plant Pathology, 70(1–3): 60–68.

Liu, D., Xin, M., Zhou, X., Wang, C., Zhang, Y. and Qin, Z. 2017. Expression and functional analysis of the transcription factor-encoding GeneCsERF004in cucumber during Pseudoperonospora cubensis and Corynespora cassiicolainfection. BMC Plant Biology, 17(1): 1–13.

Liu, X.J., Ma, Y., Shi, Y. and Ma, H.L. 2023. 2, 3-Butanediol induces brown blotch resistance in creeping bentgrass by strengthening cell wall structure and promoting lignin synthesis of precursor phenolic acid. Acta Physiologiae Plantarum, 45(3): 40.

Lorenzo, O., Chico, J.M., Saénchez-Serrano, J.J. and Solano, R. 2004. Jasmonate-Insensitive1 Encodes a MYC Transcription Factor Essential to Discriminate between Different Jasmonate-Regulated Defense Responses in Arabidopsis. The Plant Cell, 16(7): 1938–1950.

Lorenzo, O., Piqueras, R., Sánchez-Serrano, J.J. and Solano, R. 2003. Ethylene Response FACTOR1 Integrates Signals from Ethylene and Jasmonate Pathways in Plant Defense. The Plant Cell, 15(1): 165–178.

Love, A.J., Laval, V., Geri, C., Laird, J., Tomos, A.D., Hooks, M.A. and Milner, J.J. 2007. Components of Arabidopsis Defense- and Ethylene-Signaling Pathways Regulate Susceptibility to Cauliflower mosaic virus by Restricting Long-Distance Movement. Molecular plant-microbe interactions, 20(6): 659–670.

Marassatto, C.M. 2022. Induced Systemic Resistance by Beauveria Bassiana in Soybean Plants and the effects on Spodoptera frugiperda. s.l.:Doctoral dissertation, Universidade de São Paulo.

Mignolet-Spruyt, L., Xu, E., Idänheimo, N., Hoeberichts, F.A., Mühlenbock, P., Brosché, M. et al. 2016. Spreading the news: subcellular and organellar reactive oxygen species production and signalling. Journal of Experimental Botany, 67(13): 3831–3844.

Mishra, R., Nanda, S., Rout, E., Chand, S.K., Mohanty, J.N. and Joshi, R.K. 2017. Differential expression of defense-related genes in chilli pepper infected with anthracnose pathogen Colletotrichum truncatum. Physiological and Molecular Plant Pathology, 97: 1–10.

Muhammad Aslam, M., Waseem, M., Jakada, B.H., Okal, E.J., Lei, Z., Saqib, H.S.A. et al. 2022. Mechanisms of abscisic acid-mediated drought stress responses in plants. International Journal of Molecular Sciences, 23(3): 1084.

Müller, M. and Munné-Bosch, S. 2015. Ethylene response factors: a key regulatory hub in hormone and stress signaling. Plant Physiology, 169(1): 32–41.

Munkvold, G. and White, D.G. 2016. Compendium of Corn Diseases. In: American Phytopathological Society Press. s.l.:St Paul, MN.

Najafi, S., Sorkheh, K. and Nasernakhaei, F. 2018. Characterization of the APETALA2/Ethylene-responsive factor (AP2/ERF) transcription factor family in sunflower. Scientific Report, 8: 11576.

Narayan, H., Srivasatava, P., Chandra Bhatt, S., Joshi, D. and Soni, R. 2022. Plant pathogenesis and disease control. Plant Protection: From Chemicals to Biologicals, p. 95.

Nascimento, F.X., Rossi, M.J. and Glick, B.R. 2018. Ethylene and 1-Aminocyclopropane-1-carboxylate (ACC) in Plant–Bacterial Interactions. Frontiers in Plant Science, 9: 114.

Nasrullah, Ali S., Umar, M., Sun, L., Naeem, M., Yasmin, H. and Khan, N. 2022. Flooding tolerance in plants: from physiological and molecular perspectives. Brazilian Journal of Botany, 45(4): 1161–1176.

Tintor, N., Ross, A., Kanehara, K., Yamada, K., Fan, L., Kemmerling, B. et al. 2013. Layered pattern receptor signaling via ethylene and endogenous elicitor peptides during Arabidopsis immunity to bacterial infection. Proceedings of the National Academy of Sciences, 110(15): 6211–6216.

Ogata, T., Tsukahara, Y., Ito, T., Iimura, M., Yamazaki, K., Sasaki, N. et al. 2022. Cell death signalling is competitively but coordinately regulated by repressor-type and activator-type ethylene response factors in tobacco (*Nicotiana tabacum*). Plant Biology, 24(5): 897–909.

Orozco-Mosqueda, M.D.C., Santoyo, G. and Glick, B.R. 2023. Recent Advances in the Bacterial Phytohormone Modulation of Plant Growth. Plants, 12(3): 606.

Pattyn, J., Vaughan-Hirsch, J. and Poel, B.V.d. 2021. The regulation of ethylene biosynthesis: a complex multilevel control circuitry. New Phytologist, 229(2): 770–782.

Peng, J., Li, Z., Wen, X., Li, W., Shi, H., Yang, L. et al. 2014. Salt-induced stabilization of EIN3/EIL1 confers salinity tolerance by deterring ROS accumulation in Arabidopsis. PLoS Genetics, 10(10): p. e1004664.

Peng, Y., Yang, J., Li, X. and Zhang, Y. 2021. Salicylic acid: biosynthesis and signaling. Annual Review of Plant Biology, 72: 761–791.

Pieterse, C., Leon-Reyes, A., Van der Ent, S. and Van Wees, S. 2009. Networking by small-molecule hormones in plant immunity. Nature Chemical Biology, 5(5): 308–16.

Pieterse, C. M. J., Leon-Reyes, A., Ent, S. V. d. and Wees, S.C.M.V. 2009. Networking by small-molecule hormones in plant immunity. Nature Chemical Biology, 5(5): 308–316.

Poorni, K.E., Roy, M., Roy, N. and Gnanendra, T.S. 2023. Biochemical process associated with plants and beneficial microbes. In: Plant-Microbe Interaction-Recent Advances in Molecular and Biochemical Approaches. s.l.:Academic Press, pp. 73–85.

Riyazuddin, R., Bela, K., Poór, P., Szepesi, Á., Horváth, E., Rigó, G. et al. 2022. Crosstalk between the Arabidopsis Glutathione Peroxidase-Like 5 Isoenzyme (AtGPXL5) and Ethylene. International Journal of Molecular Sciences, 23(10): 5749.

Robert-Seilaniantz, A., Grant, M. and Jones, J.D. 2011. Hormone crosstalk in plant disease and defense: more than just jasmonate-salicylate antagonism. Annual Review of Phytopathology, 49: 317–343.

Rubio, V. et al. 2009. Plant hormones and nutrient signaling. Plant Molecular Biology, 69(4): 361–373.

Seymour, G.B., Østergaard, L., Chapman, N.H., Knapp, S. and Martin, C. 2013. Fruit development and ripening. Annual Review of Plant Biology, 64: 219–241.

Zhou, S., Zhang, Y.K., Kremling, K.A., Ding, Y., Bennett, J.S., Bae, J.S. et al. 2019. Ethylene signaling regulates natural variation in the abundance of antifungal acetylated diferuloylsucroses and Fusarium graminearum resistance in maize seedling roots. New Phytologist, 221(4): 2096–2111.

Shekhawat, K., Fröhlich, K., García-Ramírez, G.X., Trapp, M.A. and Hirt, H. 2023. Ethylene: A Master Regulator of Plant–Microbe Interactions under Abiotic Stresses. Cells, 12(1): 31.

Shigenaga1, A.M., Berens, M.L., Tsuda, K. and Argueso, C.T. 2017. Towards engineering of hormonal crosstalk in plant immunity. Current Opinion in Plant Biology, 38: 164–172.

Sofy, A.R., Sofy, M.R., Hmed, A.A., Dawoud, R.A., Refaey, E.E., Mohamed, H.I., El-Dougdoug, N. 2021. Molecular characterization of the Alfalfa mosaic virus infecting Solanum melongena in Egypt and the control of its deleterious effects with melatonin and salicylic acid. Plants, 10(3): 459.

Srivastava, R. and Kumar, R. 2019. The expanding roles of APETALA2/Ethylene Responsive Factors and their potential applications in crop improvement. Briefings in Functional Genomics, 18(4): 240–254.

Sun, X., Yu, G., Li, J., Liu, J., Wang, X., Zhu, G. et al. 2018. AcERF2, an ethylene-responsive factor of Atriplex canescens, positively modulates osmotic and disease resistance in Arabidopsis thaliana. Plant Science, 274: 32–43.

Tang, Q., Zheng, X.-d., Guo, J. and Yu, T. 2022. Tomato SlPti5 plays a regulative role in the plant immune response against Botrytis cinerea through modulation of ROS system and hormone pathways. Journal of Integrative Agriculture, 21(3): 697–709.

Tao, J.J., Chen, H.W., Ma, B., Zhang, W.K., Chen, S.Y. and Zhang, J.S. 2015. The role of ethylene in plants under salinity stress. Frontiers in Plant Science, 6: 1059.

Tezuka, D., Kawamata, A., Kato, H., Saburi, W., Mori, H. and Imai, R. 2019. The rice ethylene response factor OsERF83 positively regulates disease resistance to Magnaporthe oryzae. Plant Physiology and Biochemistry, 135: 263–271.

Thaler, J.S., Humphrey, P.T. and Whiteman, N.K. 2012. Evolution of jasmonate and salicylate signal crosstalk. Trends in Plant Science, 17(5): 260–270.

Tian, W., Huang, Y., Li, D., Meng, L., He, T. and He, G. 2022 . Identification of StAP2/ERF genes of potato (*Solanum tuberosum*) and their multiple functions in detoxification and accumulation of cadmium in yest: Implication for Genetic-based phytoremediation. Science of The Total Environment, 810: 152322.

Uniyal, S., Bhandari, M., Singh, P., Singh, R.K. and Tiwari, S.P. 2022. Cytokinin biosynthesis in cyanobacteria: Insights for crop improvement. Frontiers in Genetics, p. 13.

Vanderstraeten, L. and Straeten, D.V.D. 2017. Accumulation and Transport of 1-Aminocyclopropane-1-Carboxylic Acid (ACC) in Plants: Current Status, Considerations for Future Research and Agronomic Applications. Frontiers in Plant Science, 8: 38.

Vidhyasekaran, P. 2015. Jasmonate Signaling System in Plant Innate Immunity. In: In Plant Hormone Signaling Systems in Plant Innate Immunity. Dordrecht: Springer, pp. 123–194.

Vleesschauwer, D.D., Yang, Y., Cruz, C.V. and Höfte, M. 2010. Abscisic Acid-Induced Resistance against the Brown Spot Pathogen Cochliobolus miyabeanus in Rice Involves MAP Kinase-Mediated Repression of Ethylene Signaling. Plant Physiology, 152(4): 2036–2052.

Wang, F., Cui, X., Sun, Y. and Dong, C.-H. 2013. Ethylene signaling and regulation in plant growth and stress responses. Plant Cell Reports, 32(7): 1099–1109.

Wang, H., Zhao, X., Wang, Y., Li, W., Li, M., Ma, Z. et al. 2022. Exogenous carbon promotes plantlet growth by inducing ethylene signaling in grapevine. Scientia Horticulturae, 293: 110659.

Wang, J., Gu, K., Han, P., Yu, J., Wang, C., Zhang, Q. et al. 2020. Apple ethylene response factor MdERF11 confers resistance to fungal pathogen Botryosphaeria dothidea. Plant Science, 291: 110351.

Wang, K.L., Li, H. and Ecker, J.R. 2002. Ethylene Biosynthesis and Signaling Networks. The Plant Cell, 14(1): S131–S151.

Wang, L., Liu, H., Yin, Z., Li, Y., Lu, C., Wang, Q. et al. 2022. A novel guanine elicitor stimulates immunity in Arabidopsis and rice by ethylene and jasmonic acid signaling pathways. Frontiers in Plant Science, 13: 89.

Wang, L., Liu, W. and Wang, Y. 2020. Heterologous expression of Chinese wild grapevine VqERFs in Arabidopsis thaliana enhance resistance to Pseudomonas syringae pv. tomato DC3000 and to Botrytis cinerea. Plant Science, 293: 110421.

Wang, S., Han, K., Peng, J., Zhao, J., Jiang, L., Lu, Y. et al. 2019. NbALD1 mediates resistance to turnip mosaic virus by regulating the accumulation of salicylic acid and the ethylene pathway in Nicotiana benthamiana. Molecular Plant Pathology, 20(7): 990–1004.

Wei, Y., Chang, Y., Zeng, H., Liu, G., He, C. and Shi, H. 2017. RAV transcription factors are essential for disease resistance against cassava bacterial blight via activation of melatonin biosynthesis genes. Journal of Pineal Research, 64(1): e12454.

Xing, L., Di, Z., Yang, W., Liu, J., Li, M., Wang, X. et al. 2017. Overexpression of ERF1-V from Haynaldia villosa Can Enhance the Resistance of Wheat to Powdery Mildew and Increase the Tolerance to Salt and Drought Stresses. Frontiers in Plant Science, 8: 1948.

Xiong, P., Sun, C., Zhang, Y., Li, J., Liu, F., Zhu, H. et al. 2020. GhWRKY70D13 Regulates Resistance to Verticillium dahliae in Cotton Through the Ethylene and Jasmonic Acid Signaling Pathways. Frontiers in Plant Science, 11: 69.

Xu, X., Chen, Y., Li, B., Zhang, Z., Qin, G., Chen, T. et al. 2022. Molecular mechanisms underlying multi-level defense responses of horticultural crops to fungal pathogens. Horticulture Research, p. 9.

Yadav, S.K. 2010. Cold stress tolerance mechanisms in plants. A review. Agronomy for Sustainable Development, 30: 515–527.

Yang, C.L., Liang, S., Wang, H.Y., Han, L.B., Wang, F.X., Cheng, H.Q. et al. 2015. Cotton Major Latex Protein 28 Functions as a Positive Regulator of the Ethylene Responsive Factor 6 in Defense against Verticillium dahliae. Molecular Plant, 8(3): 399–411.

Yang, C., Li, W., Cao, J., Meng, F., Yu, Y., Huang, J. et al. 2017. Activation of ethylene signaling pathways enhances disease resistance by regulating ROS and phytoalexin production in rice. The Plant Journal, 89(2): 338–353.

Yang, C., Lu, X., Ma, B., Chen, S.Y. and Zhang, J.S. 2015. Ethylene Signaling in Rice and Arabidopsis: Conserved and Diverged Aspects. Molecular Plant, 8(4): 495–505.

Yang, Y., Yang, X., Guo, X., Hu, X., Dong, D., Li, G. et al. 2022. Exogenously applied methyl jasmonate induces early defense related genes in response to phytophthora infestans infection in potato plants. Horticultural Plant Journal, 8(4): 511–52.

Yan, W., Jian, Y., Duan, S., Guo, X., Hu, J., Yang, X. et al. 2023. Dissection of the Plant Hormone Signal Transduction Network in Late Blight-Resistant Potato Genotype SD20 and Prediction of Key Resistance Genes. Phytopathology, 113(3): 528–538.

Yasuda, M., Isawa, T., Satoshi Shinozaki, K.M. and Nakashita, H. 2009. Effects of colonization of a bacterial endophyte, Azospirillum sp. B510, on disease resistance in rice. Bioscience, Biotechnology, and Biochemistry, 73: 2595–2599.

Zahid, G., Iftikhar, S., Shimira, F., Ahmad, H.M. and Aka Kaçar, Y. 2023. An overview and recent progress of plant growth regulators (PGRs) in the mitigation of abiotic stresses in fruits: A review. Scientia Horticulturae, 309: 111621.

Zander, M., Thurow, C. and Gatz, C. 2014. TGA transcription factors activate the salicylic acid-suppressible branch of the ethylene-induced defense program by regulating ORA59 expression. Plant Physiology, 165(4): 1671–1683.

Zhabinskii, V.N., Khripach, N.B. and Khripach, V.A. 2015. Steroid plant hormones: effects outside plant kingdom. Steroids, 97: 87–97.

Zhang, D., Zhu, K., Shen, X., Meng, J., Huang, X., Tan, Y. et al. 2023. Two interacting ethylene response factors negatively regulate peach resistance to Lasiodiplodia theobromae. Plant Physiology, p. 279.

Zhang, J., Liao, J., Ling, Q., Xi, Y. and Qian, Y. 2022. Genome-wide identification and expression profiling analysis of maize AP2/ERF superfamily genes reveal essential roles in abiotic stress tolerance. BMC Genomics, 23(1): 1–22.

Zhang, Z., Yao, W., Dong, N., Liang, H., Liu, H. and Huang, R. 2007. A novel ERF transcription activator in wheat and its induction kinetics after pathogen and hormone treatments. Journal of Experimental Botany, 58(11): 2993–3003.

Zhao, B., Liu, Q., Wang, B. and Yuan, F. 2021. Roles of phytohormones and their signaling pathways in leaf development and stress responses. Journal of Agricultural and Food Chemistry, 69(12): 3566–3584.

Zhao, Q. and Guo, H.-W. 2011. Paradigms and Paradox in the Ethylene Signaling Pathway and Interaction Network. Molecular Plant, 4(4): 626–634.

Zhao, Y., Chang, X., Qi, D., Dong, L., Wang, G., Fan, S. et al. 2017. A Novel Soybean ERF Transcription Factor, GmERF113, Increases Resistance to Phytophthora sojae Infection in Soybean. Frontiers in Plant Science, 8: 299.

Zheng, X., Xing, J., Zhang, K., Pang, X., Zhao, Y., Wang, G. et al. 2019. Ethylene Response Factor ERF11 Activates BT4 Transcription to Regulate Immunity to Pseudomonas syringae. Plant Physiology, 180: 1132–1151.

Zhu, F., Xi, D.H., Deng, X.G., Peng, X.J., Tang, H., Chen, Y.J. et al. 2014. The Chilli Veinal Mottle Virus Regulates Expression of the Tobacco Mosaic Virus Resistance Gene N and Jasmonic Acid/Ethylene Signaling Is Essential for Systemic Resistance Against Chilli Veinal Mottle Virus in Tobacco. Plant Molecular Biology Reporter, 32(2): 382–394.

Zhu, L., Guo, J., Ma, Z., Wang, J. and Zhou, C. 2018. Arabidopsis Transcription Factor MYB102 Increases Plant Susceptibility to Aphids by Substantial Activation of Ethylene Biosynthesis. Biomolecules, 8(2): 39.

Zhu, Z., An, F., Feng, Y., Li, P., Xue, L.A.M., Jiang, Z. et al. 2011. Derepression of ethylene-stabilized transcription factors (EIN3/EIL1) mediates jasmonate and ethylene signaling synergy in Arabidopsis. Proceedings of the National Academy of Sciences, 108(30): 12539–12544.

Zwanenburg, B. and Pospíšil, T. 2013. Structure and Activity of Strigolactones: New Plant Hormones with a Rich Future. Molecular Plant, 6(1): 38–62.

5

Role of Strigolactone in the Alleviation of Biotic Stress in Plants

Marium Khatun,[1] Nor Anis Nadhirah Md Nasir,[2] Irnis Azura Zakarya[2] and A.K.M. Aminul Islam[1],*

Introduction

Plant hormones play a role in a variety of areas of plant life, including responses to biotic and abiotic stress are plant growth regulators (Llanes et al. 2018). Strigolactones (SLs) arc multifunctional molecules categorized as a new family of phytohormones that regulate a variety of plant functions (Andreo-Jimenez et al. 2015). During the previous few years, investigation into SLs have increased (Marzec and Muszynska 2015). The chemicals that are carotenoid derivatives secreted from the roots of 80% of land plants are precursor of SLs and SLs can induce a symbiotic relationship with soil AMF (Akiyama and Hayashi 2006). In 1966, SLs were first found in cotton plants as root exudates (Pandey et al. 2016). 'Strigol' was identified as the first naturally observed SLs. SLs got the name just after the discovery of the first candidate in *Striga* (Cook et al. 1966). It was previously identified as being generated in roots and stems as well as distributed through the vascular tissue (Kohlen et al. 2011). Cotton, maize, cowpea, red clover, and sorghum are among the monocots and dicots that have been recognized as SLs producers.

There are two types of SLs such as canonical SLs having a D-ring (butenolide ring) and a ABC ring (tricyclic ring) that are linked by an enol-ether bond whereas

[1] Department of Genetics and Plant Breeding, Faculty of Agriculture, Bangabandhu Sheikh Mujibur Rahman Agricultural University, Gazipur 1706, Bangladesh.
[2] School of Environmental Engineering, Universiti Malaysia Perlis (UNIMAP), Kompleks Pusat Pengajian Jejawi 3, 02600, Arau, Perlis, Malaysia.
* Corresponding author: aminulgpb@bsmrau.edu.bd

non-canonical SLs having a poorly stable structure (Al-Babili and Bouwmeester 2015, Jia et al. 2018). There are many SLs isolated from a variety of plants to date, including orobanchol, strigol, 20-epi-orobanchol, sorgolactone, sorgomol and solanacol (Koltai 2015). In cases of two model plants, e.g., Rice (*Oryza sativa*) and *Arabidopsis* (*Arabidopsis thaliana*), the activation of SL-biosynthesis genes are regulated by different plant hormones eg. salicylic acid (SA), jasmonic acid (JA), and ethylene associated with defense mechanism (Moubayidin et al. 2010). In axillary bud development stage, cytokinins (CKs) are reported to be hostile to SLs and may decrease SL production (Oldroyd 2013).

To ensure food security is the most significant task of global agriculture in a long-term way. The discovery of novel elite varieties and enhanced agricultural operations was tackled by the issue of feeding the world's rising population in fifty years ago (Gianinazzi et al. 2010). Nevertheless, with the availability of natural nutrient stocks, optimum development of these improved varieties/strategies was not possible in most soils. Due to the effects of climate change, different biotic and abiotic stresses are occurred in present field of agriculture. These stresses have a deleterious impact on yield. Plants have to regulate its lifecycle to overcome all sort of these stresses due to their sessile nature. To address these growing challenges, novel agronomic practices must be developed. However, each part such as, morphological, anatomical, physiological, and molecular plant biology must be modified practically for doing so. However, each part such as, morphological, anatomical, physiological, and molecular plant biology must be modified practically. Normally, by sensing the stress, a plant reacts to it and then transmits a signal, which is then regulated through molecular activity. Phytohormones including some signaling pathways like, calcium ions (Ca^{++}), reactive oxygen species (ROS) offer techniques for acclimation to severe environments and hence, it is a main indicators for crop function (Mishra et al. 2017). Furthermore, plant scientists are interested to do research on new metabolites which may act as phytohormones.

As an internal and external signaling molecule, SLs perform the double purpose. SLs are necessary to stimulate the interaction between plant and AMF in the root-zone area, especially in nutrition deficiency situations (Bouwmeester et al. 2003). For Orobanchaceae (root parasitic plants) and AMF (symbiotic fungus) from the phylum Glomeromycota, SLs operate as host recognition signals (Bouwmeester et al. 2007, López-Ráez et al. 2011b). SLs are also responsible to initiate Rhizobial (symbiotic microbe) association leading to nodulation in the root-zone area (Foo and Davies 2011). In 2008, the initial results indicating that aboveground plant branching were negatively regulated by SLs (Gomez-Roldan et al. 2008). Further research revealed that using a synthetic SL (GR24) can restrict axillary bud formation leading decreased branching of shoot whereas the mutants having a deficiency of SLs were shown to be fast growing (Umehara et al. 2008). SLs appear to play a ubiquitous role in growth and development of plants across a wide variety of organisms, such as monocotyledons and dicotyledons plants (Marzec and Muszynska 2015). SLs also play a vital role in host defense mechanisms that have recently been discovered (Table 1). In the present study, the importance of SLs under biotic stress is reviewed.

Table 1. Roles of strigolactones (SLs) in different plant functions (from growth and development to defense system).

Parameter	Main Functions	References
Root system Architecture	• Growth of primary roots and root hairs, e.g., crown root of *Oryza sativa*, • Prevention of adventitious root, e.g., *Solanum lycopersicum*, *Arabidopsis thaliana* and *Pisum sativum*	Rasmussen et al. 2013, Urquhart et al. 2015
Plant growth and development	• Germination of seed, • Development of seedling, • Leaf structure improvement and occurrences of senescence, • Change in the height of internode and morphology of stem •Enlargement of mesocotyl, stem, and secondary growth • Activity in reproduction	Faizan et al. 2020, Marzec and Muszynska 2015, Andreo-Jimenez et al. 2015
Environmental parameters	• Environmental parameters management, e.g., nitrogen (N) and phosphorus (P) availability, as well as intensity of light	Faizan et al. 2020
Inductor and signaling molecule	• Germination of the seeds of root parasitic plants of the genera *Striga, Orobanche, Alectra,* and *Phelipanche* through communication between plants and other organisms, e.g., bacteria or fungus	Akiyama et al. 2003, Soto et al. 2010, Yoneyama et al. 2013
Synthetic SL (GR24)	• Axillary bud growth inhibition	Mishra et al. 2017
Association of rhizobia	• Rhizosphere nodulation	Andreo-Jimenez et al. 2015
Interactions of arbuscular mycorrhizal fungi (AMF)	• Root parasitic plants management, • Uptaking photosynthetically fixed carbon and providing mineral nutrients [16–18].	Akiyama et al. 2005, Bouwmeester et al. 2007, López-Ráez et al. 2011b
Abiotic stress tolerance	Able to tolerate abiotic stress (e.g., salinity, drought and heavy metals)	Andreo-Jimenez et al. 2015
Biotic stress resistance	Plant resistance against several aboveground pathogens, and soil-borne and phytophagous insects	Liu et al. 2018, Pozo et al. 2010

Importance of Strigolactone in Perspective Biotic Stress

The majority of prior research on biotic stress was explanatory (Linderman 2000). In general, it was reported that the symbiosis between AMF and plant have many positive impacts, like promising bioagents for disease prevention strategies (Mukerji and Ciancio 2007). The extent to which the AMF colonize the roots appears to be a significant determinant of the influence of other organisms for symbiotic interactions. In the case of the mycorrhizal defense system, findings revealed that a well-established symbiosis is required before taking action against aggressor attack, with a few exceptions (Pozo et al. 2010). It was just reported that SLs play a role in the biotic stress response. SLs, on the other hand, do not offer a general role in defense system of plant, rather focusing on disease resistance only to a few pathogens. The earliest report on the role of SLs against biotic stress revealed that promoter sequences of genes were discovered associated in SL biosynthesis to include motifs recognized by TFs (transcription factors) involved in the resistant responses of several pathogens (i.e., virus, fungus, and bacterium) (Marzec 2016). In the recent research period, where multiple enhancement techniques are chosen for traditional crop management strategies, such an integrated collection is crucial. In this paper we discussed SLs biosynthesis, its implications, as well as the continuous advancement of SLs in maintaining plant growth and development under adverse environmental circumstances, particularly biotic stress.

Strigolactone Biosynthesis

Strigol, the first naturally occurring SL was a sprouting stimulator for *Striga lutea* and these substances were all together termed as strigolactones (Cook et al. 1972). Treatment of *Zea mays* with fluridone, an inhibitor of carotenoid biosynthesis was evidenced by poor SLs deposition in plants (Matusova et al. 2005). During evolutionary development, SL biosynthesis can be linked to the system's functional requirements. A wide range of plants, particularly bryophytes and algae contain the gene responsible for SL production as well as the crucial molecules, SLs have been passed through the chain of evolution for a period of time (Mishra et al. 2017). There are two sections made up of the chemical structure of SLs, such as, (i) ABC section (a lactone with tricyclic core), (ii) the D ring (a meioty of butenolide core) (Zwanenburg et al. 2009). The methyl, hydroxyl, and acetyloxyl units of A–B rings may be substituted whereas the C–D section of the SL is preserved (Xie et al. 2013). The generation of the carotenoid-derived metabolites known as natural SLs has been linked to a number of enzymes (Wakabayashi et al. 2022). In plastids, biosynthesis of SLs takes place and four enzymes are required in the procedure. Among these four, three enzymes previously identified remain functional (Fig. 1).

All trans-β-carotene are converted to 9-cis-β-carotene by an initial iron-containing enzyme, carotenoid isomerase D27 (DWARF27) (Alder et al. 2012). Two genes namely, AtD27 and D27, from two model plants, for e.g., *Arabidopsis* (Waters et al. 2012) and rice (Lin et al. 2009) were discovered that encode the first enzyme of this process. The immobile product, 9-cis-β-carotene is converted to 9-cis-β-apo-10'-carotenal with the help of carotenoid cleavage dioxygenase (CCD7) and 9-cis-β-apo-10'-carotenal to carlactone by carotenoid cleavage dioxygenase (CCD8) (Fig. 1). C19-skeleton and a C14-moiety of carlactone are analogous to the strigolactones' D-ring and the biological functions are also analogous to the strigolactones. Parasite seed germination is stimulated and shoot branching is controlled by carlactone (Scaffidi et al. 2013). In 2014, carlactone was discovered firstly in crop plants, and its significance as a potential intermediary was validated during biosynthesis of SLs. 13C-labeled carlactone is converted to [13C]-2'-epi-5-deoxystrigol and 13C-orobanchol and these newly derived products are two primary substrates of various SLs (Seto et al. 2013).

The last enzyme, monooxygenase (MAX1) is responsible for this conversion in SLs synthesis belonging to the family of cytochrome P450 (Alder et al. 2012). In *Arabidopsis*, it was first identified and works downstream of both CCD7 and CCD8, but its role was unclear for a longer period (Booker et al. 2005). Five MAX1 genes (Os01g0700900, Os01g0701500, Os01g0221900, Os06g0565100,

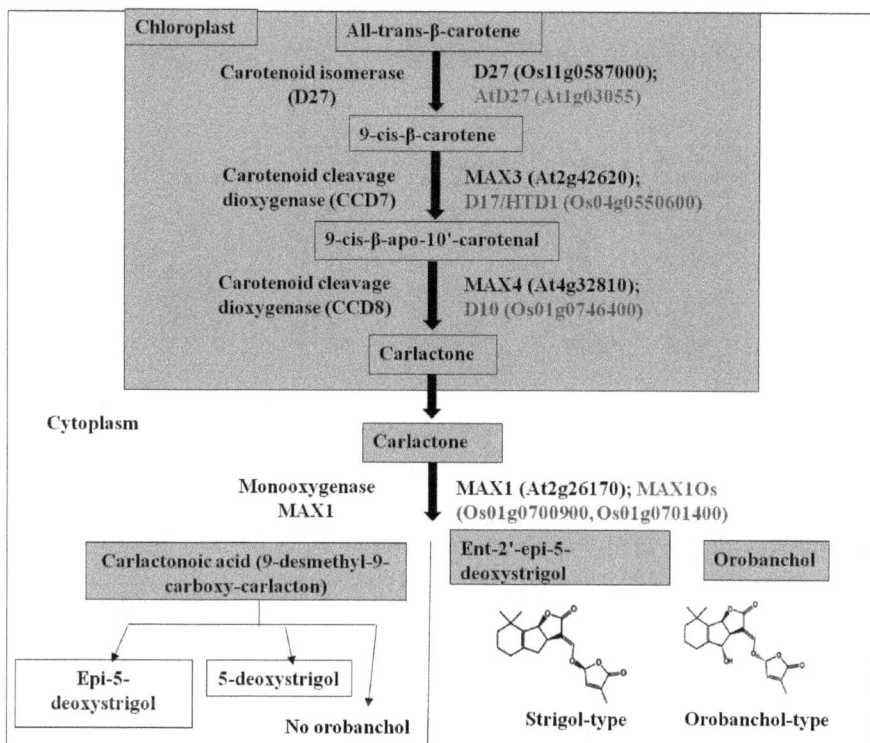

Fig. 1. Stepwise procedure and main enzymes (left side) associated in strigolactone biosynthesis. In two model plants, Arabidopsis and rice, genes encoding proteins (right side) are found in every step of the process. The two major ancestor's chemical structures of many other strigolactones, or the final product of the procedure (left side: Arabidopsis, right side: rice), are also shown (adapted from Mishra et al. 2017, Andreo-Jimenez et al. 2015, Marzec and Muszynska 2015).

and Os01g0701400) were discovered in rice. Among the five, only two MAX1 genes (Os01g0700900, Os01g0701400) were found in cultivated varieties. Natural variation of SL production associates the removal of the two genes (Cardoso et al. 2014). Ent-2'-epi-5-deoxystrigol was produced as first strigolactone from carlactone through oxidation and it was catalyzed by Os01g0700900. Orobanchol was produced with the help of the other available gene, Os01g0701400 through hydroxylation and hence, these two genes are responsible for diversity of SLs structure (Zhang et al. 2014).

Carlactone was converted to carlactonoic acid (9-desmethyl-9-carboxy-carlacton) with the help of one MAX1 gene (At2g26170) that is found in *Arabidopsis* (Abe et al. 2014). The MAX1 of *Arabidopsis* can only synthesize a little quantity of epi-5-deoxystrigol and 5-deoxystrigol but doesn't synthesize orobanchol. This phenomenon was observed when the biosynthesis of SLs in Arabidopsis was reconstituted in common tobacco (*Nicotiana benthamiana*) (Zhang et al. 2014). The relationship between strigolactone biosynthesis and its impact will be investigated under adverse environmental conditions in the following section (Fig. 2). Research on SLs revealed that there may still have some undiscovered functions in perspective

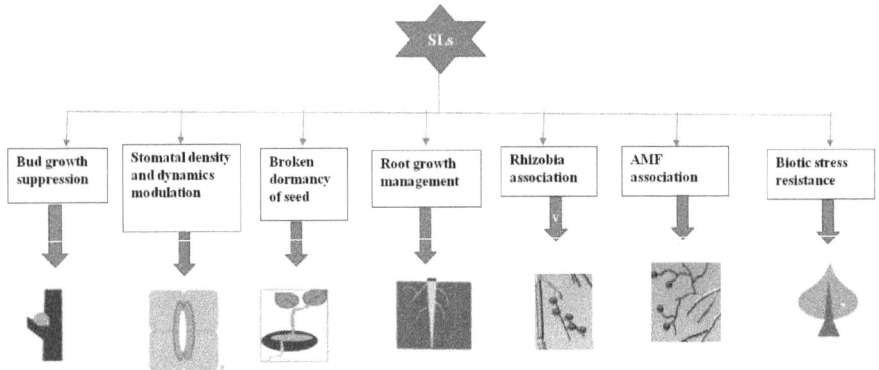

Fig. 2. Strigolactones stimulated responses in unfavorable environmental situations with a graphical view (Mishra et al. 2017).

of growth and development of plants, responses to stresses, and interaction with other phytohormones. Silico analysis especially on the genes encoding protein associated with the biosynthesis of SLs may open a new role of SLs in plants as the genetic basis of the biosynthetic pathway of SLs appears being well investigated and due to the stimuli of several stresses, enhanced formation of such hormones has been reported (Marzec and Muszynska 2015).

Symbiotic Relationship between Plant and Microbe

Plants synthesize greater portions of SLs when they are starved of nutrients. Shoot branching are suppressed and symbiosis is promoted by SLs (Umehara et al. 2008). SLs play an essential function in phosphorus (P) and nitrogen (N) deficit by modifying the architecture of root and shoot and promoting a symbiotic relationship between rhizobacteria and AMF (Marzec 2016). Fungus assure the availability of nutrients (especially, P and N) and water through the AMF symbiosis by extending hyphal growth (Mishra et al. 2017) (Fig. 3). Different types of transporter systems including nitrate, phosphate, ammonium, sulfur, carbon, zinc, protein, as well as plant aquaporin (AQ) transporters and so on are induced through AMF symbiotic relationship (Diagne et al. 2020). AMF are sometimes utilized as a biofertilizer to increase growth and yield of plants, nevertheless at a considerably lower rate compared to traditional fertilizers (Duhamel and Vandenkoornhuyse 2013). Plant's ability is affected to cope with biotic and abiotic challenges through AMF symbiotic relationship. AMF could be used not only as a biocontrol agent against various ecological pressures but also as a server to boost the nutrition of plants. To establish AMF symbiosis requires SLs (Foo et al. 2013b, Kohlen et al. 2012).

Li et al. (2018) reported on perennial ryegrass (*Lolium perenne*) by infecting the foliage of plant with a pathogen (*Bipolaris sorokiniana*) that several disease-associated enzymes including, polyphenol oxidase (PPO), catalase (CAT), and peroxidase (POD) were expressed, as well as the amount of malondialdehyde (MDA) acting as a determinant of cellular injury from peroxidation of lipids was

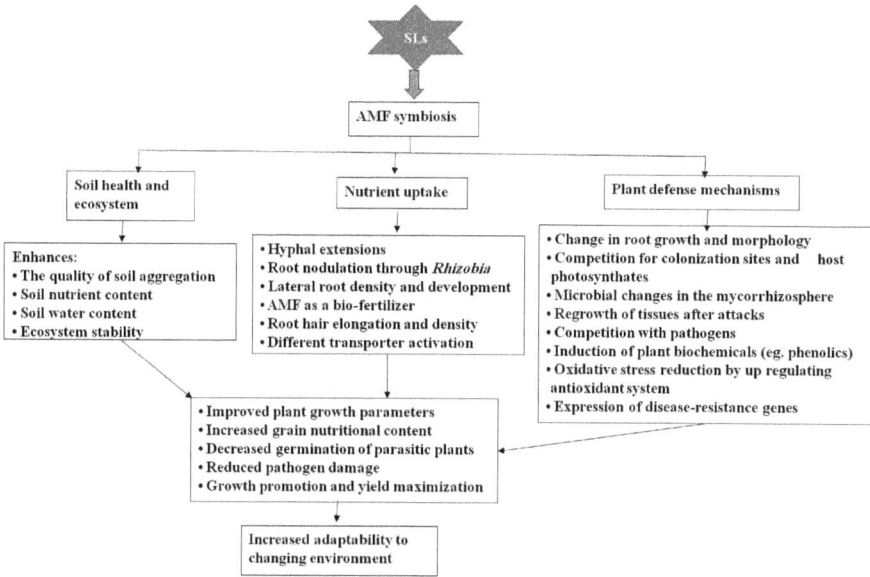

Fig. 3. Plant adaptable mechanisms induced by AMF symbiosis with plant roots under stressed environments (modified from Andreo-Jimenez et al. 2015, Liu et al. 2018, Begum et al. 2019, Diagne et al. 2020).

decreased. MDA had been linked to crop resistance against disease (Liu et al. 2014). All these diseases-associated enzymes were induced through AMF symbiosis. Plants protections enzymes such as POD and CAT can minimize stress-derived ROS (reactive oxygen species) and thus, can help plants recover from injuries incurred by abiotic and biotic challenges (Li et al. 2018). The plant ecosystem is affected positively by AMF symbiosis. Ecological stability, soil aggregation quality, plant architecture and microbial communities are also improved by symbiosis (Fig. 3) (Diagne et al. 2020). SLs serve as host detection cues for both parasite and AM fungus in the rhizosphere opens up yet another avenue for innovative management techniques. Lendzemo et al. (2005) reported maize and sorghum cultivars that witchweed (*Striga hermonthica*) infections had been reduced through AMF symbiosis. Likewise, by comparing with non-colonized plants, seed germination of *Orobanche* and *Phelipanche sp.* was reduced in AMF colonized pea, tomato, and lettuce plants (Aroca et al. 2013, Fernández-Aparicio et al. 2010, López-Ráez et al. 2011a).

A study on tomato revealed that completely established symbiosis can reduce the germination of parasitic plants due to the reduced production of SLs (López- Ráez et al. 2011a). In comparison to a plant lacking mycorrhizae, an AMF colonized plant will utilize more of such adaptive processes. Under practical situations, cultivars breeding with higher SL synthesis could be a method for increasing colonization with mycorrhiza. Natural SLs or synthetic counterparts could also be used exogenously. Small quantities of GR24 (0.1 M) had a favorable impact on the number of nodules, but greater quantities had a negative effect (De Cuyper et al. 2015).

Impact of Strigolactone on Plant Development

Phytohormones such as SLs operate as growth and development controllers. It is a new class of plant hormones that play a role in a variety of developmental stages (Davière and Achard 2016). It performs a vital function under nutrient deficient conditions in plants (especially P and N) by boosting symbiosis with AMF and nitrogen-fixing rhizobacteria as well as modifying the plant architecture of root and shoot (Fig. 4) (Marzec 2016).

Development is the primary role of SLs, like other phytohormones and SL-regulated developmental stages are dominated by their relationship with auxin (Hayward et al. 2009). The wild plants treated with GR24 exhibited higher lateral root numbers in lower phosphate condition, whereas the mutants of SLs showed lower lateral root numbers. On the contrary, GR24 suppresses both lateral root-forming capacity and outgrowth, resulting in a reduction of lateral root density in higher phosphate condition (Ruyter-Spira et al. 2011). Furthermore, in *Arabidopsis* and rice, there was a relationship between low phosphate, increased SLs, and impaired bud growth (Kohlen et al. 2011). With exception of their involvement in phosphate homeostasis maintenance, SLs were also identified as a potential pathway for controlling growth of plants in response to the availability of nitrogen (Mishra et al. 2017).

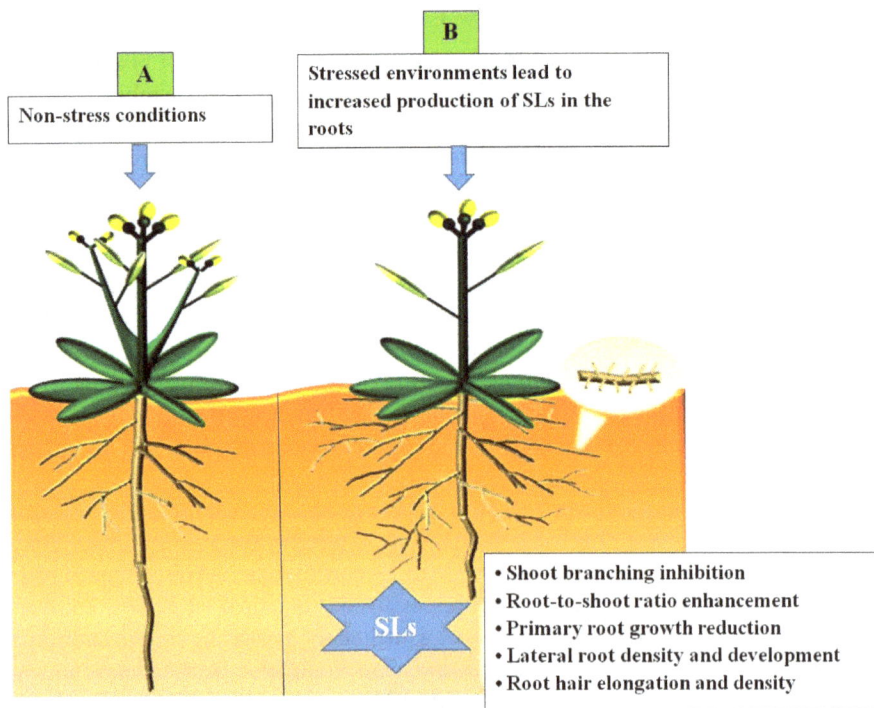

Fig. 4. Effects of strigolactones (SLs) on *Arabidopsis thaliana* (ecotype Columbia) plant growth and development (adapted from Andreo-Jimenez et al. 2015).

Role of Strigolactone against Soil-Borne Pathogens

Apart from their involvement in growth and development, SLs have also been identified as a contributor in disease resistance (Marzec 2016). The discovery of pathogen-related TF motifs in the promoters of genes involved in SL production was the first bit of proof (Torres-Vera et al. 2014). Plant mineral nutrient uptake, especially P, is increased and disease resistance particularly soil-borne diseases including *Rhizoctonia solani*, *Fusarium oxysporum*, and *Phytophthora parasitica* can be improved by mutualistic SLs induced AMF symbiosis (Liu et al. 2018). Microsymbionts are playing an increasingly important role in biotic stress regulation. AMF symbiosis has been shown in a number of studies to lessen plant pathogen damage (Table 2). Systemic protection is given against a diverse variety of aggressors and properties with systemic-acquired resistance are shared after a pathogen attack by an AMF colonization-induced protective mechanism (Nguvo and Gao 2019). After root colonization with non-pathogenic rhizobia, induced systemic resistance is also observed (Cameron et al. 2013). Based on the host plant and the system of culture, the efficiency of the relationships differs (Schüßler et al. 2001).

AMF symbiosis is thought to decrease the harmful effects due to soil diseases. Numerous experiments showed a strong decrease in the extremity of wilting and root rot occurred by fungus like *Macrophomina*, *Rhizoctonia*, *Fusarium*, or *Verticillium*, water molds (oomycetes) like *Pythium*, *Phytophthora*, and *Aphanomyces*, and bacterium like *Erwinia carotovora*, or the occurrences of the pathogens due to the presence of AMF symbiosis. Whipps (2004) produced a comprehensive assessment of the findings. Likewise, a report on mycorrhizal plants revealed that parasitic nematodes including *Meloidogyne* and *Pratylenchus* have been shown to reduce harmful impacts (Li et al. 2006). The level of security of plant strongly depends on the involvement of AMF, according to the research that examined several species of fungi or isolates (Kobra et al. 2009). Surprisingly, several findings indicated that *Glomus mosseae* had a stronger protective impact than other AMF (Ozgonen and Erkilic 2007).

Mycorrhizal plants having improved resistance to soil pathogens; many strategies act concurrently. Competitive behavior of AMF for both colonization sites and photo-synthates with the pathogens can assist plants to get host defense mechanisms (Fig. 3). *Phytophthora* was completely excluded from arbusculated cells in tomato roots, for instance (Cordier et al. 1998). The architecture and morphology of the root system have been documented to change as a result of mycorrhizal colonization (Norman et al. 1996). Although actual proof of a relationship between these alterations and pathogen infection dynamics is absent, such alterations may influence the pathogen infection dynamics. The pathogen's development may be influenced by changes in root exudation patterns (Pozo et al. 2010). The decrease of nematode attack in mycorrhizal plants is likely connected to a modification of its root exudates by AMF, according to additional root exudate investigations (Vos et al. 2012a). Nematodes are paralyzed temporarily or its penetration is decreased significantly by using mycorrhizal root exudates in mycorrhizal plants compared with non-mycorrhizal plants (Diagne et al. 2020). *Phytophthora fragariae* sporulation was greatly diminished by mycorrhizal colonization (Norman and Hooker 2000), and

Table 2. Role of strigolactones (SLs) and SLs-induced arbuscular mycorrhizal fungus (AMF) symbiosis in assisting plants in coping with a variety of biotic challenges.

Host Plants	Pathogen or Disease or Biotic stress	SLs or Induced AMF strains	Responses	References
Pepper (Capsicum annum)	Pythium Damping-Off and Root Rot (Pythium aphanidermatum)	Glomus sp.	Increasing crop plant growth and yield by lowering disease occurrence	Kumari et al. 2019
Melon (Cucumis melo)	Fusarium wilt (Fusarium oxysporum f.sp. niveum)	Funneliformis mosseae	Disease occurrence mitigation	Martínez-Medina et al. 2011
Potato (Solanum tuberosum)	PVY (potato virus Y)	Rhizophagus irregularis	Less symptoms and considerable acceleration of stem development	Thiem et al. 2014
Tobacco (Nicotiana tabacum)	TMV (tobacco mosaic virus), CGMMV (Cucumber green mottle mosaic virus)	Rhizophagus irregularis	Lower disease symptoms exhibition	Stolyarchuk et al. 2009
Maize (Zea mays)	Purple witchweed (Striga hermonthica)	Glomus margarita, G. etunicatum, Scutellospora fulgida	Reduction of Striga species occurrence	Othira 2012
Sorghum (Sorghum bicolor)	Purple witchweed (Striga hermonthica)	Funneliformis mosseae	Increased sorghum's quality	Isah et al. 2013
Soybean (Glycine max)	Charcoal rot (Macrophomina phaseolina)	Rhizophagus irregularis	Increased functional leaves and the height of the plant	Spagnoletti et al. 2020
Sugarcane (Saccharum officinarum)	Purple witchweed (Striga hermonthica)	Gigaspora margarita, Glomus etunicatum, Scutellospora fulgida,	Boosted – • growth of plant, • physiological functions, • finally, yield	Manjunatha et al. 2018
Milkvetch (Astragalus adsurgens)	Powdery mildew (Erysiphe pisi)	Funneliformis mosseae, Glomus versiforme, Claroideoglomus etunicatum	Growth of root and shoot enhancement	Liu et al. 2018
Grape-vines (Vitis vinifera)	Armillaria Root Rot (Armillaria melea)	Glomus mosseae	Plant resistance against pathogens	Nogales et al. 2009
Wheat (Triticum aestivum)	Take-all (Gaeumannomyces graminis)	AMF species	Systemic protection at the root level	Khaosaad et al. 2007

Plant	Disease/pathogen	AMF species	Effect	Reference
Tomato (Solanum lycopersicum)	Powdery mildew (Blumeria graminis f. sp. Tritici)	Rhizophagus irregularis, Funneliformis mosseae	• Reduced infection through accumulated polyphenolic chemicals, • Increased defense system against pathogens	Mustafa et al. 2016
	Wilts and blights (Phytophthora sp), Bacterial wilt (Ralstonia solanacearum)	AMF species	Systemic resistance against pathogens	Pozo et al. 2010
	Early blight (Alternaria solani)	AMF species	Plant resistance to shoot pathogens	Fritz et al. 2006, Noval et al. 2007
	Botrytis gray mold (Botrytis cinerea)	Glomus mosseae	Systemic resistance against pathogens	Jung et al. 2009
	Leaf mold (Cladosporium fulvum)	Funneliformis mosseae	Increased – • resistance to the infection of pathogen • concentration of chlorophyll • leaf photosynthesis • dry and fresh mass	Wang et al. 2018
	Wilt or root knot disease (Meloidogyne javanica)	Funneliformis mosseae	Decreased disease symptoms – • galling • reproduction of nematode • female morphological features	Siddiqui and Mahmood 1998
	Early blight (Alternaria solani), wilts and blights (Fusarium oxysporum)	Funneliformis mosseae	Reduced disease symptoms	Song et al. 2015, El-Khallal 2007
	Northern root lesion (Pratylenchus penetrans), root-knot nematode (Meloidogyne incognita)	Funneliformis mosseae	Development of defense response	Vos et al. 2012b
Banana plants (Musa acuminate)	Parasitic nematodes (Radopholus similis)	AMF species	Systemic resistance to pathogens	Elsen et al. 2008
Apple tree seedlings (Malus domestica)	Powdery mildew (Podosphaera leucotricha)	Mixture of AMF species	• Flint (trifloxystrobin) having a similar effect • Strobyon (strobilurin) having a better effect	Yousefi et al. 2011

Table 2. contd. ...

... Table 2. contd.

Host Plants	Pathogen or Disease or Biotic stress	SLs or Induced AMF strains	Responses	References
Cucumber (Cucumis sativus)	Powdery mildew (Podosphaera xanthii)	AMF species	• No consequence • Reduced foliar attack	Chandanie et al. 2006
Eggplant (Solanum melongena), Cucumber (Cucumis sativus).	Verticillium wilt (Verticillium dahlia) Angular Leaf Spot (Pseudomonas lacrymans)	Glomus versiforme	Disease symptoms reduction	Jones et al. 2011
Model plant (Arabidopsis thaliana)	Leafy gall (Rhodococcus fascians)	Treatment with GR24	Hypersensitivity of SL mutant to infection	Band et al. 2012
	Bacterial Soft Rot (Pectobacterium carotovorum) Bacterial canker or blast (Pseudomonas syringae pv. Syringae)	SL-signaling mutant	Hypersensitivity of SL mutant to infection	Oldroyd 2013
Garden tomato (Lycopersicon esculentum)	Early blight (Alternaria alternate), Gray mold (Botrytis cinerea)	SL-biosynthesis mutant slccd8	Hypersensitive to fungal attack	Torres-Vera et al. 2014
	Fusarium wilt of tomato (Fusarium oxysporum f. sp. Lycopersici)	Glomus sp.	• Decreased occurrence of disease through produced antimicrobial chemicals • Improved – • growth of plants, • photosynthetic rate, • dry matter production, • ultimately, yield	Kumari and Prabina 2019
Burclover (Medicago sp.)	Black rot (Xanthomonas campestris)	AMF species	Plant resistance to shoot pathogens	Liu et al. 2007
Barrel medic (Medicago truncatula)	Smut (Sporisorium reilianum f. sp. Zeae)	GR24 treatment	• No influence on the growth of fungus • Enhanced respiration of cell • Modified expression of gene	Jin et al. 2016
	White mold (Sclerotinia sclerotiorum) Wilt (Fusarium solani)	GR24 treatment	• Significant mycelial enhancement, • Radial expansion suppression	Fonouni-Farde et al. 2016

Plant	Disease (pathogen)	Organism / SL	Observed effect	Reference
Roses (Rosa sp.)	Anthracnos (Colletotrichum acutatum)			
	Black spot and rot (Alternaria alternate)			
	Gray mold (Botrytis cinerea)	G. mosseae	Observed positive effect	Moller et al. 2009
Chrysanthemum (Chrysanthemum sp.)	Yellow disease phytoplasmas (Phytoplasma asteris)	AMF species	Resistance to pathogen	D'Amelio et al. 2007
Pear (Pyrus communis)	Pear decline (Phytoplasma pyri)	AMF species	Phytoplasma disease tolerance	García-Chapa et al. 2004
Groundnut (Arachis hypogaea)	Southern stem rot of peanut (Sclerotium rolfsii)	Gigaspora margarita, Rhizophagus fasciculatum, Sclerocystis dussii, Acaulospora laevis, and Cucumis melo	Minimization of the harmful effects of pathogens	Diagne et al. 2020
Chickpea (Cicer arietinum L.)	Fusarium wilt (Fusarium oxysporum f.sp. ciceris)	Glomus hoi, Rhizophagus fasciculatum	Enhanced amount of phosphorus and nitrogen contents	Singh et al. 2010
Wild pea (Pisum sativum L. subsp. Elatius)	Fusarium wilt (Fusarium oxysporum f. sp. Pisi)	Various SLs (SL-biosynthesis mutant pscdd8)	In aseptic condition – • Same sensitivity between wild and mutant type • No influence on growth and development of fungi	Yu et al. 2014
Pea (Pisum sativum)	Pythium seed and seedling rot (Pythium irregulare)	Various SLs (SL-biosynthesis mutants pscdd7 and pscdd8)	In aseptic condition – • Same sensitivity between wild and mutant type • No influence on growth and development of fungi	Pimprikar et al. 2016
	Powdery mildew of pea (Erysiphe pisi var. pisi)	Glomus mosseae	Reduced – • disease occurrences • disease index from 55.2 to 28.7%	Singh et al. 2004

the chemotactic response of *Phytophthora nicotianae* zoospores was changed (Lioussanne et al. 2008). As exudation of root play a significant role in creating soil microbial populations, altered exudation can lead to altered microbial populations such as the emergence of potentially hostile organisms. Plant defenses were implicated because of the systemic nature of the generated resistance (Pozo et al. 2010).

Impact of Strigolactone on Aboveground Pathogens

The function of strigolactone in the host's defense system against bacterial infection such as *Pectobacterium carotovorum, Pseudomonas syringae,* and *Rhodococcus fascians* was revealed in a study using SL synthesis and *Arabidopsis*-signaling mutants (Stes et al. 2015). At the time of infection, *R. fascians*, which is a Gram-positive actinomycete, produces a combination of cytokinins (CKs) and leafy gall syndrome is developed in infected plants. The syndrome is characterized by the apical dominance collapse and latent axillary meristems stimulation (Band et al. 2012). Leafy gall syndrome was more evident in all SL-biosynthesis and signaling mutants compared to the wild plants. Plants cultivated on conditions containing GR24 (a synthesized homologue of SL) were unaffected by *R. fascians*, but plants grown with D2 (a SL synthesis antagonist) were more susceptible. As SLs biosynthetic genes are increased at the time of infection, the amount of SLs in plant parts is used as a determinant of resistance to *R. fascians* (Band et al. 2012).

There are fewer studies on the effects of AMF on aboveground diseases, and they appear to be unclear. Initial research linked AMF symbiosis to increased virus susceptibility (Whipps 2004), and it was widely assumed that AM plants are more vulnerable to shoot diseases. Recent research on diseases associated with various way of life, on the other hand, has shown a more complexity. Rust and powdery mildew fungi (Uromyces, Oidium, and Blumeria), the biotrophic phytopathogens, might develop better tolerance in terms of biomass and yield of plant (Pozo et al. 2010). Previous experiments on the relationship between aboveground pathogens and AMF have revealed (Table 2) that AMF can lessen the disease intensity of powdery mildew (Liu et al. 2018). The aboveground sections of over 10,000 species of plants (Glawe 2008) comprising various legume crops like standing milk-vetch (*Astragalus adsurgens*) (Nan and Li 1994) were infected by the most common phytopathogenic fungus, powdery mildews (*Erysiphe pisi*) on the globe. Researchers investigated the impact of three AMF, either alone or in conjunction, on standing milkvetch development as well as sensitivity to *E. pisi*, since AMF symbiotic relationship has been found to improve resistance against phytopathogenic diseases like powdery mildew. AMF inoculation decreased powdery mildew disease incidence or had no impact on the severity of disease, according to the few research studies that precisely evaluated AMF symbiotic effects on powdery mildew in plants. Liu and colleagues reported that yield losses incurred by powdery mildew infection of the foliage part of plants are minimized due to the existence of AMF in the rhizosphere (Liu et al. 2018).

Insect vectors transmit *Phytoplasma*, which are specific obligate parasites of the phloem tissue. Lingua et al. (2002) conducted a study on tomato and found that disease symptoms induced by a Stolbur group *phytoplasma* was reduced due to the presence of AMF symbiosis. Possible effects of insect vectors were excluded out because they were infected by grafting with contaminated scions. As a result, the defense is associated with alterations in the mycorrhizal plant's physiological state. Two major methods would be effective in aboveground mycorrhizal plant relationships. The first method would be possible alterations in the host plant's nutritional contents, as well as changes in the relationship of source-sink inside it, which could impact the plant's adaptability for stem invaders. The manipulation of defense responses is the second method (Pozo et al. 2010).

Effect of Strigolactone on Root Parasitic Plants

SLs were found to be the stimulants of root parasitic plants germination under the family Orobanchaceae, such as the genera *Orobanche, Phelipanche* (broomrapes), and *Striga* (witchweeds), prior to the discovery of their actual their function as plant hormones and signaling indicators for mutualistic relationship with microbes in the rootzone of plants (Bouwmeester et al. 2003). These obligatory parasitic weeds causing havoc on vital crops including rice, sorghum, maize, legumes, sunflower, tomato, and tobacco, are the one of the most destructive agricultural pests all over the world. Crop yields can be reduced by up to 70% as a result of them (Parker 2009). The African witchweed (also known as 'Striga'), which is primarily found in Sub-Saharan Africa, is one of the biotic limitations that has the greatest effect on production in developing nations. Owing to excessive rates of infection, these parasitic plants are a socioeconomic concern that has pushed numerous poor farmers to quit the crops (Atera et al. 2012). A possible link between AMF and its influence on parasitic plants might be developed with the identification of SLs as host recognition cues for AMF in the root-zone (Akiyama et al. 2005).

Striga germination can be inhibited or suppressed by soil microbes, especially AMF. Striga and cereal interactions can be influenced by AMF (Lendzemo et al. 2006). AMF affected the germination of Striga seeds negatively, decreased the amount of *Striga* seedlings adhering and rising, as well as pushed back Striga emergence time. AMF efficacy against *Striga* infection in protecting sugarcane, boosting crop growth, and lowering the soil *Striga* seed bank was proven in studies conducted by Manjunatha et al. (2018). As a result, the plant host's performance was improved by AMF, enabling it to survive better against *Striga* injury (Lendzemo et al. 2006). Diagne and coworkers found that mycorrhizal plants showed the reduced incidence of parasitism of ragwort leafminer (*Chromatomyia syngenesiae*) by leafminer parasite (*Diglyphus isaea*), and that the effects of three AMF species on parasitism frequencies in the laboratory were depending on the AMF species.

AMF symbiosis considerably decreased the level of parasitic plants in maize and sorghum plants in African fields afflicted with witchweed, a hemiparasite. As a result, it was proposed to use mycorrhizas for integrative parasitic weed control (López-Ráez et al. 2009). Follow-up trials under lab settings showed that AMF symbiosis reduces the effects of Striga, which is thought to be linked to a decrease in SLs synthesis (Lendzemo et al. 2007). Likewise, extracts from tomatoes cultivated with G. mosseae cause lower germination of broomrape (*Orobanche ramosa*) seeds than extracts from non-mycorrhizal plants (López-Ráez et al. 2012). Furthermore, decreased strigolactone synthesis was linked to reduced *Orobanche* susceptibility in a tomato mutant (López-Ráez et al. 2008). Overall, the reduced root parasitic plant occurrence on mycorrhizal plants appears to be due to a drop in SLs synthesis.

Role of Strigolactone against Phytophagous Insects

Studies on the influence of AMF on herbivore insects are scarce in comparison to the well-known influence on infectious nematode and fungus. The impact's size and direction are determined by the insect's feeding method and lifestyle (Koricheva et al. 2009). Several investigations under controlled and field situations have covered a wide variety of mycorrhizal plant insect interactions. Hartley and Gange (2009) determined that mycorrhizas have significant adverse impacts on rhizophagous insects in general, whereas impacts on stem-feeding insects are milder and also more varied, based on a complete analysis of the published evidence. The impact of mycorrhizal fungus on herbivorous insects differed based on the feature examined and the extent of feeding specialty, according to a meta-analysis performed by Koricheva et al. (2009). In contrast to other fungal species investigated, *Rhizoglomus intraradices*, a member of AMF, appeared to have a detrimental impact on chewing insects (Koricheva et al. 2009). There is a controversial finding on flowering plant species (*Plantago lanceolate*) found by Gange and West that AMF symbiosis enhanced leaf resistance against chewer, great tiger moth (*Arctia caja*) (Linnaeus 1758), whereas it increased the sucker, green aphid (*Mysus persicae*) performance (Gange and West 1994). Although specialized insects may improve, generalist insects are normally badly influenced by mycorrhizas. Additionally, aphids prefer mycorrhizal plants, but leaf-chewers are frequently adversely impacted by the relationship (Pozo et al. 2010).

In mycorrhizal plants, the detrimental impact on leaf-chewing insect is presumably due to their vulnerability to jasmonate-dependent defenses that are increased (Peña-Corté et al. 2004). Furthermore, mycorrhizal plants can affect herbivorous insects by enhancing the efficiency of both predators and parasitoids: for example, in tomato, the emitted volatile mixtures by mycorrhizal plants may be more appealing to aphid parasitoids compared to the emitted blends by non-mycorrhizal plants (Guerrieri et al. 2004).

Strigolactone as a Modulator Host Defense Mechanism

The formation of a successful mutualism needs a higher level of cooperation between these participants (Parniske 2008). During extensive signaling pathways, the amounts of many plant hormones (primarily ethylene (ET), salicylic acid (SA), abcisic acid (ABA), and jasmonates (JAs)) adjust the defense responses in crops (Pieterse et al. 2009). In mycorrhizal plants, these hormone levels appear to just be modified remarkably (Hause et al. 2007, López-Ráez et al. 2010), possibly altering host defense systems. In mycorrhizal roots, the plant's defensive chemicals are accumulated, but to a lesser level compared to the plant-pathogen interconnection. In mycorrhizal roots, phenylpropanoid and oxylipin metabolism, the buildup of reactive oxygen species, and certain isoforms of defense-related enzymes are activated (López-Ráez et al.

2010). Transcriptional analysis of responses of plants to AMF found considerable similarity with reactions to biotrophic pathogens (Güimil et al. 2005). Actually, AMF are comparable to biotrophic pathogens in that they are obligatory biotrophs (Paszkowski 2006). Co-inoculation between *Glomus intraradices* and *Rhizoctonia solani*, as well as suppression of pathogen-induced defense responses were observed (Guenoune et al. 2001).

Despite the fact that AMF can stimulate host defense responses in mycorrhizal mutants, only insufficient and transitory defense responses are stimulated throughout suitable AMF relationships (Liu et al. 2003). As a result, it appears that the suppression of some SA-regulated defense responses is required for AM formation. Plant defenses are modulated not only in the roots but also in the shoots during AMF development. In mycorrhizal plant leaves, insect anti-feedant chemicals are accumulated (Verhage et al. 2009), and transcriptional upregulation of defense-associated genes are occurred (Liu et al. 2007).

Suppression of some defenses has been recorded in mycorrhizal tobacco stems (Shaul et al. 1999), as has a hindrance in the system-wide development of pathogenesis-related protein 1 (PR1) after treatment with SA or synthetic analogues, as well as inhibition of some protective mechanisms (Bennett et al. 2009). The relationship with stem invaders may be affected by this manipulation. The aboveground and below-ground reciprocal influences on plants defenses is gaining increased attention in this area (Erb et al. 2009). The changed volatile profile generated by mycorrhizal plants adds another layer of complication (Rapparini et al. 2007). Volatiles may contribute in defense, such as alluring natural predators of highly harmful insects or preparing distant areas of the plants for much more effective defense action (Heil and Ton 2008). The plant will be more vulnerable to diseases that are repelled by the reactions, such as biotrophic pathogens, if AMF suppresses SA-controlled reactions. Activation of the JA signaling pathway, on the other hand, would increase the resistance of mycorrhizal plants to JA-sensible insects and necrotrophic diseases (Pozo and Azcón-Aguilar 2007). This pattern is more evident in stem relationships, when host defense modification appears to be the primary method. Because other systems are working at the same time, the significance of such a changed equilibrium will be reduced in roots, and the most general effect will be a decrease in disease (Pozo et al. 2010).

Major Constraints of Strigolactone

As SLs increase AMF symbiosis by increasing growth and branching of hyphae (Davière and Achard 2016), it is hypothesized that they can also impact harmful fungi formation. GR24 (homologue of SLs) effectively stopped six separate harmful fungi from growing radially, including parasites of foliage/stem and root. The impacts of SLs on branching of hypae became less obvious, varying between hyphal branching stimulation to repression based on the fungus. When the greatest proportion of GR24 was used, the impact was only seen in some fungus (Fonouni-Farde et al. 2016). According to another research, GR24 can improve respiration of cells without impacting growth or morphology of fungus including *Sporisorium reilianum* f. sp. zeae (pathogenic fungus) (Jin et al. 2016). Contaminated with the fungus *Alternaria alternata* and *Botrytis cinerea*, the SL biosynthesis of tomato mutant slccd8 that has a mutation in the orthologue of the MAX4 gene of *Arabidopsis* in comparison to wild-type plants, showed enhanced vulnerability to pathogens. Decreased levels of recognized plant defense hormones in affected tissues followed increased sensitivity of slccd8 to fungal diseases, along with a decrease of 30% JA and 50% SA (Oldroyd 2013). SLs, on the other hand, have not been shown to have a function in resistance to many other diseases such as *Fusarium oxysporum* and *Pythium irregulare* (Foo et al. 2016).

Infections of pea SL-biosynthesis mutants did not result in hypersensitivity against infection of fungus, and plant pathogenic treatment with several synthesized or naturally occurring SLs had no effect on growth or branching of fungus (Pimprikar et al. 2016). Almost over 30 mutant strains of *B. cinerea* were screened recently, and among these strains, only two such as, Dbcltf1 and Dbctrr1, were found to have slower response to the treatment of GR24 (Boivin et al. 2016). The findings suggested that SLs may suppress the activation of the TFs of fungus from the GATA family that are expressed by BcTLF1 and perform a key function in pathogenicity. As BcTRR1 expresses thioredoxin reductase, which is essential in management of redox system of the cells of fungus, SLs may disrupt cell redox homeostasis in *B. cinerea*. As this wild type and other mutant strains responded similarly to the treatment of GR24, it was determined that SLs have no effect on ROS generation, mitogen-activated protein kinases (MAPKs), or calcium signaling (Boivin et al. 2016). Overall, the findings suggest that SLs are not involved across all aspects of plant defense. SLs, on the other hand, appear to have a role to certain fungi and bacteria in plant resistance (Marzec 2016).

Future Prospect of Strigolactone

Plant stress management is well-known for the interplay of plant hormones with stress. The present list of options includes both traditional and newly found plant hormones, and the list is continually growing as more growth-stimulatory compounds become available. In most situations, phytohormones are first mentioned and explained under terms of production, and their function is normally treated as the second priority in stress. Development and stress-related, these are the two categories for major hormones such as ABA, ET, JA, and SA. The SL has been identified as a major potential performer in plant stress control (Marzec 2016). The biological and ecological significance of SLs in the root system is demonstrated by experimental studies. It's so fascinating and hopeful to discover new roles and functions for the many SLs in and ex planta. Because of their multifunctional character, they have a variety of possible agricultural applications. Researchers have also found variances in biological selectivity among individual SLs, while we are still learning how this occurs mechanistically (Andreo-Jimenez et al. 2015). Moreover, if it's either production or signaling is impaired, pathogen-specific plant resistance may be compromised. SL's importance in nutrient deficiency, drought, temperature, salinity, and plant defense has been acknowledged, but as a stress hormone, its status requires more research (Mishra et al. 2017).

The thing which is widely known are that synthesized homologues of SLs are a blend of at least a couple, but in most cases four, stereoisomers with distinct bioactivities, which may explain the inconsistencies in findings on the impact of GR24 on the growth and development of several pathogens (Floss et al. 2013). The dosage of GR24 or the method of application of SLs are also elements to consider (Marzec 2016). More study into the needs for distinct SLs in various biological activities should help us better understand the biological processes that take place underground. Furthermore, in order to fully utilize the possibilities of these signaling pathways in agriculture, a thorough understanding of the processes regulate SL synthesis and secretion, as well as how these are impacted by various environmental situations, is essential (Andreo-Jimenez et al. 2015).

Conclusion

Strigolactones (SLs) are multipurpose compounds that have just been defined as a novel category of plant hormones. In different to biotic and abiotic stress, it serves an important function in modulating plants growth and development. It was first discovered to promote the development of AMF symbiosis and to serve as host recognition signal as well as stimulator of the germination of parasitic weed seeds like *Striga* and *Orobanche*. Stress may have a direct impact on SLs production or signaling pathways. It forms the basis for crop persistence in the face of nutrient deficiency by increasing development-assisting soil microbes including AMF and *Rhizobia*, nitrogen-fixing bacteria's nodulation efficacy. According to existing evidence, SLs may have a role in the resistance of plants to pathogenic infections from a variety of bacterial as well as fungal groups. The synthesis of pathogenic cytokinin is a fundamental element for interacting with the SLs found in plant tissue in certain instances (e.g., after *R. fascians* infection). The molecular mechanism of SL-dependent plant-pathogen relationship remains unknown for other infections. The SL mutant's phenotype, which includes larger volume of stem branching and closure problems of stomata, could be linked to increased vulnerability to diseases. Nonetheless, the control systems governing gene expression of SL-synthesis, and also interaction among SLs as well as other defense-related plant hormones, strongly suggest that SLs play a role in biotic stress response of plants. If it is possible to make plants genetically resistant to pests, weeds, pathogens, rather than spraying chemical pesticides, weedicides, fungicides and so on; this will be the cheapest and best way to minimize the losses caused by these stresses leading to sustainable development. New research incorporating other pathogenic species and combining transcriptomic approaches alongside hormonal measures may help to clarify the function of SLs in plant resistance to certain diseases and reveal the molecular pathways behind the responses.

References

Abe, S., Sado, A., Tanaka, K., Kisugi, T., Asami, K., Ota, S. et al. 2014. Carlactone is converted to carlactonoic acid by MAX1 in Arabidopsis and its methyl ester can directly interact with AtD14 in vitro. Proc. Natl. Acad. Sci. U.S.A., 111(50): 18084–18089.

Akiyama, K., Matsuzaki, K.I. and Hayashi, H. 2005. Plant sesquiterpenes induce hyphal branching in arbuscular mycorrhizal fungi. Nature, 435(7043): 824–827.

Akiyama, K. and Hayashi, H. 2006. Strigolactones: chemical signals for fungal symbionts and parasitic weeds in plant roots. Ann. Bot. 97(6): 925–931.

Al-Babili, S. and Bouwmeester, H.J. 2015. Strigolactones, a novel carotenoid-derived plant hormone. Ann. Rev. Plant Biol., 66: 161–186.

Alder, A., Jamil, M., Marzorati, M., Bruno, M., Vermathen, M., Bigler, P. et al. 2012. The path from β-carotene to carlactone, a strigolactone-like plant hormone. Science, 335(6074): 1348–1351.

Andreo-Jimenez, B., Ruyter-Spira, C., Bouwmeester, H.J. and Lopez-Raez, J.A. 2015. Ecological relevance of strigolactones in nutrient uptake and other abiotic stresses, and in plant-microbe interactions below-ground. Plant and Soil, 394(1–2): 1–19.

Aroca, R., Ruiz-Lozano, J.M., Zamarreño, Á.M., Paz, J.A., García-Mina, J.M., Pozo, M.J. et al. 2013. Arbuscular mycorrhizal symbiosis influences strigolactone production under salinity and alleviates salt stress in lettuce plants. J. Plant Physiol., 170(1): 47–55.

Atera, E.A., Itoh, K., Azuma, T. and Ishii, T. 2012. Farmers' perspectives on the biotic constraint of Striga hermonthica and its control in western Kenya. Weed Biol. Manag., 12(1): 53–62.

Band, L.R., Úbeda-Tomás, S., Dyson, R.J., Middleton, A.M., Hodgman, T.C., Owen, M.R. et al. 2012. Growth-induced hormone dilution can explain the dynamics of plant root cell elongation. Proc. Natl. Acad. Sci. USA, 109(19): 7577–7582.

Begum, N., Qin, C., Ahanger, M.A., Raza, S., Khan, M.I., Ashraf, M. et al. 2019. Role of Arbuscular Mycorrhizal Fungi in Plant Growth Regulation: Implications in Abiotic Stress Tolerance. Front. Plant Sci., 10: 1–15.

Bennett, A.E., Bever, J.D. and Deane Bowers, M. 2009. Arbuscular mycorrhizal fungal species suppress inducible plant responses and alter defensive strategies following herbivory. Oecologia, 160(4): 771–779.

Boivin, S., Fonouni-Farde, C. and Frugier, F. 2016. How auxin and cytokinin phytohormones modulate root microbe interactions. Front. Plant Sci., 7: 1240.

Booker, J., Sieberer, T., Wright, W., Williamson, L., Willett, B., Stirnberg, P. et al. 2005. MAX1 encodes a cytochrome P450 family member that acts downstream of MAX3/4 to produce a carotenoid-derived branch-inhibiting hormone. Dev. Cell, 8(3): 443–449.

Bouwmeester, H.J., Matusova, R., Zhongkui, S. and Beale, M.H. 2003. Secondary metabolite signalling in host–parasitic plant interactions. Curr. Opin. Plant Biol., 6(4): 358–364.

Bouwmeester, H.J., Roux, C., Lopez-Raez, J.A. and Becard, G. 2007. Rhizosphere communication of plants, parasitic plants and AM fungi. Trends Plant Sci., 12(5): 224–230.

Cameron, D.D., Neal, A.L., Van Wees, S.C. and Ton, J. 2013. Mycorrhiza-induced resistance: more than the sum of its parts?. Trends Plant. Sci., 18(10): 539–545.

Cardoso, C., Zhang, Y., Jamil, M., Hepworth, J., Charnikhova, T., Dimkpa, S.O. et al. 2014. Natural variation of rice strigolactone biosynthesis is associated with the deletion of two MAX1 orthologs. Proc. Natl. Acad. Sci. U.S.A., 111(6): 2379–2384.

Cook, C.E., Whichard, L.P., Wall, M., Egley, G.H., Coggon, P., Luhan, P.A. et al. 1972. Germination stimulants. II. Structure of strigol, a potent seed germination stimulant for witchweed (*Striga lutea*). J. Am. Chem. Soc., 94(17): 6198–6199.

Cordier, C., Pozo, M.J., Barea, J.M., Gianinazzi, S. and Gianinazzi-Pearson, V. 1998. Cell defense responses associated with localized and systemic resistance to *Phytophthora parasitica* induced in tomato by an arbuscular mycorrhizal fungus. Mol. Plant-Microbe Interact., 11(10): 1017–1028.

D'Amelio R., Massa M., Gamalero E., D'Agostino G., Sampò S., Berta G. et al. 2007. Preliminary results on the evaluation of the effects of elicitors of plant resistance on chrysanthemum yellows phytoplasma infection. Bull Insectol., 60: 317–318

Davière, J.M. and Achard, P. 2016. A pivotal role of DELLAs in regulating multiple hormone signals. Mol. Plant, 9(1): 10–20.

De Cuyper, C., Fromentin, J., Yocgo, R.E., De Keyser, A., Guillotin, B., Kunert, K. et al. 2015. From lateral root density to nodule number, the strigolactone analogue GR24 shapes the root architecture of *Medicago truncatula*. J. Exp. Bot., 66(1): 137–146.

De La Noval, B., Pérez, E., Martínez, B., León, O., Martínez-Gallardo, N. and Délano-Frier, J. 2007. Exogenous systemin has a contrasting effect on disease resistance in mycorrhizal tomato (*Solanum lycopersicum*) plants infected with necrotrophic or hemibiotrophic pathogens. Mycorrhiza, 17(5): 449–460.

Diagne, N., Ngom, M., Djighaly, P.I., Fall, D., Hocher, V. and Svistoonoff, S. 2020. Roles of arbuscular mycorrhizal fungi on plant growth and performance: importance in biotic and abiotic stressed regulation. Diversity, 12(10): 1–25.

Duhamel, M. and Vandenkoornhuyse, P. 2013. Sustainable agriculture: possible trajectories from mutualistic symbiosis and plant neodomestication. Trends Plant Sci., 18(11): 597–600.

El-Khallal, S.M. 2007. Induction and modulation of resistance in tomato plants against Fusarium wilt disease by bioagent fungi (arbuscular mycorrhiza) and/or hormonal elicitors (jasmonic acid and salicylic acid): 1-Changes in growth, some metabolic activities and endogenous hormones related to defence mechanism. Aust. J. Basic Appl. Sci., 1(4): 691–705.

Elsen, A., Gervacio, D., Swennen, R. and De Waele, D. 2008. AMF-induced biocontrol against plant parasitic nematodes in Musa sp.: a systemic effect. Mycorrhiza, 18(5): 251–256.

Erb, M., Flors, V., Karlen, D., De Lange, E., Planchamp, C., D'Alessandro et al. 2009. Signal signature of aboveground-induced resistance upon belowground herbivory in maize. Plant J., 59(2): 292–302.

Faizan, M., Faraz, A., Sami, F., Siddiqui, H., Yusuf, M., Gruszka, D. et al. 2020. Role of strigolactones: Signalling and crosstalk with other phytohormones. Open Life Sci., 15(1): 217–228.

Fernández-Aparicio, M., García-Garrido, J.M., Ocampo, J.A. and Rubiales, D. 2010. Colonisation of field pea roots by arbuscular mycorrhizal fungi reduces Orobanche and Phelipanche species seed germination. Weed Res., 50(3): 262–268.

Floss, D.S., Levy, J.G., Lévesque-Tremblay, V., Pumplin, N. and Harrison, M.J. 2013. DELLA proteins regulate arbuscule formation in arbuscular mycorrhizal symbiosis. Proc. Natl. Acad. Sci. U.S.A., 110(51): 5025–5034.

Fonouni-Farde, C., Tan, S., Baudin, M., Brault, M., Wen, J., Mysore, K.S. et al. 2016. DELLA-mediated gibberellin signalling regulates Nod factor signalling and rhizobial infection. Nat. Commun., 7(1): 1–13.

Foo, E. and Davies, N.W. 2011. Strigolactones promote nodulation in pea. Planta, 234(5): 1073–1081.

Foo, E., Yoneyama, K., Hugill, C.J., Quittenden, L.J. and Reid, J.B. 2013. Strigolactones and the regulation of pea symbioses in response to nitrate and phosphate deficiency. Mol. Plant, 6(1): 76–87.

Foo, E., Blake, S.N., Fisher, B.J., Smith, J.A. and Reid, J.B. 2016. The role of strigolactones during plant interactions with the pathogenic fungus *Fusarium oxysporum*. Planta, 243(6): 1387–1396.

Fritz, M., Jakobsen, I., Lyngkjær, M.F., Thordal-Christensen, H. and Pons-Kühnemann, J. 2006. Arbuscular mycorrhiza reduces susceptibility of tomato to *Alternaria solani*. Mycorrhiza, 16(6): 413–419.

Gange, A.C. and West, H.M. 1994. Interactions between arbuscular mycorrhizal fungi and foliar-feeding insects in *Plantago lanceolata* L. New Phytol., 128(1): 79–87.

García-Chapa, M., Batlle, A., Laviña, A., Camprubí, A., Estaún, V. and Calvet, C. 2004. Tolerance increase to pear decline phytoplasma in mycorrhizal OHF-333 pear rootstock. Acta. Hort., 657: 437–441.

Gianinazzi, S., Gollotte, A., Binet, M.N., Van Tuinen, D., Redecker, D. and Wipf, D. 2010. Agroecology: the key role of arbuscular mycorrhizas in ecosystem services. Mycorrhiza 20(8): 519–530.

Glawe, D.A. 2008. The powdery mildews: a review of the world's most familiar (yet poorly known) plant pathogens. Annu. Rev. Phytopathol. 46: 27–51.

Gomez-Roldan, V., Fermas, S., Brewer, P.B., Puech-Pagès, V., Dun, E.A., Pillot, J.P. et al. 2008. Strigolactone inhibition of shoot branching. Nature, 455(7210): 189–194.

Guenoune, D., Galili, S., Phillips, D.A., Volpin, H., Chet, I., Okon, Y. et al. 2001. The defense response elicited by the pathogen *Rhizoctonia solani* is suppressed by colonization of the AM-fungus Glomus intraradices. Plant Sci., 160(5): 925–932.

Guerrieri, E., Lingua, G., Digilio, M.C., Massa, N. and Berta, G. 2004. Do interactions between plant roots and the rhizosphere affect parasitoid behaviour?. Ecol. Entomol., 29(6): 753–756.

Güimil, S., Chang, H.S., Zhu, T., Sesma, A., Osbourn, A., Roux, C. et al. 2005. Comparative transcriptomics of rice reveals an ancient pattern of response to microbial colonization. Proc. Nat. Acad. Sci., 102(22): 8066–8070.

Hartley, S.E. and Gange, A.C. 2009. Impacts of plant symbiotic fungi on insect herbivores: mutualism in a multitrophic context. Ann. Rev. Entomol., 54: 323–342.

Hayward, A., Stirnberg, P., Beveridge, C. and Leyser, O. 2009. Interactions between auxin and strigolactone in shoot branching control. Plant Physiol., 151(1): 400–412.

Heil, M. and Ton, J. 2008. Long-distance signalling in plant defence. Trends Plant Sci., 13(6): 264–272.

Isah, K.M., Kumar, N., Lagoke, S.T. and Atayese, M.O. 2013. Management of *Striga hermonthica* on sorghum (*Sorghum bicolor*) using arbuscular mycorrhizal fungi (*Glomus mosae*) and NPK fertilizer levels. Pak. J. Biol. Sci., 16(22): 1563–1568.

Jia, K.P., Baz, L. and Al-Babili, S. 2018. From carotenoids to strigolactones. J. Exp. Bot., 69(9): 2189–2204.

Jin, Y., Liu, H., Luo, D., Yu, N., Dong, W., Wang, C. et al. 2016. DELLA proteins are common components of symbiotic rhizobial and mycorrhizal signalling pathways. Nat. Commun., 7(1): 1–14.

Jones, N., Krishnaraj, P., Kulkarni, J., Patil, A., Laxmipathy, R. and Vasudeva, R. 2011. Diversity of arbuscular mycorrhizal fungi in different ecological zones of northern Karnataka. Eco. Environ. Cons., 18: 1053–1058.

Jung, S.C., García-Andrade, J., Verhage, A., Fernández, I., García, J.M., Azcón-Aguilar, C. et al. 2009. Arbuscular mycorrhiza confers systemic resistance against Botrytis cinerea in tomato through priming of JA-dependent defense responses. In Induced resistance: chances and limits, IOBC/wprs Bulletin, Working Group "Induced resistance in plants against insects and diseases." Proceedings of the meeting at Granada, Spain., pp. 8–16.

Khaosaad, T., Garcia-Garrido, J.M., Steinkellner, S. and Vierheilig, H. 2007. Take-all disease is systemically reduced in roots of mycorrhizal barley plants. Soil Biol. Biochem., 39(3): 727–734.

Kobra, N., Jalil, K. and Youbert, G. 2009. Effects of three *Glomus* species as biocontrol agents against *Verticillium*-induced wilt in cotton. J. Plant Protect Res., 49: 185–189.

Kohlen, W., Charnikhova, T., Lammers, M., Pollina, T., Tóth, P., Haider, I. et al. 2012. The tomato carotenoid cleavage dioxygenase 8 (S l CCD 8) regulates rhizosphere signaling, plant architecture and affects reproductive development through strigolactone biosynthesis. New Phytol., 196(2): 535–547.

Kohlen, W., Charnikhova, T., Liu, Q., Bours, R., Domagalska, M.A., Beguerie, S. et al. 2011. Strigolactones are transported through the xylem and play a key role in shoot architectural response to phosphate deficiency in nonarbuscular mycorrhizal host Arabidopsis. Plant Physiol., 155(2): 974–987.

Koltai, H. 2015. Cellular events of strigolactone signalling and their crosstalk with auxin in roots. J. Exp. Bot., 66(16): 4855–4861.

Koricheva, J., Gange, A.C. and Jones, T. 2009. Effects of mycorrhizal fungi on insect herbivores: a meta-analysis. Ecology, 90(8): 2088–2097.

Kumari, S.M.P. and Prabina, B.J. 2019. Protection of tomato, *Lycopersicon esculentum* from wilt pathogen, *Fusarium oxysporum* f. sp. lycopersici by arbuscular mycorrhizal fungi, *Glomus* sp. Int. J. Curr. Microbiol. Appl. Sci., 8: 1368–1378.

Kumari, S.M.P. and Srimeena, N. 2019. Arbuscular mycorrhizal fungi (AMF) induced defense factors against the damping-off disease pathogen, *Pythium aphanidermatum* in Chilli (*Capsicum annum*). Int. J. Curr. Microbiol. Appl. Sci., 8: 2243–2248.

Lendzemo, V.W., Kuyper, T.W., Kropff, M.J. and Van Ast, A.V. 2005. Field inoculation with arbuscular mycorrhizal fungi reduces *Striga hermonthica* performance on cereal crops and has the potential to contribute to integrated *Striga* management. Field Crop Res., 91(1): 51–61.

Lendzemo, V.W., Van Ast, A. and Kuyper, T.W. 2006. Can arbuscular mycorrhizal fungi contribute to Striga management on cereals in Africa?. Outlook Agric., 35(4): 307–311.

Li, H.Y., Yang, G.D., Shu, H. R., Yang, Y.T., Ye, B.X., Nishida, I. et al. 2006. Colonization by the arbuscular mycorrhizal fungus, *Glomus versiforme* induces a defense response against the root-knot nematode *Meloidogyne incognita* in the grapevine (*Vitis amurensis* Rupr.), which includes transcriptional activation of the class III chitinase gene VCH3. Plant Cell Physiol., 47(1): 154–163.

Lin, H., Wang, R., Qian, Q., Yan, M., Meng, X., Fu, Z. et al. 2009. DWARF27, an iron-containing protein required for the biosynthesis of strigolactones, regulates rice tiller bud outgrowth. Plant Cell, 21(5): 1512–1525.

Linderman, R.G. 2000. Effects of Mycorrhizas on Plant Tolerance to Diseases: Mycorrhiza-disease interactions. Arbuscular Mycorrhizas: Physiology and Function, pp. 345–365.

Lingua, G., D'Agostino, G., Massa, N., Antosiano, M. and Berta, G. 2002. Mycorrhiza-induced differential response to a yellows disease in tomato. Mycorrhiza, 12(4): 191–198.

Lioussanne, L., Jolicoeur, M. and St-Arnaud, M. 2008. Mycorrhizal colonization with *Glomus intraradices* and development stage of transformed tomato roots significantly modify the chemotactic response of zoospores of the pathogen *Phytophthora nicotianae*. Soil Biol. Biochem., 40(9): 2217–2224.

Liu, J., Blaylock, L.A., Endre, G., Cho, J., Town, C.D., VandenBosch, K.A. et al. 2003. Transcript profiling coupled with spatial expression analyses reveals genes involved in distinct developmental stages of an arbuscular mycorrhizal symbiosis. Plant Cell, 15(9): 2106–2123.

Liu, J., Maldonado-Mendoza, I., Lopez-Meyer, M., Cheung, F., Town, C.D. and Harrison, M.J. 2007. Arbuscular mycorrhizal symbiosis is accompanied by local and systemic alterations in gene expression and an increase in disease resistance in the shoots. Plant J., 50(3): 529–544.

Liu, X.L., Xi, X.Y., Shen, H., Liu, B. and Guo, T. 2014. Influences of arbuscular mycorrhizal (AM) fungi inoculation on resistance of tobacco to bacterial wilt. Tob. Sci. Technol., 49(5): 23–30.

Liu, Y., Feng, X., Gao, P., Li, Y., Christensen, M.J. and Duan, T. 2018. Arbuscular mycorrhiza fungi increased the susceptibility of *Astragalus adsurgens* to powdery mildew caused by *Erysiphe pisi*. Mycology, 9(3): 223–232.

Llanes, A., Reginato, M., Devinar, G. and Luna, V. 2018. What is known about phytohormones in halophytes? A review. Biologia, 73(8): 727–742.

López-Ráez, J.A., Charnikhova, T., Fernández, I., Bouwmeester, H. and Pozo, M.J. 2011a. Arbuscular mycorrhizal symbiosis decreases strigolactone production in tomato. J. Plant Physiol., 168(3): 294–297.

Lopez-Raez, J.A., Charnikhova, T., Mulder, P., Kohlen, W., Bino, R., Levin, I. et al. 2008. Susceptibility of the tomato mutant high pigment-2dg (hp-2dg) to *Orobanche* spp. infection. J. Agric. Food Chem. 56(15): 6326–6332.

López-Ráez, J.A., Jung, S.C., Fernandez, I., García, J.M., Bouwmeester, H. and Pozo, M.J. 2012. Mycorrhizal symbiosis as a strategy for root parasitic weed control. *IOBC-WPRS*. Granada, Spain, pp. 59–63.

López-Ráez, J.A., Matusova, R., Cardoso, C., Jamil, M., Charnikhova, T., Kohlen, W. et al. 2009. Strigolactones: ecological significance and use as a target for parasitic plant control. Pest Manag. Sci., 65(5): 471–477.

López-Ráez, J.A., Pozo, M.J. and García-Garrido, J.M. 2011b. Strigolactones: a cry for help in the rhizosphere. Botany, 89(8): 513–522.

López-Ráez, J.A., Verhage, A., Fernández, I., García, J.M., Azcón-Aguilar, C., Flors, V. et al. 2010. Hormonal and transcriptional profiles highlight common and differential host responses to arbuscular mycorrhizal fungi and the regulation of the oxylipin pathway. J. Exp. Bot., 61(10): 2589–2601.

Manjunatha, H.P., Jones Nirmalnath, P., Chandranath, H.T., Shiney, A. and Jagadeesh, K.S. 2018. Field evaluation of native arbuscular mycorrhizal fungi in the management of Striga in sugarcane (*Saccharum officinarum* L.). J. Pharm. Phytochem., 7: 2496–2500.

Martínez-Medina, A., Roldán, A. and Pascual, J. 2011. Interaction between arbuscular mycorrhizal fungi and Trichoderma harzianum under conventional and low input fertilization field condition in melon crops: growth response and Fusarium wilt biocontrol. Appl. Soil Ecol., 47(2): 98–105.

Marzec, M. 2016. Strigolactones as Part of the Plant Defence System. Trends Plant Sci., 21(11): 900–903.

Marzec, M. and Muszynska, A. 2015. *In Silico* analysis of the genes encoding proteins that are involved in the biosynthesis of the RMS/MAX/D pathway revealed new roles of strigolactones in plants. Int. J. Mol. Sci., 16(4): 6757–6782.

Matusova, R., Rani, K., Verstappen, F.W., Franssen, M.C., Beale, M.H. and Bouwmeester, H.J. 2005. The strigolactone germination stimulants of the plant-parasitic *Striga* and *Orobanche* spp. are derived from the carotenoid pathway. Plant Physiol., 139(2): 920–934.

Mishra, S., Upadhyay, S. and Shukla, R.K. 2017. The role of strigolactones and their potential cross-talk under hostile ecological conditions in plants. Front. Physiol., 7: 1–7.

Møller, K., Kristensen, K., Yohalem, D. and Larsen, J. 2009. Biological management of gray mold in pot roses by co-inoculation of the biocontrol agent *Ulocladium atrum* and the mycorrhizal fungus *Glomus mosseae*. Biol. Control., 49(2): 120–125.

Moubayidin, L., Perilli, S., Ioio, R.D., Di Mambro, R., Costantino, P. and Sabatini, S. 2010. The rate of cell differentiation controls the Arabidopsis root meristem growth phase. Curr. Biol., 20(12): 1138–1143.

Mukerji, K.G. and Ciancio, A. 2007. Mycorrhizae in the integrated pest and disease management. In General Concepts in Integrated Pest and Disease Management, pp. 245–266.

Mustafa, G., Randoux, B., Tisserant, B., Fontaine, J., Magnin-Robert, M., Sahraoui, A.L.H. et al. 2016. Phosphorus supply, arbuscular mycorrhizal fungal species, and plant genotype impact on the protective efficacy of mycorrhizal inoculation against wheat powdery mildew. Mycorrhiza, 26(7): 685–697.

Nan, Z.B. and Li, C.J. 1994. Fungal disease of pasture plants recorded in China-A check list. Pratacultural Sci. Suppl., 11: 1–160.

Nguvo, K.J. and Gao, X. 2019. Weapons hidden underneath: bio-control agents and their potentials to activate plant induced systemic resistance in controlling crop *Fusarium* diseases. J. Plant. Dis. Prot., 126(3): 177–190.

Nogales, A., Aguirreolea, J., Santa María, E., Camprubí, A. and Calvet, C. 2009. Response of mycorrhizal grapevine to *Armillaria mellea* inoculation: disease development and polyamines. Plant Soil, 317(1): 177–187.

Norman, J.R., Atkinson, D. and Hooker, J.E. 1996. Arbuscular mycorrhizal fungal-induced alteration to root architecture in strawberry and induced resistance to the root pathogen, *Phytophthora fragariae*. Plant Soil, 185(2): 191–198.

Norman, J.R. and Hooker, J.E. 2000. Sporulation of *Phytophthora fragariae* shows greater stimulation by exudates of non-mycorrhizal than by mycorrhizal strawberry roots. Mycol. Res., 104(9): 1069–1073.

Oldroyd, G.E. 2013. Speak, friend, and enter: signalling systems that promote beneficial symbiotic associations in plants. Nat. Rev. Microbiol., 11(4): 252–263.

Othira, J.O., Omolo, J.O., Wachira, F.N. and Onek, L.A. 2012. Effectiveness of arbuscular mycorrhizal fungi in protection of maize (*Zea mays* L.) against witchweed (*Striga hermonthica* Del Benth) infestation. J. Agric. Biotechnol. Sustain. Dev., 4: 37–44.

Ozgonen, H. and Erkilic, A. 2007. Growth enhancement and *Phytophthora* blight (*Phytophthora capsici* Leonian) control by arbuscular mycorrhizal fungal inoculation in pepper. Crop Prot., 26(11): 1682–1688.

Pandey, A., Sharma, M. and Pandey, G.K. 2016. Emerging roles of strigolactones in plant responses to stress and development. Front. Plant Sci., 7: 434.

Parker, C. 2009. Observations on the current status of *Orobanche* and *Striga* problems worldwide. Pest Manag. Sci., 65(5): 453–459.

Parniske, M. 2008. Arbuscular mycorrhiza: the mother of plant root endosymbioses. Nat. Rev. Micro. 6(10): 763–775.

Paszkowski, U. 2006. Mutualism and parasitism: the yin and yang of plant symbioses. Curr. Opin. Plant Biol., 9(4): 364–370.

Peña-Cortés, H., Barrios, P., Dorta, F., Polanco, V., Sánchez, C. and Sánchez, E. 2004. Involvement of jasmonic acid and derivatives in plant response to pathogen and insects and in fruit ripening. J. Plant Growth Regul., 23(3): 246–260.

Pieterse, C.M., Leon-Reyes, A., Van der Ent, S. and Van Wees, S.C. 2009. Networking by small-molecule hormones in plant immunity. Nat. Chem. Biol., 5(5): 308–316.

Pimprikar, P., Carbonnel, S., Paries, M., Katzer, K., Klingl, V. and Bohmer, M.J. 2016. A CCaMK-CYCLOPS-DELLA complex activates transcription of RAM1 to regulate arbuscule branching. Curr. Biol., 26(8): 987–998.

Pozo, M.J. and Azcón-Aguilar, C. 2007. Unraveling mycorrhiza-induced resistance. Curr. Opin. Plant Biol., 10(4): 393–398.

Pozo, M.J., Jung, S.C., López-Ráez, J.A. and Azcón-Aguilar, C. 2010. Impact of arbuscular mycorrhizal symbiosis on plant response to biotic stress: the role of plant defence mechanisms. Arbuscular Mycorrhizas: Physiology and Function, pp. 193–207.

Pozo, M.J. and Azcón-Aguilar, C. 2007. Unraveling mycorrhiza-induced resistance. Curr. Opin. Plant Biol., 10(4): 393–398.

Rapparini, F., Llusià, J. and Peñuelas, J. 2007. Effect of arbuscular mycorrhizal (AM) colonization on terpene emission and content of *Artemisia annua* L. Plant Biol., 9(S 01): e20–e32.

Rasmussen, A., Depuydt, S., Goormachtig, S. and Geelen, D. 2013. Strigolactones fine-tune the root system. Planta, 238(4): 615–626.

Ruyter-Spira, C., Kohlen, W., Charnikhova, T., Van Zeijl, A., Van Bezouwen, L., De Ruijter, N. et al. 2011. Physiological effects of the synthetic strigolactone analog GR24 on root system architecture in *Arabidopsis*: another belowground role for strigolactones?. Plant Physiol., 155(2): 721–734.

Scaffidi, A., Waters, M.T., Ghisalberti, E.L., Dixon, K.W., Flematti, G.R. and Smith, S.M. 2013. Carlactone-independent seedling morphogenesis in Arabidopsis. Plant J., 76(1): 1–9.

Schüßler, A., Schwarzott, D. and Walker, C. 2001. A new fungal phylum, the Glomeromycota: phylogeny and evolution. Mycol. Res., 105(12): 1413–1421.

Seto, Y., Sado, A., Asami, K., Hanada, A., Umehara, M. and Akiyama, K. 2014. Carlactone is an endogenous biosynthetic precursor for strigolactones. Proc. Natl. Acad. Sci., 111(4): 1640–1645.

Shaul, O., Galili, S., Volpin, H., Ginzberg, I., Elad, Y., Chet, I. et al. 1999. Mycorrhiza-induced changes in disease severity and PR protein expression in tobacco leaves. Mol. Plant-Microbe Interact., 12(11): 1000–1007.

Siddiqui, Z.A. and Mahmood, I. 1998. Effect of a plant growth promoting bacterium, an AM fungus and soil types on the morphometrics and reproduction of *Meloidogyne javanica* on tomato. Appl. Soil Ecol., 8(1–3): 77–84.

Singh, D.P., Srivastava, J.S., Bahadur, A., Singh, U.P. and Singh, S.K. 2004. Arbuscular mycorrhizal fungi induced biochemical changes in pea (*Pisum sativum*) and their effect on powdery mildew (*Erysiphe pisi*). J. Plant Dis. Prot., 266–272.

Singh, P.K., Singh, M. and Vyas, D. 2010. Biocontrol of fusarium wilt of chickpea using arbuscular mycorrhizal fungi and Rhizobium leguminosorum biovar. Caryologia, 63(4): 349–353.

Song, Y., Chen, D., Lu, K., Sun, Z. and Zeng, R. 2015. Enhanced tomato disease resistance primed by arbuscular mycorrhizal fungus. Front. Plant. Sci., 6: 786.

Soto, M.J., Fernández-Aparicio, M., Castellanos-Morales, V., García-Garrido, J.M., Ocampo, J.A., Delgado, M.J., et al. 2010. First indications for the involvement of strigolactones on nodule formation in alfalfa (*Medicago sativa*). Soil Biol. Biochem., 42(2): 383–385.

Spagnoletti, F.N., Cornero, M., Chiocchio, V., Lavado, R.S. and Roberts, I.N. 2020. Arbuscular mycorrhiza protects soybean plants against *Macrophomina phaseolina* even under nitrogen fertilization. Eur. J. Plant. Pathol., 156(3): 839–849.

Stes, E., Depuydt, S., De Keyser, A., Matthys, C., Audenaert, K., Yoneyama, K. et al. 2015. Strigolactones as an auxiliary hormonal defence mechanism against leafy gall syndrome in Arabidopsis thaliana. J. Exp. Bot., 66(16): 5123–5134.

Stolyarchuk, I.M., Shevchenko, T.P., Polischuk, V.P. and Kripka, A.V. 2009. Virus infection course in different plant species under influence of arbuscular mycorrhiza. Microbiol. Biotechnol., (7): 70–75.

Thiem, D., Szmidt-Jaworska, A., Baum, C., Muders, K., Niedojadlo, K. and Hrynkiewicz, K. 2014. Interactive physiological response of potato (*Solanum tuberosum* L.) plants to fungal colonization and Potato virus Y (PVY) infection. Acta Mycol., 49(2).

Torres-Vera, R., García, J.M., Pozo, M.J. and López-Ráez, J.A. 2014. Do strigolactones contribute to plant defence? Mol. Plant Pathol., 15(2): 211–216.

Umehara, M., Hanada, A., Yoshida, S., Akiyama, K., Arite, T., Takeda-Kamiya, N. et al. 2008. Inhibition of shoot branching by new terpenoid plant hormones. Nature, 455(7210): 195–200.

Urquhart, S., Foo, E. and Reid, J.B. 2015. The role of strigolactones in photo-morphogenesis of pea is limited to adventitious rooting. Physiol. Plant., 153(3): 392–402.

Verhage, A., García-Andrade, J., García, J.M. and Azcón-Aguilar, C. 2009. Priming plant defence against pathogens by arbuscular mycorrhizal fungi. In Mycorrhizas-functional processes and ecological impact, Springer, Berlin, Heidelberg, pp. 123–135.

Vos, C., Claerhout, S., Mkandawire, R., Panis, B., De Waele, D. and Elsen, A. 2012a. Arbuscular mycorrhizal fungi reduce root-knot nematode penetration through altered root exudation of their host. Plant Soil, 354(1): 335–345.

Vos, C.M., Tesfahun, A.N., Panis, B., De Waele, D. and Elsen, A. 2012b. Arbuscular mycorrhizal fungi induce systemic resistance in tomato against the sedentary nematode *Meloidogyne incognita* and the migratory nematode, *Pratylenchus penetrans*. Appl. Soil Ecol., 61: 1–6.

Wakabayashi, T., Shinde, H., Shiotani, N., Yamamoto, S., Mizutani, M., Takikawa, H. et al. 2022. Conversion of methyl carlactonoate to heliolactone in sunflower. Nat. Prod. Res., 36(9): 2215–2222.

Wang, Y.Y., Yin, Q.S., Qu, Y., Li, G.Z. and Hao, L. 2018. Arbuscular mycorrhiza-mediated resistance in tomato against *Cladosporium fulvum*-induced mould disease. J. Phytopathol., 166(1): 67–74.

Waters, M.T., Brewer, P.B., Bussell, J.D., Smith, S.M. and Beveridge, C.A. 2012. The Arabidopsis ortholog of rice DWARF27 acts upstream of MAX1 in the control of plant development by strigolactones. Plant Physiol., 159(3): 1073–1085.

Whipps, J.M. 2004. Prospects and limitations for mycorrhizas in biocontrol of root pathogens. Can. J. Bot., 82(8): 1198–1227.

Xie, X., Yoneyama, K., Kisugi, T., Uchida, K., Ito, S., Akiyama, K. et al. 2013. Confirming stereochemical structures of strigolactones produced by rice and tobacco. Mol. Plant, 6(1): 153–163.

Yoneyama, K., Kisugi, T., Xie, X. and Yoneyama, K. 2013. Chemistry of strigolactones: why and how do plants produce so many strigolactones. Molecular Microbial Ecology of the Rhizosphere 1: 373–379.

Yousefi, Z., Riahi, H., Khabbaz-Jolfaei, H. and Zanganeh, S. 2011. Effects of arbuscular mycorrhizal fungi against apple powdery mildew disease. Life Sci. J., 8(4): 108–112.

Yu, N., Luo, D., Zhang, X., Liu, J., Wang, W., Jin, Y. et al. 2014. A DELLA protein complex controls the arbuscular mycorrhizal symbiosis in plants. Cell Res., 24(1): 130–133.

Zhang, Y., Van Dijk, A.D., Scaffidi, A., Flematti, G.R., Hofmann, M., Charnikhova, T. et al. 2014. Rice cytochrome P450 MAX1 homologs catalyze distinct steps in strigolactone biosynthesis. Nat. Chem. Biol., 10(12): 1028–1033.

Zwanenburg, B., Mwakaboko, A.S., Reizelman, A., Anilkumar, G. and Sethumadhavan, D. 2009. Structure and function of natural and synthetic signalling molecules in parasitic weed germination. Pest Manag. Sci., 65(5): 478–491.

6

Prospective Role of Brassinosteroids in Tolerance to Biotic Stress and Plant Defence

Sayani Bandyopadhyay,[1,2] *Pratik Saha,*[1] *Anubhab Hooi,*[3]
Amitava Mondal[3,4] *and Kousik Atta*[5,6,*]

Introduction

Various microorganisms, nematodes, and insects are detrimental for plants and thus, their invasion causes huge agricultural losses in the whole world. Plants protect themselves by expressing defensive responses induced by the attacks of microbes and pests. A proper regulation of defence responses is essential for the plant's life because when they are activated, it may cause an injurious effect to the plant growth and development. The defense response of plants has an intricate and complicated mode of activation on a molecular level. Responses are often initiated with 'gene-for-gene' recognition of the pathogen. Plants carrying R genes for corresponding resistance identify pathogens by certain mechanisms like identifying certain virulence effectors produced by the pathogens. R gene–mediated resistance generally goes along

[1] Department of Agricultural Biotechnology, Faculty of Agriculture, Bidhan Chandra Krishi Viswavidyalaya, Nadia, 741252, West Bengal, India.
[2] Swami Vivekananda Institute of Modern Science, Sonarpur Station Road, Karbala, Kolkata-700103, West Bengal, India.
[3] Department of Plant Pathology, Faculty of Agriculture, Bidhan Chandra Krishi Viswavidyalaya, Nadia, 741252, West Bengal, India.
[4] Department of Plant Pathology, School of Agricultural science, JIS University, Agarpara, Kolkata 700109, West Bengal, India.
[5] Department of Plant Physiology, Faculty of Agriculture, Bidhan Chandra Krishi Viswavidyalaya, Nadia, 741252, West Bengal, India.
[6] Faculty of Agricultural sciences, GLA University, Mathura, Uttar Pradesh, 281406, India.
* Corresponding author: kousikatta1995@gmail.com

with the oxidative burst, which is the rapid generation of reactive oxygen species (ROS) (Glazebrook 2005). Phytohormones are known as the natural activators for the growth and development of plants. Along with this, phytohormones also play a vital role for protecting the plant species from various biotic stresses like the attacks of virus, bacteria, fungus, insects, and nematodes and abiotic stress like drought, submergence, temperature, salinity, etc. Plant hormones start a signalling cascade involving a number of molecular players, which lead to a model genetic pathway (Ali et al. 2014, Manghwar et al. 2022).

Steroid hormones are involved for upregulating different stress-related defence genes and the hormonal cross-talk of that hormone with other metabolic pathways, which was observed by different studies. Brassinosteroids (BRs) are a class of steroidal phytohormones which are necessary for the growth, development, and yield of plants. These phytohormones take part in the regulation of division, elongation, and differentiation of different types of cells in all over the lives of plant cell (Nawaz et al. 2017, Manghwar et al. 2022). Brassinosteroids (BRs) are key players in the regulation of growth, controlling processes like stem elongation, pollen tube development, leaf bending and unrolling, and root inhibition (Mandava 1988, Azpiroz et al. 1998, Clouse and Sasse 1998). BR was first isolated from Rape pollen (*Brassica napus*) and the hormone was named as brassins (Mitchell et al. 1970). Though around 60 compounds related to BR had been found, but brassinolide, 24-epibrassinolide, and 28-homobrassinolide are considered as the most active brassinosteroids (Anwar et al. 2018). External addition of BR for developing biotic and abiotic stress resistant plants is an innovative research approach which is a key for understanding the stress tolerance mechanisms of BRs in plants (Marková et al. 2023). Some of the crucial roles of brassinosteroid in plant species are displayed in Fig. 1.

Brassinosteroids play a role in plant adaptation to environmental stresses including heat, drought and salinity, heavy metals, and biotic stresses (Nawaz et al. 2017). BRs have also been studied in response to plant stresses like cold tolerance (Hotta et al. 1998, Li et al. 2017) and ethylene biosynthesis which is attributed for playing a crucial role in stress tolerance (Yi et al. 1999). Recent studies in the field of BR revealed that deficiency in endogenous BR causes dwarfing in plants in light conditions and a seedling similar to light- grown ones developing leaves in the absence of light (Li et al. 1996, Szekeres et al. 1996, Fujioka et al. 1997, Schumacher and Chory 2000). The application of aqueous solutions of BR containing *Lychnis iscaria* seed extract resulted in an increased tolerance of tobacco, cucumber, and tomato to viral and fungal pathogens (Roth et al. 2000). Brassinosteroids also accomplish an antagonistic role in anti-herbivory structure formation in tomato (Campos et al. 2009). Nutritional deficiency also works as a harmful environment that has a significant negative impact on plant growth, development, and crop output in addition to stressful environmental factors. Brassinosteroids take part in the regulation of the 'forging responses' of plant lateral roots in response to nutritional deprivation (Zhang 2023).

Brassinosteroids are known to be involved in response to all types of trophic pathogens, *viz*, biotrophic, hemi-biotrophic, and necrotrphic pathogens. But their mode of action to elicit tolerance and susceptibility depends primarily on the particular

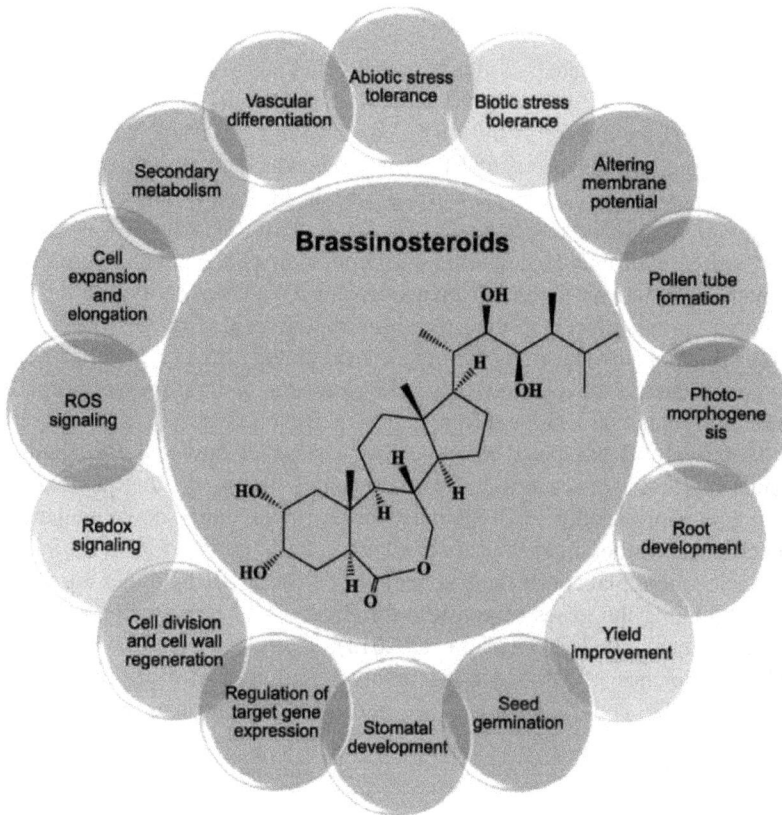

Fig. 1. Some crucial roles of BR in plants (Hafeez et al. 2021).

life cycle, infection mechanisms, and types of BR application, either exogenous or endogenous (Yu et al. 2018, Sheikhi et al. 2023). The importance of brassinosteroids along with its signalling and its interactions with various phytohormones like jasmonic Acid (JA), abscisic acid (ABA), salicylic acid (SA), auxin, gibberellin, and cytokinin will be effective to understand the disease responsive mechanism of the plants.

The study of the possible advantages and uses of brassinosteroid in boosting a plant's defence against biotic stresses, such as pests and diseases, is referred to as the prospective role of brassinosteroids for tolerance to biotic stress. Aims of studying this topic include:

1. Gaining an understanding of brassinosteroids by looking at their production, signalling pathways, mechanisms of action, and physiological, genetic and molecular basisin plants.
2. Analysing how brassinosteroids affect a plant's ability to withstand biotic stressers caused by diseases, insects, nematodes, and other pests as well as increasing awareness of how plant defence mechanisms work, which are essential for evaluating the role of brassinosteroids in horticulture and agriculture.

Signaling of Brassinosteroids

Brassinosteroids (BRs) take part in different developmental processes and responses to biotic and abiotic stresses. BRs are perceived by the BRI1 receptor and its close relatives- BRI1-Like 1 (BRL1) and BRI1-Like 3 (BRL3) (Bishop and Koncz 2002).

Endogenous cellular levels of BR, its maintenance and regulation through the processes of biosynthesis, transport, and degradation is critically important for various physiological and biological processes (Tanaka et al. 2005). When BR is applied to plants in both cases of abundance and deficit, the physiological status quo were observed to be altered. Mutations in certain plants causing BR insensitivity and deficiency have shown various defects in their physio-morphological attributes *viz,* decreased seed germination, dwarfism, dark-green and curled leaves, lowered fertility, delayed reproductive development, and development of light-grown morphology (de-etiolation) in the dark (Li and Chory, 1997, Clouse 2011).

Alternatively, it has been observed that exorbitant doses of BR application in plants increases concentration in endogenous levels of the phytohormone and triggers upregulation of BR-inactivation genes and downregulation of BR-specific biosynthetic genes. This feedback loop is crucial for maintaining BR homeostasis where plants can regulate cellular levels of BR by either upregulating or downregulating the genes responsible for BR biosynthesis (Bishop and Yokota, 2001, Tanaka et al. 2003, 2005, Zhu et al. 2013).

The percipience of BR signaling involves various cell membrane receptors which, when the conditions arise, are activated by phosphorylation cascades

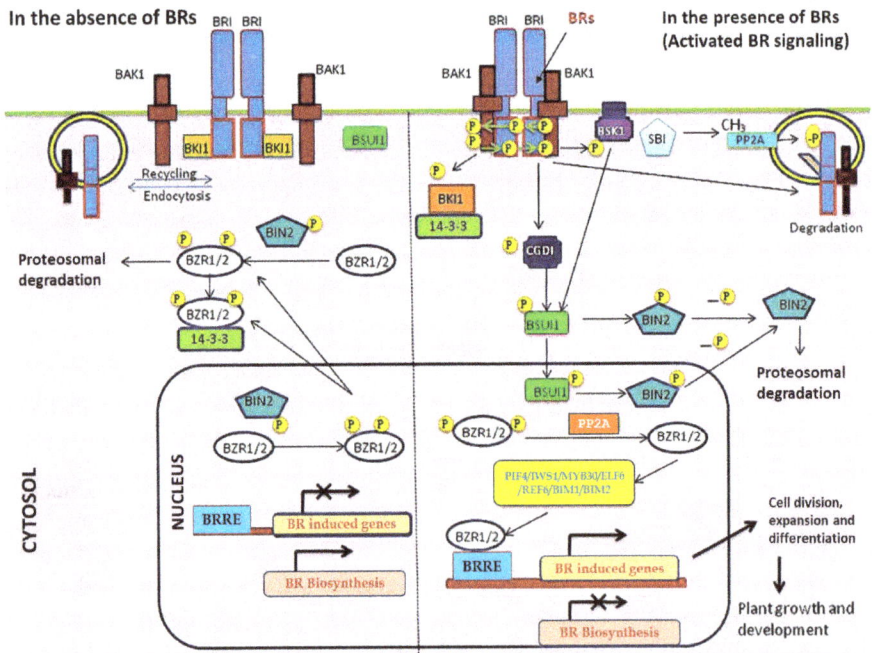

Fig. 2. Brassinosteroid-signaling pathway model in the presence and absence of Brassinosteroid (Saini et al. 2015).

(Fig. 2). These receptors are basically a group of localized leucine-rich repeat receptor kinases (LRR-RKs). They relay specific signals to control plant specific downstream transcription factors like BZR1 (BRASSINAZOLE-RESISTANT1) and BES1 (BR-INSENSITIVE-EMS-SUPPRESSOR1). These transcription factors act on promoters of multiple genes, consecutively involved in different signalling pathways (Guo et al. 2013, Wang et al. 2014, Belkhadir and Jaillais, 2015). At cellular level, the BR signalling is mainly affected by the binding affinity of BR to its receptors. It has been observed that, although both BRI1 and BRL1 have similar structural homology, BR has a higher binding affinity towards BRL1 in contrast to BRI1 due to minor differences in binding sites of the extracellular domains (She et al. 2011, 2013).

The preassembled BRI1-BAK1 receptor assembly undergoes conformational changes in their respective intercellular kinase domains discretely after conjugation with ligand. In the absence of BR, BRI1 remains inactive due to auto-inhibition in its cytoplasmic kinase domain along with BKI1 (BRI1 kinase inhibitor 1) interaction, which is an inhibitory protein (Wang et al. 2005a, Wang and Chory 2006, Jaillais et al. 2011). BIN2 (BR insensitive 2) which is a GSK3/Shaggy-like kinase, in its active state, phosphorylates BES1/BZR2 transcription factors simultaneously and inhibits BRI1 activity by protein degradation, reduced DNA binding, and Cytoplasmic retention by 14-3-3 proteins (Ye et al. 2011, Hao et al. 2013). When BR signaling needs to be activated, the BRI1 heterooligomerise with BAK1 is its co-receptor. This conjugation of BRI1 with BR in the island domain of 70 amino acids triggers conformational change in the homodimer, autophosphorylating BRI1. This promotes partial kinase activity, allowing it to dissociate from the cell membrane and interact with BAK1 (Li et al. 2002, Wang et al. 2005a, Wang et al. 2005b, Wang and Chory 2006). Concurrently, BKI1 binds with the 14-3-3 proteins, repressing their negative roles and promoting BRI1 signalling finally allowing BRI1 to interact with BAK1 transphosphorylating each other (Wang et al. 2011, Wang et al. 2008). A series of phosphorylation events are initiated by an activated BRI1 for transmitting the signal from membrane receptors to cytoplasmic regulators. These receptors are mainly downstream plasma membrane-bound RLCKs (receptor-like cytoplasmic kinases), BSKs (BR signalling kinases), and CDG1 (constitutive differential growth 1) (Tang et al. 2008, Kim et al. 2011, Sreeramulu et al. 2013).

Interactions of Brassinosteroids with Other Phytohormones

Plants are complex organisms with even more complicated cellular functions. In accordance to respond to various stresses, BR crucially plays diverse roles in interacting with various phytohormones like abscisic acid (ABA), auxin, cytokinin (CK), ethylene, gibberellins (GA), jasmonic acid (JA), polyamines (PA), and salicylic acid (SA), and also engage in plant growth and development (Choudhary et al. 2012, Gruszka 2013). Different studies that were pioneered in the field, provide knowledge about responses to several external and internal factors. Signalling components of BR interacts on a molecular level with genes and transcription factors of aforesaid plant hormones to coordinate multiple physiological functions (Saini et al. 2015).

Crosstalk with Abscisic Acid

Abscisic acid is a sesquiterpene which is responsible for various developmental, growth, and stress responses (Nambara 2017). Abscisic acid helps in the synthesis of various proteins and osmolytes, which enable plants to tolerate abiotic and biotic stresses, and acts as a general inhibitor of germination and post germination growth and metabolic activities (Srivastava 2002). ABA causes seed dormancy, stomatal closure, and adaptation to stressed conditions as observed in *Arabidopsis*.

Dehydrin (DHN), a Group II LEA protein, is known to be induced by endogenous ABA. The induced expression of dehydrin protects embryos from desiccation during late embryogenesis (Murofushi et al. 1999). Drought stress induced in wheat, characterized by ABA accumulation, was alleviated by application of 24-epibrassinolide (EBR) as a pre-sown seed treatment, invigorating growth during seed germination. EBR-treated plants showed high accumulation of low molecular weight DHNs up to 2.5 times. This indicates strong correlation between ABA and EBR, where it is implicated, that exogenous BR can regulate development of plants in water stress (Shakirova et al. 2016). BR-ABA interaction has shown hypocotyl elongation. BR-deficient and BR-insensitive mutant lines of *Arabidopsis* have been found to be more sensitive to ABA compare to the wild type. Both in cases of germination and hypocotyl elongation, the wild types outperformed the mutants where the hypocotyl was distinctly inhibited (Xue et al. 2009).

Crosstalk with Jasmonic Acid

BR interacts with JA to perform key functions in plant development responses to biotic and abiotic factors. *coi1* (Coronatine insensitive1) needed for JA responses in *Arabidopsis* are crucial in JA signaling. DWF4 is mutated to the *partially suppressing coi1* (*psc1*), encoding a vital enzyme of BR synthesis which leads to a reduction of root development. It was observed that the partial restoration of JA sensitivity by *psc1* in *coi1-2* background and the JA hypersensitivity of *psc1* in wild type *COI1* background were removed by external application of BR. We can conclude that BR plays an inimical role towards a JA signaling pathway to downregulate *COI1* and inhibits root growth and development (Ren et al. 2009).

During stressed conditions, BR tends to increase the JA levels in rice. Elevated levels of JA robustly promote thionin-encoding gene expression which encodes antimicrobial peptides. This shows possible phytohormonal interactions between BR and JA. This is further elucidated to understand the development of natural defences in tomatoes against pests. BR and JA crucially regulate trichome formation. Interaction of both hormones accumulates zgb (zingiberene) and PI-I (Proteinase inhibitor I), which compliments defence against insect herbivores. In JA-insensitive and BR-deficient mutant's *jai1-1* and *dpy*, it was clearly observed that BR restricted anti-herbivory in tomato while JA promoted it when exogenously applied. BR and JA show an antagonistic action for zgb contents, since *dpy* showed higher levels of zgb and lower levels in *jai1-1*. In *dpy* mutants, increased zingiberene biosynthesis, proteinase inhibitor expression and enhanced pubescence were observed. The importance of JA-BR was confirmed through herbivory tests using *Spodoptera frugiperda* and the tomato pest *Tutaabsoluta* where dpy leaflets showed higher mortality rates (Campos et al. 2009).

Crosstalk with Salicylic Acid (SA)

NPR1 (Non Expressor of Pathogenesis-Related Genes1), a key regulator, and WRKY70, a transcription factor, is mainly responsible for potential BR-SA crosstalk. NPR1 is a critical component that mediates stress with the help of EBR, controlling BR signalling components like BIN2 and BZR1 (Divi et al. 2010). SA is responsible for SAR (Systemic Acquired Resistance). Increased resistance has been observed in rice against *Magnaporthe grisea* and *Xanthomonas oryzae* when treated with BR. But no evidence inferred the requirement of SA in BR-mediated disease resistance, which indicated that there are independent pathways of both SA and BR in regards to disease-causing pathogens. But BR and SAR, if applied as a combined strategy, provides additive resistance against pathogens (Nakashita et al. 2003). On contrary, in rice, *Pythium graminicola* has a tendency to hijacks BR pathway and exploits it as virulence factors to induce disease. It can also be confirmed that SA and BR pathways acts antagonistically and cause the immune suppressive effect of BR. Brassinazole application on rice against *Pythium graminicola* showed reduced susceptibility caused by derepression of NPR1 and *OsWRKY45* indicating BR-mediated suppression of SA defence responses (DeVleesschauwer et al. 2012).

Crosstalk with Ethylene

Ethylene has a rather opposite effect with respect to BR. They both regulate gravitropic shoot responses. In presence of reduced ethylene signalling, there was a loss of gravitropic reorientation in *bri1-1* mutants, which indicated that loss of gravitropism was dependent on endogenous BR levels. Whereas, lack of BR enhances gravitropic reorientation in 2-day old seedlings. Ethylene and BR regulate a number of differentially expressed *AUX/IAA* (Auxin/Indole Acetic Acid) genes. In the presence of ethylene and in the absence of a BR signal, *IAA6* and *IAA9* genes were highly repressed. Ethylene downregulated *AUX/IAA* and enhances *ARF7* and *ARF19* genes to positively regulate shoot gravitropic responses. BR controls a larger set of *AUX/IAA* genes like *ARF3 and ARF12* in dark grown seedlings. Reduction in endogenous levels of BR using BRZ (Brassinosteroid synthesis inhibitor) increases the rate of reorientation in etiolated seedlings (Vandenbussche et al. 2013). Similarly, ethylene reduced a root's gravitropic responses while BR induced it (Kim et al. 2007).

Ethylene-inducing xylanase (Eix) is crucial for phytohormonal interactions between Ethylene and BR, as a response to biotic stress. Eixhas is a potential elicitor of defense-based response in plants like tobacco and tomato. LeEix2, which is a receptor of Eix, has the potential of initiating defence responses whereas, LeEix1 heterodimerizes with LeEix 2 and attenuates Eix-induced internalization and its signalling. This is put forward by binding of LeEix 1 with BAK1 which strongly points to BR-ethylene crosstalk (Bar et al. 2010).

Crosstalk with Gibberellic Acid

There are several evidences indicating antagonistic behaviour between BR-GA in defence-related response against root oomycetes *Pythium graminicola*. There was a severe spike in disease development in various GA deficient/insensitive mutants proving a positive role of GA in providing resistance against *P. graminicola*. It can be confirmed that after BR treatment and in absence of GA by application of uniconazole (a GA biosynthesis inhibitor), did not cause any additive effect (De Vleesschauwer et al. 2012). But when uniconazole was co-applied with brassinazole (a BR biosynthesis inhibitor), the induction of resistance due to brassinazole was hampered or reduced. It is also seen that exogenous GA application controls BR biosynthesis through a negative feedback loop, but promotes cell elongation by activating primary BR signalling (Tong et al. 2014). It is implied that BR signalling acts as a prerequisite for cell elongation promoted by GA. Both, in *in vitro* and *in vivo* conditions, DELLA proteins interact with BZR1 directly and discourage BZR1-DNA binding. This ultimately results in disruption in the perception of environmental cues necessary for controlling cell elongation and seedling etiolating (Bai et al. 2012, Gallego-Bartolomé et al. 2012, Li and He 2013). The transcription factor BRZ1 which when activated consequently interacts with RGA (the gal-3 repressor) to repress the BRZ1 activity. The DELLA proteins effectuate cellular elongation and physiological development by modulating GA and BR pathways and acts both as a positive and negative regulator in BR-GA crosstalk (Li et al. 2012).

Crosstalk with Auxin and Cytokinin

Crosstalk between BR and auxin is quite elusive. Plants need to maintain an optimum threshold level of BR to promote proper functioning of auxins. *BRX* (*BREVIS RADIX*), which is a regulator of cell proliferation and elongation in the root and shoot, regulates roots architecture which helps to maintain this balanced threshold. BRX is strongly induced by auxins and slightly repressed by BR, indicating it to be a focal regulator of a feedback loop (Kissoudis et al. 2014, Breda et al. 2017, Mouchel et al. 2006). It has been observed that, BRX can upregulate BR biosynthesis genes like CPD (a C-3 hydrogenase constitutive photo-morphogenesis and dwarf/ CYP90A1) and DWF4 (C-22 hydroxylase dwarf4/CYP90B1) which hints BR-auxin signalling (Tanaka et al. 2005, Mouchel et al. 2006). Auxins also help to upregulate BR biosynthesis gene expressions, since exogenous application of auxin exponentially enhanced the expression of DWF4 in *Arabidopsis*. But on reaching an optimum level of BR, the feedback loop DWF4 kicks in with the help of BRX (Chung et al. 2011, Maharjan et al. 2011).

BRs are known to have a key role in auxin transport in plants by regulating auxin-responsive genes like PIN3, PIN4 (pinformed), and auxin-resistant1/like aux1 by changing cellular localization (Hacham et al. 2011, 2012). The action on these transporter genes by repressing them indicates a pivotal role in abiotic stress regulated by BR-auxin crosstalk (Nemhauser et al. 2004). Auxin production in rice is controlled by *YUCCA* genes encoding rate limiting enzymes. It was found that *OsYUCCA* genes were downregulated during drought stress, whereas under cold stress, *OsYUCCA2, OsYUCCA3, OsYUCCA6,* and *OsYUCCA7* genes were highly upregulated and *OsYUCCA3, OsYUCCA6,* and *OsYUCCA7* showed a five-time increment in their expression under heat stress (Yamamoto et al. 2007, Du et al. 2013).

Cytokinins has an array of both direct and indirect crosstalk with BR. Auxin transport modulation indirectly helps in regulating lateral root development with the help of *PIN* genes enhanced by BR whereas, cytokinins hinders increments in auxin levels by downregulating PIN genes, which inhibits development of lateral roots (Benjamins and Scheres, 2008). BR, by means of exogenous application, enhances rooting and increase in leaf size under CKX3 (*cytokinin oxidase/dehydrogenase3*) overexpression and BRI1 ectopic expression (Vercruyssen et al. 2011). In wheat seedlings, BR application caused zeatin derivatives to accumulate in shoots and roots which were promptly reverted to control levels on BR removal. This was suggestive of an inhibition due to gene encoding CKX protein (Yuldashev et al. 2012). In rice, IPT (*isopentyl transferase*) overexpression caused increase in cytokinins, leading to increase in drought stress tolerance. The observed enhancement of cytokinins corresponds with the upregulation of different BR-related biosynthesis genes (*DWF4, DWF5, HYD1*) and BR-signaling genes (*BRI1, BZR1, BAK1, SERK1, BRH1*). The plants conferred higher grain yields and delayed drought stress responses contributing to cytokinin-BR crosstalk regulating yield and stress tolerance (Peleg et al. 2011).

Synergistic relation between cytokinin and BR was observed in *Chlorella vulgaris*, where cytokinin stimulated endogenous BR levels. Treating a culture of *Chlorella* with trans-zeatin caused considerable increase in endogenous BR levels. Accumulation of different essential organic biomolecules was boosted even more by treatment with BL and trans-zeatin simultaneously (Bajguz and Piotrowska-Niczyporuk 2014) indicating BR-CK crosstalk point. Cytokinin also interacts with BR to perpetually control the levels of ACS (the ethylene biosynthesis gene) and thereby post-transcriptionally regulate biosynthesis of ethylene. As observed in *Arabidopsis*, cytokinin is perceived by AHK (*Arabidopsis* histidine kinase) receptors which autophosphorylates to activate ARR1 (*Arabidopsis* response regulator Type B) down the lane activating ethylene biosynthesis. The target of action of both BR and CK appears to be independent of each other, although the ACS proteins are stabilized with their additive effect, indicating hormonal interaction between CK and BR (Hansen et al. 2009).

Importance of Brassinosteroids against Biotic Stresses Affecting Plants

Plants are subjected to certain invasion by multifarious microbial pathogens, nematodes, and insect. Various pathogens infect plants using different strategies. Microbes induced plant diseases restrict growth and development of plants ultimately resulting in losses in crop yield. Phytopathogens can be divided into two groups which are avirulent and virulent pathogen, based on their ability to cause infection. These pathogens can be further classified in to three categories *viz*, biotrophs, hemi-biotrophs, and necrotrophs based on their lifecycle and mode of nutrient uptake (Glazebrook 2005, Yu et al. 2018). Necrotrophic pathogens are bacterial, fungal, and oomycete species which raise the destruction of host cell and uptake nutrients from dead or decaying cells, resulting in extensive necrosis, tissue maceration, and plant rots. Biotrophic pathogens primarily enter through the wounded orifices of the plant and stomata and later spread through the intercellular spaces. They attack plant cell through formation of penetration peg, later developing into haustoria. Although they gradually infect the plant parts, they do not destroy the plant cell (Laluk and Mengiste 2010). Biotrophic pathogens can survive with their main host crops or plants in an artificial compatibility, and so their damage to the plant parts is comparatively low and not destructive (Glazebrook 2005). Necrotrophs intially kill host cells through the action of different phytotoxins produced by them and feed on them by breaking down the cells deriving nutrients. Hemibiotrophs are the pathogens which show both of the lifestyles based on the stages of life cycle of them (Pieterse et al. 2009). It is well established that brassinosteroids may able for protecting the plant parts from various attacks of the pathogens. Post-harvest application of BRs and its analogues have also been reported in some crops which is described in Table 1.

Brassinosteroids interacts with all classes of pathogens including biotroph, hemi-biotroph, and necrotroph pathogens. It has been observed in various studies that BR enhances tolerance to most of the biotrophs, though it enhances susceptibility to hemi-biotrophs and necrotrophs (Yu et al. 2018). Primary field trials in crop plants suspected that exogenously applied Brassinosteroids could make plants more resistant from a wide effect of disease infestations (Albrtcht et al. 2012). The role of BR against virus, bacteria, fungus, insects, and nematodes are described and listed in Table 2.

Table 1. Application of BRs and its equivalents after harvest.

Application	Crops	Results	References
BRs	Strawberry	Decreased weight loss, preserved color, and improved resistance to the disease *Botrytis cinerea*.	Furio et al. 2019
BRs	Pears	Cause a decrease in titratable acidity (TA) and an increase in total soluble solids (TSS).	Thapliyal et al. 2016
BL	Tomato	Increase respiration rate, ethylene generation, lycopene synthesis, soluble solids, and chlorophyll concentration, all of which contribute to ripening.	Zhu et al. 2015a,b
BRs	Jujube	Reduced blue mold rots growth.	Zhu et al. 2015a,b
24-EBR	Satsuma mandarin	Reduce disease occurrence and up the rate of weight reduction.	Zhu et al. 2015a,b
24-EBR	Eggplant	Restrictive chilling Injury Index (CI) and pulp browning.	Gao et al. 2015
BL	Peaches	Reduce *Penicillium expansum*-induced fruit degradation.	Ge et al. 2014
BL	Bell peppers	Decreased the loss of ascorbic acid and chlorophyll content more slowly.	Wang et al. 2012

Role of Brassinosteroids against Virus

Different viruses play major destructive effects in the life of the specific plant species. For surviving from against the virus attack, different stress responsive

Table 2. Effects of BR against different pathogenic organisms.

Name of Biotic stress	Plant species	Name of the pathogenic organism	Effects of BR	References
Virus	Tobacco (*Nicotiana benthamiana*)	Tobacco Mosaic Virus (TMV)	External application of BR enhanced resistance.	Deng et al. 2016
Virus	Tobacco (*Nicotiana tabacum*)	Tobacco Mosaic Virus (TMV)	BL-treated wild type plants became tolerant to the virus.	Nakashita et al. 2003
Virus	Cucumber (*Cucumis sativus*)	Cucumber Mosaic Virus (CMV)	BR level was positively correlated with CMV resistant.	Xia et al. 2009
Virus	*Arabidopsis thaliana*	Cucumber Mosaic Virus (CMV)	Resistance was enhanced due to BR levels.	Zhang et al. 2015
Virus	Barley (*Hordeum vulgare*)	Barley Stripe Mosaic Virus (BSMV)	The semi-dwarf uzu derivatives, where mutated BRI1 gene was present, showed less symptoms of the virus compared to their parents.	Ali et al. 2014
Virus	Rice (*Oryza sativa*)	Rice Stripe Virus (RSV)	Resistance was increased by the increasing level or signalling of BR and JA.	Hu et al. 2020
Virus	Rice (*Oryza sativa*)	Rice Black-Streaked Dwarf Virus (RBSDV)	BR induces susceptibility of plants to RBSDV.	He et al. 2017
Virus	Maize (*Zea mays*)	Maize Chlorotic Mottle Virus	BR biosynthesis increase the susceptibility of maize to this virus in a NO-dependent manner.	Cao et al. 2019
Bacteria	Rice (*Oryza sativa*)	*Xanthomonas oryzae*	BL induced resistance to these bacteria in rice crop.	Nakashita et al. 2003
Bacteria	Tobacco (*Nicotiana tabacum*)	*Pseudomonas syringaepv. tabaci* (Pst)	BL gave emergence to increase resistance to this bacterial pathogen in tobacco plant.	Nakashita et al.2003
Fungus	Barley (*Hordeum vulgare*)	*Fusarium culmorum*	Treatment with epiBL minimizes the severity of FHB and loss of grain yield 86% and 33%, respectively, in Lux variety.	Ali et al. 2013
Fungus	Barley (*Hordeum vulgare*)	*Fusarium culmorum*	Development of Lux and 'Akashinriki' variety in soil amended with epiBL resulted in 28% and 35% decrese in FSB symptoms, respectively.	Ali et al. 2013

Table 2. contd. ...

... Table 2. contd.

Name of Biotic stress	Plant species	Name of the pathogenic organism	Effects of BR	References
Fungus	Rice (*Oryza sativa*)	*Magnaporthe grisea*	Application of BL increased resistance against the disease.	Nakashita et al. 2003
Fungus	Tobacco (*Nicotiana tabacum*)	*Oidium* sp.	Wild type tobacco plant treated with BL became resistant to the fungal disease.	Nakashita et al. 2003
Fungus	Rice (*Oryza sativa*)	*Pythium graminicola*	Sensitivity to the diseases was increased with application of BR.	Goddard et al. 2014
Fungus	Potato (*Solanum tuberosum*)	*Phytophthora infestans*	Increase susceptibility against the disease.	Yu et al. 2018
Fungus	Cucumber (*Cucumis sativus*)	*Fusarium oxysporum*	The disease severity was reduced due to the pre-treatment of EBL.	Ding et al. 2009a
Fungus	*Brassica napus*	*Leptosphaeria maculans* and *Sclerotinia sclerotiorum*	Over-expression of *Arabidopsis* BR biosynthetic gene AtDWF4 show enhanced resistance.	Sahni et al. 2016
Insect	Tomato (*Solanum lycopersicum* L.)	*Spodoptera frugiperda* and *Tutaabsoluta*	JA–BR interaction plays important roles in the defensive mechanisms against the herbivory.	Campos et al. 2009
Nematode	*Oryza sativa*	Root-knot nematode-*Meloidogyne graminicola*	External applications of lower concentration BL causes susceptibility in the roots and external applications of higher concentration BL enforced defensive actions against RNK.	Nahar et al. 2013
Nematode	Radish (*Raphanus sativus* L.)	Root-knot nematode-*Meloidogyne incognita*	Development of the plant root along with development of RNK was enhanced in BL treated plants, compared to non-treated plants.	Ohri et al. 2004
Nematode	Tomato (*Lycopersicon esculentum* L.)	Root-knot nematode-*Meloidogyne incognita*	Augmented plant growth in both RNK inoculated susceptible and resistant varieties were observed.	Kaur et al. 2013

mechanisms were generated by the plants. Salicylic acid (SA), jasmonic acid (JA), and brassinosteroids (BRs) are the plant hormones which are popular for playing significant roles by involving in different biotic responsive mechanisms including the defence mechanism of viruses. Both the positive and negative effects of brassinosteroids (BRs) against different plant viral attacks were already observed (Nawaz et al. 2017). Though the defensive role of brassinosteroids against different biotic stresses is already known, but very limited information was found about its antiviral effects and the detail resistance mechanisms is still not clear because few research works were carried out till now about this.

The defensive responses of brassinosteroids were investigated in different studies. To understand the function of brassinosteroids in *Nicotiana benthamiana* plant against the attack of Tobacco Mosaic Virus (TMV), incorporation of both the approaches based on pharmacology and genetics were carried out. When the brassinosteroids were added externally, the plants produced resistant reactions against TMV. However, when Bikinin (inhibitor of glycogen synthase kinase-3) by which brassinosteroids signaling is activated, was applied, the plant became susceptible against TMV. NbBRI1-(BRASSINOSTEROID INSENSITIVE 1, a plasma membrane-localized receptor)-silenced plants and NbBSK1- (BR SIGNALING KINASE 1, which is phosphorylated by activated BRI1)-silenced plants stopped BR-induced resistance against TMV. Bikinin-reduced resistance against the virus was blocked by silencing of NbBES1/BZR1 (BRI1 EMS SUPPRESSOR 1 and BRASSINAZOLE RESISTANT 1), which is the crucial regulators of BR-induced transcriptional changes. In this study, different actions of BR were found against TMV. In one side, the TMV resistance was induced by BR through MEK2-SIPK cascade and RBOHB (respiratory burst oxidase homolog B) dependent ROS (reactive oxygen species) burst. On the other side, the production of RBOHB-dependent ROS was reduced by BES1/BZR1 (Deng et al. 2016).

Brassinolide (BL) is known as the major important BR, playing a significant role in hormonal regulation of growth and development of plants. The effects of BL in resistance against different biotic stresses in rice and tobacco plants were studied. The wild type tobacco (*Nicotiana tabacum*) plants, which were treated with BL showed more tolerant to tobacco mosaic virus (TMV). By analysing Brassinazole 2001, a specific BR synthesis inhibitor, and measuring BRs in the leaves infected with TMV, it was stated that the steroid hormone-mediated tolerance to the disease mechanism generate defensive responses in the tobacco plant (Nakashita et al. 2003).

Understanding of the genetical mechanisms of BR is very tough due to the strong and pleiotropic phenotypes of BR synthesis and signalling mutants which included utmost dwarfism, dark green colour of leaves, and retarded development. To understand the effect of BR in cucumber (*Cucumis sativus*), a chemical genetics method was used to manipulate the BR level. A positive correlation of BR levels with the resistance properties to cucumber mosaic virus was found (Xia et al. 2009). The potential role of BR for creating resistance against cucumber mosaic virus (CMV) was also studied in *Arabidopsis thaliana*. By manipulating the levels of BR in *Arabidopsis thaliana* plants, a positive correlation between the levels of BR and cucumber mosaic virus (CMV) resistance was observed. For BR-induced CMV resistance, BR signalling is required and this signalling controls tolerance to virus through the regulations of some antioxidative enzymes and some defence-related genes, one example is WRKY30. Two BR signalling mutants were used which were a weak allele of the BRs receptor mutant *bri1-5* (Loss-of-function mutants) and constitutive BRs response mutant *bes1-D* (BRI1 EMS SUPRESSOR 1). CMV resistance was poor in *bri1-5* and high in *bes1-D*. Upregulation of CMV tolerances by BR applications were understood in this study (Zhang et al. 2015).

Brassinosteroid receptor BRI1 plays major role in the in the brassinosteroid signalling cascade. A mutation in a conserved domain of the kinase tail of BRI1 was present in the Semi-dwarf 'uzu' barley (*Hordeum vulgare*) 2 lines, cultivars Akashinriki and Bowman. Several pathogenicity tests were carried out to test the resistant properties of uzu derivatives with their parents. The two uzu derivatives showed less symptoms of Barley Stripe Mosaic Virus (BSMV) compared to their parents. It was observed that the two uzu derivatives became tolerant to several pathogens may be because of the pleiotropic effects of BRI1 or the cascade actions of their repressed BR signalling (Ali et al. 2014).

The productivity and quality of rice plant is seriously diminished due to Rice Stripe Virus Disease (RSVD), transmitted by small brown plant hoppers (SBPH, LaodelphaxstriatellusFalle´n). Other cereal crops like wheat, barley, and maize are also affected by RSVD. In rice, it was found that if the levels or signalling of BR and jasmonic acid (JA) were enhanced then the resistance mechanism would increase. Though the susceptibility to RSV increased due to treated plants with BR or JA signaling. BR signaling pathway was inhibited by RSV infection. The increase of resistance against RSV which is conferred by BR was impaired in *OsMYC2* (a key positive regulator of JA response) knockout plants, which indicated that the active JA pathway is required for the RSV-resistant mediated by BR. *OsGSK2* (a key negative regulator of BR signaling) was also accumulated by RSV infection. *OsMYC2* was interacted and phosphorylated by *OsGSK2*, and as a result *OsMYC2* was degraded and JA-mediated RSV defence response was suppressed. This study helps to understand the cross-talk between BR and JA in response to virus infection (Hu et al. 2020).

Rice black-streaked dwarf virus (RBSDV), which is transmitted in rice, wheat, barley, and maize by the small brown planthopper (SBPH) (*Laodelphax striatellus),* causes serious damages mainly in rice and maize plant. To investigate the hormone-mediated immunity in virus defence, global gene expression of RBSDV-infected paddy was compared with the normal plants using a high-throughput-sequencing approach. In the infected rice plants, induction of the jasmonic acid (JA) pathway and reduction of the BR pathway were observed. When methyl jasmonate (MeJA) or brassinazole (BRZ) was applied, then RBSDV infection was suppressed, whereas the infection was enhanced with epibrassinolide (BL) treatment. This study revealed that the plants became resistant and susceptible to RBSDV with JA and BR, respectively. The combination MeJA and BL treatment supressed in RBSDV infection compared to a single BL treatment. Application of MeJA repressed the expression of BR biosynthesis genes, and this reduction depended on the JA co-receptor *OsCOI1*. It can be stated that, JA-mediated defence can reduce the BR-mediated susceptibility to RBSDV infection. For further understanding the negative role of BR in defence against RBSDV infection of rice plants, BR-insensitive and BR-deficient mutant were used. Both the mutant showed resistant to RBSDV infection (He et al. 2017, Zhang et al. 2019).

Maize chlorotic mottle virus (MCMV), a single-stranded positive-sense RNA virus causes malformation in the growth of maize (*Zea mays* L.) plants. In a study, global gene expressions of the virus-infected plants were compared using transcriptome sequencing. Upregulation of BR-associated genes after virus infection was seen. 2,4-epibrassinolide (BL)- the most active BR or brassinazole (BRZ), the BR biosynthesis inhibitor, was applied to the plants externally. The virus infection in BL-treated plants was higher compared with that of mock-treated and BRZ-treated plants. This result revealed that biosynthesis of BR was linked with the susceptibility to MCMV infection. To investigate further, DWARF4 (ZmDWF4, a key gene of BR biosynthesis) and nitrate reductase (ZmNR, a key gene of NO synthesis) were knocked down using virus-induced gene silencing approaches and the knock down plants showed higher resistant against MCMV compared to wild types. This study implied that BR biosynthesis enhance the susceptibility of maize to MCMV in a NO (Nitric Oxide)-dependent manner (Cao et al. 2019).

Role of Brassinosteroids against Bacteria

BRs (Brassinosteroids) have been recently found to have protective abilities against plant pathogenic bacteria based on field trial evaluations, although the molecular basis of the results is yet to be clarified since the assays do not always coincide with the field trials (Khripach et al. 2000). *Pseudomonas* sp. a rhizosphere-colonizing bacterium imparts a systemic resistance known as rhizobacteria-mediated induced systemic resistance (ISR), mediated by phytohormones like jasmonates (Knoester et al. 1999, Pieterse et al. 2000, van Wees et al. 1999). The hemibiotrophic bacteria, *Pseudomonas syringae* inject type III secretory proteins which stimulate plant infection by manipulating the host's immunity, suppressing it, and causing disease. *Pseudomonassyringae* pv. *tomato* DC3000, didn't cause host cell death in the primary stages of infection, but as the stages of infection advances, the host tissue became severely necrotic and chlorotic by producing toxins (Espinosa et al. 2004, Glazebrook, 2005, Yu et al. 2018).

In tobacco, virulence against *Pseudomonas syringae*pv. *tabaci* (*Pst*) for *Nicotiatana tabacum* cv. Xanthi nc was recorded. It was observed that the cultivar did not possess any *Pst* specific resistance gene which makes the plant compatible with the pathogen. Pathogenesis observed on the leaf was used to measure susceptibility to *Pseudomonas*. The bacterial growth was inhibited by treating it with BL, although it did not show any microbial activity. Microbial assay performed on the infected leaves by treating them with BL which indicated that BL-induced resistance was not dependent on SA biosynthesis for its development. For determining action of BL on SAR activation in tobacco, induction of PR genes and their expressions were assessed. Northern blot analysis did not detect any transcripts for PR genes in leaves. This confirmed that BL-induced resistance followed distinctly different pathway from SAR. Similar resistance was observed in case of bacterial infection caused by *Xanthomonas oryzae*pv. *oryzae* in rice. Rice cultivar (*O. sativa* cv. Aichiasahi) treated with BL conferred slight bleaching in contrast to controls treated with water which showed excessive bleaching. This provides necessary proof of BL action against bacterial infections (Nakashita et al. 2003).

In a study conducted to check resistance to *Pseudomonas syringae*pv. *tomato* DC3000 (*Pst* DC3000) infection, it was found that BES1 (BRI1-EMS-SUPPRESSOR 1), a key factor for transcription, activates BR signaling response by controlling targeted gene expressions. Prior application of BL and BRZ in *Arabidopsis* leaves infected with *Pst* DC3000 was performed to check plant-pathogen interactions and the utility of BR. It was observed that BL showed lower pathogenic population and decreased susceptibility by increasing resistance to the pathogen, whereas control had higher pathogen concentrations and BRZ escalated the advancement of the disease. Various genetically modified lines including an overexpressing BR biosynthesis *DWF4*, BR biosynthesis knockout *det2*, and *DET2OX* (Overexpressing *DET2*transgenic) lines were employed to assess immunity of plants. The biosynthetically enhanced lines conferred greater resistances against the pathogen, whereas the knockout line *det2* exhibited susceptibility towards the infection which indicated enhanced immunity conferred due to application of BR. Analysing superoxide levels in plant cell against pathogenesis conferred a surge after *Pst* DC3000 infection. Rise in levels of superoxide was seen in BL-treated, *DWF4OX* and *DET2OX* plants but lower in BRZ-treated and *det2* plants. There was an enhancement found in *PR1*, *PR2* defence-related gene expressions. This increase was observed in BR biosynthesis enhanced and BL treated lines in contrast to BRZ treated and BR deficient lines which emphasised involvement of antioxidant behaviour of BR-induced resistance to *Pst* DC3000 (Xiong et al. 2022).

In *Arabidopsis*, pathogen-induced MAPKs increase phosphorylation of BRASSINOSTEROID INSENSITIVE1-ETHYL METHANESULFONATE-SUPPRESSOR1 (BES1), which is a transcription factor implicated in the brassinosteroid (BR) signaling pathway. There was an interaction seen between BES1 and MITOGEN-ACTIVATED PROTEIN KINASE6 (MPK6). MPK6 phosphorylates BES1.bes1 loss-of-function mutants showed compromised tolerance to *Pseudomonas syringae*pv *tomato* DC3000 (Kang et al. 2015).

It was reported that plants are serving receptor like kinases (RLKs) as well as receptor like proteins for fast identification of attacking pathogen, then RLKs transfer the signals to receptors like cytoplasmic kinases (RLCKs) for activating the immune responses. It was observed that *Arabidopsis thaliana* receptor like kinase (RLK902) is one of the most important factors in resistance to the bacterial pathogen *Pseudomonas syringae*. Consequently, RLK902 immediately linked with the Brassinosteroid signaling kinase1 (BRSK1) which is a vital domain of plant immunity system (Zhao et al. 2019).

Role of Brassinosteroids against Fungus

Different fungi use different ways to infect plants. Necrotrophic infections like *Botrytis cinerea* and *Alternaria brassicicola* invade the plant through minor wounds/ fissures in the cuticle or through the stomata. They can kill host cells early in the infection process by secreting the enzymes which degrade cell walls and the lytic enzymes that cause tissue damage. To encourage host cell death, they also create phytotoxic substances like as phytotoxins and proteins. The actions of BR on these diseases appear to be pleiotropic, depending on the pathogens and plant species involved, they induce resistance, enhance vulnerability, or have no impact (Yu et al. 2018).

Dwarfism is common in BR-deficient and BR-insensitive mutants. The action of BRs in governing cell division, growth, and even differentiation is at the root of this trait (Ahmad et al., 2022). A study by Nakashita et al. (2003) on the effect of externally administered BRs on innate immunity backs this up. In small-scale disease experiments, the scientists found that BL had a beneficial but variable effect on disease resistance of tobacco and rice to different leaf pathogens. This effect was discovered to be both local and systemic. BR treatment has recently been found to protect barley from a variety of *Fusarium* infections. Exogenously applied BR, for example, has been shown to provide resistance in barley plants to a variety of fungal infections with varying trophic lifestyles. When epibrassinolide (epiBL) was applied to the heads of 'Lux' barley, it decreased the severity of *Fusarium* head blight (FHB) affected by *Fusarium culmorum* by 86% and the loss of grain yield due to FHB by 33%, respectively. Furthermore, on the Lux and 'Akashinriki' barleys, plant development in soil amended with epiBL resulted in 28% and 35% decreases in *Fusarium* seedling blight (FSB) symptoms, respectively. Based on the analysis of gene expression of these plants, it was observed that growth in epiBL amended soil stimulated the activation of genes involved in chromatin remodeling, hormone signaling, photosynthesis, and pathogenicity at the time of the early phase of FSB development (Ali et al. 2013, Kohli et al. 2019).

Overexpression of the *Arabidopsis* BR biosynthetic gene AtDWF4 in the oilseed plant *Brassica napus* showed significantly more tolerance to dehydration and heat stress, and increased resistance to necrotrophic fungal pathogens-*Leptosphaeria maculans* and *Sclerotinia sclerotiorum* (Sahni et al. 2016). Exogenously administered BR, on the other hand, had no effect on in wild-type *Arabidopsis* plants infected with the necrotrophic fungus *Alternaria brassicicola* (Yu et al. 2018). BL-induced resistance against *Magnaporthe grisea* in rice plants and *Oidium* sp. in wild type tobacco plant (Nakashita et al. 2003). BRs were found to increase sensitivity to the hemibiotrophic diseases like *Pythium graminicola* in rice, rather than increasing plant resistance (Goddard et al. 2014). By promoting mycelial growth, enhancing *Phytophthora infenstans* spore generation, BR increased the susceptibility of potato tuber tissues to the disease and decreased the immunological status of plant tissues. The immunosuppressive impact of BRs is systemic and long-lasting (Yu et al. 2018).

In potato tubers infected with *Phytophthora infenstans* yielded completely contradictory results. This indicated that whether BRs protect plants from pathogen infection or not depended on the methodology and timing of BRs addition, as well as various stimulating points of the plant or pathogen. The buildup of systemic acquired resistance (SAR) is vital in eliciting BRs-enhanced pathogen resistance in plants. In field studies with cucumber, EBR was found to have fungi-protective properties. Seed treatment with EBR, followed by foliar spraying with EBR during the flowering stage, dramatically reduced mildew growth, owing to increased POD and polyphenoloxidase activities (PPL) (Kang and Guo 2011). Independent of application mode, both root and foliar treatments of EBR significantly minimized disease severity caused by *fusarium* wilt, as well as increased plant development and reduced biomass losses. Pathogen-induced build-up of ROS, flavonoids, and phenolic compounds, as well as the activity of defence-related and ROS-scavenging enzymes, were considerably reduced by 24-epibrassinolide (EBL) treatments (Ding et al. 2009a). A study conducted on EBL-induced disease resistance in cucumber plants revealed that when EBL was applied as a foliar spray and in the root, it reduced *Fusarium* population on both root surfaces and in nutrient solution. On the other hand, it boosted the fungal and actinobacterial populations on root surfaces (Ding et al. 2009b).

Role of Brassinosteroids against Insects and Nematodes

Insects are very much detrimental for agricultural plants not only for consuming the plants, but also for being the vectors of different plant pathogens. A huge proportion of agricultural losses occur due to insect herbivory (Campos et al. 2009). Plants like tomato need large quantities of insecticides because of its high susceptibility to insect herbivory (Zalom 2002). Several defensive mechanisms against insect attack were developed by tomato plants including glandular trichomes, allelochemical's synthesis, expressions of enzymes such as proteinase inhibitors and polyphenol oxidases. Classification of trichomes was carried out on the basis of trichome stalk length/format and presence/absence of glands (Campos et al. 2009). *Zgb* (Zingiberene) is a sesquiterpene linked with tomato tolerance against different insects (Freitas et al. 2002). By studying the role of different plant hormones against insect herbivory, it was observed that the trichome density and allelochemical contents were affected by both brassinosteroids (BRs) and jasmonates (JAs) but in antithetical manner. The expressions of proteinase inhibitor, zingiberene biosynthesis, and pubescence were increased in *dpy* (the BR-deficient mutant) whereas, *jai1-1* (the JA-insensitive mutant) had the opposite results. *jai1-1* was epistatic to *dpy*, which was observed from the *dpy* X *jai1-1* double mutant.

The significance of JA–BR interaction in defence was clearly understood by the herbivory tests with the polyphagous insect-*Spodoptera frugiperda* and the tomato pest-*Tutaabsoluta*. Simultaneously, when the *S. frugiperda* selected a genotype for feeding on (two-choice assay), *dpy* and MT (Micro-Tom cultivar) did not show any statistical differences indicating that both trichomes and *zgb* are not involved in insect defensive mechanisms. For the insect, *jai1-1* and *dpy* X *jai1-1* were comparatively better than MT. Though there was no alteration for *dpy* but *dpy* X *jai1-1* gave larger caterpillars than *jai1-1* suggestibly due to the easier digestion or more nutrition provided by *dpy* X *jai1-1* to the caterpillars. In case of *T. absoluta*, mortality rates were measured rather than weight changes. In *dpy* and *jai1-1* mutants, mortality rates were comparable with MT. The wounded *dpy* mutants showed an intensified serine proteinase inhibitor, PI-I expression in comparison with MT.

Surges in mortality levels of *T. absoluta* can be attributed to high PI-I levels in *dpy* acting conjointly with increased trichome numbers and *zgb* levels, as observed in the *dpy* mutants. The normal and wounded plant did not show any PI-I expression in the *jai1-1* and *dpy* X *jai1-1* mutants because of the epistatic effect of *jai1-1* to *dpy* (Campos et al. 2009).

Systemin is known as an 18–amino acid peptide which is involved in the defensive responses against the attack of insects and wounds which induce signaling that enhances Jasmonate production, leading to expression of defensive genes. Leu-rich repeat receptor-like kinase (LRR-RLK) plays a significant role in the signaling of the plant peptide hormones- phytosulfokine and systemin. The Tomato homolog of the *Arabidopsis* bri1 (Brassinosteroid Insensitive 1), a putative LRR named tBri1, was extracted with the help of degenerate primers (Montoya et al. 2002). Sequencing homology between the putative systemin receptor, SR160 and BRI1 was found previously (Scheer and Ryan 2002). Based on the sequence analysis, the possible dual roles of tBRI1/SR160 LRR-RLK in systemin and BR signaling were understood (Montoya et al. 2002).

Declination of crop yield is caused by root-knot nematode (RKN) (*Meloidogyne* spp.) which is an obligate, sedentary endoparasite of different crops (Abad et al. 2003). The potential role of phytohormones like BR in the innate immunity of Rice against root-knot nematode (RKN)-*Meloidogyne graminicola* was evaluated along with the cross interaction with the JA pathway. Epibrassinolide (BL) was added externally in various concentrations as a source of BR. The external applications of BL at lower concentrations gave rise to susceptibility in the roots but when the concentration of BL was increased, then systemic defensive action against RNK was emboldened. Based on the q-RT-PCR results, it was revealed that BR pathway was antagonistic with JA pathway in the roots of the paddy plants. It was suggested that the balance between BR and JA pathways acts an effectual regulator of the result of the rice–RNK association (Nahar et al. 2013).

Treatment of brassinolide in *Raphanus sativus* L. seedling was carried out after which plants were infected with root-knot nematode (*Meloidogyne incognita*) for nematode's developmental studies. The plant grew better along with better development of nematodesin the treated plants, compared to un-treated plants (Ohri et al. 2004). The potency of 28-homobrassinolide (HBl) was observed in one resistant and one susceptible varieties of tomato (*Lycopersicon esculentum*) five days after RNK (*M. incognita*) inoculation. It was observed that the plant growth which was truncated in susceptible varieties due to nematode invasion, regained its physical stature after exogenous BR application. The resistant varieties did not show any adverse physiological growth since nematodes could not invade the roots. Presowing application of HBl seed treatment in the resistant plant moreover increased the plant growth. Furthermore, RNK inoculation in resistant variety showed enhanced specific activities complemented by the HBl treatment (Kaur et al. 2013).

Conclusion

Plants face several stresses and diseases although their life cycle owing to abiotic, edaphic, and biotic factors. Owing to the recent peculiarities of weather conditions causing myriad of stresses and disease, dependence on chemical supplements, pesticides and herbicides to boost crop yield is exponentially increasing. This causes undue strain on the ecosystem. The utility of phytohormones like brassinosteroids, and its familiar homologs for protecting the plants from biotic stresses was observed by several researchers which were mainly based on observing the tolerance either by exogenous applications of BR or overexpressing any genes of BR.

Various other phytohormones like JA, SA, ABA, gibberelins, auxins, and cytokinins work directly in tandem or sometime individually with BRs to alleviate stresses. Applications of BR in several plants showed resistance and/or increased tolerance to several viruses like TMV, CMV, RSV, etc. Similarly, increased resistance was observed against bacterial pathogens like *Xanthomonas oryzae* and *Pseudomonas syringae* pv. *tabaci* (*Pst*), and fungal pathogens like *Fusarium* sp, *Magnaporthe grisea*, *Oidium spp.*, *Leptosphaeria maculans and Sclerotinia sclerotiorum*. Effects of BR were also studied in different insects and RNK. Sometimes, applications of BR also induced susceptibility against various plant pathogens. Although the detailed mechanisms of BR imparting resistance are yet to be unveiled, it is pertinent to pursue the effects of BR against different microorganisms and pests in different plant species for protecting the plants against the biotic stresses.

References

Abad, P., Favery, B., Rosso, M.N. and Castagnone-Sereno, P. 2003. Root-knot nematode parasitism and host response: molecular basis of a sophisticated interaction. Mol. Plant Pathol., 4(4): 217–224.

Ahmad, A., Shahzadi, I., Akram, W., Yasin, N.A., Khan, W.U. and Wu, T. 2022. Plant Proteomics and Metabolomics Investigations in Regulation of Brassinosteroid. In: M.T.A., M. Yusuf, F. Qazi and A. Ahmad (Eds.), Brassinosteroids Signalling. Springer, Singapore, pp. 17–45.

Albrecht, C., Boutrot, F., Segonzac, C., Schwessinger, B., Gimenez-Ibanez, S., Chinchilla, D. et al. 2012 Brassinosteroids inhibit pathogen-associated molecular pattern-triggered immune signaling dependent of the receptor kinase BAK1. Proc. Natl. Acad. Sci., 109: 303–308.

Ali, S.S., Gunupuru, L.R., Kumar, G.B., Khan, M., Scofield, S., Nicholson, P. et al. 2014. Plant disease resistance is augmented in uzu barley lines modified in the brassinosteroid receptor BRI1. BMC Plant Biol., 14(1): 1–15.

Ali, S.S., Kumar, G.S., Khan, M. and Doohan, F.M. 2013. Brassinosteroid enhances resistance to fusarium diseases of barley. Phytopathology, 103(12): 1260–1267.

Anwar, A., Liu, Y., Dong, R., Bai, L., Yu, X. and Li, Y. 2018. The physiological and molecular mechanism of brassinosteroid in response to stress: a review. Biol. Res., 51: 1–15

Azpiroz, R., Wu, Y., LoCascio, J.C. and Feldmann, K.A. 1998. An Arabidopsis brassinosteroid-dependent mutant is blocked in cell elongation. Plant Cell., 10: 219–230.

Bai, M.Y., Shang, J.X., Oh, E., Fan, M., Bai, Y., Zentella, R., Sun, T. and Wang, Z 2012. Brassinosteroid, gibberellin and phytochrome impinge on a common transcription module in Arabidopsis. Nat. Cell Biol., 14: 810–817.

Bajguz, A. and Piotrowska-Niczyporuk, A 2014. Interactive effect of brassinosteroids and cytokinins on growth, chlorophyll, monosaccharide and protein content in the green alga Chlorella vulgaris (Trebouxiophyceae). Plant Physiol. Biochem., 80: 176–183.

Bar, M., Sharfman, M., Ron, M. and Avni, A. 2010. BAK1 is required for the attenuation of ethylene-inducing xylanase (Eix)-induced defense responses by the decoy receptor LeEix1. Plant J., 63: 791–800.

Belkhadir, Y. and Jaillais, Y. 2015. The molecular circuitry of brassinosteroidsignaling. New Phytol., 206, 522–540.

Benjamins, R. and Scheres, B. 2008. Auxin: the looping star in plant development. Annu. Rev. Plant Biol., 59: 443–465.

Bishop, G.J. and Koncz, C. 2002. Brassinosteroids and plant steroid hormone signaling. Plant Cell., 14: S97–S110.

Bishop, G.J. and Yokota, T. 2001. Plants steroid hormones, brassinosteroids: current highlights of molecular aspects on their synthesis/metabolism, transport, perception and response. Plant Cell Physiol., 42(2): 114–120.

Breda, A.S., Hazak, O. and Hardtke, C.S. 2017. Phosphosite charge rather than shootward localization determines OCTOPUS activity in root protophloem. Proc. Natl. Acad. Sci., 114(28): E5721–E5730.

Campos, M.L., De Almeida, M., Rossi, M. L., Martinelli, A. P., Litholdo, J.C.G., Figueira, A. et al. 2009. Brassinosteroids interact negatively with jasmonates in the formation of anti-herbivory traits in tomato. J. Exp. Bot., 60(15): 4347–4361.

Cao, N., Zhan, B. and Zhou, X. 2019. Nitric oxide as a downstream signaling molecule in brassinosteroid-mediated virus susceptibility to Maize chlorotic mottle virus in maize. Viruses, 11(4): 368.

Choudhary, S.P., Yu, J.Q., Yamaguchi-Shinozaki, K., Shinozaki, K. and Tran, L.S.P. 2012. Benefits of brassinosteroid crosstalk. Trends Plant Sci., 17(10): 594–605.

Chung, Y., Maharjan, P.M., Lee, O., Fujioka, S., Jang, S., Kim, B. et al. 2011. Auxin stimulates DWARF4 expression and brassinosteroid biosynthesis in Arabidopsis. Plant J., 66(4): 564–578.

Clouse, S.D. 2011. Brassinosteroid signal transduction: from receptor kinase activation to transcriptional networks regulating plant development. Plant Cell., 23: 1219–1230.

Clouse, S.D. and Sasse, J.M. 1998. BRASSINOSTEROIDS: Essential Regulators of Plant Growth and Development. Annu. Rev. Plant Physiol. Plant Mol. Biol., 49: 427–451.

De Vleesschauwer, D., Van Buyten, E., Satoh, K., Balidion, J., Mauleon, R., Choi, I.R. et al. 2012. Brassinosteroids antagonize gibberellin–and salicylate-mediated root immunity in rice. Plant Physiol., 158: 1833–1846.

Deng, X.G., Zhu, T., Peng, X.J., Xi, D.H., Guo, H., Yin, Y. et al. 2016. Role of brassinosteroidsignaling in modulating Tobacco mosaic virus resistance in *Nicotiana benthamiana*. Sci. Rep., 6(1): 1–14.

Ding, J., Shi, K., Zhou, Y.H. and Yu, J.Q. 2009a. Effects of root and foliar applications of 24-epibrassinolide on Fusarium wilt and antioxidant metabolism in cucumber roots. Hortscience, 44(5): 1340–1345.

Ding, J., Shi, K., Zhou, Y.H. and Yu, J.Q. 2009b. Microbial community responses associated with the development of *Fusarium oxysporum* f. sp. *cucumerinum* after 24-epibrassinolide applications to shoots and roots in cucumber. Eur. J. Plant Pathol., 124: 141–150.

Divi, U.K. and Krishna, P. 2009. Brassinosteroid: a biotechnological target for enhancing crop yield and stress tolerance. New Biotechnol., 26: 131–136.

Du, H., Liu, H. and Xiong, L. 2013. Endogenous auxin and jasmonic acid levels are differentially modulated by abiotic stresses in rice. Front. Plant Sci., 4: 397.

Espinosa, A. and Alfano, J.R., 2004. Disabling surveillance: Bacterial type III secretion system effectors that suppress innate immunity. Cell Microbiol., 6: 1027–1040.

Freitas, J.A., Maluf, W.R., Cardoso, M.D.G., Gomes, L.A.A. and Bearzotti, E. 2002. Inheritance of foliar zingiberene contents and their relationship to trichome densities and whitefly resistance in tomatoes. Euphytica., 127(2): 275–287.

Fujioka, S., Li, J., Choi, YH., Seto, H., Takatsuto, S., Noguchi, T. et al. The Arabidopsis deetiolated2 mutant is blocked early in brassinosteroid biosynthesis. Plant Cell., 9: 1951–1962.

Furio, R., Salazar, S., Martínez-Zamora, G., Coll, Y., Hael-Conrad, V. and Díaz Ricci, J. 2019. Brassinosteroids promote growth, fruit quality and protection against Botrytis on Fragaria x ananassa. Eur. J. Plant Pathol. https ://doi.org/10.1007/s1065 8-019-01704 -3.

Gallego-Bartolomé, J., Mingueta, E.G., Grau-Enguixa, F., Abbasa, M., Locascioa, A., Thomas, S.G. et al. 2012. Molecular mechanism for the interaction between gibberellin and brassinosteroidsignaling pathways in *Arabidopsis*. Proc. Natl. Acad. Sci. USA, 109: 13446–13451.

Gao, H., Kang, L., Liu, Q., Cheng, N., Wang, B. and Cao, W. 2015. Effect of 24-epibrassinolide treatment on the metabolism of eggplant fruits in relation to development of pulp browning under chilling stress. J. Food Sci. Technol., 52: 3394–3401.

Ge, Y., Li, C., Tang, R., Sun, R. and Li, J. 2014. Effects of postharvest brassinolide dipping on quality parameters and antioxidant activity in peach fruit. In: Proceedings of III international symposium on postharvest pathology: using science to increase food availability, pp. 377–384.

Glazebrook, J. 2005. Contrasting mechanisms of defense against biotrophic and necrotrophic pathogens. Annu. Rev. Phytopathol., 43: 205–227.

Goddard, R., Peraldi, A., Ridout, C. and Nicholson, P. 2014. Enhanced disease resistance caused by BRI1 mutation is conserved between *Brachypodiumdistachyon* and barley (Hordeum vulgare). Mol. Plant-Microbe Interact., 27(10): 1095–1106.

Gruszka, D. 2013. The brassinosteroidsignaling pathway—New key players and interconnections with other signaling networks crucial for plant development and stress tolerance. Int. J. Mol. Sci., 14(5): 8740–8774.

Guo, H., Li, L., Aluru, M., Aluru, S. and Yin, Y. 2013. Mechanisms and networks for brassinosteroid regulated gene expression. Curr. Opin. Plant Biol., 16: 545–553.

Hacham, Y., Holland, N., Butterfield, C., Ubeda-Tomas, S., Bennett, M.J., Chory, J. et al. 2011. Brassinosteroid perception in the epidermis controls root meristem size. Development, 138(5): 839–848.

Hacham, Y., Sela, A., Friedlander, L. and Savaldi-Goldstein, S. 2012. BRI1 activity in the root meristem involves post-transcriptional regulation of PIN auxin efflux carriers. Plant SignalingBehav., 7(1): 68–70.

Hafeez, M.B., Zahra, N., Zahra, K., Raza, A., Khan, A., Shaukat, K. et al. 2021. Brassinosteroids: Molecular and physiological responses in plant growth and abiotic stresses. Plant Stress, 2: 1–7.

Hansen, M., Chae, H.S. and Kieber, J.J. 2009. Regulation of ACS protein stability by cytokinin and brassinosteroid. Plant J., 57(4): 606–614.

Hao, J., Yin, Y. and Fei, S.Z. 2013. Brassinosteroidsignaling network: implications on yield and stress tolerance. Plant Cell Rep., 32(7): 1017–1030.

He, Y., Zhang, H., Sun, Z., Li, J., Hong, G., Zhu, Q. et al. 2017. Jasmonic acid-mediated defense suppresses brassinosteroid-mediated susceptibility to Rice black streaked dwarf virus infection in rice. New Phytol., 214(1): 388–399.

Hotta, Y., Tanaka, T., Bingshan, L., Takeuchi, Y. and Konnai, M. 1998. Improvement of cold resistance in rice seedlings by 5-aminolevulinic acid. J. Pesticide Sci., 23: 29–33.

Hu, J., Huang, J., Xu, H., Wang, Y., Li, C., Wen, P. et al. 2020. Rice stripe virus suppresses jasmonic acid-mediated resistance by hijacking brassinosteroidsignaling pathway in rice. PLoSpathog., 16(8): e1008801.

Jaillais, Y., Hothorn, M., Belkhadir, Y., Dabi, T., Nimchuk, Z.L., Meyerowitz, E.M. et al. 2011. Tyrosine phosphorylation controls brassinosteroid receptor activation by triggering membrane release of its kinase inhibitor. Genes Dev., 25(3): 232–237.

Kang, S., Yang, F., Li, L., Chen, H., Chen, S. and Zhang, J. 2015. The Arabidopsis transcription factor Brassinosteroid Insensitive1-Ethyl Methanesulfonate-Suppressor1 is A Direct Substrate of Mitogen-Activated Protein Kinase6 and regulates immunity. Plant Physiol., 167(3): 1076–1086.

Kang, Y.Y., Guo, S.R., 2011. Role of brassinosteroids on horticultural crops. pp. 269–288. In: S. Hayat, S. and A. Ahmad (Eds.). Brassinosteroids: A Class of Plant Hormone. Springer, Dordrecht.

Kaur, R., Ohri, P. and Bhardwaj, R. 2013. Effect of 28-homobrassinolide on susceptible and resistant cultivars of tomato after nematode inoculation. Plant Growth Regul., 71(3): 199–205.

Khripach, V., Zhabinskii, V. and de Groot, A. 2000. Twenty years of brassinosteroids: steroidal plant hormones warrant better crops for the XXI century. Ann. Bot., 86(3): 441–447.

Kim, T.W., Guan, S., Burlingame, A.L. and Wang, Z.Y. 2011. The CDG1 kinase mediates brassinosteroid signal transduction from BRI1 receptor kinase to BSU1 phosphatase and GSK3-like kinase BIN2. Mol. Cell., 43(4): 561–571.

Kim, T.W., Lee, S.M., JOO, S.H., Yun, H.S., Lee, Y.E.W., Kaufman, P.B. et al. 2007. Elongation and gravitropic responses of Arabidopsis roots are regulated by brassinolide and IAA. Plant, Cell Environ., 30(6): 679–689.

Kissoudis, C., van de Wiel, C., Visser, R.G. and van der Linden, G. 2014. Enhancing crop resilience to combined abiotic and biotic stress through the dissection of physiological and molecular crosstalk. Front. Plant Sci., 5: 207.

Knoester, M., Pieterse, C.M., Bol, J.F. and Van Loon, L.C. 1999. Systemic resistance in Arabidopsis induced by rhizobacteria requires ethylene-dependent signaling at the site of application. Mol. Plant-Microbe Interact., 12(8): 720–727.

Kohli, S.K., Bali, S., Khanna, K., Bakshi, P., Sharma, P., Sharma, A., Verma, V. et al. 2019. A current scenario on role of Brassinosteroids in plant defense triggered in response to biotic challenges. pp. 367–388. In: S. Hayat, M. Yusuf, R. Bhardwaj and A. Bajguz (Eds.). Brassinosteroids: Plant growth and development. Springer, Singapore.

Laluk, K. and Mengiste, T. 2010. Necrotroph attacks on plants: Wanton destruction or covert extortion? Arabidopsis Book., 8: e0136.

Li, J. and Chory, J. 1997. A putative leucine-rich repeat receptor kinase involved in brassinosteroid signal transduction. Cell., 90: 929–938.

Li, J., Wen, J., Lease, K.A., Doke, J.T., Tax, F.E. and Walker, J.C. 2002. BAK1, an Arabidopsis LRR receptor-like protein kinase, interacts with BRI1 and modulates brassinosteroidsignaling. Cell., 110(2): 213–222.

Li, J.M., Nagapal, P., Vitart, V., McMorris, T.C. and Chory, J. 1996. A role for brassinosteroids in light-dependent development of Arabidopsis. Science, 272: 398–401.

Li, Q.F., Wang, C., Jiang, L., Li, S., Sun, S.S. and He, J.X. 2012. An interaction between BZR1 and DELLAs mediates direct signalling crosstalk between brassinosteroids and gibberellins in Arabidopsis. Sci. Signal., 5: ra72.

Li, Q.F. and He, J.X. 2013. Mechanisms of signaling crosstalk between brassinosteroids and gibberellins. Plant Signal. Behav., 8: e24686.

Li, H., Ye, K., Shi, Y., Cheng, J., Zhang, X. and Yang, S. 2017. BZR1 positively regulates freezing tolerance via CBF-dependent and CBF-independent pathways in Arabidopsis. Mol. Plant., 10: 545–559.

Maharjan, P.M., Schulz, B. and Choe, S. 2011. BIN2/DWF12 antagonistically transduces brassinosteroid and auxin signals in the roots of Arabidopsis. J. Plant Biol., 54(2): 126–134.

Mandava, N. 1988. Plant growth-promoting brassinosteroids. Annu. Rev. Plant Physiol. Plant Mol. Biol., 39: 23–52.

Marková, H., Tarkowská, D., Čečetka, P., Kočová, M., Rothová, O. and Holá, D. 2023. Contents of endogenous brassinosteroids and the response to drought and/or exogenously applied 24-epibrassinolide in two different maize leaves. Frontiers in Plant Science, 14: 1848.

Manghwar, H., Hussain, A., Ali, Q. and Liu, F. 2022. Brassinosteroids (BRs) Role in Plant Development and Coping with Different Stresses. Int. J. Mol. Sci., 23(3): 1012.

Mitchell, J., Mandava, N., Worley, J.F., Plimmer, J.R. and Smith, M.V. 1970. Brassins—a New Family of Plant Hormones from Rape Pollen. Nature., 225: 1065–1066.

Montoya, T., Nomura, T., Farrar, K., Kaneta, T., Yokota, T. and Bishop, G.J. 2002. Cloning the tomato curl3 gene highlights the putative dual role of the leucine-rich repeat receptor kinase tBRI1/SR160 in plant steroid hormone and peptide hormone signaling. Plant Cell., 14(12): 3163–3176.

Mouchel, C.F., Osmont, K.S. and Hardtke, C.S. 2006. BRX mediates feedback between brassinosteroid levels and auxin signalling in root growth. Nature., 443(7110): 458–461.

Murofushi, N., Yamane, H., Sakagami, Y., Imaseki, H.,Kamiya, Y., Iwamura, H. et al. 1999. Plant Hormones. pp. 19–136. In: D. Barton, K. Nakanishi and O. Meth-Cohn (Eds.). Comprehensive Natural Products Chemistry, Pergamon.

Nahar, K., Kyndt, T., Hause, B., Höfte, M. and Gheysen, G. 2013. Brassinosteroids suppress rice defense against root-knot nematodes through antagonism with the jasmonate pathway. Mol. Plant-Microbe Interact., 26(1): 106–115.

Nakashita, H., Yasuda, M., Nitta, T., Asami, T., Fujioka, S., Arai, Y. et al. 2003. Brassinosteroid functions in a broad range of disease resistance in tobacco and rice. The Plant Journal., 33(5): 887–898.

Nambara, E. 2017. Abscisic Acid. pp. 361–366. In: B. Thomas, B.G. Murray and D.J. Murphy (Eds.). Encyclopedia of Applied Plant Sciences (Second Edition), Academic Press.

Nawaz, F., Naeem, M., Zulfiqar, B., Akram, A., Ashraf, M.Y., Raheel, M. et al. 2017. Understanding brassinosteroid-regulated mechanisms to improve stress tolerance in plants: a critical review. Environ. Sci. Pollut. Res., 24: 15959–15975.

Nemhauser, J.L., Mockler, T.C., Chory, J. and Dangl, J. 2004. Interdependency of brassinosteroid and auxin signaling in Arabidopsis. PLoS Biol., 2(9): e258.

Ohri, P., Kaur, S., Khurma, U.R., Bhardwaj, R. and Sohal, S.K. 2004. Effect of brassinolide on development of *Meloidogyne incognita* in *Raphanus sativus* L. plants. Indian J. Nematol., 34(2): 165–168.

Peleg, Z., Reguera, M., Tumimbang, E., Walia, H. and Blumwald, E. 2011. Cytokinin-mediated source/sink modifications improve drought tolerance and increase grain yield in rice under water-stress. Plant Biotechnol., J. 9: 747–758.

Pieterse, C.M., Leon-Reyes, A., Van der Ent, S. and Van Wees, S.C. 2009. Networking by small-molecule hormones in plant immunity. Nat. Chem. Biol., 5: 308–316.

Pieterse, C.M., Van Pelt, J.A., Ton, J., Parchmann, S., Mueller, M.J., Buchala, A.J. et al. 2000. Rhizobacteria-mediated induced systemic resistance (ISR) in Arabidopsis requires sensitivity to jasmonate and ethylene but is not accompanied by an increase in their production. Physiol. Mol. Plant Pathol., 57(3): 123–134.

Ren, C., Han, C., Peng, W., Huang, Y., Peng, Z., Xiong, X. et al. 2009. A leaky mutation in DWARF4 reveals an antagonistic role of brassinosteroid in the inhibition of root growth by jasmonate in *Arabidopsis*. Plant Physiol., 151: 1412–1420.

Roth, U., Friebe, A. and Schnabl, H. 2000. Resistance induction in plants by a brassinosteroid containing extract of *Lychnis viscaria* L. Zeitschrift für Naturforschung C., 55: 552–559.

Sahni, S., Prasad, B.D., Liu, Q., Grbic, V., Sharpe, A., Singh, S.P. et al. 2016. Overexpression of the brassinosteroid biosynthetic gene DWF4 in Brassica napus simultaneously increases seed yield and stress tolerance. Sci. Reports., 6(1): 1–14.

Saini, S., Sharma, I. and Pati, P.K., 2015. Versatile roles of brassinosteroid in plants in the context of its homoeostasis, signaling and crosstalks. Front. Plant Sci., 6: 950.

Scheer, J.M. and Ryan Jr, C.A., 2002. The systemin receptor SR160 from *Lycopersicon peruvianum* is a member of the LRR receptor kinase family. Proc. Natl. Acad. Sci., 99(14): 9585–9590.

Schumacher, K. and Chory, J. 2000. Brassinosteroid signal transduction: still casting the actors. Curr. Opi. Plant Biol., 3: 79–84.

Shakirova, F., Allagulova, C., Maslennikova, D., Fedorova, K., Yuldashev, R., Lubyanova, A. et al. 2016. Involvement of dehydrins in 24-epibrassinolide-induced protection of wheat plants against drought stress. Plant Physiol. Biochem., 108: 539–548.

She, J., Han, Z., Kim, T.W., Wang, J., Cheng, W., Chang, J. et al. 2011. Structural insight into brassinosteroid perception by BRI1. Nature., 474: 472–476.

She, J., Han, Z., Zhou, B. and Chai, J. 2013. Structural basis for differential recognition of brassinolide by its receptors. Protein Cell., 4: 475–482.

Sheikhi, S., Ebrahimi, A., Heidari, P., Amerian, M.R., Rashidi-Monfared, S. and Alipour, H. 2023. Exogenous 24-epibrassinolide ameliorates tolerance to high-temperature by adjusting the biosynthesis of pigments, enzymatic, non-enzymatic antioxidants, and diosgenin content in fenugreek. Sci. Rep., 13: 6661.

Sreeramulu, S., Mostizky, Y., Sunitha, S., Shani, E., Nahum, H., Salomon, D. et al. 2013. BSK s are partially redundant positive regulators of brassinosteroidsignaling in Arabidopsis. Plant J. 74(6): 905–919.

Srivastava, L.M. 2002. CHAPTER 10—Abscisic Acid. In: Srivastava, L.M. (Ed.). Plant Growth and Development. Academic Press, pp. 217–232.

Szekeres, M., Nemeth, K., Koncz-Kalman, Z., Mathur, J., Kauschmann, A., Altmann, T. et al. 1996. Brassinosteroids rescue the deficiency of CYP90, a cytochrome P450, controlling cell elongation and deetiolation in Arabidopsis. Cell., 85: 171–182.

Tanaka, K., Asami, T., Yoshida, S., Nakamura, Y., Matsuo, T. and Okamoto, S. 2005. Brassinosteroid homeostasis in Arabidopsis is ensured by feedback expressions of multiple genes involved in its metabolism. Plant Physiol., 138(2): 1117–1125.

Tanaka, K., Nakamura, Y., Asami, T., Yoshida, S., Matsuo, T. and Okamoto, S. 2003. Physiological roles of brassinosteroids in early growth of Arabidopsis: brassinosteroids have a synergistic relationship with gibberellin as well as auxin in light-grown hypocotyl elongation. J. Plant Growth Regul., 22(3): 259–271.

Tang, W., Kim, T.W., Oses-Prieto J.A., Sun, Y., Deng, Z., Zhu, S. et al. 2008. BSKs mediate signal transduction from the receptor kinase BRI1 in Arabidopsis. Science, 321(5888): 557–60.

Thapliyal, V.S., Rai, P.N. and Bora, L. 2016. Influence of pre-harvest application of gibberellin and brassinosteroid on fruit growth and quality characteristics of pear (*Pyrus pyrifolia* (Burm.) Nakai) cv. Gola. J. Appl. Natl. Sci., 8(4): 2305–2310.

Tong, H., Xiao, Y., Liu, D., Gao, S., Liu, L., Yin, Y. et al. 2014. Brassinosteroid regulates cell elongation by modulating gibberellin metabolism in rice. Plant Cell., 26: 4376–4393.

van Wees, S., Luijendijk, M., Smoorenburg, I., Van Loon, L.C. and Pieterse, C.M. 1999. Rhizobacteria-mediated induced systemic resistance (ISR) in Arabidopsis is not associated with a direct effect on expression of known defense-related genes but stimulates the expression of the jasmonate-inducible gene Atvsp upon challenge. Plant Mol. Biol., 41(4): 537–549.

Vandenbussche, F., Callebert, P., Zadnikova, P., Benkova, E. and Van Der Straeten, D. 2013. Brassinosteroid control of shoot gravitropism interacts with ethylene and depends on auxin signaling components. Am. J. Bot., 100: 215–225.

Vercruyssen, L., Gonzalez, N., Werner, T., Schmülling, T. and Inzé, D. 2011. Combining enhanced root and shoot growth reveals cross talk between pathways that control plant organ size in Arabidopsis. Plant Physiol., 155(3): 1339–1352.

Wang, Q., Ding, T., Gao, L., Pang, J. and Yang, N. 2012. Effect of brassinolide on chilling injury of green bell pepper in storage. Sci. Hortic., 144: 195–200.

Wang, H., Yang, C., Zhang, C., Wang, N., Lu, D., Wang, J. et al. 2011. Dual role of BKI1 and 14-3-3 s in brassinosteroidsignaling to link receptor with transcription factors. Dev. Cell., 21(5): 825–834.

Wang, W., Bai, M.Y. and Wang, Z.Y. 2014. The brassinosteroidsignaling network-a paradigm of signal integration. Curr. Opin. Plant Biol., 21: 147–153.

Wang, X. and Chory, J. 2006. Brassinosteroids regulate dissociation of BKI1, a negative regulator of BRI1 signaling, from the plasma membrane. Science, 313(5790): 1118–1122.

Wang, X., Goshe, M.B., Soderblom, E.J., Phinney, B.S., Kuchar, J.A., Li, J. et al. 2005b. Identification and functional analysis of in vivo phosphorylation sites of the Arabidopsis Brassinosteroid-Insensitive1 receptor kinase. Plant Cell., 17(6): 1685–1703.

Wang, X., Kota, U., He, K., Blackburn, K., Li, J., Goshe, M.B. et al. 2008. Sequential transphosphorylation of the BRI1/BAK1 receptor kinase complex impacts early events in brassinosteroidsignaling. Dev. Cell., 15(2): 220–235.

Wang, X., Li, X., Meisenhelder, J., Hunter, T., Yoshida, S., Asami, T. et al. 2005a. Autoregulation and homodimerization are involved in the activation of the plant steroid receptor BRI1. Dev. Cell., 8(6): 855–865.

Xia, X.J., Wang, Y.J., Zhou, Y.H., Tao, Y., Mao, W.H., Shi, K. et al. 2009. Reactive oxygen species are involved in brassinosteroid-induced stress tolerance in cucumber. Plant Physiol., 150(2): 801–814.

Xiong, J., Wan, X., Ran, M., Xu, X., Chen, L. and Yang, F. 2022. Brassinosteroids Positively Regulate Plant Immunity via BRI1-EMS-SUPPRESSOR 1-Mediated Glucan Synthase-Like 8 Transcription. Front. Plant Sci., 13: 854899–854899.

Xue, L., Du, J., Yang, H., Xu, F., Yuan, S. and Lin, H. 2009. Brassinosteroids Counteract Abscisic Acid in Germination and Growth of Arabidopsis. Zeitschrift für Naturforschung C., 64(3–4): 225–230.

Yamamoto, Y., Kamiya, N., Morinaka, Y., Matsuoka, M. and Sazuka, T. 2007. Auxin biosynthesis by the YUCCA genes in rice. Plant Physiol., 143(3): 1362–1371.

Ye, H., Li, L. and Yin, Y. 2011. Recent Advances in the Regulation of BrassinosteroidSignaling and Biosynthesis Pathways. J. Integr. Plant Biol., 53(6): 455–468.

Yi, H.C., Joo, S., Nam, K.H., Lee, J.S., Kang, B.G. and Kim, W.T. 1999. Auxin and brassinosteroid differentially regulate the expression of three members of the 1-aminocyclopropane-1carboxylate synthase gene family in mung bean (*Vigna radiata* L.). Plant Mol. Biol., 41: 443–454.

Yu, M.H., Zhao, Z.Z. and He, J.X. 2018. Brassinosteroidsignaling in plant–microbe interactions. Int. J. Mol. Sci., 19(12): 4091.

Yuldashev, R., Avalbaev, A., Bezrukova, M., Vysotskaya, L., Khripach, V. and Shakirova, F. 2012. Cytokinin oxidase is involved in the regulation of cytokinin content by 24-epibrassinolide in wheat seedlings. Plant Physiology and Biochemistry, 55: 1–6.

Zalom, F.G. 2002. Pests, endangered pesticides and processing tomatoes. Acta Hortic., 613: 223–233.

Zhang, D. 2023. Hormonal control of plant stress responses: Brassinosteroids and gibberellin. Front. Plant Sci., 14: 1–3.

Zhang, DW., Deng, XG., Fu, FQ. and Lin, H.H. 2015. Induction of plant virus defense response by brassinosteroids and brassinosteroidsignaling in *Arabidopsis thaliana*. Planta., 241: 875–885.

Zhang, H., He, Y., Tan, X., Xie, K., Li, L., Hong, G. et al. 2019. The dual effect of the brassinosteroid pathway on rice black-streaked dwarf virus infection by modulating the peroxidase-mediated oxidative burst and plant defense. Mol. Plant-Microbe Interact., 32(6): 685–696.

Zhao, Y., Wu, G., Shi, H. and Tng, D. 2019. Receptor-Like Kinase 902 associates with and phosphorylates Brassinosteroid-Signaling Kinase1 to regulate plant immunity. Molecular Plant., 12(1): 59–70.

Zhu, F., Yun, Z., Ma, Q., Gong, Q., Zeng, Y., Xu, J. et al. 2015a. Effects of exogenous 24-epibrassinolide treatment on postharvest quality and resistance of Satsuma mandarin (*Citrus unshiu*). Postharvest Biol. Technol., 100: 8–15.

Zhu, T., Tan, W-R., Deng, X-G., Zheng, T., Zhang, D-W. and Lin, H-H. 2015b. Effects of brassinosteroids on quality attributes and ethylene synthesis in postharvest tomato fruit. Postharvest Biol. Technol., 100: 196–204.

Zhu, W., Wang, H., Fujioka, S., Zhou, T., Tian, H., Tian, W. and Wang, X. 2013. Homeostasis of brassinosteroids regulated by DRL1, a putative acyltransferase in Arabidopsis. Mol. Plant., 6(2): 546–558.

7

Jasmonic Acid
Application for Growth Improvement and Protection against Biotic Stress

Yachana Jha,[1,]* *Heba I. Mohamed,*[2] *M.T. El-Mahdy*[3] *and Mona F.A. Dawood*[4]

Introduction

Plants are the only organism on earth with the potential to utilize sunlight, water, mineral nutrients, and carbon dioxide to carry out all their functions. But plants, being living organisms, have interactions with several abiotic and biotic factors as well as having to survive in an ever-changing natural environment (Wang et al. 2020). The plethora of environmental fluctuations is one of the main causes of reduced crop productivity due to the plant's limited ability to attain its full genetic potential in such conditions. One such environmental condition is the frequent interaction of plants with a diverse group of pathogenic microorganisms (Rassa et al. 2016). There is a vast range of microbial interactions encountered throughout the lifetime of the plants, and these interactions may be deleterious or beneficial, depending on whether the interaction is symbiotic or pathogenic (Jha et al. 2014). The underground root system of the plants continuously exudes important nutrients like carbon containing primary and secondary metabolites, mineral ions, water, mucilage, and enzymes,

---

[1] N. V. Patel College of pure and Applied Sciences, CVM University, V. V. Nagar, Anand (Gujarat), India.

[2] Biological and Geological Science Depertament. Faculty of Education, Ain Shams University, Cairo, Egypt.

[3] Department of Pomology, Faculty of Agriculture, Assiut University, Assiut 71526, Egypt.

[4] Botany and Microbiology Department, Faculty of Science, Assiut University, Assiut, 71516, Egypt.

* Corresponding author: yachanajha@ymail.com

which initiate diverse interactions of the plant root with different microorganisms (Walker et al. 2023).

Plant-microbe interaction is one of the important communications that characterizes the root zone below ground. Among these interactions, when the interaction of pathogenic microbes with plants takes place, it will be counted as the main biotic stressor (Jha 2019a). A two-way communication process needs to be established among the plant and pathogen upon plant-pathogen interactions, where plants develop the potential to recognize the pathogen for the activation of its defense system. Additionally, the pathogen develops the potential to modulate the plant metabolism to generate a favorable condition for its establishment and growth (Rojas et al. 2014).

The group of biotic stressors for plants includes fungi, bacteria, viruses, insects, nematodes, and herbivores. All these biotic stressors are mobile, but plants, being sessile organisms, require multilevel response mechanisms to resist, mitigate, or recover from the adverse effects of biotic stressors (Gimenez et al. 2018). Plants lack a well-defined immune system like animals and totally depend on systemic signals produced or the innate immunity of individual cells during infection (Jha 2022). Plant pathogens are classified on the basis of pathogen lifestyle into biotrophic, necrotrophic, and hemibiotrophic groups, which apply different mechanisms for infecting and sustaining the host plant (Rajarammohan 2021).

Based on the microbe-induced mechanism, plants regulate their innate immune system and opt for a specific response against the pathogen. Plants develop a complicated immune response strategy to protect themselves from pathogen challenge (Anderson 2023). The majority of pathogens have been blocked by the front-line physical barriers of the plant, such as rigid cell walls, antimicrobial secondary metabolites, and waxy cuticles (Jha 2019b). After successfully breaking these physical barriers, pathogens encounter an efficient immune system of the plant, which induces them to check the progression of pathogen colonization in the plant (Vasvi et al. 2021). Plants involve many fundamental biological tools to perceive such interaction and categorized it as helpful or deleterious to activate specific responses or signals. One major signaling biomolecule that exists in plants is known as plant hormones, which underlie numerous responses according to environmental signals. There are about nine natural plant hormones that have been reported to act in plants according to environmental signals. These phytohormones are ethylene, abscisic acid, gibberellins, auxin, cytokinin, brassinosteroid, strigolactone, salicylic acid, and jasmonic acid (Jha 2020). Among these phytohormones, jasmonic acid is omnipresent in higher plants, so it needs higher attention to explore its action on plant defense mechanisms and stress responses (Wang et al. 2020). The aim of this chapter is to explore the role of jasmonic acid in the induction of plant defense responses to overcome biotic stress.

Biotic Stress and Plant Defense Responses

Plants continuously encounter different types of biotic stresses and are able to protect themselves through diverse mechanisms at the molecular, biochemical, and morphological levels, which remain coordinated by different signaling pathways

(Akbar et al. 2023). Biotic stresses in plants are commonly caused by bacteria, fungi, viruses, nematodes, insects, pests, and herbivores, primarily as pathogens, pests, or parasites. Among these, only bacteria, fungi, viruses, and nematodes are considered pathogens and responsible for plant diseases (Nazarov et al. 2020). The fungi have different modes of infection, as the group of fungi that will produce toxin to kill the host for utilizing the cell content as a source of nutrients is nectrotrophs, or the group of fungi that will not kill the host but feed on living host cells is biotrophs (Shuping and Eloff 2017).

Fungus, along with bacteria, is responsible for causing many types of symptoms in plants, like cankers, leaf spots, and vascular wilts, and also infecting diverse plant parts (Jha 2022a). Viruses not only cause local lesions but also cause systemic damage, resulting in malformations, wilting, chlorosis, and stunting, although viruses rarely kill their hosts (Hayano-Saito and Hayashi 2020). Nematodes feed on plant parts to uptake plant cell content as nutrients, and parasitic nematodes attack plant roots to cause soil-borne diseases (Bahadur 2021). Nematodes are responsible for causing nutrient deficiency in plants, which causes wilting or stunting in infected plants (Ahmad et al. 2021, 2022). While pests like insects and mites cause damage to plants by piercing, sucking, feeding, and lying eggs, they also act as vectors for the transmission of plant viruses (Qiu et al. 2023).

Jasmonates (JA and methyl jasmonates (MeJA) are lipid-derived bioactive chemicals that are formed through the oxylipin biosynthetic pathway. These molecules play a significant role in enhancing plant tolerance to necrotrophic pathogens, herbivores, mechanical injury, and various environmental stresses (Eng et al. 2021). From seed germination to fruit ripening and finally to senescence, JA affects a number of cellular and developmental processes (Ruan et al. 2019). Additionally, JA modifies plant physiology and activates a number of signaling pathways and antioxidant defense mechanisms that are intimately linked to increasing plant immunity and giving tolerance to stress (Wang et al. 2020).

The biochemically connected jasmonic acid pathway reduces reactive oxygen species (ROS) and malondialdehyde (MDA) content caused by biotic stress; together, they both synergistically activate the plant's enzymatic and non-enzymatic antioxidant mechanisms. Antioxidant enzymes that are upregulated by JA mutually include catalase (CAT), peroxidase (POX), superoxide dismutase (SOD), ascorbate peroxidase (APX), glutathione peroxidase (GPX), and glutathione reductase (GR), whereas glutathione, ascorbate, phenolic compounds, flavonoids, α-tocopherol, soluble sugar, and amino acids are non-enzymatic antioxidants that are produced by JA together during stress for scavenging ROS (Wang et al. 2019, 2021).

All these biotic stresses are collectively responsible for massive crop loss and productivity (Ahmad et al. 2022). Plants developed a complex immune system to deal with such biotic stresses for their survival. Plants possess a passive front line of defense mechanisms such as physical barriers of specialized trichomes, thick cuticles, and waxes to check the entry of pathogens or chemical barriers to prevent pathogens or insects from attacking plants (Arya 2021). The second level of plant defense mechanisms includes pathogen recognition for the induction of plant defense responses. Pathogen recognition in plants takes place by cell surface receptors called pattern recognition receptors, which are the site for binding of pathogens on the plant cell surface with the help of receptor-compatible molecules known as pathogen-associated molecular patterns (Greenwood et al. 2023). Upon interaction of pathogenic molecules with this receptor, they will induce plant defense to check their progression in plant cells. After that, the second level of plant defense has been carried out with the help of plant resistance proteins, which will recognize specific pathogen effectors called Avr proteins for the efficient and effective activation of plant defense mechanisms (Jha 2019c).

The plant defense mechanisms activated upon the entry of a plant pathogen generally activate hypersensitive responses, which include programmed cell death in the infected and surrounding cells to block the spreading of infection or localized clearance of the pathogen. The hypersensitive responses are mediated by resistance genes that encode proteins, and activation of the hypersensitive response in plants results in increased resistance to a broad range of pathogens for an extended period of time (Balint-Kurti 2019). This incident is classified as plant systemic resistance and is divided into systemic acquired resistance (SAR) and induced systemic resistance (ISR). Systemic resistance is induced by non-pathogenic microbes, and systemic acquired resistance is induced by pathogenic microbes. In systemic acquired resistance, accumulation of salicylic acid takes place to trigger the production of pathogenic-related proteins, while non-pathogenic microbes induce plant immunity by inducing systemic resistance (Jha and Subramanian 2012). The activation of systemic acquired resistance by non-pathogenic microbes is ethylene and jasmonic acid (JA) pathway-dependent. The activation of both salicylic acid and jasmonic acid signaling pathways by non-pathogenic microbes indicates the existence of a complex multidirectional signaling pathway required for the activation of systemic resistance in plants.

Jasmonic Acid Biosynthesis in Plant

The stress-related responses and plant growth have been regulated by the establishment of signaling networks among the plant hormones. Jasmonic acid, its isoleucine conjugate (JA-Ile), and its methyl ester (MeJA) are derivatives of fatty acids and are jointly known as jasmonates, which play a significant role in the growth and developmental processes of plants and are considered stress-related hormones (Lemos et al. 2016). Jasmonic acid is an endogenous chemical produced by all higher plants at three distinct sites, that is, chloroplast, peroxisome, and cytoplasm, by opting for the hexadecane pathway by using hexadecatrienoic acid as a precursor or the octadecane pathway by using α-linolenic acid as a precursor (Fig. 1). The biosynthesis of jasmonic acid started in the chloroplasts with the release of polyunsaturated fatty acids like α-linolenic acid with the help of lipid hydrolyzing enzymes from the chloroplast membranes (Ali et al. 2020).

Upon the release of α-linolenic acid from galactolipids, the synthesis of 12-oxo-phytodienoic acid (12-OPDA) or deoxymethylated dienic acid (dn-OPDA) from polyunsaturated fatty acids takes place in the chloroplast with the help of a set of enzymes, including galactolipase, 13-lipoxygenase (LOX), allene oxide synthase, and allene oxide cyclase. Then the final intermediate 12-oxo-phytodienoic acid has been exported to the peroxisomes from the chloroplast, where it is finally converted

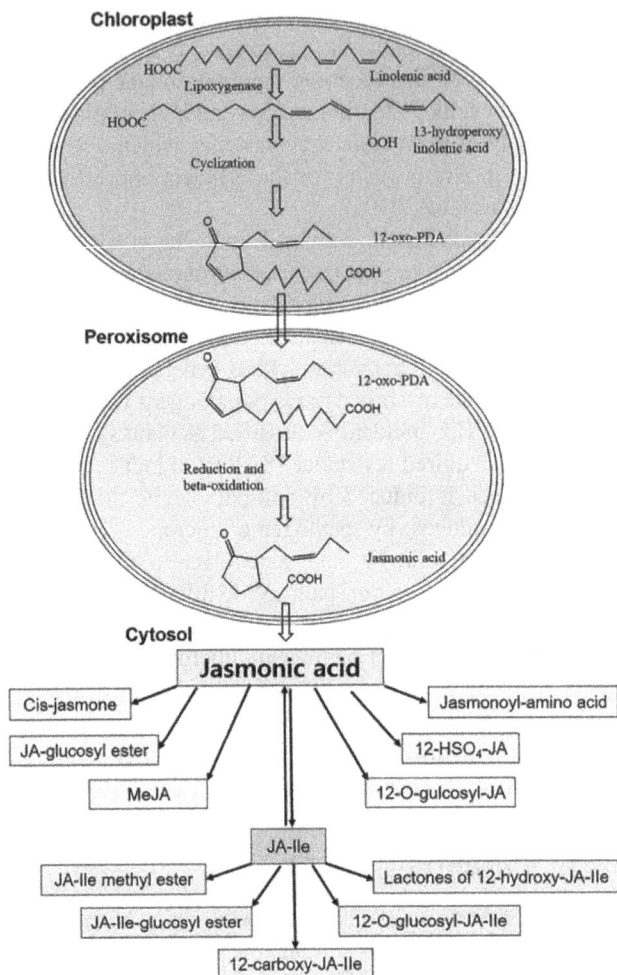

Fig. 1. Biosynthesis, metabolism, transport of jasmonic acid, and jasmonic acid signaling response pathway in plant cells (Ali et al. 2020).

into (+)-7-*iso*-JA by the help of enzyme 12-oxo-phytodienoic acid reductase (OPR), and then by the three consecutive beta oxidation reactions by using multi-functional enzymes of the peroxisome, that is, acyl-CoA oxidase and l-3-ketoacyl-CoA thiolase, it will form diverse intermediate (Wasternack and Hause 2019). The final product of peroxisome (+)-7-*iso*-JA by general epimerization forms a more stable *trans* configuration, which undergoes different modifications to produce diverse JA derivatives in the cytoplasm. In the cytoplasm, JA is catabolized for the production of different structures by different chemical reactions, such as MeJA, JA-Ile, *cis-jasmone,* and 12-hydroxyjasmonic acid (Gareth 2020).

Jasmonic Acid Signaling in Plant

Induction of jasmonic acid in plants takes place naturally as a signaling molecule for a vast range of physiological activities and under stress. The jasmonic acid signal is possibly transduced by the activation of receptors that interact with it. The defense responses in plants are stimulated by both jasmonic acid and 12-oxo-phytodienoic acid, while the plant's physiological activity is stimulated only by jasmonic acid. It indicates that a minimum of two transducer pathways exist for jasmonic acid signaling in plants. The jasmonic acid precursor 12-oxo-phytodienoic acid, the free jasmonic acid, the methyl ester of jasmonic acid, that is, MeJA, and conjugates of jasmonic acid are considered to be the most active forms of jasmonic acid in plants, and it is the only proven ligand for jasmonic acid signaling in plants (Fig. 2) (Ruan et al. 2019).

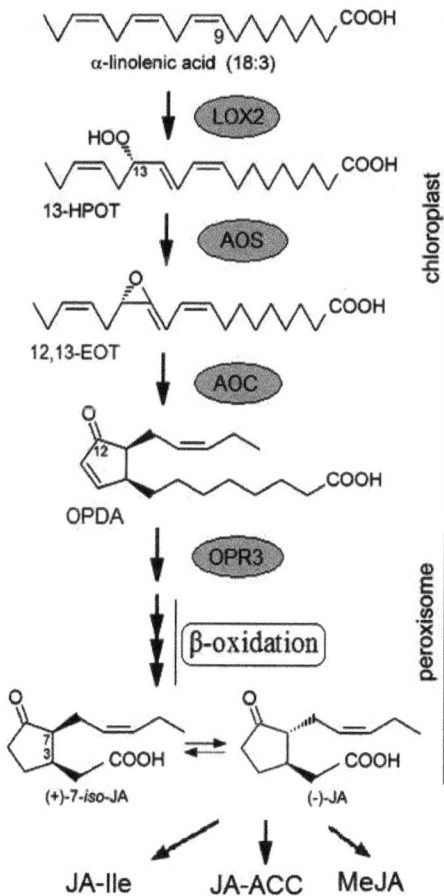

Fig. 2. Jasmonic acid mediated signaling in plant for growth and resistance against stress (Ruan et al. 2019).

Jasmonic acid and the methyl ester of jasmonic acid are recognized as the most bioactive forms in plants, as they are present in abundance and participate in several physiological activities. The primary signal transduction process for jasmonic acid perception is based on a basic helix-loop helix domain containing transcription factors (TFs). Jasmonic acid perception, the signal transduction process, assembles on the basic-helix-loop helix domain of transcription factors. The basic-helix-loop helix transcription factors form a complex with the promoter designated as G-box of the jasmonate-responsive gene (Zhang et al. 2011). Jasmonate-response is under negative regulation by jasmonate-zim-domain (JAZ) proteins, which act as repressors to regulate the expression of the jasmonate-responsive gene. Jasmonate-zim-domain repressors control jasmonic acid response under stress conditions, and jasmonate-zim-domain interacts with coronatine insensitive 1 (COI1) (Fig. 3). This complex has been included in the co-reception of biologically active jasmonic acid. So, the binding of basic-helix-loophelix-related transcription factors is repressed by homo- or heterodimers of jasmonate-zim-domain proteins (Zhao et al. 2018).

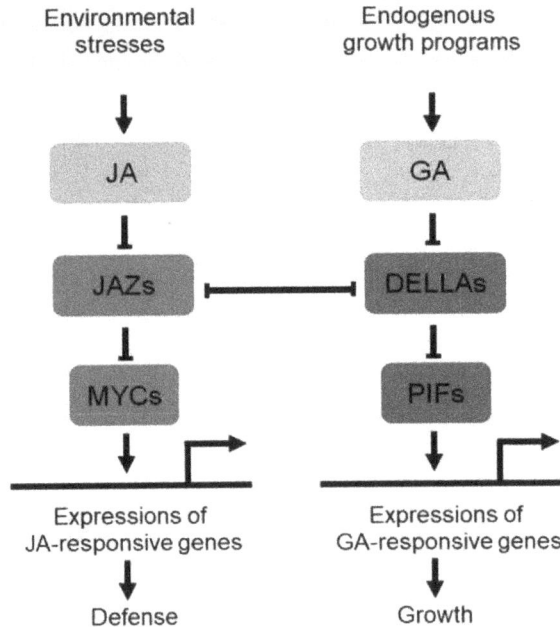

Fig. 3. Jasmonic acid mediated effect on plant through different responsive gene (Jang et al. 2020).

Cross-talk between JA and Other Plant Hormone Signaling Pathways

Plants act as sources of nutrition for several other terrestrial organisms on earth. So, the herbivores and pathogens continuously attack the plant for food and niche, which constantly imposes a negative effect on plant survival. During evolution, for the adaptation of plants to such conditions, plants develop an amazing range of specialized plant defense mechanisms, which may be repellant, toxic, or anti-nutritional for the pathogens. Other indirect defense mechanisms are also developed by the plant, such as chemical compounds produced by the plant that attract natural enemies of such plant pathogens (Bali et al. 2019).

The coordination mechanism of such local and systemic defense responses in plants provides the potential for the plant to develop resistance against a broad spectrum of pathogen attacks, collectively known as induced immunity (Jha et al. 2021). The plant-pathogen interaction has been mediated by a cell surface-localized pattern recognition receptor, which specifically interacts with the pathogenic foreign proteins known as pathogen-associated molecular patterns. Another method for induced resistance is based on the efficiency of plant intracellular resistance proteins to recognize and counteract the toxic protein or molecule induced by the plant pathogen in the plant. It also includes various other plants signaling components like the production of ROS, activation of MAP kinase pathways, calcium-dependent signaling cascades, etc. (Jha 2022b).

However, the effectiveness of the induction depends on the strength and timing of the induction and decides the fate of plant-pathogen interaction. But the advantage of such induction under mild infection for plants is that once it induces, it will remain effective against a broad range of plant pathogens. Among all such players in plants, the phytohormone network is a central chemical signaling molecule that will induce the chemical and morphological defense activity of the plant under adverse conditions, including pathogen attacks. In plants, the diverse danger signals converge on the immune-promoting effects of two major defense hormones, jasmonic acid and salicylic acid (Pacifici et al. 2015).

Role of Jasmonic Acid in Plant Growth and Development

Many aspects of plant growth and development are influenced by jasmonic acid. The jasmonic acid modulates many physiological activities of the plants, like inhibiting the germination of seeds and activating abscission and senescence of leaves after sensing adverse environmental conditions (Jarocka-Karpowicz and Markowska 2021). The leaf senescence response induced by jasmonic acid in a COI1-dependent manner includes the degradation of important chloroplast enzymes and proteins like RUBISCO, the loss of chlorophyll pigments, and the accumulation of new proteins, which activate the inactivation of functional chloroplast enzymes (Sarwat et al. 2013). The large subunit of RUBISCO enzyme has been coded by the chloroplast gene, and the cleavage of the transcript for RUBISCO enzyme has been initiated by jasmonic acid; in this way, jasmonic acid inhibits the translation of the large subunit of RUBISCO enzyme (Jha and Mohamed 2022).

Jasmonic acid-mediated leaf senescence takes place by upregulating the expression of senescence-associated genes and genes for enzymes responsible for chlorophyll catabolic activity, as well as by downregulating photosynthesis-related genes (Table 1). Stomatal functions are necessary for photosynthesis, which is the site of gaseous exchange, and its opening has been regulated by pairs of guards. Stomata also regulate water loss through transpiration and play a critical role in plant immunity to pathogens (Jha et al. 2022). The mechanism behind the jasmonic acid-mediated closure of stomata is that methyl-jasmonic acid interacts with coronatine insensitive1

Table 1. Effect of jasmonic acid in plant for sustains growth under biotic stress.

Effect of Jasmonic Acid	Category of Response	Reference
Root Elongation	Growth	Yang et al. 2017
Root Hair development		Xu et al. 2019
Stamen Development		Acosta and Przybyl 2019
Leaf Senescence		Abid et al. 2019
Wounding Response	Biotic	Lee et al. 2020
Pathogen defense		Gupta et al. 2020
Anthocyanin Accumulation		Ting et al. 2014
Flowering Repression		Browse and Wallis 2019
Herbivorous Resistance		Ricardo et al. 2016

for the activation of plasma membrane H^+-ATPase and activates the influx of Ca^{2+}, efflux of Cl^-/H^+, and generation of ROS (Zhao et al. 2023). Simultaneously, it will also activate outward K^+ channels for the efflux of K^+, which causes a loss of turgor in guard cells and results in stomatal closure.

Jasmonic acid competitive antagonist coronatine-O-methyloxime can block the interaction of JAZs and COI (Fig. 4), which antagonizes the inhibitory effect of coronatine on primary root growth and enhances the growth of the primary root of the plants (Yang et al. 2017). Jasmonic acid also has a role in the formation of lateral and adventitious roots. Jasmonic acid promotes lateral root formation by upregulating the expression of the ethylene-associated amphiphilic repression domain, which binds to and activates the promoters of the auxin biosynthetic genes. In contrast, jasmonic

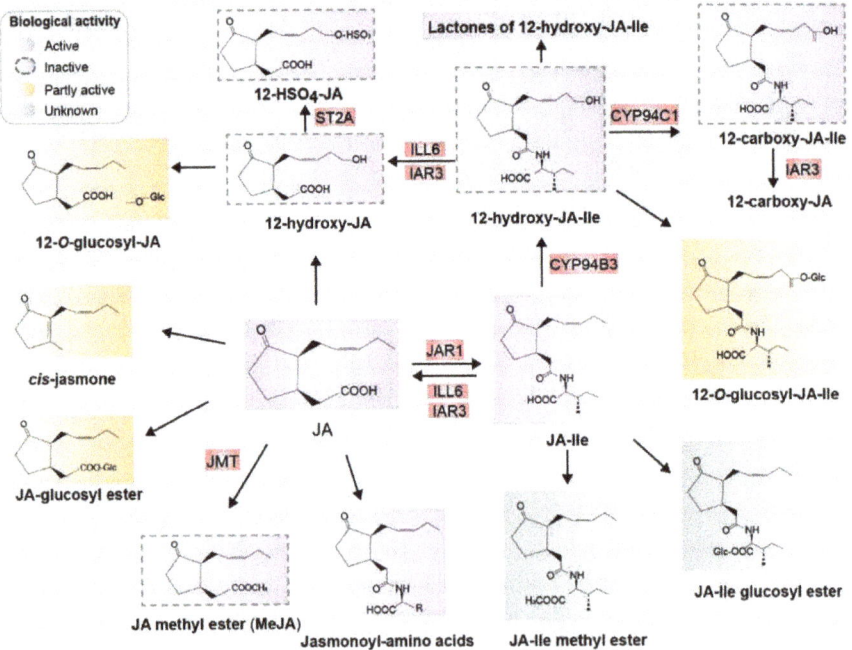

Fig. 4. Major jasmonates arising from the metabolic conversion of JA (Wang et al. 2021).

acid negatively regulates adventitious root formation in the coronatine-mediated cascade. There are reports showing that enhanced accumulation of jasmonate has been observed in actively dividing and growing plant cells and tissues like developing flowers, seed pods, tuber formation, hypocotyl hooks, and tendril coiling (Siddiqi and Husen 2019). OPDA is an endogenous hormone responsible for tendril coiling owing to its dramatic increase during the initiation and progression of coiling.

At the same time, jasmonic acid also has an inhibitory effect on plant growth under adverse environmental conditions. Jasmonic acid represses leaf expansion by inhibiting the activity of the mitotic cyclin CycB1,2 and cell division rather than by affecting cell size. The coronatine-insensitive 1-jasmonate-zim-domain cascade mediates the jasmonic acid-induced inhibition of the leaf expansion gene. Jasmonic acid inhibits flowering in plants under stress to delay the transition of the plant from vegetative to reproductive growth. The jasmonic acid showed its inhibitory effect on plant flowering, which is mediated through the COI1-JAZ cascade. The repressing effect of jasmonic acid on plant growth parameters is a suitable strategy for the enhanced survival of the plant in natural environments, which allows plants to divert the metabolite for acquiring the plant and enhances focus on defending against various stresses. In this way, jasmonic acid's inhibitory effect on plant growth parameters directly promotes plant defense responses (Montejano-Ramírez et al. 2020).

Jasmonic Acid-Triggered Immunity

It is a multi-level process where the highly conserved core of the jasmonic acid module has the ability to recognize diverse range of effectors molecules for the induction of a specific defense pattern to chemical diversity and pathogen species interaction to overcome pathogen attack (Campos ct al. 2014). The main role of jasmonic acid in induced immunity has three major activities: first, upon pathogen attack or tissue injury, the rapid synthesis of jasmonic acid with its receptor-active derivative like jasmonoyl-L-isoleucine takes place in the plant, followed by a systemic response in the entire plant as per the affected tissue and types of eliciting signals (Fragoso et al. 2014). The second expression of nearly all major proteins and classes of secondary metabolites has been promoted by jasmonic acid, which includes pathogenesis-related proteins, anti-nutritional proteins, phenylpropanoids, terpenoids, alkaloids, phenolics, and amino acid derivatives (De Geyter et al. 2012, De Vleesschauwer et al. 2013, Jha and Subramanian 2021). Third, the development of morphological structures like the formation of resin ducts, nectaries, and glandular trichomes is also promoted by the jasmonic acid, which secretes highly diverse chemical compounds having a direct or indirect role in plant defense against pathogens (Campos et al. 2014). Jasmonic acid-triggered immunity is more effective for necrotrophic or tissue-consuming pathogens, whereas biotrophic pathogens are subjected to the effects of salicylic acid-triggered immunity (Kou et al. 2021).

Jasmonic acid promotes resistance towards a wide range of pathogens in both monocot, dicot, and gymnosperm species. Jasmonic acid-mediated signal transduction is often coupled with the perception of danger signals by the cell surface receptors and the expression of jasmonic acid-responsive defense genes (Campos et al. 2014). The jasmonic acid-mediated signal transduction pathway activation depends on the formation of a conserved core domain for the formation of a complex with jasmonic acid-Ile. The unique feature of jasmonic acid-Ile-mediated signal transduction pathway-triggered immune responses is the endogenously rapidly reversible accumulation of immune signals in vegetative tissues, a potent site for pathogen attack (Marquis et al. 2022). The synthesis of jasmonic acid starts in the plastids with the C18 linolenic acid precursor, which gets converted into a functional intermediate, cyclic 12-oxo-phytodienoic acid. Then, by subsequent reduction and oxidation, cyclic 12-oxo-phytodienoic acid gets converted into jasmonic acid in the peroxisome. However, jasmonic acid is conjugated with Ile in the cytosol for the formation of jasmonic acid-ile. Many genes required for jasmonic acid-Ile biosynthesis are coordinately upregulated by the jasmonic acid signaling pathway (Wang et al. 2021). Jasmonic acid has a crosstalk with other plant hormones to stimulate a complex signaling network in plants and regulate plant hormone signaling pathways, which is a central mechanism of plant stress response. Jasmonic acid never acts independently but always acts in coordination with other plant hormones in a complex signaling network. There are several cross-talks of jasmonic acid with different phytohormones like auxin, abscisic acid, salicylic acid, ethylene, brassinosteroids, and gibberellins in different signaling pathways.

Jasmonic acid induced biosynthesis of auxin in plant under biotic stress

The growth and development of plants under stress are coordinatedly regulated by jasmonic acid and auxin phytohormones (Yang et al. 2019). This signaling pathway is carried out with the help of coronatine insensitive1 and jasmonate-zim-domain, which will act as activators. After the induction of auxin hormone in the plant, the induction of auxin-ARF signaling takes place to induce the synthesis of jasmonic acid in plants under stress. Endogenous jasmonic acid induction in plants activates the expression of the auxin synthase gene to sustain the growth of plants under stress (Xu et al. 2020). So, jasmonic acid modulates the expression of the jasmonate-zim-domain for the modulation of the biosynthesis of auxin specifically in the plant root, which is the most potent site for pathogenic interaction.

The endogenous jasmonic acid induction initiated the formation of coronatine insensitive1 and the jasmonate-zim-domain complex, resulting in the inactivation of the jasmonate-zim-domain, which will allow the activation of the gene for the induction of flower development. While effectors of auxin signaling pathways are responsible for regulating the endogenous level of jasmonic acid to induce the growth of petal and stamen (Acosta and Przybyl 2019). Thus, the coordinated auxin and jasmonic acid signaling pathways are able to modulate the two important growth parameters of the plant under biotic stress.

Jasmonic acid induced coordination with abscisic acid in plant under biotic stress

Jasmonic acid and abscisic acid signaling pathways are responsible for regulating plant defense and growth under stressful conditions (Chen et al. 2016). This signaling pathway has several participants, like the jasmonate-zim-domain of jasmonic acid and abscisic acid receptor pyrabactin resistance-1-like proteins, which are responsible for modulating metabolic reprogramming through coordination with the jasmonic acid signaling pathway. The complex of jasmonate-zim-domain and abscisic acid receptor pyrabactin resistance 1-Like proteins have developed the potential for transcriptional activation of the expression of jasmonic acid-responsive genes to provide resistance against pathogens and insects attacks (Zhang et al. 2020).

The endogenous abscisic acid production will block the jasmonate-zim-domain of jasmonic acid to modulate the binding of different effectors of the abscisic acid to carry out specific physiological activity of the plant under stress. Thus, the interaction between jasmonic acid and abscisic acid in the signaling pathway by different effectors, especially to coordinate the balance between plant growth and defense is responsible for supervising elicitor-induced reprogramming of the plant metabolic pathway to sustain plant growth under biotic stress (Martin et al. 2020). Therefore, jasmonic acid and abscisic acid signaling pathways coordinately regulate plant responses to pathogen infection while antagonizing plant growth and development to sustain plant growth in stressful environments.

Jasmonic acid-induced coordination with salicylic acid in plants under biotic stress

Salicylic acid is an important plant defense hormone that helps in the growth and development of the plant under different environmental conditions, modulates important physiological activities of the plant, and induces a better tolerance and defense response in plants under biotic stress conditions (Luo et al. 2019). It induces system-acquired resistance in plants to overcome the adverse effects of environmental stress. Salicylic acid is a component of system-acquired resistance, which is responsible for broad-spectrum resistance against biotrophic pathogens, while induced systemic resistance is associated with potentiated expression of the jasmonate-responsive gene and is active against necrotrophic pathogens (Ricardo et al. 2016).

Jasmonic acid synthesis in plants antagonizes the accumulation of salicylic acid by modulating a series of transcriptional factors, and these transcriptional factors act as repressors for the salicylic acid biosynthesis gene. The interaction of salicylic acid and jasmonic acid signaling pathways involves several components like redox regulators like glutathione, thioredoxin, mitogen-activated protein kinase, etc. Mitogen-activated protein kinase positively regulates effectors of the salicylic acid signaling pathway and negatively regulates effectors of the jasmonic acid-responsive genes. Mitogen-activated protein kinase is one of the main components of the JA and SA signaling pathways, which coordinately induce resistance against necrotrophic or biotrophic pathogens (Huang et al. 2023). Salicylic acid is responsible for the activation of early defence-related gene expression, while jasmonic acid is responsible for the activation of late defence-related gene expression in infected plants.

Jasmonic acid induced coordination with ethylene in plant under biotic stress

Jasmonic acid and ethylene synergistically or coordinately modulate plant biotic stress responses by involving several effectors in both pathways. The jasmonate-zim-domain of the jasmonic acid signaling pathway and ethylene insensitive 3 or its homologue of the ethylene signaling pathway are involved in this interaction of the two phytohormones (Wang et al. 2021). In this interaction, the jasmonate-zim-domain is the site for the binding of a repressor for the transcription of ethylene insensitive3 of the ethylene signaling pathway as well as the activator binding site to promote the transcription of the plant defense in expression and develop resistance against necrotrophic pathogen infection (Gupta et al. 2020). So, the jasmonic acid signaling pathways coordinately interact with the ethylene signaling pathway to regulate plant defense responses against necrotrophic pathogens by modulating the expression of plant defense proteins.

Jasmonic acid induced coordination with brassinosteroids in plants under biotic stress

The phytohormone brassinosteroids plays a significant role in plant development and growth. Brassinosteroids are uniformly distributed in all growing plant tissue and present in higher concentrations in pollen, fruits, and seeds (Zhang et al. 2019). So, the brassinosteroids is responsible for the regulation of a broad range of plant growth responses, including the development of roots, differentiation of vascular tissues, elongation of hypocotyls, transition of floral activity, development of the anther, and pollen grain maturation (An et al. 2023). So, brassinosteroids hormone is responsible for the induction of plant growth, while jasmonic acid inhibits plant growth. The interaction between brassinosteroids and jasmonic acid signaling pathways is responsible for the establishment of a balance between defense response and plant growth (Yang et al. 2017).

The concentration of brassinosteroids in the plant has significantly affected the physiological activity of the plant. For example, brassinosteroids in high concentration are responsible for the activation of BR signaling cascades including brassinosteroids-related transcriptional factors, brassinosteroids-related receptors, and brassinosteroids-related kinases for the regulation of plant defense responses, while in low concentration they stimulate the expression of genes for early or late biosynthesis of brassinosteroids and the accumulation of anthocyanin to induce plant defense responses (Nawaz et al. 2017). The coordination in the two pathways is like endogenous jasmonic acid inhibiting the biosynthesis of brassinosteroids and endogenous brassinosteroids inhibiting the biosynthesis of jasmonic acid to establish coordination between plant growth and defense response under biotic stress (Yang et al. 2019).

Jasmonic acid induced coordination with gibberellins in plant under biotic stress

The naturally existing plant hormone gibberellins are recognized as essential plant growth regulator responsible for the induction of plant cell division and elongation for plant growth. The gibberellins and jasmonic acid signaling pathways antagonistically and coordinately modulate defense response and plant growth, but the plant defense response is established at the cost of plant growth (Browse and Wallis 2019). This interaction between gibberellins and jasmonic acid has taken place with the help of effectors like the jasmonate-zim-domain of the jasmonic acid pathway and many transcriptional factors of the gibberellin biosynthesis gene to coordinate the biosynthesis of these two phytohormones and the growth and defense response of the plant under biotic stress (Chai et al. 2022).

Jasmonic Acid Regulates Plant Response to Biotic Stress

Plants regularly get exposed to many biotic stresses like pathogenic microbes, parasitic microbes, herbivores, and insects, and to sustain their existence, plants require not only to maintain growth but also to develop resistance against all such pathogenic interactions (Table 2). Plants, being sessile, develop multilayer, sophisticated response mechanisms to acquire resistance, recover, and mitigate the adverse effects of biotic stress. When plants are affected by biotic stress, they receive information about the stress signals and respond accordingly (Jha 2018a). The fundamental regulator molecule, which directly senses and transmits adverse environmental signals for successive activation of specific plant defense responses, is naturally produced by the plant and is called a phytohormone. Plant hormones are non-toxic organic chemical compounds naturally produced by the plant in very low concentrations. They are of different types and act as signaling networks that regulate stress responses. Among all these hormones, jasmonic acid is an important hormone that plays a very important role in developing resistance against various biotic stresses, mainly pathogens or insects.

Role of jasmonic acid against insect as biotic stress

Plants being sessile have constantly faced threats from a multitude of insects and pests, responsible for severe crop loss globally (Jha and Mohamed 2022b). At the same time, due to its inability to escape, it also developed highly sophisticated defense systems to protect it from all such attacks. Defence against such insect attack has induced many molecular signals in plants, resulting in the induction of jasmonic acid. Plants are continuously attacked by several different types of insects, which directly feed on plant parts and are responsible for the rapid increase in endogenous jasmonic acid levels (Salazar-Mendoza et al. 2023). Jasmonic acid is a lipid-derived plant hormone that plays a major role in regulating plant defensive responses against insect attack. Invasion of plants by insects results in the elicitation of a wide variety of chemical signals, which are recognized by the plants for the induction of immune responses (Jha 2019d). Some chemical toxins injected by insects or produced by

Table 2. Effect of jasmonic acid against bacterial, fungal, insects, viral, and nematode infections.

Biotic Stress	Effect of Jasmonic Acid	Reference
Insects	Accumulation of defensive proteins, which will affect the digestive physiology of insect.	Jang et al. 2020
Insects	By stimulating the defense signaling pathway, SA and JA therapy reduces the growth of insects, pests, and bugs while also enhancing plant growth, development, and yield attributes.	Stella de Freitas et al. 2019
Insects	Increased synthesis of harmful proteins and induced plant defense mechanisms.	Zhang et al. 2015
Insects	Additionally, trypsin protease inhibitor (TrypPI), POD, and PPO activation are all dependent on COI1.	Ye et al. 2012
Insects	Following the consumption of xanthotoxin-containing diets, carboxylesterase (CarE) activities, glutathione-S-transferase (GSTs) activities, and cytochrome mono-oxygenases (P450s) content increased more. Larvae exposed to MeJA fumigation also displayed higher enzyme activity in a dose-dependent manner, with lower and medium concentrations of MeJA producing higher detoxification enzyme activities than higher concentrations of MeJA. Additionally, MeJA enhanced the growth of larvae fed on diets with lower (0.05%) doses of xanthotoxin and the control diet without toxins.	Chen et al. 2023
Herbivores	OsAOS1 and OsAOS2 both participate in JA biosynthesis brought on by herbivores.	Zeng et al. 2021
Herbivores	Treatment of soybean plants with sodium nitroprusside and methyl jasmonate might improve their capacity to fend off infection by cotton leaf worms. In comparison to sodium sodium nitroprusside, methyl jasmonate was more efficient.	Mohamed et al. 2016
Herbivores	Treatment of soybean plants with sodium nitroprusside and methyl jasmonate would be effective enhancing the ability of these plants to resists cotton leaf worm infection and increased secondary metabolites and antioxidants responsible for plant defense.	Mohamed et al. 2021
Herbivores	As a direct signaling molecule, jasmonic acid significantly increases the defense responses to *Spodoptera exigua* infestation in the plants under study; nevertheless, methyl jasmonate may have a secondary effect by causing JA buildup. In both test plants, where its level is significantly lower than JA and MeJA, JA-L-Ile demonstrated a less effective function in defense responses to *S. exigua* attack.	Al-Zahrani et al. 2020
Fungi	Modulate plant signaling molecule to activate plant defense responses.	He et al. 2018
Fungi	Together, SA and JA serve a critical role in activating the plant's defense mechanisms. They also stimulate growth, signaling pathways, and TFs that reduce pathogenic infection.	Luo et al. 2019
Fungi	Together, SA and JA upregulate the activity of enzymes like CAT, PAL, and PPO. They also both activate the MAPK, WRKY, Lexyl2, and atpA pathways, which reduce ROS and MDA levels.	Wang et al. 2019
Fungi	Many signaling genes associated with JA were expressed more frequently to increase resistance. BnLOX2, BnAOS, and BnPDF are all included.	Wang et al. 2012
Fungi	Significantly reduced lipid peroxidation and H_2O_2 concentration levels	Brenya et al. 2020

Table 2. contd. ...

... Table 2. contd.

Biotic Stress	Effect of Jasmonic Acid	Reference
Fungi	TaJAZ1 overexpression increased TaPR1/2 expression, a pathogenesis-related gene, to protect against Bgt.	Jing et al. 2019
Bacteria	Jasmonic acid inhibit salicylic acid mediated plant defense to suppress bacterial infection by downregulating pathogen associated molecular pattern.	Gupta et al. 2020
Nematode	Jasmonic acid by modulating the activity and expression of antioxidative enzymes restrict the nematode infection.	Martínez-Medina et al. 2017
Nematode	Together, SA and JA increase the expression of defence-related genes, such as LRR receptor-like serine/threonine-protein kinase, serine/threonine-protein kinase, and enzymatic antioxidant, which enhance plant immunity against nematodes.	Du et al. 2020
Virus	Jasmonic acid induce specific transcription factor to initiate the ARGONAUTE 18-mediated host defense network in plant against virus.	Zhang et al. 2016
Virus	Exogenous application of MeJA is effective for inducing MYMIV tolerance in *Vigna mungo*, as evidenced by the increased expression of the defence marker genes lipoxygenase and phenylalanine ammonia-lyase and the decreased expression of *V. mungo* Yellow Mosaic India Virus (MYMIV) coat-protein encoding gene after MYMIV infection.	Chakraborty and Basak 2018

wounded host plants cells themselves are categorized as herbivore-associated molecular patterns and damage-associated molecular patterns, respectively (Gao et al. 2023).

Perception of insect attack by plants activates a series of events such as the induction of calcium waves and electrical signals, accompanied by a burst of jasmonic acid accumulation. Upon mechanical wounding or insect feeding, there is instant accumulation of jasmonic acid, not only limited to wounded local sites but systemically induced in the entire plant by the transport of precursors for the synthesis of jasmonic acid from the infected site to the entire plant (Li et al. 2023). The defensive proteins accumulated in the plant after herbivore attack will affect the digestive physiology of insects for the protection of the plant from herbivorous insects, and all defensive proteins get accumulated by jasmonic acid (Jang et al. 2020).

Role of jasmonic acid against fungi as biotic stress

Plants face extensive survival pressure from several types of plant pathogens, such as fungi. Plant pathogenic fungi are one of the most common causes of crop loss due to plant diseases and are responsible for public health risks. Plants are mostly infected by necrotropic and hemibiotropic fungi, which are responsible for activating JA in plants for their protection. But some hemibiotropic fungi have the ability to hydrolyze the JA produced by the plant (Wang et al. 2020).

Endophytic fungal infection is not able to activate a strong hypersensitive response but is able to induce the production and accumulation of different types of plant secondary metabolites in infected host cells (Jha and Yadav 2021). However, biotropic fungal infections in plants cause rapid production of jasmonic acid and the expression of defense-related genes. Jasmonic acid is a well-known plant signaling molecule responsible for plant defense responses in plants in response to fungal elicitors or microbial infections. While infection of necrotropic fungi requires suppression of jasmonic acid-mediated defense in plants, achieved by the production of cell wall degrading enzymes, host-specific toxins, or acting as an elicitor of the activation of the SA pathway in plants.

Role of jasmonic acid against bacteria as biotic stress

Plants continuously get affected by bacterial pathogens, which is responsible for the remarkable crop loss. The level of infection caused by the bacteria in plants depends on several factors, including disease-susceptible cultivars, environmental status, and the availability of nutrients. Bacterial plant pathogens like *P. syringae,* by acting on the jasmonic acid-mediated signaling pathway and/or by disturbing the JA-SA balance, effectively disrupt hormonal equilibrium in the plant cell to establish infection in the plant (Ke et al. 2019). Many necrotropic pathogens produce the polyketide toxin coronatine, which is responsible for regulating jasmonic acid-mediated plant pathogen interaction as well as promoting plant bacterial infection by down-regulating the induction of pathogen-associated molecular pattern immunity. The pathogen-associated molecular pattern immunity induces a defense response, and stomatal closure is a common mechanism in plants upon infection (Jha 2019e). Coronatine can interact with JAZ and is responsible for the activation of the JA signaling pathway to inhibit salicylic acid-mediated plant defense and promote bacterial infection by regulating pathogen-associated molecular patterns (Ding et al. 2022).

Role of jasmonic acid against virus

Plants are constantly challenged by many pathogens during their lives, including viruses, which have adverse effects on plants. Plants, to sustain their survival, develop multilevel defense mechanisms to overcome the adverse effects of all such interactions. Viral interaction with the plant can disrupt hormonal pathways and establish itself in the plant system. Plant hormones play a significant role in plant defense against viruses (Shang et al. 2011). The plant hormone jasmonic acid generally modulates plant growth and stimulates defense against necrotrophic pathogens, but with ethylene, it usually activates induced systemic resistance against viruses. The viral receptor protein directly interacts with the ZIM domain of the jasmonic acid signaling pathway to establish infection in the host plant, as reported by Oka et al. (2013). JA-mediated plant resistance against viral infection is via microRNAs as reported by Zhang et al. (2016), and JA signaling is up-regulated during viral infection. JA-induced MYB transcription factor initiates the ARGONAUTE 18-mediated host defense network in virally infected plants to overcome the adverse effects of infection.

Conclusion

Jasmonic acid is generally recognized as a stress hormone produced by the plant under stress, such as biotic stress: during pathogen infection, being affected by herbivores, or biotic stress like ultraviolet radiation or wounding. It equally efficiently acts as a signaling compound responsible for regulating plant cellular defense and development activity. Many plant physiological activities, like root growth, leaf senescence, and stamen growth, are directly regulated by jasmonic acid, as are the production and accumulation of different metabolites like terpenoids and phytoalexins, which are also modulated by the plant hormone jasmonic acid. All such regulation by jasmonic acid has been carried out by the JA receptor and different types of effector molecules like JAZ-domain proteins, CORONATINE INSENSITIVE 1, and central repressors. So, induction and production of jasmonic acid in plants have very versatile effects on plant growth and defense against different types of stresses.

References

Acosta, I.F. and Przybyl, M. 2019. Jasmonate Signaling during Arabidopsis Stamen Maturation. Plant Cell Physiol., 60(12): 2648–2659.

Ahmad, G., Khan, A., Khan, A.A., Ali, A. and Mohhamad, H.I. 2021. Biological control: a novel strategy for the control of the plant parasitic nematodes. Antonie van Leeuwenhoek, 114(7): 885–912.

Ahmad, G., Amir, K.H.A.N., Ansari, S., Elhakem, A., Rokayya, S.A.M.I. and Mohamed, H.I. 2022. Management of root-knot nematode infection by using fly ash and Trichoderma harzianum in Capsicum annum plants by modulating growth, yield, photosynthetic pigments, biochemical substances, and secondary metabolite profiles. Notulae Botanicae Horti Agrobotanici Cluj-Napoca, 50(1): 12591–12591.

Ali, M.S. and Baek, K.H. 2020. Jasmonic Acid Signaling Pathway in Response to Abiotic Stresses in Plants. Int. J. Mol. Sci., 21(2): 621.

Al Zahrani, W., Bafeel, S.O. and El-Zohri, M. 2020. Jasmonates mediate plant defense responses to *Spodoptera exigua* herbivory in tomato and maize foliage. Plant Signal Behave., 15(5): 1746898.

Akbar, M.U., Aqeel, M., Shah, M.S., Jeelani, G., Iqbal, N., Latif, A. et al. 2023. Molecular regulation of antioxidants and secondary metabolites act in conjunction to defend plants against pathogenic infection. S. Afr. J. Bot., 161: 247–257.

An, S., Liu, Y., Sang, K., Wang, T., Yu, J., Zhou, Y. et al. 2023. Brassinosteroid signaling positively regulates abscisic acid biosynthesis in response to chilling stress in tomato. J. Integr. Plant Biol., 65(1): 10–24.

Anderson, J.C. 2023. Ill Communication: Host Metabolites as Virulence-Regulating Signals for Plant-Pathogenic Bacteria. Annu. Rev. Phytopathol., 10: 1146.

Arya, G.C., Sarkar, S., Manasherova, E., Aharoni, A. and Cohen, H. 2021. The Plant Cuticle: An Ancient Guardian Barrier Set Against Long-Standing Rivals. Front. Plant Sci., 12: 663165.

Bahadur, A. 2021. Nematodes Diseases of Fruits and Vegetables Crops in India. In: Cristiano, C. and T.E. Kaspary (Eds.). Nematodes—Recent Advances, Management and New Perspectives. Intech Open.

Bali, S., Jamwal, V.L., Kaur, P., Kohli, S.K., Ohri, P., Gandhi, S.G. et al. 2019. Role of P-type ATPase metal transporters and plant immunity induced by jasmonic acid against lead (Pb) toxicity in tomato. Ecotoxicol. Environ. Saf., 174: 283–294.

Balint-Kurti, P. 2019. The plant hypersensitive response: concepts, control and consequences. Mol. Plant Pathol., 20(8): 1163–1178.

Brenya, E., Chen, Z.-H., Tissue, D., Papanicolaou, A. and Cazzonelli, C.I. 2020. Prior exposure of *Arabidopsis* seedlings to mechanical stress heightens jasmonic acid-mediated defense against necrotrophic pathogens. BMC Plant Biol., 20: 548.

Browse, J. and Wallis, J.G. 2019. *Arabidopsis* Flowers Unlocked the Mechanism of Jasmonate Signaling. Plants (Basel)., 8(8): 285.

Chakraborty, N. and Basak, J. 2018 Exogenous application of methyl jasmonate induces defense response and develops tolerance against mungbean yellow mosaic India virus in Vigna mungo. Funct. Plant Biol., 46(1): 69–81.

Chai, Z., Fang, J., Huang, C., Huang, R., Tan, X., Chen, B. et al. 2022. A novel transcription factor, ScAIL1, modulates plant defense responses by targeting DELLA and regulating gibberellin and jasmonic acid signaling in sugarcane. J. Exp. Bot., 73(19): 6727–6743.

Chen, X., Jiang, W., Tong, T., Chen, G., Zeng, F., Jang, S. et al. 2021. Molecular Interaction and Evolution of Jasmonate Signaling With Transport and Detoxification of Heavy Metals and Metalloids in Plants. Front Plant Sci., 12: 665842.

Chen, H.Y., Hsieh, E.J., Cheng, M.C., Chen, C.Y., Hwang, S.Y. and Lin, T.P. 2016. ORA47 (octadecanoid-responsive AP2/ERF-domain transcription factor 47) regulates jasmonic acid and abscisic acid biosynthesis and signaling through binding to a novel cis-element. New Phytol., 211(2): 599–613.

Chen, L., Song, J., Wang, J., Ye, M., Deng, Q., Wu, X. et al. 2023. Effects of Methyl Jasmonate Fumigation on the Growth and Detoxification Ability of Spodoptera litura to Xanthotoxin. Insects 14: 145.

Campos, M.L., Kang, J.H., and Howe, G.A. 2014. Jasmonate-triggered plant immunity. J. Chem. Ecol. 40: 657–675.

Damián, Balfagón., Soham, Sengupta., Aurelio, Gómez-Cadenas., Felix, B. Fritschi., Rajeev, K. Azad., Ron, Mittler et al. 2019. Jasmonic Acid Is Required for Plant Acclimation to a Combination of High Light and Heat Stress. Plant Physiol., 181: 1668–1682.

De Geyter, N., Gholami, A., Goormachtig, S. and Goossens, A. 2012. Transcriptional machineries in jasmonate-elicited plant secondary metabolism. Trends. Plant Sci. 17: 349–359.

De Vleesschauwer, D., Gheysen, G. and Höfte, M. 2013. Hormone defense networking in rice: tales from a different world. Trends Plant Sci., 18: 555–565

Ding, L.N., Li, Y.T., Wu, Y.Z., Li, T., Geng, R., Cao, J. et al. 2022. Plant disease resistance-related signaling pathways: recent progress and future prospects. Int. J. Mol. Sci., 23(24): 16200.

Du, C., Shen, F., Li, Y., Zhao, Z., Xu, X., Jiang et al. 2020. Effects of salicylic acid, jasmonic acid and reactive oxygen species on the resistance of *Solanum peruvianum* to *Meloidogyne incognita*. Sci. Hortic., 273: 109649.

Eng, F., Marin, J.E., Zienkiewicz, K., Gutiérrez-Rojas, M., Favela-Torres, E. and Feussner, I. 2021. Jasmonic acid biosynthesis by fungi: derivatives, first evidence on biochemical pathways and culture conditions for production. Peer J., 9: e10873.

Fragoso, V., Rothe, E., Baldwin, I.T. and Kim, S.G. 2014. Root jasmonic acid synthesis and perception regulate folivore-induced shoot metabolites and increase *Nicotiana attenuata* resistance. New Phytol., 202: 1335–1345.

Gao, J., Tao, T., Arthurs, S.P., Hussain, M., Ye, F. and Mao, R. 2023. Saliva-Mediated Contrasting Effects of Two Citrus Aphid Species on Asian Citrus Psyllid Feeding Behavior and Plant Jasmonic Acid Pathway. Insects., 14(8): 672.

Gareth, Griffiths. 2020. Jasmonates: biosynthesis, perception and signal transduction. Essays Biochem, 64(3): 501–512.

Gimenez, E., Salinas, M. and Manzano-Agugliaro, F. 2018. Worldwide research on plant defense against biotic stresses as improvement for sustainable agriculture. Sustainability, 10(2): 391.

Gupta, A., Bhardwaj, M. and Tran, LP. 2020. Jasmonic Acid at the Crossroads of Plant Immunity and *Pseudomonas syringae* Virulence. Int. J. Mol. Sci., 21(20): 7482.

Greenwood, J.R., Zhang, X. and Rathjen, J.P. 2023. Precision genome editing of crops for improved disease resistance. Curr. Biol., 33(11): R650–R657.

Hayano-Saito, Y. and Hayashi, K. 2020. Stvb-i, a Rice Gene Conferring Durable Resistance to Rice stripe virus, Protects Plant Growth From Heat Stress. Front Plant Sci., 11: 519.

He, Y., Han, J. and Liu, R. 2018. Integrated transcriptomic and metabolomic analyses of a wax deficient citrus mutant exhibiting jasmonic acid-mediated defense against fungal pathogens. Hortic. Res., 5: 43.

Hu, Y., Jiang, Y., Han, X., Wang, H., Pan, J. and Yu, D. 2017. Jasmonate regulates leaf senescence and tolerance to cold stress: crosstalk with other phytohormones. J. Exp. Bot., 68(6): 1361–1369.

Huang, P.C., Tate, M., Berg-Falloure, K.M., Christensen, S.A., Zhang, J., Schirawski, J. et al. 2023. A non-JA producing oxophytodienoate reductase functions in salicylic acid-mediated antagonism with jasmonic acid during pathogen attack. Mol. Plant Pathol., 24(7): 725–741.

Jang, G., Yoon, Y. and Choi, Y.D. 2020. Crosstalk with Jasmonic Acid Integrates Multiple Responses in Plant Development. Int. J. Mol. Sci., 21: 305.

Jarocka-Karpowicz, I. and Markowska, A. 2021. Therapeutic Potential of Jasmonic Acid and Its Derivatives. Int. J. Mol. Sci., 22: 8437.

Jha, Y. and Subramanian, R.B. 2012. Isolation of root associated bacteria from the local variety of rice GJ-17 World Res. J. Geoinformatics, 1: 21–26.

Jha, Y. and Subramanian, R.B. 2014. Identification of plant growth promoting rhizobacteria from *Suaeda nudiflora* plant and its effect on maize. Indian Journal of Plant Protection 42(4): 422–429.

Jha, Y. 2022a Plant-Microbe Interactions in the Pedosphere Necessary for Plant to Overcome Various Stresses. In: B. Giri, R. Kapoor, Q.S. Wu and A. Varma (Eds.). Structure and Functions of Pedosphere. Springer, Singapore.

Jha, Y. 2018a. Effects of salinity on growth physiology, accumulation of osmo-protectant and autophagy-dependent cell death of two maize varieties. Russ. Agri. Sci., 44(2): 124–130.

Jha, Y. 2019a. Higher induction of defense enzymes and cell wall reinforcement in maize by root associated bacteria for better protection against *Aspergillus niger*. J. Plant Prot. Res., 59(3): 341–349.

Jha, Y. 2019b. Mineral Mobilizing Bacteria mediated regulation of secondary metabolites for proper photosynthesis in maize under stress, in edited Book Photosynthesis Productivity and Environmental Stress, 197–213.

Jha, Y. 2019c. Endophytic Bacteria as a Modern Tool for Sustainable Crop Management Under Stress. In: B. Giri, R. Prasad, Q.S. Wu and A. Varma (Eds.). Biofertilizers for Sustainable Agriculture and Environment. Soil Biology, vol. 55. Springer, Cham.

Jha, Y. 2019d. The Importance of Zinc-Mobilizing Rhizosphere Bacteria to the Enhancement of Physiology and Growth Parameters for Paddy under Salt-Stress Conditions. Jordan J. Biol. Sci., 12(2): 167–173.

Jha, Y. 2019e. Endophytic Bacteria-Mediated Regulation of Secondary Metabolites for the Growth Induction in *Hyptissuaveolens* Under Stress. In: D. Egamberdieva and A. Tiezzi (Eds.). Medically Important Plant Biomes: Source of Secondary Metabolites. Microorganisms for Sustainability, vol 15. Springer, Singapore.

Jha, Y. 2020. Plant Microbiomes with Phytohormones Attribute for Plant Growth and Adaptation Under the Stress Conditions, Agriculture, Microorganisms for Sustainability, 85–103.

Jha, Y. 2022. Enhanced cell viability with induction of pathogenesis related proteins against *Aspergillus niger* in maize by endo-rhizospheric bacteria. Jordan J. Biol. Sci., 15(1): 139–147.

Jha, Y. 2022b. Nitric Oxide Cross-talk with Phytohormones Vis-à-Vis Photosynthetic Regulation Under Extreme Environments. In: M.A. Ahanger and P. Ahmad (Eds.). Nitric Oxide in Plants.

Jha, Y., Dehury, B., Kumar, S.P.J., Chaurasia, A., Singh, U.B., Yadav, M.K. et al. 2022. Delineation of molecular interactions of plant growth promoting bacteria induced β-1,3-glucanases and guanosine triphosphate ligand for antifungal response in rice: a molecular dynamics approach. Mol. Biol. Rep., 49(4): 2579–2589.

Jha, Y. and Mohamed, H.I. 2022. Plant Secondary Metabolites as a Tool to Investigate biotic Stress Tolerance in Plants: A Review. Gesunde Pflanzen., 74: 771–790.

Jha, Y. and Mohamed, H.I. 2023. Inoculation with Lysinibacillus fusiformis strain YJ4 and Lysinibacillus sphaericus strain YJ5 Alleviates the Effects of Cold Stress in Maize Plants. Gesunde Pflanzen, 75: 77–95.

Jha, Y., Kulkarni, A. and Subramanian, R.B. 2021. Psychrotrophic Soil Microbes and Their Role in Alleviation of Cold Stress in Plants. In: A.N. Yadav (Ed.). Soil Microbiomes for Sustainable Agriculture. Sustainable Development and Biodiversity, vol. 27. Springer, Cham.

Jha, Y. and Subramanian, R.B. 2021. Use of Halo-tolerant Bacteria to Improve the Bioactive Secondary Metabolites in Medicinally Important Plants under Saline Stress. Microbiomes of Extreme Environments, 74–86.

Jha, Y. and Yadav, A.N. 2021. *Piriformospora indica*: Biodiversity, Ecological Significances, and Biotechnological Applications for Agriculture and Allied Sectors. In: A.M. Abdel-Azeem, A.N. Yadav, N. Yadav and Z. Usmani (Eds.). Industrially Important Fungi for Sustainable Development. Fungal Biology. Springer, Cham.

Jing, Y., Liu, J., Liu, P., Ming, D. and Sun, J. 2019. Overexpression of *TaJAZ1* increases powdery mildew resistance through promoting reactive oxygen species accumulation in bread wheat. Sci. Rep., 9: 5691.

Ke, Y., Kang, Y., Wu, M., Liu, H., Hui, S., Zhang, Q. et al. 2019. Jasmonic acid-involved OsEDS1 signaling in rice-bacteria interactions. Rice., 12: 1–12.

Kou, M.Z., Bastías, D.A., Christensen, M.J., Zhong, R., Nan, Z.B. and Zhang, X.X. 2021. The Plant Salicylic Acid Signaling Pathway Regulates the Infection of a Biotrophic Pathogen in Grasses Associated with an *Epichloë* Endophyte. J. Fungi (Basel)., 7(8): 633.

Lee, H.J., Park, J.S. and Shin, S.Y. 2020. Submergence deactivates wound-induced plant defense against herbivores. Commun. Biol., 3: 651.

Lemos, M., Xiao, Y., Bjornson, M., Wang, J.Z., Hicks, D., Souza, A.D. et al. 2016. The plastidial retrograde signal methyl erythritol cyclopyrophosphate is a regulator of salicylic acid and jasmonic acid crosstalk. J. Exp. Bot., 67(5): 1557–1566.

Li, T., Feng, M., Chi, Y., Shi, X., Sun, Z., Wu, Z. et al. 2023. Defensive Resistance of Cowpea Vigna unguiculata Control Megalurothrips usitatus Mediated by Jasmonic Acid or Insect Damage. Plants., 12(4): 942.

Liu, S., Tian, Y., Jia, M., Lu, X., Yue, L., Zhao, X. et al. 2020. Induction of Salt Tolerance in *Arabidopsis thaliana* by Volatiles From *Bacillus amyloliquefaciens* FZB42 via the Jasmonic Acid Signaling Pathway. Front. Microbiol., 11: 562934.

Luo, J., Xia, W., Cao, P., Xiao, Z.A., Zhang, Y., Liu, M. et al. 2019. Integrated transcriptome analysis reveals plant hormones jasmonic acid and salicylic acid coordinate growth and defense responses upon fungal infection in poplar. Biomolecules, 9(1): 12.

Martínez-Medina, A., Fernandez, I., Lok, G.B., Pozo, M.J., Pieterse, C.M.J., and Van Wees, S.C.M. 2017. Shifting from priming of salicylic acid- to jasmonic acid-regulated defenses by *Trichoderma* protects tomato against the root knot nematode *Meloidogyne incognita*. New Phytol., 213: 1363–1377.

Marquis, V., Smirnova, E., Graindorge, S., Delcros, P., Villette, C., Zumsteg, J. et al. 2022. Broad-spectrum stress tolerance conferred by suppressing jasmonate signaling attenuation in Arabidopsis JASMONIC ACID OXIDASE mutants. Plant J., 109: 856–872.

Martin, R.L., Le Boulch, P., Clin, P., Schwarzenberg, A., Yvin, J.C., Andrivon, D. et al. 2020. A comparison of PTI defense profiles induced in Solanum tuberosum by PAMP and non-PAMP elicitors shows distinct, elicitor-specific responses. Plos One., 15(8): e0236633.

Mohamed, H.I., Haleem, A.B.D.E.L., Mohammed, M.A. and Mogazy, A.M. 2016. Effects of plant defense elicitors on soybean (Glycine max L.) growth, photosynthetic pigments, osmolyts and lipid components in response to cotton worm (Spodoptera littoralis) infestation. Bangladesh J. Bot., 45(3): 597–604.

Mohamed, H.I., Haleem, M.A., Mohamed, N.M., Ashry, N.A., Zaky, L.M. and Mogazy, A.M. 2021. Comparative effectiveness of potential elicitors of soybean plant resistance against Spodoptera Littoralis and their effects on secondary metabolites and antioxidant defense system. Gesunde Pflanzen., 73(3): 273–285.

Montejano-Ramírez, V., García-Pineda, E. and Valencia-Cantero, E. 2020. Bacterial compound N, N-dimethylhexadecylamine modulates expression of iron deficiency and defense response genes in Medicago truncatula independently of the jasmonic acid pathway. Plants., 9(5): 624.

Nawaz, F., Naeem, M., Zulfiqar, B., Akram, A., Ashraf, M.Y., Raheel, M. et al. 2017. Understanding brassinosteroid-regulated mechanisms to improve stress tolerance in plants: a critical review. Environ Sci Pollut Res., 24: 15959–15975.

Nazarov, P.A., Baleev, D.N., Ivanova, M.I., Sokolova, L.M., and Karakozova, M.V. 2020. Infectious Plant Diseases: Etiology, Current Status, Problems and Prospects in Plant Protection. Acta Naturae., 12(3): 46–59.

Oka, K., Kobayashi, M., Mitsuhara, I. and Seo, S. 2013. Jasmonic acid negatively regulates resistance to Tobacco mosaic virus in tobacco. Plant Cell Physiol., 54: 1999–2010.

Ono, K., Kimura, M., Matsuura, H., Tanaka, A. and Ito, H. 2019. Jasmonate production through chlorophyll a degradation by Stay-Green in Arabidopsis thaliana. J. Plant Physiol., 238: 53–62.

Pacifici, E., Polverari, L. and Sabatini, S. 2015. Plant hormone cross-talk: The pivot of root growth. J. Exp. Bot., 66: 1113–1121.

Qiu, C., Zeng, J., Tang, Y., Gao, Q., Xiao, W. and Lou, Y. 2023. The Fall Armyworm, *Spodoptera frugiperda* (Lepidoptera: Noctuidae), Influences *Nilaparvata lugens* Population Growth Directly, by Preying on Its Eggs, and Indirectly, by Inducing Defenses in Rice. Int. J. Mol. Sci., 24(10): 8754.

Raíssa, Mesquita, Braga., Manuella, Nóbrega. Dourado. and Welington, Luiz. Araújo. 2016. Microbial interactions: ecology in a molecular perspective. Braz. J. Microbiol., 47: 86–98.

Rajarammohan, S. 2021. Redefining Plant-Necrotroph Interactions: The Thin Line Between Hemibiotrophs and Necrotrophs. Front. Microbiol,. 12: 673518.

Ricardo, A.R. Machado., Mark, McClure., Maxime, R. Hervé., Ian, T. Baldwin. and Matthias, Erb. 2016. Benefits of jasmonate-dependent defenses against vertebrate herbivores in nature. eLife 5: e13720.

Rojas, C.M., Senthil-Kumar, M., Tzin, V. and Mysore, K.S. 2014. Regulation of primary plant metabolism during plant-pathogen interactions and its contribution to plant defense. Front Plant Sci., 10; 5: 17.

Ruan, J., Zhou, Y., Zhou, M., Yan, J., Khurshid, M., Weng, W. et al. 2019. Jasmonic Acid Signaling Pathway in Plants. Int. J. Mol Sci., 20(10): 2479.

Salazar-Mendoza, P., Bento, J.M.S., Silva, D.B., Pascholati, S.F., Han, P. and Rodriguez-Saona, C. 2023. Bottom-up effects of fertilization and jasmonate-induced resistance independently affect the interactions between tomato plants and an insect herbivore. J. Plant Interact. 18(1): 2154864.

Sarwat, M., Naqvi, A.R., Ahmad, P., Ashraf, M. and Akram, N.A. 2013. Phytohormones and microRNAs as sensors and regulators of leaf senescence: assigning macro roles to small molecules. Biotechnol. Adv., 31(8): 1153–1171.

Shang, J., Xi, D.H., Xu, F., Wang, S.D., Cao, S., Xu, M.Y. et al. 2011. A broad-spectrum, efficient and nontransgenic approach to control plant viruses by application of salicylic acid and jasmonic acid. Planta., 233: 299–308.

Shuping, D.S.S. and Eloff, J.N. 2017. The use of plants to protect plants and food against fungal pathogens: A review. African Journal of Traditional, Complementary and Alternative Medicines., 14(4): 120–127.

Siddiqi, K.S. and Husen, A. 2019. Plant response to jasmonates: current developments and their role in changing environment. Bull. Natl. Res. Cent. 43: 153.

Stella de Freitas, T.F., Stout, M.J. and Sant'Ana, J. 2019. Effects of exogenous methyl jasmonate and salicylic acid on rice resistance to Oebalus pugnax. Pest. Manag. Sci., 75(3): 744–752.

Ting, Li., Kun-Peng, Jia., Hong-Li, Lian., Xu, Yang., Ling, Li. and Hong-Quan, Yang. 2014. Jasmonic acid enhancement of anthocyanin accumulation is dependent on phytochrome A signaling pathway under far-red light in Arabidopsis. Biochem. Biophy. Res. Commun., 454: 78–83.

Ullah, A., Akbar, A. and Yang, X. 2019. Chapter 7 - Jasmonic acid (JA)-mediated signaling in leaf senescence. In: Sarwat, M., Tuteja, N., editors. *Senescence Signaling and Control in Plants.* Academic Press; Cambridge, MA, U.S.A.; 111–123.

Vasvi, Chaudhry., Paul, Runge., Priyamedha, Sengupta., Gunther, Doehlemann., Jane, E. Parker. and Eric, Kemen. 2021. Shaping the leaf microbiota: plant–microbe–microbe interactions. J. Exp. Bot., 72: 36–56.

Wang, Y., Mostafa, S., Zeng, W. and Jin, B. 2021. Function and Mechanism of Jasmonic Acid in Plant Responses to Abiotic and Biotic Stresses. Int. J. Mol Sci., 22: 8568.

Wang, J., Song, L., Gong, X., Xu, J. and Li, M. 2020. Functions of jasmonic acid in plant regulation and response to abiotic stress. Int. J. Mol. Sci., 21(4): 1446.

Wang, Z., Tan, X., Zhang, Z., Gu, S., Li, G. and Shi, H. 2012. Defense to *Sclerotinia sclerotiorum* in oilseed rape is associated with the sequential activations of salicylic acid signaling and jasmonic acid signaling. Plant Sci., 184: 75–82.

Wang, Q., Chen, X., Chai, X., Xue, D., Zheng, W., Shi, Y. et al. 2019. The involvement of jasmonic acid, ethylene, and salicylic acid in the signaling pathway of *Clonostachys rosea*-induced resistance to gray mold disease in tomato. Phytopathol., 109(7): 1102–1114.

Walker, T.S., Bais, H.P., Grotewold, E. and Vivanco, J.M. 2003. Root exudation and rhizosphere biology. Plant Physiol., 132(1): 44–51.

Wasternack, C. and Hause, B. 2019 The missing link in jasmonic acid biosynthesis. Nat. Plants, 5: 776–777.

Xu, P., Zhao, P.X., Cai, X.T., Mao, J.L., Miao, Z.Q. and Xiang, C.B. 2020. Integration of Jasmonic Acid and Ethylene Into Auxin Signaling in Root Development. Front. Plant Sci., 11: 271.

Yang, Z.B., He, C., Ma, Y., Herde, M. and Ding, Z. 2017. Jasmonic Acid Enhances Al-Induced Root Growth Inhibition. Plant Physiol., 173(2): 1420–1433.

Yang, J., Duan, G., Li, C., Liu, L., Han, G., Zhang, Y. et al. 2019. The crosstalks between jasmonic acid and other plant hormone signaling highlight the involvement of jasmonic acid as a core component in plant response to biotic and abiotic stresses. Front Plant Sci., 10: 1349.

Wang, Z., Wong, D.C.J., Wang, Y., Xu, G., Ren, C., Liu, Y. et al. 2021. GRAS-domain transcription factor PAT1 regulates jasmonic acid biosynthesis in grape cold stress response. Plant Physiol., 186(3): 1660–1678.

Ye, M., Luo, S.M., Xie, J.F., Li, Y.F., Xu, T., Liu, Y. et al. 2012. Silencing *COI1* in rice increases susceptibility to chewing insects and impairs inducible defense. PLoS ONE., 7: e36214.

Yoshiaki, Ueda., Shahid, Siddique. and Michael, Frei. A. 2015 Novel Gene, OZONE-RESPONSIVE APOPLASTIC PROTEIN1, Enhances Cell Death in Ozone Stress in Rice. Plant Physiol., 169: 873–889.

Yuan, M., Huang, Y. and Ge, W. 2019. Involvement of jasmonic acid, ethylene and salicylic acid signaling pathways behind the systemic resistance induced by *Trichoderma longibrachiatum* H9 in cucumber. BMC Genomics, 20: 144.

Zeng, J., Zhang, T., Huangfu, J., Li, R. and Lou, Y. 2021. Both allene oxide synthases genes are involved in the biosynthesis of herbivore-induced jasmonic acid and herbivore resistance in rice. Plants., 10: 442.

Zhao, W., Huang, H., Wang, J., Wang, X., Xu, B., Yao, X. et al. 2023. Jasmonic acid enhances osmotic stress responses by MYC2-mediated inhibition of protein phosphatase 2C1 and response regulators 26 transcription factor in tomato. The Plant Journal, 113(3): 546–561.

Zhang, H., Hedhili, S., Montiel, G., Zhang, Y., Chatel, G., Pré, M. et al. 2011. The basic helix-loop-helix transcription factor CrMYC2 controls the jasmonate-responsive expression of the ORCA genes that regulate alkaloid biosynthesis in *Catharanthus roseus*. Plant J., 67(1): 61–71.

Zhang, Y.T., Zhang, Y.L., Chen, S.X., Yin, G.H., Yang, Z.Z., Lee, S. et al. 2015. Proteomics of methyl jasmonate induced defense response in maize leaves against Asian corn borer. BMC Genom., 16: 224.

Zhang, W., Zhu, K., Wang, Z., Zhang, H., Gu, J., Liu, L. et al. 2019. Brassinosteroids function in spikelet differentiation and degeneration in rice. J. Integr. Plant Boil., 61(8): 943–963.

Zhang, Q., Dai, W. and Wang, X. 2020. Elevated CO_2 concentration affects the defense of tobacco and melon against lepidopteran larvae through the jasmonic acid signaling pathway. Sci. Rep., 10: 4060.

Zhao, F., Li, G. and Hu, P. 2018. Identification of basic/helix-loop-helix transcription factors reveals candidate genes involved in anthocyanin biosynthesis from the strawberry white-flesh mutant. Sci. Rep., 8: 2721.

8

Antimicrobial Potential of Salicylic Acid in Plants

Parteek Prasher[1],* and *Mousmee Sharma*[2]

Introduction

Salicylic acid is a plant hormone associated with maintaining plant immunity and vital processes in plants such as thermogenesis, floral induction, root initiation, seed germination, stomata closure, and response to the various types of biotic/abiotic stress (Lefevere et al. 2020). Salicylic acid is known to switch the innate immune response in plants which effectively counters the pathogenic attack in plants. Apart from this, salicylic acid plays an indirect role in regulating the plant growth and development such as providing properties such as drought resistance and disease resistance (War et al. 2011). The pathogens require host tissues for their survival and progression. Salicylic acid is known to activate the various components of plant defence system mainly through non-expressor of PR1 (NPR1) protein. The inactive form of cytoplasmic NPR1 is mainly oligomeric and has constitutive function in plant cells. However, on plant infection, it gets activated resulting in a higher production of salicylic acid that eventually alters the redox potential of cell (Hayat et al. 2012). Apparently, the oligomeric form of NPR1 is transformed to bioactive monomeric form that on entering the nucleus displays interactions with TGA proteins, which results in the expression of salicylic acid-dependent genes associated with pathogenesis (Vincent et al. 2022). Mainly, the plants where salicylate hydroxylase (NahG) is overexpressed are unable to bear the beneficial effects of salicylic acid during pathogenic attack and possess a poor systemic acquired resistance (Lubbers et al. 2021). Loaded with antimicrobial potential, the salicylic acid surprisingly does not pose any toxic effect on the endophytes, also regarded as plant-friendly microbes

[1] Department of Chemistry, University of Petroleum & Energy Studies, Energy Acres, Dehradun 248007, India.

[2] Department of Chemistry, Uttaranchal University, Arcadia Grant, Dehradun 248007, India.

* Corresponding author: parteekchemistry@gmail.com

that play an essential role in mediating the fixation of essential nutrients and metal ions via symbiosis (Benjamin et al. 2022). Figure 1 represents the various biological role of salicylic acid in plants. In this review, we have presented the antimicrobial profile of salicylic acid and the mechanism of microbial inhibition thereof. In this chapter, we discuss about the potential antimicrobial applications of salicylic acid.

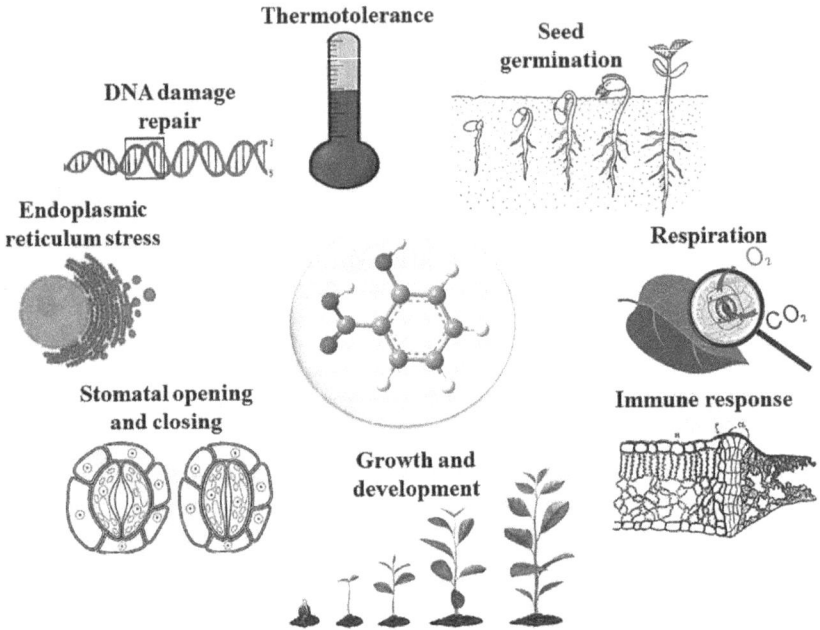

Fig. 1. Biological functions of salicylic acid in plants (source or if it own).

Biosynthesis of Salicylic Acid in Plants

The biosynthesis of salicylic acid occurs mainly from chorismate via isochorismate synthase (ICS) and phenylalanine ammonia-lyase (PAL) pathways out of which the former is a major contributor (> 90%) for the salicylic acid synthesis. The ICS pathway initiates with the conversion of chorismate to isochorismate in the presence of isochorismate synthase enzyme (Lefevere et al. 2020). This synthetic step takes place in plants in plastids from where the isochorismate is further transferred to the cytosol. Eventually, the enzyme isochorismate synthase conjugates to L-glutamate resulting in the formation of isochorismate-9-glutamate in the presence of amidotransferase enzyme located in the cytosol. The decomposition of ICS-glutamate results in the subsequent formation of salicylic acid and 2-hydroxy-acryloyl-N-glutamate (Chen et al. 2009). The contribution of PAL pathway for salicylic acid synthesis is limited to less than 10% that occurs entirely in the cytosol. Phenylalanine ammonia-lyase enzyme converts phenylalanine to trans-cinnamic acid that is further converted to salicylic acid via benzaldehyde and ortho-coumaric acid intermediates (Zhang et al. 2021). The salicylic acid exists in free and inactive forms namely salicylic acid glucoside and salicylic acid glucose ester. These forms mainly accumulate in cell vacuoles where they generate active form on hydrolysis (Mishra et al. 2021). The pathogenic attack on plant triggers the levels of both forms of salicylic acid that provides systemic acquired resistance to the host plant. Furthermore, the methylation of salicylic acid causes the formation of volatile forms referred as methyl salicylate that plays a role in enhancing the membrane permeability (Ogawa et al. 2006). Furthermore, the methyl salicylate plays a leading role in regulating the plant-insect interactions and signalling of systemically acquired resistance (Shine et al. 2016). Eventually, the salicylic acid formed via ICS and PAL pathways undergoes hydroxylation to form 2,3-dihydrobenzoic acid. Figure 2 illustrates the biosynthetic pathways for salicylic acid formation in plants.

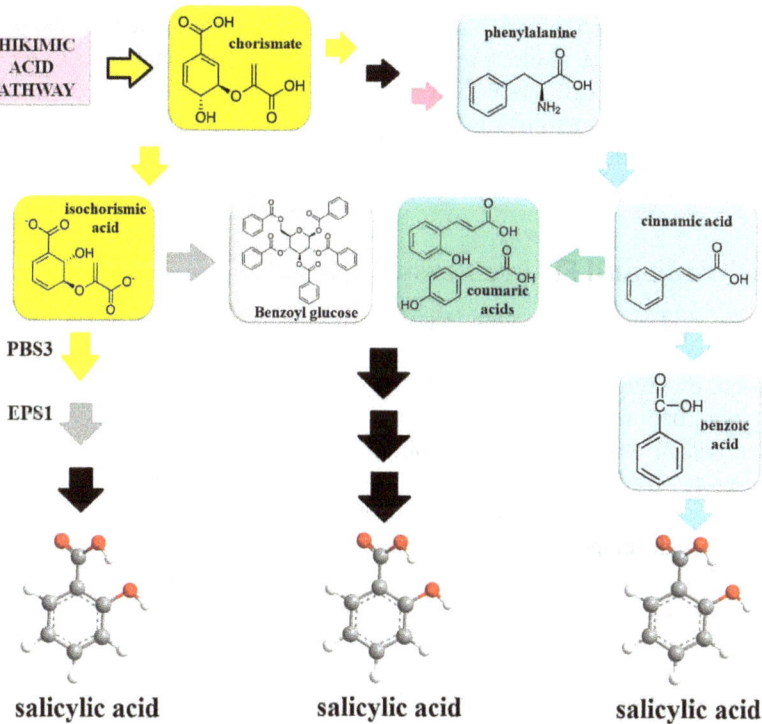

Fig. 2. Various pathways for the biosynthesis of salicylic acid in plants.

Antimicrobial Activity of Salicylic Acid in Plants

Salicylic acid is a potent molecule that induces the development of resistance in plants against various biotic and abiotic stresses. However, it may undergo glycosylation that limits its efficacy. Furthermore, it exhibits phytotoxicity that challenges its developments as plant protecting molecule. These limitations are overcome by chemical modification or by the synthesis of analogues of salicylic acid due to the presence of carboxylic and hydroxyl functional groups that enable its derivatization. Research group led by Joshi et al. (2022) have reported the triazine-based dendrimers of salicylic acid possessing an enhanced antimicrobial activity against *Staphylococcus aureus*. Similarly, the functional analogues of salicylic acid such as nicotinic acid derivatives, salicylate or benzoate compounds, β-aminobutyric acid, neonicotinoid compounds, pyrimidine derivatives, and derivatives based on thiadiazole, pyrazole, and thiazole heterocycles have been reported (Faize et al. (2018). The derivatives of salicylic acid have been reported to exhibit antifungal activity against *Rhizoctonia solani, Botrytis cinerea, Alternaria alternata, Phytophtora infestans, Fusarium germinarum,* and *Fusarium culmorum* by inhibiting the mycelial growth. Reaction of salicylic acid with several 2-bromoalkanoates in weakly basic medium result in the regioselective derivatization at the carboxyl group of salicylic acid (Wodnicka et al. (2017). Salicylic acid offers the production of phenolic compounds as secondary metabolites in the cell suspension cultures of *Thevetia peruviana* that play an important role in imparting the microbial resistance to the plants (Mendoza et al. (2018). Table 1 illustrates selected examples of the antimicrobial effect of salicylic acid.

Conclusion and Future Perspectives

Salicylic acid, a phytohormone is involved in the regulation of stress response in plants. The stress could be biotic or abiotic, where the biotic stress refers to the invasion by microbes to cause infection. The microbial invasion switches on deleterious pathways in the host plant, such as redox stress, and the altered expression of antioxidant enzymes and metabolites. These factors switch the production of phyto-protective phenolic compounds, of which salicylic acid is a major molecule. Salicylic acid does not exert any toxic effect on plant-friendly microbes, which further supports its development as plant protecting agent. However, the glycosylation of salicylic acid may be a major challenge to show an optimal bioactivity to which the chemical modification serves as a workable alternative. Functionalization of salicylic acid at the carboxyl terminal provides a range of analogies that exhibit enhanced antimicrobial potency with minimal cytotoxicity on plant cells. Further work needs to be done to develop salicylic acid-based molecules for protecting the host plant against the microbial invasion, without causing much alteration in the original biocidal profile of salicylic acid.

Table 1. Antimicrobial activity of salicylic acid.

Plant	Target microbe	Antimicrobial effect	Ref.
Ruta graveolens	*Staphylococcus aureus, Pseudomonas aeruginosa, Klebsiella pneumoniae, Candida albicans, Escherichia coli*	Modulation in the isoprenoid metabolism	Attia et al. 2018
Persea americana	*Phytophthora cinnamomi, Aspergillus*	Disruption of plasma membrane, ion leakage, and alteration in activity of enzymes associated with the biosynthesis of plasma membrane	Sanchez et al. (2014)
Solanum lycopersicum	*Rhizopus stolonifer*	Disruption in membrane integrity and disruption of mitochondrial function	Kong et al. (2019)
Malus pumila	*Penicillium expansum*	Inhibition of the germination of phytopathogen	Neto et al. (2016)
Citrus sinensis var. Moro	*Penicillium digitatum* Sacc.	Antioxidant activity of salicylic acid	Aminifard et al. (2013)
Amphipterygium adstringens	*Streptococcus mutans* and *Porphyromonas gingivalis*	Inhibition of microbial growth	Cruz et al. (2011)
Solanum tuberosum L.	*Dickeya solani*	Significant reduction in the symptoms of microbial infection	Czarkowski et al. (2015)
Ginkgo biloba	Vancomycin-resistant *Enterococcus faecalis* CDC-286	Inhibition of microbial growth	Choi et al. (2009)
Oryza sativa	*Pseudomonas aeruginosa,* and *Rhizoctonia solani*	Accumulation of peroxidases in plant	Saikia et al. (2006)
Rosmarinus officinalis L.	*Bacillus cereus, Escherichia coli, Pseudomonas aeruginosa, Micrococcus flavus, Listeria monocytogenes*	Stimulation of antioxidant enzymes	Esawi et al. (2017)
Solanum lycopersicum	*Fusarium solani, Rhizopus stolonifer*	Inhibition of fungal mycelial growth	Kong et al. (2016)

Table 1. contd. ...

... Table 1. contd.

Plant	Target microbe	Antimicrobial effect	Ref.
Arabidopsis	*Pseudomonas syringae*	Age-related resistance	Cameron et al. (2004)
Mangifera indica	*Colletotrichum gloeosporioides*	Inhibition of postharvest anthracnose	He et al. (2017)
Asparagus officinalis	*Fusarium oxysporum* f. sp. *Asparagi*	Development of resistance against fungal pathogenesis	He et al. (2005)
Arabidopsis thaliana	*Pseudomonas aeruginosa*	Repression of virulence factors and diminished transcription of exoproteins	Prithiviraj et al. (2005)
Citrus reticulata Blanco	*Penicillium digitatum, Penicillium italicum*	Enhanced expression of the genes associated with plant defence	Moosa et al. (2021)
Solanum lycopersicum	*Trichoderma harzianum*	Intensified plant defence system against fungal wilting	AL-surhanee et al. (2022)
Phoenix dactylifera L.	*Nigrospora* spp.	Inhibition of growth and toxicity, reduction in toxin production	Abbas et al. (2019)
Arabidopsis thaliana	*Streptomyces* sp. *136, Chryseobacterium* sp. *8, Pseudomonas* sp. *50, Escherichia coli, Bacillus* sp. *125, Brevundimonas* sp. *374*	Modulation of bacterial colonization	Lebeis et al. (2015)
Solanum lycopersicum	*Geotrichum candidum, Trichothecium roseum*	Inhibition of the mycelium growth	Ghazanfar et al. (2019)
Stevia rebaudiana	*Rhizoctonia solani, Pythium aphanidermatum, Xanthomonas campestris, Pseudomonas syringae, Cercospora kikuchi, Phomopsis* spp., *Fusarium* spp. *Pythium myriotylum, Pythium debaryanum*	Inhibition of microbial growth	Sedghi et al. (2013)
Tagetes erecta Linn.	*Pseudomonas aeruginosa, Staphylococcus aureus, Bacillus subtilis, Klebsiella pneumoniae, Candida albicans, Acinetobacter baumanii*	Inhibition of microbial growth	Devika et al. (2014)

Microchaete Sp. NCCU-342	*Pseudomonas aeruginosa*	Antioxidant activity and induction of resistance against pathogen	Naaz et al. (2021)
Solanum lycopersicum	*Ralstonia solanacearum, Bacillus methylotrophicus, Bacillus subtilis*	Inhibition of bacterial wilt disease	Almoneafy et al. (2013)
Solanum lycopersicum	*Botrytis cinerea*	Decrease in the severity of infection	Salem et al. (2020)
Phaseolus vulgaris	*Pseudomonas savastanoi* pv. *phaseolicola*	Inhibition of culture growth and Halo blight disease	Cooper et al. (2022)
Citrus sinensis	*Alternaria alternata, Penicillium digitatum*	Disruption in microbial cell wall, shrinkage of cytoplasm, leakage of organelles	Shoala et al. (2021)
Oryzae sativa	*Xanthomonas oryzae* pv. *Oryzae*	Increased production of ROS, modification in cell wall, structural alternations in essential proteins	Thanh et al. (2017)
Capsicum annuum	*Fusarium oxysporium*	Prevention of wilt disease caused by fungus	Yousif et al. (2018)
Oryzae sativa	*Magnaporthe grisea*	Increased production of antioxidant enzymes	Daw et al. (2008)
Vitis vinifera	*Eutypa lata*	Fungicidal effect	Amborabe et al. (2003)
Arabidopsis	*Dickeya dadantii*	Increased tolerance of plant towards microbial infection, overexpression of ATG8 proteins	Rigault et al. (2021)

References

Abbas, M.H. 2019. *In Vitro* Antifungal Activity of Different Plant Hormones on the Growth and Toxicity of *Nigrospora* spp. on Date Palm (*Phoenix dactylifera* L.). Open Plant Sci. J., 10: 10–20.

Almoneafy, A.A., Ojaghian, M.R., Xu, S.-F., Ibrahim, M., Xie, G.-L., Shi, Y. et al. 2013. Synergistic effect of acetyl salicylic acid and DL-Beta-aminobutyric acid on biocontrol efficacy of *Bacillus* strains against tomato bacterial wilt. Trop. Plant Pathol., 38: 102–113.

AL-surhanee, A.A. 2022. Protective role of antifusarial eco-friendly agents (*Trichoderma* and salicylic acid) to improve resistance performance of tomato plants. Saudi J. Biol. Sci., 29: 2933–2941.

Amborabe, B.E., Lessard, P.F., Chollet, J.F. and Roblin, G. 2002. Antifungal effects of salicylic acid and other benzoic acid derivatives towards *Eutypa lata*: structure–activity relationship. Plant Physiol. Biochem., 40: 1051–1060.

Attia, E.Z., El-Baky, R.M.A., Desoukey, S.Y., Mohamed, M.A.H., Bishr, M.M. and Kamel, M.S. 2018. Chemical composition and antimicrobial activities of essential oils of *Ruta graveolens* plants treated with salicylic acid under drought stress conditions. Future J. Pharm. Sci., 4: 254–264.

Benjamin, G., Pandharikar, G. and Frendo, P. 2022. Salicylic Acid in Plant Symbioses: Beyond Plant Pathogen Interactions. Biology (MDPI). 11: Article 861.

Cameron, R.K. and Zaton K. 2004. Intercellular salicylic acid accumulation is important for age-related resistance in Arabidopsis to *Pseudomonas syringae*. Physiol. Mol. Plant Pathol., 65: 197–209.

Chen, Z., Zheng, Z., Huang, J., Lai, Z. and Fan, B. 2009. Biosynthesis of salicylic acid in plants. Plant Signal Behav., 4: 493–496.

Choi, J.G., Jeong, S.I., Ku, C.S., Sathishkumar, M., Lee, J.J., Mun, S.P. and Kim, S.M. 2009. Antibacterial activity of hydroxyalkenyl salicylic acids from sarcotesta of Ginkgo biloba against vancomycin-resistant Enterococcus. Fitoterapia., 80: 18–20.

Cooper, B., Campbell, K.B. and Garrett, W.M. 2022. Salicylic Acid and Phytoalexin Induction by a Bacterium that Causes Halo Blight in Beans. Mol. Physiol. Plant Pathol., 112: 1766–1775.

Cruz, B.E.R., Esturau, N., Nieto, S.S., Romero, I., Juarez, I.C. and Cruz, J.F.R. 2011. Isolation of the new anacardic acid 6-[16'Z-nonadecenyl]-salicylic acid and evaluation of its antimicrobial activity against *Streptococcus mutans* and *Porphyromonas gingivalis*., 25: 1282–1287.

Czarkowski, R., Wolf, J.M., Krolicka, A., Ozymko, Z., Narajczyk, M., Kaczynska, N. et al. 2015. Salicylic acid can reduce infection symptoms caused by Dickeya solani in tissue culture grown potato (Solanum tuberosum L.) plants. Eur. J. Plant Pathol., 141: 545–558.

Daw, B.D., Zhang, L.H. and Wang, Z.Z. 2008.

Devika, P. and Koilpillail. 2014. Antimicrobial Activity Study of Flavonoids and Salicylic Acid Extracted from Tagetes Erecta Linn. Nanobio. Pharm. Technol., 493–497.

Esawi, M.A., Elansary, H.O., Shanhorey, N.A., Abdel-Hamid, A.M.E., Ali Hm and Elshikh, M.S. 2017. Salicylic Acid-Regulated Antioxidant Mechanisms and Gene Expression Enhance Rosemary Performance under Saline Conditions. Front. Physiol. 8: Article 716.

Faize, L. and Faize, M. 2018. Functional analogues of salicylic acid and their use in crop protection. Agronomy (MDPI). 8: Article 5.

Ghazanfar, M.U., Raza, W., Iqbal, Z., Ahmad, S., Qamar, M.I. and Hussain, M. 2019. Antifungal activity by resistance inducing salicylic acid and plant extracts against postharvest rots of tomato. Int. J. Biosci., 14: 264–274.

Hayat, S., Irfan, M., Wani, A., Nasser, A. and Ahmad, A. 2012. Salicylic acids. Plant Signal Behav., 7: 93–102.

He, C.Y. and Wolyn, D.J. 2005. Potential role for salicylic acid in induced resistance of asparagus roots to *Fusarium oxysporum* f. sp. *asparagi*. Plant Pathol., 54: 227–232.

He, J., Ren, Y., Chen, C., Liu, J. and Pei, Y. 2017. Defense Responses of Salicylic Acid in Mango Fruit Against Postharvest Anthracnose, Caused by *Colletotrichum gloeosporioides* and its Possible Mechanism. J. Food Saf. 37: Article e12294.

Joshi, M. and Dashora, K. 2022. Development of triazine-based dendrimer of salicylic acid with increased antibacterial activity. J. Adv. Sci. Res., 13: 327–332.

Kong, J., Xie, Y.-F., Guo, Y.-H., Cheng, Y.-L., Qian, H. and Yao, W.-R. 2016. Biocontrol of postharvest fungal decay of tomatoes with a combination of thymol and salicylic acid screening from 11 natural agents. LWT-Food Sci. Technol., 72: 215–222.

Kong, J., Zhang, Y., Ju, J., Xie, Y., Guo, Y., Cheng, Y., Qian, H., Quek, S.Y. and Yao, W. 2019. Antifungal effects of thymol and salicylic acid on cell membrane and mitochondria of *Rhizopus stolonifer* and their application in postharvest preservation of tomatoes. Food Chem., 285: 380–388.

Lebeis, S.L., Paredes, S.H., Lundberg, D.S., Breakfield, N., Gehring, J., McDonald, M. et al. 2015. Salicylic acid modulates colonization of the root microbiome by specific bacterial taxa. Science. 349: Article 6250.

Lefevere, H., Bauters, L. and Gheysen, G. 2020. Salicylic acid biosynthesis in plants. Front. Plant Sci. 11: Article 338.

Lubbers, R.J.M., Dilokpimol, A., Visser, J., Hilden, K.S., Makela, M.R. and Vries, R.P. 2021. Discovery and functional analysis of a salicylic acid hydroxylase form *Aspergillus niger*. Appl. Environ. Microbiol., 87: Article e02701-20.

Mendoza, D., Cuaspud, O., Arias, J.P., Ruiz, O. and Arias, M. 2018. Effect of salicylic acid and methyl jasmonate in the production of phenolic compounds in plant cell suspension cultures of *Thevetia peruviana*. Biotechnol. Rep., 19: Article e00273.

Mishra, A.K. and Baek, K.-H. 2021. Salicylic acid synthesis and metabolism: a divergent pathway for plants and bacteria. Biomolecules (MDPI). 11: Article 705.

Moosa, A., Farzand, A., Sahi, S.T., Khan, S.A., Aslam, M.N. and Zubair, M. 2021. Salicylic acid and *Cinnamomum verum* confer resistance against *Penicillium* rot by modulating the expression of defence linked genes in *Citrus reticulata* Blanco. Postharvest Biol. Technol. 181: Article 111649.

Naaz, H., Afzal, B., Sami, N., Yasin, D., Khan, N.J. and Fatma, T. 2021. Salicylic Acid Induced Physiological Responses of Microchaete Sp. NCCU-342 Under PQ Stress. DOI: 10.21203/rs.3.rs-839759/v1.

Neto, A.C., Luiz, C., Maraschin, M. and Piero, R.M.D. 2016. Efficacy of salicylic acid to reduce *Penicillium expansum* inoculum and preserve apple fruits. Int. J. Food Microbiol., 221: 54–60.

Ogawa, D., Nakajima, N., Seo, S., Mitsuhara, I., Kamada, H. and Ohashi, Y. 2006. The phenylalanine pathway is the main route of salicylic acid biosynthesis in *Tobacco mosaic virus*-infected tobacco leaves. Plant Biotechnol., 23: 395–398.

Prithiviraj, B., Bais, H.P., Weir, T., Suresh, B., Najarro, E.H., Dayakar, B.V. et al. 2005. Down Regulation of Virulence Factors of *Pseudomonas aeruginosa* by Salicylic Acid Attenuates Its Virulence on *Arabidopsis thaliana* and *Caenorhabditis elegans*. Infect. Immun., 73: 5319–5328.

Rigault, M., Citerne, S., Daubresse, C.M. and Dellagi, A. 2021. Salicylic acid is a key player of Arabidopsis autophagy mutant susceptibility to the necrotrophic bacterium *Dickeya dadantii*. Sci. Rep., 11: Article 3624.

Saikia, R., Kumar, R., Arora, D.K., Gogoi, D.K. and Azad, P. 2006. *Pseudomonas aeruginosa* inducing rice resistance against *Rhizoctonia solani*: Production of salicylic acid and peroxidases. Fol. Microbiol., 51: 375–380.

Salem, E.A. and Shafea, Y.M.E. 2020. Assessment of Salicylic Acid Protective Role Against *Botrytis Cinerea* Induced Damage in Tomato, In Various Storage Periods. Adv. Biotechnol. Microbiol., 15: Article 555923.

Salicylic acid enhances antifungal resistance to Magnaporthe grisea in rice plants. Australasian Plant Pathol., 37: 637–644.

Sanchez, G.R., Mercado, E.C. and Pineda, E.G. 2014. Avocado roots treated with salicylic acid produce phenol-2,4-bis (1,1-dimethylethyl), a compound with antifungal activity. J. Plant Physiol., 171: 189–198.

Sedghi, M. and Toluie, S.G. 2013. Influence of Salicylic Acid on the Antimicrobial Potential of Stevia (Stevia rebaudiana Bertoni, Asteraceae) Leaf Extracts against Soybean Seed-Borne Pathogens. Tropical J. Pharm. Res., 12: 1035–1038.

Shine, M.B., Yang, J.-W., El-Habbak, M., Nagyabhyru, P., Fu, D.-Q., Navarre, D. et al. 2016. Cooperative functioning between phenylalanine ammonia lyase and isochorismate synthase activities contributes to salicylic acid biosynthesis in soybean. New Phytol., 212: 627–636.

Shoala, T., Monir, G.A. and Amin, B.H. 2021. Effects of Salicylic Acid in The Normal and Nano Form Against Selected Fungi That Infect Citrus Trees (Citrus sinensis). Int. J. Sci. Res. Sustain. Dev., 4: 1–14.

Thanh, T.L., Thumanu, K., Wongkaew, S., Boonkerd, N., Teaumroong, N., Phansak, P. et al. 2017. Salicylic acid-induced accumulation of biochemical components associated with resistance against *Xanthomonas oryzae* pv. *oryzae* in rice. J. Plant Interact., 12: 108–120.

Vincent, S.A., Ebertz, A., Spanu, P.D. and Devlin, P.F. 2022. Salicylic Acid-Mediated Disturbance Increases Bacterial Diversity in the Phyllosphere but Is Overcome by a Dominant Core Community. Front. Microbiol., 13: Article 809940.

War, A.R., Paulraj, M.G., War, M.Y. and Ignacimuthu, S. 2011. Role of salicylic acid in induction of plant defense system in chickpea (*Cicer arietinum* L.). Plant Signal Behav., 6: 1787–1792.

Wodnicka, A., Huzar, E., Krawczyk, M. and Kwiecien, H. 2017. Synthesis and antifungal activity of new salicylic acid derivatives. Polish J. Chem. Technol., 19: 143–148.

Yousif, D.Y.M. 2018. Effects Sprayed Solution of Salicylic Acid to Prevent of Wilt Disease Caused by Fussarium oxysporium. IOP Conf. Ser. 1003, Article 012001.

Zhang, H., Huang, Q., Yi, L., Song, X., Li, L., Deng, G. et al. 2021. PAL-mediated SA biosynthesis pathway contributes to nematode resistance in wheat. The Plant J., 107: 698–712.

9

Roles of Gibberellins in Plant Defense against Biotic and Abiotic Stress

Murad Muhammad,[1,2,*] *Abdul Basit,*[3] *Aqsa Arooj,*[4] *Gopal Dixit,*[5]
Muhammad Majeed,[6] *Dwaipayan Sinha,*[7] *Heba I. Mohamed*[8]
and Wen-Jun Li[1,9]

Introduction

A group of plant hormones known as gibberellins is crucial for controlling multiple stages of plant development (Hedden and Sponsel 2015). Gibberellins are categorized by their enantiomeric (ent) structure rather than biological action. Avar et al. (2015) categorize them as ent-gibberellin-ringed cyclic diterpenes. C20 GAs are those that have all 20 carbon atoms, while C19 GAs are those that lack one carbon atom

[1] State Key Laboratory of Desert and Oasis Ecology, Key Laboratory of Ecological Safety and Sustainable Development in Arid Lands, Xinjiang Institute of Ecology and Geography, Chinese Academy of Sciences, Urumqi, 830011, PR China.
[2] University of Chinese Academy of Sciences, Beijing 100049, PR China.
[3] Department of Horticulture Science, Kyungpook National University, Daegu 41566, Korea.
[4] Key Laboratory of Urban Pollutant Conversion, Institute of Urban Environment, Chinese Academy of Sciences, No. 1799 Jimei Road, Xiamen City, Fujian 361021, China.
[5] Department of Botany, Upadhi PG College (MJP Rohilkhand University), Pilibhit, India.
[6] Department of Botany, University of Gujrat, Hafiz Hayat Campus Gujrat-50700, Punjab Pakistan.
[7] Department of Botany, Government General Degree College, Mohanpur, Paschim Medinipur, West Bengal, India.
[8] Department of Biological and Geological Sciences, Faculty of Education, Ain Shams, University, Cairo, 11341, Egypt.
[9] State Key Laboratory of Biocontrol, Guangdong Provincial Key Laboratory of Plant Resources and Southern Marine Science and Engineering Guangdong Laboratory (Zhuhai), School of Life Sciences, Sun Yat-Sen University, Guangzhou 510275, PR China.
* Corresponding author: Muradbotany1@uop.edu.pk

(Sponsel 2016). The GAs that affect higher plants biologically are C19 compounds. A group of closely similar tetracyclic diterpenoid acids known as gibberellins serves an essential function as plant growth hormones. An individual subscription number, GAn, is assigned to each recognized gibberellin, where n generally corresponds to the discovery order. Gibberellic acid (GA) is the first gibberellin with a recognized structural identity. The abundance of known gases is due to the variety of the ent-gibberellin ring system (Toner et al. 2021). Due to this diversity, cells can undergo various structural changes. Gibberellins are produced by higher plants, fungi, and bacteria, and their chemical structure is based on diterpenoid acids that contain isoprene residues (Jan et al. 2021).

Gibberellins play a vital role in various stages of plant growth, including seed germination, stem lengthening, blooming, and the postponement of senescence in several organs (Ali and Baloch 2020). Regulating the quantity of root hairs also influences how well nutrients are absorbed. Transcriptional regulators suppress GA responses called DELLA proteins, which diterpenoids like GAs bind to and activate. The GID1 (GA insensitive dwarf1) receptor carries out the perception of active forms of GA. Because DELLA protein levels are reduced at high GA concentrations, repression of GA responses is diminished (Nelson and Steber 2016). The genera *Azospirillum, Arthrobacter, Agrobacterium, Azotobacter, Pseudomonas, Bacillus, Flavobacterium, Acinetobacter, Clostridium, Micrococcus, Rhizobium, Burkholderia, Xanthomonas* are some examples of Gram-positive and negative bacteria that can synthesize gibberellin (Kour et al. 2019). It is well known that PGPBs with the ability to produce gibberellins promote plant growth and that this promotion is frequently associated with other advantageous physiologic traits of plants, interactions with other hormones, and control of the plant's overall hormonal balance (Salazar-Cerezo et al. 2018). The fact that GA is a mobile signal in plants is essential for several processes involved in plant development and adaptive growth. According to the research (Binenbaum et al. 2018, Muhammad et al. 2021), plants regulate their GA signaling on various levels, including biosynthesis, metabolism, and perception.

Gibberellins are crucial in nearly every aspect of plant life, including growth and development. They regulate various molecular activities, including creating cell membranes, RNA, proteins, cell walls, and shedding. They regulate growth and development at the organ level, start germination, promote flowering, and promote the formation of internodes (stem nodes) (Tassadduq et al. 2022). The hydrolysis of starch to sugar by gibberellins is thought to boost growth by reducing the water potential inside the cell (Castro-Camba et al. 2022). More water can now enter the cell, which causes it to enlarge. They may also hinder development (Rademacher 2016). A subset of these GAs mediates developmental and environmental cues as endogenous plant growth and development regulators. Some GAs have biological action in higher plants. They promote cell elongation and/or division to aid organ growth (Barbosa and Dornelas 2021, Gantait et al. 2015). In addition, they cause developmental changes between stages, such as the vegetative and reproductive stages, the juvenile and adult growth phases, and the latent and germination stages of a seed.

According to several studies, gibberellins aid plants in warding off viral infections by preventing the reproduction and transmission of some viruses (Allie et al. 2014, Carr et al. 2019, Shimura and Pantaleo 2011). These hormones have been demonstrated to activate immune signaling pathways and gene expression related to plant defense responses (Sinha et al. 2023) and the production of antiviral proteins. Gibberellins have been demonstrated to increase plant health and productivity by boosting their immune systems and reducing the severity and spread of viral infections (Alazem and Lin 2015, Qiong et al. 2022). Gibberellins have also been demonstrated to have antibacterial effects against various pathogens, including bacteria and fungi. They can stop the spread of these diseases by interfering with cellular processes or preventing the synthesis of necessary molecules (Toner et al. 2021). Gibberellins prevent cell wall development, which stunts bacterial growth (Shah et al. 2021). These hormones can prevent mycelial growth and spore germination in the case of fungi, protecting plants against infection (Chakraborty et al. 2020, Goswami et al. 2016). Gibberellins' antiviral and antibacterial properties are critical for plant defense and crop production. Using gibberellins' potential might create new approaches to treating viral and microbial illnesses in agricultural contexts. Traditional chemical pesticides may no longer be required if these hormones are used to create ecologically friendly plant protection solutions (Ghosh and Chakraborty 2021).

This chapter investigates gibberellins' ability to inhibit plant viral and bacterial growth. Specifically, we will investigate the hormonal signaling pathways and changes in gene expression that underlie these responses. This chapter also aims to show how gibberellins can be used instead of harmful synthetic pesticides and antiviral agents to protect crops and guarantee agricultural sustainability.

Biosynthesis of Gibberellin

Gibberellins (GAs) are tetracyclic, diterpenoid-containing endogenous plant growth regulators (Wang and Deng 2014). Gibberellins are produced via the terpenoid route and involve three cellular locations namely plastid, endoplasmic reticulum, and cytoplasm (Rizza and Jones 2019). Higher plant GA biosynthesis can be separated into three stages: (i) production of ent-kaurene from geranyl geranyl diphosphate (GGDP) in proplastids, (ii) conversion of ent-kaurene to GA12 via cytochrome P450 monooxygenases, and (iii) creation of C20- and C19-GAs in the cytoplasm (Sun 2008). GGDP is a common predecessor to GAs, carotenoid compounds, and chlorophylls (Ilahy et al. 2019, Hedden et al. 2020). It is transformed to ent-kaurene, the GA pathway's first dedicated intermediary. GAs are diterpenoids produced from trans-geranylgeranyl diphosphate (GGPP), which is cyclized to the tetracyclic hydrocarbon source ent-kaurene via ent-copalyl diphosphate in a pair of steps (Keeling et al. 2010). Ent-kaurene is synthesised in plants primarily in plastids via the methylerythritol 4-phosphate (MEP) pathway, with some contribution from the mevalonic acid (MVA) pathway, which is probably reliant on the influx of isoprenoid intermediaries of GGPP synthesis into the plastids from the cytosol (Davidson et al. 2004, Bajguz and Piotrowska-Niczyporuk, 2023). The two-step conversion of GGPP to ent-kaurene begins with proton-initiated cyclization to the dicyclic ent-copalyl diphosphate (CPP), which is catalyzed by ent-copalyl diphosphate synthase (CPS), a type-II diterpene cyclase. This type of enzyme has a conserved DXDD pattern, with the middle aspartate giving a proton to start the reaction, and a water molecule linked to histidine and asparagine acting as the catalytic base to accept a proton and stop the activity (Köksal et al. 2014, Lemke et al. 2019). In the second stage of the path, ent-kaurene oxidase (KO) and ent-kaurenoic acid oxidase (KAO) catalyze stepwise oxidation followed by ring contraction to create GA12 (Hedden and Sponsel 2015). The committed step in GA-specific biosynthesis is the ring contraction catalyzed by the cytochrome P450 (CYP) ent-kaurenoic acid oxidase (KAO) (Wang et al. 2012). In higher plants, physiologically active gibberellins are biosynthesized from GA12 via two principal parallel processes, the early-non-hydroxylation pathway and the early-13-hydroxylation pathway, which result in GA4 (66) and GA1, respectively. Although gibberellin 13-hydroxylase has not been found in higher plants, these steps are catalyzed by 2-oxoglutarate-dependent dioxygenases (Toyomasu and Sassa 2010). In higher plants, physiologically active gibberellins are biosynthesized from GA12 via two principal parallel processes, the early-non-hydroxylation pathway and the early-13-hydroxylation pathway, which result in GA4 and GA1, respectively. Although gibberellin 13-hydroxylase has not been found in higher plants, these steps are catalyzed by 2-oxoglutarate-dependent dioxygenases (Toyomasu and Sassa 2010). By oxidizing C-20, Gibberellin 20-oxidase (GA20ox) changes GA12 and GA53 into GA9 and GA20, respectively. *Arabidopsis* was the first plant to exhibit the γ-lactone formation type GA20ox, which results in bioactive gibberellins (Lo et al. 2008, Toyomasu and Sassa 2010).

Signaling and Mode of Action of GA

Ueguchi-Tanaka et al. (2005) used the gibberellin insensitive dwarf1 (gid1) mutant of rice to successfully identify the GA receptor, which marked a major step forward in our understanding of the signaling mechanism for this hormone. While *Arabidopsis* has three GID1 genes (GID1a, GID1b, and GID1c), rice only has one. To totally eliminate the GA response in Arabidopsis, all three GID1 genes must be deleted (Schwechheimer, 2012). The proteins they express have strong potential as GA receptors, according to the available information. All three *Arabidopsis* GID1 proteins displayed GA binding activity in *E. coli* at affinities similar to GID1. Second, the GA-insensitive dwarf phenotype of the gid1-1 mutant was restored in rice through the amplification of specific *Arabidopsis* GID1 genes. Finally, yeast two-hybrid studies showed that GID1a-1c interacted with the five *Arabidopsis* DELLAs in a GA-dependent manner (Nakajima et al. 2006). DELLAs are a class of plant-specific TFs that have been shown to have a GRAS domain at their C-termini and a variable N-terminus. In contrast to other members of the GRAS family, DELLAs have a unique N-terminal region (Wu et al. 2011, Aoyanagi et al. 2020). The N-terminal GA-perception region of DELLAs is essential for binding to GID1, while the C-terminal GRAS domain mostly contributes to GA repression through interactions with other regulatory proteins (Du et al. 2017, Blázquez et al. 2020, Xue et al. 2022). N-terminal domains are characterized by polymeric Ser/Thr/Val (poly S/T/V) motifs and the DELLA and VHYNP domains, both of which are required for GID1 binding (Willige et al., 2007, Xue et al., 2022). The C-terminal domain of GID1 undergoes a conformational shift upon binding GA due to increased extracellular GA levels (Harberd et al. 2009). In order to create a binding domain (exposed hydrophobic surfaces) for DELLA, the synthesis of GA-GID1 causes a conformational change in the N-Ex of GID1. DELLAs play a pivotal role as the key regulatory nodes in this GA signaling transduction system. Five growth-inhibiting proteins, RGA-LIKE 1 (RGL1), RGL2, and RGL3 REPRESSOR OF ga1-3 (RGA), GA-INSENSITIVE (GAI), and RGA, make up the transcriptional regulators of the GRAS family, known as DELLAs (Bao et al. 2019).

SLR1 in rice (Hirano et al. 2012), GIBBERELLIC ACID INSENSIVE(Peng et al. 1999), REPRESSOR-OF-ga1-3, and RGA-LIKE1 (RGL1), RGL2, and RGL3 in *Arabidopsis* (Achard and Genschik 2009), are all DELLA growth repressors that the soluble GID1 proteins interact with after hormone attachment. The C-terminal domain of DELLA proteins undergoes a conformational change when GA binds to its receptor GID1 protein, promoting the formation of the GA-GID1-DELLA complex. As a result, DELLA is polyubiquitinated by the E3 ubiquitin ligase SCFSLY1/GID2 and degraded by the 26S proteasome (Xue et al. 2022). Flowering locus T (FT) is transcriptionally triggered by GA in leaves to induce blooming, especially during long days (LDs) (Fukazawa et al. 2021).

Compared to plants with defective GA signaling or the GA-deficient mutant ga1-3, FT production is greatly enhanced in plants treated with exogenous GA or in mutants with continuously active GA (Galvo et al. 2012). Gibberellin also regulates hypocotyl elongation, another way. Both the DELLA protein GA INSENSITIVE (GAI) and the DELLA protein REPRESSOR OF ga1-3 (RGA) act as repressors of the GA signaling pathway and prevent hypocotyl extension when exposed to red light (de Leucas et al. 2008, Feng et al. 2008). Since these DELLA proteins are degraded by the proteasome in the presence of white light, their inhibitory effect on hypocotyl elongation is no longer present (King et al. 2001).

As a visual indicator of plant stress, anthocyanin accumulation has been reported (Li and Ahammed 2023). The expression of genes-encoding anthocyanins in higher plants is controlled by the conserved MBW complex, which consists of the MYB, bHLH, and WD40 subunits. By binding competitively to bHLH and MYB/ bHLH, respectively, MYBL2 and the JAZ family of proteins inhibit MBW activity (Xu et al. 2015). Under normal conditions or in the presence of GA, DELLAs are degraded to produce MYBL2/JAZs, which prevent the formation and function of MBW complex and turn off the expression of anthocyanin biosynthesis genes. To the contrary, DELLAs assemble and sequester MYBL2/JAZs in response to abiotic stresses, leading to the formation of the MBW complex, which in turn stimulates anthocyanin synthesis and ultimately aids in stress tolerance (Xie et al. 2016, Muhammad et al. 2021). Gibberellins (GAs) play an important role in plant development by stimulating organ growth and triggering phase transitions at critical points. Evidence for the involvement of long-day (LD) and biennial plants in flower initiation is well-established, and our understanding of the mechanisms behind floral induction is expanding (Mutasa-Göttgens and Hedden 2009). The *Arabidopsis* DELLA-BRM-NF-YC module formation reveals the involvement of BRAHMA (BRM), a core subunit of the chromatin-remodeling SWItch/sucrose nonfermentable (SWI/SNF) complex that regulates gene expression and is involved in numerous biological processes. DELLA, BRM, and NF-YC are all transcription factors that interact with one another, specifically, DELLA proteins improve the physical contact between BRM and NF-YC proteins. By preventing NF-YCs from binding to SOC1, an essential floral integrator gene, flowering is stifled. On the other hand, DELLA proteins facilitate BRM binding to SUPPRESSOR OF OVEREXPRESSION OF CONSTANS1 (SOC1). Early flowering is caused by the GA-induced degradation of DELLA proteins, which in turn disrupts the DELLA-BRM-NF-YC module, prevents BRM from inhibiting NF-YCs, and decreases BRM's DNA-binding capacity. Phosphorus (P) is the second-most important macronutrient for plants, and its deficiency can stunt their development. Root hair elongation and anthocyanin accumulation are just a couple of the adaptively relevant responses to Pi loss in plants, and research shows that Pi deficiency decreases bioactive GA levels, which causes DELLA accumulation.

Overview of the Importance Of Antiviral and Antimicrobial Effects in Plants

The antiviral and antibacterial capabilities of plants significantly impact the plants' overall health, growth, and ability to endure. An explanation of the relevance of these results is presented as follows:

Disease Resistance

Plants are subject to constant exposure to a diverse range of microbial and viral pathogens, which have the potential to impede their growth, reduce productivity, and induce various diseases (Miller et al. 2017). The antiviral and antimicrobial properties aid in mitigating the severity and transmission of diseases by supporting plants in their defense against pathogens. The results above contribute to plants' overall welfare and viability by enhancing plant resistance (Huang et al. 2016).

Crop protection

Crop plants are vulnerable to harm by microbes and viruses, which can reduce yields and have other adverse economic effects. Because of its antiviral and antibacterial capabilities, this compound can be used as a replacement for or in addition to chemical pesticides (Divekar et al. 2022). Utilizing plants' natural defensive systems can lessen reliance on synthetic pesticides and create more sustainable crop protection techniques (Pathania et al. 2021).

Enhanced plant immunity

The ability of plants to fight against bacterial and viral diseases depends on their immune system. These actions can activate genes linked to defense systems, improve immunological signaling pathways, and trigger the creation of antimicrobial substances. The immune system of plants is strengthened by their antiviral and antibacterial capabilities, increasing their resistance to illnesses and their chances of survival (Das et al. 2022, Fang and Ramasamy 2015).

Sustainable agriculture

Antiviral and antibacterial properties in plant protection are consistent with sustainable agriculture principles. When opposed to using only chemical treatments, the employment of naturally occurring defensive systems can lessen the adverse effects of pesticide use on the ecosystem. Furthermore, it can aid in maintaining beneficial creatures and biodiversity, resulting in a more environmentally responsible and well-balanced agricultural system (Domonkos et al. 2021).

Description and Classification of GA in Plants

Gibberellins are a subclass of plant hormone compounds from the tetracyclic diterpenoid class. These substances originate from the mevalonic acid pathway, which goes through several enzymatic steps and finally produces molecules that are precursors to gibberellins (Fig.1) (Wahab et al. 2022).

The precursor molecules are then put through several enzyme processes, such as oxidations, reductions, and other changes, to become active gibberellins. The fungus *Gibberella fujikuroi*, which was discovered to produce substances that encouraged the elongation of rice seedlings, gave gibberellins their name (Table 1) (Rademacher 1994). Gibberellins were first discovered in these compounds. Since then, gibberellins have been found in various plant species, and it has been shown that they can control a wide range of factors involved in a plant's growth and development (Binenbaum et al. 2018). Up to this point, 136 gibberellins have been identified and divided into several classes based on their biological functions and unique structural makeup features. Gibberellic acid (GA_3), the most well-known and thoroughly studied gibberellin, is the standard gibberellin for the GA group. According to Hernandez-Garca et al. (2002), GA1, GA4, GA7, GA19, GA20, and GA24 are other important gibberellins. These gibberellins come in various chemical configurations and play multiple roles in plant development and growth. The total number of gibberellins is 136. However, only gibberellin GA3 and a combination of gibberellins GA4 and GA7 are used in

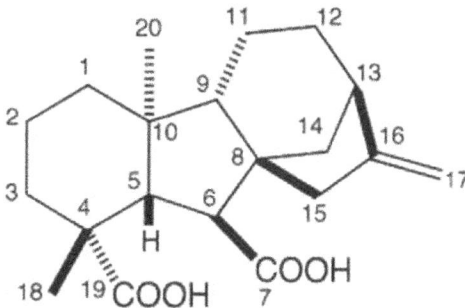

Fig. 1. Basic gibberellane structure, with numbering of constituent carbon atoms as indicated (Peters 2013).

Table 1. Gibberellins are produced by microorganisms.

Gibberellin	Microorganism*					
GA$_1$	G			P	R	A
GA$_2$	G					
GA$_3$	G		N			A
GA$_4$	G	S		P	R	
GA$_7$	G					
GA$_9$	G	S		P	R	
GA$_{10}$	G					
GA$_{11}$	G					
GA$_{12}$	G			P		
GA$_{13}$	G	S				
GA$_{14}$	G	S				
GA$_{15}$	G	S		P		
GA$_{16}$	G					
GA$_{20}$	G			P	R	
GA$_{24}$	G	S		P		
GA$_{25}$	G	S		P		
GA$_{20}$	G	S				
GA$_{36}$	G	S				
GA$_{37}$	G					
GA$_{40}$	G					
GA$_{41}$	G					
GA$_{42}$	G					
GA$_{47}$	G					
GA$_{54}$	G					
GA$_{55}$	G					
GA$_{56}$	G					
GA$_{57}$	G					
GA$_{78}$	G					
GA$_{82}$				P		

*G = *Gibberella fijikuroi*
S = *Sphaceloma manihoticola* and *forther species* R = *Rhizobiumphaseoli* P = *Phaeosphaeria* sp.
A = *Azospirillum lipoferum* and *A. brasilense* , N = *Neurospora crassa*

commercial applications (Cui et al. 2020). A few gibberellins, including GA1, GA3, GA4, and GA7 (Rademacher 2015), are thought to have intrinsic biological activity in higher plants. This is the case regardless of the number of GAs. One of the most often utilized plant growth hormones is GA3. More recent research has focused on additional GA hormones, including GA4 and GA7, because of their unique effects on plants. Some research (Hedden and Thomas 2012, Niu et al. 2015, Qian et al. 2018) suggests that GA4+7 can successfully promote fruit set and improve cucumber, pear, and apple fruit size.

Qian et al. (2018) found that compared to pollinated fruit, fruit treated with GA4+7 developed more sugar and fewer organic acids. As a result, utilizing GA4+7 as a farming treatment for horticulture crops has been demonstrated to raise production and income substantially. However, the application of GA4+7 in cereal crops is restricted, and the physiological mechanism by which exogenous GA4+7 regulates maize grain filling remains elusive (Cui et al. 2020). As a result, we have no idea how GA4+7 will affect the density of maize grains at this time.

The Role of GA in Plant Growth and Development

Gibberellins are a plant hormone class with many roles in plant growth and development. Stem elongation is a well-known function of gibberellins. They stimulate cell division and growth at stem internodes, increasing plant stature (Toungos 2018). Plants like rice and wheat, where their function may be more pronounced, rely heavily on gibberellins for proper stem elongation. For dormant seeds to start the germination process, gibberellins must be present. Hydrolytic enzymes, such as amylases, which break down starch and other storage components in the seed, are produced at a higher rate (Bilal et al. 2018). This is done so the seedlings can receive the nutrients they need to flourish.

Gibberellins eliminate several physiological and environmental barriers to germination (Rafiullah et al. 2021). Gibberellins have a role in controlling flowering or the progression of a plant from its vegetative to its reproductive state. They affect the initiation and culmination of flower development and the expansion of floral structures. Nkhata et al. (2018) found that gibberellins regulate flowering through interactions with auxins and cytokinins. Fruit growth and development require gibberellins, a type of plant hormone. They cause the fruit to grow in size by fostering cellular replication and multiplication inside the fruit. Ripening isn't the only thing that gibberellins affect, they also affect sugar buildup and various other qualitative traits, such as the maturation of color. Research demonstrates that gibberellins can occasionally slow fruit ripening. For instance, citrus plantations use gibberellic acid (GA) to modify flowering and fruiting and to prevent or alleviate the severity of some physiological issues caused by environmental factors (Oleska et al. 2020).

The chemical must be used at the correct times and amounts to be effective. Since GA promotes cell division and elongation, it has been used for years to control certain horticultural crops' flowering and fruit development (Bisht et al. 2018). The substance was given to the plants, and that did the trick. There are four main benefits to using GA on citrus, as stated by Garmendia et al. (2019). Among these are the amelioration of albedo breakdown, the lessening of watermark (especially on *Imperial mandarin*), the lessening of oleocellosis, and the enhancement of rind quality.

The mesocarp or albedo (the layer of white internal rind) separates from the exocarp or flavedo (the layer of orange or yellow external rind), causing the rind to develop creases and often cracks, this process is known as albedo breakdown. One term for this phenomenon is 'albedo breakdown.' It manifests itself in refined, deeply excavated grooves in the rind, which, when they cross over, give the impression of a lumpy, mushy fruit. This is a severe issue since the pressure inside the container could cause the fruit to explode (Ali et al. 2022). Combining GA with up to three calcium (Ca) sprays is indicated for the maximum level of control. Maintaining a healthy weight and drinking enough water is also crucial. High phosphorus (P) levels are associated with thinner rinds are more prone to developing albedo. Increases in nitrogen (N) and potassium (K) concentrations have been observed in denser rints. Producing economically acceptable yields in the areas prone to the condition may need tolerance of rinds with a larger thickness. Moisture stress, whether from a shortage of moisture or an excess of moisture, dramatically increases the occurrence and severity of albedo (Pal et al. 2016, Uthman and Garba 2023).

The growth of leaves is affected by gibberellins because they stimulate cell division and elongation. They accomplish this by changing the proportions of the plant's leaves, which in turn modifies the plant's anatomy. The natural aging process of leaves, called senescence, is regulated by gibberellins (Yang et al. 2020). The plant hormone gibberellin is crucial in developing tubers in plants like potatoes. They affect the initiation of underground storage structures, or tubers, and their subsequent development. Gibberellins encourage cell division in the stolon, the stem-like structure from which tubers emerge. Food-storing tuber plants have stolons. (Kpczynski 2018).

Antiviral Effects of Gibberellins

Infectious diseases of plants pose significant challenges to agricultural production and the safety of food supplies. Gibberellins have been known for a long time for their functions in the growth and development of plants. However, recent research (Li et al. 2022, Narayanan and Glick 2023) has shown that gibberellins also exhibit antiviral features. According to Nkhata et al. (2018), for researchers to develop novel methods for battling viral infections in plants, they need to understand how gibberellins suppress the proliferation of viruses and activate innate immunity. Gibberellins, found in plants, have been shown to have antiviral effects, which bodes well for developing new methods to control viral infections in agricultural settings. Gibberellins have two possible mechanisms of action, the first of which is the direct inhibition of viral replication, and the second is the stimulation of the body's innate immune response (Hedden 2020). Both of these mechanisms suggest that gibberellins have the potential to be utilized as natural antiviral drugs.

Mechanism of action of GA in inhibiting viral growth

The antiviral properties of gibberellins have been associated with many processes. Starting with generating viral RNA or proteins, gibberellins can stop viral replication directly. This interference with the viral replication machinery reduces viral load and delays viral dissemination within plant cells (Allie et al. 2014). Additionally, gibberellins influence the expression of genes associated with defense implicated in antiviral responses. The expression of genes encoding pathogenesis-related (PR) proteins, which are known to have an antiviral effect, has been demonstrated to be stimulated by them. Gibberellins also activate the signaling pathways for salicylic acid (SA) and jasmonic acid (JA), which are essential for plant defense against viral infections (Carr et al. 2019).

Gibberellins significantly strengthen a plant's inherent defenses against viral infections. Antiviral agents such as reactive oxygen species (ROS) and phytoalexins are produced due to their activation (Yang et al. 2022). Gibberellins trigger systemic acquired resistance (SAR), a potent immune response that protects against viruses (Shah et al. 2014).

Gibberellins have been shown to have antiviral activities in various plant-virus systems across several investigations (Calil and Fontes 2017, Carr et al. 2019, Gupta et al. 2021, Zhao and Li 2021). To give you an example, researchers found that administering gibberellic acid (GA3) to tobacco plants that were infected with cucumber mosaic virus (CMV) led to a considerable reduction in the amount of viral replication as well as a less severe manifestation of the infection's symptoms. Previous research has shown that gibberellins can successfully prevent the buildup of Tomato yellow leaf curl virus (TYLCV) in tomato plants, hence increasing the plants' resistance to this harmful viral infection, it has been established by tests in which gibberellins were administered to tomato plants. (Sofy et al. 2020, Sun et al. 2016).

Antimicrobial Effects of Gibberellins

Plant-associated bacteria produce gibberellins, which regulate the host plant's internal activities, including growth and development (Keswani et al. 2022). These processes include stem elongation, germination, flowering, and cell expansion. The gibberellin-insensitive dwarf1 (GID1) protein is a nuclear-localized receptor for GA phytohormones, via which they exercise their impact in plants. The binding and interaction of bioactive GA with the GID1 protein leads to allosteric changes in GID1. This means it will likely interact with DELLA growth repressor proteins like gibberellin insensitive (Davière and Achard 2016), generated in response to environmental stresses like osmotic or thermal stress. This contact causes a conformational change in the DELLA protein, making it more amenable to degradation via ubiquitin-mediated pathways (Keswani et al. 2022). Plant health and agricultural output are both seriously threatened by microbial infections. Despite their long-held association with the mechanisms that promote plant growth, gibberellins have recently been scrutinized for their effectiveness as antibacterial agents. Understanding how gibberellins inhibit microbial multiplication and generate systemic acquired resistance in plants is crucial for developing effective techniques for controlling microbial infections ((Kamal et al. 2015).

Mechanism of action of GA in inhibiting microbial growth

There are many ways in which gibberellins (GAs) restrict microbial growth. GAs first interfere with DNA replication, transcription, and translation, all vital metabolic processes in microbial organisms. This disruption impairs the microbial cells' normal functioning, growth, and reproduction (Ismail et al. 2021). Furthermore,

GAs change the shape and function of microbial cell membranes, which allows more cellular contents to leak out. This diminished membrane integrity further hinders microbial survival and growth (Harman et al. 2021). In addition, GAs encourage the production of phytoalexins and antimicrobial peptides in plants. Phytoalexins are secondary metabolites with direct antibacterial activity, while antimicrobial peptides are small proteins with broad-spectrum antimicrobial properties (Meena et al. 2020). In addition, GAs activate salicylic acid (SA), jasmonate (JA), and ethylene (ET) signaling pathways, all of which regulate plant defenses against microbial infections. Defense-related genes are elevated, and antimicrobial compounds are generated when these pathways function. In addition to influencing signaling pathways involved in plant defense responses, GAs also affect the levels of reactive oxygen species (ROS) and nitric oxide (NO). Gibberellins function in many ways to reduce microbial growth and promote the plant's defense against microbial illnesses (Figueiredo et al. 2016, Chen et al. 2018, Khan et al. 2023, Meena et al. 2020, Yang et al. 2015).

GA-Induced Systemic Acquired Resistance in Plants

Gibberellins substantially influence systemic acquired resistance, a defense mechanism that provides long-lasting and comprehensive immunity against microbial infections. By activating defense-related genes and pathways, gibberellin causes plants to produce pathogenesis-related (PR) proteins and other defensive compounds (Jain and Khurana 2018). Gibberellins increase the synthesis of nitric oxide (NO) and reactive oxygen species (ROS), which are secondary mediators in the signaling pathways associated with systemic acquired resistance. These signaling molecules activate systemic defense responses in healthy plant tissues, safeguarding the plant against a broad spectrum of microbial pathogens (Verma et al. 2020).

Gibberellins have been shown in various plant-microbe interactions to possess antibacterial capabilities, as proven by many studies (Abdelsattar et al. 2023). It has been established that treating tomato and rice plants with gibberellic acid (GA$_3$) reduces the formation and growth of bacterial illnesses such as *Pseudomonas syringae* and *Xanthomonas* spp. One example is treating tomato plants with GA3 is similar to treating rice plants with GA$_3$. Similarly, it has been demonstrated that gibberellins can prevent the spread of fungi such as *Botrytis cinerea* and *Fusarium spp.* in *Arabidopsis* and strawberry plants (Walid and Ibrahim 2019).

The use of gibberellins in plant protection necessitates, among other things, the following fundamental components: GAs can combat various microbial ailments, including bacterial and fungal infections. They can lessen disease frequency and severity by preventing pathogenic bacteria's growth and development (Naqvi 2019). GAs interfere with important bacterial metabolic activities, cause cell membrane damage, and encourage the formation of antimicrobial peptides and phytoalexins, among other antimicrobial chemicals. GAs are helpful in disease management due to their antibacterial properties (Paul and Roychoudhury 2021). GAs trigger a component of plant systemic acquired resistance. SAR is a defense mechanism that offers comprehensive and continuing protection against various diseases (Klessig et al. 2018). Plants exposed to GAs produce pathogenesis-related (PR) proteins and other defense-related substances by activating defense-related genes. This activation improves the plant's ability to combat microbial diseases, even in unaffected areas of the plant (Jain and Khurana 2018).

Gibberellins can improve several plant defensive systems, including generating defense-related chemicals, cell wall fortification, and defense-related communication pathways. They promote the production of phytoalexins, which are antibacterial chemicals produced by plants in response to pathogen attacks (McNeil et al. 2018). Plant defense is facilitated by reactive oxygen species (ROS) and nitric oxide (NO), both of which are affected by GAs (Prakash et al. 2019). Gibberellins are a natural alternative to synthetic pesticides for plant protection. When GAs are obtained from plants, the environment and non-target animals are considered safe in their vicinity (Costa et al. 2019). They are less hazardous than synthetic pesticides, lowering the risk of pesticide residues and environmental damage. Using GAs for plant protection allows farmers to practice sustainable agriculture by reducing their dependency on synthetic pesticides (Meena et al. 2020). Gibberellins can be used in integrated pest management (IPM) programs, emphasizing various approaches to reduce pests and illnesses. They can be used with other strategies like resistant crop varieties, biological control agents, and cultural norms. Pest and disease control techniques can be made more sustainable and efficient by introducing GAs into IPM programs (Sharma and Bakshi 2022).

Use of GA as a Natural Alternative to Synthetic Pesticides

As an alternative to synthetic pesticides, gibberellic acids (GAs) are increasingly used to protect plants. They can treat various diseases and insects because of their antibacterial and insecticidal capabilities. By hindering the spread of bacteria and fungi, GAs can lessen the occurrence and impact of illness (Sansinenea, 2019). They can also delay the development of insects, which could reduce the overall pest population. GAs help to improve plant health and decrease agricultural losses by targeting pests and diseases (Yatoo et al. 2021). GAs are a non-toxic alternative to synthetic pesticides, one of their main benefits. The environmental impact of GAs made from plants is minimal (Sharma et al. 2019). They don't stick around as long and don't harm anything besides the pest they're meant to kill, including beneficial insects, pollinators, and natural predators. GAs can promote ecological harmony and biodiversity in agricultural systems while lessening the environmental impact of conventional pesticides (Garraway 2020).

The concepts of sustainable agriculture are compatible with the usage of GAs. Farmers can lessen their reliance on synthetic pesticides and the overall quantity of chemicals used in agricultural systems by switching to natural alternatives like GAs. This method increases agricultural sustainability, protects biodiversity, and saves money in the long run (Puglia et al. 2021). Also, additional components of an IPM (Integrated Pest Management) strategy can be incorporated into pest management plans if GAs are used. To optimize the efficiency and long-term viability of pest and disease control, it is advisable to employ genetic algorithms in conjunction with biological control agents, cultural approaches, and resistant crop varieties (El-Hack et al. 2018).

Pros and Cons of Using GA in Plant Protection

Gibberellins (GAs) provide a lot of advantages for plant defense methods. GAs provide a non-chemical substitute for farmers who want to wean themselves off chemical pesticides. Plant-based GAs are thought to be environmentally beneficial and support sustainable farming methods. Compared to synthetic pesticides, they are less dangerous for use around non-target organisms such as beneficial insects, pollinators, and natural predators. This helps to maintain the farmland's natural balance and biodiversity (Shukla 2020). The capacity of GAs to pinpoint specific diseases and pests is one advantage. They efficiently control some pest populations by stopping the growth of insects and limiting the transmission of infectious diseases. With less impact on beneficial species, this tailored method enables farmers to focus on certain pests or illnesses (Lahlali et al. 2022). Systemic acquired resistance (SAR) is another potential adverse effect of GAs on plants. This defense process strengthens the plant's resistance to several diseases and offers ongoing, all-encompassing security. GAs can lessen the need for pesticide applications by enhancing the plant's natural resistance to pests (Chatterjee and Chattopadhyay 2023, Sansinenea, 2019). Sustainable agricultural methods can coexist with the usage of GAs for plant protection. Synthetic pesticide usage is decreased, reducing environmental contamination and dangers related to environmental residues. Farmers can contribute to the long-term ecological viability of agricultural systems and the conservation of natural resources by employing natural alternatives like GAs (An et al. 2022).

A variety of rules and considerations apply to the use of GAs for plant protection. The target organism and environmental conditions can impact how effective GAs are. If GAs are ineffective against particular organisms, other pest and virus control strategies may be required (Lahlali et al. 2022). Because environmental factors like temperature, humidity, and plant type can also impact the effectiveness of GAs, it is essential to carefully consider the proper application settings (Farhangi et al. 2023). Additionally, GAs could need to be used at higher concentrations than synthetic pesticides, which could increase costs and labor requirements (Behera et al. 2021). It cannot be easy to create GAs that have adequate distribution and coverage. The poor potency of GAs compared to some synthetic pesticides is another drawback. Even though these methods have shown potential in combating particular illnesses and pests, they might only be helpful when applied in concert (Sharma et al. 2023).

Application of Gibberellins in Plant Protection against Abiotic Stress

Gibberellins (GAs), which alter plant physiology and defense processes in many ways, are now recognized as possible plant protection strategies (El-Maraghy et al. 2021). In addition to several advantages, using GAs to manage plant diseases and pests improves sustainable agricultural methods (Fig. 2) (Harman et al. 2021). It is becoming increasingly clear that the growth hormones of the gibberellin (GA) class play a crucial role in the adaptation to abiotic stress. Growth inhibition in plants has been linked to reduced GA levels and signaling in response to a variety of conditions (Table 2). These challenges include exposure to cold, salt, and osmotic stress.

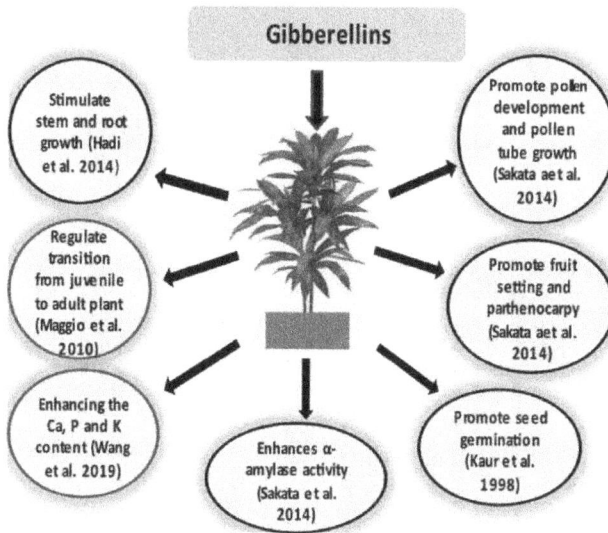

Fig. 2. Gibberellins control a wide variety of plant processes, as depicted in this diagram (Niharika et al. 2021).

Future Perspectives on the Use of GA in Plant Protection

The potential of gibberellins (GAs) for plant protection is only beginning to be explored. New research results and plans for GA application in plant protection are presented here. Ongoing studies aim to improve GAs by creating more potent formulations and delivery technologies (Trivedi et al. 2021). Encapsulation methods and nanoformulations enhance the stability, targeted dispersion, and prolonged release of GAs. These developments can potentially enhance the efficiency and effectiveness of GA application, leading to better plant protection results (Tleuova et al. 2020).

Further studies will focus on determining the most effective combinations of GAs with other natural substances, biocontrol agents, or synthetic pesticides (Anuar et al. 2023) to maximize each treatment's effectiveness. When GAs are used with good management practices, they can be much more effective and have a more comprehensive range of activity against diseases and pests. Combining them can boost pest control and reduce our reliance on synthetic pesticides (Lahlali et al. 2022), which is the goal of integrated pest management. Targeted use of GA has become possible with the development of precision agricultural technologies (Camara et al. 2018). Site-specific software has the potential to maximize GA use, reduce waste, and lessen environmental consequences because it relies on real-time monitoring and data-driven decision-making (Chamara et al. 2022). The optimal time, dose, and administration of GAs can be achieved with the help of remote sensing, imaging technologies, and sensor-based systems, as reported by Liao et al. (2020).

Biotechnology and genetic engineering can potentially boost the effectiveness of GA production and plant protection (Shelton et al. 2020). Plants' tolerance to pests and diseases can be enhanced by increasing endogenous GA levels, which can be achieved by manipulating GA biosynthesis pathways. Additionally, genetic

Table 2. Effects of various stresses and gibberellic acid's (GA) mitigating effects on diverse plant species.

Stress	Plants	Effects	Conc. Of GA	Mitigating effect by GA	References
Heavy metal (nickel)	*Triticum aestivum*	Reduced plant height, increased electrolyte leakage, decreased antioxidant enzyme activity, dried fresh weight, and decreased root length and MDA chlorophyll concentration.	10^{-6} M and 10^{-8} M	Reduced electrolyte leakage and increased plant height, fresh weight, dry weight, antioxidant content, enzyme activity, and chlorophyll content.	Siddiqui et al. 2011
Heavy metal (nickel)	*Glycine max*	Antioxidant enzyme activity, chlorosis, chlorophyll content, and oxidative stress all cause a decrease in dry weights.	0.05 mM	Growth, dry weights of the roots and shoots, chlorophyll content, and activity of the antioxidant enzymes all increased.	Saeidi-Sar et al. 2007
Heavy metal cadmium, molybdenum	*Hordeum vulgare*	Reduce alkaline phosphatase, acid, radicles, protein α-amylase activity, and sugar content.	0.5 μM	Acid and alkaline phosphatase activity and sugar and protein content in radicles were elevated.	Amri et al. 2016
Heavy Metal Chromium	*Pisum sativum*	Decreased protein, NR activity, nitrogen concentration, and seed germination	10 and 100 μM	Increased protein content, nitrogen content, NR activity, seed germination, and plant length	Gangwar et al. 2011
Heavy metal Cadmium	*Arabidopsis Thaliana*	A reduction in root and shoot growth, chlorosis, a rise in MDA levels, and elevated lipid peroxidation.	5 μM	Improved root and shoot development, less lipid peroxidation, and lower MDA concentrations	Zhu et al. 2012
Heavy Metal Cadmium	*Parthenium Hysterophorus*	Reduced biomass and plant growth	10^{-9}, 10^{-7} and 10^{-5} M	Increased biomass and plant growth	Hadi et al. 2014
Heavy metal Cadmium	*Brassica napus*	Less fresh and dry weight, chlorophyll, seed germination, and more MDA.	50 μM	The amount of chlorophyll and the fresh and dry weight both increased	Meng et al. 2009
Heavy Metal Nickel	*Vigna radiata*	Reduced fresh weight, dry weight, shoot and root length, chlorophyll content, photosynthetic pigments, and oxidative stress are all signs of stress.	10^4M	Expanded shoots and roots, increased fresh weight and dried weight, and produced more photosynthetic pigments and chlorophyll.	Ali et al. 2015
Heavy Metal Copper	*Pisum sativum*	Disrupted cellular homeostasis and induced oxidation.	1 μM	Thioredoxin/ferredoxin systems under control	Ben Massoud et al. 2018

Stress	Species	Observation	Concentration	Result	Reference
Low temp stress	*Solanum lycopersicum*	Proline, CAT, MDA, POD, and electrolyte leakage increased.	0.2 mmol L^{-1}	Decreased levels of malondialdehyde (MDA), peroxide dismutase (POD), proline, and catalase. Reduced loss of electrolytes.	Ding et al. 2015
Low temp stress	*Oryza aestivum*	Results in a sharp fall in the amount of developed pollen and the seed germination rate.	10 μM	Enhanced seed germination rate and encouraged maturation of pollen.	Sakata et al. 2014
Low temp stress	*Anacardium occidentale*	The declined postharvest quality	180 mg L^{-1}	Enhanced postharvest quality without affecting the physicochemical factors by reducing both mass and firmness loss	Souza et al. 2016
Low temp stress	*Potato tubers*	The sweetening of potatoes induced by the cold during postharvest storage	-	We have demonstrated a reduction in dormancy, increased reducing sugars, and increased sprouting.	Xie et al. 2018
Low temp stress	*Solanum lycopersicum*	Chilling injury (CI) in fruits severed	0.5 mM	Decrease in the chilling injury (CI) index	Zhu et al. 2016
Allelopathic stress (Juglone)	*Lepidium sativum*	Reductions in germinating seeds' viability, plant growth (including root and shoot development), wet and dry weight	1 mM	Improvements in both wet and dry seed weight, root and shoot development, and germination rates.	Terzi and Kocacaliskan 2009
Salt stress	*Triticum aestivum*	There was an increase in proline concentration but a decrease in leaf area, photosynthetic pigments, carbohydrates, and proteins.	100 mg L^{-1} 10 ml of 100 ppm	The amount of protein, carbohydrates, and photosynthetic pigments in the leaves increased.	Shaddad et al. 2013
Salt stress	*Brassica juncea*	Decreases in RWC, photosynthetic leaf area, shoot and root length, and fresh and dry weight.	75 ml pot^{-1}, conc. 75 mg l^{-1}	Increased fresh and dry weight, shoot and root length, Leaf RWC, reduction in chlorophyll content, and increases in photosynthetic pigment. Electrolyte leakage was also reduced.	Ahmad 2010
Salt stress	*Beta vulgaris*	The parameters and seed germination percentage dropped, but the MGT rose.	0,100, and 200 ppm	Mean germination time (MGT) they were reduced, while seed germination and germination percentage parameters increased.	Kandil et al. 2014

Table 2. contd. ...

... Table 2. contd.

Stress	Plants	Effects	Conc. Of GA	Mitigating effect by GA	References
Salt stress	*Abelmoschus esculentus*	Reduced the number of growth characteristics	0.1 mM	Increased Ca, Fe, K, Ca, and Mg, concentrations, as well as osmoprotectants	Wang et al. 2019
Salt stress	*Zea mays*	RWC, dry weight, chlorophyll, and leaf electrolyte loss all increased.	50 ppm and 100 ppm	Electrolyte leakage was reduced when leaf RWC, dry weight, and chlorophyll contents increased.	Tuna et al. 2008
Salt stress	*Phaseolus vulgaris*	There was a decrease in the leaf area, fresh weight, photosynthetic pigments, dry weight protein, activity, antioxidant enzyme, anthocyanin content, and hydrogen peroxide (H2O2) content. At the same time, there was a rise in MDA levels.	0.05 mM	Increases occurred in these variables as well: fresh and dry weight, leaf area, dry weight, and photosynthetic pigments, as well as protein anthocyanin content.	Saeidi-Sar et al. 2013
Salt stress	*Glycine max*	Reduced chlorophyll, root, and shoot length, fresh and dry biomass, and endogenous ABA, SA, and JA levels.	0.5 μM, 1.0 μM, and 5.0 μM	Increased amounts of isoflavones, endogenous SA, ABA, and JA, as well as chlorophyll content, root and shoot length, and dry and fresh biomass plant	Hamayun et al.
Salt stress	*Oryza sativa*	Reduced concentrations of certain lipids	10 μM	Increased lipid biosynthesis	Liu et al. 2018
Drought stress	*Zea mays*	The length of the cob, the seed weight, the diameter of the cob, the diameter of the stem, the chlorophyll index of the leaves, and the length of the internodes all decreased.	50, 100, and 150 mg^1	The plant grew taller, the stem was wider, the leaves were greener (higher SPAD values), the seeds were heavier, and the cob was wider and longer.	Akter et al. 2014
Drought stress	*Zea mays*	Increased electrolyte leakage and relative water content in the leaf, larger root: shoot ratio, lower total chlorophyll concentration, lower dry weight.	25 and 50 mg L^1	Electrolyte leakage decreased, while leaf relative water content, dry weight, and total chlorophyll content increased.	Kaya et al. 2006

modification techniques allow adding the GA synthesis genes to vulnerable crop kinds, imparting intrinsic resilience (Jogawat et al. 2021). Climate change poses significant risks to plant health and can potentially affect the dynamics of pests and diseases. The development of ways to help plants become more resistant to the effects of climate change is likely to be the focus of future research on using GAs for plant protection. GAs can potentially improve agricultural output and lessen the impact of abiotic stresses on plant health in various settings (Divekar et al. 2022). Regulatory conformity and market acceptance will become increasingly crucial as GAs become widely used as plant protection tools. Future work will address regulatory concerns, guarantee product safety, and evaluate the possible effects of GAs on the environment and human health (Mittal et al. 2020). Natural gas's market adoption and acceptance can be increased by educating farmers, stakeholders, and consumers on its benefits and safety (Hofmann et al. 2020).

Conclusion

The plant hormones gibberellins (GAs) have been shown to offer great potential for application in sustainable farming and pest control. Thanks to their antiviral and antibacterial properties, plants can use GAs as powerful weapons against microbial and viral illnesses. GAs reduce viral protein synthesis, block viral replication, and boost the plant's natural defenses against pathogens. A further aspect contributing to the plant's overall resilience is the capacity of the plant's defensive mechanisms to elicit systemic acquired resistance (SAR). Using GAs for crop protection and agriculture has numerous benefits. They reduce the need for chemical-based treatments by serving as a natural alternative to synthetic pesticides. Since GAs are less toxic, they are less likely to harm non-target creatures, which is good for maintaining ecological balance. Targeting specific diseases and parasites and activating SAR provides plants with a robust and long-lasting defense. Farmers may encourage environmentally responsible farming practices and lessen the environmental damage caused by synthetic pesticides by using GAs in integrated pest management (IPM) programs. When it comes to protecting plants with GAs, however, a few more factors need to be considered. The organism it is meant to treat and the existing environmental circumstances affect how well GA works.

More study is needed to determine the ideal composition, dose, and application methods for the best possible effect. We must also address regulatory frameworks and market acceptance to ensure GAs may be utilized in agriculture safely and appropriately. It's not hard to see how gibberellins could be helpful in various contexts. However, it is challenging to get economic value from the product due to the high cost of production. This is why only a tiny subset of crop species can benefit from using gibberellins in industrial agriculture. Several studies have been conducted on the topic of approaches to lessen GA_3 production costs, with an emphasis on dietary factors that influence GA_3 production.

Several methods, such as fermentation, can boost gibberellin microbial production rates and use more efficient strains. Reduced fermentation costs can be achieved using any strategy that increases production speed. Genetic modification to increase gibberellin production could make it more cost-effective. If successful, this strategy for lowering production costs for gibberellin should increase the production of the compound. Consumer preferences are expected to drive growth in the worldwide gibberellins market over the next few years as more people seek healthier options and experiment with new flavors.

References

Abd El-Hack, M.E., Alagawany, M., Elrys, A.S., Desoky, E.S.M., Tolba, H.M., Elnahal, A.S. et al. 2018. Effect of forage Moringa oleifera L.(moringa) on animal health and nutrition and its beneficial applications in soil, plants and water purification. Agriculture, 8(9): 145.

Abdelsattar, A.M., Abdelsattar, A.M., Elsayed, A., El-Esawi, M.A. and Heikal, Y.M. 2023. Enhancing Stevia rebaudiana growth and yield through exploring beneficial plant-microbe interactions and their impact on the underlying mechanisms and crop sustainability. Plant Physiol. Biochem., 198: 107673.

Achard, P. and Genschik, P. 2009. Releasing the brakes of plant growth: how GAs shutdown DELLA proteins. J. Exp. Bot., 60(4): 1085–1092.

Ahmad, P. 2010. Growth and antioxidant responses in mustard (Brassica juncea L.) plants subjected to combined effect of gibberellic acid and salinity. Agron. Soil Sci., 56(5): 575–88.

Akter, N., Rafiqul Islam, M., Abdul Karim, M. and Hossain, T. 2014. Alleviation of drought stress in maize by exogenous application of gibberellic acid and cytokinin. J Crop Sci. Biotech., 17: 41–8.

Alazem, M. and Lin, N.S. 2015. Roles of plant hormones in the regulation of host–virus interactions. Mol. Plant Pathol., 16(5): 529–540.

Ali, M.A., Asghar, H.N., Khan, M.Y., Saleem, M., Naveed, M. and Niazi, N.K. 2015. Alleviation of nickel-induced stress in mungbean through application of gibberellic acid. Int. J. Agric. Biol., 17(5).

Ali, S. and Baloch, A.M. 2020. Overview of sustainable plant growth and differentiation and the role of hormones in controlling growth and development of plants under various stresses. Recent pat. food Nutr. Amp, Agric., 11(2): 105–114.

Ali, S., Khan, A.S., Anjum, M.A., Ejaz, S., Khaliq, G., Hussain, S., Ahmad, R. and Saleem, M.S. 2022. Pre-and Postharvest Physiological Disorders of Citrus Fruit. Citrus Production, 371–390.

Allie, F., Pierce, E. J., Okoniewski, M. J., and Rey, C. 2014. Transcriptional analysis of South African cassava mosaic virus-infected susceptible and tolerant landraces of cassava highlights differences in resistance, basal defense and cell wall associated genes during infection. BMC Genom., 15: 1–30.

Amri, B., Khamassi, K., Ali, M.B., da Silva, J.A.T. and Kaab, L.B.B. 2016. Effects of gibberellic acid on the process of organic reserve mobilization in barley grains germinated in the presence of cadmium and molybdenum. S. Afr. J. Bot., 106: 35–40.

An, C., Sun, C., Li, N., Huang, B., Jiang, J., Shen, Y., Wang, C., Zhao, X., Cui, B., Wang, C. and Li, X. 2022. Nanomaterials and nanotechnology for the delivery of agrochemicals: strategies towards sustainable agriculture. J. Nanobiotechnology, 20(1): 1–19.

Anuar, M.S.K., Hashim, A.M., Ho, C.L., Wong, M.Y., Sundram, S., Saidi, N.B. and Yusof, M.T. 2023. Synergism: biocontrol agents and biostimulants in reducing abiotic and biotic stresses in crop. World J. Microbiol. Biotechnol., 39(5): 123.

Aoyanagi, T., Ikeya, S., Kobayashi, A. and Kozaki, A. 2020. Gene Regulation via the Combination of Transcription Factors in the INDETERMINATE DOMAIN and GRAS Families. Genes, 11(6), 613.

Bajguz, A., & Piotrowska-Niczyporuk, A. 2023. Biosynthetic Pathways of Hormones in Plants. Metabolites,. 13(8): 884.

Barbosa, N.C.S. and Dornelas, M.C. 2021. The roles of gibberellins and cytokinins in plant phase transitions. Trop. Plant Biol., 14: 11–21.

Behera, B., Das, T.K., Raj, R., Ghosh, S., Raza, M.B. and Sen, S. 2021. Microbial consortia for sustaining productivity of non-legume crops: prospects and challenges. Agric. Res. J., 10: 1–14.

Ben Massoud, M., Karmous, I., El Ferjani, E. and Chaoui, A. 2018. Alleviation of copper toxicity in germinating pea seeds by IAA, GA3, Ca and citric acid. J Plant Interact., 13(1): 21–9.

Bilal, L., Asaf, S., Hamayun, M., Gul, H., Iqbal, A., Ullah, I., Lee, I.J. and Hussain, A. 2018. Plant growth promoting endophytic fungi Asprgillus fumigatus TS1 and Fusarium proliferatum BRL1 produce gibberellins and regulates plant endogenous hormones. Symbiosis, 76: 117–127.

Binenbaum, J., Weinstain, R. and Shani, E. 2018. Gibberellin localization and transport in plants. Trends Plant Sci., 23(5): 410–421.

Bisht, T.S., Rawat, L., Chakraborty, B. and Yadav, V. 2018. A Recent Advances in Use of Plant Growth Regulators (PGRs) in Fruit Crops—A Review.

Blázquez, M.A., Nelson, D.C. and Weijers, D. 2020. Evolution of plant hormone response pathways. Annual Review of Plant Biology, 71: 327–353.

Calil, I.P. and Fontes, E.P. 2017. Plant immunity against viruses: antiviral immune receptors in focus. Ann. Bot., 119(5): 711–723.

Camara, M.C., Vandenberghe, L.P., Rodrigues, C., de Oliveira, J., Faulds, C., Bertrand, E. and Soccol, C.R. 2018. Current advances in gibberellic acid (GA 3) production, patented technologies and potential applications. Planta, 248: 1049–1062.

Carr, J.P., Murphy, A.M., Tungadi, T. and Yoon, J.Y. 2019. Plant defense signals: Players and pawns in plant-virus-vector interactions. Plant Sci., 279: 87–95.

Castro-Camba, R., Sánchez, C., Vidal, N. and Vielba, J.M. 2022. Plant development and crop yield: The role of gibberellins. Plant J., 11(19): 2650.

Ćavar, S., Zwanenburg, B. and Tarkowski, P. 2015. Strigolactones: occurrence, structure, and biological activity in the rhizosphere. Phytochem. Rev., 14: 691–711.

Chakraborty, M., Hasanuzzaman, M., Rahman, M., Khan, M.A.R., Bhowmik, P., Mahmud, N.U. et al. 2020. Mechanism of plant growth promotion and disease suppression by chitosan biopolymer. Agriculture, 10(12): 624.

Chamara, N., Islam, M.D., Bai, G.F., Shi, Y. and Ge, Y. 2022. Ag-IoT for crop and environment monitoring: Past, present, and future. Agric. Syst., 203: 103497.

Chatterjee, S. and Chattopadhyay, S. (Eds.). 2023. Nucleic Acid Biology and its Application in Human Diseases, Springer Nature.

Chen, M., Zhou, S., Zhu, Y., Sun, Y., Zeng, G., Yang, C. et al. 2018. Toxicity of carbon nanomaterials to plants, animals and microbes: Recent progress from 2015-present. Chemosphere, 206: 255–264.

Costa, J.A.V., Freitas, B.C.B., Cruz, C.G., Silveira, J. and Morais, M.G. 2019. Potential of microalgae as biopesticides to contribute to sustainable agriculture and environmental development. J. Environ. Sci. Health A ., Part B, 54(5): 366–375.

Cui, W., Song, Q., Zuo, B., Han, Q. and Jia, Z. 2020. Effects of gibberellin (GA4+ 7) in grain filling, hormonal behavior, and antioxidants in high-density maize (Zea mays L.). Plants 9(8): 978.

Das, P.P., Singh, K.R., Nagpure, G., Mansoori, A., Singh, R.P., Ghazi, I.A. et al. 2022. Plant-soil-microbes: A tripartite interaction for nutrient acquisition and better plant growth for sustainable agricultural practices. Environ. Res., 214: 113821.

Davidson, S.E., Smith, J.J., Helliwell, C.A., Poole, A.T. and Reid, J.B. 2004. The pea gene LH encodes ent-kaurene oxidase. Plant Physiol., 134(3): 1123–1134.

Davière, J.M. and Achard, P. 2016. A pivotal role of DELLAs in regulating multiple hormone signals. Mol Plant, 9(1): 1.–20.

de Lucas, M., Davière, J.M., Rodríguez-Falcón, M., Pontin, M., Iglesias-Pedraz, J.M., Lorrain, S. et al. 2008. A molecular framework for light and gibberellin control of cell elongation. Nature, 451(7177): 480–484.

Ding, Y., Sheng, J., Li, S., Nie, Y., Zhao, J., Zhu, Z. et al. 2015. The role of gibberellins in the mitigation of chilling injury in cherry tomato (Solanum lycopersicum L.) fruit. Postharvest Biol Technol., 101: 88–95.

Divekar, P.A., Narayana, S., Divekar, B.A., Kumar, R., Gadratagi, B.G., Ray, A. et al. 2022. Plant secondary metabolites as defense tools against herbivores for sustainable crop protection. Int. J. Mol. Sci., 23(5): 2690.

Domonkos, M., Tichá, P., Trejbal, J. and Demo, P. 2021. Applications of cold atmospheric pressure plasma technology in medicine, agriculture and food industry. Appl. Sci., 11(11): 4809.

Du, R., Niu, S., Liu, Y., Sun, X., Porth, I.A.Y. and Li, W. 2017. The gibberellin GID1-DELLA signalling module exists in evolutionarily ancient conifers. Sci. Rep., 7: 16637.

El-Maraghy, S.S., Tohamy, A.T. and Hussein, K.A. 2021. Plant protection properties of the Plant GrowthPromoting Fungi (PGPF): Mechanisms and potentiality. Curr. Res. Environ. Appl. Mycol 11(1): 391–415.

Ezquer, I., Salameh, I., Colombo, L. and Kalaitzis, P. 2020. Plant cell walls tackling climate change: Insights into plant cell wall remodeling, its regulation, and biotechnological strategies to improve crop adaptations and photosynthesis in response to global warming. Plant J. 9(2): 1–27.

Fang, Y. and Ramasamy, R.P. 2015. Current and prospective methods for plant disease detection. Biosens., 5(3): 537–561.

Farhangi, H., Mozafari, V., Roosta, H.R., Shirani, H. and Farhangi, M. 2023. Optimizing growth conditions in vertical farming: enhancing lettuce and basil cultivation through the application of the Taguchi method. Sci. Rep., 13(1): 6717.

Feng, S., Martinez, C., Gusmaroli, G., Wang, Y., Zhou, J., Wang, F. et al. 2008. Coordinated regulation of Arabidopsis thaliana development by light and gibberellins. Nature, 451 (7177): 475–479.

Figueiredo, M.D.V.B., Bonifacio, A., Rodrigues, A.C. and de Araujo, F.F. 2016. Plant growth-promoting rhizobacteria: key mechanisms of action. Microbial-Mediated Induced Systemic Resistance in Plants, 23–37.

Fukazawa, J., Ohashi, Y., Takahashi, R., Nakai, K. and Takahashi, Y. 2021. DELLA degradation by gibberellin promotes flowering via GAF1-TPR-dependent repression of floral repressors in Arabidopsis. The Plant Cell., 33(7): 2258–2272.

Galvão, V.C., Horrer, D., Küttner, F. and Schmid, M. 2012. Spatial control of flowering by DELLA proteins in Arabidopsis thaliana. Development. 139(21): 4072–4082.

Gangwar, S., Singh, V.P., Srivastava, P.K. and Maurya, J.N. 2011. Modification of chromium (VI) phytotoxicity by exogenous gibberellic acid application in Pisum sativum (L.) seedlings. Acta Physiol. Plant., 33: 1385–97.

Gantait, S., Rani Sinniah, U., Ali, N. and Chandra Sahu, N. 2015. Gibberellins-a multifaceted hormone in plant growth regulatory network. Curr. Protein Pept. Sci., 16(5): 406–412.

Garmendia, A., Beltran, R., Zornoza, C., Garcia-Breijo, F.J., Reig, J. and Merle, H. 2019. Gibberellic acid in Citrus spp. flowering and fruiting: A systematic review. PloS One, 14(9): e0223147.

Garraway, J.L. 2020. Insecticides, Fungicides and Herbicides. Biotechnology-The Science and the Business, CRC Press, 497–514.

Ghosh, D. and Chakraborty, S. 2021. Impact of viral silencing suppressors on plant viral synergism: a global agro-economic concern. Appl. Microbiol. Biotechnol., 105: 6301–6313.

Goswami, D., Thakker, J.N. and Dhandhukia, P.C. 2016. Portraying mechanics of plant growth promoting rhizobacteria (PGPR): a review. Cogent Food Agric., 2(1): 1127500.

Gupta, N., Reddy, K., Bhattacharyya, D. and Chakraborty✉, S. 2021. Plant responses to geminivirus infection: guardians of the plant immunity. Virol. J., 18(1): 143.

Hadi, F., Ali, N. and Ahmad, A. 2014. Enhanced phytoremediation of cadmium-contaminated soil by Parthenium hysterophorus plant: effect of gibberellic acid (GA3) and synthetic chelator, alone and in combinations. Bioremediat J., 18(1): 46–55.

Hamayun, M., Khan, S.A., Khan, A.L., Shin, J.H., Ahmad, B., Shin, D.H. et al. 2010. Exogenous gibberellic acid reprograms soybean for higher growth and salt stress tolerance. J. Agric. Food Chem. 58(12): 7226–32.

Harberd, N.P., Belfield, E. and Yasumura, Y. 2009. The angiosperm gibberellin-GID1-DELLA growth regulatory mechanism: how an "inhibitor of an inhibitor" enables flexible response to fluctuating environments. The Plant Cell, 21(5): 1328–1339.

Harman, G., Khadka, R., Doni, F. and Uphoff, N. 2021. Benefits to plant health and productivity from enhancing plant microbial symbionts. Front. Plant Sci., 11: 610065.

Hedden, P. 2020. The current status of research on gibberellin biosynthesis. Plant Cell Physiol., 61(11): 1832–1849.

Hedden, P. 2020. The Current Status of Research on Gibberellin Biosynthesis. Plant & cell physiology, 61(11): 1832–1849.

Hedden, P. and Sponsel, V. 2015. A century of gibberellin research. J. Plant Growth Regul., 34: 740–760.

Hedden, P. and Thomas, S.G. 2012. Gibberellin biosynthesis and its regulation. Biochem. J., 444(1): 11–25.

Hedden, P. and Sponsel, V. A. 2015. Century of Gibberellin Research. J. Plant Growth Regul., 34: 740–760.

Hernández-García, J., Briones-Moreno, A. and Blázquez, M.A. 2021, January. Origin and evolution of gibberellin signaling and metabolism in plants. Seminars in Cell and Developmental Biology, Elsevier.

Hirano, K., Kouketu, E., Katoh, H., Aya, K., Ueguchi-Tanaka, M., & Matsuoka, M. (2012). The suppressive function of the rice DELLA protein SLR1 is dependent on its transcriptional activation activity. Plant J Title(s). 71(3), 443–453.

Hofmann, T., Lowry, G.V., Ghoshal, S., Tufenkji, N., Brambilla, D., Dutcher, J.R. et al. 2020. Technology readiness and overcoming barriers to sustainably implement nanotechnology-enabled plant agriculture. Nat. Food., 1(7): 416–425.

Huang, J., Yang, M. and Zhang, X. 2016. The function of small RNAs in plant biotic stress response. J. Integr. Plant Biol., 58(4): 312–327.

Ilahy, R., Tlili, I., Siddiqui, M.W., Hdider, C. and Lenucci, M.S. 2019. Inside and Beyond Color: Comparative Overview of Functional Quality of Tomato and Watermelon Fruits. Front. Plant Sci., 10: 769.

Ismail, M.A., Amin, M.A., Eid, A.M., Hassan, S.E.D., Mahgoub, H.A., Lashin, I. et al. 2021. Comparative Study between exogenously applied plant growth hormones versus metabolites of microbial endophytes as plant growth-promoting for Phaseolus vulgaris L. Cells, 10(5): 1059.

Jain, D. and Khurana, J.P. 2018. Role of pathogenesis-related (PR) proteins in plant defense mechanism. Mol. Plant Pathol., 265–281.

Jan, R., Asaf, S., Numan, M., Lubna and Kim, K.M. 2021. Plant secondary metabolite biosynthesis and transcriptional regulation in response to biotic and abiotic stress conditions. J. Agron. 11(5): 968.

Jogawat, A., Yadav, B., Chhaya, Lakra, N., Singh, A.K. and Narayan, O.P. 2021. Crosstalk between phytohormones and secondary metabolites in the drought stress tolerance of crop plants: a review. Physiol. Plant., 172(2): 1106–1132.

Kamal, R., Gusain, Y.S., Kumar, V. and Sharma, A.K. 2015. Disease management through biological control agents: An eco-friendly and cost effective approach for sustainable agriculture-A Review. Agric. Rev., 36(1): 37–45.

Kandil, A.A., Sharief, A.E., Abido, W.A.E. and Awed, A.M. 2014. Effect of gibberellic acid on germination behaviour of sugar beet cultivars under salt stress conditions of Egypt. Sugar Tech., 16: 211–21.

Kaur, S., Gupta, A.K. and Kaur, N. 1998. Gibberellin A3 reverses the effect of salt stress in chickpea (Cicer arietinum L.) seedlings by enhancing amylase activity and mobilization of starch in cotyledons. Plant Growth Regul., 26: 85–90.

Kaya, C., Levent Tuna, A. and Alfredo, A.A. 2006. Gibberellic acid improves water deficit tolerance in maize plants. Acta Physiol Plant., 28: 331–7.

Keeling, C.I., Dullat, H.K., Yuen, M., Ralph, S.G., Jancsik, S. and Bohlmann, J. 2010. Identification and functional characterization of monofunctional ent-copalyl diphosphate and ent-kaurene synthases in white spruce reveal different patterns for diterpene synthase evolution for primary and secondary metabolism in gymnosperms. Plant Physiol., 152(3): 1197–1208.

Kępczyński, J. 2018. Induction of agricultural weed seed germination by smoke and smoke-derived karrikin (KAR 1), with a particular reference to Avena fatua L. Acta Physiol. Plant, 40: 1–10.

Keswani, C., Singh, S.P., García-Estrada, C., Mezaache-Aichour, S., Glare, T.R., Borriss, R. et al. 2022. Biosynthesis and beneficial effects of microbial gibberellins on crops for sustainable agriculture. J. Appl. Microbiol., 132(3): 1597–1615.

Khan, M., Ali, S., Al Azzawi, T.N.I. and Yun, B.W. 2023. Nitric Oxide Acts as a Key Signaling Molecule in Plant Development under Stressful Conditions. Int. J. Mol. Sci., 24(5): 4782.

King, K.E., Moritz, T. and Harberd, N.P. 2001. Gibberellins are not required for normal stem growth in Arabidopsis thaliana in the absence of GAI and RGA. Genetics, 159(2): 767–776.

Klessig, D.F., Choi, H.W. and Dempsey, D.M.A. 2018. Systemic acquired resistance and salicylic acid: past, present, and future. MPMI, 31(9): 871–888.

Köksal, M., Potter, K., Peters, R.J. and Christianson, D.W. 2014. 1.55Å-resolution structure of ent-copalyl diphosphate synthase and exploration of general acid function by site-directed mutagenesis. Biochim Biophys Acta., 1840(1): 184–190.

Kour, D., Rana, K.L., Yadav, A.N., Yadav, N., Kumar, V., Kumar, A. et al. 2019. Drought-tolerant phosphorus-solubilizing microbes: biodiversity and biotechnological applications for alleviation of drought stress in plants. Plant growth promoting rhizobacteria for sustainable stress management: Volume 1: Rhizobacteria in abiotic stress management, 255–308.

Lahlali, R., Ezrari, S., Radouane, N., Kenfaoui, J., Esmaeel, Q., El Hamss, H. et al. 2022. Biological control of plant pathogens: A global perspective. Microorganisms, 10(3): 596.

Lemke, C., Potter, K.C., Schulte, S. and Peters, R.J. 2019. Conserved bases for the initial cyclase in gibberellin biosynthesis: from bacteria to plants. The Biochemical Journal, 476(18): 2607–2621.

Li, L., Zhang, H., Yang, Z., Wang, C., Li, S., Cao, C. et al. 2022. Independently evolved viral effectors convergently suppress DELLA protein SLR1-mediated broad-spectrum antiviral immunity in rice. Nat. Commun., 13(1): 6920.

Li, Z. and Ahammed, G.J. 2023. Hormonal regulation of anthocyanin biosynthesis for improved stress tolerance in plants. Plant Physiol. Biochem. PPB, 201: 107835.

Liao, Z., Zhou, Q. and Gao, B. 2022. Electrochemical Microneedles: Innovative Instruments in Health Care. Biosensors, 12(10): 801.

Liu, X., Wang, X., Yin, L., Deng, X. and Wang, S. 2018. Exogenous application of gibberellic acid participates in up-regulation of lipid biosynthesis under salt stress in rice. Theoretical Exp Plant Physi., 30: 335–45.

Lo, S.F., Yang, S.Y., Chen, K.T., Hsing, Y.I., Zeevaart, J.A., Chen, L.J. et al. 2008. A novel class of gibberellin 2-oxidases control semidwarfism, tillering, and root development in rice. The Plant Cell., 20(10): 2603–2618.

Maggio, A., Barbieri, G., Raimondi, G. and De Pascale, S. 2010. Contrasting effects of GA 3 treatments on tomato plants exposed to increasing salinity. J. Plant Growth Regul., 29: 63–72.

McNeil, M.D., Bhuiyan, S.A., Berkman, P.J., Croft, B.J. and Aitken, K.S. 2018. Analysis of the resistance mechanisms in sugarcane during Sporisorium scitamineum infection using RNA-seq and microscopy. PloS One, 13(5): e0197840.

Meena, M., Swapnil, P., Divyanshu, K., Kumar, S., Harish, Tripathi, Y.N. et al. 2020. PGPR-mediated induction of systemic resistance and physiochemical alterations in plants against the pathogens: Current perspectives. J. Basic Microbiol., 60(10): 828–861.

Meena, R.S., Kumar, S., Datta, R., Lal, R., Vijayakumar, V., Brtnicky, M. et al. 2020. Impact of agrochemicals on soil microbiota and management: A review. Land, 9(2): 34.

Meng, H., Hua, S., Shamsi, I.H., Jilani, G., Li, Y. and Jiang, L. 2009. Cadmium-induced stress on the seed germination and seedling growth of Brassica napus L., and its alleviation through exogenous plant growth regulators. J. Plant Growth Regul., 58: 47–59.

Miller, R.N.G., Costa Alves, G.S. and Van Sluys, M.A. 2017. Plant immunity: unravelling the complexity of plant responses to biotic stresses. Ann. Bot., 119(5): 681–687.

Mittal, D., Kaur, G., Singh, P., Yadav, K. and Ali, S.A. 2020. Nanoparticle-based sustainable agriculture and food science: Recent advances and future outlook. Front. Nanotechnol., 2: 579954.

Muhammad, M., Badshah, L., Shah, A.A., Shah, M.A., Abdullah, A., Bussmann, R.W. et al. 2021. Ethnobotanical profile of some useful plants and fungi of district Dir Upper, Tehsil Darora, Khyber Pakhtunkhwa, Pakistan. ERA., 21: 1–15.

Muhammad, M., Badshah, L., Shah, A.A., Shah, M.A., Abdullah, A., Bussmann, R.W. et al. 2021. Ethnobotanical profile of some useful plants and fungi of district Dir Upper, Tehsil Darora, Khyber Pakhtunkhwa, Pakistan. Ethnobot. Res. Appl., 21: 1–15.

Mutasa-Göttgens, E. and Hedden, P. 2009. Gibberellin as a factor in floral regulatory networks. Exp. Bot., 60(7): 1979–1989.

Nakajima, M., Shimada, A., Takashi, Y., Kim, Y. C., Park, S. H., Ueguchi-Tanaka, M. et al. 2006. Identification and characterization of Arabidopsis gibberellin receptors. Plant J Title(s), 46(5): 880–889.

Naqvi, S.A.H. 2019. Bacterial leaf blight of rice: An overview of epidemiology and management with special reference to Indian sub-continent. Pak. J. Agric. Res., 32(2): 359.

Narayanan, Z. and Glick, B.R. 2022. Secondary metabolites produced by plant growth-promoting bacterial endophytes. Microorganisms, 10(10): 2008.

Nelson, S.K. and Steber, C.M. 2016. Gibberellin hormone signal perception: down-regulating DELLA repressors of plant growth and development. Annu. Rev. Plant Biol., Volume 49: Gibberellins, The: 153–188.

Niharika, Singh, N.B., Singh, A., Khare, S., Yadav, V., Bano, C. and Yadav, R.K. 2021. Mitigating strategies of gibberellins in various environmental cues and their crosstalk with other hormonal pathways in plants: a review. Plant Mol. Biol., 39: 34–49.

Niu, Q., Wang, T., Li, J., Yang, Q., Qian, M. and Teng, Y. 2015. Effects of exogenous application of GA 4+ 7 and N-(2-chloro-4-pyridyl)-N′-phenylurea on induced parthenocarpy and fruit quality in Pyrus pyrifolia 'Cuiguan'. J. Plant Growth Regul., 76: 251–258.

Nkhata, S.G., Ayua, E., Kamau, E.H. and Shingiro, J.B. 2018. Fermentation and germination improve nutritional value of cereals and legumes through activation of endogenous enzymes. Food Sci. Nutr., 6(8): 2446–2458.

Oleńska, E., Małek, W., Wójcik, M., Swiecicka, I., Thijs, S. and Vangronsveld, J. 2020. Beneficial features of plant growth-promoting rhizobacteria for improving plant growth and health in challenging conditions: A methodical review. Sci. Total Environ., 743: 140682.

Pal, P., Yadav, K., Kumar, K. and Singh, N. 2016. Effect of Gibberellic Acid and Potassium Foliar Sprays on Productivity and Physiological and Biochemical Parameters of Parthenocarpic Cucumber cv. 'Seven Star F'. J. Hortic. Res., 24(1): 93–100.

Pathania, A., Singh, L. and Sharma, P.N. 2021. Host Plant Resistance: An Eco-Friendly Approach for Crop Disease Management. Microbial Biotechnology in Crop Protection, Springer, 395–449.

Paul, A. and Roychoudhury, A. 2021. Go green to protect plants: repurposing the antimicrobial activity of biosynthesized silver nanoparticles to combat phytopathogens. Nanotechnol. Environ. Eng., 6(1): 10.

Peng, J., Richards, D.E., Hartley, N.M., Murphy, G.P., Devos, K.M., Flintham, J.E. et al. 1999. 'Green revolution' genes encode mutant gibberellin response modulators. Nature, 400: 256–261.

Peters, R.J. 2013. Gibberellin phytohormone metabolism. Isoprenoid synthesis in plants and microorganisms: New Concepts and Experimental Approaches, 233–49.

Prakash, V., Singh, V.P., Tripathi, D.K., Sharma, S. and Corpas, F.J. 2019. Crosstalk between nitric oxide (NO) and abscisic acid (ABA) signalling molecules in higher plants. Environ. Exp. Bot., 161: 41–49.

Puglia, D., Pezzolla, D., Gigliotti, G., Torre, L., Bartucca, M.L. and Del Buono, D. 2021. The opportunity of valorizing agricultural waste, through its conversion into biostimulants, biofertilizers, and biopolymers. Sustainability, 13(5): 2710.

Qian, C., Ren, N., Wang, J., Xu, Q., Chen, X. and Qi, X. 2018. Effects of exogenous application of CPPU, NAA and GA4+ 7 on parthenocarpy and fruit quality in cucumber (Cucumis sativus L.). Food Chem., 243: 410–413.

Qiong, T.A.N.G., Zheng, X.D., Jun, G.U.O. and Ting, Y.U. 2022. Tomato SlPti5 plays a regulative role in the plant immune response against Botrytis cinerea through modulation of ROS system and hormone pathways. J. Integr. Agric., 21(3): 697–709.

Rademacher, W. 1994. Gibberellin formation in microorganisms. Plant Growth Regul., 15: 303–14.

Rademacher, W. 2015. Plant growth regulators: backgrounds and uses in plant production. J. Plant Growth Regul., 34: 845–872.

Rademacher, W. 2016. Chemical regulators of gibberellin status and their application in plant production. Annu. Plant Rev., Volume 49: Gibberellins, The: 359–404.

Rafiullah, F.I.W., Basit, A., Ullah, I., Sajid, M., Shah, S.T., Ahmad, I. et al. 2021. Vegetative Growth and Root Development of Euonymus Japonica Cuttings is Influenced by Various Concentrations of Indole Butyric Acid (IBA) and Transplantation Dates. Ann. Romanian Soc. Cell Biol., 948–959.

Rizza, A. and Jones, A.M. 2019. The makings of a gradient: spatiotemporal distribution of gibberellins in plant development. Curr. Opin. Plant Biol., 47: 9–15.

Saeidi-Sar, S., Khavari-Nejad, R.A., Fahimi, H., Ghorbanli, M. and Majd, A. 2007. Interactive effects of gibberellin A 3 and ascorbic acid on lipid peroxidation and antioxidant enzyme activities in Glycine max seedlings under nickel stress. Russ. J Plant Physiol., 54: 74–9.

Sakata, T., Oda, S., Tsunaga, Y., Shomura, H., Kawagishi-Kobayashi, M., Aya, K. et al. 2014. Reduction of gibberellin by low temperature disrupts pollen development in rice. Plant Physiol., 164(4): 2011–9.

Salazar-Cerezo, S., Martínez-Montiel, N., García-Sánchez, J., Pérez-y-Terrón, R. and Martínez-Contreras, R.D. 2018. Gibberellin biosynthesis and metabolism: A convergent route for plants, fungi and bacteria. Microbiol. Res., 208: 85–98.

Sansinenea, E. 2019. Bacillus spp.: As plant growth-promoting bacteria. Secondary metabolites of plant growth promoting rhizomicroorganisms: Discovery and Applications, 225–237.

Schwechheimer, C. 2012. Gibberellin signaling in plants—the extended version. Front. Plant Sci., 2: 107.

Shaddad, M.A.K., Abd El-Samad, H.M. and Mostafa, D. 2013. Role of gibberellic acid (GA3) in improving salt stress tolerance of two wheat cultivars. Int. J. Plant Physiol. Biochem., 5(4): 50–7.

Shah, J., Chaturvedi, R., Chowdhury, Z., Venables, B. and Petros, R.A. 2014. Signaling by small metabolites in systemic acquired resistance. Plant J., 79(4): 645–658.

Shah, S.M., Shah, A.A., Muhammad, M., Abdussalam, M., Rehman, K.U. and Khan, H. 2021. Phytodiversity And Ecological Features Of Weed Species Of Sufaid Sung, Peshawar. WSSP., 27(2): 227.

Sharma, B., Tiwari, S., Kumawat, K.C. and Cardinale, M. 2023. Nano-biofertilizers as bio-emerging strategies for sustainable agriculture development: Potentiality and their limitations. Sci. Total Environ., 860: 160476.

Sharma, B., Vaish, B., Monika, Singh, U.K., Singh, P. and Singh, R.P. 2019. Recycling of organic wastes in agriculture: an environmental perspective. Int. J. Environ. Res., 13: 409–429.

Sharma, N. and Bakshi, A. 2022. 15 Integrated Pest Management through genetically modified crops. Genetically Modified Crops and Food Security: Commercial, Ethical and Health Considerations.

Shelton, A.M., Long, S.J., Walker, A.S., Bolton, M., Collins, H.L., Revuelta, L. et al. 2020. First field release of a genetically engineered, self-limiting agricultural pest insect: evaluating its potential for future crop protection. Front. Bioeng. Biotechnol., 482.

Shimura, H. and Pantaleo, V. 2011. Viral induction and suppression of RNA silencing in plants. Biochim. Biophys. Acta Gene Regul. Mech., 1809(11–12): 601–612.

Shukla, A. 2020. Chapter-5 Biopesticides and Agricultural Pest Management. Recent Trends, 21: 97.

Siddiqui, M.H., Al-Whaibi, M.H. and Basalah, M.O. 2011. Interactive effect of calcium and gibberellin on nickel tolerance in relation to antioxidant systems in Triticum aestivum L. Protoplasma., 248: 503–11.

Sinha, D., Odoh, U.E., Ganguly, S., Muhammad, M., Chatterjee, M., Chikeokwu, I. and Egbuna, C. 2023. Phytochemistry, history, and progress in drug discovery. Phytochemistry, Computational Tools and Databases in Drug Discovery, Elsevier, 1–26.

Sofy, A.R., Dawoud, R.A., Sofy, M.R., Mohamed, H.I., Hmed, A.A. and El-Dougdoug, N.K. 2020. Improving regulation of enzymatic and non-enzymatic antioxidants and stress-related gene stimulation in Cucumber mosaic cucumovirus-infected cucumber plants treated with glycine betaine, chitosan and combination. Mol., 25(10): 2341.

Souza, K.O., Viana, R.M., de Siqueira Oliveira, L., Moura, C.F.H. and Miranda, M.R.A. 2016. Preharvest treatment of growth regulators influences postharvest quality and storage life of cashew apples. Sci. Horticult., 209: 53–60.

Sponsel, V.M. 2016. Signal achievements in gibberellin research: the second half-century. Annu. Rev. Plant Biol., Volume 49: Gibberellins, The: 1–36.

Sun, T.P. 2008. Gibberellin metabolism, perception and signaling pathways in Arabidopsis. The Arabidopsis Book, 6: e0103.

Sun, W.J., Lv, W.J., Li, L.N., Yin, G., Hang, X., Xue, Y. et al. 2016. Eugenol confers resistance to Tomato yellow leaf curl virus (TYLCV) by regulating the expression of SlPer1 in tomato plants. N Biotechnol., 33(3): 345–354.

Tassadduq, S.S., Akhtar, S., Waheed, M., Bangash, N., Nayab, D.E., Majeed, M. et al. 2022. Ecological Distribution Patterns of Wild Grasses and Abiotic Factors. Sustainability, 14(18): 11117.

Terzi, I. and Kocaçalışkan, I. 2009. Alleviation of juglone stress by plant growth regulators in germination of cress seeds. Sci. Res. Essays., 4(5): 436–9.

Tleuova, A.B., Wielogorska, E., Talluri, V.P., Štěpánek, F., Elliott, C.T. and Grigoriev, D.O. 2020. Recent advances and remaining barriers to producing novel formulations of fungicides for safe and sustainable agriculture. JCR, 326: 468–481.

Toner, P., Nelson, D., Rao, J.R., Ennis, M., Moore, J.E. and Schock, B. 2021. Antimicrobial properties of phytohormone (gibberellins) against phytopathogens and clinical pathogens. Access Microbiol., 3(10).

Toungos, M.D. 2018. Plant growth substances in crop production: A Review. IJIABR, 6: 1–8.

Trivedi, P., Mattupalli, C., Eversole, K. and Leach, J.E. 2021. Enabling sustainable agriculture through understanding and enhancement of microbiomes. New Phytol., 230(6): 2129–2147.

Tuna, A.L., Kaya, C., Dikilitas, M. and Higgs, D. 2008. The combined effects of gibberellic acid and salinity on some antioxidant enzyme activities, plant growth parameters and nutritional status in maize plants. Environ Exp. Bot., 62(1): 1–9.

Ueguchi-Tanaka, M., Ashikari, M., Nakajima, M., Itoh, H., Katoh, E., Kobayashi, M. et al. 2005. Gibberellin Insensitive Dwarf1 encodes a soluble receptor for gibberellin. Nature, 437(7059): 693–698.

Uthman, A. and Garba, Y. 2023. Citrus Mineral Nutrition and Health Benefits: A Review. Citrus Research-Horticultural and Human Health Aspects.

Verma, N., Tiwari, S., Singh, V.P. and Prasad, S.M. 2020. Nitric oxide in plants: an ancient molecule with new tasks. J. Plant Growth Regul., 90: 1–13.

Wahab, A., Abdi, G., Saleem, M.H., Ali, B., Ullah, S., Shah, W. et al. 2022. Plants' physio-biochemical and phyto-hormonal responses to alleviate the adverse effects of drought stress: A comprehensive review. Plants, 11(13): 1620.

Wahab, A., Muhammad, M., Munir, A., Abdi, G., Zaman, W. et al. 2023. Role of Arbuscular Mycorrhizal Fungi in Regulating Growth, Enhancing Productivity, and Potentially Influencing Ecosystems under Abiotic and Biotic Stresses. Plants., 12(17): 3102.

Waheed, A., Li, C., Muhammad, M., Ahmad, M., Khan, K.A., Ghramh, H.A. et al. 2023. Sustainable Potato Growth under Straw Mulching Practices. Sustainability, 15(13): 10442.

Walid, N. and Ibrahim, M.M. 2019. EFFECT OF Trichoderma harzianum AND Aneurinobacillus migulanus Applications On Auxin, Gibberellic Acid and Abscisic Acid Content Of Gladiolus Corms Infected With Fusarium oxysporum f. sp. Gladioli. Future, 3: 44–53.

Wang, Q., Hillwig, M.L., Wu, Y. and Peters, R.J. 2012. CYP701A8: a rice ent-kaurene oxidase paralog diverted to more specialized diterpenoid metabolism. Plant Physiol., 158(3): 1418–1425.

Wang, Y. and Deng, D. 2014. Molecular basis and evolutionary pattern of GA-GID1-DELLA regulatory module. Mol. Genet. Genom., 289(1): 1–9.

Wang, Y.H., Zhang, G., Chen, Y., Gao, J., Sun, Y.R., Sun, M.F. et al. 2019. Exogenous application of gibberellic acid and ascorbic acid improved tolerance of okra seedlings to NaCl stress. Acta Physiol. Plant., 41: 1–0.

Wang, Y.H., Zhang, G., Chen, Y., Gao, J., Sun, Y.R., Sun, M.F. et al. 2019. Exogenous application of gibberellic acid and ascorbic acid improved tolerance of okra seedlings to NaCl stress. Acta Physiol Plant., 41: 1–0.

Willige, B.C., Ghosh, S., Nill, C., Zourelidou, M., Dohmann, E.M., Maier, A. et al. 2007. The DELLA domain of GA INSENSITIVE mediates the interaction with the GA INSENSITIVE DWARF1A gibberellin receptor of Arabidopsis. The Plant Cell, 19(4): 1209–1220.

Wu, J., Kong, X., Wan, J., Liu, X., Zhang, X., Guo, X. et al. 2011. Dominant and pleiotropic effects of a GAI gene in wheat results from a lack of interaction between DELLA and GID1. Plant Physiol., 157(4): 2120–2130.

Xie, Y., Onik, J.C., Hu, X., Duan, Y. and Lin, Q. 2018. Effects of (S)-carvone and gibberellin on sugar accumulation in potatoes during low temperature storage. Molecules., 23(12): 3118.

Xie, Y., Tan, H., Ma, Z. and Huang, J. 2016. DELLA Proteins Promote Anthocyanin Biosynthesis via Sequestering MYBL2 and JAZ Suppressors of the MYB/bHLH/WD40 Complex in Arabidopsis thaliana. Mol Plant. 9(5): 711–721.

Xu, W., Dubos, C. and Lepiniec, L. 2015. Transcriptional control of flavonoid biosynthesis by MYB-bHLH-WDR complexes. Trends Plant Sci., 20(3): 176–185.

Xue, H., Gao, X., He, P. and Xiao, G. 2022. Origin, evolution, and molecular function of DELLA proteins in plants. Crop J., 10(2): 287–299.

Yang, C., Dolatabadian, A. and Fernando, W.D. 2022. The wonderful world of intrinsic and intricate immunity responses in plants against pathogens. Can. J. Plant Pathol., 44(1): 1–20.

Yang, S., Zhang, K., Zhu, H., Zhang, X., Yan, W., Xu, N. et al. 2020. Melon short internode (CmSi) encodes an ERECTA-like receptor kinase regulating stem elongation through auxin signaling. Hortic. Res., 7.

Yang, Y.X., J Ahammed, G., Wu, C., Fan, S.Y. and Zhou, Y.H. 2015. Crosstalk among jasmonate, salicylate and ethylene signaling pathways in plant disease and immune responses. Curr. Protein Pept. Sci., 16(5): 450–461.

Yatoo, A.M., Ali, M.N., Baba, Z.A. and Hassan, B. 2021. Sustainable management of diseases and pests in crops by vermicompost and vermicompost tea. A Review. ASD, 41: 1–26.

Zhang, C., Jian, M., Li, W., Yao, X., Tan, C., Qian, Q. et al. 2023. Gibberellin signaling modulates flowering via the DELLA-BRAHMA-NF-YC module in Arabidopsis. The Plant Cell., 35(9): 3470–3484.

Zhao, S. and Li, Y. 2021. Current understanding of the interplays between host hormones and plant viral infections. PLoS Pathog., 17(2): e1009242.

Zhu, X.F., Jiang, T., Wang, Z.W., Lei, G.J., Shi, Y.Z., Li, G.X. et al. 2012. Gibberellic acid alleviates cadmium toxicity by reducing nitric oxide accumulation and expression of IRT1 in Arabidopsis thaliana. J. Hazard. Mater., 239: 302–7.

Zhu, Z., Ding, Y., Zhao, J., Nie, Y., Zhang, Y., Sheng, J. et al. 2016. Effects of postharvest gibberellic acid treatment on chilling tolerance in cold-stored tomato (Solanum lycopersicum L.) fruit. Food Bioprocess Technol., 9: 1202–9.

10

Plant Phenolics and Oxidative Stress

Parteek Prasher[1],* and *Mousmee Sharma*[2]

Introduction

Phenolics represent aromatic compounds containing a benzene ring having hydroxyl groups with a main role in providing defense to the plants from oxidative stress (Dai et al. 2010, Lopes et al. 2023). The phenolics are also known to play an essential role in plant physiology by providing the scaffolding support leading to the structural integrity of the plant (Bhattacharya et al. 2010, Tungmunnithum et al. 2018, Shahidi et al. 2015). The phenolics secreted by injured plants provide antimicrobial properties. Some of the plant phenolics act as chemical agents for deterring the herbivores and pathogens and in allelopathy (Chenier et al. 2013, Zhang et al. 2023). The plant phenolics are synthesized mainly by the shikimate pathway and phenylpropanoid pathway. Figure 1 represents the categories of plant-derived phenolics (Cosme et al. 2020). The synthesized polyphenols accumulated in the subepidermal layers are exposed to pathogen attack and mechanical stress (Tyagi et al. 2020, Wallis et al. 2020). The polyphenolic chemicals and pigments secreted by the plants attract pollinators, symbiotic microbes, and fruit-scattering animals that cause seed dispersal. Intake of polyphenols in humans occurs from the nuts, fruits, seeds, and vegetables is highly beneficial in the mitigation of oxidative stress and related ailments, including ageing and age-related disorders (Lin et al. 2016, Balasundram et al. 2006, Cheynier et al. 2012). The major categories of phenolic compounds in plants include lignans, flavonoids, isoflavonoids, flavones, stilbenes, coumarins, and phenolic acids (Friedman et al. 2000, Maqsood et al. 2014, Zhang et al. 2022). Besides the normal antioxidant function, the plant polyphenols and their

---

[1] Department of Chemistry, University of Petroleum & Energy Studies, Energy Acres, Dehradun 248007, India.
[2] Department of Chemistry, Uttaranchal University, Arcadia Grant, Dehradun 248007, India.
* Corresponding author: parteekchemistry@gmail.com

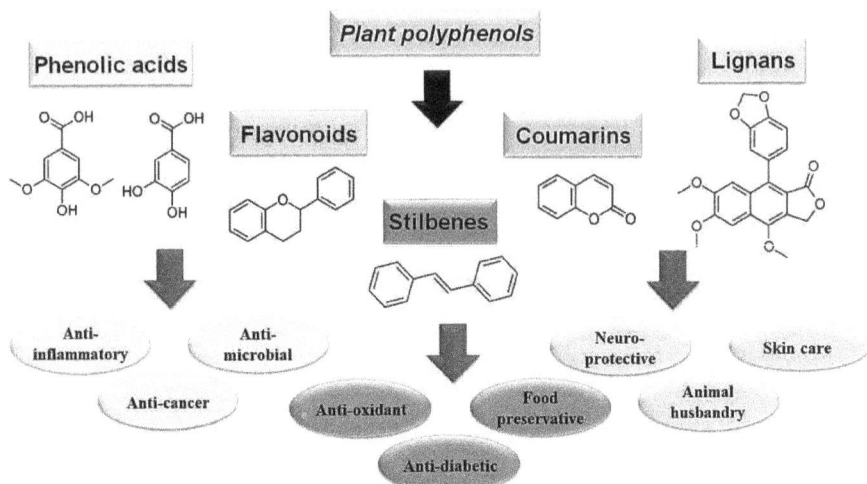

Fig. 1. Broad classification and health benefits of phenolics derived from plants.

chemically synthesized derivatives present several other health benefits including anti-inflammatory, antimicrobial, anticancer, anti-neurodegenerative, cardioprotective, and immunity boosting properties (Albuquerque et al. 2021, Anantharaju et al. 2016, Kopjar et al. 2014, Lv et al. 2021, Sun and Shahrajabian 2023) as shown in Fig. 1. Therefore, the plant-derived polyphenols cater to the pressing need for the development of nature-derived chemicals as green therapeutics with potential health effects. Table 1 presents the various categories of plant phenolics. In this chapter, we present an account of the antioxidant properties of plant phenolics and their effect in maintaining an optimal health.

The production of phenolic compounds in plants occurs as a result of the secondary metabolism in plants that results in the formation of medicinally important natural products such as alkaloids, terpenoids, and phenolics such as lignins, coumarins, and chalcones as the immediate products (Kumar et al. 2019, Goufo et al. 2014). Further biochemical transformations result in the formation of flavonones, dihydroflavonols, and flavonols that play in leading role in the regulation of stress response in the plants (Lattanzio et al. 2013, Mandal et al. 2010). Eventually, the flavone classes of compounds are metabolized to yield isoflavones, pterocarpans, coumestans, anthocyanins, and condensed tannins (Thilakarathna et al. 2013, Walle et al. 2004).

These metabolites are mainly responsible for providing the color to flowers and for enabling the plant to defend against invading microbes (Yang et al. 2021, Murota et al. 2018). The polyphenolic such as coumestans exhibit anticancer properties and hold high medicinal value. It also serves as phytoalexin that comes into contact with invading parasites and inhibits their growth and development on the plants (Nehybova et al. 2014, Jeandet et al. 2014). Besides their isolation from the plant sources, the naturally occurring phenolic compounds have also been chemically synthesized and modified by introducing different substituents to further tune the therapeutic profile and to explore the potential healing effects (Bhattacharya et al. 2010, Direito et al. 2021). Figure 2 represents the flowchart for the synthesis of phenolics as secondary metabolites.

Table 1. Types of plant phenolics.

Category of Phenolic	Carbon backbone	No. of C atoms
Phenol	C_6	6
Phenolic acid	C_6-C_1	7
Lignin, melanin, flavolans	$[C_6$-$C_3]_n$ OR $[C_6]_n$ OR $[C_6$-C_3-$C_6]_n$	N
Lignan and neolignan	$[C_6$-$C_3]_2$	18
Biflavonoids	$[C_6$-C_3-$C_6]_2$	30
Xanthones, stilbenes	$[C_6$-C_1-$C_6]$	13
Anthraquinone	$[C_6$-C_2-$C_6]$	14
Naphthoquinone	$[C_6$-$C_4]$	10
Coumarin, isocoumarin, chromone	$[C_6$-$C_3]$	9
Phenylacetic acid, phenyl propenes, hydroxy cinnamic acids	$[C_6$-$C_2]$	8
Kaviol A, colpol	$[C_6$-C_4-$C_6]$	16
Amentoflavone	$[C_6$-C_3-$C_6]_2$	30
Mangiferin	$[C_6$-C_1-$C_6]$	13
Gallic, salicylic acid	$[C_6$-$C_1]$	7
Resveratrol, emodin	$[C_6$-C_2-$C_6]$	14
Acetophenones, tyrosine derivatives, phenyl acetic acids	$[C_6$-$C_2]$	8
Chalconoids, flavonoids, isoflavonoids, neoflavonoids	$[C_6$-C_3-$C_6]$	15
Catechol, hydroquinone, 2,6-dimethoxybenzoquinone	$[C_6$-$C_1]$	7

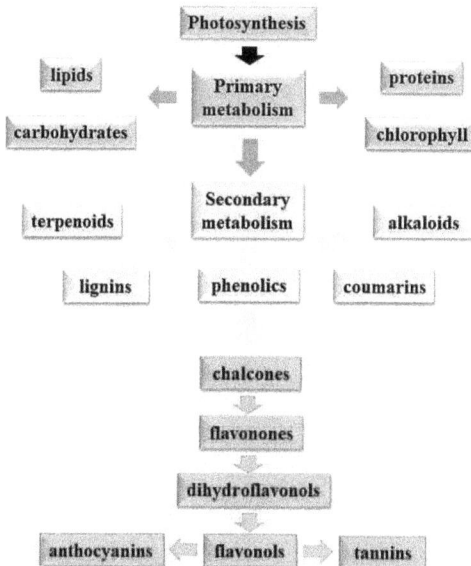

Fig. 2. Synthesis of phenolic phytochemicals as a part of the secondary metabolism in plants.

Biosynthesis of Plant Phenolics

The biosynthesis of phenolic compounds in plants is induced by cuts or injury or the production of reactive oxygen species (ROS) to biosynthesize lignin as a stress response (Marchiosi et al. 2020, Rehman et al. 2012, Kandar et al. 2021). The stress response or injury induces the oxidative decomposition of sugar to erythrose-4-phosphate and phosphoenolpyruvate that serve as the precursor for the synthesis of phenolics via a shikimic acid pathway and phenylpropanoid pathway (Moreno et al. 2012). The biosynthesis of phenolic compounds mainly occurs via the shikimic acid pathway and acetic acid pathway (Seigler et al. 1998, Vickery et al. 1981). In the former, phosphoenolpyruvate and erythrose-4-phosphate combine to synthesize 3-dehydroquinate whose subsequent dehydration with shikimate dehydrogenase results in the formation of 3-dehydroshikimic acid (Wilson et al. 1998). Further reduction of 3-dehydroshikimic acid with NADPH provides shikimic acid, which is converted to chorismic acid that undergoes Claisen rearrangement to form prephenic acid (Weinstein et al. 1962, Sanchez et al. 2019). This product gets converted to the amino acid 'tyrosine' after several steps, which forms the precursor for the biosynthesis of phenolic compounds. Figure 3 represents the synthesis of phenolics by the shikimic acid pathway.

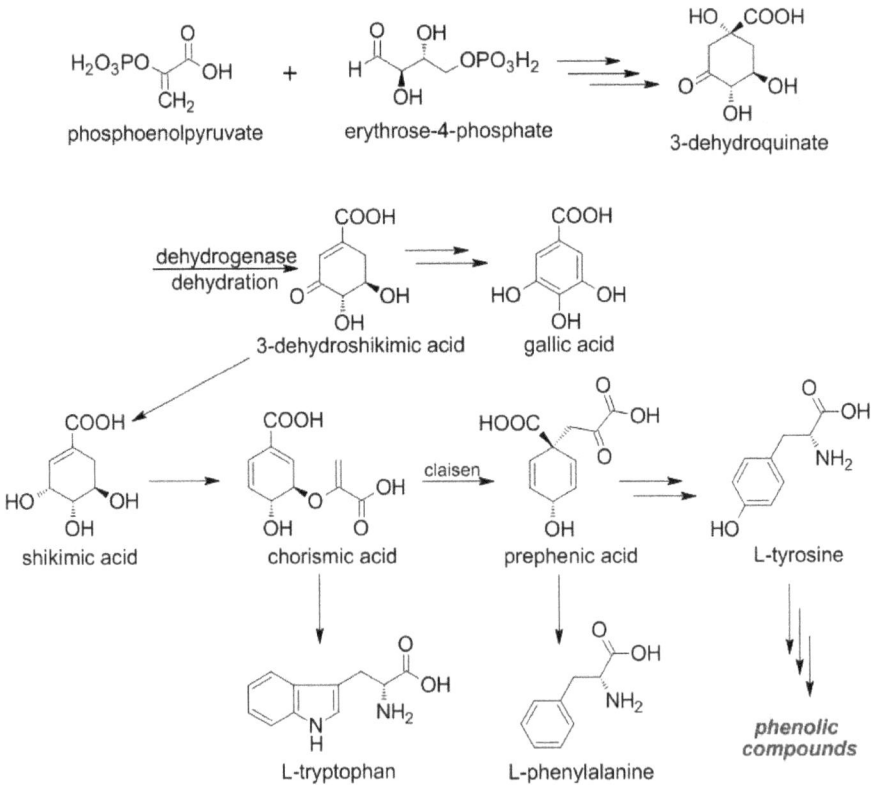

Fig. 3. Biosynthesis of phenolic compounds via the shikimic acid pathway.

The phenylpropanoid pathway is another essential biosynthesis route for obtaining phenolics. This pathway is quite like the shikimic acid pathway till the L-phenylalanine stage, which in the next step undergoes deamination in the presence of phenylalanine ammonia lyase enzyme to yield cinnamic acid (Sharma et al. 2019, Grace et al. 2000). Furthermore, the hydroxylation and conversion to Coenzyme A results in the biosynthesis of *p*-coumaroyl Coenzyme A. This final compound is the precursor for the further synthesis of various phenolic compounds (Kumar et al. 2020, Zhang et al. 2021). Figure 4 represents the synthesis of phenolics by the phenylpropanoid pathway.

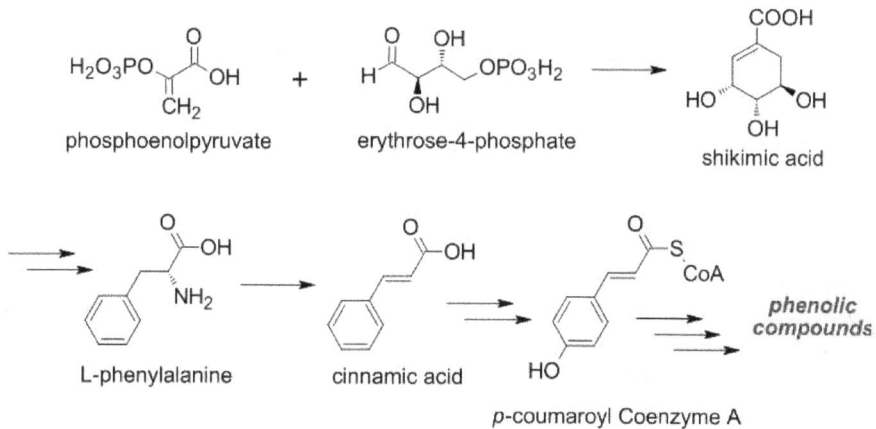

Fig. 4. Biosynthesis of phenolic compounds via the phenylpropanoid pathway.

Functions of Phenolic Compounds

The phenolic compounds manage the stress conditions in plant under the effect of external or internal stimuli (Samec et al. 2021, Isah et al. 2019, Tak et al. 2020). As such, the mechanical injury which is a type of external stimulus triggers the biosynthesis of phenolates in the plant cell wall to avoid further damage to the plant (Hu et al. 2022, Rhodes et al. 2019, Chen et al. 2018). Similarly, deficiency of water acts as a stress condition to plants, which eventually causes suberization in roots (Silva et al. 2021, Aroca et al. 2012, Feng et al. 2016). The deficiency of elements such as iron causes a surge in glucose and sucrose levels that results in the production of plant phenolics (Michalak et al. 2006). The lignification in the plants is caused as a result of the accumulation of phenolics in response to the stress caused by deficiency of essential elements and plant nutrients such as sulfur and potassium. Other than these factors, the metal accumulation, biomagnification, and heavy metal stress serve as inducers of stress response in plants (Antonova et al. 2012). Nevertheless, the physical factors including excessive heat and cold, UV irradiation, and heat shock also play an important role in mediating stress response in plants, resulting the production of phenolics (Salar et al. 2013, Calderon et al. 2015).

Plant phenolics play an important role in growth regulation of plants starting from the stage of seed germination (Lannucci et al. 2013, Williams et al. 2017).

The germinating seeds contain phenolic growth inhibitors in their outer membranes that provide protection against premature seed germination and during unfavorable environmental conditions due to the enzyme catalyzed formation of tannins and lignins in the seed coat (Tarzi et al. 2012, Reigosa et al. 1999). However, during stratification, phenolic compounds present in the seed coat decrease tremendously (Mala et al. 2021, Marambe et al. 1992). Overall, these factors render the seed coat water impervious, but the availability of atmospheric oxygen is unaffected. It can be inferenced from these observations that the seed coat phenolics prevent premature germination by regulating the exchange of water and gas through the seed and provide protection to the developing embryo from harsh environmental conditions (Alkaltham et al. 2020, Amoros et al. 2006). The plant phenolics play both growth inhibitory and growth promoting role in the vegetative parts of plant depending on their interactions with enzymes and essential proteins.

The plant phenolics containing multiple hydroxyl groups are known to regulate the activity of Indole-3-acetic acid IAA-oxidase (Lee et al. 1982). The phenolics containing adjacent 2/3-hydroxyl groups play an inhibitory role towards IAA-oxidase. Conversely, the phenolics containing a single hydroxyl group stimulate the activity of IAA-oxidase, hence, promoting the plant's growth (Lee et al. 1980).

Phenolics such as phytoalexins are secondary metabolites that offer a antimicrobial property to the plant in response to biotic stimulus caused by viruses, fungi, and bacteria pathogens (Jeandet et al. 2015, Ghimire et al. 2017, Othman et al. 2019). The plants that possess resistance towards microbial attacks have naturally higher content of phenolics such as o-coumaric acid, p-coumaric acid, ferulic acid, catechol, and protocatechuic acid (Ota et al. 2011). The enzymes such as phenylalanine ammonia-lyase, peroxidase, and polyphenol oxidase play an important role in the synthesis of phenolics and therefore contribute towards microbial resistance (Ngadze et al. 2012, Vanitha et al. 2009). The wounded plant parts develop an immunity mechanism where the secretion and production of phenolics increases with a second injury to the plant (Hu et al. 2022, Dehghanian et al. 2022, Mandal et al. 2010, Bhattacharya et al. 2010, Savatin et al. 2014, Dzialo et al. 2016).

Similar observation has been reported with mechanically wounded plants infested with microbial disease that show an increased production of phenolics. Some plants have inherently higher content of phenolics depending on their genetics that evolved over the years. In addition, the plants utilize polyphenolic compounds such as proanthocyanidins, phenolic acids, flavonoids, and anthocyanins for exerting the antioxidant activity by balancing the effect of reactive oxygen species (Pandey et al. 2009). Figure 5 illustrates the functions of plant phenolics.

Oxidative Stress in Plants

As the microbial cells invade plants, they face a variety of stress response such as oxidative stress and chemical defense by the host through phytochemicals. The phenolic compounds form a major category of phytochemicals that induce a stress response in plants. These are encountered by the invading microbes on the leaf surface and in the rhizosphere (Zhang et al. 2021, Slobodnikova et al. 2016). The phenolates induce a variety of toxicity to the invading microbes such as degradation

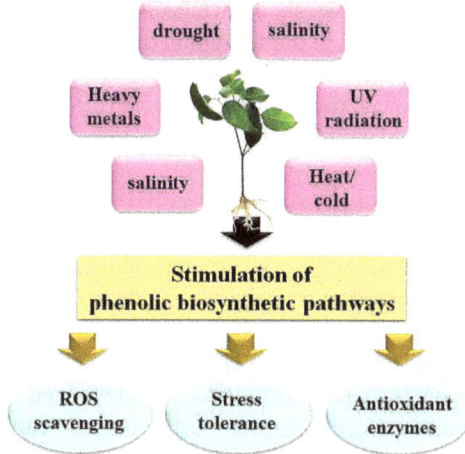

Fig. 5. Functions of phenolics in plants.

of microbial enzymes that inflicts significant damage to the metabolic machinery of host microbe. Regulation of the redox response by phenolates is of paramount importance (Oulahal et al. 2022). Phenolates, also termed as antioxidants are reported to trigger cellular responses that effectively counter the oxidative stress induced in the plant by invading pathogens (Tako et al. 2020).

Hydrogen peroxide is the most significant contributor to redox stress by exhibiting a dual role. It not only participates in signal transduction at low concentrations, while the higher levels of hydrogen peroxide exert a toxic effect on the host cells. The optimal levels of hydrogen peroxide in the foliage occur as approximately 1 µmol/g while it appears as 10 µM in peroxisomes (Cheeseman et al. 2006, Costa et al. 2010). In plants, the optimal levels of hydrogen peroxide are essential for the growth and development of cells. It also mediates the oxidation of phenolates and polymerization of lignin in the cell wall of plant cells (Smirnoff et al. 2019). The altered level of hydrogen peroxide (up to 10 µM in plants) however exerts toxicity and plays a major role in the apoptosis and ageing. The toxicity of hydrogen peroxide in plants is caused by the deactivation of essential enzymes by oxidation of cysteine residues to methionine that alters the catalytic activity of enzymes (Kurutas et al. 2016). Similarly, the hydrogen peroxide reacts with ferric and cupric ions present as essential elements in plant cells and converts them into the toxic hydroxide radical which potentially damages the DNA molecules. The exposure of plant cells to hydrogen peroxide may result in the generation of superoxide ions owing to the activation of NOX enzymes (Nita et al. 2016). Lastly, the presence of hydrogen peroxide at a concentration of 10 µM has been found to inhibit the fixation of carbon dioxide in Calvin cycle that adversely affects the process of photosynthesis.

The oxidative stress in photosynthetic plants is mainly caused by singlet molecular oxygen due to the presence of chlorophyll that serves the function of photosensitizer, thereby generating this reactive species in leaves (Foyer et al. 2018). During the process of photosynthesis, the absorption of light in chlorophyll occurs in light-harvesting complexes that have a long-lived excited state for converting the excitatory energy to electrochemical potential via charge separation (Dmitrieva et al.

2020). However, the excited triplet state of chlorophyll can supply sufficient energy to the molecular oxygen leading to the formation of a singlet molecular oxygen that serves as the reactive oxygen species which degrades the activity of D1 protein thereby causing a loss in the activity of photosystem II (Sharma et al. 2012). These events result in bleaching of chlorophyll pigment and loss of photosynthesis in plants (Tripathy et al. 2012).

Superoxide anion radicals generated normally in the cells by a single-electron reduction of molecular oxygen is rapidly converted to hydrogen peroxide by superoxide dismutase enzyme that regulates the intracellular building up of this reactive oxygen species (Tsujimoto et al. 1993, Dionisi et al. 1975). Under normal conditions, these reactive oxygen species customarily combine to another molecule of superoxide anion radical in acidic pH, resulting in the formation of oxygen molecule and hydrogen peroxide. However, the unregulated levels of superoxide anion radical trigger a series of ROS reactions directly, via enzymatic processes, or by metal catalysis (Palma et al. 2019, Sachdev et al. 2021).

Hydroxyl radical is another potent oxidant having a short half-life, positive redox potential of +2V, and a considerable affinity towards a range of biomolecules. This species is reported to oxidize DNA, lipids, amino acids, proteins, and sugars resulting in permanent damage to the plant and inducing genetical instability to the plant (Chimi et al. 1991). The hydroxyl radical is generated from hydrogen peroxide in the presence of iron and copper ions via a Fenton reaction (Ozyurek et al. 2008). Due to the absence of hydroxyl radical eliminating enzymes, the cells adopt various mechanisms to maintain iron homeostasis (Burkitt et al. 2000). The presence of ferritin and metallothioneins that can store iron, copper, and zinc is direct evidence in support of this occurrence.

Source of Reactive Oxygen Species in Plants

The free radicals that serve as reactive oxygen species are formed as a by-product of various biological processes and aerobic metabolic reactions. These redox processes occur in organelles such as peroxisomes, cells wall, chloroplast, mitochondria, endoplasmic reticulum, plasma membrane, and apoplast. As a result of these processes in the mentioned organelles, the free radicals that are formed include superoxide ions, singlet oxygen, hydrogen peroxide, and hydroxyl radical (Hasanuzzaman et al. 2020). The chief source for the generation of ROS is photosynthesis and various components associated with it such as photosystem I and II, electron transport chain, and photorespiration. These components generate singlet oxygen and superoxide ions by Mehler reaction. The reduction of molecular reduction to generate ROS in the form of hydrogen peroxide and superoxide radical anion occurs due to the electron leakage in mitochondrial complex I and III (Moucheshi et al. 2014).

Similarly, hydrogen peroxide is formed as a by-product during the glycolate pathway in the process of photorespiration in peroxisomes. This event leads to the intracellular production of hydrogen peroxide when the glycolate generated in the stroma of chloroplast gets oxidized. The enzymes such as respiratory burst oxidase homolog (BROH) catalyze the electron transfer from cytoplasmic NAD(P) H to molecular oxygen during stress conditions, thereby generating a superoxide anion radical, which is a reactive oxygen species (Garg et al. 2009). The RBOHs are expressed mainly in nucleus, cell wall, vacuoles, and mitochondria. Similarly, during the adverse abiotic conditions, the class III peroxidases from apoplasts serve as a source of oxidative burst by producing ROS such as hydroxyl radical and hydrogen peroxide. Figure 6 indicates the various ROS sources in plants (Huang et al. 2019).

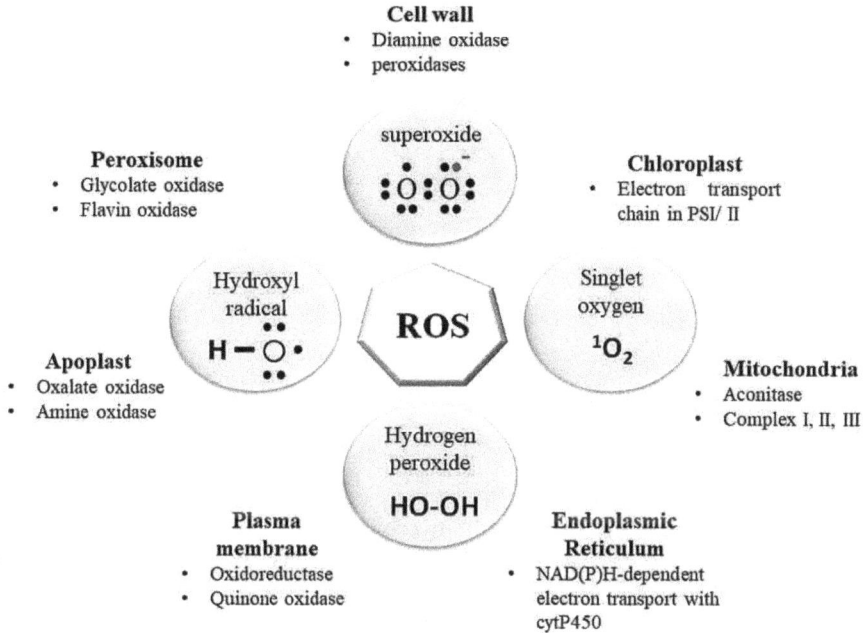

Cell wall
- Diamine oxidase
- peroxidases

Peroxisome
- Glycolate oxidase
- Flavin oxidase

superoxide

Chloroplast
- Electron transport chain in PSI/ II

Hydroxyl radical

ROS

Singlet oxygen
$^1O_2$

Apoplast
- Oxalate oxidase
- Amine oxidase

$H - O \cdot$

Mitochondria
- Aconitase
- Complex I, II, III

Hydrogen peroxide

HO-OH

Plasma membrane
- Oxidoreductase
- Quinone oxidase

Endoplasmic Reticulum
- NAD(P)H-dependent electron transport with cytP450

Fig. 6. Sources of reactive oxygen species in plants.

Phenolic Compounds and Redox Stress

The antioxidant potential of phenolics is related to their unique structure that contains an aromatic ring with hydroxyl or methoxy substituents that can trap the free radicals. Phenolics can efficiently donate electrons or hydrogen atoms owing to their ability to stabilize the phenol radical. The phenolics that contain *o*-dihydroxy substituents participate in the complexation with metal ions that prevents the formation of reactive oxygen species. The phenolics are also known to capture singlet oxygen and inhibit lipid peroxidation by effectively trapping the alkoxy radicals formed in lipids. The phenolates reportedly modify the peroxidation kinetics due to a decrease in membrane fluidity and amendment in the lipid bilayer that discourages the diffusion of free radicals and the subsequent peroxidation reactions.

Tannins as Antioxidants

Tannins are naturally occurring polyphenolic phytochemicals with a high molecular weight and several phenolic groups (Fig. 7). Tannins can be divided in three groups: condensed tannins also known as proanthocyanidins, phlorotannins predominantly

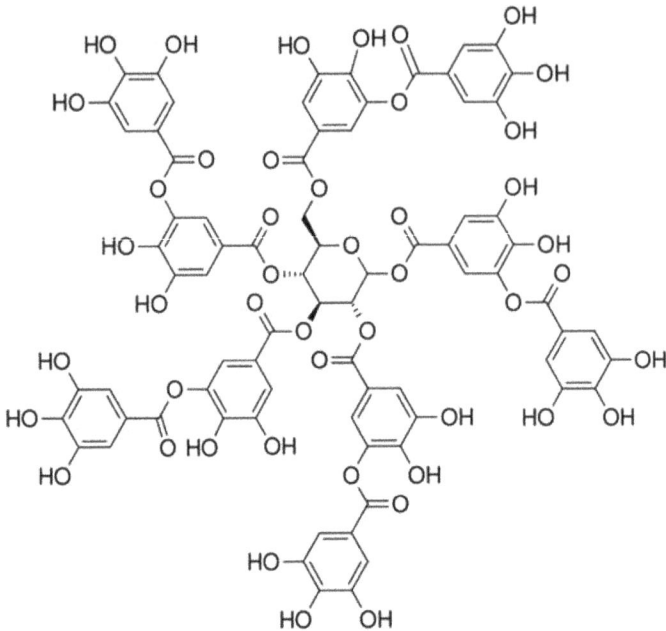

Fig. 7. Basic structure of tannin.

present in marine brown algae, and hydrolysable tannins. Tannins containing multiple phenolic groups are heavily polymerized due to which they serve as effective antioxidants. In some plants, tannins are inducted due to the biotic and abiotic stress where they control the antioxidative damage to the plant in response to UV-B radiation and drought conditions (Das et al. 2020). Condensed tannins are the end-product of the flavonoid pathway, and they contribute to the survival of plants by absorbing harmful radiation which may contribute towards the production of free radicals (Jiang et al. 2020). The condensed tannins also serve as antifeedants, biocidals, and maintain the nutrient cycle by contributing towards ecophysiological roles.

Gu et al. (2008) reported antioxidant activity of tannins derived from persimmom pulp by fractionation with polysulfone ultrafiltration membrane, having a molecular weight of 10,000 Da. Both low and high molecular weight tannins are obtained by this method Gu et al. (2008). Thiolysis degradation of the latter provided products such as gallocatechin, epicathechin-3-*O*-gallate, and epigallocathechin-3-*O*-gallate. The antioxidant activity of persimmon tannins obtained by this method was checked via 2-deoxyribose oxidation system, salicylic acid system for scavenging hydroxyl radicals. The scavenging of superoxide anions and lipid peroxidation of linoleic acid was evaluated for further confirming the antioxidant profile. Notably, the high molecular weight tannins displayed commendable antioxidant activity as compared to their low molecular weight counterpart in a dose dependent manner. The antioxidant potential of tannins is also reported in ruminants. The condensed tannins are mainly oligomers that possess a high molecular weight and have a poor bioavailability. These factors restrict the antioxidant potential of condensed tannins in living animals. The source of condensed tannins used as nutritional strategy for ruminants include legumes, plants, and trees. In addition, these sources also provide benefits as regulators of ruminal biohydrogenation, which has a positive impact on the fatty acid profile of meat and milk fats.

Moreover, these agents are known to reduce bloating, minimize methane emissions, ameliorate the oxidation stability of dietary metabolites, prevent the degradation of proteins, and check the development of internal parasites. The antioxidant activity of condensed tannins also arises from their chelation with transition metals and the inhibition of pro-oxidative enzymes (Soldado et al. 2021). The accumulation of condensed tannins in circulation and their subsequent buildup in tissues exerts a direct antioxidant action. The bioavailability of condensed tannins and their metabolites in circulation largely depends on their molecular weight, chemical structure, stereochemical configuration, degree of polymerization, and number of hydrophilic hydroxyl groups as evidenced by the absorption and transportation of only tetramers and oligomers to the various organs. Based on their stereochemical configuration, the absorption rate of condensed tannins varies in the order (–)-epicatechin > (+)-epicatechin = (+)-catechin > (–)-catechin. Hydrolyzable tannins also display a remarkable antioxidant potential and inhibit the activity of tyrosinase and hyaluronidase. The tannins isolated from the aqueous ethanolic extract of the leaves of *Eucalyptus globulus* contained four gallotannins and five ellagitannins that exerted antioxidant potential. Notably, the hydrolysable tannins displayed a remarkable antioxidant activity like superoxide dismutase and displayed a potent inhibition of tyrosinase and hyaluronidase (Sugimoto et al. 2009). The inducible antioxidant potential of condensed tannins on hybrid poplar has been reported by Gourlay et al. (2019). Condensed tannins bear significant antioxidant potential *in vitro* in both high light and during nitrogen deficiency. The hybrid transgenic poplar plants further possessed 50-times higher levels of condensed tannins as compared to the control that resulted in considerably higher antioxidant activity.

Chalcones as Antioxidants

The plant-derived chalcones present commendable antioxidant potential owing to the presence of phenolic groups. Chemically designed chalcones have been reported to further tune their antioxidant profile. Pyrazine derivatives of chalcones containing alkyl substituent on the pyrazine ring were reported to exert the radical scavenging property by supporting the cell growth exposed to hydrogen peroxide (Stepanic et al. 2019). The radical scavenging activity of licorice chalcones, and naturally occurring chalcones such as non-prenylated butein and prenylated xanthohumol are well known. Structure activity relationship analysis of chalcones has screened the presence of free -OH groups at C2' position in ring A and catechol group in ring B. Similarly, the presence of α, β double bond is an important feature of chalcones for exerting the free radical scavenging activity. Chalcones also serve as the starting material for the biosynthesis of flavonoids that have a robust antioxidant profile.

Sivakumar et al. (2011) designed chalcone derivatives, investigated their structure activity relationship analysis, and studied their antioxidant activity via DPPH-scavenging assay, superoxide radical-scavenging, reducing potential, and hydrogen peroxide scavenging studies. The chalcone derivatives with methoxy and methyl thionyl substituents at para position of A ring and hydroxyl group at the para position of B ring are best suited for displaying optimal antioxidant activity. Sokmen et al. (2016) reported the antioxidant profile of curcuminoids and chalcones isolated from *Curcuma longa*. The antioxidant properties of isolated chalcones were ascertained by DPPH and β-carotene/ linoleic acid assay where a correlation between free radical scavenging potency and concentration was found. The chalcones with four phenolic groups displayed optimal antioxidant ability. Besides naturally occurring chalcones the chemically synthesized chalcones are also reported to have potent antioxidant activity. Chalcone fatty acid esters synthesized by reaction of chalcone with palmitic and stearic acid by electrophilic and Michael addition reaction have been shown to exert superior antioxidant potential as compared to the commonly used antioxidant 'ascorbic acid' (Lahsasni et al. 2014).

The antioxidant potential arises due to the substituent pattern on aryl moieties, where the hydroxyl group plays a significant role due to its facile conversion to phenoxy radical via hydrogen atom transfer mechanism. The location of hydroxyl group on B ring at ortho- and para-positions exhibited strong antioxidant activity and the hydroxyl group on A ring attached to *p*-coumaric acid and carbonyl group at *o*-position exhibited antioxidant potential due to their potency to absorb UV rays. Furthermore, the presence of unsaturation at C3 position that connects A ring with B results in conjugation that further ameliorates antioxidant potential and has a photooxidation tendency. The absence of conjugated system by the loss of carbon-carbon double bond and presence of electron donating substituent at 5' position retains the antioxidation and photooxidation ability. Furthermore, the B ring derived from *p*-coumaric acid enables the trapping of peroxy radical (Daikatsu et al. 2008). Rationally designed chalcone derivatives containing amide functionality at the B ring were synthesized via Claisen-Schmidt reaction and were reported to possess antioxidant activities by acting as free radical scavengers. Due to this activity, the reported analogues showed inhibitory potential against lipopolysaccharide induced NF-kB activation. Furthermore, these chalcone derivatives were known to inhibit the proinflammatory mediators such as TNF-α, IL-1β, IL-6, and PGE2 (Chu et al. 2016).

Chalcone derivatives substituted with halogens at the 4-position of B ring and methoxy or hydroxyl group at 3,4-position of theA ring were reported by Wang et al. (2019) to possess antioxidant properties. These derivatives provided defense against hydrogen peroxide induced oxidative damage in the cells and played a significant role in the scavenging of free radicals. Furthermore, these chalcone derivatives exerted neuroprotective and cytoprotective potential that proved worthy for application as pharmaceutically active ingredients. The radical scavenging activity of chalcones has been elucidated with the help of density function theory and time-dependent density function theory. Hydrogen transfer mechanism, sequential proton loss electron transfer, and electron transfer followed by proton transfer mechanisms explain the radical scavenging properties of chalcones. The chalcones reportedly adopt a fully planar conformation in their neutral, anionic, and cationic forms and the 2'-OH groups on chalcone play an important role in the stabilization of phenolic radicals by forming intramolecular hydrogen bonds.

Importantly, the presence of electron donating substituents on B-ring plays a significant role in exerting the antioxidant activity to the chalcones. Of all the free radical scavenging mechanisms, the hydrogen transfer mechanism plays an important role due to its thermodynamic favorability in gas phase and in nonpolar environment. The sequential proton loss electron transfer is a preferred mechanism in the polar medium (Xue et al. 2019). Ammaji et al. (2022) reported the antioxidant activity of chalcone analogies substituted with hydroxy and chloro substituents synthesized via the Claisen-Schmidt reaction. The free radical scavenging activity of test compounds was tested with DPPH and MABA free radical assays that indicated 1.8-times higher activity of fluorine and methoxy substituted chalcones as compared to the unsubstituted molecule. The activity and bioavailability of test compounds further improves with the presence of electron releasing groups at the ortho position. The test compounds displayed drug-likeliness based on their physicochemical parameters that further supports their development as leads. However, these molecules surpassed Lipinksi Rule of Five and showed high absorption in the gastrointestinal tract.

Flavones as Antioxidants

Flavones have been known to possess antioxidant and antiradical properties. The studies have shown that the presence of hydroxyl group at 2',3',4' positions on the B ring of flavone offers radical scavenging activity as indicated in DPPH assay. Further investigations by MDA test showed inhibitory effect on oxidation of lipids in tissues. The mechanism of these events was elucidated to find that the inhibition of xanthine oxidase leads to antioxidant property. The reported flavones acted as competitive inhibitors of the enzyme where C-2 and C-6 hydroxyl group in flavones accommodates in the binding pocket where C-7 hydroxyl group on xanthine accommodates normally. However, the protection of hydroxyl group with methoxy substituent becomes unfavorable for antioxidant activity due to the loss of radical scavenging activity exerted by the hydroxyl group (Cotelle et al. 1996). Naturally occurring flavones berberisinol, oleanolic acid, gallic acid, 8-oxoberberine, berberine, and β-sitosterol have been isolated from *Berberis baluchistanica* were found to be enriched with antioxidant activity which was evaluated via the DPPH free radical scavenging assay (Parvez et al. 2019).

In addition to the reactive oxidation species, advanced glycation end-products and metalloproteinases play a leading role in the degenerative processes in tissues. Antioxidant flavones such as chrysin, apigenin, and luteolin, in addition to flavonols such as quercetin, kaempferol, and mirycetin have been known to offer antioxidant properties and inhibit the production of metalloproteinases and advanced glycation end-products. Notably, the antioxidant activity increased in proportion to the number of hydroxyl groups. The inhibition of metalloproteinases by flavones and flavonols are related to the total polar surface area of molecule and logP values (Ronsisvalle et al. 2020). Ali et al. (2019) performed DFT analysis on flavones and flavonols to ascertain their antioxidant activity. Reportedly, the flavonol molecules containing hydroxyl group at 3-position on B ring are conformationally non-planar, while the resulting radicals and flavones showed a planar conformation. The free radical was stabilized by low bond dissociation energy. Further, the conversion of nonplanar flavonols to planar radicals and their association with resonance towards the carbonyl group served as a driving force. The formation of hydrogen bonding between ortho hydroxyl group, 3-OH group with catecholic radical, and resonance towards the carbonyl functionality of nearby pyrone oxygen contributed towards free radical scavenging. The antioxidant activity done experimentally corroborated with the DFT results. The flavonoids extracted from alfalfa displayed commendable *in vitro* antioxidant activity as demonstrated via DPPH free radical assay. The antioxidant potential arises due to inhibition of superoxide dismutase and glutathione peroxidase (Chen et al. 2020).

Flavonoids and phenolics isolated from *Ammoides atlantica* have been reported to display antioxidant properties (Benteldjoune et al. 2019). The antioxidant molecules obtained from the isolates were identified as hydroxybenzoic acids, flavones, flavonols, lignan, and hydroxycinnamic acid which showed radical scavenging activity in DPPH assay.

Conclusion

The plant phenolics bear an established profile as redox stress controlling agents caused due to various biotic and abiotic stimuli. The plant phenolics are also known to modulate the activity of antioxidant enzymes that results in effective management of oxidative stress and diseases caused by this event. The free radical scavenging activity of plant phenolics has been ascertained due to the generation of phenolic radicals. The free radical scavenging activity of plant phenolics has been utilized for the management of diseases such as cancer, inflammation, and ischemia. The antioxidant profile is further utilized in the design and development of synthetic molecules that bear similar properties as that of phenolics for achieving radical scavenging properties.

References

Albuquerque, B.R., Heleno, S.A., Oliveira, M.B.P.P., Barros, L. and Ferreira, I.C.F.R. 2021. Phenolic compounds: current industrial applications, limitations and future challenges. Food Funct., 12: 14–29.

Ali, H.M. and Ali, I.H. 2019. Structure-antioxidant activity relationships, QSAR, DFT calculation, and mechanisms of flavones and flavonols. Med. Chem. Res., 28: 2262–2269.

Alkaltham, M.S., Salamatullah, A.M., Ozcan, M.M., Uslu, N. and Hayat, K. 2021. The effects of germination and heating on bioactive properties, phenolic compounds and mineral contents of green gram seeds. LWT., 134: Article 110106.

Ammaji, S. Masthanamma, R.R. Bhandare, S. Annadurai and A.B. Shaik. 2022. Antitubercular and antioxidant activities of hydroxy and chloro substituted chalcone analogues: Synthesis, biological and computational studies. Arab. J. Chem., 15: Article 103581.

Amoros, M.L.L., Hernandez, T. and Estrella, I. 2006. Effect of germination on legume phenolic compounds and their antioxidant activity. J. Food Comp. Anal., 19: 277–283.

Anantharaju, P.G., Gowda, P.C., Vimalambike, M.G. and Madhunapantula, S.R.V. 2016. An overview on the role of dietary phenolics for the treatment of cancers. Nutrition J. 15, Article 99.

Antonova, G.F., Varaksina, T.N., Zheleznichenko, T.V. and Stasova, V.V. 2012. Changes in phenolic acids during maturation and lignification of scots pine xylem. Russian J. Dev. Biol., 43: 199–208.

Aroca, R., Porcel, R. and Lozano, J.M.R. 2012. Regulation of root water uptake under abiotic stress conditions. J. Exp. Botany, 63: 43–57.

Balasundram, N., Sundram, K. and Samman, S. 2006. Phenolic compounds in plants and agri-industrial by-products: Antioxidant activity, occurrence, and potential uses. Food Chem., 99: 191–203.

Benteldjoune, M., Boudiar, T., Backouche, A., Contreras, M.M., Sanchez, J.L., Bensouici, C. et al. 2019. Antioxidant activity and characterization of flavonoids and phenolic acids of *Ammoides atlantica* by RP–UHPLC–ESI–QTOF–MS^n. J. Natural Fibers., 35: 1639–1643.

Bhattacharya, A., Sood, P. and Citovsky, V. 2010. The roles of plant phenolics in defence and communication during Agrobacterium and Rhizobium infection. Mol. Plant Pathol., 11: 705–719.

Burkitt, M.J. and Duncan, J. 2000. Effects of trans-Resveratrol on Copper-Dependent Hydroxyl-Radical Formation and DNA Damage: Evidence for Hydroxyl-Radical Scavenging and a Novel, Glutathione-Sparing Mechanism of Action. Arch. Biochem. Biophys., 381: 253–262.

Calderon, M.G., Ferrer, T.P., Mrazova, A., Balang, P.P., Vilkova, M., Delgado, C.M.P. et al. 2015. Modulation of phenolic metabolism under stress conditions in a Lotus japonicus mutant lacking plastidic glutamine synthetase. Front. Plant Sci., 6, Article 760.

Cheeseman, J.M. 2006. Hydrogen peroxide concentrations in leaves under natural conditions. J. Exp. Botany., 57: 2435–2444.

Chen, Q., Lu, X., Guo, X., Liu, J., Liu, Y., Guo, Q. et al. 2018. The specific responses to mechanical wound in leaves and roots of Catharanthus roseus seedlings by metabolomics. J. Plant Interact., 13: 450–460.

Chen, S., Li, X., Liu, X., Wang, N., An, Q., Ye, X.M. et al. 2020. Investigation of Chemical Composition, Antioxidant Activity, and the Effects of Alfalfa Flavonoids on Growth Performance. 2020, Article 8569237.

Cheynier, V., Comte, G., Davies, K.M., Lattanzio, V. and Martens, S. 2013. Plant phenolics: Recent advances on their biosynthesis, genetics, and ecophysiology. Plant Physiol. Biochem., 72: 1–20.

Cheynier, V. 2012. Phenolic compounds: from plants to foods. Phytochem. Rev., 11: 153–177.

Chimi, H., Cillard, J., Cillard, P. and Rahmani, M. 1991. Peroxyl and hydroxyl radical scavenging activity of some natural phenolic antioxidants. J. Amer. Oil Chem. Soc., 68: 307–312.

Chu, J. and Guo, C.-L. 2016. Design and Discovery of Some Novel Chalcones as Antioxidant and Anti-Inflammatory Agents via Attenuating NF-κB. Arch. Pharm., 349: 63–70.

Cosme, P., Rodriguez, A.B., Espino, J. and Garrido, M. 2020. Plant Phenolics: Bioavailability as a Key Determinant of Their Potential Health-Promoting Applications. Antioxidants (MDPI). 9, Article 1263.

Costa, A., Drago, I., Behera, S., Zottini, M., Pizzo, P., Schroeder, J.I. et al. 2010. H$_2$O$_2$ in plant peroxisomes: an *in vivo* analysis uncovers a Ca^{2+}-dependent scavenging system. Plant J. 62: 760–772.

Cotelle, N., Bernier, J.-L., Catteau, J.-P., Pommery, J., Wallet, J.-C. and Gaydou, E.M. 1996. Antioxidant activity of hydroxy-flavones. Free Radical Biol. Med., 20: 35–43.

Dai, J. and Mumper, R.J. 2010. Plant Phenolics: Extraction, Analysis and Their Antioxidant and Anticancer Properties. Molecules (MDPI). 15: 7313–7352.

Daikatsu, Y. and Sato, T. 2008. Antioxidant and photooxidant activity of chalcone derivatives. J. Japan Petroleum Inst., 51: 298–308.

Das, A.K., Islam, N., Faruk, O., Ashaduzzaman, M. and Dungani, R. 2020. Review on tannins: Extraction processes, applications and possibilities. South Af. J. Botany., 135: 58–70.

Dehghanian, Z., Habibi, K., Dehghanian, M., Aliyar, S., Lajayer, B.A., Astatkie, T. et al. 2022. Reinforcing the bulwark: unravelling the efficient applications of plant phenolics and tannins against environmental stresses. Heliyon. 8, Article e09094.

Dionisi, O., Galeotti, T., Terranova, T. and Azzi, A. 1975. Superoxide radicals and hydrogen peroxide formation in mitochondria from normal and neoplastic tissues. Biochim. Biophys. Acta. 403: 2920300.

Direito, R., Rocha, J., Sepodes, B. and Figueira, M.E. 2021. Phenolic Compounds Impact on Rheumatoid Arthritis, Inflammatory Bowel Disease and Microbiota Modulation. Pharmaceutics (MDPI)., 13: Article 145.

Dmitrieva, V.A., Tyutereva, E.V. and Voitsekhovskaja, O.V. 2020. Singlet Oxygen in Plants: Generation, Detection, and Signaling Roles. Int. J. Mol. Sci. (MDPI)., 21: Article 3237.

Dzialo, M., Mierziak, J., Korzun, U., Preisner, M., Szopa, J. and Kulma, A. 2016. The potential of plant phenolics in prevention and therapy of skin disorders. Int. J. Mol. Sci. (MDPI)., 17: Article 160.

Savatin, D.V., Gramegna, G., Modesti, V. and Cervone, F. 2014. Wounding in plant tissue: the defense of a dangerous passage. Front. Plant Sci., 5: Article 470.

Feng, W., Lindner, H., Robbins, N.E. and Dinneny, J.R. 2016. Growing Out of Stress: The Role of Cell- and Organ-Scale Growth Control in Plant Water-Stress Responses. Plant Cell., 28: 1769–1782.

Foyer, C.H. 2018. Reactive oxygen species, oxidative signaling and the regulation of photosynthesis. Environ. Exp. Bot., 154: 134–142.

Friedman, M. and Jurgens, H.S. 2000. Effect of pH on stability of plant phenolic compounds. J. Agric. Food. Chem., 48: 2101–2110.

Garg, N. and Manchanda, G. 2009. ROS generation in plants: Boon or bane? Plant Biosys., 143: 81–96.

Ghimire, B.K., Seong, E.S., Yu, C.Y., Kim, S.H. and Chung, I.M. 2017. Evaluation of phenolic compounds and antimicrobial activities in transgenic Codonopsis lanceolata plants via overexpression of the γ-tocopherol methyltransferase (γ-tmt) gene. South Af. J. Botany., 109: 25–33.

Goufo, P. and Trindade, H. 2014. Rice antioxidants: phenolic acids, flavonoids, anthocyanins, proanthocyanidins, tocopherols, tocotrienols, γ-oryzanol, and phytic acid. Food Sci. Nutrition., 2: 75–104.

Gourlay, G. and Constabel, C.P. 2019. Condensed tannins are inducible antioxidants and protect hybrid poplar against oxidative stress. Tree Physiol., 39: 345–355.

Grace, S.C. and Logan, B.A. 2000. Energy dissipation and radical scavenging by the plant phenylpropanoid pathway. Phil. Trans. Royal Soc., 355: 1499–1510.

Gu, H.-F., Li, C.-M., Xu, Y.-J., Hu, W.-F., Chen, M.-H. and Wan, Q.-H. 2008. Structural features and antioxidant activity of tannin from persimmon pulp. Food Res. Int., 41: 208–217.

Hasanuzzaman, M., Bhuyan, M.H.M.B., Parvin, K,, Bhuiyan, T.F., Anee, T.I. Nahar, K. et al. 2020. Regulation of ROS Metabolism in Plants under Environmental Stress: A Review of Recent Experimental Evidence. Int. J. Mol. Sci. (MDPI). 21, Article 8695.

Hu, W., Sarengaowa, Guan, Y. and Feng, K. 2022. Biosynthesis of Phenolic Compounds and Antioxidant Activity in Fresh-Cut Fruits and Vegetables. Front. Microbiol., 13, Article 906069.

Hu, W., Sarengowa, Guan, Y. and Feng, K. 2022. Biosynthesis of Phenolic Compounds and Antioxidant Activity in Fresh-Cut Fruits and Vegetables. Front. Microbiol., 46: 1029–1056.

Huang, H., Ullah, F., Zhou, D.-X., Yi, M. and Zhao, Y. 2019. Mechanisms of ROS Regulation of Plant Development and Stress Responses. Front. Plant Sci., 10: Article 800.

Isah, T. 2019. Stress and defense responses in plant secondary metabolites production. Biol. Res., 52: Article 39.

Jeandet, P., Hebrard, C., Deville, M.A., Cordelier, S., Dorey, S., Aziz, A. et al. 2014. Deciphering the Role of Phytoalexins in Plant-Microorganism Interactions and Human Health. Molecules (MDPI)., 19: 18033–18056.

Jeandet, P. 2015. Phytoalexins: Current Progress and Future Prospects. Molecules (MDPI)., 20: 2770–2774.

Jiang, Y., Zhang, H., Qi, X. and Wu, G. 2020. Structural characterization and antioxidant activity of condensed tannins fractionated from sorghum grain. J. Cereal. Sci., 92: Article 102918.

Kandar, C.C. 2021. Secondary Metabolites from Plant Sources. In: Pal, D. and Nayak, A.K. (Eds.). Bioactive Natural Products for Pharmaceutical Applications. Advanced Structured Materials, vol 140. Springer, Cham.

Kopjar, M., Orsolic, M. and Pilizota, V. 2014. Anthocyanins, Phenols, and Antioxidant Activity of Sour Cherry Puree Extracts and their Stability During Storage. Int. J. Food Properties, 17: 1393–1405.

Kumar, N. and Goel, N. 2019. Phenolic acids: Natural versatile molecules with promising therapeutic applications. Biotechnol. Rep., 24: Article e00370.

Kumar, S., Abedin, M.M., Singh, A.K. and Das, S. 2020. Role of Phenolic Compounds in Plant-Defensive Mechanisms. In: Lone, R., Shuab, R. and Kamili, A. (Eds.). Plant Phenolics in Sustainable Agriculture. Springer, Singapore.

Kurutas, E.B. 2016. The importance of antioxidants which play the role in cellular response against oxidative/nitrosative stress: current state. Nutr. J. 15: Article 71.

Lahsasni, S.A., Korbi, F.H.A. and Aljaber, N.A.A. 2014. Synthesis, characterization and evaluation of antioxidant activities of some novel chalcones analogues. Chem. Cent. J. 8: Article 32.

Lannucci, A., Fragasso, M., Platani, C. and Papa, R. 2013. Plant growth and phenolic compounds in the rhizosphere soil of wild oat (*Avena fatua* L.). Front. Plant Sci. 4: Article 509.

Lattanzio, V. 2013. Phenolic Compounds: Introduction. In: Ramawat, K. and Mérillon, J.M. (Eds.). Natural Products. Springer, Berlin, Heidelberg.

Lee, T.T., Starratt, A.N. and Jevnikar, J.J. 1982. Regulation of enzymic oxidation of indole-3-acetic acid by phenols: Structure-Activity Relationships., 21: 517–523.

Lee, T.T. 1980. Effects of phenolic substances on metabolism of exogenous indole-3-acetic acid in maize stems. Physiol. Plant., 50: 107–112.

Lin, D., Xiao, M., Zhao, J., Li, Z., Xing, B., Li, X. et al. 2016. An Overview of Plant Phenolic Compounds and Their Importance in Human Nutrition and Management of Type 2 Diabetes. Molecules (MDPI). 21: Article 1374.

Lopes, M., Sanches-Silva, A., Castilho, M., Cavaleiro, C. and Ramos, F. 2023. Halophytes as source of bioactive phenolic compounds and their potential applications. Critical Reviews in Food Science and Nutrition, 63(8): 1078–1101.

Lv, Q.-Z., Long, J.-T., Gong, Z.-F., Nong, K.-Y., Liang, X.-M., Qin, T. et al. 2021. Current State of Knowledge on the Antioxidant Effects and Mechanisms of Action of Polyphenolic Compounds. Natural Product Commun., 16: 1–13.

Mala, M., Norrizah, J.S. and Azani, S. 2021. *In vitro* seed germination and elicitation of phenolics and flavonoids in *in vitro* germinated *Trigonella foenum graecum* plantlets. Biocatal. Agric. Biotechnol., 32: Article 101907.

Mandal, S.M., Chakraborty, D. and Dey, S. 2010. Phenolic acids act as signaling molecules in plant-microbe symbioses. Plant Signal Behav., 5: 359–368.

Mandal, S.M., Chakraborty, D. and Dey, S. 2010. Phenolic acids as signaling molecules in plant-microbe symbioses. Plant Signal. Behav., 5: 359–368.

Maqsood, S., Benjakul, S., Abushelaibi, A. and Alam, A. 2014. Phenolic Compounds and Plant Phenolic Extracts as Natural Antioxidants in Prevention of Lipid Oxidation in Seafood: A Detailed Review. Comprehensive Rev. Food Sci. Food Saf., 13: 1125–1140.

Marambe, B. and Ando, T. 1992. Phenolic acids as potential seed germination inhibitors in animal-waste composts. Soil Sci. Plant Nut., 38: 727–733.

Marchiosi, R., Santos, W.D., Constantin, R.P., Lima, R.B., Soares, A.R., Teixeira, A.F. et al. 2020. Biosynthesis and metabolic actions of simple phenolic acids in plants. Phytochem. Rev., 19: 865–906.

Michalak, A. 2006. Phenolic Compounds and Their Antioxidant Activity in Plants Growing under Heavy Metal Stress. Polish J. Environ. Studies., 15: 523–530.

Moreno, A.B., Benavides, J., Zevallos, L.C. and Velazquez, D.A.J. 2012. Plants as Biofactories: Glyphosate-Induced Production of Shikimic Acid and Phenolic Antioxidants in Wounded Carrot Tissue. J. Agric. Food Chem., 60: 11378–11386.

Moucheshi, A.S., Shekoofa, A. and Pessarakli, M. 2014. Reactive Oxygen Species (ROS) Generation and Detoxifying in Plants. J. Plant Nutrition., 37: 1573–1585.

Murota, K., Nakamura, Y. and Uehara, M. 2018. Flavonoid metabolism: the interaction of metabolites and gut microbiota. Biosci. Biotechnol. Biochem., 82: 600–610.

Nehybova, T., Smarda, J. and Benes, P. 2014. Plant Coumestans: Recent Advances and Future Perspectives in Cancer Therapy. Anti-cancer Agent Med. Chem., 14: 1351–1362.

Ngadze, E., Icishahayo, D., Coutinho, T.A. and Waals, J.E. 2012. Role of Polyphenol Oxidase, Peroxidase, Phenylalanine Ammonia Lyase, Chlorogenic Acid, and Total Soluble Phenols in Resistance of Potatoes to Soft Rot. Am. Pathophysiol. Soc., 96: 186–192.

Nita, M. and Grzybowski, A. 2016. The Role of the Reactive Oxygen Species and Oxidative Stress in the Pathomechanism of the Age-Related Ocular Diseases and Other Pathologies of the Anterior and Posterior Eye Segments in Adults. 2016. Article 3164734.

Ota, A., Abramovic, H., Abram, V. and Ulrih, N.P. 2011. Interactions of p-coumaric, caffeic and ferulic acids and their styrenes with model lipid membranes. Food Chem., 125: 1256–1261.

Othman, L., Sleiman, A. and Massih. 2019. Antimicrobial activity of polyphenols and alkaloids In middle eastern plants. Front. Microbiol., 10, Article 911.

Oulahal, N. and Degraeve, P. 2022. Phenolic-Rich Plant Extracts with Antimicrobial Activity: An Alternative to Food Preservatives and Biocides? Front. Microbiol., 12: Article 753518.

Ozyurek, M., Bektasoglu, B., Guclu, K. and Apak, R. 2008. Hydroxyl radical scavenging assay of phenolics and flavonoids with a modified cupric reducing antioxidant capacity (CUPRAC) method using catalase for hydrogen peroxide degradation. Anal. Chim. Acta., 616: 196–206.

Palma, J.M., Gupta, D.K. and Corpas, F.J. 2019. Hydrogen Peroxide and Nitric Oxide Generation in Plant Cells: Overview and Queries. In: Gupta, D., Palma, J. and Corpas, F. (Eds.). Nitric Oxide and Hydrogen Peroxide Signaling in Higher Plants. Springer, Cham.

Pandey, K.B. and Rizvi, S.I. 2009. Plant polyphenols as dietary antioxidants in human health and disease. 2: 270–278.

Parvez, S., Saeed, M., Ali, M.S., Fatima, I., Khan, H. and Ullah, I. 2019. Antimicrobial and Antioxidant Potential of Berberisinol, a New Flavone from *Berberis baluchistanica*. Chem. Natural Compd., 55: 247–251.

Rehman, F., Khan, F. and Badruddin, S. 2012. Role of Phenolics in Plant Defense Against Insect Herbivory. In: Khemani, L., Srivastava, M. and Srivastava, S. (Eds.). Chemistry of Phytopotentials: Health, Energy and Environmental Perspectives. Springer, Berlin, Heidelberg.

Reigosa, M., Souto, X. and Gonźlez, L. 1999. Effect of phenolic compounds on the germination of six weeds species. Plant Growth Regulation, 28: 83–88.

Rhodes, J., Michael and Wooltorton, L.S.C. 2019. The Biosynthesis of Phenolic Compounds in Wounded Plant Storage Tissues. Biochemistry of wounded plant tissues, edited by Günter Kahl, Berlin, Boston: De Gruyter, pp. 243–286.

Ronsisvalle, S., Panarello, F., Longhitano, G., Siciliano, E.A., Montenegro, L. and Panico, A. 2020. Natural flavones and flavonoids: relationships among antioxidant activity, glycation and metalloproteinase inhibition. Cosmetics (MDPI)., 7: Article 71.

Sachdev, S., Ansari, S.A., Ansari, M.I., Fujita, M. and Hasanuzzaman, M. 2021. Abiotic Stress and Reactive Oxygen Species: Generation, Signaling and Defense Mechanisms. 2021. Abiotic Stress and Reactive Oxygen Species: Generation, Signaling, and Defense Mechanisms. Antioxidants (MDPI). 10: Article 277.

Salar, R.K., Certik, M., Brezova, V., Brlejova, M., Hanusova, V. and Breierova, E. 2013. Stress influenced increase in phenolic content and radical scavenging capacity of Rhodotorula glutinis CCY 20-2-26. 3 Biotech, 3: 53–60.

Samec, D., Karalija, E., Sola, I., Bok, V.V. and Sondi, B.S. 2021. The Role of Polyphenols in Abiotic Stress Response: The Influence of Molecular Structure. Plants (MDPI). 10, Article 118.

Santos-Sánchez, NormaCOronado, R.S., Carlos, B.H. and Canongo, C.V. 2019. Shikimic Acid Pathway in Biosynthesis of Phenolic Compounds. In: Plant Physiological Aspects of Phenolic Compounds, edited by Marcos Soto-Hernández, Rosario García-Mateos, Mariana Palma-Tenango. London: IntechOpen, 10.5772/intechopen.83815

Seigler, D.S. 1998. Shikimic Acid Pathway. In: Plant Secondary Metabolism. Springer, Boston, MA.

Shahidi, F. and Ambigaipalan P. 2015. Phenolics and polyphenolics in foods, beverages and spices: Antioxidant activity and health effects—A review. J. Funct. Food., 18: 820–897.

Sharma, A., Shahzad, B., Rahman, A., Bhardwaj, R., Landi, M. and Zheng, B. 2019. Response of Phenylpropanoid Pathway and the Role of Polyphenols in Plants under Abiotic Stress. Molecules (MDPI). 24, Article 2452.

Sharma, P., Jha, A.B., Dubey, R.S. and Pessarakli, M. 2012. Reactive Oxygen Species, Oxidative Damage, and Antioxidative Defense Mechanism in Plants under Stressful Conditions. J. Botany. 2012. Article 217037.

Silva, N.D.G., Murmu, J., Chabot, D., Hubbard, K., Ryser, P., Molina, I. et al. 2021. Root Suberin Plays Important Roles in Reducing Water Loss and Sodium Uptake in Arabidopsis thaliana. Metabolites (MDPI). 11, Article 735.

Sivakumar, P.M., Prabhakar, P.K. and Doble, M. 2011. Synthesis, antioxidant evaluation, and quantitative structure–activity relationship studies of chalcones. Med. Chem. Res., 20: 482–492.

Slobodnikova, L., Fialova, S., Rendekova, K., Kovac, J. and Mucaji, P. 2016. Antibiofilm activity of plant polyphenols. 21, Article 1717.

Smirnoff, N. and Arnaud, D. 2019. Hydrogen peroxide metabolism and function in plants. New Phytol., 221: 1197–1214.

Sökmen, M. and Khan, M.A. 2016. The antioxidant activity of some curcuminoids and chalcones. Inflammopharmacol., 24: 81–86.

Soldado, D., Bessa, R.J.B. and Jeronimo, E. 2021. Condensed Tannins as Antioxidants in Ruminants-Effectiveness and Action Mechanisms to Improve Animal Antioxidant Status and Oxidative Stability of Products. Animals (MDPI). 11, Article 3243.

Stepanic, V., Matijasic, M., Horvat, T., Verbanac, D., Chlupacova, M.K., Saso, L. et al. 2019. Antioxidant Activities of Alkyl Substituted Pyrazine Derivatives of Chalcones-*In Vitro* and *In Silico* Study. Antioxidants. 8, Article 90.

Sugimoto, K., Nakagawa, K., Hayashi, S., Amakura, Y., Yoshimura, M., Yoshida, T. et al. 2009. Hydrolyzable Tannins as Antioxidants in the Leaf Extract of *Eucalyptus globulus* Possessing Tyrosinase and Hyaluronidase Inhibitory Activities. Food Sci. Technol. Res., 15: 331–336.

Sun, W. and Shahrajabian, M.H. 2023. Therapeutic potential of phenolic compounds in medicinal plants—Natural health products for human health. Molecules, 28(4): 1845.

Tak, Y. and Kumar, M. 2020. Phenolics: A Key Defence Secondary Metabolite to Counter Biotic Stress. In: Lone, R., Shuab, R. and Kamili, A. (Eds.). Plant Phenolics in Sustainable Agriculture. Springer, Singapore.

Tako, M., Kerekes, E.B., Zambrano, C., Kotogan, A., Papp, T., Krisch, J. and Vagvolgyi, C. 2020. Plant Phenolics and Phenolic-Enriched Extracts as Antimicrobial Agents against Food-Contaminating Microorganisms. Antioxidants (MDPI). 9: Article 165.

Tarzi, B.G., Gharachorloo, M., Baharinia, M. and Mortazavi, S.A. 2012. The Effect of Germination on Phenolic Content and Antioxidant Activity of Chickpea. Iran J. Pharm. Res., 11: 1137–1143.

Thilakarathna, S.H. and Rupasinghe, H.P.V. 2013. Flavonoid Bioavailability and Attempts for Bioavailability Enhancement. Nutrients (MDPI), 5: 3367–3387.

Tripathy, B.C. and Oelmuller, R. 2012. Reactive oxygen species generation and signaling in plants. Plant Signal Behav., 7: 1621–1633.

Tsujimoto, Y., Hashizume, H. and Yamazaki, M. 1993. Superoxide radical scavenging activity of phenolic compounds. Int. J. Biochem., 25: 491–494.

Tungmunnithum, D., Thongboonyou, A., Pholboon, A. and Yangsabai, A. 2018. Flavonoids and Other Phenolic Compounds from Medicinal Plants for Pharmaceutical and Medical Aspects: An Overview. Medicines (MDPI), 5: Article 93.

Tyagi, K., Shukla, P., Rohela, G.K., Shabnam, A.A. and Gautam, R. 2020. Plant Phenolics: Their Biosynthesis, Regulation, Evolutionary Significance, and Role in Senescence. In: Lone, R., Shuab, R. and Kamili, A. (Eds.). Plant Phenolics in Sustainable Agriculture. Springer, Singapore.

Vanitha, S.C., Niranjana, S.R. and Umesha, S. 2009. Role of Phenylalanine Ammonia Lyase and Polyphenol Oxidase in Host Resistance to Bacterial Wilt of Tomato. J. Pathophysiol., 157: 552–557.

Vickery, M.L. and Vickery, B. 1981. Shikimic Acid Pathway Metabolites. In: Secondary Plant Metabolism. Palgrave, London.

Walle, T. 2004. Absorption and metabolism of flavonoids. Free Rad. Biol. Med., 36: 829–837.

Wallis, C.M. and Galarneau, E.R.A. 2020. Phenolic Compound Induction in Plant-Microbe and Plant-Insect Interactions: A Meta-Analysis. Front. Plant Sci., 6: 221–227.

Wang, J., Huang, L., Cheng, C., Li, G., Xie, J., Shen, M. et al. 2019. Design, synthesis and biological evaluation of chalcone analogues with novel dual antioxidant mechanisms as potential anti-ischemic stroke agents. Acta Pharm. Sinica B., 9: 335–350.

Weinstein, L.H., Porter, C.A. and Laurencot, H.J. 1962. Role of the Shikimic Acid Pathway in the Formation of Tryptophan in Higher Plants: Evidence for an Alternative Pathway in the Bean. Nature, 194: 205–206.

Williams, R.D. and Hoagland, R.E. 2017. The effects of naturally occurring phenolic compounds on seed germination. Weed Sci., 30: 206–212.

Wilson, D., Patton, S., Florova, G., Hale, V. and Reynolds, K.A. 1998. The shikimic acid pathway and polyketide biosynthesis. J. Ind. Microbiol. Biotech., 20: 299–303.

Xue, Y., Liu, Y., Zhang, L., Wang, H., Luo, Q., Chen, R. et al. 2019. Antioxidant and spectral properties of chalcones and analogous aurones: Theoretical insights. Int. J. Quant. Chem., 119. Article e25808.

Yang, G., Hong, S., Yang, P., Sun, Y., Wang, Y., Zhang, P. et al. 2021. Discovery of an ene-reductase for initiating flavone and flavonol catabolism in gut bacteria. Nature Commun. 12, Article 790.

Zhang, J. and Sun, X. 2021. Recent advances in polyphenol oxidase-mediated plant stress responses. Phytochem. 181, Article 112588.

Zhang, Q., Yang, W., Liu, J., Liu, H., Lv, Z., Zhang, C. et al. 2021. Postharvest UV-C irradiation increased the flavonoids and anthocyanins accumulation, phenylpropanoid pathway gene expression, and antioxidant activity in sweet cherries (*Prunus avium* L.). Postharvest Biol. Technol., 175, Article 111490.

Zhang, Y., Cai, P., Cheng, G. and Zhang, Y. 2022. A Brief Review of Phenolic Compounds Identified from Plants: Their Extraction, Analysis, and Biological Activity. Natural Product Commun. DOI: 10.1177/1934578X211069721

Zhang, Y., Li, Y., Ren, X., Zhang, X., Wu, Z. and Liu, L., 2023. The positive correlation of antioxidant activity and prebiotic effect about oat phenolic compounds. Food Chemistry, 402: 134231.

11

Brassinosteroid

A Stress-Reliever Molecule for Plants under Abiotic Stress

Naresh Kumar,[1] Charu Lata,[2,]* Gurpreet kaur,[3]
Anshul Sharma Manjul[2] and Sonu[2]

Introduction

Plant sensitivity to various abiotic stresses poses a significant threat to increasing agricultural crop yields (Zhang et al. 2022). These stresses, including salt, drought, and temperature fluctuations, account for more than half of the global decline in food crop productivity, resulting in economic losses of $14–19 million (Kaur et al. 2022). Future projections indicate that climate change will exacerbate the frequency and severity of these challenges (Kumar et al. 2018). Therefore, enhancing plant tolerance to abiotic stresses is an urgent global priority to ensure long-term food security for the growing population. Multiple strategies have been employed to achieve resistance to abiotic stresses in plants, ranging from conventional breeding methods to the latest genome editing techniques (Ahmad et al. 2021, Kaur et al. 2022), However, finding a practical and universally applicable solution to this problem remains elusive due to the advantages and disadvantages associated with each strategy. Traditional breeding and marker-assisted selection are cost-effective and socially acceptable approaches for achieving desired outcomes (Wani et al. 2018). Nonetheless, they are time-consuming and limited by reproductive barriers and restricted genetic diversity. Modern technologies, such as genetic engineering and genome editing, have shown promise in enhancing plant stress tolerance (Kaur et al. 2022). However, they present ethical challenges related to germplasm cross-

---
[1] Eternal University, Baru Sahib, Himachal Pradesh, India.
[2] ICAR-Indian Institute of Wheat and Barley Research, Regional Station, Shimla, India.
[3] ICAR-Central Soil Salinity Research Institute, Karnal, India.
* Corresponding author: charusharmabiotech@gmail.com

contamination, potential health risks, and public acceptance issues. Thus, researchers are compelled to explore environmentally friendly techniques that can yield desired results more efficiently.

Plant growth regulators, naturally present in trace amounts in all plants, play a crucial role in regulating various morphological, physiological, biochemical, and molecular characteristics (Janowska and Andrzejak 2023, Zahid et al. 2023). Extensive research has demonstrated their efficacy in agriculture for enhancing crop growth and productivity (Manghwar et al. 2022). Recently, there has been increased interest in their ability to regulate abiotic stress responses in plants. This discussion delves into the details of brassinosteroids (BRs), including their chemical structure, occurrence, and distribution in the plant kingdom. Furthermore, we highlight the interplay between BRs and physiological, biochemical, and molecular processes under abiotic stress conditions. Lastly, we examine their role in plant signaling mechanisms.

Brassionosteroids

Brassinosteroids are a recently discovered group of plant hormones, joining the ranks of cytokinins, auxins, gibberellins, ethylene, and abscisic acid, thereby expanding the spectrum of phytohormones in the plant kingdom. Considered growth-promoting hormones, BRs play a crucial role in governing various physiological processes essential for normal plant growth and development. Found ubiquitously throughout the plant kingdom, BRs are approximately 70 poly-hydroxylated sterol derivatives. Interestingly, despite their plant origin, BRs share structural similarities with animal steroid hormones, which are renowned for their roles in embryonic and post-embryonic development as well as adult homeostasis regulation. Like animal hormones, BRs regulate the expression of numerous genes involved in cell division, differentiation, intricate metabolic pathways, and overall developmental processes that contribute to morphogenesis. Notably, BRs significantly influence plant growth, particularly flowering and cell expansion (Clouse 2011). The versatile role of BRs in plant growth and development has garnered recognition in the scientific community (Nazir et al. 2019, Wei and Li 2016). Plant-specific BR ligands directly interact with cell surface receptors known as BRASSINOSTEROID INSENSITIVE 1 (BRI1), which are leucine-rich repeat receptor kinases. These receptors, along with BRI1 ASSOCIATED RECEPTOR KINASE (BAK1), initiate signaling in the cytoplasm through phosphorylation cascades. The BSU1 protein is phosphorylated in this cascade, which also causes the BRASSINOSTEROID-INSENSITIVE 2 (BIN2) proteins to be degraded by the proteasome. Inactivation of BIN2 enables the activation of BES1/BZR1, which translocate to the nucleus and regulates the expression of target genes in plants (Hafeez et al. 2021). Furthermore, BRs exhibit crosstalk with other plant hormones, concurrent with these signaling cascades (Kour et al. 2021, Mubarik et al. 2021, Planas-Riverola et al. 2019). Figure 1 illustrates several fundamental roles of BRs in plants.

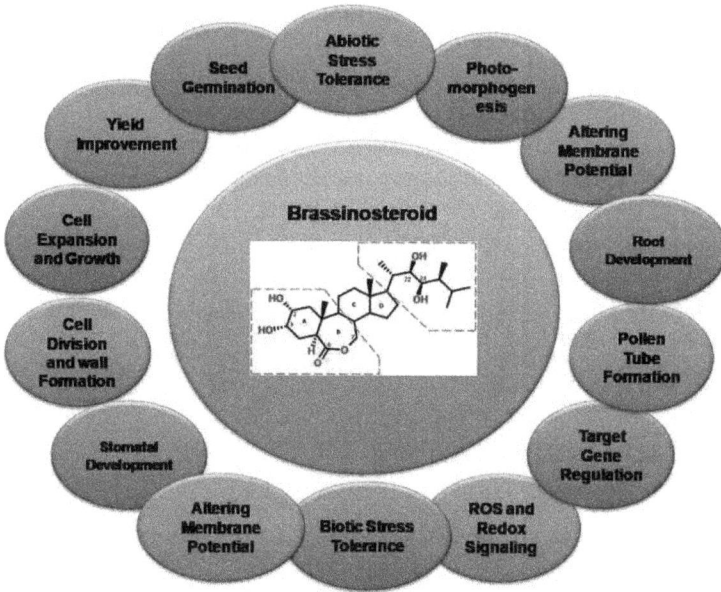

Fig. 1. Depicts the vital roles of brassinosteroids in plants.

History of BR Discovery

The discovery of brassinosteroids as plant growth promoting compounds with steroid properties occurred independently through research conducted at the United States Department of Agriculture (USDA) and Nagoya University in Japan (Yokota 1999). The USDA team, led by Mitchell, conducted a thirty-year-long analysis of organic pollen extracts from various plant species in search of new plant hormones (Mitchell et al. 1970). The most active growth-promoting substance was identified in the pollen of *Brassica napus* and named 'brassins.' Mandava (1988) conducted a bioassay on bean second-internode, demonstrating the significant effects of brassinosteroids on cell elongation and division. Spraying brassinosteroids on young seedlings of radishes, leafy vegetables, and potatoes resulted in increased yields. Mitchell and Gregory (1972) attributed hormonal status to 'brassins' based on these initial findings, as they observed measurable growth regulation when small amounts of these specific translocatable organic compounds were applied to other plants. After extensive efforts by various USDA laboratories, the chemical nature of the active component of brassins was discovered. Pilot-plant solvent extractions of 227 kg of *B. napus* pollen collected from bees (Steffens 1991), followed by extensive column chromatography, led to the identification of a steroidal lactone through single crystal X-ray analysis, which was named brassinolide (Grove et al. 1979). In subsequent years, the need for extensive plant extraction procedures was eliminated, and brassinolide (BL) and its stereoisomer, 24-epiBL, was chemically synthesized. Research then focused on exploring the physiological effects of BR in various biological systems, as well as testing its application in greenhouse and field settings to enhance crop yield during the 1980s (Cutler et al. 1991). Japanese research groups in the 1990s unraveled the biosynthetic pathway of BRs from common membrane sterols and identified endogenous BRs in *Arabidopsis* tissues (Shozo Fujioka et al. 1997, Shozo Fujioka and Sakurai 1997). Through mutant studies, genetic and biochemical evidence established that BRs are crucial for normal plant growth and development. Research on BRs has revealed that they regulate a wide range of physiological and morphological responses in plants. These include leaf bending and epinasty, stem elongation, activation of proton pumps, induction of ethylene biosynthesis, synthesis of nucleic acids and proteins, activation of photosynthesis, and regulation of carbohydrate assimilation and allocation (Bajguz and Hayat 2009). Additionally, BRs have been found to provide protection to plants against various biotic and abiotic stresses, such as high temperatures, drought, salt, and heavy metals (Bajguz and Hayat 2009, Sasse 2003).

Chemical Structure of BRs

The structure of BRs was determined through spectroscopic analysis including EI-MS, FAB-MS, NMR, and X-ray diffraction. The first structure that was determined was (22R, 23R, 24S)-2α, 3α, 22, 23-tetrahydroxy-24-methyl-B-homo-7-oxa-5α-cholestan-6-one, and it is known as brassinolide (BL) (see Fig. 2(A)). Yokota et al. (1982) isolated another BR called castasterone (CS) from the insect galls of chestnut (*Castanea crenata*). The structure of CS was provided by Yokota (1999) as (22R,23R,24S)-2α,3α,22,23-tetrahydroxy-24-methyl-5α-cholestan-6-one (see Fig. 2(B)).

The natural structure of BRs is derived from the 5α-cholestane skeleton, and the type and position of functionality in the A/B rings and the side chain contribute to structural variations (Yokota 1997). Oxidation and reduction reactions during biosynthesis lead to these structural modifications. Naturally, there are 70 BRs, divided into 65 free compounds and 5 conjugated compounds. These BRs can be categorized as C27, C28, or C29 consisting of 10, 38, or 16 compounds, respectively, depending on the alkyl substitutions in the side chain. Depending on the cholestane side chain, BRs are further categorized into 11 types with different substituents at C-23, C-24, and C-25 (Bajguz 2011). Among these compounds, brassinolide is the most active and well-known BR compound.

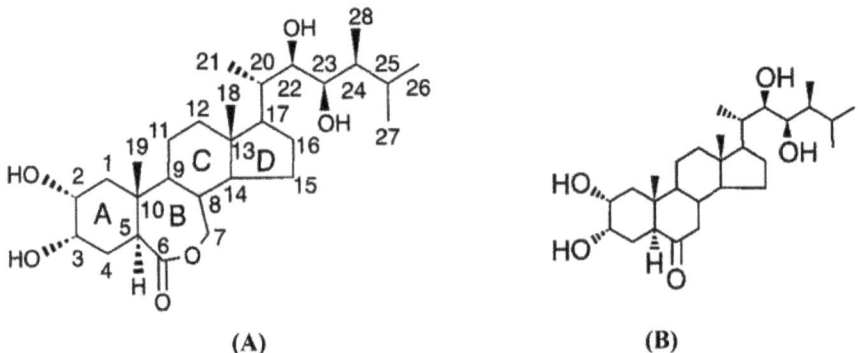

(A)   (B)

Fig. 2. Structures of Brassinosteroids: (A) Brassinolide, (B) Castasterone.

Occurrence and Distribution of BRs in the Plant Kingdom

Precise examination of a plant species by GC-MS or LC confirmed the presence of BRs in all cases under study, including 1 pteridophyte, 1 bryophyte, 3 algae, 6 gymnosperms and 53 angiosperms (41 dicots and 12 monocots) (Bajguz 2011, Fujioka 1999). The level of endogenous BRs varies across plant tissues as presented in Table 1.

The highest levels of BRs were reported in pollen and immature seeds. Immature tissue shows greater physiological response to BRs, that's why young, growing shoots contain higher BR levels than mature tissue (Table 1).

Table 1. Distribution of endogenous levels of BR reported in different tissue types in a range of plant species.

S.No.	Species	Tissue	Quantity (ng/g f.w.)
	Arabidopsis thaliana	Shoot, Seed, Silique	0.11–5.4
	Brassica campestris	Seed, Sheath	0.00013–0.094
	Brassica napus	Pollen	> 100
	Raphanus sativus	Seed	0.3–0.8
	Vicia faba	Pollen, Seed	5–628
	Helianthus annuus	Pollen	21–106
	Pisum sativum seed	Shoot	0.164–3.13
	Pinus thunbergii	Pollen	89
	Solanum lycopersicum	Shoo	0.029–1.69
	Cupressus arizonica	Pollen	1.0–6,400
	Equisetum arvense	Strobilus	0.152–0.349
	Catharanthus roseus	Cultured Cells	0.047–4.5
	Oryza sativa	Shoot	0.0084–0.0136
	Hydrodictyon reticulatum	Green Algae	0.3–0.4
	Citrus sinensis	Pollen	36.2
	Theasinensis	Leaves	0.001–0.02
	Zea mays pollen	Shoot	2.0–120
	Typha latifolia	Pollen	68
	Lilium elegans	Pollen	1.0–50

Information adopted from (Bajguz 2011, Clouse 2011, Fujioka 1999)

Role of Brassinosteroids against Abiotic Stress

Literature survey describes that tolerance towards any stress (biotic or abiotic) in plants is not restricted to one mechanism or gene only. The same is true with the role of brassinosteroids in providing tolerance against various stress in plants (Table 2) and has been explained in the below mentioned sections.

Table 2. Role of exogenous brassinosteroids in plants under major abiotic stresses.

Abiotic stress	Brassinosteroids	Role of Brassinosteroids under Abiotic Stress	References
Salt stress	Brassinolide	Regulate the uptake and transport of essential ions, such as potassium (K^+) and sodium ($Na+$), reduce oxidative stress and enhance osmotic regulation	Mu et al. 2022, Rattan et al. 2022, Vikram et al. 2022
	Homobrassinolide	Enhances the expression of H^+-ATPase and Na^+/H^+ antiporters, which help plants maintain proper ion homeostasis and reduce the accumulation of toxic ions, particularly sodium (Na^+), Modulate the antioxidant defense	
	Epibrassinolide	Modulate the activity of ion transporters and maintain ion homeostasis, enhance antioxidant defense, promote osmotic adjustment	
Drought stress	Brassinolide	Reduce transpirational water loss and improve water use efficiency, enhance antioxidant defense, modulate the expression of genes involved in drought response	Ahmad Lone et al. 2021, Castañeda-Murillo et al. 2022
	Homobrassinolide	Improve photosynthesis and water use efficiency, regulate stomatal conductance, promote the accumulation of osmolytes, such as proline and sugars	
	Epibrassinolide	Enhance water uptake efficiency, regulate stomatal conductance, activate antioxidant defense	
Heat stress	Brassinolide	Regulate the expression of HSPs, modulate the antioxidant defence system, maintain the membrane stability	Kothari and Lachowiec 2021, Yanhua et al. 2023
	Homobrassinolide	Induce the expression of HSPs, enhance the antioxidant defense, improve photosynthesis	
	Epibrassinolide	Regulate the HSPs expression, enhance the antioxidant defense, promote the osmotic adjustment	
Cold stress	Brassinolide	Induce the expression of cold responsive gene, LEA and dehydrins, regulate the osmotic adjustment and modulate the antioxidant defense	Ahammed et al. 2020, Lado et al. 2023, Ramirez and Poppenberger 2020
	Homobrassinolide	Increase the expression of cold-responsive genes such as COR, promote membrane stability, induce osmotic adjustment	
	Epibrassinolide	Enhance the expression of dehydrins and COR proteins, regulate the antioxidant defense and reduce oxidative damage, promote the osmotic adjustment	

Table 2. contd. ...

... Table 2. contd.

Abiotic stress	Brassinosteroids	Role of Brassinosteroids under Abiotic Stress	References
Heavy metal stress	Brassinolide	Enhance the expression of metal detoxification enzymes such as metallothionines and phytochelatins, regulate the ion transport and homeostasis	Ahammed et al. 2020, Basit et al. 2022, Ren et al. 2023
	Homobrassinolide	Induce the expression of metallothioneins, glutathione-S-transferases (GSTs), and ATP-binding cassette (ABC) transporters	
	Epibrassinolide	Regulate the metal transport and sequestration, stimulate the phytochelatin	

Salt stress

Under salty conditions, the application of BRs improves seed germination by enhancing ethylene biosynthesis, leading to an increased expression of CsACO1 and CsACO2 (Wei et al. 2015). BRs also modulate putrescine metabolism, improving the germination process and enhancing salt tolerance. There is an interaction between BRs and Gibberellic acid (GA) that regulates cell elongation, controlling plant height during salt stress (Tong et al. 2014). BRs have the ability to modify cell wall carbohydrate content, regulate cell division and enlargement, thereby improving plant growth under salt stress (Ashraf et al. 2010). Furthermore, BRs can slow down chlorophyll degradation and the degradation of proteins associated with the light-harvesting complex, restore optimal levels of chlorophyll pigments, enhance light capture efficiency, and regulate transcription and translation processes involved in photosynthetic pigment formation during salinity.

Additionally, BRs application enhances the photosynthetic rate by increasing the activity of carbonic anhydrase, RuBisCo, and other key enzymes involved in the Calvin cycle. BRs are also involved in the maintenance and development of stomata and chloroplast structure, leading to improved stomatal conductance and CO_2 intake under salinity (Fariduddin et al. 2014). BRs promote photosynthetic capacity by enhancing the repair and stabilization process of the D1 protein. The overall photosynthetic state and efficiency, represented by PSII (φPSII), increase with the application of BRs. BRs have been shown to reduce the content of malondialdehyde (MDA), a product of lipid peroxidation, under salinity.

Furthermore, BRs treatment enhances the activity of various enzymes associated with the antioxidative defense system, including superoxide dismutase (SOD), catalase (CAT), ascorbate peroxidase (APX), monodehydroascorbate reductase (MDHAR), glutathione reductase (GR), glutathione peroxidase (GPX), peroxidase (POD), and polyphenol oxidase (PPO), under salinity. BR treatment mitigates the inhibitory effects of salinity on germination by increasing glutathione accumulation. BRs have also been reported to enhance protein concentration, which is associated with increased translation rate and Hill reaction efficiency, among other factors (Vikram et al. 2022).

Application of BRs increases the levels of proline, betaine, and soluble sugars, which play important roles in salinity tolerance mechanisms, including osmoregulation. Salinity affects nitrogen metabolism by interfering with the absorption and xylem loading of nitrate ions, resulting in decreased NR activity in leaves. However, BRs application accelerates the activity of nitrate reductase (NR), nitrite reductase (NiR), glutamine synthase, and glutamate synthase in plants experiencing salinity. BRs also enhance nodule nitrogenase activity. Plant growth and development are influenced by the concentration of ions, and BRs application regulates the uptake of various ions such as Na^+, K^+, Ca^{2+}, Mg^{2+}, N and P under salinity (Soni et al. 2021). BRs application maintains the ionic balance by reducing Na^+ concentration and restoring K^+ levels. BRs also restrict the transport of Na+ ions and are involved in wall loosening of epicotyl and hypocotyl, which helps maintain plant biomass under saline environments (Vikram et al. 2022).

Drought Stress

When drought-stressed plants are treated with BRs, their water content and water potential increase, thereby improving plant viability under low water potential conditions. BRs aid in alleviating water stress by enhancing relative water content (RWC) (Ahmad Lone et al. 2021). Leaf wilting is overcome, and plant growth is improved with BRs treatment under drought conditions. BRs application also maintains root nodulation and nitrogenase activity in plants experiencing drought stress. Drought tolerance is closely associated with the accumulation of abscisic acid (ABA), and BR application can enhance ABA levels (Singh and Roychoudhury 2023). Additionally, BRs treatment increases zeatin content in drought-stressed plants. BRs application enhances the maximum quantum yield of photosystem II (PS II), the activity of ribulose-1,5-bisphosphate carboxylase/oxygenase, and the photosynthetic capacity (Castañeda-Murillo et al. 2022). It also improves light utilization efficiency, gaseous exchange processes, and related traits under drought stress.

Under drought conditions, BR application alters the expression of genes encoding both structural and regulatory proteins. For instance, increased transcript levels of BnCBF5 and BnDREB (two key drought-responsive genes) induced by BRs partly contribute to enhanced drought tolerance (Hossain et al. 2022). BRs application in drought-experiencing plants leads to increased content of soluble sugars and proline. BRs reduce the levels of reactive oxygen species (ROS) and malondialdehyde (MDA), protect against membrane injury, and increase the activities of antioxidative enzymes during drought. BRs application helps plants maintain optimum biomass under drought conditions, and increased biomass can lead to higher acid invertase activity in young leaves. Furthermore, BRs assist in seed setting under drought stress conditions, thereby increasing seed yield (Kazemi et al. 2023).

Heavy Metals

Brassinosteroids improve seed germination and germination percentage under heavy metal stress. They have the capacity to reduce metal load and promote plant growth and development, resulting in improved plant length and weight. BRs also reduce the absorption of heavy metals from the surrounding environment (Kapoor et al. 2022). Additionally, BRs enhance the biosynthesis of metal chelators such as phytochelatins (PC), which function in metal detoxification. By alleviating damage to the reaction centers and O_2^- evolving complexes, BRs eliminate the toxic effects of heavy metals on photochemical pathways and ensure efficient electron transport (Holá 2022). BRs can modulate the activity of proteins and membrane enzymes, preventing the loss of proteins, monosaccharides, and other sugars. They also increase the biosynthesis of proline and glycine betaine. Under heavy metal stress, BRs likely maintain the modified cell redox state by regulating the activities of enzymes such as SOD, CAT, AAO, GST, PPO as well as enzymatic and non-enzymatic antioxidants involved in the Halliwell-Asada cycle. BRs help re-establish the redox potential by reducing MDA levels and maintain membrane stability index (Kapoor et al. 2022). Furthermore, BRs increase the number of nodules, the activity of the NR enzyme, the content of leghemoglobin, and nitrogen during heavy metal stress. BRs mitigate the harmful effects of heavy metals on plant growth and enhance biomass. Moreover, under heavy metal stress, BRs improve biological yield, fruit yield, and fruit quality (Rajewska et al. 2016, Vardhini 2016).

Heat Stress

The application of BRs can mitigate the harmful effects of heat in plants. BRs play a role in upregulating genes associated with plant growth and development, cell elongation, cell wall modification enzymes, auxin responsive factors, and transcription factors responsible for mitigating heat stress. BRs help restore the expression of developmental proteins that are suppressed under heat stress and induce higher levels of protective proteins, primarily heat shock proteins (HSPs). HSPs aid in the proper folding, intracellular distribution, and degradation of proteins. BRs also enhance the synthesis of osmolytes to combat heat stress. During heat stress, the secondary messenger Ca^{2+} plays a crucial role and is necessary for the induction of TFs, HSPs, and osmolytes in response to heat stress. BRs are known to increase cytosolic Ca^{2+} concentration by binding to the BR receptor, resulting in the opening of Ca^{2+} ion channels in the plasma membrane (Zhao et al. 2013). BRs also induce biosynthetic pathways for phytohormones such as ABA and ethylene, which act as secondary messengers and activate the transcription of genes producing heat shock proteins at the molecular level (Kothari and Lachowiec 2021). Exogenous application of BRs minimizes the accumulation of reactive oxygen species and increases the activity of antioxidant enzymes such as SOD, POX, and APX during heat stress. BRs pretreatment during heat stress improves the contents of photosynthetic pigments, net CO_2 assimilation rate, stomatal conductance, photochemical activity of PSI, and water-use efficiency (Hafeez et al. 2021, Kothari and Lachowiec 2021). Plants treated with BRs during heat stress produce almost similar biomass as plants under optimal conditions and increasing grain yield is closely correlated with increasing biomass.

Signal Transduction of BRs under Abiotic Stress

In the realm of agricultural advancement, the presence of abiotic stresses poses significant limitations, leading to a notable focus on utilizing BRs in stress conditions to enhance crop productivity (Table 2). Over the past twenty years, extensive research has provided compelling evidence that supports the involvement of several key components within BR signaling in mitigating abiotic stress, thereby improving stress response and tolerance (Khan et al. 2022). Various scientists have reported on the signaling cascade of BRs, highlighting their effectiveness in mitigating various abiotic stresses such as heat stress, salt stress, drought stress, pesticide stress, cold stress, photo-oxidative stress, heavy metal stress, and submergence stress (Kumari and Hemantaranjan 2018, 2019, Ahanger et al. 2020, Mubarik et al. 2021, Sharma et al. 2015, Eremina et al. 2017, Xia et al. 2009, Zhong et al. 2020, Peleng et al, 2011). Current studies have predominantly focused on the model plant *Arabidopsis thaliana*, providing recent insights into the signaling and transduction of BRs under abiotic stresses.

Brassinosteroid response is triggered by BRI1, a steroid receptor that interacts with BR ligand and is located on the cell membrane. BRI1, along with its homologs BRL1 and BRL3, belongs to the leucine-rich repeat receptor-like kinase (LRR-RLK) family and is located on the plasma membrane. The extracellular domains of BRI1, BRL1, and BRL2 perceive BR molecules at nanomolar concentrations. Detailed atomic structures of BRI1 and BRL1 in complex with BL have provided insights into the structural basis of BR perception at the cell surface. Both receptors contain 25 and 24 units of LRRs, respectively, exhibiting a similar fundamental framework with a horseshoe-like structure in their extracellular domains, which plays a crucial role in BR binding. However, BRL1 exhibits higher affinity for BR compared to BRI1, attributed to minor structural differences in their BR binding pockets (Saini et al. 2015). Subsequently, a cascade of cytoplasmic signaling is triggered, leading to the activation of BR-associated gene expression and the regulation of cellular processes (Fig. 3).

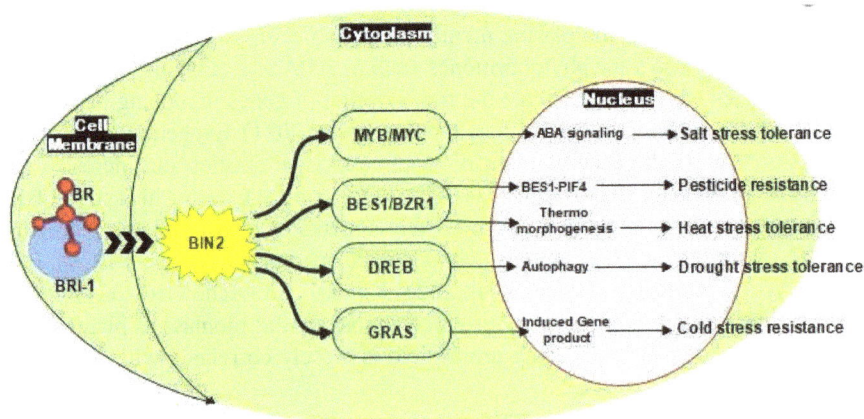

Fig. 3. Schematic illustration depicts the signal transduction of brassinosteroids under abiotic stress through several genes.

Specific stress-related signaling pathways have been observed in primary roots as an adaptive response to abiotic stresses (Planas-Riverola et al. 2019). BRs play a crucial role in managing environmental challenges and promoting balanced growth processes, either independently or through cross-talk with other hormones (Planas-Riverola et al. 2019). Several signaling cascades within plant cells are known to participate in the BR-mediated response to abiotic stresses. Fàbregas et al. (2018), reported an increase in osmo-protectant production as an adaptive mechanism mediated by BR signaling, Zou et al. (2018) identified the activation of the antioxidant system, while Ye et al. (2017) reported the induction of stress-responsive transcripts. Recent studies have revealed that under drought stress and starvation, BR signaling, particularly through BIN2, interacts with the autophagy pathway to regulate normal plant growth and survival. Under heat stress, reduced availability of BES1 homodimers leads to de-repression of BR biosynthesis, followed by the activation of thermo-morphogenesis through the BES1-PIF4 complex.

Cui et al. (2012) provided evidence for the significance of BR signaling mediated by BRI1 in conferring salt stress tolerance in *Arabidopsis*. In their study, they found that the BR-insensitive mutant bin2-1 exhibited increased sensitivity to salinity stress, which correlated with the suppression of stress-responsive gene induction (Zeng et al. 2010). In transgenic *Arabidopsis thaliana*, BR treatment stimulates the ZmBES1/BZR1-5 transcription factor, which controls drought and salt tolerance by binding to E-box regions and inducing the expression of stress-related downstream genes (Sun et al. 2020).

Ding et al. (2021) observed the upregulation of AnBES1/BZR1-3 and BZR1-5 under drought and salinity-induced stress conditions in *Ammopiptanthus nanus*. Additionally, under salt treatment in *Arabidopsis thaliana*, brassinolide induces the expression of a multimeric protein (iron sequestering protein) (Sharma et al. 2013). Xia et al. (2009) reported that, manipulation of BR biosynthesis genes has shown promising results in overcoming stress conditions. When HYDROXYSTEROID DEHYDROGENASE1 and AtDWF4, which are BR biosynthesis genes, were overexpressed in *Arabidopsis* mutant plants, they showed enhanced tolerance to abiotic stresses compared to wild-type plants (Divi and Krishna 2010, Li and Jin 2007).

Crosstalk between BRs and Physiological and Biochemical Process under Abiotic Stress

Seed germination

Soil moisture and partial pressure of oxygen are important features for seed germination and reduction in these parameters due to various abiotic stresses leads to germination failure (Khalid et al. 2019). Water stress delayed germination of Radish (*Raphanus sativus* L.) seeds but supplementation of BRs (EBL and HBL) improved the seed germination (Mahesh et al. 2013). EBL helped the *Arabidopsis* seeds to germinate under sodium chloride salinity by influencing the activity of diacylglycerol kinase (DGKs) enzyme. DGKs enzymes are central to the complex lipid signaling network which produces phosphatidic acid, a key lipid messenger (Derevyanchuk et al. 2019). Aluminium (Al) toxicity significantly decreased seed germination energy, germination percentage, germination index, and vigour index but seeds primed with BRs (EBL) significantly resist the effect of Al toxicity on these germination traits (Basit et al. 2022).

Plant Growth

Nutrients, water, and energy are required by the plants for cell proliferation and growth. Abiotic stress hinders these processes which reduces the growth of plants (Kaur et al. 2021, Mann et al. 2019, Sanwal et al. 2019). BRs regulate various aspects related to plant growth and development namely, expansion, cell division, stem cell maintenance, vascular development, elongation of different cell types, floral transition, hypocotyl elongation, root growth, shoot growth, stomata patterning, pollen tube growth, seed germination, xylem formation, and photomorphogenesis. Na^+ stress reduced growth of root and shoot in *Eucalyptus urophylla* plants, but the plants sprayed with EBL showed an increment in the various parameters related to root and shoot anatomy directly related to growth (de Oliveira et al. 2019). *Solanum lycopersicum* seedlings sprayed with EBL resulted in the amelioration of sodium chloride mediated decline in shoot length and dry weight (Ahanger et al. 2020). Leaf growth parameters, namely, leaf area, leaf dry weight, specific leaf area and equivalent water thickness were highly affected by drought stress and lowest values were reported in the plants of *Solanum quitoense* Lam not sprayed with brassinosteroid analogue DI-31. Moreover, plant growth was also least affected in the plants spayed with this brassinosteroid analogue (Castañeda-Murillo et al. 2022).

Photosynthesis and Related Traits

The process of photosynthesis is badly hampered when the plant is exposed to any kind of stress and the primary reason being the lowering of chlorophyll pigments (Kaur et al. 2021). BRs helps in the restoration of these pigments in plants experiencing abiotic stress (Houimli et al. 2010, Pereira et al. 2019). Brassinosteroids lowers the chlorophyll catabolism by reducing the chlorophyllase activity under abiotic stresses (Bartwal and Arora 2020). Under low temperature, application of BRs enhanced the chlorophyll biosynthesis (M. Zhao et al. 2019). Brassinosteroids, EBL, and HBL, helped the potato plants maintain optimum chlorophyll levels under salinization (Efimova et al. 2018). Similarly, EBL helped soybean plants to maintain optimum levels of chlorophyll under water deficient conditions (Pereira et al. 2019).

Abiotic stress generates water deficient conditions and plants lowers down the stomatal conductance to avoid water loss due to transpiration. This process affects CO_2 assimilation which further affects the photosynthesis process. Reduction in transpiration rate, stomatal conductance, and CO_2 assimilation rate has been reported due to various abiotic stresses (Singh and Thakur 2018). Application of BRs increased CO_2 fixation rate, stomatal conductance, and transpiration rate and hence, the photosynthetic process. The increment in the photosynthetic rate by the application of BRs cannot be attributed to sugar-signal-induced feedback as BRs also enhanced the content of sugars and increased CO_2 assimilation provided more carbohydrate for metabolism and export to sink. Instead, BRs increased Rubisco activity which increased the activity of other Calvin-cycle enzymes which enhanced the photosynthesis rate (Hafeez et al. 2021, Yu et al. 2004). Brassinosteroids, EBL, and 7,8-Dihydro-8α-20-Hydroxyecdysome (DHECD), enhanced the transpiration rate, stomatal conductance, and CO_2 assimilation rate in the rice plantlets exposed to heat stress (Thussagunpanit et al. 2015). Under cold stress, *Capsicum annuum* L. leaves were subjected to foliar application of EBL, resulting in increased photosynthesis and the development of well-defined thylakoid membranes in the chloroplasts. Moreover, proteomics data revealed that various differentially upregulated proteins were associated with the photosynthetic electron transfer chain and calvin cycle (Jie Li et al. 2022).

Water Relations

Abiotic stress (especially salinity and drought) causes reduction in the water content of plants which causes water stress and it is characterized by relative water content (RWC), leaf water potential, stomatal closure, and poor growth. Supplementation of BRs helps the plants to thrive under such conditions (Fariduddin et al. 2014). Rice grown in drought showed low values for leaf water potential, osmotic potential, pressure potential, and relative water content over the controls, but application of BRs, improved these water-related attributes (Farooq et al. 2009). Salinity exposure to maize plants reduced the leaf water potential and osmotic potential but exogenously applied EBL helped the plants to maintain optimum water and osmotic potential required to grow under saline conditions (Kaya et al. 2018).

RWC reduction due to drought is a common phenomenon, but chilli plants treated with BRs (EBL and DHECD (7,8-dihydro-8α-20- hydroxyecdysone)) facing drought maintained optimum water levels (Khamsuk et al. 2018). Similarly, in *Brassica juncea* (L.) Czern. plants exposed to Cd toxicity maintained optimum RWC supplemented with EBL through foliar application (Alam et al. 2020).

Osmoregulation

Plants drew water from the soil when water potential of the plant is less than that of the soil but various abiotic stress components lower down the osmotic potential of the soil and to survive and reproduce under such conditions, plants accumulate various cellular compatible solutes commonly known as osmolytes or osmoprotectants (Ozturk et al. 2021). BRL3 (vascular-enriched member of the BR receptor family) might regulate the biosynthesis of various osmolytes. BR-mediated biosynthesis of osmoprotectant, glycine betaine, is due to the better performance of the enzyme betaine aldehyde dehydrogenase (BADH), which converts choline to glycine betaine. BRs also modulates the expression of genes encoding polyamine synthesis enzymes (Sharma et al. 2019). Proline, soluble proteins, and soluble sugars increased in grapevine supplemented with EBL exposed to chilling stress (Xi et al. 2013). Proline content increased in potato under salinity stress but more increment was observed in the plants supplemented with brassinolide (BL) (Hu et al. 2016). BRs treatment increased total osmolyte content, proline, glycine betaine, and mannitol in maize seedlings exposed to salt stress (Rattan et al. 2020). In *Brassica rapa* L., application of BRs leads to increased synthesis of osmoprotectants, glycine betaine, proline, and soluble sugars in drought-stressed plants (Ahmad Lone et al. 2021).

Antioxidants

During stressful conditions, electrons with high energy state are transferred to the molecular oxygen and reactive oxygen species (ROS) are generated, which targets the biomolecules. Excess ROS production disrupts the homeostasis as the balance between the ROS production and its scavengers, that is, antioxidants, is disrupted which causes oxidative stress. Non-enzymatic (ascorbic acid, glutathione, α tocopherol, carotenoids, proline, phenolic compounds, dehydrins, annexins, sugars, cysteine, methionine, flavonoids and polyamines), and enzymatic (Superoxide dismutase (EC 1.15.1.1), catalase (EC 1.11.1.6), ascorbate peroxidase (EC1.11.1.11), monodehydroascorbate reductase (EC 1.6.5.4), dehydroascorbate reductase (EC 1.8.5.1), glutathione reductase (EC 1.6.4.2), guaiacol peroxidase (EC1.11.1.9), glutathione peroxidase (EC 1.11.1.9), and glutathione S-transferase (EC 2.5.1.18)) components of the antioxidative defense system help the plant to escape the effects of oxidative stress (Ahmad et al. 2010, Soares et al. 2019, Soni et al. 2021).

BRs viz., EBL, and HBL pretreatment, ameliorate the oxidative stress by Cd toxicity in radish seedlings because enhancement in the expression of *FeSOD, Cu/ ZnSOD, MnSOD, CAT1, CAT2*, and *CAT3* genes was investigated (Sharma et al. 2018). Application of EBL enhanced the activity of antioxidative enzymes (SOD, CAT, APX, and POX) in salt stress on mung bean seedlings (Sharma et al. 2013). The application of EBL resulted in a reduction in the concentration of reactive oxygen species (ROS), including O_2^- and H_2O_2, as well as improved membrane stability traits. This effect was associated with decreased levels of malondialdehyde (MDA) and electrolyte leakage (EL), which can be indicative of cell damage. Furthermore, the enhanced activity of antioxidative enzymes was found to be intrinsically linked to these beneficial effects (Ribeiro et al. 2019). Salt stress enhanced the synthesis of MDA but less content was observed in the salt stressed maize seedlings pre-treated with BRs (EBL and HBL). Moreover, this pretreatment enhanced the activities of enzymatic antioxidants, POX, CAT, MDHAR, and DHAR. The authors reported that BRs-mediated salt stress tolerance, might involve a complex pathway, regulating the plant defense system by activating the transcription factors BZR1/BES1 (Rattan et al. 2020).

Exogenous application of EBL reduced the oxidative stress in tomato plants exposed to NaCl salinity as low levels of superoxide, H_2O_2, MDA, and electrolyte leakage (EL) were monitored by the investigators. This reduction was due to the upregulation of the antioxidative defence system (Ahanger et al. 2020). Foliar application of EBL improved the AsA-GSH Cycle in brown mustard (*Brassica juncea* (L.) Czern.) under Cd toxicity (Alam et al. 2020). Grape seedlings exposed to Cadmium (Cd) phytotoxicity and treated with EBL, showed low accumulation of oxidative stress indicators, MDA, and H_2O_2 due to the enhanced activity of antioxidants. EBL treatment induced enhancement in the activity of SOD and POX enzymes. Moreover, supplementation of EBR also increased the activity of the enzymes (APX, MDHAR, DHAR, and GR) involved in the H_2O_2 scavenging AsA-GSH (Ascorbate-glutathione) cycle and also the content of non-enzymatic antioxidants (ascorbate and glutathione) involved in this pathway (Li et al. 2022).

Ionic Content

Ions are required by the plants to carry out the various metabolic processes, but excess or inadequate level of ions disrupts the ionic homeostasis. The ionic phase of the salinity stress arises due to the excess concentration of Na^+ ions within the plant (Kaur et al. 2021). Potassium is considered as essential in plants as it play vital roles in various processes. When there is high Na^+ concentration, Na^+ ions compete with the K^+ binding sites due to same ionic radii. In tomato, NaCl salinity increased the Na^+ and decreased the K^+ content but this effect was ameliorated in the plants treated with EBL (Ahanger et al. 2020). EBL decreased the absorption and accumulation of Na^+ ions in the *Eucalyptus urophylla*. Moreover, salinity-exposed plants also showed increments in P, K, Mg, Zn, Cu, and Mo after being treated with EBL (de Oliveira et al. 2019). Application of BRs (HBL and EBL), reduced the Na^+ accumulation and significantly increased the K^+ ions over the control in maize seedlings. BRs reduces the transportation of Na^+ ions into the plant and improved the K^+/Na^+ ratio (Rattan et al. 2020).

Apart from NaCl stress, BRs also helps the plants to combat the phytotoxicity created by several metal ions. The application of EBR through spraying on cowpea plants exposed to Cd resulted in a significant reduction in Cd concentrations and an increase in nutrient contents across all tissues (Santos et al. 2018). In *Arabidopsis thaliana*, EBL alleviated the stress created by antimony (Sb) (Wu et al. 2019). Also, EBL application confronts the rice plants from the Al phytotoxicity. Plants sprayed with this BR showed more root-shoot length, and fresh and dry weight compared to the plants in Al toxic conditions. Moreover, Al uptake by the plant was also reduced due to BR application (Basit et al. 2022).

Crosstalk between BRs and Molecular Processes under Abiotic Stress

Abiotic stress tolerance is a complex phenomenon influenced by multiple genes (Basit et al. 2021). The capacity of brassinosteroids to regulate these genes, thereby mitigating the effects of abiotic stress, has been extensively studied for the past two decades. These genes can be categorized into further two classes that are explained below.

Regulatory genes

The regulatory genes influenced by brassinosteroids in response to abiotic stress include various kinases, phosphatases, and transcription factors that play important roles in phosphorylation, dephosphorylation, and gene expression regulation. Genetic and molecular studies have demonstrated that BRs directly or indirectly control these regulatory proteins for abiotic stress adaptation (Kour et al. 2021).

Mitogen-activated protein kinases (MAPKs), BIN2 kinase, and BSU1 phosphatase are examples of kinases and phosphatases involved in BR-mediated signaling in plants. Transcriptome studies have revealed that BRs exhibit a synergistic effect in stimulating the expression of MAPK pathway genes under stress conditions. This finding suggests the involvement of BRs in abiotic stress responses (Kong et al. 2021). BRs also control the growth of stomata through interactions with MAP kinase genes.

PP2A is a phosphatase protein with a dual role in BR signaling. It interacts with the transcription factor BZR1 to induce BR-responsive genes and also dephosphorylates the BRI1 receptor, thereby turning off the BR pathway (Manghwar et al. 2022). Several transcription factors, including WRKY, GRAS, MYB/MYC, DREB, bZIP, NPR, and NAC, are regulated by BRs and are involved in the modulation of gene expression. WRKY transcription factors, known for their involvement in abiotic stress response, function either upstream or downstream of various BR-mediated signaling pathways (Madhunita and Oelmüller 2014). When coupled with the transcription factor BES1, transcription factors including WRKY46, WRKY54, and WRKY70 have been demonstrated to provide tolerance to drought stress (Manghwar et al. 2022, Shi et al. 2018).

The GRAS and MYB/MYC transcription factors have also been implicated in BR signaling pathways, as demonstrated by genetic and microarray studies (Bartwal and Arora 2020). In response to stress, the BR-responsive transcription factor BZR1 effectively modulates the expression of MYB transcription factors, which regulate lignin production (Kaur and Pati 2019). The DREB family of transcription factors plays a major role in the reprogramming of genes involved in abiotic stress responses at the transcript level. BR application under stress conditions enhances the expression of multiple DREB transcription factors, leading to the accumulation of stress-relieving protective proteins (Xie 2018). bZIP transcription factors are phylogenetically conserved and participate in various plant activities. In response to stress, it has been demonstrated that they combine BR signaling and endoplasmic reticulum stress signaling pathways.

NAC transcription factors constitute one of the largest plant TF families and were initially thought to govern abiotic stress adaptation through interaction with other phytohormones such as ABA (Jia et al. 2018). However, a previous study indicated that NAC transcription factors have a negative impact on BR production in *Arabidopsis* (Chung et al. 2014). NPR (nonexpresser of pathogenesis-related genes) transcription factors are primarily involved in biotic stress adaptation but also play a recognized role in abiotic stress tolerance (Sharma et al. 2017). They have been discovered to mediate the relationship between salicylic acid and temperature and salinity stress tolerance.

Functional Genes

Abiotic stress affects a variety of genes in plants and many of these genes, including lectins, HSPs, LEA proteins, and genes related to the PGR metabolism, are modulated by BRs. Lectins, sugar-binding proteins, play a crucial role in innate immune responses and their involvement in abiotic stress signaling pathways has been recognized. BR application increases the formation of agglutinin of wheat germ in response to drought stress through an ABA (Avalbaev et al. 2020). They have also been shown to upregulate the mRNA expression of JAC-LEC1-3 in response to abiotic stress (Kaur and Pati 2019). Conversely, BR treatment reduced the salt stress-induced expression of the mannose-binding lectin (Kaur and Pati 2019), but the precise involvement of BR in lectin modulation remains unknown. Heat shock proteins (HSPs), another group of genes, play a crucial role in the initial defense against environmental stress. BRs stimulate the production of HSPs in the cell's powerhouses during stress. In Arabidopsis, the BR-regulated transcription factor BES1 interacts with HSP90 protein to regulate BR levels (Albertos et al. 2022).

The multifunctional protein family LEA (late embryogenesis abundant) is widely known for its work in abiotic stress tolerance (Mertens et al. 2018). In response to abiotic stress, BRs directly regulate the expression of several LEA genes (that is rd29 and erd10) (Hussan et al. 2022). Dehydrins, the most important proteins in the LEA family, are synthesized in wheat upon BR pretreatment, aiding in drought and heavy metal stress relief (Bartwal and Arora 2020). Additionally essential for stress tolerance is the adjustment of cytoskeleton components in response to varied environmental stressors (Lin et al. 2014). BRs are potentially involved in modulating the dynamics of cytoskeleton proteins (Kaur and Pati 2019).

BRs have also been linked to the regulation of the expression of lipocalins, abscisic acid stress ripening (ASR)-like protein, ferritin, and TUD1 (Kaur and Pati 2019). Additionally, it is becoming clearer how BRs affect the regulation of genes involved in abiotic stress adaption via micro-RNA (Sharma et al. 2017).

Crosstalk between BRs and other Phyto-hormones under Abiotic Stress

In addition to their primary role in growth and developmental processes, BRs have been identified to interact with various phytohormones, playing a vital role in managing abiotic stress. BRs engage in crosstalk with phytohormones such as ethylene, gibberellins (GA), abscisic acid (ABA), auxin, cytokinin (CK), salicylic acid (SA), and jasmonic acid (JA) in response to the perception of abiotic stress. This interplay between BRs and other phytohormones regulates numerous target genes involved in plant growth and developmental processes under abiotic stress conditions. Plants adapt to different environmental stresses by modulating the levels of stress-responsive phytohormones. Recent studies on BRs have highlighted their regulatory roles in controlling the functions of other hormones under abiotic stress. The response of BRs to different factors, signaling cascades, BR components crosstalk with other phytohormones, genes and transcription factors, all contribute to their control of abiotic stress (Kaur et al. 2022). Understanding the intricate mechanisms of BR signaling and their interconnections with other phytohormones is of great importance for improving modern agriculture. In this section, we shed light on the interplay between BRs and other phytohormones under abiotic stress.

Brassinosteroid Crosstalk with Auxin

Brassinosteroids and auxin are considered as 'master regulators,' having additive impacts on plant development and growth because of their molecular interaction, which is critical in controlling numerous elements of plant life (Chaiwanon and Wang 2015). In the model plant *Arabidopsis*, there is feedback inhibition of DWF4 by BR. Treatment with auxin enhances the expression of DWF4, which in turn triggers the synthesis of BR through the auxin-induced BRX gene. Phosphorylation by BIN2 reduces the inhibitory effect of auxin on ARF2, leading to the improvement in the activities of ARF promoter which leads to the enhancement in the auxin synthesis and expression of BR-regulated genes (Vert et al. 2008). Several studies have reported the involvement of IAA/auxin genes in auxin synthesis which is regulated by BR (Bashri et al. 2022). However, a comprehensive understanding of the interaction between auxin and BR in regulating stress responses still requires further investigation.

Brassinosteroid Crosstalk with Cytokinin

Compounds derived from adenine that regulates plant growth processes under abiotic stress conditions are known as cytokinins (CKs). Two key enzymes, namely isopentenyltransferases (IPTs) and CK oxidase/dehydrogenases (CKXs), play crucial roles in the synthesis and suppression of CKs, respectively (Yasin et al. 2022). It has been discovered by Vercruyssen et al. (2011) that the responses mediated by brassinosteroids are targeted by these two enzymes. When BR is applied exogenously, it causes the overexpression of CKX3 and the production of BRI1, which lengthens the leaves and roots in response to abiotic stresses. Several studies have demonstrated that under abiotic stress conditions, there is communication between brassinosteroids and cytokinins that promotes the formation of anthocyanins (Yuan et al. 2015). These studies provide evidence for the crosstalk between BR and CKs, suggesting their involvement in enhancing crop yield.

Brassinosteroid Crosstalk with Ethylene

Brassinosteroids and ethylene have a reciprocal regulation on various signaling and metabolic systems that are crucial for plant growth and development. Additionally, both of these phytohormones are well known for their involvement in plant adaptation to diverse abiotic stress (Shahzadi et al. 2022). Through the activation of ACO and ACS gene expression of ethylene biosynthesis, BRs promote the synthesis of ethylene. Several studies have reported the induction of ethylene production and reactive oxygen species (ROS) generation by BRs, subsequently leading to an increase in alternative oxidase (AOX) capacity. Elevated AOX activity helps in mitigating excessive ROS production and preventing oxidative damage in plant cells during various abiotic stress conditions. For instance, cucumber seedlings treated with BL (the most active BR) before exposure to salt, drought, and cold stress exhibited enhanced ethylene biosynthesis and an increased capacity of the alternative oxidase pathway (Wei et al. 2015).

The biosynthesis of ethylene is regulated by BRs in a dose-dependent manner, either negatively or positively. In their study, Guo et al. (2019) provided evidence that the exogenous application of brassinosteroids promoted banana ripening. They observed that BR treatment enhanced the expression of key ripening-related genes, namely MaACS1, MaACO13, and MaACO14, which are involved in the regulation of ethylene production. This finding suggests that BR plays a role in regulating banana ripening by influencing ethylene biosynthesis.

Brassinosteroid Crosstalk with Gibberellins

Brassinosteroids and gibberellins (GAs) are important plant hormones that play significant roles in plant growth and developmental processes, as well as in controlling abiotic stresses. The interaction between BRs and GAs has been revealed through associated gene complexes, indicating interactions at both the protein and DNA levels (Abbas et al. 2022, Yusuf et al. 2022). The interaction between BZR1/BES1 proteins and DELLA proteins in *Arabidopsis* facilitates signaling crosstalk between BRs and GAs, regulating cell elongation under both normal and stress conditions. Under heavy metal stress, the interaction between DELLA protein and BRZ1 leads to an increase in the levels of ROS-scavenging enzymes (Achard and Genschik 2009).

Direct interactions between BZR1 and DELLA prevent BZR1 from attaching to DNA. This prevention of BZR1 causes the disruption of signal transduction essential for the optimal seedling etiolation. BZR1's transcriptional activity is inhibited by the DELLA gene, while DELLA is degraded by GA. Both GA and BR affect the expression of the BRZ1 gene, and the interaction of BES1/BRZ1 promotes GA production, leading to enhanced degradation of DELLA (Ross and Quittenden 2016).

Brassinosteroid Crosstalk with Abscisic Acid

Abscisic acid plays a significant role in plant protection, particularly under various stress conditions. The interaction between BR and abscisic acid primarily revolves around the gene expression regulation and variation in the activities of proteins. During abiotic stress, a complex is formed through interactions involving histone deacetylase19 (HDA19), TOPLESS (TPL)/TOPLESS-RELATED (TPR) proteins, and BRI1-EMS suppressor1 (BES1). This complex disrupts the promoter region (E-box) and leads to the suppression of the expression of the ABA insensitive 3 (ABI3) gene in the presence of brassinosteroids. This molecular mechanism highlights the role of the complex in modulating the response to abiotic stress by regulating gene expression related to abscisic acid (ABA) signaling. Furthermore, the phosphorylation of the ABI5 transcription factor by BIN2 leads to the upregulation of ABA-related gene expression and self-stimulation of SnRK2s genes and kinase activity (Wang et al. 2018). When BZR1 TF binds to the G-box promoter of ABI5, it results in the deactivation of ABI3 and ABI5 gene expression, subsequently decreasing the stress response by downregulating ABA-regulated gene expression.

In addition to their interactions with other phytohormones such as jasmonic acid, salicylic acid, and polyamines, brassinosteroids also interact with these plant hormones under various abiotic and biotic stresses, thereby enhancing the plants' tolerance capacity under extreme conditions. However, there is a knowledge gap that requires attention in order to untie the intricate interplay among hormones, transcription factors, and several metabolites, in order to fully understand the comprehensive interactive mechanism of phytohormones under abiotic stresses.

Conclusion

Plant growth regulators have proven to be highly effective in enhancing plant vigor when confronted with abiotic stress conditions. The literature has exaggerated the stress-alleviating advantages of BRs. These compounds are ecofriendly and alter a range of stress-related responses, thereby promoting the production of high-yield agriculture crops resilient to abiotic stress. While their importance in governing numerous cellular and molecular systems is acknowledged, additional understanding is necessary to grasp the intricacies of these complex mechanisms.

References

Abbas, H.M.K., Askri, S.M.H., Ali, S., Fatima, A., Qamar, M.T. ul, Xue, S.-D. et al. 2022. Mechanism Associated with Brassinosteroids Crosstalk with Gibberellic Acid in Plants. pp. 101–115. In: Brassinosteroids Signalling: Intervention with Phytohormones and Their Relationship in Plant Adaptation to Abiotic Stresses. Springer.

Achard, P. and Genschik, P. 2009. Releasing the brakes of plant growth: How GAs shutdown DELLA proteins. Journal of Experimental Botany, 60(4): 1085–1092.

Ahammed, G.J., Li, X., Liu, A. and Chen, S. 2020. Brassinosteroids in plant tolerance to abiotic stress. Journal of Plant Growth Regulation, 39(4): 1451–1464.

Ahanger, M.A., Mir, R.A., Alyemeni, M.N. and Ahmad, P. 2020. Combined effects of brassinosteroid and kinetin mitigates salinity stress in tomato through the modulation of antioxidant and osmolyte metabolism. Plant Physiology and Biochemistry, 147: 31–42.

Ahmad Lone, W., Majeed, N., Yaqoob, U. and John, R. 2021. Exogenous brassinosteroid and jasmonic acid improve drought tolerance in *Brassica rapa* L. genotypes by modulating osmolytes, antioxidants and photosynthetic system. Plant Cell Reports, 1–15.

Ahmad, M., Ali, Q., Hafeez, M.M. and Malik, A. 2021. Improvement for biotic and abiotic stress tolerance in crop plants. Biological and Clinical Sciences Research Journal, 2021(1).

Ahmad, P., Jaleel, C.A., Salem, M.A., Nabi, G. and Sharma, S. 2010. Roles of enzymatic and nonenzymatic antioxidants in plants during abiotic stress. Critical Reviews in Biotechnology, 30(3): 161–175.

Alam, P., Kaur Kohli, S., Al Balawi, T., Altalayan, F.H., Alam, P., Ashraf, M. et al. 2020. Foliar application of 24-Epibrassinolide improves growth, ascorbate-glutathione cycle, and glyoxalase system in brown mustard (*Brassica juncea* (L.) Czern.) under cadmium toxicity. Plants, 9(11): 1487.

Albertos, P., Dündar, G., Schenk, P., Carrera, S., Cavelius, P., Sieberer, T. et al. 2022. Transcription factor BES1 interacts with HSFA1 to promote heat stress resistance of plants. The EMBO Journal, 41(3): e108664.

Ashraf, M., Akram, N.A., Arteca, R.N. and Foolad, M.R. 2010. The physiological, biochemical and molecular roles of brassinosteroids and salicylic acid in plant processes and salt tolerance. Critical Reviews in Plant Sciences, 29(3): 162–190.

Avalbaev, A., Bezrukova, M., Allagulova, C., Lubyanova, A., Kudoyarova, G., Fedorova, K. et al. 2020. Wheat germ agglutinin is involved in the protective action of 24-epibrassinolide on the roots of wheat seedlings under drought conditions. Plant Physiology and Biochemistry, 146: 420–427.

Bajguz, A. 2011. Brassinosteroids–occurrence and chemical structures in plants. Brassinosteroids: a class of plant hormone, 1–27.

Bajguz, A. and Hayat, S. 2009. Effects of brassinosteroids on the plant responses to environmental stresses. Plant Physiology and Biochemistry, 47(1): 1–8.

Bartwal, A. and Arora, S. 2020. Brassinosteroids: molecules with myriad roles. Co-Evolution of Secondary Metabolites, 869–895.

Bashri, G., Fatima, A., Singh, S. and Prasad, S.M. 2022. Interplay of Brassinosteroids and Auxin for Understanding of Signaling Pathway. pp. 137–154. In: Brassinosteroids Signalling: Intervention with Phytohormones and Their Relationship in Plant Adaptation to Abiotic Stresses. Springer.

Basit, F., Liu, J., An, J., Chen, M., He, C., Zhu, X., et al. 2021. Brassinosteroids as a multidimensional regulator of plant physiological and molecular responses under various environmental stresses. Environmental Science and Pollution Research, 28(33): 44768–44779.

Basit, F., Liu, J., An, J., Chen, M., He, C., Zhu, X. et al. 2022. Seed priming with brassinosteroids alleviates aluminum toxicity in rice via improving antioxidant defense system and suppressing aluminum uptake. Environmental Science and Pollution Research, 29(7): 10183–10197.

Castañeda-Murillo, C.C., Rojas-Ortiz, J.G., Sánchez-Reinoso, A.D., Chávez-Arias, C.C. and Restrepo-Díaz, H. 2022. Foliar brassinosteroid analogue (DI-31) sprays increase drought tolerance by improving plant growth and photosynthetic efficiency in lulo plants. Heliyon, 8(2): e08977.

Chaiwanon, J. and Wang, Z.-Y. 2015. Spatiotemporal brassinosteroid signaling and antagonism with auxin pattern stem cell dynamics in *Arabidopsis* roots. Current Biology, 25(8): 1031–1042.

Chung, Y., Kwon, S.I. and Choe, S. 2014. Antagonistic regulation of *Arabidopsis* growth by brassinosteroids and abiotic stresses. Molecules and Cells, 37(11): 795.

Clouse, S.D. 2011. Brassinosteroids. The *Arabidopsis* Book/American Society of Plant Biologists, 9.

Cui, F., Liu, L., Zhao, Q., Zhang, Z., Li, Q., Lin, B. et al. 2012. *Arabidopsis* ubiquitin conjugase UBC32 is an ERAD component that functions in brassinosteroid-mediated salt stress tolerance. The Plant Cell, 24(1): 233–244.

Cutler, H.G., Yokota, T. and Adam, G. 1991. Brassinosteroids: chemistry, bioactivity, and applications. ACS Publications.

de Oliveira, V.P., Lima, M.D.R., da Silva, B.R.S., Batista, B.L. and da Silva Lobato, A.K. 2019. Brassinosteroids confer tolerance to salt stress in *Eucalyptus urophylla* plants enhancing homeostasis, antioxidant metabolism and leaf anatomy. Journal of Plant Growth Regulation, 38(2): 557–573.

Derevyanchuk, M., Kretynin, S., Kolesnikov, Y., Litvinovskaya, R., Martinec, J., Khripach, V. et al. 2019. Seed germination, respiratory processes and phosphatidic acid accumulation in *Arabidopsis* diacylglycerol kinase knockouts–The effect of brassinosteroid, brassinazole and salinity. Steroids, 147: 28–36.

Ding, L., Guo, X., Wang, K., Pang, H., Liu, Y., Yang, Q. et al. 2021. Genome-wide analysis of BES1/BZR1 transcription factors and their responses to osmotic stress in *Ammopiptanthus nanus*. Journal of Forest Research, 26(2): 127–135.

Divi, U.K. and Krishna, P. 2010. Overexpression of the brassinosteroid biosynthetic gene AtDWF4 in Arabidopsis seeds overcomes abscisic acid-induced inhibition of germination and increases cold tolerance in transgenic seedlings. Journal of Plant Growth Regulation, 29: 385–393.

Efimova, M.V., Khripach, V.A., Boyko, E.V., Malofii, M.K., Kolomeichuk, L.V., Murgan, O.K. et al. 2018. The priming of potato plants induced by brassinosteroids reduces oxidative stress and increases salt tolerance. In Doklady Biological Sciences. Springer, 478: 33–36.

Fàbregas, N., Lozano-Elena, F., Blasco-Escámez, D., Tohge, T., Martínez-Andújar, C., Albacete, A. et al. 2018. Overexpression of the vascular brassinosteroid receptor BRL3 confers drought resistance without penalizing plant growth. Nature Communications, 9(1): 4680.

Fariduddin, Q., Yusuf, M., Ahmad, I. and Ahmad, A. 2014. Brassinosteroids and their role in response of plants to abiotic stresses. Biologia Plantarum, 58(1): 9–17.

Farooq, M., Wahid, A. and Basra, S.M.A. 2009. Improving water relations and gas exchange with brassinosteroids in rice under drought stress. Journal of Agronomy and Crop Science, 195(4): 262–269.

Fujioka, S. 1999. Natural occurrence of brassinosteroids in the plant kingdom. Brassinosteroid: Steroidal Plant Hormones, 21–45.

Fujioka, Shozo, Li, J., Choi, Y.-H., Seto, H., Takatsuto, S., Noguchi, T. et al. 1997. The *Arabidopsis* deetiolated2 mutant is blocked early in brassinosteroid biosynthesis. The Plant Cell, 9(11): 1951–1962.

Fujioka, Shozo and Sakurai, A. 1997. Brassinosteroids. Natural Product Reports, 14(1): 1–10.

Grove, M.D., Spencer, G.F., Rohwedder, W.K., Mandava, N., Worley, J.F., Warthen Jr, J.D. et al. 1979. Brassinolide, a plant growth-promoting steroid isolated from Brassica napus pollen. Nature, 281(5728): 216–217.

Guo, Y., Shan, W., Liang, S., Wu, C., Wei, W., Chen, J. et al. 2019. MaBZR1/2 act as transcriptional repressors of ethylene biosynthetic genes in banana fruit. Physiologia Plantarum, 165(3): 555–568.

Hafeez, M.B., Zahra, N., Zahra, K., Raza, A., Khan, A., Shaukat, K. et al. 2021. Brassinosteroids: Molecular and physiological responses in plant growth and abiotic stresses. Plant Stress, 2, 100029.

Holá, D. 2022. Brassinosteroids and primary photosynthetic processes. pp. 59–104. In: Brassinosteroids in Plant Developmental Biology and Stress Tolerance. Elsevier.

Hossain, A., Venugopalan, V.K., Rahman, M.A., Alam, M.J., Al-Mahmud, A., Islam, M.A. et al. 2022. Physiological, biochemical, and molecular mechanisms of plant steroid hormones brassinosteroids under drought-induced oxidative stress in plants. pp. 99–130. In: Emerging Plant Growth Regulators in Agriculture. Elsevier.

Houimli, S.I.M., Denden, M. and Mouhandes, B.D. 2010. Effects of 24-epibrassinolide on growth, chlorophyll, electrolyte leakage and proline by pepper plants under NaCl-stress. EurAsian Journal of BioSciences, 4.

Hu, Y., Xia, S., Su, Y., Wang, H., Luo, W., Su, S. et al. 2016. Brassinolide increases potato root growth in vitro in a dose-dependent way and alleviates salinity stress. BioMed Research International, 2016.

Hussan, S. ul, Rather, M.A., Dar, Z.A., Jan, R., Dar, Z.M., Wani, M.A. et al. 2022. Decoding the Enigma of Drought Stress Tolerance Mechanisms in Plants and its Application in Crop Improvement. pp. 339–368. In: S.S. Mahdi and R. Singh (Eds.). Innovative Approaches for Sustainable Development: Theories and Practices in Agriculture. Cham: Springer International Publishing. https://doi.org/10.1007/978-3-030-90549-1_22

Janowska, B. and Andrzejak, R. 2023. Plant growth regulators for the cultivation and vase life of geophyte flowers and leaves. Agriculture, 13(4): 855.

Jia, D., Gong, X., Li, M., Li, C., Sun, T. and Ma, F. 2018. Overexpression of a novel apple NAC transcription factor gene, MdNAC1, confers the dwarf phenotype in transgenic apple (*Malus domestica*). Genes, 9(5): 229.

Kapoor, D., Bhardwaj, S., Gautam, S., Rattan, A., Bhardwaj, R. and Sharma, A. 2022. Brassinosteroids in plant nutrition and heavy metal tolerance. pp. 217–235. In: Brassinosteroids in Plant Developmental Biology and Stress Tolerance. Elsevier.

Kaur, G., Sanwal, S.K., Sehrawat, N., Kumar, A., Kumar, N. and Mann, A. 2021. Assessing the effect of salinity stress on root and shoot physiology of chickpea genotypes using hydroponic technique. Indian Journal of Genetics and Plant Breeding, 81(04): 92–95.

Kaur, N. and Pati, P.K. 2019. Harnessing the potential of brassinosteroids in abiotic stress tolerance in plants. pp. 407–423. In: Brassinosteroids: Plant Growth and Development. Springer.

Kaur, N., Saini, S., Marothia, D. and Pati, P.K. 2022. Crosstalk of Reactive Oxygen Species and Brassinosteroids in Plant Abiotic Stress Mitigation. pp. 59–64. In: Jasmonates and Brassinosteroids in Plants. CRC Press.

Kaur, N., Sharma, S., Hasanuzzaman, M. and Pati, P.K. 2022. Genome Editing: A Promising Approach for Achieving Abiotic Stress Tolerance in Plants. International Journal of Genomics, 2022.

Kaya, C., Aydemir, S., Akram, N.A. and Ashraf, M. 2018. Epibrassinolide application regulates some key physio-biochemical attributes as well as oxidative defense system in maize plants grown under saline stress. Journal of Plant Growth Regulation, 37(4): 1244–1257.

Kazemi, S., Rafati Alashti, M. and Khodabin, G. 2023. Evaluation of the Effect of Foliar Application of Brassinosteroid on Physiological Characteristics and Yield of Rapeseed Genotypes Under Late-Season Drought Stress. Journal of Crops Improvement, 25(1): 111–126.

Khalid, M.F., Hussain, S., Ahmad, S., Ejaz, S., Zakir, I., Ali, M.A. et al. 2019. Impacts of abiotic stresses on growth and development of plants. pp. 1–8. In: Plant Tolerance to Environmental Stress. CRC Press.

Khamsuk, O., Sonjaroon, W., Suwanwong, S., Jutamanee, K. and Suksamrarn, A. 2018. Effects of 24-epibrassinolide and the synthetic brassinosteroid mimic on chili pepper under drought. Acta Physiologiae Plantarum, 40(6): 1–12.

Khan, M.T.A., Yusuf, M., Akram, W. and Qazi, F. 2022. Signal transduction of brassinosteroids under abiotic stresses. pp. 1–16. In: Brassinosteroids Signalling: Intervention with Phytohormones and Their Relationship in Plant Adaptation to Abiotic Stresses. Springer.

Kong, Q., Mostafa, H.H., Yang, W., Wang, J., Nuerawuti, M., Wang, Y. et al. 2021. Comparative transcriptome profiling reveals that brassinosteroid-mediated lignification plays an important role in garlic adaption to salt stress. Plant Physiology and Biochemistry, 158: 34–42.

Kothari, A. and Lachowiec, J. 2021. Roles of Brassinosteroids in Mitigating Heat Stress Damage in Cereal Crops. International Journal of Molecular Sciences, 22(5): 2706.

Kour, J., Kohli, S.K., Khanna, K., Bakshi, P., Sharma, P., Singh, A.D. et al. 2021. Brassinosteroid signaling, crosstalk and, physiological functions in plants under heavy metal stress. Frontiers in Plant Science, 12: 608061.

Kumar, P., Tokas, J., Kumar, N., Lal, M. and Singal, H.R. 2018. Climate change consequences and its impact on agriculture and food security. International Journal of Chemical Studies, 6(6): 124–133.

Lado, J., Rey, F. and Manzi, M. 2023. Phytohormones and Cold Stress Tolerance. pp. 207–226. In: Plant Hormones and Climate Change. Springer.

Li, B.-B., Fu, Y.-S., Li, X.-X., Yin, H.-N. and Xi, Z.-M. 2022. Brassinosteroids alleviate cadmium phytotoxicity by minimizing oxidative stress in grape seedlings: Toward regulating the ascorbate-glutathione cycle. Scientia Horticulturae, 299: 111002.

Li, Jianming and Jin, H. 2007. Regulation of brassinosteroid signaling. Trends in Plant Science, 12(1): 37–41.

Li, Jie, Sohail, H., Nawaz, M.A., Liu, C. and Yang, P. 2022. Physiological and proteomic analyses reveals that brassinosteroids application improves the chilling stress tolerance of pepper seedlings. Plant Growth Regulation, 96(2): 315–329.

Lin, F., Qu, Y. and Zhang, Q. 2014. Phospholipids: molecules regulating cytoskeletal organization in plant abiotic stress tolerance. Plant Signaling & Behavior, 9(5): e28337.

Madhunita, B. and Oelmüller, R. 2014. WRKY transcription factors: jack of many trades in plants. Plant Signaling and Behavior, 9(2).

Mahesh, K., Balaraju, P., Ramakrishna, B. and Rao, S.S.R. 2013. Effect of brassinosteroids on germination and seedling growth of radish (*Raphanus sativus* L.) under PEG-6000 induced water stress. American Journal of Plant Sciences.

Mandava, N.B. 1988. Plant growth-promoting brassinosteroids. Annual Review of Plant Physiology and Plant Molecular Biology, 39(1): 23–52.

Manghwar, H., Hussain, A., Ali, Q. and Liu, F. 2022. Brassinosteroids (BRs) role in plant development and coping with different stresses. International Journal of Molecular Sciences, 23(3): 1012.

Mann, A., Kaur, G., Kumar, A., Sanwal, S.K., Singh, J. and Sharma, P.C. 2019. Physiological response of chickpea (*Cicer arietinum* L.) at early seedling stage under salt stress conditions. Legume Research: An International Journal, 42(5).

Mertens, J., Aliyu, H. and Cowan, D.A. 2018. LEA proteins and the evolution of the WHy domain. Applied and Environmental Microbiology, 84(15): e00539–18.

Mitchell, J.W., Mandava, N.B., Worley, J.F., Plimmer, J.R., Smith and M.V. Brassins. 1970. A new family of plant hormones from rape pollen. Nature (London) 225: 1065–1066.

Mitchell, J.W. and Gregory, L.E. 1972. Enhancement of overall plant growth, a new response to brassins. Nature New Biology, 239(95): 253–254.

Mu, D., Feng, N., Zheng, D., Zhou, H., Liu, L., Chen, G. et al. 2022. Physiological mechanism of exogenous brassinolide alleviating salt stress injury in rice seedlings. Scientific Reports, 12(1): 20439. https://doi.org/10.1038/s41598-022-24747-9

Mubarik, M.S., Khan, S.H., Sajjad, M., Raza, A., Hafeez, M.B., Yasmeen, T. et al. 2021. A manipulative interplay between positive and negative regulators of phytohormones: A way forward for improving drought tolerance in plants. Physiologia Plantarum, 172(2): 1269–1290.

Nazir, F., Hussain, A. and Fariduddin, Q. 2019. Interactive role of epibrassinolide and hydrogen peroxide in regulating stomatal physiology, root morphology, photosynthetic and growth traits in *Solanum lycopersicum* L. under nickel stress. Environmental and Experimental Botany, 162: 479–495.

Ozturk, M., Turkyilmaz Unal, B., García-Caparrós, P., Khursheed, A., Gul, A. and Hasanuzzaman, M. 2021. Osmoregulation and its actions during the drought stress in plants. Physiologia Plantarum, 172(2): 1321–1335.

Pereira, Y.C., Rodrigues, W.S., Lima, E.J.A., Santos, L.R., Silva, M.H.L. and Lobato, A.K.S. 201.
    Brassinosteroids increase electron transport and photosynthesis in soybean plants under water
    deficit. Photosynthetica, 57(1): 181–191.
Planas-Riverola, A., Gupta, A., Betegón-Putze, I., Bosch, N., Ibañes, M. and Caño-Delgado, A.I. 2019.
    Brassinosteroid signaling in plant development and adaptation to stress. Development, 146(5):
    dev151894.
Rajewska, I., Talarek, M. and Bajguz, A. 2016. Brassinosteroids and response of plants to heavy metals
    action. Frontiers in Plant Science, 7: 629.
Ramirez, V.E. and Poppenberger, B. 2020. Modes of brassinosteroid activity in cold stress tolerance.
    Frontiers in Plant Science, 11: 583666.
Rattan, A., Kapoor, D., Ashish, Kapoor, N., Bhardwaj, R. and Sharma, A. 2022. Involvement of
    brassinosteroids in plant response to salt stress. pp. 237–253. In: Ahammed, G.J., Sharma, A. and
    Yu, J. (Eds.). Brassinosteroids in Plant Developmental Biology and Stress Tolerance. Academic
    Press.
Rattan, A., Kapoor, D., Kapoor, N., Bhardwaj, R. and Sharma, A. 2020. Brassinosteroids regulate
    functional components of antioxidative defense system in salt stressed maize seedlings. Journal of
    Plant Growth Regulation, 39(4): 1465–1475.
Ren, Y., Li, X., Liang, J., Wang, S., Wang, Z., Chen, H. et al. 2023. Brassinosteroids and gibberellic acid
    actively regulate the zinc detoxification mechanism of Medicago sativa L. seedlings. BMC Plant
    Biology, 23(1): 1–13.
Ribeiro, D.G. dos, S., Silva, B.R.S. da and Lobato, A.K. da S. 2019. Brassinosteroids induce tolerance to
    water deficit in soybean seedlings: contributions linked to root anatomy and antioxidant enzymes.
    Acta Physiologiae Plantarum, 41(6): 1–11.
Ross, J.J. and Quittenden, L.J. 2016. Interactions between Brassinosteroids and Gibberellins: Synthesis or
    Signaling? The Plant Cell, 28(4): 829–832.
Santos, L.R., Batista, B.L. and Lobato, A.K.S. 2018. Brassinosteroids mitigate cadmium toxicity in
    cowpea plants. Photosynthetica, 56(2): 591–605.
Sanwal, S.K., Kaur, G. and Mann, A. 2019. Response of Okra (Abelmoschus esculentus L.) Genotypes to
    Salinity Stress in Relation to Seedling Stage.
Sasse, J.M. 2003. Physiological actions of brassinosteroids: an update. Journal of Plant Growth
    Regulation, 22: 276–288.
Shahzadi, I., Ahmad, A., Noreen, Z., Akram, W., Yasin, N.A. and Khan, W.U. 2022. Brassinosteroid and
    Ethylene-Mediated Cross Talk in Plant Growth and Development. pp. 117–136. In: Brassinosteroids
    Signalling: Intervention with Phytohormones and Their Relationship in Plant Adaptation to Abiotic
    Stresses. Springer.
Sharma, A., Shahzad, B., Kumar, V., Kohli, S.K., Sidhu, G.P.S., Bali, A.S. et al. 2019. Phytohormones
    regulate accumulation of osmolytes under abiotic stress. Biomolecules, 9(7): 285.
Sharma, Indu, Sharma, A., Pati, P. and Bhardwaj, R. 2018. Brassinosteroids reciprocates heavy metals
    induced oxidative stress in radish by regulating the expression of key antioxidant enzyme genes.
    Brazilian Archives of Biology and Technology, 61.
Sharma, Isha, Kaur, N. and Pati, P.K. 2017. Brassinosteroids: a promising option in deciphering remedial
    strategies for abiotic stress tolerance in rice. Frontiers in Plant Science, 8: 2151.
Sharma, Isha, Kaur, N., Saini, S. and Pati, P.K. 2013. Emerging dynamics of brassinosteroids research.
    Biotechnology: Prospects and Applications, 3–17.
Sharma, V., Kumar, N., Verma, A. and Gupta, V.K. 2013. Exogenous Application of Brassinosteroids
    Ameliorates Salt-Induced Stress in Mung Bean Seedlings. LS: International Journal of Life
    Sciences, 2(1): 7–13.
Shi, W.-Y., Du, Y.-T., Ma, J., Min, D.-H., Jin, L.-G., Chen, J. et al. 2018. The WRKY transcription factor
    GmWRKY12 confers drought and salt tolerance in soybean. International Journal of Molecular
    Sciences, 19(12): 4087.
Singh, A. and Roychoudhury, A. 2023. Abscisic acid in plants under abiotic stress: crosstalk with major
    phytohormones. Plant Cell Reports, 1–14.
Singh, J. and Thakur, J.K. 2018. Photosynthesis and abiotic stress in plants. pp. 27–46. In: Biotic and
    abiotic stress tolerance in plants. Springer.

Soares, C., Carvalho, M.E., Azevedo, R.A. and Fidalgo, F. 2019. Plants facing oxidative challenges—A little help from the antioxidant networks. Environmental and Experimental Botany, 161: 4–25.

Soni, S., Kumar, A., Sehrawat, N., Kumar, N., Kaur, G., Kumar, A. et al. 2021. Variability of durum wheat genotypes in terms of physio-biochemical traits against salinity stress. Cereal Research Communications, 49(1): 45–54.

Steffens, G.L. 1991. US Department of Agriculture Brassins Project: 1970–1980. ACS Publications.

Sun, L., Feraru, E., Feraru, M.I., Waidmann, S., Wang, W., Passaia, G. et al. 2020. PIN-LIKES coordinate brassinosteroid signaling with nuclear auxin input in Arabidopsis thaliana. Current Biology, 30(9): 1579–1588.

Thussagunpanit, J., Jutamanee, K., Kaveeta, L., Chai-arree, W., Pankean, P., Homvisasevongsa, S. et al. 2015. Comparative effects of brassinosteroid and brassinosteroid mimic on improving photosynthesis, lipid peroxidation, and rice seed set under heat stress. Journal of Plant Growth Regulation, 34(2): 320–331.

Tong, H., Xiao, Y., Liu, D., Gao, S., Liu, L., Yin, Y. et al. 2014. Brassinosteroid regulates cell elongation by modulating gibberellin metabolism in rice. The Plant Cell, 26(11): 4376–4393.

Vardhini, B.V. 2016. Brassinosteroids are potential ameliorators of heavy metal stresses in plants. Plant Metal Interaction, 209–237.

Vercruyssen, L., Gonzalez, N., Werner, T., Schmülling, T. and Inzé, D. 2011. Combining Enhanced Root and Shoot Growth Reveals Cross Talk between Pathways That Control Plant Organ Size in Arabidopsis. Plant Physiology, 155(3): 1339–1352.

Vert, G., Walcher, C.L., Chory, J. and Nemhauser, J.L. 2008. Integration of auxin and brassinosteroid pathways by Auxin Response Factor 2. Proceedings of the National Academy of Sciences, 105(28): 9829–9834.

Vikram, Pooja, Sharma, J. and Sharma, A. 2022. Role of Brassinosteroids in plants responses to salinity stress: A review. Journal of Applied and Natural Science, 14(2): 582–599.

Wang, H., Tang, J., Liu, J., Hu, J., Liu, J., Chen, Y. et al. 2018. Abscisic Acid Signaling Inhibits Brassinosteroid Signaling through Dampening the Dephosphorylation of BIN2 by ABI1 and ABI2. Molecular Plant, 11(2): 315–325. https://doi.org/10.1016/j.molp.2017.12.013

Wani, S.H., Choudhary, M., Kumar, P., Akram, N.A., Surekha, C., Ahmad, P. et al. 2018. Marker-assisted breeding for abiotic stress tolerance in crop plants. pp. 1–23. In: Biotechnologies of Crop Improvement, Volume 3. Springer.

Wei, L.-J., Deng, X.-G., Zhu, T., Zheng, T., Li, P.-X., Wu, J.-Q. et al. 2015. Ethylene is involved in brassinosteroids induced alternative respiratory pathway in cucumber (*Cucumis sativus* L.) seedlings response to abiotic stress. Frontiers in Plant Science, 6: 982.

Wei, Z. and Li, J. 2016. Brassinosteroids regulate root growth, development, and symbiosis. Molecular Plant, 9(1): 86–100.

Wu, C., Li, F., Xu, H., Zeng, W., Yu, R., Wu, X. et al. 2019. The potential role of brassinosteroids (BRs) in alleviating antimony (Sb) stress in Arabidopsis thaliana. Plant Physiology and Biochemistry, 141: 51–59.

Xi, Z., Wang, Z., Fang, Y., Hu, Z., Hu, Y., Deng, M. et al. 2013. Effects of 24-epibrassinolide on antioxidation defense and osmoregulation systems of young grapevines (*V. vinifera* L.) under chilling stress. Plant Growth Regulation, 71(1): 57–65.

Xia, X.-J., Huang, L.-F., Zhou, Y.-H., Mao, W.-H., Shi, K., Wu, J.-X. et al. 2009. Brassinosteroids promote photosynthesis and growth by enhancing activation of Rubisco and expression of photosynthetic genes in Cucumis sativus. Planta, 230: 1185–1196.

Xie, Z. 2018. Functions and mechanisms of AP2/ERF and MYB family transcription factors in Brassinosteroid-regulated plant growth and stress responses (PhD Thesis). Iowa State University.

Yanhua, C., Yaliang, W., Huizhe, C., Jing, X., Yikai, Z., Zhigang, W. et al. 2023. Brassinosteroids Mediate Endogenous Phytohormone Metabolism to Alleviate High Temperature Injury at Panicle Initiation Stage in Rice. Rice Science, 30(1): 70–86.

Yasin, N.A., Shah, A.A., Ahmad, A. and Shahzadi, I. 2022. Cross talk between brassinosteroids and cytokinins in relation to plant growth and developments. pp. 171–178. In: Brassinosteroids Signalling: Intervention with Phytohormones and Their Relationship in Plant Adaptation to Abiotic Stresses. Springer.

Ye, H., Liu, S., Tang, B., Chen, J., Xie, Z., Nolan, T.M. et al. 2017. RD26 mediates crosstalk between drought and brassinosteroid signalling pathways. Nature Communications, 8(1): 14573.

Yokota, T. 1999. The history of brassinosteroids: discovery to isolation of biosynthesis and signal transduction mutants. Brassinosteroids, 1–20.

Yokota, Takao. 1997. The structure, biosynthesis and function of brassinosteroids. Trends in Plant Science, 2(4): 137–143.

Yokota, Takao, Arima, M. and Takahashi, N. 1982. Castasterone, a new phytosterol with plant-hormone potency, from chestnut insect gall. Tetrahedron Letters, 23(12): 1275–1278.

Yu, J.Q., Huang, L.F., Hu, W.H., Zhou, Y.H., Mao, W.H., Ye, S.F. et al. 2004. A role for brassinosteroids in the regulation of photosynthesis in *Cucumis sativus*. Journal of Experimental Botany, 55(399): 1135–1143.

Yuan, L.B., Peng, Z.H., Zhi, T.T., Zho, Z., Liu, Y., Zhu, Q. et al. 2015. Brassinosteroid enhances cytokinin-induced anthocyanin biosynthesis in *Arabidopsis* seedlings. Biologia Plantarum, 59(1): 99–105.

Yusuf, M., Khan, M.T.A., Faizan, M., Khalil, R. and Qazi, F. 2022. Role of brassinosteroids and its cross talk with other phytohormone in plant responses to heavy metal stress. pp. 179–201. In: Brassinosteroids Signalling: Intervention with Phytohormones and Their Relationship in Plant Adaptation to Abiotic Stresses. Springer.

Zahid, G., Iftikhar, S., Shimira, F., Ahmad, H.M. and Kaçar, Y.A. 2023. An overview and recent progress of plant growth regulators (PGRs) in the mitigation of abiotic stresses in fruits: A review. Scientia Horticulturae, 309, p.111621.

Zeng, H., Tang, Q. and Hua, X. 2010. Arabidopsis brassinosteroid mutants det2-1 and bin2-1 display altered salt tolerance. Journal of Plant Growth Regulation, 29: 44–52.

Zhang, H., Zhu, J., Gong, Z. and Zhu, J.-K. 2022. Abiotic stress responses in plants. Nature Reviews Genetics, 23(2): 104–119.

Zhao, M., Yuan, L., Wang, J., Xie, S., Zheng, Y., Nie, L. et al. 2019. Transcriptome analysis reveals a positive effect of brassinosteroids on the photosynthetic capacity of wucai under low temperature. BMC Genomics, 20(1): 1–19.

Zhao, Y., Qi, Z. and Berkowitz, G.A. 2013. Teaching an Old Hormone New Tricks: Cytosolic Ca^{2+} Elevation Involvement in Plant Brassinosteroid Signal Transduction Cascades. Plant Physiology, 163(2): 555–565.

Zou, L.-J., Deng, X.-G., Zhang, L., Zhu, T., Tan, W.-R., Muhammad, A. et al. 2018. Nitric oxide as a signaling molecule in brassinosteroid-mediated virus resistance to Cucumber mosaic virus in *Arabidopsis thaliana*. Physiologia Plantarum, 163(2): 196–210.

12

The Regulatory Function of Polyamines in Plant Abiotic Stress

Asmaa M. Mogazy

Introduction

The global population increment combined with a reduction of arable land in response to human activities and climate change are growing challenges to improve crop productivity to achieve food security and address human needs. Abiotic stresses like drought, chilling, freezing, and salinity limited plant productivity induce almost 70% of yields loss. Climate changes and stressful conditions cause cellular water content rapid decline, ROS generation, and cell membrane destruction that signal plants response, accumulating enzymatic and non-enzymatic antioxidants and several water-soluble compounds with low molecular weights known as compatible solutes (Gupta and Huang 2014, FAO 2019). Among these compounds, polyamines (PAs) is a nitrogenous compound which when accumulated, plays protective role in plant defence mechanism against abiotic stress. PAs are biological active component exhibiting abilities to drive plant tolerance to stressful circumstances. These compounds exist in free or conjugated forms, and become active independently or by interacting with other secondary metabolites to alleviate stress responses (Tsaniklidis et al. 2020, Filippo et al. 2023). In response to abiotic stress and/or nutrient deficiency, Pas' content increases multi-folds (Chen et al. 2019). Polyamines' role in plant protection is supported by several studies that revealed osmolytes accumulation in stressed plants after exogenous application with polyamines (Liu et al. 2007, Isa 2019). Also, transcriptomic studies showed that the activity of genes wasresponsible

Biological and Geological Sciences Department, Faculty of Education, Ain Shams University, Cairo, 11341, Egypt.
Email: asmaa_mahmoud@edu.asu.edu.eg

for PA formation, which is triggered under stressful environmental conditions (Marco et al. 2011).

PAs showed physiological functions include stabilizing membranes, macromolecules protection, free radical chelating, regulating cell division and elongation, flowering by interacting with plant growth regulators, and organizing DNA replication (Qi et al. 2010, Childs et al. 2017, Isa 2019). Their cationic nature results in a high affinity to interact strongly with cellular component with a negative charge like nucleic acids (DNA and RNA) proteins and phospholipids (Groppa and Benavides 2008, Tyagi et al. 2022). PAs electrostatic properties are responsible for enzyme activity regulation and manging replication and transcription during cell division process (Mustafavi et al. 2018). Gilad and Gilad (2003) originated the term 'polyamine stress response' (PSR), which refers to the common reaction induced in organisms subjected to stressful stimuli. PAs are important for cell protection and survival when faced with abiotic stressors since they works as endogenous growth regulators and modulators for plant physiological responses (Chen et al. 2019). It is now known universally that polyamines are synthesized and can be isolated from different organisms, not only plants. Generally, prokaryotes synthesize putrescine and spermidine (Javier and Portugal 2023).

Spermine (Spm) was the first recorded PAs by Antonie van Leeuwenhoek in 1677 in aging human sperms. In 1885 Ludwig Brieger identified other two compounds: putrescine (Put) and cadaverine (Cad). The chemical composition of spermidine (Spd), spermine and putrescine was revealed by Harold Ward Dudley and his colleagues in 1920s (Tyagi et al. 2022). Putrescine (diamine), spermidine (triamine), and spermine (tetramine) are the most common PAs that were recognized in organisms like bacteria and various plants, less common polyamines thermospermine (Tspm, tetra-amine), and Cad, 1,3-diaminopropane are synthesized by prokaryotes such as archaea, diatoms, and some plants but not recognized in either animals or bacteria (Javier and Portugal 2023). PA distribution in plants correlates with their function and differs depending on plant species, stage, organ, and tissue, Put and Spm are found at higher concentrations in tobacco plant young leaves and apical meristem than mature leaves while Spd distribution showed a contrary pattern. Also, Spd and Spm are formed in shoot and apical meristems, while Put is synthesized in plant underground organs (Li et al. 2020). In carrots, Put is localized in cell cytoplasm, while Spm is localized in the cell wall (Cai et al. 2006).

Climatic changes make plants suffer from osmotic stress, membrane oxidation, and ionic stress, causing intensive crop damage and sever decline in productivity (Gong et al. 2020, Javeed et al. 2021). That's why implementing new approaches for plant food production and adaptation to environmental changes to achieve food security is a growing need. Regulation of plant metabolites formation is a promising approach to develop plants that are climate-smart, multiple stress-tolerant, and highly productive (Farooq et al. 2022). The chapter focused on the impact of accretion polyamine by plants challenging abiotic stress in strengthening plant tolerance to such stressors.

Polyamine Metabolic Pathways in Plant

As discussed earlier, the level of polyamines concentration increased when the plant is subjected to either a single or combination of stresses. Previous studies revealed that abiotic stress could stimulate polyamines synthesis while the mechanism of polyamines synthesis in plant cells needed to be revealed (Chen et al. 2019). Polyamines comprise a low molecular weight aliphatic component containing nitrogen. The biosynthetic pathway of polyamines (Fig. 1) begins with amino acids and is catalysed with a net of enzymes, the genes of these enzymes was detected in different plant species. Ornithine (Orn) and/or arginine (Arg) decarboxylation was revealed as the first route in polyamine biosynthesis. Decarboxylation of Arg resulted in intermediate compound agmatine (Agm) by the action of arginine-decarboxylase (ADC) (Docimo et al. 2012). Agm then loses NH_3, resulting in diamine putrescine. Arginine conversion into putrescine is divided into three steps depends on three different enzymes: arginine decarboxylase (ADC), agmatine iminohydrolase (AIH), and N-carbamoyl putrescine amidohydrolase (CPA) sequentially. On the contrary,

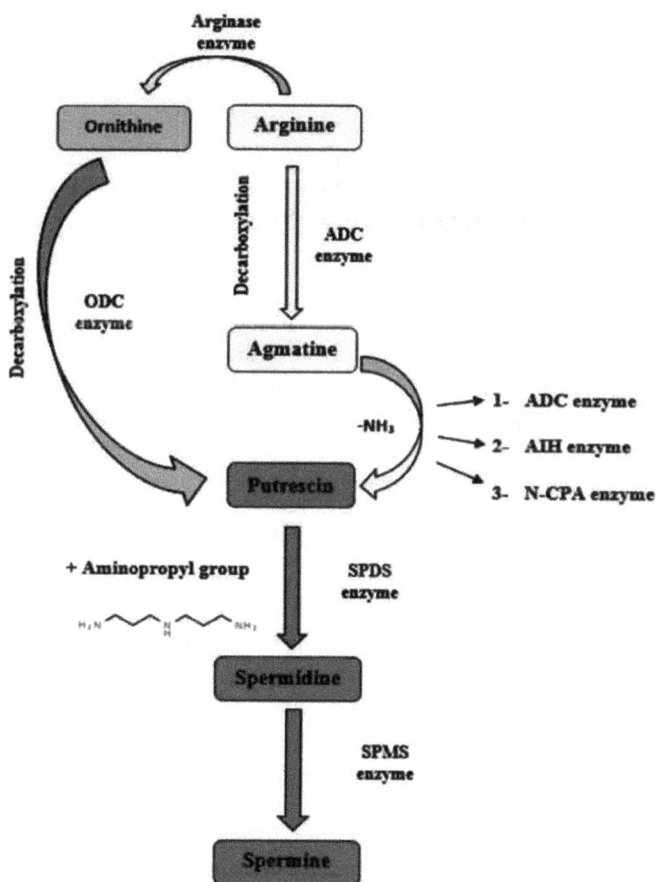

Fig. 1. Biosynthetic pathway of polyamines in plants.

ornithine conversion needs only ornithine decarboxylase enzyme (ODC) activity to form putrescine (Pegg 2016). Aminopropyl group are consecutively added to putrescine, forming higher polyamine spermidine, and spermine. The conversion process catalysed by the action of spermidine synthase (SPDS) and spermine synthase (SPMS) enzymes and Tspm-synthase (TSPMS), respectively. The added aminopropyl group is derived from methionine after its converting into S-adenosylmethionine (SAM) by the action of methionine adenosyltransferase (MAT); at this point, SAM acts as an aminopropyl donor when decarboxylated with S-adenosylmethionine decarboxylase (SAMDC). Some plants like in *Arabidopsis* and some *Brassicaceae* members recorded the absence of gene-coding ODC. Thus, polyamines synthesis following only the ADC pathway in these plants indicates that the ornithine pathway is not the main route for polyamines synthesis (Hanfrey et al. 2001).

According to previous studies, ten genes coding six enzymes responsible for polyamine biosynthesis were characterized and cloned: two coding genes for ADC (*ADC1* and *ADC2*), a single coding gene for AIH and CPA separately, two coding genes for SPDS (*SPDS1* and *SPDS2*) and SPMS (*SPMS* and *ACL5*) each and four coding genes for SAMDC (*SAMDC* 1–4) (Knott et al. 2007, Chen et al. 2019). In all plant tissues *ADC1* is constitutively expressed while abiotic stressed plants showed *ADC2* is exposed (Aktar et al. 2021). Cellular polyamines content is regulated by its degradation into pyrroline, ammonia, and H_2O_2 by diamine oxidase (DAO) and polyamine oxidase (PAO) enzymes activity (Shao et al. 2022). Primary polyamines like putrescine degradation take place by DAO activity which needs Cu^+ as cofactor while PAO catalyse the oxidation of Spd, Spm. Polyamine oxidation resulted in Pyrroline, ammonia, and H_2O_2 (Shao et al. 2022).

Polyamines-Derived Plant Growth and Development

Previous studies provided evidence for the importance of polyamines in plant growth through different life cycle stages (Table 1). Polyamines (Put, Spd, and Cad) accumulated in germinated seeds at early stages of germination process (Movahed et al. 2012). External treatment by polyamines enhanced embryo development of (*Vitis vinifera*) and vegetative growth characters of *Gerbera jamesonii* (Jiao et al. 2018, Saeed et al. 2019); in addition, the vegetative growth and flowering of tobacco plants accelerated after Spm treatment (Zhu et al. 2020). Further, it was found that PAs plays role in pollen grain development and regulation of fertilization (Fellenberg et al. 2009, Fincato et al. 2012, Bokvaj et al. 2015). Many reviews observed the effect of exogenous PAs treatment on fruit formation and ripening; the content of Put increased combined with the decline of Spm and Spd levels in *Musa acuminata*, peach during ripening (*Prunus persica*), and cherry tomato (*S. lycopersicum* var. cerasiforme) (Borges et al. 2019, Tsaniklidis et al. 2016). In addition, many studies revealed relationship between PAs and plant leaf senescence. Providing plants with PAs externally was shown to delay oat and leaf senescence (Mizrahi et al. 1989). This effect could be linked with the effect of PAs on decreasing H_2O_2, scavenging ROS, stabilizing membrane structure, and increasing antioxidants (Mo et al. 2020).

Table 1. The role of polyamines in plant growth.

Plant	Polyamine	Effect	Reference
Sugarcane	PUT	Enhance embryo development	Reis et al. 2016
Decalepis Hamiltonii	PUT	Enhance rooting	Matam et al. 2017
Wheat	Spm and Spd	Enhance plant shoot and root growth	Pála et al. 2019
Oryza sativa	PUT	Enhance root length	Lee 1997
Poncirus trifoliata	PUT	Enhance root length	Wu et al. (2010)
Tomato	PUT	Enhance shoot growth	Yadav et al. (2019)
Antirrhinum majus	PUT	Enhance plant height and fresh weight	Badawy et al. 2015
Dendrobium nobile	PUT and Spd	Promote flowering with buds' increment	Li et al. 2014
Arabidopsis thaliana	Spd	Elevate flowering	Applewhite et al. 2010
Wheat	Spd and Spm	Increase grain weight and grain filling	Liu et al. 2013
Antirhinum majus	PUT	Increase number of f inflorescences	Iman et al. 2018
Wheat	PUT	Decrease ethylene content and enhance plant growth	Yu et al. 2016

Polyamine and Plant Response to Abiotic Stress

When plants are exposed to either single or combination of multiple abiotic stress an alleviation in primary polyamines content was recorded (Table 2). Several studies were conducted to figure out the role of polyamines in mitigating plant response to abiotic stress. Role of polyamines in increasing plant tolerance to a wide range of abiotic stress may be attributed to their acidity neutralizing and cell membrane protective ability (Tyagi et al. 2022).

Heat stress

In response to climatic changes, plants may suffer from heat waves resulting in yield reduction, plant destruction, and affecting food security (Raza et al. 2021). Polyamines contribute to many plants' physiological process of modulating plant thermos-tolerance; however, the pattern of plant response to polyamine action differ among plants. Alfalfa heat stress tolerance was associated with high Spd leaves content as reported by (Yang and Yang 2002). Heat-stressed tobacco plants tends to accrete proline and PAs (Cvikrova et al. 2012). Similar results were observed in pigeon pea as gene expression of polyamine biosynthesis increased in plants after experiencing heat stress (Ramakrishna et al. 2021).

Çetinbaş-Genç (2020) studies spotted on the importance of exogenous Put application in enhancing tea plants tolerance to heat stress where ROS scavenging system was activated, and pollen grains activity increased as well. Amooaghaie and Moghym (2011) revealed that, PAs might bind Ca^{2+} ions with membrane phospholipids since heatstressed soybeans plants that previously treated with Put, Spd, and Spm showed heat shock protein alteration that affected root and hypocotyl

Table 2. Changes in polyamines content in plants subjected to different abiotic stress.

Plant	Stress	Polyamine	Reference
Lotus tenuis	Drought	PUT	Espasandin et al. 2014
Oryza sativa	Drought	PUT	Capell et al. 2004
Arabidopsis thaliana	Drought	PUT	Alcázar et al. 2010
Agrostis stolonifera	Drought	Spm, Spd	Krishnan and Merewitz 2017
Thymus vulgaris L.	Drought	PUT	Mohammadi et al. 2018
Cherry tomato (*Lycopersicon esculentum*)	Salinity/ drought	Spd	Fariduddin et al. 2018
Solanum melongena	Salinity	PUT, Spm, Spdm	Prabhavathi and Rajam 2007
Nicotiana tabacum	Salinity	PUT, Spdm, Spm	Kumria and Rajam (2002)
Cucumis sativus	NaCl	Spd	Wu et al. 2018
Arabidopsis thaliana	Low temperature	PUT, Spm	Tiburcio et al. 2011
Mango (*Mangifera indica* cv. Tommy Atkins)	Low temperature	PUT	Gonzalez-Aguilar et al. 2000
Nicotiana tabacum	Low temperature	PUT, Spm, Spdm	Zhao et al. 2010
Lemon (*Citrus limon* L. Burm,cv. Verna)	Heat	Put, Spm	Valero et al. 1998
Wheat (*T. aestivum* L.)	Heavy metal (Al)	PUT	Yu et al. 2016
Soybean (*Glycine max*)	Heavy metal(Cd)	PUT	Chmielowska-Bak et al. 2013
Wheat (*Triticum aestivum* L.)	$CdCl_2$	PUT	Benavides et al. 2018
Sunflower (*Helianthus annuus*)	$CuCl_2$	PUT	Faisal et al. 2018
Maize (*Z. mays* L.)	Heavy metal (Cr)	Spm	Naz et al. 2021

growth improvement and cell membrane stabilization as lower electrolyte leakage and a decline of MDA content was detected. It was claimed that polyamines play role in protecting cellular membrane by binding to phospholipids which reflects photosynthetic apparatus protection under stress (Alcázar et al. 2006a, Chen et al. 2019). Moreover, plants treated with putrescine recorded a decreasing of chlorophyll loss, decreasing of thylakoids damage, and increment of PSII activity (Zhang et al. 2009). Also, much of the evidence referred to the fact that polyamines contribute to free radical scavenging and act as antioxidants thus protecting stressed plants from further damage (Groppa and Benavides 2008).

According to Sang et al. (2017), tomato seedlings that received Spd exogenous treatment showed higher expression for genes coding cellular defence proteins under heat stress. The PA treatment affected chlorophyll and photosynthesis efficiency, retaining the plant water content and enhancing stomata conductivity in wheat plants under heat stress (Jing et al. 2019).

Cold stress

Low temperature in a plant's early growing stages threatened seed germination and seedling growth. In addition, cold cause pollen sterility and flowering delay, affecting yield attributes in plants (Anwar et al. 2021). Previous studies spotted on the connection between chilling stress and PAs production (Todorova et al. 2015). An increment in Put content was reported in bell pepper when storing them at chilling conditions while Spd and Spm remain at low levels (Serrano et al. 1997).

According to Cuevas et al. (2008), the putrescine level increased in *Arabidopsis* after chilling stress to increase plant tolerance. Put-treated fennel seeds showed higher germination percentage and seedling development compared with untreated seeds under low temperature conditions (Mustafavi et al. 2015). Same results were reported in tomato and *Stevia rebaudiana* stressed plants (Song et al. 2014, Moradi Peynevandi et al. 2018). Sun et al. (2020) claimed exogenous application of Put could extremely reduce damage effect of chilling stressed *Anthurium andraeanum* plants by modulating membrane stability, reducing MDA, increasing proline content, and scavenging ROS. Put treatment associated with profuse production of osmolytes, proline, glycine betaine, and ascorbic acid in plants after low temperature stress (Abbasi et al. 2019). Studies on mung bean seedlings treated with Spd showed less injury in response to cold stress. Spd modulated the ascorbate–glutathione pathway to reduce oxidative stress in plants cell and elimination of ROS (Nahar et al. 2015). In rice plants, the Spd treatment enhanced plant tolerance to low temperature stress through alpha-amylase activity elevation and improving antioxidants' capacity (Sheteiwy et al. 2017).

Supporting the role of exogenous polyamines treatment in enhancing plants' tolerance to low temperature stress, Jankovska-Bortkevič et al. (2020) reported an increment in proline content associated with a decline in ethylene (ET) emission in oilseed rape under cold stress after exogenous PAs (Put, Spd and Spm) application.

Drought stress

Global warming and climatic changes affected crop plants productivity negatively due to the reduction of water availability and planted areas across the world in response to water shortage (Tyagi et al. 2022, Ali et al. 2022). The ability of PAs in manipulating plant drought stress was explained in previous studies in the light of regulating plant osmolytes production and maintaining water transport by controlling stomatal closure (Chen et al. 2019). Many studies were established to study the effect of providing drought-stressed plants with PAs externally to control the effect of drought on plants. The results revealed that Put-treated thyme plants showed higher water content in plants leaves, less cell injury, and more dry matter accumulation as compared with stressed untreated plants (Mohammadi et al. 2018).

Similarly, Sanchez-Rodriguez et al. (2016) claimed that tomato plants subjected to drought stress showed an accumulation of PAs that induced more accumulation in osmoprotectants and ROS reduction. Many reviews observed the effect of Spm especially on manipulating plant drought stress; wheat and cherry tomato plants that subjected to Spm showed drought tolerance associated with ROS scavenging and enhanced plant productivity (Liu et al. 2016).

Exogenous foliar treatment of wheat, alfalfa (*Medicago sativa*), onion (*Allium fistulosum*) plants with Put, Spd, and Spm decreased the negative effect of waterlogging stress by increasing antioxidant content in stressed plants and decreasing oxidative damage (Du et al. 2018, Zhang et al. 2019).

Salt Stress

Salinity stress is a major abiotic stress affecting plants growth and crop productivity. Salinity-stressed plants showed cell membrane injury, reduction of photosynthesis efficiency, production of free radicals, and reduced enzyme activity that affect plant growth negatively (Chen et al. 2019). Plants responded to salinity stress by alerting physiological and biochemical process aiming at antioxidants. Pas and hormones accumulation increased antioxidant activity as ROS chelating took place (Raza et al. 2022). Salinity-stressed *Arabidopsis* plants a showed significant increment in Put content as a natural response to alleviate drastic salinity stress, in addition, plants with mutation reflects Put production deficit failed to adapt salinity stress unless plants were provided with Put externally (Urano et al. 2004). Exogenous application of spermidine for salinity-stressed rice plants caused recovery from damaged plasma membrane (Liu et al. 2011).

Heavy Metals

Heavy metals (HV) contamination soil increased through passed decays after the Industrial Revolution (Ali et al. 2022b). Some heavy metals play role in plants nutrition like copper (Cu), zinc (Zn), and iron (Fe) while other heavy metals are toxic for plant growth like lead (Pb), cadmium (Cd), mercury (Hg), and arsenic (As) (Ali et al. 2021). According to Serrano-Martinez and Casas (2011), heavy metals severely affected the growth of plants by causing oxidative stress and cell injury. Theapplication of PAs to heavy metal-stressed plants increased antioxidant production and regulated ROS scavenge process.

These results were supported as tolerance of HM stressed plants increased after exogenous Put treatment. Treating frogbit (*Hydrocharis dubia* (Bl.)) plant with Spd eliminate adverse toxicity of cadmium (Yang et al. 2013). Other plant species showed results align with previous finding such as *Ulva lactuca*, *Helianthus annuus*, *Triticum aestivum*, and rice (Kumar et al. 2010, Groppa et al. 2008, Groppa et al. 2007, Roychoudhury et al. 2012). Sunflower and wheat plants stressed with high concentrations of copper metal showed Put and Spm content elevation (Groppa et al. 2003).

Assumed Mechanisms of Polyamines in Mitigating Plant Tolerance to Abiotic Stress

Many studies suggested the role of polyamine metabolism in regulating other cellular molecules metabolism under stress (Fig. 2). Among these molecules, ET and nitric

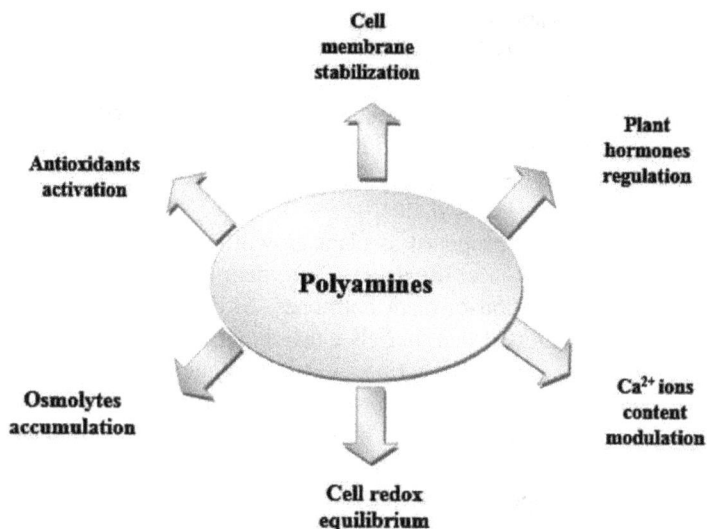

Fig. 2. Physiological role and signaling of polyamines in plants under abiotic stress.

oxide (NO) had gained a lot of attention as they share a common precursor needed for PAs biosynthesis: L-Arg (Recalde et al. 2021). One of the very primitive plant stress responses is H_2O_2 production due to PAs oxidation, that signals cell death in case of high concentration production (Quan et al. 2008). NO production can modify stress protein activity and protein-protein crosstalk controlling plant stress response (Tanou et al. 2013).

According to Chen et al. (2019), NO-contributed S-nitrosylated proteins level increased in response to PAs in *Arabidopsis* plants. Additionally, low concentrations of H_2O_2 could modulate stomatal closure during water shortage stress through stimulating the expression of genes coding nitric oxide synthase (NOS) that modulate abscisic acid (ABA) synthesis (Neill et al. 2008).

Many previous studies revealed the stimulation role of PAs for antioxidants synthesis. Among these studies, a proteome analysis revealed the effect of Spd and Put treatment in regulating nucleoside diphosphate kinase (NDPK) as well as enzymatic antioxidant content in bermuda grass (Shi et al. 2013). The link between PAs and NO-ET metabolism was also suggested by other studies, PAs (Spd and Spm) plant treatment induced rapidly NO production drift, indicating NO mediating effect in polyamine metabolism (Hussain et al. 2011, Chen et al. 2019). Rosales et al. (2012) found that PAs contribute to nitrate reductase (NR) activity regulation that modulate NO levels in plants under stress.

According to Montilla-Bascon et al. (2017) findings, barley plants possess genes responsible for NO oxidizing into NO^{3-} producing less NO showed PAs (Spd) accumulation after being stressed by drought. NO synthesis reduction caused rising in PAs levels accompanied with reduction in ET biosynthesis. According to Filippou et al. (2013), the application of sodium nitroprusside (SNP- NO donor) on *Medicago tranculata* leaves modulates Pas' metabolism. Many studies focused on the regulatory role of H_2O_2 in PAs and NO metabolism, Groppa et al. (2003)

claimed that exogenous application of Spd and Spm enhanced NO and H_2O_2 contents significantly in tomato plants during chilling stress. However, application of the H_2O_2 inhibitor decreased Spd impact on NO accumulation, indicating H_2O_2 involvement in the PA-induced NO production. Treating temperature stressed plants with NO stimulated PAs production that involved in stomata closure, restoring redox balance in the cell and regulating the melatonin level to resist heat damage (Adamipour et al. 2020, Jahan et al. 2019, Wang et al. 2016).

Polyamines first were proposed as plant growth regulators due to their effect on plant development (Chen et al. 2019). Then, many studies focused PAs effect in up and/or down regulation of plant hormones, Bitrian et al. (2012), Milhinhos and Miguel (2013) suggested that there is either synergistically or antagonistically crosstalk between PAs and plant hormones. It was noticed that PAs effect could counteract the effect of ET showing a protective role against plant senescence and fruit ripening (Nambeesan et al. 2008). This effect could be derived from the fact that S -adenosylmethionine (SAM) is the substrate for both (PAs and ET) biosynthesis, creating a kind of competition between ET or PAs synthesis (Lasanajak et al. 2014). There is an alternative relationship between PAs and ABA; both ABA and PAs interacted to regulate stressed plants' physiological and biochemical process, adapting stressors such as Ca^{2+} homeostasis, redox equilibrium, NO modulation (Pal et al. 2018). Put treatment influence the expression of the gene responsible for ABA synthesis (9-cis-epoxycarotenoid dioxygenase3). On the other hand, ABA treatment alerts S-adenosyll-methionine synthetase 1 (SAM1), SAM3, spermidine synthase (SPDS3), and peroxisomal polyamine oxidase (PAO) genes expressions in plants grown under unfavourable circumstances (Chen etal. 2019). *Arabidopsis* plant with ABA-deficient aba2 mutants showed a decline in the ABA level accompanied with progressive decline in the formation of PAs (Alcázar et al. 2006b).

The effect of PAs and ABA as active molecules in elevating plant resistance to abiotic stress was revealed by many studies since tolerant plants subjected to environmental challenges showed an accumulation of both components (Anwar et al. 2021). Also, an interaction between conjugated PAs and jasmonic acid (JA) has been observed in many plants such as mango and apples (Gonzalez-Aguilar et al. 2000, Yoshikawa et al. 2007). The treatment of methyl jasmonate plants was associated with increasing Spd and Spm contents at low-temperature stress in tolerant plants. Additionally, the interaction between PAs and gibberellic acid (GA) was revealed by (Alcazar et al. 2005). An antagonistic effect between Put and GA has been demonstrated as Put accumulation interferes with GA biosynthesis pathway in the final step that blocks GA formation in transgenic *Arabidopsis* (Alcazar et al. 2005). On the contrary, salicylic acid (SA) interacted with polyamines positively; it was found that various plant species such as *Arabidopsis*, citrus, tomato, bamboo, and asparagus that treated with SA showed Spd, Spm, and Put levels rise (Chen et al. 2019).

Many studies were conducted in past decades investigating the role of poly amines in plant development and protection depending on the cross-talk between Pas and other cellular active molecules. However more studies are needed to fill knowledge gaps about how PAs could signal plant response to stress and to find out the connection between PAs and other plant cellular molecules pathways; genome editing is a useful approach for this purpose. A comprehensive picture of effect of the gene's expression manipulation is associated with PAs synthesis under unfavourable conditions. Further biochemical and molecular collaborative studies are needed to reveal the cross-talk between PAs and cellular active molecules to develop crops' adaptation ability to future stress and to achieve food security (Tyagi et al. 2023).

New Insights for Abiotic Stress Tolerance in Plant Depending on PAs

One of the main problems relates to understanding the role of PAs in plant resistance to abiotic stress. This aligns with Hussain et al. (2011)'s observation on Pas as the "mysterious modulator of stress response in plants". Numerous studies summarized the expected role of PAS as compatible solutes with amino acids (proline and glycine betaine), scavenging ROS, protecting cellular macromoleculars like DNA and RNA, stimulating antioxidant production, and regulating of ion channels and ABA (Minocha et al. 2014). A comprehensive image about the role of PAs in plant resistance was suggested depending on four types of studies:, (i) enhancing the level of PAs in transgenic plants increased their tolerance to abiotic stressors, (ii) elevation of the level of PAs in stressed plants which is accompanied with enhancing their tolerance, (iii) plants with mutants of PAs biosynthesis showed more abiotic stress sensitivity, (iv) exogenous application of PAs to plants raise their resistance to stressful conditions. In the next paragraphs some findings from recent studies will be summarized.

Majumdar et al. (2013) claimed enzymes responsible for Put synthesis content increased in the early stages of plant stress; this observation aligns with (Urano et al. 2004, Guo et al. 2014) as gene expression for PAs synthesis enzymes was increased where two or more copies of these genes were recognized in stressed plants. In addition, genes responsible for Spm biosynthesis enzymes expression were increased *A. thaliana* plants after stress treatment. On the other hand, *Arabidopsis* plants with mutation of ADC referred to Put production reduction showed high sensitivity to low temperature stress (Cuevas et al. 2008).

Put production by drought-stressed *Lotus tenuis* plants was shown to regulate genes corresponding to drought stress resistance (Espasandin et al. 2014). Several genes responsible for polyamine biosynthesis enzymes encoding like ADC, ODC, SAMDC, or SPDS were implemented on plants, improving stress tolerance (Liu et al. 2007). Transgenic rice *ADC* gene showed high Put production during drought as compared with wild plants; the Put then transformed into Spm which further increased the plant's tolerance (Capell et al. 2004, Peremarti et al. 2009). In addition, both wild and transgenic drought-stressed plants showed the same response while transgenic plants recovered faster, compared towild plants (Peremarti et al. 2009). Introducing the *SPDS* (spermidine synthase) gene in *Arabidopsis* and sweet potato from *Cucurbita ficifolia* showed overexpression along with abiotic stress tolerance (Kasukabe et al. 2004, Wi et al. 2006). According to Qi et al. (2010) introducing the S-adenosylmethionine synthetase (SsSAMS*2)* gene in tobacco plants from *Suaeda* *salsa* induced rapid increment in polyamines that affected the rate of photosynthetic and biomass positively. Same results were reported by Wen et al. (2008) as overexpression of polyamines synthesis genes in pear (*Pyrus communis L.*) plants induced abiotic stress tolerance.

Similar results were observed in transgenic eggplants (Prabhavathi and Rajam 2007). According to He et al. (2008), transgenic *Pyrus communis* plants showed an overexpression of SPDS genes coding Spd compared to wild plants. Further accumulation was recorded in both plants after subjecting salinity stress. Antioxidants exhibited an accumulation while MDA and H_2O_2 decreased in transgenic plants which indicate less injury due to salinity. Mohapatra et al. (2009) established an experiment comparing control and transgenic cells of poplar (*Populus nigra* x *maximowiczii*) that differ in their polyamins content; the comparison involved antioxidant enzymes synthesis as well as some cellular molecules. Cells that were transformed with β-glucuronidase (GUS) named control cells, high putrescine (HP) cells were transformed with a mouse ornithine decarboxylase (mODC) gene. Results revealed that HP cells produced high levels of soluble proteins and H_2O_2 and lower level of antioxidants enzymes. The study revealed the protective role of Put in scavenging ROS.

Nanotechnology and Polyamines

The integration between nanotechnology and plant biology considered a new approach that may have recently been used for improving crop productivity and to achieve food security. Nano-fertilizers has gained great attention as many studies revealed that nano-fertilizers application enhanced soil quality, enhance plant growth, and increase plant resistance to biotic and abiotic stress (Tyagi et al. 2023). Nano particles (NPs) used as bio-stimulants to improve plant response to stressful conditions altering physiological process and inducing chemical changes in plant cells (Jalil and Ansari 2019, Rajput et al. 2021). Many reporters revealed that NPs such as zinc oxide, titanium oxide, and iron oxide could activate genes related to antioxidant enzymes production reducing stress injury (Rani et al. 2020, Rostamizadeh et al. 2020). Other reports revealed the connection between Nps application and enhancing PAs levels in plants (Mushtaq et al. 2020).

Faiz et al. (2022) reported the accumulation of PAs in seedlings that were previously permitted with NPs. Under salinity stress, the growth and productivity of *Chili pepper* plants enhanced due to chitosan nanoparticles encapsulated spermine application that led to accumulation of active component in treated plants as compared to untreated plants (Ramadan et al. 2022). Koleva et al. (2022) showed that the application of iron and silicon oxide nanoparticles regulates physiochemical process, inducing stress gene expression increment accompanied with Put and Spd content enhancement in *Phaseolus vulgaris* under cadmium stress. Engineered nanoparticles could enhance the content of PAs in stressed plants that could suggest an appropriate approach in achieving global food security.

Conclusion

Polyamines could be considered as group of organic molecules that is ubiquitous in all plant cells. The importance of PAs in signaling plant response to adverse effect of abiotic stress was revealed by many studies. Plants that suffer from PAs content deficiency are characterized by reduction in growth traits, yield, and to a lesser degree of abiotic stress tolerance, indicating the role of PAs in enhancing plant growth, productivity, and stress tolerance. Putrescine, spermidine, and spermine had gained much attention as they are the most common PAs in plants; the effect of exogenous treatment with these PAs in alleviating abiotic stress on plants was supported by many studies as well as the connection between them and many other cellular molecules production. PAs contribute to antioxidants, plant hormones, ion contents modulation, and stabilizing cell plasma membrane. PA application on plants could offer a potentially successful approach towards achieving global food security.

References

Abbasi, N.A., Ali, I., Hafiz, I.A., Alenazi, M.M. and Shafiq, M. 2019. Effects of Putrescine Application on Peach Fruit During Storage. Sustain., 11(7): 2013–29. doi:10.3390/su11072013.

Adamipour, N., Khosh-Khui, M., Salehi, H., Razi, H., Karami, A. and Moghadam, A. 2020. Regulation of Stomatal Aperture in Response to Drought Stress Mediating with Polyamines, Nitric Oxide Synthase and Hydrogen Peroxide in *Rosa canina* L. Plant Signal Behav., 15(9): 1790844.

Aktar, F., Islam, M.S., Milon, M.A.-A., Islam, N. and Islam, M.A. 2021. Polyamines: An Essentially Regulatory Modulator of Plants to Abioticstress Tolerance: A review. Asian J. Appl. Sci., 9: 195–204.

Alcázar, R., Cuevas, J.C., Patrón, M., Altabella, T. and Tiburcio, A.F. 2006a. Involvement of Polyamines in Plant Response to Abiotic Stress. Physiol. Plant., 128: 448.

Alcazar, R., Cuevas, J.C., Patron, M., Altabella, T. and Tiburcio, A.F. 2006b. Abscisic Acid Modulates Polyamine Metabolism Under Water Stress in *Arabidopsis thaliana*. Physiol. Plant, 128: 448–455. https://doi. org/ 10. 1111/j.1399-3054. 2006. 00780.x.

Alcázar, R., Garcia-Martinez, J.L., Cuevas, J.C., Tiburcio, A.F. and Altabella, T. 2005. Overexpression of ADC2 in Arabidopsis Induces Dwarfism and Late-Flowering Through GA Deficiency. Plant J., 43: 425–436. https:// doi. org/ 10.1111/j. 1365-313X. 2005. 02465.x.

Alcázar, R., Planas, J., Saxena, T., Zarza, X., Bortolotti, C., Cuevas, J. et al. 2010. Putrescine Accumulation Confers Drought Tolerance in Transgenic Arabidopsis Plants Over-Expressing the Homologous Arginine Decarboxylase 2 Gene. Plant Physiol. Biochem., 48: 547–552.doi: 10.1016/j.plaphy.2010.02.002.

Ali, S., Tyagi, A. and Bae, H. 2021. Ionomic Approaches for Discovery Of Novel Stress-Resilient Genes In Plants. Int. J. Mol. Sci., 22: 7182.

Ali, S., Tyagi, A., Park, S., Mir, R.A., Mushtaq, M., Bhat, B., Al-Mahmoudi. H. and Bae, H. 2022. Deciphering the Plant Microbiome to Improve Drought Tolerance: Mechanisms and Perspectives. Environ. Exp. Bot., 27: 104933.

Amooaghaie, R. and Moghym, S. 2011. Effect of Polyamines on Thermotolerance and Membrane Stability of Soybean Seedling. Afr. J. Biotechnol., 10(47): 9673–9676. doi:10.5897/ajb10.2446.

Anwar, M.P., Khalid, M.A.I., Islam, A.M., Yeasmin, S., Sharif, A., Hadifa, A. et al. 2021. Potentiality of Different Seed Priming Agents to Mitigate Cold Stress of Winter Rice Seedling. Phyton., 90(5): 1491–1506. doi:10.32604/phyton.2021.015822.

Applewhite, P.B., Kaur-Sawhney, R. and Galston, A.W. 2010. A Role For Spermidine in The Bolting and Flowering of Arabidopsis. Physiol. Plant, 108: 314–320. 10.1034/j.1399-3054.2000.108003314.x.

Badawy, E.S.M., Kandil, M.M., Habib, A.M. and El-Sayed, I.M. 2015. Influence of Diatomite, Putrescine and Alpha-Tocopherol on Some Vegetative Growth and Flowering of *Antirrhinum majus* L. Plants. J. Hortic. Sci. Ornam. Plants, 7: 7–18.

Benavides, M.P., Groppa, M.D., Recalde, L. and Verstraeten, S.V. 2018. Effects of polyamines on cadmium- and copper-mediated alterations in wheat (*Triticum aestivum* L.) and sunflower (*Helianthus annuus* L,) seedling membrane fluidity. Arch. Biochem. Biophys., 654: 27–39. doi: 10.1016/j.abb.2018.07.008.

Bitrian, M., Zarza, X., Altabella, T., Tiburcio, A.F. and Alcazar, R. 2012. Polyamines Under Abiotic Stress: Metabolic Crossroads and Hormonal Crosstalks in Plants. Metabolites, 2(3): 516–528.

Bokvaj, P., Hafidh, S. and Honys, D. 2015. Transcriptome Profiling of Male Gametophyte Development in *Nicotiana tabacum*. Genomics Data 3: 106–111. https://doi. org/10.1016/j. gdata. 2014.12.002

Borges, C.V., Belin, M.A.F., Amorim, E.P., Minatel, I.O, Monteiro, G.C., Gomez, H.G. et al. 2019. Bioactive Amines Changes During the Ripening and Thermal Processes of Bananas and Plantains. Food Chem. 298: 125020. https://doi.org/10. 1016/j.foodc hem. 2019.125020

Cai, Q., Zhang, J., Guo, C. and Al, E. 2006. Reviews of The Physiological Roles and Molecular Biology of Polyamines in Higher Plants. J. Fujian. Inst. Educ., 7: 118–124. https://doi. org/10.3969/j.issn.1673-9884.2006.10.039.

Capell, T., Bassie, L. and Christou, P. 2004. Modulation of the Polyamine Biosynthetic Pathway in Transgenic Rice Confers Tolerance to Drought Stress. Proc. Natl. Acad. Sci. USA, 101: 9909–9914.

Capell, T., Bassie, L. and Christou, P. 2004. Modulation of the Polyamine Biosynthetic Pathway in Transgenic Rice Confers Tolerance to Drought Stress. Pronate. Acad. Sci. USA, 101: 9909–9914. doi:10.1073/pnas.0306974101.

Çetinbaş-Genç, A. 2020. Putrescine Modifies the Pollen Tube Growth Of Tea (*Camellia sinensis*) By Affecting Actin Organization and Cell Wall Structure. Protoplasma, 257(1): 89–101. doi:10.1007/s00709-019-01422-x.

Chen, D., Shao, Q., Yin, L., Younis, A. and Zheng, B. 2019. Polyamine function in plants: metabolism, regulation on development, and roles in abiotic stress responses. Front Plant Sci., 9: 1945. https://doi.org/10. 3389/fpls. 2018. 01945.

Childs, C., Holdsworth, R.E., Christopher, A.L., Jackson, M., Anzocchi, T., Walsh, J.J. 2017. Introduction to the geometry and growth of normal faults. Geol. Soc., 439: 1–9.

Chmielowska-Bak, J., Lefevre, I., Lutts, S. and Deckert, J. 2013. Short term signaling responses in roots of young soybean seedlings exposed to cadmium stress. J. Plant Physiol., 170: 1585–1594. https://doi.org/10.1016/j.jplph.2013.06.019.

Cona, A., Cenci, F., Cervelli, M., Federico, R., Mariottini, P., Moreno, S. 2003. Polyamine Oxidase, a Hydrogen Peroxide-producing Enzyme, is up-regulated by light and down-regulated by auxin in the outer tissues of the maize mesocotyle. Plant Physiol., 131: 803–813.

Cona, A., Rea, G., Angelini, R., Federico, R. and Tavladoraki, P. 2006. Functions of Amine Oxidases in Plant Development and Defence. Trends Plant Sci., 11: 80–88.

Cuevas, J.C., López-Cobollo, R., Alcázar, R., Zarza, X., Koncz, C., Altabella, T. et al. 2008. Putrescine is Involved in Arabidopsis freezing Tolerance and Cold Acclimation by Regulating Abscisic Acid Levels in Response to Low Temperature. Plant Physiol., 148: 1094–1105 10.1104/pp.108.122945

Cvikrová, M., Gemperlová, L., Dobrá, J., Martincová, O., Prásil, I.T., Gubis, J. et al. (2012). Effect of Heat Stress on Polyamine Metabolism in Proline-Overproducing Tobacco Plants. Plant Sci., 182: 49–58. doi: 10.1016/j.plantsci.2011.01.016.

Danaee, E. and Abdossi, V. 2019. Phytochemical and Morphophysiological Responses in Basil (*Ocimum basilicum* L.) Plant to Application of Polyamines. J. Med. Plants, 18: 125–133.

Docimo, T., Reichelt, M., Schneider, B., Kai, M, Kunert, G, Gershenzon, J. et al. 2012. The first step in the biosynthesis of cocaine in *Erythroxylum coca*: the characterization of arginine and ornithine decarboxylases. Plant Mol. Biol., 78: 599–615. https://doi.org/10.1007/s11103- 012-9886-1.

Du, H.Y., Liu, D.X., Liu, G.T., Liu, H.P., Kurtenbach, R. 2018. Relationship between polyamines and anaerobic respiration of wheat seedling root under waterlogging stress. Russ. J. Plant Physiol., 65: 874–881. https://doi.org/10.1134/S1021 44371 80600 55.

Espasandin, F.D., Maiale, S.J., Calzadilla, P., Ruiz, O.A. and Sansberro, P.A. 2014. Transcriptional Regulation of 9-cis-epoxycarotenoid Dioxygenase (NCED) Gene by Putrescine Accumulation Positively Modulates ABA Synthesis and Drought Tolerance in *Lotus tenuis* Plants. Plant Physiol. Biochem., 76: 29–35. doi: 10.1016/j.plaphy.2013.12.018.

Faisal, A., Ibrahim, M.F.M. and Shehata, A. 2018. Exogenous Applied Putrescine Elevate Drought Tolerance of Sunflower Plants By Modifying of Some Physio-Biochemical Parameters. Arab. Univ. J. Agric. Sci., Special Issue, 26(2A).

Faiz, S., Yasin, N.A., Khan, W.U., Shah, A.A., Akram, W., Ahmad, A. et al. 2022. Role of Magnesium Oxide Nanoparticles in The Mitigation of Lead-Induced Stress in *Daucus carota*: Modulation in Polyamines and Antioxidant Enzymes. Int. J. Phytoremediation, 24(4): 364–372.

FAO. 2019. Agricultural Data. Food and Agriculture Organization of The United Nations. Available at: http://www.fao.org/faostat/en/#data/QC (Accessed 29/01/2019).

Fariduddin, Q., Khan, T.A., Yusuf, M., Aafaqee, S.T. and Khalil, R.R.A.E. 2018. Ameliorative Role of Salicylic Acid and Spermidine in The Presence of Excess Salt in *Lycopersicon sculentum*. Photosynthetica, 56: 750–762. https://doi. org/ 10.1007/s11099-017-0727-y.

Fellenberg, C., Bottcher, C. and Vogt, T. 2009. Phenylpropanoid Polyamine Conjugate Biosynthesis in *Arabidopsis thaliana* Flower Buds. Phytochemistry, 70: 1392–1400. https://doi.org/10.1016/j. phytochem. 2009. 08. 010.

Filippo, B., Marotta, G., Rosini M. and Minarini, A. 2023. Polyamine–Drug Conjugates: Do They Boost Drug Activity? Molecules, 28(11): 4518. https://doi.org/10.3390/molecules28114518.

Filippou, P., Antoniou, C. and Fotopoulos, V. 2013. The Nitric Oxide Donor Sodium Nitroprusside Regulates Polyamine and Proline Metabolism in Leaves of *Medicago truncatula* Plants. Free Radic. Biol. Med., 56: 172–183.

Fincato, P., Moschou, P.N., Ahou, A., Angelini, R., Roubelakis-Angelakis, K.A., Federico, R. et al. 2012. The Members of *Arabidopsis Thaliana* PAO Gene Family Exhibit Distinct Tissue- and Organ-Specific Expression Pattern During Seedling Growth and Flower Development. Amino. Acids, 42: 831–841. https://doi.org/10.1007/s00726-011-0999-7.

Gilad, G.M. and Gilad, V.H. 2003. Overview of the Brain Polyamine-Stress-Response: Regulation, Development, and Modulation by Lithium and Role in Cell Survival. Cell Mol. Neurobiol., 23: 637–649. doi:10.1023/A:1025036532672.

Gong, X., Liu, Y., Huang, D., Zeng, G., Liu, S., Tang, H. et al. 2016. Effects of Exogenous Calcium and Spermidine on Cadmium Stress Moderation and Metal Accumulation in *Boehmeria nivea* (L.) Gaudich. Environ. Sci. pollut. Res., 23(9): 8699–708. doi:10.1007/s11356-016-6122-6.

Gonzalez-Aguilar, G.A., Fortiz, J., Cruz, R., Baez, R. and Wang, C.Y. 2000. Methyl Jasmonate Reduces Chilling Injury and Maintains Postharvest Quality of Mango Fruit. J. Agric. Food Chem., 48: 515–519. https:// doi. org/10.1021/jf990 2806.

Groppa, M.D. and Benavides, M.P. 2008. Polyamines and Abiotic Stress: Recent Advances. Amino Acids 34: 35–45.

Groppa, M.D., Benavides, M.P. and Tomaro, M.L. 2003. Polyamine Metabolism in Sunflower and Wheat Leaf Discs Under Cadmium or Copper Stress. Plant Sci., 164: 293–299. https://doi.org/10.1016/S0168-9452(02) 00412-0.

Groppa, M.D., Tomaro, M.L. and Benavides, M.P. 2007. Polyamines and Heavy Metal Stress: The Antioxidant Behaviour of Spermine in Cadmium and Copper-Treated Wheat Leaves. Biometals, 20: 185–195. https://doi.org/10.1007/s10534-006-9026-y.

Groppa, M.D., Zawoznik, M.S., Tomaro, M.L. and Benavides, M.P. 2008. Inhibition of Root Growth and Polyamine Metabolism In Sunflower (*Helianthus annuus*) Seedlings Under Cadmium And Copper Stress. Biol. Trace. Elem. Res., 126: 246–256. https://doi.org/10.1007/s12011-008-8191-y.

Guo, Z., Tan, J., Zhuo, C., Wang, C., Xiang, B. and Wang, Z. 2014. Abscisic Acid, H_2O_2 and Nitric Oxide Interactions Mediated Cold-Induced S-Adenosylmethionine Synthetase in Medicago Sativa Subsp. Falcata That Confers Cold Tolerance Through Up-Regulating Polyamine Oxidation. Plant Biotechnol. J. Epub ahead of print]. 10.1111/pbi.12166.

Gupta, B. and Huang, B. 2014. Mechanism of Salinity Tolerance in Plants: Physiological, Biochemical, and Molecular Characterization. Int. J. Genom, Article ID 701596, 18 pp. Doi:10.1155/2014/701596

Hanfrey, C., Sommer, S., Mayer, M.J., Burtin, D., Michael, A.J. et al 2001. Arabidopsis Polyamine Biosynthesis: Absence of Ornithine Decarboxylase and The Mechanism of Arginine Decarboxylase Activity. Plant J., 27: 551–560. https://doi.org/10.1046/j.1365-313X.2001.01100.x.

He, L., Ban, Y., Inoue, H., Matsuda, N., Liu, J. and Moriguchi, T. 2008. Enhancement of Spermidine Content And Antioxidant Capacity in Transgenic Pear Shoots Overexpressing Apple Spermidine Synthase in Response To Salinity And Hyper osmosis. Phytochem., 69: 2133–41.

Hussain, S.S., Ali, M., Ahmad, M. and Siddique, K.H.M. 2011. Polyamines: Natural and Engineered Abiotic and Biotic Stress Tolerance in Plants. Biotechnol. Adv., 29: 300–311.doi:10.1016/j. biotechadv.2011.01.003.

Iman, M.E.S., Kandil, M.M., Badawy, M.E.S., Abdalla, M.A.E.F., Mahgoub, H.M. and Habib, M.A. 2018. Effect of Diatomite, Putrescine and Alpha-Tocopherol on Flower Characters and Anatomical Flower Bud Structure of *Antirhinum majus* L. Plant. Middle East J. Agric. Res., 7: 1747–1755.

Isah, T. 2019. Stress and Defense Responses in Plant Secondary Metabolites Production. Biol Res 52, 39. https://doi.org/10.1186/s40659-019-0246-3.

Jahan, M.S., Shu, S., Wang, Y., Chen, Z., He, M., Tao, M. et al. 2019. Melatonin Alleviates Heat-Induced Damage of Tomato Seedlings by Balancing Redox Homeostasis and Modulating Polyamine and Nitric Oxide Biosynthesis. BMC Plant Biol., 19(1): 1–16.

Jalil, S.U. and Ansari, M.I. 2019. Nanoparticles and Abiotic Stress Tolerance in Plants: Synthesis, Action, and Signaling Mechanisms. pp. 549–561. In: Iqbal, M., Khan, R., Reddy, P.S. (Eds.). Plant Signaling Molecules. Woodhead Publishing, Sawston.

Jankovska-Bortkevič, E., Gavelienė, V., Šveikauskas, V., Mockevičiūtė, R., Jankauskienė, J., Todorova, D. et al. 2020. Foliar Application of Polyamines Modulates Winter Oilseed Rape Responses to Increasing Cold. Plants, 9: 179–194. doi:10.3390/plants9020179.

Javeed, H.M.R., Ali, M., Skalicky, M., Nawaz, F., Qamar, R., Rehman, A. et al. 2021. Lipoic Acid Combined with Melatonin Mitigates Oxidative Stress and Promotes Root Formation and Growth in Salt-Stressed Canola Seedlings (*Brassica napus* L.). Molecules, 26(11): 3147–3172. doi:10.3390/ molecules26113147.

Javier, S. and Portugal, F. 2023. Putrescine Detected in Strains of *Staphylococcus aureus*. Pathogens, 12(7): 881. https://doi.org/10.3390/pathogens12070881.

Jiao, Y., Li, Z., Xu, K., Yurui Guo, Chen Zhang and Tiemei Li. 2018. Study on Improving Plantlet Development and Embryo Germination Rates in *In vitro* Embryo Rescue of Seedless Grapevine. New Zeal J. Crop Hortic. Sci. https:// doi. org/10.1080/01140 671. 2017.13383 01.

Jing, J.G., Guo, S.Y., Li, Y.F. and Li, W.H. 2019. Effects of Polyamines on Agronomic Traits and Photosynthetic Physiology of Wheat Under High Temperature Stress. Photosynthetica., 57(4): 912–920. doi:10.32615/ps.2019.104.

Kasukabe, Y., He, L., Nada, K., Misawa, S., Ihara, I. and Tachibana, S. 2004. Overexpression of Spermidine Synthase Enhances Tolerance to Multiple Environmental Stresses and Up-Regulates the Expression of Various Stress Regulated Genes in Transgenic *Arabidopsis thaliana*. Plant Cell Physiol., 45: 712–722.

Knott, J.M., Römer, P. and Sumper, M. 2007. Putative Spermine Synthases From *Thalassiosira pseudonana* and *Arabidopsis thaliana* Synthesize Thermo Spermine Rather Than Spermine. FEBS Lett., 581: 3081–3086.

Koleva, L., Umar, A., Yasin, N.A., Shah, A.A., Siddiqui, M.H. and Alamri, S. 2022. Iron Oxide and Silicon Nanoparticles Modulate Mineral Nutrient Homeostasis and Metabolism in Cadmium-Stressed *Phaseolus vulgaris*. Front Plant Sci., 13: 806781. doi:10.3389/fpls.2022.806781. PMID: 35386669, PMCID: PMC8979000.

Krishnan, S. and Merewitz, E.B. 2017. Polyamine Application Effects on Gibberellic Acid Content in Creeping Bentgrass During Drought Stress. J. Am. Soc. Horticult. Sci., 142: 135–142. doi:10.21273/ JASHS03991-16.

Kumria, R. and Rajam, M.V. 2002. Ornithine Decarboxylase Transgene in Tobacco Affects Polyamines, in Vitro-Morphogenesis and Response to Salt Stress. J. Plant Physiol., 159: 983–990. doi:10.1078/0176-1617-00822.

Lasanajak, Y., Minocha, R., Minocha, S.C., Goyal, R., Fatima, T., Handa, A.K. et al. 2014. Enhanced Flux of Substrates Into Polyamine Biosynthesis But Not Ethylene in Tomato Fruit Engineered With Yeast S -Adenosylmethionine Decarboxylase Gene. Amino. Acids, 46(3): 729–742.

Lee, T.M. 1997. Polyamine Regulation of Growth and Chilling Tolerance of Rice (*Oryza sativa* L.) Roots Cultured *in Vitro*. Plant Sci., 122: 111–117.

Li, C., Pei, Z.X. and Gan, L.Y. 2014. Effects of Photoperiod on Flowering and Polyamine Contents of Nobile-Type Dendrobium. Zhiwu Shengli Xuebao/Plant Physiol. J., 50: 1167–1170.

Li, J., Meng, Y., Wu, X. and Sun, Y. 2020. Polyamines and related signaling pathways in cancer. Cancer Cell Int. 20: 539. https://doi.org/10.1186/s12935-020-01545-9.

Liu, J., Nada, K., Pang, X., Honda, C., Kitashiba, H. and Moriguchi, T. 2006. Role of Polyamines in Peach Fruit Development and Storage. Tree Physiol., 26: 791–798. https://doi.org/10.1093/treep hys/26.6.791.

Liu, J.H., Kitashiba, H., Wang, J., Ban, Y. and Moriguchi, T. 2007. Polyamines and Their Ability to Provide Environmental Stress Tolerance to Plants. Plant Biotechnol., 24: 117–126.

Liu, Q., Nishibori, N., Imai, I. and Hollibaugh, J.T. 2016a. Response of Polyamine Pools in Marine Phytoplankton to Nutrient Limitation and Variation in Temperature and Salinity. Mar Ecol Prog Ser 544: 93–105. https://doi. org/10.3354/meps11583.

Liu, Y., Liang, H., Lv, X., Liu, D., Wen, X. and Liao, Y. 2016b. Effect of Polyamines on The Grain Filling of Wheat Under Drought Stress. Plant Physiol. Biochem., 100: 113–129. https://doi.org/10.1016/j. plaphy. 2016. 01. 003.

Liu, Y., Gu, D., Wu, W. and Al, E. 2013. The Relationship Between Polyamines and Hormones in The Regulation of Wheat Grain Filling. PLoS One. 29,8(10): e78196. doi:10.1371/journal.pone.0078196. PMID: 24205154, PMCID: PMC3812141.

Liu, Y., Zuo, Z. and Hu, J. 2010. Effects of Exogenous Polyamines on Growth and Drought Resistance of Apple Seedlings. J. Northwest for Univ, 25: 000039–000042.

Majumdar, R., Shao, L., Minocha, R., Long, S. and Minocha, S.C. 2013. Ornithine: the Overlooked Molecule in Regulation of Polyamine Metabolism. Plant Cell Physiol., 54: 990–1004. doi:10.1093/pcp/pct053.

Marco, F., Alcázar, R., Tiburcio, A.F. and Carrasco, P. 2011. Interactions Between Polyamines and Abiotic Stress Pathway Responses Unraveled by Transcriptome Analysis of Polyamine Overproducers. Omi. A. J. Integr. Biol., 15: 775–781. doi: 10.1089/omi.2011.0084.

Matam, P. and Parvatam, G. 2017. Putrescine and Polyamine Inhibitors in Culture Medium Alter inVitro Rooting Response of *Decalepis Hamiltonii* Wight & Arn. Plant Cell Tissue Organ Cult. (PCTOC), 128: 273–282.

Milhinhos, A. and Miguel, C. 2013. Hormone Interactions in Xylem Development: A Matter of Signals. Plant Cell Rep., 32(6): 867–883.

Minocha, R., Majumdar, R., Minocha, S.C. 2014. Polyamines and Abiotic Stress in Plants: A Complex Relationship. Front Plant Sci. May 5,5:175. doi: 10.3389/fpls. 00175. PMID: 24847338, PMCID: PMC4017135.

Mizrahi, Y., Applewhite, P.B. and Galston, A.W. 1989. Polyamine Binding to Proteins in Oat and *Petunia protoplasts*. Plant Physiol., 91: 738–743. https://doi.org/10.1104/pp.91.2.738.

Mo, A., Xu, T., Bai, Q. et al. 2020. FaPAO$_5$ Regulates Spm/Spd Levels as a Signaling During Strawberry Fruit Ripening. Plant Direct, 4: e00217. https://doi.org/10.1002/pld3.217.

Mohammadi, H., Ghorbanpour, M. and Brestic, M. 2018. Exogenous Putrescine Changes Redox Regulations and Essential Oil Constituents in Field-Grown *Thymus Vulgaris* L. Under Well-Watered and Drought Stress Conditions. Ind. Crops Prod., 122: 119–132. doi:10.1016/j.indcrop.2018.05.064.

Mohapatra, S., Minocha, R., Long, S. and Minocha, S.C. 2009. Putrescine Overproduction Negatively Impacts the Oxidative State of Poplar Cells in Culture. Plant Physiol. Biochem., 47: 262–71.

Montilla-Bascon, G., Rubiales, D., Hebelstrup, K.H., Mandon, J., Harren, F.J.M., Cristescu, S.M. et al. 2017. Reduced Nitric Oxide Levels During Drought Stress Promote Drought Tolerance in Barley and Is Associated With Elevated Polyamine Biosynthesis. Sci. Rep., 7: 1–15. https://doi. org/10.1038/s41598-017-13458-1.

Moradi, Peynevandi, K., Razavi, S.M. and Zahri, S. 2018. The Ameliorating Effects of Polyamine Supplement on Physiological and Biochemical Parameters of *Stevia rebaudiana* Bertoni Under Cold Stress. Plant Prod. Sci. doi:10.1080/1343943X.2018.1437756.

Movahed, N., Eshghi, S., Tafazoli, E. and Jamali, B. 2012. Effects of Polyamines on Vegetative Characteristics, Growth, Flowering and Yield of Strawberry ("Paros" and 'Selva'). Acta Hortic 926: 287–294. https://doi.org/10.17660/actah ortic.2012.926.39.

Mushtaq, T., Shah, A.A., Akram, W. and Yasin, N.A. 2020. Synergistic Ameliorative Effect of Iron Oxide Nanoparticles and *Bacillus Subtilis* S4 Against Arsenic Toxicity in *Cucurbita moschata*: Polyamines, Antioxidants, and Physiochemical Studies. Int. J. Phytorem., 22(13): 1408–1419.

Mustafavi, S.H., Naghdi-Badi, H., Sękara, A., Mehrafarin, A., Janda, T., Ghorbanpour, M. et al. (2018). Polyamines and Their Possible Mechanisms Involved in Plant Physiological Processes and Elicitation of Secondary Metabolites. Acta Physiol. Plant, 40: 102–120. doi:10.1007/s11738-018-2671-2.

Mustafavi, S.H., Shekari, F. and Abbasi, A. 2015. Putrescine Improve Low Temperature Tolerance of fennel (*Foeniculum vulgare* Mill.) seeds. Cercet. Agron. Mold., 48(1): 69–76. doi:10.1515/cerce-2015-0018.

Nahar, K., Hasanuzzaman, M., Alam, M.M. and Fujita, M. 2015. Exogenous Spermidine Alleviates Low Temperature Injury in Mung Bean (*Vigna radiata* L.) Seedlings by Modulating Ascorbate-Glutathione and Glyoxalase Pathway. Int. J. Mol. Sci., 16(12): 30117–32. doi:10.3390/ijms161226220.

Nambeesan, S., Handa, A.K. and Mattoo, A.K. 2008. Polyamines and Regulation of Ripening and Senescence. In: Paliyath, G., Murr, D.P., Handa, A.K. and Lurie, S. (Eds.). Postharvest Biology and Technology of Fruits, Vegetables and Flowers. Wiley, Hoboken, pp. 319–340.

Naz, R., Sarfraz, A., Anwar, Z., Yasmin, H., Nosheen, A., Keyani, R. et al. 2021. Combined Ability of Salicylic Acid and Spermidine to Mitigate The Individual and Interactive Effects of Drought and Chromium Stress in Maize (*Zea mays* L.). Plant Physiol. Biochem., 159: 285–300. https://doi.org/10.1016/j.plaphy.2020.12.022.

Neill, S., Barros, R., Bright, J., Desikan, R., Hancock, J., Harrison, J. et al. 2008. Nitric Oxide, Stomatal Closure, and Abiotic Stress. J Exp Bot 59: 165–176. https://doi.org/10.1093/jxb/erm293.

Pál, M., Ivanovska, B., Oláh, T., Tajti, J., Hamow, K.Á, Szalai, G et al. 2019. Role of polyamines in plant growth regulation of Rht wheat mutants. Plant Physiol Biochem. 2019 Apr, 137: 189–202. doi: 10.1016/j.plaphy.2019.02.013. Epub 2019 Feb 19. PMID: 30798173.

Pal, M., Tajti, J., Szalai, G., Desikan, R., Hancock, J. and Harrison, J. 2018. Interaction of Polyamines, Abscisic Acid and Proline Under Osmotic Stress in The Leaves of Wheat Plants. Sci. Rep., 8: 1–12. https://doi.org/10.1038/s41598- 018-31297-6.

Pegg, A.E. 2016. Functions of Polyamines in Mammals. J. Biol. Chem., 291(21): 14904–12. doi: 10.1074/jbc. R116.731661.

Peremarti, A., Bassie, L., Christou, P. and Capell, T. 2009. Spermine Facilitates Recovery from Drought but Does Not Confer Drought Tolerance In Transgenic Rice Plants Expressing *Datura stramonium* S-Adenosylmethionine Decarboxylase. Plant Mol. Biol., 70: 253–264.

Prabhavathi, V.R. and Rajam, M.V. 2007. Polyamine Accumulation in Transgenic Egg Plant Enhances Tolerance to Multiplea Biotic Stresses and Fungal Resistance. Plant Biotechnol., 24: 273–282. doi:10.5511/plantbiotechnology.24.273.

Qi, Y.C., Wang, F.F., Zhang, H. and Liu, W.Q. 2010. Overexpression of Suadea Salsa S-Adenosylmethionine Synthetase Gene Promotes Salt Tolerance in Transgenic Tobacco. Acta Physiol. Plant, 32: 263–269.

Quan, L.J., Zhang, B., Shi, W.W., Li, H.Y. (2008). Hydrogen Peroxide in Plants: A Versatile Molecule of The Reactive Oxygen Species Network. J. Integr. Plant Biol., 50: 2–18.https://doi.org/10.1111/j.1744-7909.2007.00599.x.

Rajput, V.D., Minkina, T., Kumari, A., Singh, V.K., Verma, K.K., Mandzhieva, S. et al. 2021. Coping With the Challenges of Abiotic Stress in Plants: New Dimensions in The Field Application of Nanoparticles. Plants, 10(6): 1221.

Ramadan, M.E., El-Saber, M.M., Abdelhamid, A.E. and El-Sayed, A.A. 2022. Effect of Nano chitosan Encapsulated Spermine on Growth, Productivity and Bioactive Compounds of Chili Pepper (*Capsicum annuum* L.) Under Salinity Stress. Egypt. J. Chem., 65: 187–98.

Ramakrishna, G., Kaur, P., Singh, A., Yadav, S.S, Sharma, S. et al. 2021. Comparative Transcriptome Analyses Revealed Different Heat Stress Responses in Pigeonpea (*Cajanus cajan*) and Its Crop Wild Relatives. Plant Cell Rep., 40: 881–898. https://doi.org/10.1007/s00299-021-02686-5.

Rani, P., Kaur, G., Rao, K.V., Singh, J. and Rawat, M. 2020. Impact of Green Synthesized Metal Oxide Nanoparticles on Seed Germination and Seedling Growth of *Vigna radiata* (Mung Bean) and *Cajanus cajan* (Red Gram). J. In. org. Organo met Polym. Mater, 30(10): 4053–4062.

Raza, A., Tabassum, J., Fakhar, A.Z., Sharif, R., Chen, H., Zhang, C. et al. 2022. Smart Reprograming of Plants Against Salinity Stress Using Modern Biotechnological Tools. Crit. Rev. Biotechol., 12: 1–28.

Raza, A., Tabassum, J., Kudapa, H. and Varshney, R.K. 2021. Can Omics Deliver Temperature Resilient Ready-To-Grow Crops? Crit. Rev. Biotechol., 41(8): 1209–1232.

Recalde, L., Gomez Mansur, N.M., Cabrera, A.V., Matayoshi, C.L., Gallego, S.M, Groppa, M.D. et al. 2021. Unravelling Ties in the Nitrogen Network: Polyamines and Nitric Oxide Emerging As Essential Players In Signalling Roadway. Ann. Appl. Biol., 178(2): 192–208.

Reis, R.S., Vale, E.M., Heringer, A.S. and Al, E. 2016. Putrescine Induces Somatic Embryo Development and Proteomic Changes in Embryogenic Callus of Sugarcane. J. Proteomics 130: 170–179. doi: 10.1016/j.jprot.2015.09.029.

Rosales, E.P., Iannone, M.F., Groppa, M.D. and Benavides, M.P. 2012. Polyamines Modulate Nitrate Reductase Activity in Wheat Leaves: Involvement of Nitric Oxide. Amino. Acids 42: 857–865.

Rostamizadeh, E., Iranbakhsh, A., Majd, A., Arbabian, S. and Mehregan, I. 2020. Green Synthesis of Fe_2O_3 Nanoparticles Using Fruit Extract of *Cornus mas* L. and Its Growth-Promoting Roles in Barley. J Nanostructure Chem., 10(2): 125–130.

Roychoudhury, A., Basu, S. and Sengupta, D.N. 2012. Antioxidants and Stress-Related Metabolites in The Seedlings of Two Indica Rice Varieties Exposed To Cadmium Chloride Toxicity. Acta Physiol. Plant, 34: 835–847. https://doi.org/10.1007/s11738-011-0881-y.

Saeed, A.A.J.M., Abdulhadi, M.D. and Salih, S.M. 2019. Response of Gerbera (*Gerbera jamesonii*) cv. "Great Smoky Mountains" to Foliar Application of Putrescine, Spermidine and Salicylic Acid. In: IOP Conference Series: Earth and Environmental Science. P 012067.

Sanchez-Rodriguez, E., Romero, L. and Ruiz, J.M. 2016. Accumulation on Free Polyamines Enhanced Antioxidant Response in Fruit of Grafting Tomato Plants Under Water Stress. J Plant Physiol 190: 72–78. https://doi.org/10.1016/j.jplph.2015. 10. 010.

Sang, Q., Shan, X., An, Y., Shu, S., Sun, J. and Guo, S. 2017. Proteomic Analysis Reveals the Positive Effect of Exogenous Spermidine in Tomato Seedlings' Response To High-Temperature Stress. Front. Plant Sci., 8: 120–134. doi: 10.3389/fpls.2017.00120.

Serrano, M., Martinez-Madrid, M.C., Pretel, M.T., Romojaro, F. and Riquelme, F. 1997. Modified Atmosphere Packaging Minimizes Increases in Putrescine and Abscisic Acid Levels Caused by *Chilling injury* in Pepper Fruit. J. Agric. Food Chem., 45: 1668–1672. https://doi.org/10.1021/jf960866h.

Serrano, M., Pretel, M.T., Martinez-Madrid, M.C., Romojaro, F., Riquelme, F.. 1998. CO_2 Treatment of Zucchini Squash Reduces Chilling-Induced Physiological Changes. J. Agric. Food Chem., 46: 2465–2468. https://doi.org/ 10.1021/jf970 864c.

Serrano-Martinez, F. and Casas, J.L. 2011. Effects of Extended Exposure to Cadmium and Subsequent Recovery Period on Growth, Antioxidant Status and Polyamine Pattern In: *In Vitro* Cultured Carnation. Physiol Mol Biol Plants, 17: 327–338. https://doi.org/10.1007/s12298-011-0081-7.

Shao, J., Huang, K., Batool, M., Idrees, F., Afzal, R., Haroon, M. et al. 2022. Versatile Roles of Polyamines in Improving Abiotic Stress Tolerance of Plants. Front Plant Sci., 13: 1003155. doi:10.3389/fpls.2022.1003155. PMID: 36311109, PMCID: PMC9606767.

Sheteiwy, M., Shen, H., Xu, J., Guan, Y., Song, W. and Hu, J. 2017. Seed Polyamines Metabolism Induced by Seed Priming with Spermidine and 5- Aminolevulinic Acid for Chilling Tolerance Improvement in Rice (*Oryza sativa* L.) seedlings. Environ. Exp. Bot., 137: 58–72. doi: 10.1016/j.envexpbot.2017.02.007.

Shi, H., Ye, T. and Chan, Z. 2013. Comparative Proteomic and Physiological Analyses Reveal The Protective Effect of Exogenous Polyamines in The Bermuda Grass (*Cynodon dactylon*) Response to Salt and Drought Stresses. J. Proteome. Res. 12: 4807–4829.

Song, Y., Diao, Q., and Qi, H. (2014). Putrescine Enhances Chilling Tolerance of Tomato (*Lycopersicon esculentum* Mill.) Through Modulating Antioxidant Systems. Acta Physiol. Plant. 36, 3013–3027. doi: 10.1007/s11738-014-1672-z.

Sun, X., Yuan, Z., Wang, B., Zheng, L., Tan, J., and Chen, F. 2020. Physiological and Transcriptome Changes Induced by Exogenous Putrescine in Anthurium Under Chilling Stress. Bot. Stud. 61(1), 28–39. doi: 10.1186/s40529-020-00305-2.

Tanou, G., Ziogas, V., Belghazi, M., Christou, A., Filippou, P., Job, D. et al. 2013. Polyamines Reprogram Oxidative and Nitrosative Status and The Proteome of Citrus Plants Exposed to Salinity Stress. Plant Cell Environ. DOI:10.1111/pce.12204.

Tiburcio, A.F., Altabella, A.T. and Ferrando, M.A. 2011. Plant Having Resistance to Low-Temperature Stress and Method of Production Thereof. EP20090 793978.

Todorova, D., Katerova, Z., Alexieva, V. and Sergiev, I. 2015. Polyamines–Possibilities for Application to Increase Plant Tolerance and Adaptation Capacity to Stress. Genet. Plant Physiol., 5(2): 123–144.

Tsaniklidis, G., Kotsiras, A., Tsafouros, A., Roussos, A.P., Aivalakis, G., Katinakis, P. et al. 2016. Spatial and Temporal Distribution of Genes Involved in Polyamine Metabolism During Tomato Fruit Development. Plant Physiol. Biochem. 100: 27–36. https://doi.org/10.1016/j.plaphy.2016. 01.001.

Tsaniklidis, G., Pappi, P., Tsafouros, A., Charova, S.N., Nikoloudakis, N., Roussos, P.A. et al. 2020. Polyamine Homeostasis in Tomato Biotic/Abiotic Stress Cross-Tolerance. Gene, 727: 144230–144240. doi:10.1016/j.gene.2019.144230.

Tyagi, A., Ali, S., Ramakrishna, G., Singh, A., Park, S., Mahmoudi, H. et al. 2023. Revisiting the Role of Polyamines in Plant Growth and Abiotic Stress Resilience: Mechanisms, Crosstalk, and Future Perspectives. J. Plant Growth Regul, 42: 5074–5098. https://doi.org/10.1007/s00344-022-10847-3.

Urano, K., Yoshiba, Y., Nanjo, T., Ito, T., Yamaguchi-Shinozaki, K. and Shinozaki, K. 2004. Arabidopsis Stress-Inducible Gene for Arginine Decarboxylase Atadc2 Is Required for Accumulation Of Putrescine in Salt Tolerance. Biochem. Biophys. Res. Commun., 313: 369–375 10.1016/j.bbrc.2003.11.119.

Valero, D., Martinez-Romero, D., Serrano, M. and Riquelme, F. 1998. Postharvest Gibberellin and Heat Treatment Effects on Polyamines, Abscisic Acid and Firmness in Lemons. J. Food Sci., 63: 611–615. https://doi.org/10.1111/j.1365-2621.1998.tb157 96.x.

Wang, Y., Luo, Z., Mao, L. and Ying, T. 2016. Contribution of Polyamines Metabolism and GABA Shunt to Chilling Tolerance Induced by Nitric Oxide in Cold-Stored Banana Fruit. Food Chem., 197: 333–339. https://doi.org/10.1016/j.foodc hem.2015.10.118.

Wen, X.P., Pang, X.M., Matsuda, N., Kita, M., Inoue, M., Hao, Y.J. et al. 2008. Overexpression of the Apple Spermidine Synthase Gene in Pear Confers Multiple Abiotic Stress Tolerance by Altering Polyamine Titers. Transgenic. Res., 17: 251–263.

Wi, S.J., Kim, W.T. and Park, K.Y. 2006. Overexpression of Carnation S-Adenosylmethionine Decarboxylase Gene Generates A Broad Spectrum Tolerance to Abiotic Stresses in Transgenic Tobacco Plants. Plant Cell Rep., 25: 1111–1121.

Wu, J., Shu, S., Li, C., Sun, J. and Guo, S. 2018. Spermidine-mediated Hydrogen Peroxide Signaling Enhances the Antioxidant Capacity of Salt-stressed Cucumber Roots. Plant Physiol. Biochem., 128: 152–162. doi:10.1016/j.plaphy.2018.05.002.

Wu, Q.S., Zou, Y.N. and He, X.H. 2010. Exogenous Putrescine, Not Spermine or Spermidine, Enhances Root Mycorrhizal Development and Plant Growth of Trifoliate Orange (*Poncirus trifoliata*) Seedlings. Int. J. Agric. Biol., 12: 576–580.

Yadav, V., Singh, N.B., Singh, H., Singh, A. and Hussain, I. 2019. Putrescine Affects Tomato Growth and Response of Antioxidant Defense System Due to Exposure to Cinnamic Acid. Int. J. Veg. Sci., 25: 259–277.

Yang, H.Y., Shi, G.X., Li, W.L. and Wu, W.L. 2013. Exogenous Spermidine Enhances *Hydrocharis dubia* Cadmium Tolerance. Russ. J. Plant Physiol., 60: 770–775. https://doi.org/10.1134/S1021 44371 3060162.

Yang, Y. and Yang, X. 2002. Effect of Temperature on Endogenous Polyamines Content of Leaves in Chinese Kale (*Brassica alboglabra* bailey) Seedlings. Hua Nan Nong Ye Da Xue Xue Bao = J South China Agric. Univ., 23: 9–12.

Yoshikawa, H., Honda, C. and Kondo, S. 2007. Effect of Low-Temperature Stress on Abscisic Acid, Jasmonates, and Polyamines in Apples. Plant Growth Regul., 52: 199–206. https://doi.org/10.1007/s10725-007-9190-2.

Yu, Y., Jin, C., Sun, C., Wang, J., Ye, Y., Zhou, W. et al. 2016. Inhibition of Ethylene Production by Putrescine Alleviates Aluminium-Induced Root Inhibition in Wheat Plants. Sci. Rep., 6: 1–10. https://doi.org/10.1038/srep1 8888.

Zhang, Q., Liu, X., Zhang, Z., Liu, N., Li, D., Hu, L. et al. 2019. Melatonin Improved Waterlogging Tolerance in Alfalfa (*Medicago sativa*) by Reprogramming Polyamine and Ethylene Metabolism. Front Plant Sci., 10: 4. https://doi.org/10. 3389/fpls.2019. 00044.

Zhang, R.H., Li, J., Guo, S.R. and Tezuka, T. 2009. Effects of Exogenous Putrescine on Gas Exchange Characteristics and Chlorophyll Fluorescence of NaCl-Stressed Cucumber Seedlings. Photosynth Res., 100: 155–162.

Zhao, Y., Du, H., Wang, Z. and Huang, B. 2010. Identification of Proteins Associated with Water-Deficit Tolerance in C4 Perennial Grass Species, *Cynodon dactylon* × *Cynodon transvaalensis* and *Cynodon dactylon*. Physiol. Plant., 141: 40–55. doi:10.1111/j.1399-3054.2010.01419.x.

Zhu, H., Tian, W., Zhu, X., Tang, X, Wu, L., Hu, X. et al. 2020. Ectopic Expression of GhSAMDC1 Improved Plant Vegetative Growth and Early Flowering Through Conversion of Spermidine to Spermine in Tobacco. Sci. Rep. 10: 1–10. https://doi.org/10.1038/s41598-020-71405-z.

13

Karrikin

Regulation of Biosynthesis, Hormonal Crosstalk, and Its Role In Biotic and Abiotic Stress Adaptation

Mona F.A. Dawood,[1],* Heba I. Mohamed,[2] Yachana Jha[3] and M.T. El-Mahdy[4]

Introduction

Smoke was deduced to instigate the germination of many species from various families (Dixon et al. 2009) where KARs were reported as the major germination promoting compound found in smoke (Antala et al. 2020). So that, the germination of smoke-responsive species is very likely enhanced by KARs and Karrikins (KARs) are defined as a family of closely related small organic molecules that are prepaed via flaring the plant materials with a structurally substituted butenolide moiety (Nelson et al. 2012). KARs are promising new plant growth regulators, and an as yet unidentified endogenous molecules perceived through the same signaling pathway as a potential new phytohormone (Antala et al. 2020). KAR1 has been demonstrated to be active at very low concentrations, in the range of 10^{-10}–10^{-7} M. It was found to be neither toxic nor genotoxic at 3×10^{-10}–10^{-4} M (Light et al. 2009), and thus,

[1] Botany and Microbiology Department, Faculty of Science, Assiut University, Assiut, 71516, Egypt.
[2] Biological and Geological Science Department, Faculty of Education, Ain Shams University, Cairo, 11341, Egypt.
[3] N. V. Patel College of pure and Applied Sciences, CVM University, V. V. Nagar, Anand (Gujarat), India.
[4] Department of Pomology, Faculty of Agriculture, Assiut University, Assiut 71526, Egypt.
   Email: marwa.refaat@agr.aun.edu.eg
* Corresponding author: Mo_fa87@aun.edu.eg

it is safe for animals and humans. KAR1 is produced from d-xylose during fire, in small amounts (Flematti et al. 2015). It has been reported that KARs receptor is present in all phylogenetic taxa of plants, including mosses, liverworts, or green algae (Douglas et al. 2017). Indeed, many studies reported that KARs elevated germination of dicotyledonous, as monocotyledonous plants belonging to different plant life forms—Annuals, perennials, woody plants, and different importance for people—Weeds or agricultural and horticultural crops. Besides the contribution of KARs as apromoting agent to release seed dormancy, their roles in reducing hypocotyl elongation, and instigating, cotyledon expansion and seedling vigor as reviewed by Yang et al. (2019). KAR1 was shown to stimulate seed germination in fire-prone environments (Merritt et al. 2006), in hemi- and holo-parasitic seeds (Daws et al. 2007), and in several Australian *Asteraceae* species (Merritt et al. 2006). Likewise, seeds of some crop plants such as lettuce, tomato, okra, bean, maize, and rice responded positively to KAR1 (Kulkarni et al. 2011). The compound is able to increase both the rate and percentage of seed germination, it also improves seedling growth. Moreover, and importantly, KAR1 allows seeds to germinate at sub- and supra-optimal temperatures, and also at a low water potential (Light et al. 2009).

The smoke water obtained by different methods, needs to be dissolved in various ratios due to the variable content of germination stimulators and inhibitors (Gupta et al. 2019). Although smoke water has a lower price compared to KARs, using KARs is more convienct for agriculture and plant discipline than aerosol smoke and smoke-water (Nelson et al. 2012). Another way for the application of KARs to agriculturally soils can be indirectly through biochar, especially the KARs have been identified recently (Kochanek et al. 2016). In addition to KARs being an important constituent for biochar, it confers other advantages such as improving nutrient sorption and water holding capacity of the soil, as well as carbon sequestration (Mona et al. 2019). Nevertheless, KARs content of biochar, like in the case of smoke, is dependent on charred material and the technology of pyrolysis. The individual plant species response relates to its sensitivity to KARs and inhibitory compounds within the biochar (Kochanek et al. 2016). Due to a huge demand for the compound in research and on account of its potential application in agriculture, several methods of synthesis using d-xylose or other compounds as substrates have been developed (Yang et al. 2019). Recently there is a focus on the potential functions of KARs in mediating abiotic stress endurance in plants. In this chapter, we will briefly cover the history of KARs discovery, signaling networks, and functioning. Furthermore, the crosstalks between KARs with various phytohormnes were also addressed. In addition, the potential mechanisms by which KARs normalize the plant's adaption to abiotic stresses were highlighted.

History of Discovery and Major Sources

It has been known for many years that seeds of some fire-followers were stimulated to germinate not by the heat of a fire but by the chemicals it produced (Nelson et al. 2012). De Lange and Boucher (1990) stated that smoke have the same effect as fire, as can water in which char was soaked or through which smoke was passed, producing 'smoke water'. Plant-derived smoke functions as a potential bioactivator that promotes seed germination and plant growth (Khatoon et al. 2020). Plant-derived smoke from fynbos plant material was developed as a germination stimulator in the 1990s, which sparked the interest of numerous scientists and researchers in smoke ecology and its use in plant growth and ontogensis (De Lange and Boucher 1990, Light 2018). Application of plant-derived smoke to various fynbos plant species in the Asteraceae, Ericaceae, Restionaceae, and Protaceae family exhibits exceptional responses toward seed germination, according to a different experiment conducted by Brown (1993). The combustion of plant tissue by smoke, which occurs naturally, aids in the regrowth of vegetation in ecosystems that are vulnerable to wildfire. However, burning plant matter and soaking the smoke in distilled water can be used to produce smoke water artificially (De Lange and Boucher 1990). According to Staden et al. (2000), the smoke produced from burning Passerina vulgaris and Themeda triandra L., was collected in two litres of distilled water using a vacuum system, whereas various plants have been employed for producing smoke water in the past. Studies claimed that seed germination of nearly different plant species could be influenced by smoke-infused water (Dixon et al. 2009) and the same response for many agricultural and horticultural crops (Kępczynski 2018). The smoke contains a variety of biologically active substances that may be dissolved in water. These compounds exhibit amazing responses toward seed germination and seedling growth when this diluted solution is administered to dormant seeds or seedlings (Singh et al. 2023). With the passage of time, numerous studies observed that plant-derived smoke acted as an enhancer of seed germination for both areas that are prone to fire and those that are not (Staden et al. 2000, Jefferson et al. 2014). It has been mentioned that out of 80 genera, 1200 species were shown to have seed-germination accelerated by smoke in various environments (Dixon et al. 2009). After learning that smoke derived from plants could be a potent tool for seed germination, scientists conducted numerous studies to identify the components of smoke that were essential for seed germination or seedling development. According to gas chromatography and mass spectroscopy/atomic spectroscopy data, smoke created from plants contained 71 different phytochemicals (Baldwin et al. 1994). A number of research groups around the world separated thousands of compounds in smoke water into many portions by liquid chromatography and studied every fraction for seed-sprouting behavior (Baldwin et al. 1994, Flematti et al. 2004). Effective fractions were further fractionated and assayed, led to the isolation of an active compound, which was confirmed through chemical synthesis (Flematti et al. 2004).

Although the positive role of smoke in germination of dormant and non-dormant seeds, and in seedling growth, has been extensively studied since 1990, only in 2004 a germination-active compound, 3-methyl-2H-furo[2,3-c]pyran-2-one was identified in plant-derived smoke (Staden et al. 2004) and burnt cellulose (Flematti et al. 2004). Initially, this compound was termed butenolide, later on, to distinguish the butenolide present in smoke from other butenolides, it was entitled karrikinolide or karrikin-1 (KAR1) (Flematti et al. 2009), derived from the word "karrik" meaning smoke in the Australian Aboriginal language (Kępczyński 2018). We now understand that a group of butenolide compounds isolated from smoke, the first member of which was identified independently by two researchers' teams, plays a major role in germination promotion (Antala et al. 2019). This compound contains a specific type of lactone known as a butenolide fused to a pyran ring with the systematic name 3-methyl-2H-furo[2,3-c]pyran-2-one. Subsequently, several closely related compounds were discovered in smoke and collectively referred to as 'karrikins'. It is common in biology to add the '-in' suffix to denote a group of related molecules, such as 'auxin, cytokinins' in plants. Fire and smoke play important roles in Aboriginal culture in Australia, and the karrikin name acknowledges that fact. The original compound identified is often referred to as 'karrikinolide': the '-olide' suffix indicates that it is a lactone (Flematti et al. 2015). The karrikins are abbreviated to KARs and numbered in order of their identification in smoke. The first karrikin discovered was KAR1, also known as karrikinolide. Since karrikins can be produced by burning sugars such as xylose, the pyran ring of karrikins is probably derived from such pyranose sugars (Flematti et al. 2015). In addition to KARs, cyanohydrins were identified as a germination signal, which can stimulate the germination of some KAR-insensitive species (Flematti et al. 2009).

The Photochemistry of Karrikins

Karrikins (KAR1) was the first compound recognized as the highly active constituent present in smoke water and the pyrolysis of carbohydrates, like cellulose and simple sugars (glucose, xylose) also considered as another source of karrikins. Thus, there is no known natural source for karrikins other than the pyrolysis of carbohydrates (Flemmati et al. 2011). The molecular identity of KARs as butenolide (3-methyl-2H-furo [2, 3-c] pyran-2-one) was confirmed by spectrometry detction (Flemati et al. 2004). Nuclear magnetic resonance (NMR) spectroscopy was used to determine the structure of KAR1 and to confirm its presence in plant-derived smoke (Staden et al. 2004). Even though just six KARs so far showed physiological activity in plants, almost 50 analogs of KAR1 with different substitutions have been synthesized (De Cuyper et al. 2017). The differences between the six known KARs are based on methyl substitutions. These KARs are described as KAR1 to KAR6 (Waters et al. 2014). The studies on karrikin structure–activity relationship demonstrated that the methyl group at C-3 is important for biological activity: introduction of methyl at C-4 or C-7 reduces the activity, introduction at C-5 being tolerated well (Nelson et al. 2012). Subsequent studies led to detection of five KAR1 analogs, KAR2–KAR6, in smoke (Flematti et al. 2009). Concentrations of these compounds were much lower than those of KAR1. Thus, KAR1 was considered to be the major factor responsible for the stimulatory effect of smoke on seed germination. The comparison of KAR1 effects on seed germination with those exerted by other karrikins revealed different, plant species-specific responses to these compounds (Waters et al. 2014).

Only C, H, and O are present in the two-ring structures of KARs. The pure KARs have a melting point of 118–119°C, and they are the substances of crystalline character. KARs can be quickly dissolved in carbon-based solvents and moderately in water. The primary function of KAR1 is to enhance germination and seedling growth. These substances are created in the ecosystem by incomplete plant combustion caused by fire, and they can ingest soil particles and persist in the ecosystem for a number of years (Flematti et al. 2015). These soil-bound lactones also aid in promoting the germination of fire-ephemeral seeds and speeding up the development of their community tolerance. In a recent study, the UHPLC-ESI (+)-MS/MS (Ultra High Performance Liquid Chromatography-Tandem Mass Spectroscopy) approach showed that karrikins might be quantified by standard method using smoke solutions. However, they also discovered that KAR2 was present in greater amounts than KAR1 and that variations in the composition of karrikins might have an impact on smoke's ability to promote germination (Hrdli˘cka et al. 2019). Additionally, various other constituents extracted from plant-derived smoke might also be used as potent activators, such as cyanohydrin, which has been proved to be essential for defence against a variety of plant diseases and pests (Flematti et al. 2013). It is interesting to note that additional butenolides that have adverse effects on seed germination are also found in plant-derived smoke. One of these, KAR1, is a powerful germination promotor, whereas 3, 4, 5-trimethylfuran-2(5H)-one, suppresse the seed germination in addition to lowering the activity of KAR1 (Light et al. 2010). This substance was identified as hydroquinone using 1D and 2D NMR techniques. This hydroquinone functions as a powerful at low concentrations, but exhibits inhibitory effects at higher quantities (Singh et al. 2023).

Perception and Signaling Mechanism of Karrikins

Although KARs have been known about from more than 15 years ago, the real mechanism of perception still unclear. However, it does not mean that there is no information regarding KARs signaling cascade, which begins by the sensing of KARs, and then translated to varios morphological and physiological responses. Two genes, KAI2 (KARRIKIN INSENSITIVE2) and MAX2 (MORE AXILLARY GROWTH2), are essential for the action mechanism of karrikin in the signaling pathway (Waters et al. 2014). KAI2, a protein that is a paralogue of D14 protein (perceives strigolactone), presents karrikin in seeds. KAI2 is a α/β hydrolase family and act as a karrikin receptor and an enzyme work for the signaling process (Guo et al. 2013). The receptor KAI2 has a catalytic triad of Ser95-His246-Asp217 amino acid residues in which the amino acid Ser95 accounts for a nucleophilic attack on the butenolide ring of karrikin (Waters et al. 2014). While, because of its susceptibity to many karrikins and some chemicals, KAI2 displays replication and heterogenity in weedy ephemerals (Sun et al. 2020). An F-box protein known as MAX2 (more axillary growth growth2) was initially known in a forward genetic detection of the kai2 mutant (Nelson et al. 2011). When KAI2 detects karrikin, it joins forces with MAX2 protein to form a complex that later helps degrade repressors like SMAX1 and SMAX2 by forming the SCF complex (Nelson et al. 2011, Guo et al. 2013, Stanga et al. 2016). In the process of degradation, the SCF complex binds to the ubiquitin bodies and uses 26 proteasomal complexes to break down the target proteins (Stirnberg et al. 2007). Receptor KAI2 is important for cotyledon expansion, shortening petioles, and leaves to achieve wild type size, anthocyanins, and chlorophylls' accumulation and enhanced expression of CHLOROPHYLL A/B BINDING PROTEIN 3 and CHALCONE SYNTHASE, which are light-responsive genes (Sun et al. 2011). According to a study by Sun et al. (2016), treating *Arabidopsis* with KARs causes an up-regulation of the d14-like 2 (DLK2) gene, which demonstrates the close closeness with KAR1 and SL signalling. A member of an eight-member family of proteins, SMXL (Suppressor of

MAX2-Like) shares sequence similarities with HSP101 (HEAT SHOCK PROTEIN 101), a protein that blocks the karrikin signalling pathway in plants. It is interesting to note that KAR can degrade when exposed to UV radiation. However, additional smoke-derived aromatic chemicals that serve as UV ray absorbers might be able to stop its destruction (Scaffidi et al. 2012). KAR1 can therefore remain active in plant soil for about 7 years as a result of these "organic sunblockers" (Flematti et al. 2015). There is evidence that a variety of plant species, including annuals, weedy plants in agriculture, and herbs, respond to KAR1 activity (Nelson et al. 2012). These plant species exhibit physiological modifications that have been ingested into their genomic systems and are exploited as a regenerative tool by fire ephemerals in the post-fire ecosystem. The perception and signaling mechanism of karrikins are represented in Fig. 1.

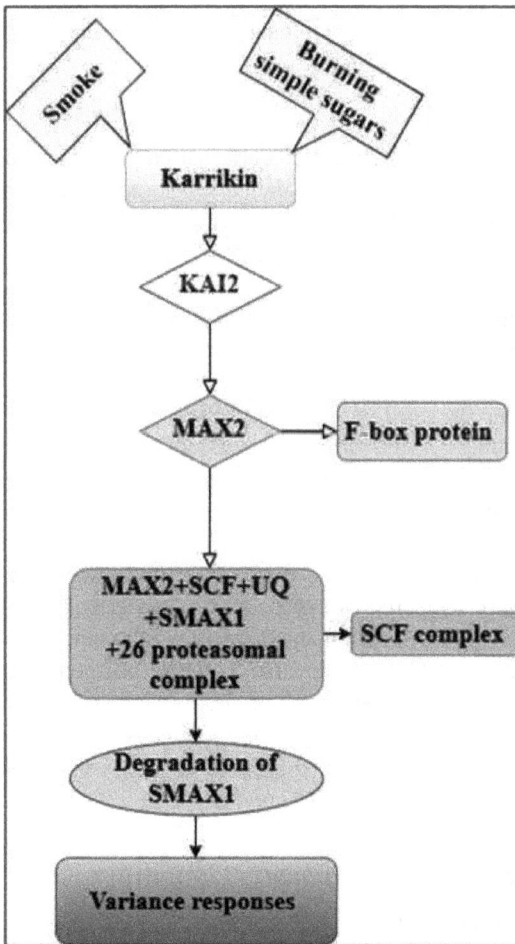

Fig. 1. Perception and signaling mechanism of Karrikins. KAI2 (KARRIKIN INSENSITIVE2), MAX2 (MORE AXILLARY GROWTH2), UB (ubiquitin bodies), SMAX1 (SUPPRESOR MORE AXILLARY GROWTH1).Functions of KARs in Plant Development

Expoliting of Storage Reserves during Seed Germination

Seeds respond to various interior and external environmental factors like internal storage reserves, temperature, air, oxygen, H_2O, light, shade, or dark to establish their suitability to germination process. These signals may act as promotors or inhibitors of germination. The germination signals related to smoke water have paid much attention due to their important roles on the germination of many seeds of plants. Wildfire smoke contains certain potent bioactive compounds including KAR1, trimethylbutenolide (TMB), and smoke-water (SW). In particular, KAR may modulate the early seed germination by influencing the activity of hydrolase and the amounts of lipids, protein, carbohydrate, and starch during seed germination. In this regard, Gupta et al. (2019) found that SW and KAR1 promote seed germination of lettuce through reducing ABA level and promoting the hydrolase activity, however, the KAR1-related compound TMB inhibits the germination of lettuce seeds by increasing ABA and inhibiting cytokinins under dark conditions.

Seed Germination

KARs have been found to stimulate seed germination. KARs are found in the smoke arising from fire and can then promote the germination of seeds on the ground after the fire (Soundappan et al. 2015). KARs was found to interact with light signals to stimulate seed germination (Waters et al. 2014, Meng et al. 2016). KARs can postpone soybean seed germination under shad stress, but not in the dark or under white light, by regulating the biosynthesis of ABA and GA (Stirnberg et al. 2002). The primary dormancy of *Arabidopsis* seeds can be overcome by KARs as it perceives KARs quickly and sensitively. KARs are an potenial enhancer of seed germination, but they do not cope the need or signalling of GAs. Thus, during pre-germination, the levels of GAs and ABA of *Arabidopsis* seeds do not alter when exposed to KARs (Nelson et al. 2009). KAR2 stimulates germination of *Arabidopsis* seeds under favorable conditions, but it can decrease germination when exposed to osmotic potential or at high temperature (Wang et al. 2018). However, KAR1 enhanced the germination establishment and quality of tef under elevated temperature and low osmotic pressure (Ghebrehiwot et al. 2008). Jain et al. (2006) reported KAR1 positively increased the germination and promotion of tomato seedling development under temperature stress. On the other hand, KAR2 postponed the germination of soybean via increasing ABA biogenesis and degeneration of gibberellin (Meng et al. 2016). Thus, further studies needed to examine the alteration of germinating seeds of different species and the impact of KARs on biochemical alterations during seed imbibition and post-imbibition must be conducted.

Photomorphogenesis and Hypocotyl Elongation

In seedlings, hypocotyl lengthening is a skotomorphogenic reaction that is needed for driving seedlings toward the soil so they may receive the sunlight. There are various environmental and physiological conditions that can affect this elongation (Wang et al. 2018). KARs play a key function in preventing hypocotyl extension in *Arabidopsis* under red light, however these symptoms disappeared in the MAX2 mutant (Nelson et al. 2011, Scaffidi et al. 2014). Further, Wang et al. (2020) showed that inhibitors like SMXL2 might be polyubiquitylated by karrikin to control the hypocotyl elongation of the seedlings. KAR2 is the most potential KAR in the improvement of germination and repression of hypocotyl lengthening of *Arabidopsis* (Nelson et al. 2009, 2010). Inhibition of hypocotyl elongation and cotyledon expansion are a light-dependent effect due to KARs exposure. Under continuous red light, KARs were observed to positively increase the contents of chlorophyll a and b in *Arabidopsis thaliana* (Nelson et al. 2009). KARs alone regulate germination and hypocotyl elongation of plants, but when KARs combined with SLs aid in the regulation of leaf morphology in *Arabidopsis* (Antala et al. 2020) Regarding transcription-related to ligt absorption anf photomorphogensis, the bZIP transcription factor family includes HY5 "(LONG HYPOCOTYL 5)", which binds to the G-box (conserved six base-pair sequence) in the upstream promoter region of gene that controls the response to light (Chattopadhyay et al. 1998). HY5 is essential for reprogramming the genes that can further modify the molecular process underlying photomorphogenesis. These transcription factors (TFs) enhanced photomorphogenesis by acting downstream of various photoreceptors. However, some workers have demonstrated that karrikin modulates seedling photomorphogenesis independently of the HY5-dependent signaling mechanism. The KAI2 mutant had a prolonged hypocotyl phenotype like that of the max2 mutant of *Arabidopsis*. In addition, the combined mutant HY5/KAI2 had a longer hypocotyl than that of the single mutants HY5 or KAI2 (Waters and Smith 2013).

Root System Development

KAR signaling mediates important root traits like the density of lateral roots, including rhizome growth route, root straightness, and root hair development. Some of the root architecture characters, which had been previously reported for SLs are indeed mediated by KARs or by the crosstalk between SLs and KARs (Yang et al. 2019). KARs are responsible for hair root development, the direction of root growth, root diameter, and root waving. Villaécija-Aguilar et al. (2019) deduced that both KARs and SLs influence the density of lateral roots. Although KARs non significantly affect the primary root length of *Arabidopsis* (Villaécija-Aguilar et al. 2019), KAR1 application positively effect the roots of rice, tomato, okra, bean, maize, and carrot root (Jain and Staden 2006, Jain er al. 2006, Kulkarni et al. 2006, 2007, Ghebrehiwot et al. 2008, Demir et al. 2012). Kulkarni et al. (2006) observed that KARs enhanced the root extent and the number of lateral roots in rice plants. Under cobalt stress, dilution of smoke 1:500, effectively increased the number of adventitious roots, lateral roots, and adventitious-root length in *Ipomoea marguerite* (Aslam et al. 2014). According to the findings of Zhong et al. (2020), under ambient circumstances, smoke produced by plants controlled the ornithine-synthesis pathways, which in turn lengthened the roots and promoted the hypocotyl growth. Otori et al. (2021) revealed that plant-derived smoke improved the root growth under flooding stress by triggering the ascorbate/glutathione cycle, increasing the synthesis of energy-related molecules, and scavenging the ROS. It has previously been discussed that karrikin played a significant role in root development in KAI2-dependent manner. In *Lotus japonicas*, the SMAX1 mutant had small primary roots and extended the root hairs in contrast to wild-type plants (Carbonnel et al. 2020). Thus, in the presence of plant-derived smoke and karrikin, KAI2 and "MAX2" formed a complex with SMAX1, degraded the repressors, and stimulated root elongation and root hair growth, root skewing and waving, adventitious root density, and lateral root density (Swarbreck et al. 2019, Swarbreck et al. 2020).

Leaf Morphogenesis

KARs are involved in the regulation of leaf shape. It has been reported that the leaf aspect ratio of the *smxl6/7/8* triple mutant of *Arabidopsis* was increased, and the *max2* leaf phenotypes were restored by mutation of *SMXL6/7/8* in comparison with the wild type (Guo et al. 2019). Although no effect on the petiole of KAR2 and SMAX1 was detected, reduced blade width and length were observed in "*smax1max2* and *smax1smxl6,7max2*" mutant plants compared with "*max2* or *smxl6,7max2*" mutant plants. The phenological analysis of *Arabidopsis* leaves suggested that "*smxl6/7/8, smax1*" and their respective receptor mutants may perform opposite functions in leaf morphology (Umehara et al. 2008, Guo et al. 2019).

Shoot Branching

To date, there is no studies conforming the modulation role of KARs to plant branching. Any production of KAR signaling-linked mutants may aid to explore whether KAR functions in regulating plant branching (Yang et al. 2019).

Mycorrhizal symbiosis

A recent study shows that GmMAX2-mediated SL and KAR signaling are involved in regulating soybean–rhizobia interaction and nodulation through interactions with auxin and JA hormones (Waters et al. 2012). "DWARF14-LIKE (D14L)" is homologous to KAI2 and co-participates in the regulation of MAX2-dependent KAR signaling pathways (Kagiyama et al. 2013, Gutjahr et al. 2015). The *hebiba* mutant lost its ability to respond to arbuscular mycorrhizal fungi in rice, and D14L may be responsible for this loss of symbiosis, suggesting that KAR receptor complexes are involved in the perception of arbuscular mycorrhizal fungi in rice (Jia et al. 2014). Also, D14L, which is an *Arabidopsis* KAI2 analog in rice, is necessary for the initiation of colonization events by arbuscular mycorrhizal fungi, but KAR2 was not effective in colonization enhancement of wild-type roots by arbuscular mycorrhizal fungi (Gutjahr et al. 2015). Whether other KARs play some role in plant-fungi symbiosis or what another signal is perceived by KAI2 is unclear uptill now.

Photosynthetic Yield

Photosynthesis is a vital physiological process that enhance the plant output, but negatively impacted by different abiotic stresses. Under harsh conditions, the integrity of the chloroplast membrane is degraded, impairing the physiological and biochemical systems. Normally, the chloroplast contains PSI and PSII activity, which collects light energy by the thylakoid membrane. However, under stress, the thylakoid membrane swells, which reduces the effectiveness of the chloroplast (Goussi et al. 2018). The PSII activity, stomatal conductance, and the photochemical quenching coefficient of *Isatis indigotica* seedling increased by smoke water besides stimulating the effectiveness of light-harvesting complex which recommends the importance of smoke-water in increasing the performance of photosynthesis (Zhou et al. 2013). The application of karrikin mitigates cadmium stress by increasing intercellular, CO_2 concentration, stomatal conductivity, transpirational rate, and photosynthetic yield (Shah et al. 2020). According to proteomic reports, immunoblot analysis revealed that application of plant derived smoke improve the photosynthesis mechanism via increasing the production of RuBisCO activase and RuBisCO large/small subunits, which were decreased during flooding stress (Komatsu et al. 2022). In addition, plant-derived smoke also improves the amount of photosynthetic pigments, chlorophyll a and b, that were decreased under flooding stress (Komatsu et al. 2022). Sharifi and Bidabadi (2020) reported that using of karrikin joined with calcium, enhanced the chlorophyll fluorescence values (Fv/Fm) in black cumin seeds under salinity stress. KARs enhanced the chlorophyll level in Arabidopsis, tef, and carrot (Ghebrehiwot et al. 2013, Nelson et al. 2010, Akeel et al. 2019). KARs not only influence the chlorophyll content, but also enhance net photosynthesis ratedue to increasing stomatal conductance and higher intercellular CO_2 level, which was found in KAR1-treated carrot plants (Akeel et al. 2019). However, foliar application of KAR1 on amaranth caused a reduction in chlorophylls content (Ngoroyemoto et al. 2019). The mechanism behind the KARs influence on chlorophyll concentration and photosynthesis is, for now, unknown, but the method of application may be decisive. The general functions of karrikin for plants were represented in Fig. 2.

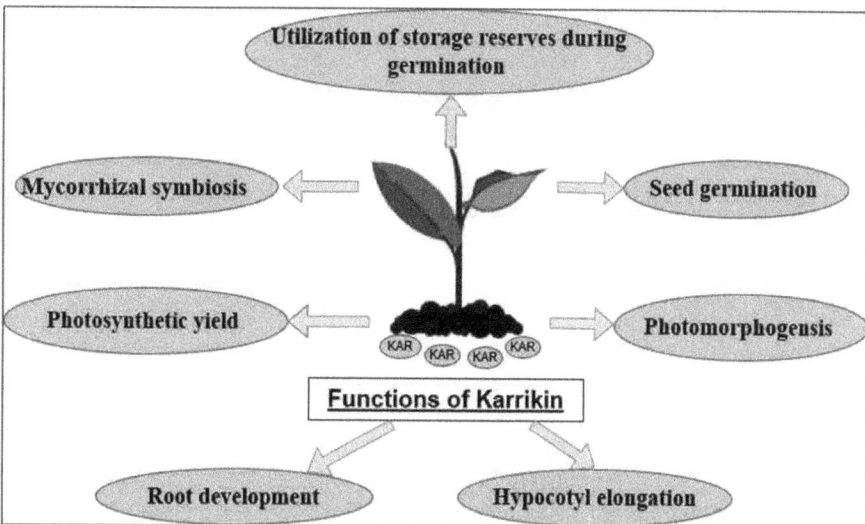

Fig. 2. The major functions of karrikin for plants.

The Interaction between Karrikin and Phytohormones

KAR Crosstalk with Auxin

Auxin is a key factor in regulating plant growth and development. Auxin is synthesized mainly at the tops of branches and in young leaves and is transported downward in the main stem by the polar auxin transport stream (PATS) (Wang et al. 2014). The "TRANSPORT INHIBITOR RESISTANT 1/AUXIN F-BOX (TIR1/AFB)" protein recognizes auxin, and together they form a co-receptor complex through E3 ubiquitin ligase with Aux/IAA protein, which acts as a repressor of the transcription of auxin-regulated genes. Aux/IAA interacts with the transcription factor "AUXIN RESPONSE FACTOR (ARF)" and is ubiquitinated and degraded in an ARF-dependent manner (Szemenyei et al. 2008). During auxin signal response, "TPL (TOPLESS) interacts with IAA12/BODENLOS (IAA12/BDL)" through an ethylene-responsive element binding factor amphiphilic repression (EAR) motif (Brewer et al. 2009). The interactions between auxin and KAR signaling are still unclear. KARs reduce the level of endogenous auxin by inhibiting the expression of IAA-responsive genes, which promotes seed germination (Yamaguchi et al. 2008). Meng et al. (2016) denoted that the activity of karrikin might suppress the transcription of the *IAA1* response gene, lowering intracellular levels of IAA, thus karrikin promotes seed germination in *Arabidopsis*. Plant-derived smoke enhanced the IAA and cytokinin accumulation in cells, promoting embryonic cell multiplication, and differentiation (Mathnoom and Al-Timmen 2020). In *Arabidopsis*, Karrikin played a significant role by regulating auxin accumulation and polar auxin transport, which might contributes to hypocotyl shade-response (Xu et al. 2022). These studies suggested a possible connection between the karrikin and IAA in mediating their physiological function in reducing abiotic stresses.

KAR crosstalk with gibberellin

Gibberellic acid (GA) is a crucial regulator of plant growth and development. Defects in GA synthesis and signaling lead to many defective phenotypes, such as inhibited germination, delayed root growth and flowering, male sterility, dwarfing, reduced rate of seed setting, and increased tiller buds (Lo et al. 2008, Silverstone et al. 1997). In general, GA participates in the regulation of seed germination by KAR signaling. KARs promote germination of dormant *Arabidopsis* seeds, and the stimulation of such germination is partly dependent on "DELLA" proteins. KAR1 can partially restore the inhibited seed germination of the GA-insensitive *sleepy1* mutant (Waters et al. 2014). Application of karrikin to *Arabidopsis* resulted in an increase in the expression of GA anabolic-genes including "GA-3-oxidase 1 (GA3ox1) and GA3ox2" (Nelson et al. 2010). KAR1 and GA3 collaborate to convert the ABA into phaseic acid, promoting the germination of Avena fatua seeds (Kępczynski 2018). The effect of karrikin regarding seed-germination enhancement was reversed by a number of GA3 inhibitors, including paclobutrazol, ancymidol, and flurprimidol. However, the administration of GA3 could change these adverse effects (Cembrowska-Lech and Kepczynski 2017, Kepczynski et al. 2013). Exogenous GA3, applied with or without smoke solution, promoted the caryopses germination, however ACC had a tendency to help germination of Avena fatua seeds only in the presence of smoke solution (Kepczynski et al. 2006).

KAR Crosstalk with Abscisic Acid

The indigenous phytohormone ABA, which belongs to the isoprenoids class (terpenoids), is a global plant growth inhibitor. ABA is derived from isopentenyl precursor (a 5-C compound). Dormant seeds face severe ABA stress. According to multiple studies, ABA inhibits the biosynthetic pathways that produce GA in an antagonistic manner, preventing plant development and growth (Cutler et al. 2010). KARs negatively regulate seed germination by inhibiting GA synthesis and promoting ABA synthesis in soybean (Stirnberg et al. 2002). CYP707As, involved in abscisic acid catabolism, are effectors newly discovered to be involved in KAR responses in *Arabidopsis* and parasitic plants (Matilla 2000). However, how CYP707As link ABA and KAR signaling remains unclear. According to the study conducted by Kamran et al. (2013), on the combined effects of ABA and smoke-saturated water, plant-derived smoke reduced the negative effects of ABA and boosted the lettuce seedling vigour as well as germination rates. In the *Arabidopsis* plant, the KAI2 protein (a paralog of D14) is crucial for improving the drought- tolerance mechanism. In a transcriptomic-analysis study, Soos et al. (2009) demonstrated that maize seedlings exposed to plant-derived smoke modulated the expression of genes involved in stress and ABA signalling, leading to an increase in seedling vigour. However, the "MAX2" protein participated in the polyubiquitylation of protein repressors during karrikin signaling by creating a complex with the KAI2 receptor and repressors such SMAX1 (Li and Tran 2015, Kleman and Matusova 2023). Furthermore, a different microarray-investigation, comparing the wild type and max2 mutant of the *Arabidopsis* plant under circumstances of dehydration, also demonstrated that the max2 mutant downregulated the expression of genes associated with dehydration-stress and ABA (Ha et al. 2014). These results suggested that karrikin might imitate the ABA in abiotic stress tolerance. Findings related to genetics and proteomics also suggested the role of the kai2 and karrikin mutants in the tolerance mechanism to extreme stress-conditions. Recent research on the *Sapium sebiferum* plant suggested that the action of karrikin might reduce the expression of a number of ABA-signalling genes, including "SNF1-RELATED PROTEIN KINASE2.3, SNF1-RELATED PROTEIN KINASE2.6, ABI3, and ABI5". This suggested that karrikin might improve the stress-tolerance mechanism in plants (Shah et al. 2020).

KAR Crosstalk with Ethylene

Ethylene is a plant growth regulator, released from ethephon or it can be constructed from exogenous precursor, ACC (1-aminocyclopropane-1-carboxylic acid). However, when combined with karrikin, exogenous administration of ethylene or its precursor ACC might be able to break the dormancy of *Avena fatua* caryopses. However, ethylene or its precursor (ACC) inhibitors, such AVG (aminoethoxyvinylglycine), AIB (-aminoisobutyric acid), and $CoCl_2$, were unable to prevent the impact of karrikin (Kepczynski and Van Staden 2012). It is interesting to note that karrikin alone was insufficient for the production of ethylene, as some endogenous quantities of ethylene might be necessary for dormancy release in response to karrikin (Kepczynski and Van Staden 2012). Ethylene affected the root growth and development by regulating the biosynthesis of auxin and its transport-dependent local auxin distribution (Ruzicka et al. 2007). Karrikin regulated the root development, but its entire signalling cascade has not yet been investigated (Swarbreck 2021). A transcriptome investigation of the karrikin in root revealed that karrikin in combination with ACS7 (ACC Synthase 7), the rate-limiting enzyme in the ethylene production process, that increase root lengthening. Application of karrikin to *Lotus japonicas*, the expression of *ACS7* rises in a KAI2-dependent pathway, promoting the root growth, karrikin signalling model "(KAI2-MAX2-SMAX1)", thereby, increased the growth of lateral roots and ethylene production (Carbonnel et al. 2020). In dormant seedlings of *Avena fatua*, KAR1 induced the ethylene biosynthesis via upregulating the expression of genes related with ethylene-receptor genes, ACS, ACO (ethylene-metabolism genes), and ethylene-receptor genes. Therefore, such an increased expression of ACS, ACO and ethylene-receptor genes might be an important factor that might play a significant role in dormancy release of KAR1 treated plants (Sami et al. 2021). Sami et al. (2021) observed that the KAR1-induced ethylene biosynthesis during the dormancy release was accredited to the regulation of chief ethylene-metabolism genes (ethylene-receptor genes, ACS, and ACO). Such an increased-expression of key ethylene biosynthetic genes (ACS and ACO) and ethylene-receptor genes was considered to be possibly responsible for the increased ethylene-increase in KAR1-treated plants.

KAR Crosstalk with Cytokinin

It has been evidenced that KAR is involved in regulating CK is that KAR1 and smoke water are able to raise the levels of cytokinins in *Eucomis autumnalis* and *Spinacia oleracea* L plants. For example, SW and KAR1 increase significantly the levels of ciszeatin, dihydrozeatin, and isopentenyladenine of CKs and then yield a greater number of leaves in spinach plants (Aremu et al. 2016). Meanwhile, SW and KAR1 treatments also accumulate higher concentrations of isoprenoid-type CKs in the aerial organs of *Eucomis autumnalis* (Du et al. 2018). These data indicate that the crosstalk between KAR and CK also plays an important role in plant growth and development.

KAR Crosstalk with Salicylic Acid

Recent study showed that the contents of free phenolic acids and salicylic acids were increased significantly in spinach plants treated with SW and KAR1, implying that there is a crosstalk possibility between KAR and salicylic acid (Aremu et al. 2016).

Gathering together, in the crosstalk between plant hormones, contribution in the regulation of a biological process does not mean that the phytohormones are jointly reliant on each other to manage this process.

KARs Assist Plant's Adaption to Stress Conditions

The sessility of plants in their natural zones has a well-recognized impact on plants survival and performance during their life cycle. Therefore, they are greatly threatened by challenging environmental stressors such as water scarcity, temperature fluctuations, soil salinity, light, heavy metals, and flooding. On the hand, plants have advanced captivating mechanisms empowering rapid detection of uncontrolled ecological changes associated with favorably superior molecular responses, leading to significant phenotypic and physiological cascades. Noteworthy, recent explorations have shown that KARs not only stimulate seedlings growth but also have been implemented in defending plants against threating stresses to a remarkable extent. Till the moment, promotive efforts have been conducted to address the specific contribution role of KARs in withstanding unstable circumstances.

Drought Stress

Drought which are caused by the lack of precipitation has been recently developed to be a matter of great concern threating sustainable agriculture practices and food security worldwide (Sallam et al. 2019, Moursi et al. 2021). Globally, a novel report has exposed that the total harvest loss generated by drought stress is around 7% in important crops (Lesk et al. 2016). Recently, literatures are rich with the worth of KARs in minimizing cumulative drought stress-impact on plants through varied applications. Exogenic supplement of KARs so far has been examined on multiple plant species under abiotic factors such as drought and have proved effective role in improving plant stress tolerance particularly during seed germination and early seedling establishment stages (Antala et al. 2022, Singh et al. 2023). Although, the drought stress response is a convoluted process in plants and the precise mechanism by which KARs standardize plant stress acclimatization is sometimes vague; a number of research studies depicted the positive role of KARs and its responsive mechanism toward drought stress tolerance even at low concentrations. The potential mechanisms of KARs induced-drought stress adaption can be summarized in: (1) regulate the imbalance between antioxidants and oxidants to inhibit the oxidative impairment, (2) repair offended cell membrane, (3) enhance photosynthetic robustness, and (4) instigate stress corresponded genes. For example, various plant species exhibited a remarkable enhancement in their growth under drought stress (PEG 6000) in the presence of KARs (10 μM) such as "*Trachyspermum copticum, Foeniculumvulgare*, and *Cuminum cyminum*". In this context, KARs mainly stimulated germination quality, seedling potency, and radicle elongation in varying degrees between species comparing to control plants (MousaviNik et al. 20 16), which could be ascribed to the control of lipid peroxidation and the rapid programming of antioxidant detoxifying system (Sunmonu et al. 2016). Modulation of antioxidant capacity and regulation of related genes expression due to KARs application has also been demonstrated in many works. KARs in nanomolar concentration (1nM) was found to improve seed germination of *Sapium sebiferum* and subsequent ontogenesis under drought stress via regulating redox homeostasis when upregulated ABA signaling and stress regulatory genes (Shah et al. 2020). Moreover, under extended drought stress, the addition of 100 nM KAR_1 to *Agrostis stolonifera* improved water deficient withstand by modulating the antioxidant machinery capacity and stress signaling genes implemented in chlorophyll catabolism (Tan et al. 2023). KAR_1 upregulated leaf water content, chlorophyll synthesis, proline pool, and lowered MDA level as well as triggered the antioxidant responsive genes. In a triplicate study of stress conducted by Moyo et al. (2022), seeds of *Brassica napus* were co-treated with SW and KAR_1 and were subjected to (1) diverse osmotic potential concentrations in the range of 0 up to 0.73 MPa initiated by PEG 6000, (2) temperatures alterations (10 to 40°C), and (3) different light conditions. The experimental results indicated that treated seeds absorbed high quantity of water in a short time (1 h) after treatment thus exhibited better seed germination and quality in terms of seedlings biomass, length, and root/shoot height. Moreover, increased seed germination by 50% under heat stress and increased germination parameters as well under constant light conditions. On the other hand, genetic analyses exploiting suitable model systems such as mutants of *Arabidopsis thaliana* (*Arabidopsis*) lacking KARs receptors have expanded our knowledge on KARs signaling and role during oxidative stress. For instance, '*kai2* (KARRIKIN INSENSITIVE2)' and '*max2* (MORE AXILLARY GROWTH2)' mutant lines which are characterized by the same phenotype of elongated hypocotyl and cotyledonary petioles, are highly considered in this concept. Several KARs receptor genes of Strigolactones (SLs) and KARs such as D14 (DWARF14) and *kai2* perform parallel and interplay with the next receptor *max2* to assist plants cope with abiotic stressors (Guo et al. 2013, Waters et al. 2013, 2014, Smith and Li 2014, Yao et al. 2016). It has been proved that the kARs-*kai2* signaling network induced stress resistance by suppressing sprouting in *Arabidopsis* under tense environments (Wang et al. 2018). The behavior of *Arabidopsis* mutants (*kai2*) was noted under water-deficient circumstances. Mutant plants displayed hypersensitivity to water shortage by displaying changes in ABA sensitivity, failure in stomatal function, disturbance in anthocyanin synthesis, and elevated membrane damage (Li et al. 2017). Transcriptomic data indicated that *kai2* and ABA signaling interaction stimulated the cuticle development, stomatal function, cell membrane steady, and anthocyanin biogenesis, which all together contributed to control sensitivity to drought stress. In transcriptomic studies, DEGs analysis of the drought resistant *Phormium tenax* showed that "response to karrikin" is one of the most expressed nodes followed by drought stress (Bai, et al. 2017). The role of D14 and *kai2* on drought stress signaling was illustrated. Transcription analysis obtained by *kai2* and D14 mutants pointed out that both receptors have important regulation role during stress, revealing interplay between hormonal signal and biochemical metabolism. *kai2* stimulated photosynthesis efficiency and metabolism of trehalose and glucosinolates, while D14 showed a different signaling pathway by affecting cytokinin and brassinosteroid upregulation (Li et al. 2020). In another study, the intricate function of *KAR UPREGULATED F-BOX 1* (*kuf1*) receptor in drought regulation was investigated by Tian et al. (2022). The authors argued that the receptor negatively implemented in the regulation of drought-tolerance mechanism when *kuf1* mutant individuals showed higher resistance to drought than wild-type ones. This corresponded to the regulation of stomatal closure, lipid metabolism, and ABA responses as shown by responsive genes in transcriptomic analysis. Furthermore, another study indicated that SUPPRESSOR of MAX2 1 (SMAX1) and SMAX1-LIKE2 (SMXL2) negatively involved in dehydration combat in *Arabidopsis smax1 smxl2 (s1,2)* double-mutants, which showed improved performance under drought by regulating the mechanism of cuticle development (Feng et al. 2022).

Salinity Stress

Salinity is one of the most prevailing stressors faced by higher plants. It is badly intimating oxidative damage in plants and affecting physiological dynamics and metabolism (Dawood et al. 2022a, Sheteiwy et al. 2022, Abeed et al. 2023). Currently, it affects around third of cultivated lands around the world and causes unavoided loss to strategic crops (El-Mahdy et al. 2022). From germination till maturation, salinity can reduce seed sprouting rate, weaken chlorophyll synthesis, accelerate premature senescence, and may cause fatality after protracted vulnerability to excessive salt levels (Zhu 2002, Hewedy et al. 2023). KARs were stated in several scientific papers to play a sophisticated role upon exposure to salt stress, however the role of KARs in the regulation of salt tolerance is complicated. For instance, priming seeds of *Ceratotheca triloba* with SW and KAR_1 showed a positive influence on germination sturdiness and remodeled the detrimental impacts of salinity, drought, and low temperature stress (Masondo et al. 2018). In the same way, black cumin seeds treated with KARs showed better NaCl tolerance than untreated plants via decreasing oxidative stress and improving net yield (Sharifi and Shirani Bidabadi 2020). It was showed that KARs along with calcium organized H_2O_2 and MDA over-accumulation, enriched chlorophyll biogenesis, increased unsaturated fatty acids pool and, stimulated the activity of antioxidant enzymes (Sharifi and Shirani Bidabadi 2020). In wheat, KARs modulated salinity-induced negative influence by regulating the hormonal signal and responsive genes between gibberellins and abscisic acid biosynthesis, hence, improved seedlings germination and biomass. Co-exposure to salinity and KARs resulted in enhanced GSH/GSSG balance, motivated the activities of enzymatic antioxidants, and maintained the redox and K^+/Na^+ homeostasis (Shah et al. 2021). kai2 mutants exhibited delay in germination and early growth submission in comparison to with wild type, proposing that kai2 has a vital impact on supporting seed sprouting under salinity (Mostafa et al. 2022). Similar outcomes were stated by (Kochanski et al. 2023), downregulation of kai2 receptor in *Arabidopsis* showed morphological aberrations and biomass reductions due to oxidative damage effect and toxicity by increased sodium ion accumulation. The contribution role of SMXLs gene family in salinity and osmotic tolerance was ascertained. This evident from the study supported by (Fu et al. 2022), who found upregulation in 'GmSMXL2.1, GmSMXL6.1, GmSMXL7.1, and GmSMXL8.1' under NaCl stress and downregulated under PEG. A further analysis revealed high expression of GmKAI2, GmD14, and GmMAX2, denoting that the 'GmMAX2-moderated KAR/SL' signaling pathways during salt stress (Fu et al. 2022).

Similarly, plant-derived smoke solution, among important sources of KARs, showed a positive response in restraining oxidative stress caused by salinity. Priming rice seeds with SW improved germination ratio, chlorophyll content, and stabilized ion homeostasis (Jamil et al. 2014). Matching results was reported in maize (Waheed et al. 2016). Plant-derived smoke (PDS) has been noted to recover germination and biomass capacity in wheat via improving the induction of superoxide dismutase and peroxidase enzymes along with regulating germination responsive genes such as *TaSAM, TaPHY, TaBGU* (Hayat et al. 2021). Likewise in wheat, low concentrations (10 ppm and 25 ppm) of PDS solution enhanced germination and overall growth upon exposure to salinity. It mainly modulated water relations and increased production of some osmoprotectants such as proline, hence, added stability under stress (Shabir et al. 2022). SW (0.1%) boosted the expression levels of *CAT*, and *Cu/Zn-SOD, TaDREB1, TaWRKY2, and TaWRKY19 genes* in salt-affected plants, in addition to elevating proline and phenolics production level (Çatav et al. 2021).

Temperature Stress

Fluctuations in temperatures have harmful effects on plant survival and ontogensis and triggers great consequences for the agricultural sector (Sallam et al. 2018, Dawood et al. 2020). To adapt with cold or heat stress, plants developed a set of signaling cascades to intermediate ROS homeostasis. Cold resilient plants have settled a protective system called cold acclimation (Thomashow 1999, Shi and Yang 2014, El-Mahdy et al. 2018). Early reports demonstrated that seed primed with KARs or its derivates compounds considerably developed better vigorous seedlings than the untreated seeds by exposure to high temperature stress (Jain et al. 2007, Ghebrehiwot et al. 2008). In this regard, *kai2* mutants of Arabidopsis at morphological and physiological levels showed sever damage effects and hypersensitivity to heat stress compared to wild type plants (Abdelrahman et al. 2022). Furthermore, gene ontology and genome enrichment analysis revealed a repression in important pathways such as heat shock protein binding associated with downregulation in their gene expressions (Abdelrahman et al. 2022). On the other hand, Zhao et al. (2012) transcriptomic analysis of differentially expressed genes on *Chorispora bungeana* plantlets under cold temperatures, indicated that the node 'response to KAR' was a common node in gene set enrichment results under chilling stress between *Chorispora bungeana* and *Arabidopsis* (Zhao et al. 2012). Moreover, chilling tolerant species have altered the expression level of ABA-associated gene expression, which regulated stomata function and stabilized redox homeostasis (Hong et al. 2012, Shi et al. 2012, Jurczyk et al. 2019). Upon freezing exposure (−20°C), kai2 mutants had severe damage impact on controls (Shah et al. 2021). Overexpression of Sapium sebiferum KAI2 (SsKAI2) revealed better resistance to cold stress by accumulating more proline, soluble sugars, and proteins as one of the effective strategies to cope with cold stress (El-mahdy et al. 2018), which led to less membrane damage and ROS mediation. Results also exhibited significant expression of chilling related genes such as cold shock proteins and C-REPEAT BINDING FACTORS and boosted ABA sensitivity and linked genes. In tomato seedlings, the interplay among KAR, SLs, and ABA toward chilling response was investigated via KAI2-, MAX1-, SnRK2.5-silenced, or cosilenced plant materials. Data revealed that SW and KAR had a role in enhancing tomato growing and net biomass under critical low temperatures by mediating nutritional accumulation, photosynthesis outputs, ROS controlling, and CBF transcriptional activation (Liu et al. 2023).

Light Stress

Under shade conditions or inconsistent light accessibility, the light captured by phytochromes receptors often showed inhibitory signaling pathway during photosynthesis, resulting in low-light stress and intensify the generation of FR-enriched environment (El-Keblawy and Gairola 2017). In crowded plant comminutes, light intensity modulates seed developmental progress and agronomic attributes. Simultaneously, it causes disturbance in phenological, physiological, and biochemical pathways, triggering sensitivity to various environmental stresses (Wit et al. 2013). Plants growing under low-light stress showed diminution in photosynthesis rate and growth abnormality syndromes. However, KARs have been displayed in earlier studies to mitigate the low-light disorders in plants by enhancing light signaling pathway and modulating photomorphogenesis (Nelson et al. 2010). KAR2 a members in KARs family was found to obstruct elongation of hypocotyls (Waters et al. 2012). kai2 mutant was originally known as "*HYPOSENSITIVE TO LIGHT (HTL)*" (Sun and Ni 2011). kai2 and max2 *Arabidopsis* mutants stimulated hypocotyl elongation, signifying that KARs and KLs are expressively implemented in the shade response (Xu et al. 2022). At the same time, a hypocotyl elongation assay in the double mutant of *hy5kai2* exhibited prolonged growth contrary to *hy5* and *kai2* mutants, which is similar to that of *hy5max2* (Waters and Smith 2013). This highlights that kai2 signaling pathway is elaborated in the regulation of seedling morphogenesis autonomously of *HY5*. Moreover, application of KARs have been reported to interrelate with auxin to eradicate the low-light stress. Low-light stress modifies hypocotyl extension by stimulating auxin expression and transport (Cole et al. 2011). Auxin gathering in the hypocotyl upsurges considerably under low-light conditions (Procko et al. 2014). This is further confirmed by *Arabidopsis* IAA deficient mutants, which showed endurable stress in response to shade conditions (Petrasek and Friml 2009). Moreover, several analyses have exposed that MAX2 can hinder IAA transport by downregulating the transcript level of the PIN gene that controls IAA transport (Shen et al. 2012). In contrary, in a recent investigation by Hamon-Josse (2022), *KAI2* was shown to be essential for the alterations in PIN protein abundance, which was not prerequisite for the light-mediated changes in PIN gene expression. KARs also affect hormones network, regulating seed dormancy and constant sprouting. Meng et al. (2016) demonstrated that KARs retarded the germination of soybean seeds by disturbing the dynamic balance between abscisic acid and gibberellin biosynthesis, which further ascertained by gene expression profiles. A preliminary study also showed KARs stimulated the expression of gibberellin metabolic genes GA3ox1 and GA3ox2 that involved in seed development (Nelson et al. 2010).

Heavy Metal Stress

Overaccumulation of heavy metals in plant tissues can inhibit seedling establishment and subsequent growth and disturb a wide range of metabolic pathways that eventually may cause plant fatality (Elazab et al. 2021, Mourad et al. 2021, Dawood et al. 2023, 2022b,c). Recently, from 2020, novel scientific evidences proved the cardinal contribution of KAR_1 in moderating the plant response under array of heavy metal stressors like cadmium, boron, and lead. In Brussels sprouts, co-treated seeds with 10^{-5} M, 10^{-7} M, and 10^{-10} M solution of KAR_1 and Cd (5 mg/L), showed promoted germination frequency and biomass production compared to Cd treatment only. Moreover, KAR_1 in this study relieved Cd toxicity by declining oxidative injury and advancing antioxidant system (Shah et al. 2020). Under *in vitro* conditions, Sardar et al. (2021) indicated that Cd toxicity was suppressed in *Coriandrum sativum L.* by KAR1 addition (10^{-6} M) and showed enhancement in metal tolerance index. In a complex study, KAR_1 was tested to modulate triplicate stressors, Cd joined with BDE-28 (2,4,4'-Tribromodiphenyle, a persistent organic pollutant) and heat stress (Ahmad et al. 2021). KAR_1 positively reduced contaminants content, improved growth parameters, and activated antioxidant guard system. KAR_1 also exhibited significant effect on modulating boron toxicity in wheat seedlings via stimulating the activities of glutathione reductase, and antioxidant enzymes which contributed to efficient removal of boron accumulation (Küçükakyüz and Çatav et al. 2021). Apart from the rare publications in this regard, Ibrahim et al. (2022) reported that the addition of smoke solution, isolated from a wild lemongrass, had a progressive role in enhancing arsenic and mercury stress in wheat.

Application of KARs in Plant Tissue Culture

The implementation of biotechnological approaches, such as plant tissue culture, in the market need will benefit tackle some complicated bottlenecks to advance output, improve disease sensitivity, analysis of functional genomics, and genetic improvement. In the recent era, with the increasing requirement of elite genotypes with superior characteristics, *in vitro* techniques became the adopted method for the desired clonal multiplication of plant material in a short period. Thus, developing protocols to improve micropropagation of plants needs to be optimized. A key progress in tissue culture applications was direct morphogenesis by modulating media formation exploiting diverse stimulants in addition to different levels and combinations of components. SW and KARs are among powerful auxiliary additives were reported to improve different growth aspects under *in vitro* conditions. This could greatly help to achieve a steady supply of economically profitable plant species in the near future.

Seed Germination

In plant tissue culture systems, various factors integrate to affect seed germination and morphogenesis in plants. Seed germination is specially considered as the first and most significant physiological period in the plant whole life. Therefore, optimizing germination consistency and robustness is an energetic influencer in the commercial production of high-quality seeds. In additional to the potential role of SW and KARs in releasing dormancy and stimulating germination, their role was also studied as prospective additives can positively impact *in vitro* germination and subsequent growth. low concentrations of KARs have been noted to be highly effectual in stimulation of *in vitro* germination across diverse species, especially valuable ornamentals such as orchids. Examples of *in vitro* response of seeds to SW and KARs are presented in Table 1.

Table 1. Summary of the impact of smoke water (SW) and KARs on *in vitro* plant germination.

Concentration	Impact	species	Reference
SW 10%	Induced high germination rate and differentiation of protocorms	*Vanda parviflora*	Malabadi et al. 2008
SW (1:500)	Elevated germination frequency	*Aloe Ferox*	Bairu et al. 2009
SW (10^{-5})	Increased germination vigor and number of regenerated shoots	**Salvia stenophylla**	Musarurwa et al. 2010)
SW 10%	Improved germination capacity and full recovery.	*Xenikophyton smeeanum*	Malabadi et al. 2012
SW 10%	Enhanced germination and regeneration and acclimatation	**Pholidota pallida**	Muglund et al. 2012
SW 10%	Improved germination and whole recovery of plantlets	**Oberonia ensiformis**	Malabadi et al. 2012
SW (1, 5, and 10%).	Inhibited growth and development of protocorms	*Arundinella hirta*, *Microstegium japonicum*, *Miscanthus sinensis*, *Paspalum thunbergii Kunth ex Steud.*, and *Themeda triandra*	Teixeira Da Silva et al. 2013
SW (1: 250) KAR1 (10^{-7}, 10^{-8}, 10^{-9}).	SW Augmented germination rate index and development rate index, while KAR_1 has no effect	Ansellia africana	Papenfus et al. 2016
SW (0–1 mg/L)	Increased germination rate in short time	*Pterostylis despectans*	Ritmejerytė et al. 2018
SW 10%	Caused the optimal regeneration rate	*Aloe arborescens*	Espinosa-Leal and Garcia-Lara 2020
KAR_2 (5 or 10 M).	Improved germination ratio (5 or 10 μM)	*Balsamorhiza sagittata* and *B. deltoidea*	Monthony 2020
SW (1:1000, 1:1250)	Empowered germination capacity and robustness	*Sceletium tortuosum*	Sreekissoon et al. 2021

Somatic embryogenesis

In breeding programs, somatic embryogenesis is one of the most modern tools to produce large numbers of selected germplasm. Somatic embryogenesis occupies prominence importance in transformation studies. Moreover, it provides a systematic model to elucidate the precise developmental changes from early to mature embryonic phases and study the mechanisms controlling the series of events that appears during zygotic embryo initiation (Heringer et al. 2018). The aptitude of cells to morphogenetically differentiate into embryos relies on its totipotency, where plant cells are capable of generating new individuals from single or gathered somatic cell. To optimize cells regeneration and differentiation during embryogenesis procedure, which is a main goal in several studies, exogenic substances such as SW and KARs have been reported to increase regeneration proficiency. For instance, in an early study, Senaratna et al. (1999). Found that pretreating hypocotyl of *Pelargonium hortorum* with SW-increased embryo differentiation. Moreover, SW was sufficient to improve regeneration ability in the critical stage of cotyledonary stage, which lead to increase the number of germinated embryos in three cultivars of *Pinus wallichiana* (Malabadi and Nataraja 2007). Ghazanfari et al. (2012) noted that SW was proved to boost plantlet regeneration capacity as well as the height of roots and shoots of newly formed plantlets of *Brassica napus*. KAR1 was also found to positively impact embryogenesis rate in *Baloskion tetraphyllum* (Ma et al. 2006). The secondary embryogenesis was also exhibited to be improved when SW applied prior or during germination and enhanced the formation of the secondary embryogenesis (Abdollahi et al. 2012).

KARs Promote the Production of Secondary Metabolites

Plants often generate an expansive variety of specialized and valuable secondary metabolites that differ in their pathway construction and development. These metabolites are not fundamental for the plant growth, but they impact some metabolic routes as signalling molecules during vital physiological and biochemical reactions. Moreover, these naturally compounds are excessively produced in higher plants as a principal strategy implemented in contesting environmental stress factors. Plants which acquire therapeutic metabolites with beneficial pharmacological effects are highly regarded as economically important plants used in medications, flavors, fragrances, nutraceutical, insecticides, dyes, etc. On the other hand, secondary metabolites have complicated structures that sometimes challenging to be reproduced synthetically. Thus, phytochemical dynamic molecules are more frequently isolated from their biological sources.

In response to the immense requirement for accelerated the production rate of secondary metabolites in plants by the pharmaceutical and other industrial applications, numerous approaches have been manipulated to progress their synthesis and abundance in plant cells. Among these approaches, tissue culture platforms combined with distinctive additives in culture media can enhance or modulate the large-scale generation of secondary metabolites in several plant organs. SW and KARs are attractive plant-derived bioactive constituents, and have been well documented to act as elicitors and trigger important metabolic pathways to stimulate the *in vitro* productivity of some central secondary metabolites such as phenols and flavonoids. Importantly, SW and KARs are recommended nontoxic molecules, safe to be utilized for enhancing bioactive compounds in in vitro culture systems. Those published reports discussing the role of SW and KAR1in scaling up important secondary metabolites by utilizing *in vitro* plant tissue culture procedures are displayed and summarized in Table 2.

Table 2. Summary of the impact of smoke water and KARs on bioactive compounds under *in vitro* conditions.

Plant	Impact on bioactive compounds	The most effective dose	Reference
Isatis indigotica Fort.	SW augmented the content of indigo in plant shoots	SW (1:500)	Zhou et al. 2011
Musa spp. AAA cultivar 'Williams')	SW and KAR_1 improved the accumulation of total phenolic in leaves and roots	SW (1:500 dilutions) $KAR_1 (1.0 \times 10^{-19})$	Aremu et al. 2012
Aloe arborescens Mill	SW increased phenolics, flavonoids, and iridoid content as well as improved antioxidant activity	SW (1: 500)	Amoo et al. 2013
Tulbaghia ludwigiana and *T. violacea*	SW boosted the accumulation of phenolics, flavonoids and tannins only in *T. ludwigiana* plantlets	SW (1:500)	Aremu et al. 2014
Salvia miltiorrhiza	SW and KAR_1 stimulated the production of salvianolic and rosmarinic acids in hairy roots through modulating phenolic biogenesis pathway	SW (1:500) and $KAR_1 (10^{-9} M)$	Zhoue et al. 2018
Eucomis autumnalis subspecies autumnalis	SW and KAR_1 increased the accumulation of 16 types of phytochemicals including coumaric acid, eucomic acid, derivatives of hydroxybenzoic and hydroxycinnamic acids, in addition to flavonoids	SW (1:500 and 1:1500) $KAR_1 (1.0 \times 10^{-19})$	Aremu et al. 2019
Salvia miltiorrhiza	KAR_1 activated the signaling pathway of nitric oxide and jasmonic acid to promote the synthesis of endogenous tanshinone I (T-I)	$KAR_1 (10^{-9} M)$	Zhoue et al. 2019
Salvia miltiorrhiza	KAR_1 promoted the production of some molecules including (Ca^{2+})–calmodulin (CaM), brassinolideand Jasmonic acid, and salvianolic acid b	$KAR_1 (10^{-9} M)$	Sun et al. 2020
Salvia miltiorrhiza	KAR_1 modulated the production level of jasmonic acid, hydrogen peroxide, and salvianolic acid b.	$KAR_1 (10^{-9} M)$	Sun et al. 2020
Salvia miltiorrhiza	The proteins corresponded to the biosynthesis of salvianolic acids and lignin were regulated by SW and karrikinolide in the hairy roots.	SW (1:1000) $KAR_1 (10^{-9} M)$	Sun et al. 2021
Salvia miltiorrhiza	KAR_1 upregulated the expression of genes corresponded to salvianolic acid biosynthesis	—	Duan et al. 2022

Conclusions

KARs are relatively simple active molecules affecting several physiological and morphological features of different species. The KAR-responsive species enabled to study signaling cascade of KAR perception. Analysis of mutants shows that receptor KAI2 in complex with F-box protein MAX2 can degrade repressors SMAX1 and SMXL2, which release the number of genes from repression that stimulates germination and cause morphological responses of aboveground and belowground organs. KARs can also stimulate the germination and seedling growth of several crops under optimal and suboptimal conditions. Therefore, more studies under field conditions, are needed to discover possible benefits of KARs use in agriculture nowadays.

References

Abdelrahman, M., Mostofa, M.G., Tran, C.D., El-Sayed, M., Li, W., Sulieman, S. et al. 2022. The karrikin receptor karrikin insensitive2 positively regulates heat stress tolerance in Arabidopsis thaliana. Plant and Cell Physiol., 63: 1914–1926.

Abdollahi, M.R., Ghazanfari, P., Corral-Martínez, P., Moieni, A. and Seguí-Simarro, J.M. 2012. Enhancing secondary embryogenesis in Brassica napus by selecting hypocotyl-derived embryos and using plant-derived smoke extract in culture medium. Plant Cell Tissue Organ. Cult., 110: 307–315.

Abeed, A.H., AL-Huqail, A.A., Albalawi, S., Alghamdi, S.A., Ali, B., Alghanem, S.M. et al. 2023. Calcium nanoparticles mitigate severe salt stress in Solanum lycopersicon by instigating the antioxidant defense system and renovating the protein profile. S. Afr. Stat. J., 161: 36–52.

Ahmad, A., Shahzadi, I., Mubeen, S., Yasin, N.A., Akram, W., Khan, W.U. et al. 2021. Karrikinolide alleviates BDE-28, heat and Cd stressors in Brassica alboglabra by correlating and modulating biochemical attributes, antioxidative machinery and osmoregulators. Ecotoxicology and Environmental Safety, 213: 112047.

Akeel, A., Khan, M.M.A., Jaleel, H. and Uddin, M. 2019. Smoke-saturated Water and Karrikinolide Modulate Germination, Growth, Photosynthesis and Nutritional Values of Carrot (Daucus carota L.). Journal of Plant Growth Regulation., 38: 1387–1401.

Amoo, S.O., Aremu, A.O. and Van Staden, J. 2013. Shoot proliferation and rooting treatments influence secondary metabolite production and antioxidant activity in tissue culture-derived Aloe arborescens grown ex vitro. Plant Growth Regulation, 70: 115–122.

Antala, M. 2022. Physiological roles of karrikins in plants under abiotic stress conditions. pp. 193–204. In Emerging Plant Growth Regulators in Agriculture. Academic Press.

Antala, M., Sytar, O., Rastogi, A. and Brestic, M. 2019. Potential of karrikins as novel plant growth regulators in agriculture. Plants., 9(1): 43.

Aremu A.O., Plackova L., Novak, O., Stirk, W.A., Dolezal, K. and Van Staden, J. 2016. Cytokinin profiles in ex vitro acclimatized Eucomis autumnalis plants pre-treated with smoke-derived karrikinolide. Plant Cell Reports., 35: 227–238.

Aremu, A.O., Masondo, N.A. and Van Staden, J. 2014. Smoke–water stimulates secondary metabolites during in vitro seedling development in Tulbaghiaspecies. S. Afr. J. Bot., 91: 49–52.

Aremu, A.O., Bairu, M.W., Finnie, J.F. and Van Staden, J. 2012. Stimulatory role of smoke–water and karrikinolide on the photosynthetic pigment and phenolic contents of micropropagated 'Williams' bananas. Plant Growth Regul., 67: 271–279.

Aremu, A.O., Masondo, N.A., Gruz, J., Doležal, K. and Van Staden, J. 2019. Potential of smoke-water and one of its active compounds (karrikinolide, KAR1) on the phytochemical and antioxidant activity of Eucomis autumnalis. Antioxidants, 8: 611.

Aslam, M.M., Akhter, A., Jamil, M., Khatoon, A., Malook, I. and Rehman, S.U. 2014. Effect of Plant-derived smoke solution on root of Ipomoea marguerite cuttings under Cobalt Stress. Journal of Bio-Molecular Sciences, 2: 6–11.

Bai, Z.Y., Wang, T., Wu, Y.H., Wang, K., Liang, Q.Y., Pan, Y.Z. et al. 2017. Whole-transcriptome sequence analysis of differentially expressed genes in Phormium tenax under drought stress. Scientific Reports, 7: 41700.

Bairu, M.W., Kulkarni, M.G., Street, R.A., Mulaudzi, R.B. and Van Staden, J. 2009. Studies on seed germination, seedling growth, and *in vitro* shoot induction of Aloe ferox Mill., a commercially important species. HortScience, 44: 751–756.

Baldwin, I.T., Staszak-Kozinski, L. and Davidson, R. 1994. Up in smoke: I. Smoke-derived germination cues for postfire annual, Nicotiana attenuata torr. Ex. Watson. Journal of Chemical Ecology, 20: 2345–2371.

Brewer, P.B., Dun, E.A., Ferguson, B.J., Rameau, C. and Beveridge, C.A. 2009. Strigolactone Acts Downstream of Auxin to Regulate Bud Outgrowth in Pea and Arabidopsis. Plant Physiology, 150: 482–493.

Brown, N.A.C. 1993. Promotion of germination of fynbos seeds by plant-derived smoke. New Phytologist., 123: 575–583.

Carbonnel, S., Das, D., Varshney, K., Kolodziej, M.C., Villa´ecija-Aguilar, J.A. and Gutjahr, C. 2020. The karrikin signaling regulator SMAX1 controls Lotus japonicus root and root hair development by suppressing ethylene biosynthesis. Proceedings of the National Academy of Sciences of the United States of America, 117: 21757–21765.

Çatav, Ş.S., Surgun-Acar, Y. and Zemheri-Navruz, F. 2021. Physiological, biochemical, and molecular responses of wheat seedlings to salinity and plant-derived smoke. S. Afr. Stat. J. 139: 148–157.

Cembrowska-Lech, D. and KepczyDski, J. 2017. Plant-derived smoke induced activity of amylases, DNA replication and 2-tubulin accumulation before radicle protrusion of dormant *Avena fatua* L. caryopses. Acta Physiologia Plantarum., 39: 1–12.

Chattopadhyay, S., Ang, L.H., Puente, P., Deng, X.W. and Wei, N. 1998. Arabidopsis bZIP protein HY5 directly interacts with light-responsive promoters in mediating light control of gene expression. Plant Cell., 10: 673–683.

Cole, B., Kay, S.A. and Chory, J. 2011. Automated analysis of hypocotyls growth dynamics during shade avoidance in Arabidopsis. Plant J., 65: 991–1000.

Dawood, M.F., Moursi, Y.S., Amro, A., Baenziger, P.S. and Sallam, A. 2020. Investigation of heat-induced changes in the grain yield and grains metabolites, with molecular insights on the candidate genes in barley. Agronomy, 10: 1730.

Dawood, M.F., Sofy, M.R., Mohamed, H.I., Sofy, A.R. and Abdel-Kader, H.A. 2023. N-or/and P-deprived Coccomyxa chodatii SAG 216–2 extracts instigated mercury tolerance of germinated wheat seedlings. Plant and Soil, 483(1–2): 225–253.

Dawood, M.F., Sofy, M.R., Mohamed, H.I., Sofy, A.R. and Abdel-kader, H.A. 2022. Hydrogen sulfide modulates salinity stress in common bean plants by maintaining osmolytes and regulating nitric oxide levels and antioxidant enzyme expression. J. Soil Sci. Plant Nutr., 22: 3708–3726.

Dawood, M.F., Tahjib-Ul-Arif, M., Sohag, A.A.M. and Abdel Latef, A.A.H. 2022. Fluoride mitigates aluminum-toxicity in barley: morpho-physiological responses and biochemical mechanisms. BMC Plant Biology, 22(1): 1–17.

Daws, M.I., Davies, J., Pritchard, H.W., Brown, N.A. and Van Staden, J. 2007. Butenolide from plant-derived smoke enhances germination and seedling growth of arable weed species. Plant Growth Regulation, 51: 73–82.

De Cuyper, C., Struk, S., Braem, L., Gevaert, K., De Jaeger, G. and Goormachtig, S. 2017. Strigolactones, karrikins and beyond. Plant, Cell, and Environment., 40: 1691–1703.

De Lange, J.H. and Boucher, C. 1990. Autecological studies on *Audouinia capitata* (Bruniaceae). I. Plant-derived smoke as a seed germination cue. South African Journal of Botany, 156: 700–703.

Demir, I., Ozuaydin, I., Yasar, F. and van Staden, J. 2012. Effect of smoke-derived butenolide priming treatment on pepper and salvia seeds in relation to transplant quality and catalase activity. South African Journal of Botany, 78: 83–87.

Dixon, K.W., Merrit, D.J. and Flematti, G.R. 2009. Karrikinolide—A Phytoreactive Compound Derived from Smoke with Applications in Horticulture, Ecological Restoration and Agriculture. Acta Horticulture, 813: 155–170.

Douglas, R.B., Rothfels, C.L., Stevenson, D.W.D., Graham, W.S., Wong, G.K.-S., Nelson, D.C. et al. 2017. Evolution of strigolactone receptors by gradual neo-functionalization of KAI2 paralogues. BMC Plant Biology, 15: 52.

Downes, K.S., Light, M.E., Pošta, M., Kohout, L. and van Staden, J. 2013. Comparison of germination responses of *Anigozanthos flavidus* (Haemodoraceae), *Gyrostemon racemiger* and *Gyrostemon ramulosus* (Gyrostemonaceae) to smoke-water and the smoke-derived compounds karrikinolide (KAR1) and glyceronitrile. Annals of Botany, 111: 489–97.

Du, H., Huang, F., Wu, N., Li, X., Hu, H. and Xiong, L. 2018. Integrative regulation of drought escape through ABA-dependent and-independent pathways in rice. Molecular Plant., 11(4): 584–597.

Duan, W., Sun, H., Ding, W., Zhou, J., Fang, L. and Guo, L. 2022. Transcriptomic analysis of Salvia miltiorrhiza hairy roots in response to smoke-isolated karrikinolide (KAR1) using RNA-seq. S Afr J. Bot., 150: 768–778.

Elazab, D.S., Abdel-Wahab, D.A. and El-Mahdy, M.T. 2021. Iron and zinc supplies mitigate cadmium toxicity in micropropagated banana (Musa spp.). PCTOC, 145: 367–377.

El-Keblawy, A. and Gairola, S. 2017. Dormancy regulating chemicals alleviate innate seed dormancy and promote germination of desert annuals. J. Plant Growth Regul., 36: 300–311.

El-Mahdy, M.T., Youssef, M. and Eissa, M.A. 2018. Impact of *in vitro* cold stress on two banana genotypes based on physio-biochemical evaluation. S. Afr. Stat. J., 119: 219–225.

El-Mahdy, M.T., Youssef, M. and Elazab, D.S. 2022. *In vitro* screening for salinity tolerance in pomegranate (Punica granatum L.) by morphological and molecular characterization. Acta Physiol. Plant, 44: 27.

Espinosa-Leal, C.A. and Garcia-Lara, S. 2020. *In vitro* germination and initial seedling development of krantz aloe by smoke-saturated water and seed imbibition. HortTechnology, 30: 619–623.

Feng, Z., Liang, X., Tian, H., Watanabe, Y., Nguyen, K.H., Tran, C.D. et al. 2022. SUPPRESSOR of MAX2 1 (SMAX1) and SMAX1-LIKE2 (SMXL2) negatively regulate drought resistance in Arabidopsis thaliana. Plant and Cell Physiology, 63: 1900–1913.

Flematti, G.R., Dixon, K.W. and Smith, S.M. 2015. What are karrikins and how were they 'discovered' by plants? BMC Plant Biology, 13(1): 1–7.

Flematti, G.R., Ghisalberti, E.L., Dixon, K.W. and Trengove, R.D. 2004. A compound from smoke that promotes seed germination. Science, 305: 977.

Flematti, G.R., Ghisalberti, E.L., Dixon, K.W. and Trengove, R.D. 2009. Identification of alkyl substituted 2 H-furo [2, 3-c] pyran-2-ones as germination stimulants present in smoke. Journal of Agriculture and Food Chemistry, 57: 9475–9480.

Flematti, G.R., Waters, M.T., Scaffidi, A., Merritt, D.J., Ghisalberti, E.L., Dixon, K.W. et al. 2013. Karrikin and cyanohydrin smoke signals provide clues to new endogenous plant signaling compounds. Molecular Plant., 6: 29–37.

Flematti, G.R., Scaffidi, A., Dixon, K.W., Smith, S.M. and Ghisalberti, E.L. 2011. Production of the seed germination stimulant karrikinolide from combustion of simple carbohydrates. Journal of Agriculture and Food Chemistry, 59: 1195–1198.

Fu, X., Wang, J., Shangguan, T., Wu, R., Li, S., Chen, G. et al. 2022. SMXLs regulate seed germination under salinity and drought stress in soybean. Plant Growth Regul., 96: 397–408.

Ghazanfari, P., Abdollahi, M.R., Moieni, A. and Moosavi, S.S. 2012. Effect of plant-derived smoke extract on in vitro plantlet regeneration from rapeseed (Brassica napus L. cv. Topas) microspore-derived embryos. Int. J. Plant Prod., 6: 1735–6814.

Ghebrehiwot, H., Kulkarni, M.G., Bairu, M. and van Staden, J. 2013. Plant-derived aerosol-smoke and smoke solutions influence agronomic performance of traditional cereal crop, tef. Experimental Agriculture, 49: 244–255.

Ghebrehiwot, H., Kulkarni, M.G., Kirkman, K.P. and van Staden, J. 2008. Smoke-Water and a Smoke-Isolated Butenolide improve germination and seedling vigour of *Eragrostis tef* (Zucc.) Trotter under high temperature and low osmotic potential. Journal of Agronomy and Crop Science, 194: 270–277.

Goussi, R., Manaa, A., Derbali, W., Cantamessa, S., Abdelly, C. and Barbato, R. 2018. Comparative analysis of salt stress, duration and intensity, on the chloroplast ultrastructure and photosynthetic apparatus in *Thellungiella salsuginea*. Journal of Photochemistry and Photobiology B, Biology, 183: 275–287.

Guo, Y.X., Zheng, Z.Y., La Clair, J.J., Chory, J. and Noel, J.P. 2013. Smoke-derived karrikin perception by the alpha/beta-hydrolase KAI2 from Arabidopsis. Proceedings of the National Academy of Sciences of the United States of America, 110: 8284–8289.

Gupta, S., Hrdlicka, J., Ngoroyemoto, N., Nemahunguni, N.K., Gucký, T., Novák, O. et al. 2019. Preparation and Standardisation of smoke-water for seed germination and plant growth stimulation. Journal of Plant Growth Regulation., 1–8.

Gupta, S., Plačková, L., Kulkarni, M.G., Doležal, K. and Van Staden, J. 2019. Role of smoke stimulatory and inhibitory biomolecules in phytochrome-regulated seed germination of *Lactuca sativa*. Plant Physiology, 181(2): 458–470.

Gutjahr, C., Gobbato, E., Choi, J., Riemann M., Johnston M.G., Summers W. et al. 2015. Rice perception of symbiotic arbuscular mycorrhizal fungi requires the karrikin receptor complex. Science, 350: 1521–1524.

Ha, C.V., Leyva-Gonzalez, M.A., Osakabe, Y., Tran, U.T., Nishiyama, R., Watanabe, Y. and Tran, L.S.P. 2014. Positive regulatory role of strigolactone in plant responses to drought and salt stress. Proceedings of the National Academy of Sciences, 111: 851–856.

Hamon-Josse, M., Villaécija-Aguilar, J.A., Ljung, K., Leyser, O., Gutjahr, C. and Bennett, T. 2022. KAI2 regulates seedling development by mediating light-induced remodelling of auxin transport. New Phytologist, 235: 126–140.

Hayat, N., Afroz, N., Rehman, S., Bukhari, S.H., Iqbal, K., Khatoon, A. et al. 2021. Plant-derived smoke ameliorates salt stress in wheat by enhancing expressions of stress-responsive genes and antioxidant enzymatic activity. Agronomy, 12: 28.

Heringer, A.S., Santa-Catarina, C. and Silveira, V. 2018. Insights from proteomic studies into plant somatic embryogenesis. Proteomics, 18: 1700265.

Hewedy, O.A., Elsheery, N.I., Karkour, A.M., Elhamouly, N., Arafa, R.A., Mahmoud, G.A.E. et al. 2023. Jasmonic acid regulates plant development and orchestrates stress response during tough times. Environmental and Experimental Botany, 208: 105260.

Hong, Z., Lan, N., Liu, Y., Wang, Y., Zhang, A., Tan, M. et al. 2012. The C2H2-type zinc finger protein ZFP182 is involved in abscisic acid-induced antioxidant defense in rice. J. Integr. Plant Biol., 54: 500–510. doi: 10.1111/j.1744-7909.2012.01135.x

Hrdli˘cka, J., Gucký, T., Novak, O., Kulkarni, M., Gupta, S., van Staden, J. et al. 2019. Quantification of karrikins in smoke water using ultra-high performance liquid chromatography–tandem mass spectrometry. Plant Methods, 15: 1–12.

Ibrahim, M., Nawaz, S., Iqbal, K., Rehman, S., Ullah, R., Nawaz, G., et al. 2022. Plant-derived smoke solution alleviates cellular oxidative stress caused by arsenic and mercury by modulating the cellular antioxidative defense system in wheat. Plants, 11: 1379.

Jain, N., Kulkarni, M.G. and van Staden, J.A. 2006. Butenolide, isolated from smoke, can overcome the detrimental effects of extreme temperatures during tomato seed germination. Plant Growth Regulation., 49: 263–267.

Jain, N. and Van Staden, J. 2007. The potential of the smoke-derived compound 3-methyl-2H-furo [2, 3-c] pyran-2-one as a priming agent for tomato seeds. Seed Science Research, 17: 175–181.

Jain, N. and van Staden, J.A. 2006. Smoke-derived butenolide improves early growth of tomato seedlings. Plant Growth Regulation., 50: 139–148.

Jamil, M., Kanwal, M., Aslam, M.M., Khan, S.U., Malook, I., Tu, J. et al. 2014. Effect of plant-derived smoke priming on physiological and biochemical characteristics of rice under salt stress condition. Aust. J. Crop Sci. 8: 159–170.

Jefferson, L., Pennacchio, M. and Havens-Young, K. 2014. Ecology of Plant-Derived Smoke: Its Use in Seed Germination. Oxford University Press.

Jia, K.P., Luo, Q., He, S.B., Lu, X.D. and Yang, H.Q. 2014. Strigolactone-regulated hypocotyl elongation is dependent on cryptochrome and phytochrome signaling pathways in Arabidopsis. Molecular Plant., 7: 528–540.

Jurczyk, B., Grzesiak, M., Pociecha, E., Wlazło, M. and Rapacz, M. 2019. Diverse stomatal behaviors mediating photosynthetic acclimation to low temperatures in Hordeum vulgare. Front. Plant Sci., 9: 1963. doi: 10.3389/fpls.2018.01963

Kagiyama, M., Hirano, Y., Mori, T., Kim, S.Y., Kyozuka, J., Seto, Y. et al. 2013. Structures of D14 and D14L in the strigolactone and karrikin signaling pathways. Genes Cells., 18: 147–160.

Kamran, M., Imran, Q., Khatoon, A., Lee, I. and Rehman, S. 2013. Effect of plant extracted smoke and reversion of abscisic acid stress on lettuce. Pakistan Journal of Botany, 45: 1541–1549.

Kępczynski, J. 2018. Induction of agricultural weed seed germination by smoke and smoke-derived karrikin (KAR1), with a particular reference to *Avena fatua* L. Acta Physiologia Plantarum, 40: 1–10.

Kępczynski, J., Białecka, B., Light, M.E. and Van Staden, J. 2006. Regulation of Avena fatua seed germination by smoke solutions, gibberellin A 3 and ethylene. Plant Growth Regulation, 49: 9–16.

Kępczynski, J. and Van Staden, J. 2012. Interaction of karrikinolide and ethylene in controlling germination of dormant *Avena fatua* L. caryopses. Plant Growth Regulation, 67: 185–190.

Khatoon, A., Rehman, S. U., Aslam, M. M., Jamil, M. and Komatsu, S. 2020. Plant-derived smoke affects biochemical mechanism on plant growth and seed germination. International Journal of Molecular Science, 21:7760.

Kleman, J. and Matusova, R. 2023. Strigolactones: current research progress in the response of plants to abiotic stress. Biologia, 78: 307–318.

Kochanek, J., Long, R.L., Lisle, A.T. and Flematti, G.R. 2016. Karrikins identified in biochars indicate post-fire chemical cues can influence community diversity and plant development. PLoS ONE., 11: e0161234.

Kochanek, J., McGregor, G., Tryggestad, K.A., Flematti, G.R., Wang, Y. and Krisantini, S. 2023. The Role of Karrikins in Addressing Pressing Global Challenges: Plant Abiotic Stress Tolerance, Biodiversity Restoration, and Karrikin Commercial Realities. In Strigolactones, Alkamides and Karrikins in Plants. CRC Press, pp. 249–271.

Komatsu, S., Yamaguchi, H., Hitachi, K., Tsuchida, K., Rehman, S.U. and Ohno, T. 2022. Morphological, biochemical, and proteomic analyses to understand the promotive effects of plant-derived smoke solution on wheat growth under flooding stress. Plants, 11(11): 1508.

Kulkarni, M.G., Ascough, G.D. and van Staden, J. 2007. Effects of foliar applications of smoke-water and a smoke-isolated butenolide on seedling growth of okra and tomato. HortScience, 42: 179–182.

Kulkarni, M.G., Light, M.E. and Van Staden, J. 2011. Plant-derived smoke: old technology with possibilities for economic applications in agriculture and horticulture. South African Journal of Botany, 77: 972–979.

Kulkarni, M.G., Sparg, S.G., Light, M.E. and van Staden, J. 2006. Stimulation of Rice (*Oryza sativa* L.) Seedling Vigour by Smoke-water and Butenolide. Journal of Agronomy Crop Science, 192: 395–398.

Lesk, C., Rowhani, P. and Ramankutty, N. 2016. Influence of extreme weather disasters on global crop production. Nature, 529: 84–87.

Li, W., Nguyen, K.H., Chu, H.D., Ha, C.V., Watanabe, Y., Osakabe, Y. et al. 2017. The karrikin receptor KAI2 promotes drought resistance in Arabidopsis thaliana. PLoS Genetics, 13: e1007076.

Li, W., Nguyen, K. H., Chu, H. D., Watanabe, Y., Osakabe, Y., Sato, M. et al. 2020. Comparative functional analyses of DWARF14 and KARRIKIN INSENSITIVE 2 in drought adaptation of Arabidopsis thaliana. The Plant Journal 103: 111–127.

Li, W. and Tran, L.S.P. 2015. Are karrikins involved in plant abiotic stress responses? Trends of Plant Science, 20: 535–538.

Light, M.E, Daws, M.I. and Van Staden, J. 2009. Smoke-derived butenolide: towards understanding its biological effects. South African Journal of Botany, 75: 1–7.

Light, M.E. 2018. Special collection of articles on 'smoke ecology and applications of plant-derived smoke'. South African Journal of Botany, 115: 217–218.

Light, M.E., Burger, B.V., Staerk, D., Kohout, L. and Van Staden, J. 2010. Butenolides from plant-derived smoke: natural plant-growth regulators with antagonistic actions on seed germination. J. Nat. Prod. 73: 267–269.

Liu, M., Shan, Q., Ding, E., Gu, T. and Gong, B. 2023. Karrikin increases tomato cold tolerance via strigolactone and the abscisic acid signaling network. Plant Sci., 332: 111720.

Lo, S.F., Yang, S.Y., Chen, K.T., Hsing, Y.L., Zeevaart, J.A.D., Chen, L.J. and Yu, S.M. 2008. A Novel Class of Gibberellin 2-Oxidases Control Semidwarfism, Tillering, and Root Development in Rice. Plant Cell., 20: 2603–2618.

Ma, G.-H., Bunn, E., Dixon, K. and Flematti, G. 2006. Comparative enhancement of germination and vigor in seed and somatic embryos by the smoke chemical 3-methyl-2H-furo [2,3-C] pyran-2-one in Baloskion tetraphyllum (Restionaceae). *In Vitro* Cell. Dev. Biol., 42: 305–308.

Malabadi, R.B. and Nataraja, K. 2007. Smoke-saturated water influences somatic embryogenesis using vegetative shoot apices of mature trees of Pinus wallichiana AB Jacks. J. Plant. Sci., 2: 45–53.

Malabadi, R.B., Meti, N.T., Mulgund, G.S., Nataraja, K. and Kumar, S.V. 2012. Smoke saturated water promoted in vitro seed germination of an epiphytic orchid Oberonia ensiformis (Rees) Lindl. Research in Plant Biology, 2.

Malabadi, R.B., Teixeira da Silva, J.A. and Mulgund, G.S. 2008. Smoke-saturated water influences in vitro seed germination of Vanda parviflora Lindl. Seed Science and Biotechnology, 2: 65–69.

Masondo, N.A., Kulkarni, M.G., Finnie, J.F. and Van Staden, J. 2018. Influence of biostimulants-seed-priming on Ceratotheca triloba germination and seedling growth under low temperatures, low osmotic potential and salinity stress. Ecotoxicol. Environ. Saf., 147: 43–48.

Mathnoom, S.N. and al-timmen, W.M.A. 2020. The effect of smoke water extract on endogenous phytohormones of *cucumis sativus* L. seeds exposed to salt stress. Plant Cell Biotechnology and Molecular Biology, 1–11.

Matilla, A.J. 2000. Ethylene in seed formation and germination. Seed Scientific Research, 10: 111–126.

Meng, Y., Chen, F., Shuai, H., Luo, X., Ding, J., Tang, S. et al. 2016. Karrikins delay soybean seed germination by mediating abscisic acid and gibberellin biogenesis under shaded conditions. Scientific Reports, 6: 22073.

Merritt, D.J., Kristiansen, M., Flematti, G.R., Turner, S.R., Ghisalberti, E.L., Trengove, R.D. et al. 2006. Effects of a butenolide present in smoke on light-mediated germination of Australian Asteraceae. Seed Scientific Research, 16: 29–35.

Mona, S., Rachna, B., Deepak, B., Bala, K. and Nisha, R. 2019. Biochar for Reclamation of Saline Soils. pp. 451–466. In: Giri, B. and Varma, A. (Eds.). Microorganisms in Saline Environments: Strategies and Functions, Springer: Cham, Germany.

Monthony, A.S., Baethke, K., Erland, L.A. and Murch, S.J. 2020. Tools for conservation of Balsamorhiza deltoidea and Balsamorhiza sagittata: Karrikin and thidiazuron-induced growth. *In Vitro* Cellular & Developmental Biology-Plant, 56: 398–406.

Mostofa, M.G., Abdelrahman, M., Rahman, M.M., Tran, C.D., Nguyen, K.H., Watanabe, Y. et al. 2022. Karrikin receptor KAI2 coordinates salt tolerance mechanisms in arabidopsis thaliana. Plant and Cell Physiology, 63: 1927–1942.

Mourad, A.M., Amin, A.E.E.A.Z. and Dawood, M.F. 2021. Genetic variation in kernel traits under lead and tin stresses in spring wheat diverse collection. Environmental and Experimental Botany, 192: 104646.

Moursi, Y.S., Dawood, M.F., Sallam, A., Thabet, S.G. and Alqudah, A.M. 2021. Antioxidant Enzymes and Their Genetic Mechanism in Alleviating Drought Stress in Plants. pp. 233–262. In: Organic Solutes, Oxidative Stress, and Antioxidant Enzymes Under Abiotic Stressors. CRC Press.

MousaviNik, M., Jowkar, A. and RahimianBoogar, A. 2016. Positive effects of karrikin on seed germination of three medicinal herbs under drought stress. Iran Agric Res., 35: 57–64.

Moyo, M., Amoo, S.O. and Van Staden, J. 2022. Seed priming with smoke water and karrikin improves germination and seedling vigor of Brassica napus under varying environmental conditions. Plant Growth Regul., 97: 315–326.

Mulgund, G.S., Meti, N.T., Malabadi, R.B., Nataraja, K. and Kumar, S.V. 2012. Smoke promoted in vitro seed germination of Pholidota pallida Lindl. Research in Plant Biology, 2: 24–29.

Musarurwa, H.T., Van Staden, J. and Makunga, N.P. 2010. *In vitro* seed germination and cultivation of the aromatic medicinal Salvia stenophylla (Burch. ex Benth.) provides an alternative source of α-bisabolol. Plant Growth Regulation, 61: 287–295.

Nelson, D.C., Flematti, G.R., Riseborough, J.A., Ghisalberti E.L., Dixon K.W. and Smith S.M. 2010. Karrikins enhance light responses during germination and seedling development in Arabidopsis thaliana. Proc. Natl. Acad. Sci. USA, 107: 7095–7100.

Nelson, D.C., Flematti, G.R., Ghisalberti, E.L., Dixon, K.W. and Smith, S.M. 2012. Regulation of seed germination and seedling growth by chemical signals from burning vegetation. Annual Review of Plant Biology, 63: 107–130.

Nelson, D.C., Flematti, G.R., Riseborough, J.A., Ghisalberti, E.L., Dixon, K.W. and Smith, S.M. 2010. Karrikins enhance light responses during germination and seedling development in *Arabidopsis thaliana*. Proceedings of the National Academy of Sciences of the United States of America, 107: 7095–7100.

Nelson, D.C., Risenborough, J.-A., Flematti, G.R., Stevens, J., Ghisalberti, E.L., Dixon, K.W. et al. 2009. Karrikins discovered in smoke trigger arabidopsis seed germination by a mechanism requiring gibberellic acid synthesis and light. Plant Physiology, 149: 863–873.

Nelson, D.C., Scaffidi, A., Dun, E.A., Waters, M.T., Flematti, G.R., Dixon, K.W. et al. 2011. F-box protein MAX2 has dual roles in karrikin and strigolactone signaling in *Arabidopsis thaliana*. Proceedings of the National Academy of Sciences of the United States of America, 108: 8897–8902.

Ngoroyemoto, N., Gupta, S., Kulkarni, M.G., Finnie, J.F. and van Staden, J. 2019. Effect of organic biostimulants on the growth and biochemical composition of *Amaranthus hybridus* L. South African Journal of Botany. 124: 87–93.

Otori, M., Murashita, Y., Ur Rehman, S. and Komatsu, S. 2021. Proteomic study to understand promotive effects of plant-derived smoke on soybean (*Glycine max* L.) root growth under flooding stress. Plant Molecular Biology Reporter, 39: 24–33.

Papenfus, H.B., Naidoo, D., Pošta, M., Finnie, J.F. and Van Staden, J. 2016. The effects of smoke derivatives on *in vitro* seed germination and development of the leopard orchid Ansellia africana. Plant Biology, 18: 289–294.

Petrášek, J. and Friml, J. 2009. Auxin transport routes in plant development. Development, 136: 2675–2688.

Procko, C., Crenshaw, C.M., Ljung, K., Noel, J.P. and Chory, J. 2014. Cotyledon- generated auxinis required for shade-induced hypocotyls growth in Brassica rapa. Plant Physiol., 165: 1285–1301.

Ritmejerytė, E., Obvintseva, A. and Huynh, T. 2018. The effect of smoke derivatives and carbon utilisation on symbiotic germination of the endangered Pterostylis despectans (orchidaceae). Lankesteriana [online]. 18, n. 3.

Ruzicka, K., Ljung, K., Vanneste, S., Podhorska, R., Beeckman, T., Friml, J. et al. 2007. Ethylene regulates root growth through effects on auxin biosynthesis and transport-dependent auxin distribution. Plant and Cell., 19: 2197–2212.

Sallam, A., Alqudah, A.M., Dawood, M.F., Baenziger, P.S. and Börner, A. 2019. Drought stress tolerance in wheat and barley: advances in physiology, breeding and genetics research. Int. J. Mol. Sci., 20: 3137.

Sallam, A., Amro, A., El-Akhdar, A., Dawood, M. F., Kumamaru, T. and Stephen Baenziger, P. 2018. Genetic diversity and genetic variation in morpho-physiological traits to improve heat tolerance in Spring barley. Mol. Biol. Rep., 45: 2441–2453.

Sami, A., Zhu, Z.H., Zhu, T.X., Zhang, D.M., Xiao, L.H., Yu, Y. and Zhou, K.J. 2021. Influence of KAR1 on the plant growth and development of dormant seeds by balancing different factors. International Journal Environmental Science Technology, 2021: 1–10.

Sardar, R., Ahmed, S. and Yasin, N.A. 2021. Seed priming with karrikinolide improves growth and physiochemical features of Coriandrum sativum under cadmium stress. Environmental Advances 5: 100082.

Scaffidi, A., Waters, M.T., Bond, C.S., Dixon, K.W., Smith, S.M., Ghisalberti, E.L. et al. 2012. Exploring the molecular mechanism of karrikins and strigolactones. Bioorganic and Medicinal Chemistry Letters, 22: 3743–3746.

Scaffidi, A., Waters, M.T., Sun, Y.K., Skelton, B.W., Dixon, K.W., Ghisalberti, E.L. et al. 2014. Strigolactone hormones and their stereoisomers signal through two related receptor proteins to induce different physiological responses in Arabidopsis. Plant Physiology, 165: 1221–1232.

Senaratna, T., Dixon, K., Bunn, E. and Touchell, D. 1999. Smoke-saturated water promotes somatic embryogenesis in geranium. Plant Growth Regul., 28: 95–99.

Shabir, S., Ilyas, N., Asif, S., Iqbal, M., Kanwal, S. and Ali, Z. 2022. Deciphering the role of plant-derived smoke solution in ameliorating saline stress and improving physiological, biochemical, and growth responses of wheat. J. of Plant Growth Regul., 41: 2769–2786.

Shah, A.A., Khan, W.U., Yasin, N.A., Akram, W., Ahmad, A., Abbas, M. et al. 2020. Butanolide alleviated cadmium stress by improving plant growth, photosynthetic parameters and antioxidant defense system of Brassica oleracea. Chemosphere, 261: 127728.

Shah, F.A., Ni, J., Yao, Y., Hu, H., Wei, R. and Wu, L. 2021. Overexpression of karrikins receptor gene Sapium sebiferum KAI2 promotes the cold stress tolerance via regulating the redox homeostasis in Arabidopsis thaliana. Front. Plant Sci., 12: 657960.

Shah, F.A., Wei, X., Wang, Q., Liu, W., Wang, D., Yao, Y. et al. 2020. Karrikin improves osmotic and salt stress tolerance via the regulation of the redox homeostasis in the oil plant *Sapium sebiferum*. Frontiers in Plant Science, 11: 216.

Sharifi, P. and Shirani Bidabadi, S. 2020. Protection against salinity stress in black cumin involves karrikin and calcium by improving gas exchange attributes, ascorbate–glutathione cycle and fatty acid compositions. SN Applied Sciences, 2: 1–14.

Sheteiwy, Mohamed, S., Zaid Ulhassan, Weicong Qi, Haiying Lu, Hamada AbdElgawad, Tatiana Minkina, et al. 2022. Association of jasmonic acid priming with multiple defense mechanisms in wheat plants under high salt stress. Frontiers in Plant Science, 13(2022): 886862.

Shi, Y. and Yang, S. 2014. ABA regulation of the cold stress response in plants. pp. 337–363. In: D.P. Zhang (Ed.). Abscisic Acid: Metabolism, Transport and Signaling. (Dordrecht: Springer). doi: 10.1007/978-94-017-9424-4_17

Silverstone, A.L., Chang, C.W., Krol, E. and Sun, T.P. 1997. Developmental regulation of the gibberellin biosynthetic gene GA1 in *Arabidopsis thaliana*. Plant Journal, 12: 9–19.

Singh, S., Uddin, M., Khan, M.M.A., Chishti, A.S., Singh, S. and Bhat, U.H. 2023. The role of plant-derived smoke and karrikinolide in abiotic stress mitigation: An Omic approach. Plant Stress, 2023: 100147.

Smith, S.M. and Li, J. 2014. Signalling and responses to strigolactones and karrikins. Curr. Opin. Plant Biol., 21: 23–29.

Soos, V., Sebestyen, E., Juhasz, A., Pinter, J., Light, M.E., Van Staden, J. et al. 2009. Stress-related genes define essential steps in the response of maize seedlings to smoke-water. Functional and Integrative Genomics, 9: 231–242.

Soundappan, I., Bennett, T., Morffy, N., Liang, Y., Stanga, J.P., Abbas, A. et al. 2015. SMAX1-LIKE/D53 Family Members Enable Distinct MAX2-Dependent Responses to Strigolactones and Karrikins in Arabidopsis. Plant Cell., 27: 3143–3159.

Sreekissoon, A., Finnie, J.F. and Van Staden, J. 2021. Effects of smoke water on germination, seedling vigour and growth of Sceletium tortuosum. S. Afr. J. Bot., 139: 427–431.

Staden, J.V., Brown, N.A., Jager, A.K. and Johnson, T.A. 2000. Smoke as a germination cue. Plant Species Biology, 15: 167–178.

Stanga, J.P., Morffy, N. and Nelson, D.C. 2016. Functional redundancy in the control of seedling growth by the karrikin signaling pathway. Planta, 243: 1397–1406.

Stirnberg, P., van de Sande, K. and Leyser, H.M.O. 2002. MAX1 and MAX2 control shoot lateral branching in Arabidopsis. Development, 129: 1131–1141.

Stirnberg, P., Furner, I.J. and Ottoline Leyser, H.M. 2007. MAX2 participates in an SCF complex which acts locally at the node to suppress shoot branching. Plant Journal, 50: 80–94.

Sun, H., Xu, Z., Zhou, J. and Guo, L. 2020. Karrikins-induced accumulation of Salvianolic acid B is regulated by Jasmonic acid and hydrogen peroxide in Salvia miltiorrhiza. S. Afr. J. Bot., 130: 371–374.

Sun, X.D. and Ni, M. 2011. Hyposensitive to Light, an Alpha/Beta Fold Protein, Acts Downstream of Elongated Hypocotyl 5 to Regulate Seedling De-Etiolation. Molecular Plant, 4: 116–126.

Sun, Y.K., Flematti, G.R., Smith, S.M. and Waters, M.T. 2016. Reporter gene-facilitated detection of compounds in Arabidopsis leaf extracts that activate the karrikin signaling pathway. Frontiers in Plant Science, 7: 1799.

Sun, Y.K., Yao, J., Scaffidi, A., Melville, K.T., Davies, S.F., Bond, C.S. and Waters, M.T. 2020. Divergent receptor proteins confer responses to different karrikins in two ephemeral weeds. Nature Communication, 11: 1264.

Sunmonu, T.O., Kulkarni, M.G. and VanStaden, J. 2016. Smoke-water, karrikinolide and gibberellic acid stimulate growth in bean and maize seedlings by efficient starch mobilization and suppression of oxidative stress. S. Afr. Stat. J., 102: 4–11

Swarbreck, S.M. 2021. Phytohormones interplay: Karrikin signalling promotes ethylene synthesis to modulate roots. Trends in Plant Science, 26: 308–311.

Swarbreck, S.M., Guerringue, Y., Matthus, E., Jamieson, F.J. and Davies, J.M. 2019. Impairment in karrikin but not strigolactone sensing enhances root skewing in Arabidopsis thaliana. Plant Journal. 98: 607–621.

Swarbreck, S.M., Mohammad-Sidik, A. and Davies, J.M. 2020. Common components of the strigolactone and karrikin signaling pathways suppress root branching in Arabidopsis. Plant Physiology, 184: 18–22.

Szemenyei, H., Hannon, M. and Long, J.A. 2008. TOPLESS mediates auxin-dependent transcriptional repression during Arabidopsis embryogenesis. Science, 319: 1384–1386.

Tan, Z.Z., Wang, Y.T., Zhang, X.X., Jiang, H.Y., Li, Y., Zhuang, L.L. et al. 2023. Karrikin1 Enhances Drought Tolerance in Creeping Bentgrass in Association with Antioxidative Protection and Regulation of Stress-Responsive Gene Expression. Agronomy, 13: 675.

Teixeira Da Silva, J.A. 2013. Smoke-saturated water from five grasses growing in Japan inhibits in vitro protocorm-like body formation in hybrid cymbidium. Journal of Plant Development, 20.

Thomashow, M.F. 1999. PLANT COLD ACCLIMATION: freezing tolerance genes and regulatory mechanisms. Annu. Rev. Plant Physiol. Plant Mol. Biol., 50: 571–599.

Tian, H., Watanabe, Y., Nguyen, K.H., Tran, C.D., Abdelrahman, M., Liang, X. et al. 2022. KARRIKIN UPREGULATED F-BOX 1 negatively regulates drought tolerance in Arabidopsis. Plant Physiol., 190: 2671–2687.

Umehara, M., Hanada, A., Yoshida, S., Akiyama, K., Arite, T., Takeda-Kamiya. N. et al. 2008. Inhibition of shoot branching by new terpenoid plant hormones. Nature, 455: 195–200.

Van Staden, J., Jäger, A.K., Light, M.E. and Burge, B.V. 2004. Isolation of the major germination cue from plant-derived smoke. South African Journal of Botany, 70: 654–659.

Villaécija-Aguilar, J.A., Hamon-Josse, M., Carbonnel, S., Kretschmar, A., Schmidt, C., Dawid, C. et al. 2019. SMAX1/SMXL2 regulate root and root hair development downstream of KAI2-mediated signaling in Arabidopsis. PLoS Genetics, 15: e1008327.

Waheed, M.A., Jamil, M., Khan, M.D., Shakir, S.K. and Rehman, S.U. 2016. Effect of plant-derived smoke solutions on physiological and biochemical attributes of maize (Zea mays L.) under salt stress. Pak. J. Bot., 48: 1763–1774.

Wang R.H. and Estelle M. 2014. Diversity and specificity: Auxin perception and signaling through the TIR1/AFB pathway. Curr. Opin. Plant Biol., 21: 51–58.

Wang, L., Waters, M.T. and Smith, S.M. 2018. Karrikin-KAI2 signalling provides Arabidopsis seeds with tolerance to abiotic stress and inhibits germination under conditions unfavourable to seedling establishment. New Phytologist, 219: 605–618.

Wang, L., Xu, Q., Yu, H., Ma, H., Li, X., Yang, J. et al. 2020. Strigolactone and karrikin signaling pathways elicit ubiquitination and proteolysis of SMXL2 to regulate hypocotyl elongation in Arabidopsis. Plant and Cell, 32: 2251–2270.

Waters, M.T., Nelson, D.C., Scaffidi, A., Flematti, G.R., Sun, Y.K.M., Dixon, K.W. et al. 2012. Specialisation within the DWARF14 protein family confers distinct responses to karrikins and strigolactones in Arabidopsis. Development, 139: 1285–1295.

Waters, M.T., Scaffidi, A., Sun, Y.K., Flematti, G.R. and Smith, S.M. 2014. The karrikin response system of Arabidopsis. Plant Journal., 79: 623–631.

Waters, M.T. and Smith, S.M. 2013. KAI2- and MAX2-mediated responses to karrikins and strigolactones are largely independent of HY5 in Arabidopsis seedlings. Molecular Plant, 6: 63–75.

Waters, M.T., Nelson, D.C., Scaffidi, A., Flematti, G.R., Sun, Y.K., Dixon, K.W. et al. 2012. Specialisation within the dwarf14 protein family confers distinct responses to karrikins and strigolactones in Arabidopsis. Development, 139: 1285–1295.

Wit, M., Spoel, S.H., Sanchez-Perez, G.F., Gommers, C.M., Pieterse, C.M., Voesenek, L.A. et al. 2013. Perception of low red:far-red ratio compromises both salicylic acid-and jasmonic acid-dependent pathogen defences in Arabidopsis. Plant J., 75: 90–103.

Xu, P., Jinbo, H. and Cai, W. 2022. Karrikin signaling regulates hypocotyl shade avoidance response by modulating auxin homeostasis in Arabidopsis. New Phytologist, 236: 1748–1761.

Yamaguchi, S. 2008. Gibberellin metabolism and its regulation. Annual Review of Plant Biology, 59: 225–251.

Yang, T., Lian, Y. and Wang, C. 2019. Comparing and contrasting the multiple roles of butenolide plant growth regulators: strigolactones and karrikins in plant development and adaptation to abiotic stresses. International Journal of Molecular Sciences, 20(24): 6270.

Yao, R.F., Ming, Z.H., Yan, L.M., Li S.H., Wang, F., Ma, S. et al. 2016. DWARF14 is a non-canonical hormone receptor for strigolactone. Nature, 536: 469–473.

Zhao, Z., Tan, L., Dang, C., Zhang, H., Wu, Q. and An, L. 2012. Deep-sequencing transcriptome analysis of chilling tolerance mechanisms of a subnival alpine plant, Chorispora bungeana. BMC Plant Biology, 12: 1–17.

Zhong, Z., Kobayashi, T., Zhu, W., Imai, H., Zhao, R., Ohno, T. et al. 2020. Plantderived smoke enhances plant growth through ornithine-synthesis pathway and ubiquitin-proteasome pathway in soybean. Journal of Proteomics, 221: 103781.

Zhou, J., Fang, L., Wang, X., Guo, L. and Huang, L. 2013. Effects of smoke-water on photosynthetic characteristics of *Isatis indigotica* seedlings. Sustainable Agriculture Research, 2: 526-2016-3796.

Zhou, J., Ran, Z., Xu, Z., Liu, Q., Huang, M., Fang, L. et al. 2018. Effects of smoke-water and smoke-isolated karrikinolide on tanshinones production in salvia miltiorrhiza hairy roots. S. Afr. J. Bot., 119: 265–270.

Zhou, J., Van Staden, J., Guo, L.P. and Huang, L.Q. 2011. Smoke-water improves shoot growth and indigo accumulation in shoots of Isatis indigotica seedlings. S. Afr. J. Bot., 77: 787–789.

Zhou, J., Xu, Z.X., Sun, H. and Guo, L.P. 2019. Smoke-isolated karrikins stimulated tanshinones biosynthesis in Salvia miltiorrhiza through endogenous nitric oxide and jasmonic acid. Molecules, 24: 1229.

Zhu, J.K. 2002. Salt and drought stress signal transduction in plants. Annu. Rev. Plant Biol., 53: 247–273. doi:10.1146/annurev.arplant.53.091401.143329

14

Effect of Phytohormones on Nodulation and Nitrogen Fixation

Ayesha Khan,[1] *Abdul Basit,*[2,]* *Inayat Ullah,*[3] *Nihal Gören Sağlam*[4] and *Heba I. Mohamed*[5]

Introduction

Nitrogen fixation occurs through association of leguminous crops (FaFaCuRo clade; Fa-bales, Fagales, Cucurbitales, Rosales) (Raza et al. 2020) with rhizobia bacteria (nitrogen fixing) by making nodules (specialized organ for bacteria living and fixing nitrogen) (Desbrosses and Stougaard 2011, Ferguson et al. 2010, Oldroyd 2013). Due to such an association, legumes (almost 60 million years ago) were able to absorb a higher amount of CO_2 from the air than other crops due to the presence of nitrogen (Sprent 2007). This symbiotic relationship gets more particular from Rhizobium bacteria entry through injured points of roots to enter through contagion filaments (without lateral roots), which led to nodule development, which enhances nitrogen fixing in auspicious environments (Sprent 2008). The combination of several functions led to nodule development, such as injury of the host crop providing an entry point, nutrients, and metabolite interchange in storage organs (due to confined gene appearance), nitrogenase action (in less oxygen areas), and native easing of shielding reactions (Breakspear et al. 2014, Clarke et al. 2014, Benezech et al. 2020). Nodule formation occurs by various elastic progressive series (Fig. 1).

[1] Department of Horticulture, The University of Agriculture, 25120, Peshawar, Pakistan.
[2] Department of Horticultural Science, Kyungpook National University, 41566 Daegu, South Korea.
[3] Department of Agricultural Mechanization and Renewable Energy Technologies, Faculty of Crop Production Sciences, The University of Agriculture Peshawar, Pakistan.
[4] Istanbul University, Faculty of Science, Department of Biology, Istanbul, Turkey.
[5] Department of Biological and Geological Sciences, Faculty of Education, Ain Shams, University, Cairo, 11341, Egypt.
* Corresponding author: abdulbasithort97@gmail.com, abdulbasit97_lily@knu.ac.kr

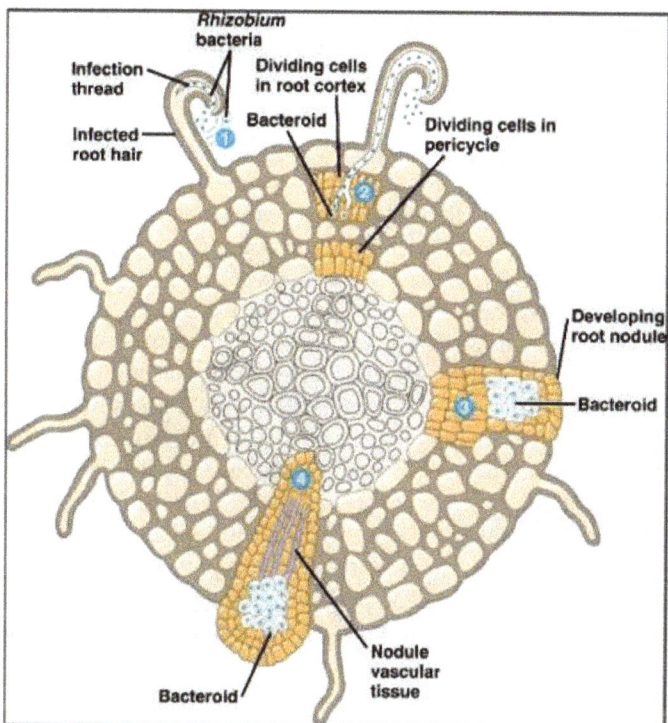

Fig. 1. Nodule formation.

Bacterial nitrogenase (in the root nodule) converts N_2 into an available form inside plants (RNS, root nodule symbiosis). While nitrogen and phosphate nutrients and water uptake by plants occur through mycorrhizae, fungi, and plant root associations (Hawkins et al. 2000, Hodge et al. 2001, Govindarajulu et al. 2005, Guether et al. 2009). As a macronutrient, nitrogen has a great role in crop yield, for which certain nitrogenous fertilizers have been applied, but more usage has led to water pollution (Vance 2001).

Such pollution could be minimized by the usage of biological nitrogen fixing bacteria in fields (add 40 million N per year) (Herridge et al. 2008), which also provides food (having more protein) while facilitating other crops and enhances soil N assets (Oldroyd et al. 2011). Due to lower input costs and climatic pollution, biological nitrogen fixation (BNF) is very successful (Udvardi and Poole 2013).

Nitrogen has two sources: synthetic and biological. Symbiotic associations (mycorrhizae and rhizobia) lessen plant nutritional (N and carbon) deficiencies. Association of actinorhizal plants with Frankia (filamrnous bacteria) (Gherbi et al. 2008) and arbuscular mycorrhizae (AM) association with other plants (angiosperms, gymnosperms, pteridophytes, and certain bryophytes) led to arbuscules (tree-shaped subcellular organ) development, which causes nutrients to interchange (Harrison 2005, Wang and Qiu 2006). Transfers of molecular indicators (between host crop and microbe) along with phytohormones make the association stronger, such as an AM start via plant strigolactones (Gomez-Roldan et al. 2007). RNS has developed from AM association by studying various aspects (mutual association, infection progression, genetics, and hormonal control) (Bonfante and Genre 2008).

Phytohormones play a significant role in initiating signals during stress and controlling meristematic activities during nodule formation (Dobrev et al. 2005, Javid et al. 2011), in addition to their involvement in its growth and development. Nodule formation and nitrogen fixation are also regulated by PGPRs (plant growth-promoting rhizobacteria) and hormones (Lucas-Gracia et al. 2004, Naz et al. 2009). Symbiosis gets efficient when rhizobia change hormone breakdown (Spaepen et al. 2007) along with its analysis (Boiero et al. 2007).

The presence of hormones (auxin, cytokinin, gibberellin, and ABA) has been highly observed in plants with infected roots rather than normal, although their production source is unknown, whether host crop or bacteria, due to the lack of information on the availability of responsible genes for auxin and cytokinin in plants to biosynthesize. Phytohormones are also processed in rhizobia, where nodule formation cannot be affected by sudden changes in their respective genes responsible for nodule development (Brett and Mathesisus 2014). The role of rhizobia strains in cytokinin is under study. Lipochitin-oligosaccharide signals (Nod factors) get processed by flavonoids secreted by host plant roots (Oldroyd 2013) and develop nodules on roots (cortex and pericycle) through cell division and hormone alterations (production, sensitivity, transference, and its buildup) (Desbrosses and Stougaard 2011, Ding and Oldroyd 2009, Mathesius 2008) and sometimes without the involvement of rhizobia (Giraud et al. 2007).

All the events are similar to general events that occur in plant growth, that is, cell division, formation of organs, reaction towards unfavourable conditions, etc. Various hormones have a key role in nodule development (*Medicago truncatula* = indeterminate and *Lotus japonicus* + *Glycine max* = determinate), along with plant growth and regulation, which need to be identified. Differentiated nodules develop from the binding of pericycle and root (outer cortex) cells, which is called determinate, while pericycle and cortex (inner) fusion leads to extended (indeterminate) nodule development (Ferguson et al. 2010, Soyano et al. 2021) (Fig. 2). This chapter describes the association of host plants with microbes, their signal interchange, and their control by phytohormones (endogenous and exogenous).

Contribution of Plant Hormones

Among all plant hormones (Zhao 2010, Kiseleva et al. 2012, Chandler et al. 2015), auxin (cell elongation) and cytokinin (cell differentiation and division) have a key role in systematic and applied fields, such as their observation during nodule development (Desbrosses and Stougaard, 2011, Mukherjee and Ané 2011, Ferguson and Mathesius 2014).

These hormones control all events (from bacteria identification to nodule ageing) of nodule and bacteroid development. All hormones (ethylene = negatively, auxin and cytokinin = positively) work differently in different conditions (Mortier et al. 2014, Mathesius 2008, Penmetsa et al. 2008), such as infective filament formation, which is negatively regulated by cytokinin and ethylene (Murray et al. 2007). Hormones regulate nodulation in the model plant *Arabidopsis thaliana* and other plant species

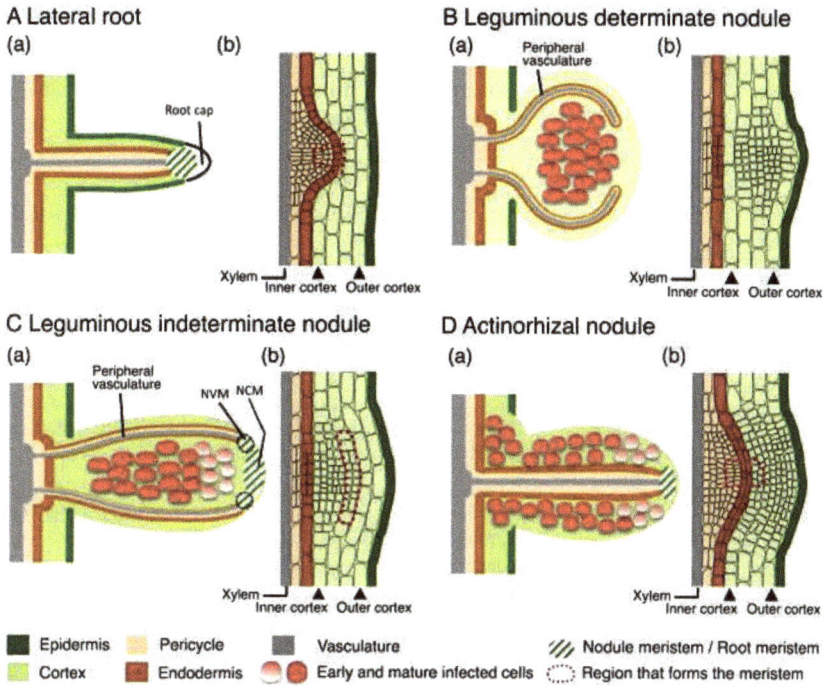

Fig. 2. Diagrammatic representation of the growth patterns of lateral roots and root nodules in legumes and actinorhizal plants (a). (A) Lateral roots emerge through formative division of multiple cells in the pericycle and endodermis (b), and they are distinguished by a central vascular system (a). (B, C) Leguminous root nodules are made up primarily of cells from the parental root's cortex and are distinguished by peripheral vasculatures (a). In the outer and inner root cortex of legumes, cortical cell division begins, resulting in the formation of determinate (Bb) and indeterminate (Cb) nodules, respectively. The meristem is still present at the growing nodules' apex in indeterminate nodules (Ca and b). The meristem area is divided into two different domains. The peripheral vasculature (Ca) is where the nodule vascular meristem (NVM) is found. Rhizobia can infect new cells produced by the nodule central meristem (NCM). The ambiguous nodules' apical to basal portions gradually get infected (Ca). After the development of nodule primordia and the expansion of infected cells, the growth of determinate nodules is stopped (Ba). (D) A central vasculature that develops from the parental pericycle and endodermis distinguishes actinorhizal nodules (a). Cell division follows infection in the cortex (b), endodermis, and root pericycle. A meristem made up of endodermal and cortical cells is formed by pericycle-derived cells in the growing nodule primordia (b). These cells are surrounded by multiple layers of divided endodermal cells. Cortical cells that were split during the nodule are infected.

(Table 1). All three hormones work together to promote nodule placement on roots (Ferguson and Mathesius 2014). Figure 3 shows that plant hormones regulate various events (enhanced adjacent root density and nodule formation) in response to variation in ABA sensitivity (in legumes and non-leguminous crops) (Liang and Harris 2005).

Role of Hormones in Symbiosome and Arbuscule Development

Certain structures of arbuscules and symbiosomes in AM and RNS exchange nutrients and are formed using various host plant substances. Host roots (cortex cells)

Table 1. The main plant hormones role in nodules and roots development in both leguminous and non-leguminous plant species (Bensmihen 2015).

Hormones	Nodules development	Rooting in other crops (*Arabidopsis thaliana* and Dicots)	Rooting in leguminous crops
Cytokinin	Initiate nodules	Inhibit side roots development	Inhibits lateral roots formation
Ethylene	-	Limit major + adjacent roots	Same as auxin but also limits major root size
Jasmonic acid	Initiate nodules via nitrogen fixation	Develop side roots but limit major roots growth	Inhibit major roots size + enhance adjacent roots development
Gibberellic acid	Nodulation at optimum amount	Downstream adjacent roots	Reduced adjacent roots (dwarf pea variant)
Auxin	Indeterminate nodules (low quantity)	Lateral roots (low quantity) + inhibition (high quantity)	Lateral roots (low quantity) + inhibition (high quantity)
Abscisic acid	Regulate nitrogen fixation+nodules development	Negativel influence roots primodia + lateral roots development + meristematic activity	Limits lateral roots (high dose) + promote (low dose) + enhance meristematic activity
Nitric oxide	Aging of nodules + prompt nodulation + contagion	Promote lateral roots growth	Adjacent root primordial
Brassinosteroids	Develop nodulation	Positively regulate lateral roots + apical meristematic activity	Down regulate adjacent roots + major root size

also facilitate symbiotic microbes and efficient nodules. In arbuscule formation, various hormones (found in arbuscule cells) play a major role, such as the presence of genes and enzymes responsible for JA analysis (Hause et al. 2002, Isayenkov et al. 2005), and their higher concentration (due to arbuscules collapsing) occurs in mycorrhizal roots (Vierheilig 2004). Hormones controlling GA synthesis (found in *Bradyrhizobium japonicum*) also regulate symbiosis (Patra and Mandal 2022), while rhizobia or host crops control its concentration (in pre- and post-bacteroid distinction). Nitric oxide reductase lowers the inhibitory responses of nitric oxide, acting as a hormone and inhibitor of nitrogenase activity through Bradyrhizobium bacteroids (Meakin et al. 2006).

Efficient nodule development in indeterminate legumes also occurs through hormones (NO and auxin) (Pii et al. 2007). ENOD genes (MtENOD40-1 and MtENOD40-2) in *Medicago truncatula* get synthesized by cytokinin for bacteroid advancement (Wan et al. 2007) and are observed in mycorrizhal roots that have their positive control. Symbiosome development, bacteroid distinction, nodule formation under an ethylene-independent response, and MtRR4 (the major cytokinin reaction controller) appearance have been controlled by EFD (the transcriptional controller) (Vernié et al. 2008, Gonzalez-Rizzo et al. 2006). Another plant hormone is ABA, which inhibits ethylene synthesis and prones *Solanum lycopersicum* plants to AMF contamination and arbuscule formation (using Sitiens tomato variants having low

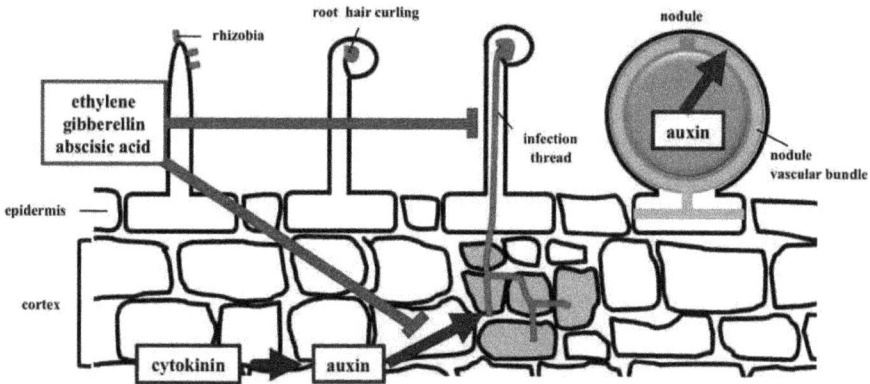

Fig. 3. Effects of phytohormones on leguminous plants' nodulation process. Cortical cell division requires auxin and cytokinin, and auxin functions downstream of cytokinin signalling. Rhizobial infection and cortical cell division are inhibited by ethylene, gibberellin, and abscisic acid.

ABA quantities) (Herrera-Medina et al. 2007). More arbuscules led to higher auxin concentrations in G. intraradices-inoculated roots (Jentschel et al. 2007). Arbuscule formation and phosphate carriers (StPT3 and StPT4) in S*olanum tuberosum* get controlled by a signaling molecule called lysophosphatidylcholine (Drissner et al. 2007).

Plant Hormones as Negative Regulator

Pea plants without nodules due to ethylene (nodule inhibitor) could be recovered by using aminoethoxyvinylglycine (Serova et al. 2019). Autoregulation, derived from the shoot, is another hormone's inhibitory action where abundant nodules develop in its absence. Autoregulation during grafting (navigating graft union) is provoked by root-initiated systemic transporters (Umehara et al. 2008). Autoregulatory factors were once considered ABA. *In vitro* studies of pea root (explant cortex) showed that auxin in media facilitates cell division (in the pericycle), while both hormones (auxin and cytokinin) lead to nodule formation. Factor is a substitute for cytokinin in the root stele. This stele factor could be disabling by Nod + rhizobia, and mitogenic Nod factors (in normal *Pisum sativum* not supernodulating) could be considered to enhance autoregulation (Wu et al. 2021).

Morphology of Root Nodules under Phytohormones

The role of hormones like auxin and cytokinin has been observed in nodule appearance during symbiotic association (Oldroyd et al. 2011, Murray 2011), and lateral root appearance has its own significance where they influence hormone activities along response initiation (Ferreira and Kieber 2005). *In vitro* media, these hormone concentrations should be balanced for appropriate output. Genes and proteins (nodules, ENOD2, ENOD11, ENOD12, ENOD40) causing early nodule development to get analyzed by cytokinin due to rhizobium, show endogenous cytokinin presence, and facilitate cortex cell growth (Mathesius et al. 2000). Nodules

develop under the influence of auxin inhibitors (pseudonodules developers) and
ENOD40 (Mathesius et al. 2000). The absence of NOD factors activates cytokinin to
develop nodules (in leguminous plants). A sudden change in histidine kinase causes
nodulation (Tirichine et al. 2007), while its absence in *Medicago sativa* led to fewer
nodules in quantity (Gonzalez-Rizzo et al. 2006) but not its power of infection (Murray
et al. 2007). In such situations, endogenous cytokinin develops nodule structure and
also contagious filament development (Gage and Margolin 2000, Murray, 2011, Liu
et al. 2015). Auxin increases gibberellin (GB) formation and cooperates with BS
(brassinosteroids) (Ross et al. 2000, Kim et al. 2000). GB and BS have a great role in
nodule and lateral root formation, where their absence causes less nodule formation
in *Pisum sativum* than in the wild form, which was recovered by supplying GB from
outside (Ferguson et al. 2005).

Plant Hormones Involved in Nodule Organogenesis

Phytohormones regulate plant growth, development, organ formation, and the
sensing of foreign signals (Ferguson and Mathesius 2014). Figure 4 shows plant
hormones role in nodulation.

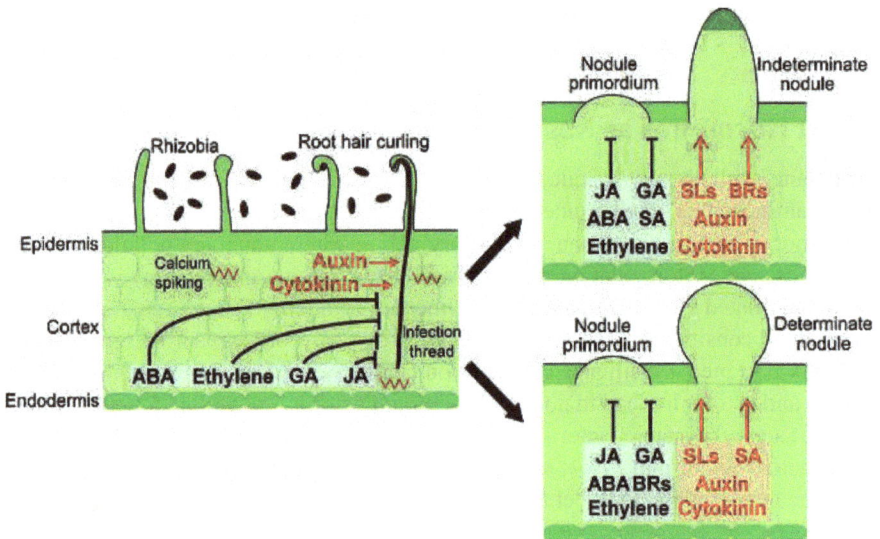

Fig. 4. Role of hormones in nodulation.

Auxin

Auxin (natural = IAA and synthetic = IBA, NAA, and 2,4-D) is found in apical
meristematic tissues and flows beside the phloem and roots. It causes cell elongation
and apical dominance to suppress lateral shoot growth. Auxin moves from younger
leaves (Lomax et al. 2001) and gathers in the apical tissues of the root to start division
(Ljung et al. 2001). Cell division and distinction (Moubayidin et al. 2009) occur
by cytokinin, whose action is stopped by exogenous auxin usage (Nordstrom et al.

2004). Although both hormones in proper balance can support each other's synthesis (Chandler and Werr 2015).

Auxin biosynthesis, metabolism, and signaling

Tryptophan is the precursor of IAA (indole-3-acetic acid) synthesis (Zhao 2012, Korasick et al. 2013), as auxin is involved in almost every event of plant growth (Teale et al. 2006). Tryptophan is transferred into IPA (indole-3-pyruvic acid) through the action of Tryptophan Amidotransferase and its associated proteins (Stepanova et al. 2008), and through the YUCCA protein, it is transferred into free indole acetic acid (Zhao et al. 2001). Sensitivity of TRANSPORT INHIBITOR RESISTANT1 (TIR1) indicates AFB receptors initiate signaling path of auxin due to intracellular auxin, but it could be disabled via many (oxidation, amino acid, and carbohydrate conjugation) methods (Dharmasiri et al. 2005a, 2005b, Parry et al. 2009). ARF (AUXIN RESPONSE FACTOR) gets stopped in case of TIR/AFB disabling (absence of IAA). Aux/IAA efficiency (due to Skp1-Cullin-F-box formation upon IAA sensitivity by TIR1/AFB receptors) gets destroyed (proteasome-dependent), and resultant gene workibility is controlled (negatively and positively) by dimers (made from ARFs that come out of Aux/IAA control) fused with auxin reactive substances (Lin et al., 2020). Nodule organ formation occurs by auxin, which has scientific understanding currently (Kohlen et al. 2018).

Auxin in nodule development

Auxin has a key role in infection and nodule formation. The appearance of the Enod gene and factors stopping auxin flow cause nodule formation (Rightmyer and Long 2011), and auxin reacts in two ways (LjLHK1 and LJNIN) (Suzaki et al. 2012). MtLBD16 appears due to MtNIN-dependent induction of MtYuc2 (*Medicago truncatula*) (Schiessl et al. 2019). Improper nodule development of the Cre1 variant (*Medicago truncatula*) could be recovered by the use of auxin transport inhibitors (Ng et al. 2015). Efficient analysis and flow of auxin occur during nitrogen fixation in root hairs (Breakspear et al. 2014, Nadzieja et al. 2018). Asl18a inside roots is controlled by LjNIN via intron cis factors and auxin (Soyano et al. 2019). NF-YA1 (*L. japonicus*) controls Shi/Sty genes (known as Yuc appearance controllers) too in *Arabidopsis* (Hossain et al. 2016).

The main role of these elements, along with LBD16 and SHI/STYs (auxin-detecting points), is to act as symbiotic messengers, activating auxin (for cell division, side root formation, and nodulation). Auxin is also detected through binding IAA to GH3s to act in developmental events. GH3s show significance in metabolic response due to auxin detection in nodules and contagious progression (van Noorden et al. 2007, Singh et al. 2014a, Breakspear et al. 2014). Auxin elements (IBA-Ala, IAA-Asp, and IAA-Glc) hydrolyze in nodulation by MtIAR33 and MtIAR34 (Campanella et al. 2008). Nodule development becomes low due to lower auxin flow (in the case of *M. truncatula Rhizobiium* injection), which is why auxin flow is more important than its metabolism (van Noorden et al. 2006). Variation in auxin flow (due to gravitropic roots) causes nodule disposition in *Lotus japonicus* (Nadzieja et al. 2019), and its exogenous usage (ingress inhibitor) causes lower nodule development (*Medicago truncatula*) before rhizobia spread (Roy et al. 2017). Lax2 variants that synthesise less auxin led to lower nodules than wild species (Roy et al. 2017), while auxin importers (MtLAX1 and MtLAX2) could be easily observed at the early nodule developmental level (de Billy et al. 2001, Roy et al. 2017).

Auxin transformation and symbiotic detection get difficult due to the PIN family (Ng et al. 2020). Contagious root hairs have great auxin indication (cortex and pericycle of young and apex of mature nodules) (Pacios-Bras et al. 2003, Suzaki et al. 2012, Breakspear et al. 2014), as researchers found it in *Lotus japonicus* and *Medicago truncatula* (pDR5 and pGH3: Gus and DII-Venus) (Takanashi et al. 2011, Suzaki et al. 2012, Guan et al. 2013, Breakspear et al. 2014, Franssen et al. 2015, Nadzieja et al. 2019).

Interaction of auxin and other signaling pathways during nodule organogenesis

Auxin and cytokinin stabilizes each other (El-Showk et al. 2013) during developmental events (especially nodulation), and cytokinin (LHK1/CRE1-dependent) enhances auxin synthesis. MtPin appearance and protein buildup based on MtCRE1 that occurs due to reticence of polar auxin transport (upon rhizobium contagion) (Plet et al. 2011) led to lower nodule formation (in Cre1 plants) and could be recovered by NPA or TIBA application. PAT stops the flavonoids buildup (MtCRE1-dependent) (Ng et al. 2015). Auxin gathered due to CK signaling (MtCRE1-dependent that prompts MtNin, MtSty, MtLbd16, MtYuc2, and MtYuc8) (Schiessl et al. 2019). Recovery of miR160 (sense auxin and CK) action and appearance during nodule development has been achieved by the usage of these hormones (Turner et al. 2013, Nizampatnam et al. 2015). Signals initiated after rhizobium contagion by LjLHK1 inhibit auxin buildup (epidermal cells in *L. japonicus*) and genes (LjLog and LjCyp735a) that synthesize CK (Nadzieja et al. 2018). Although both hormones work together while developing nodules.

Cytokinins

Cytokinin (kinetin, zeatin, and BAP) is present in all types of creatures, where it produces at the root apex, moves along the xylem (passively and nonpolarly), and initiates shoot formation to suppress apical dominance (caused by auxin). The combined effect of both hormones causes cell distinction but also decreases each other's activity (Kurep and Smalle 2022). Initiation of signals in cytokinin starts through two mechanisms: the ARR sensing controller and the histidine kinase receptor, which activate associated genes (Ferreira and Kieber 2005). Nematode causes cytokinin to produce cecidia, while cytokinin also forms nodules but limits side root development (Lohar et al. 2004). Isopentenyl transferases (cytokinin enzymes) enhance root size (triple *Arabidopsis thaliana*) (Ioio et al. 2007), while lowering cell division at fast-growing points. Cytokinin acts in root nodulation (shape and structure) and plant growth processes. Reduced cytokinin in transgenic plants (due to the cytokinin oxidase/dehydrogenase gene) results in altered appearances (more lateral, accessory, and meristematic roots) (Werner et al. 2003).

Cytokinin synthesis, metabolism, and signaling

CKs help in nodulation and root hair contagion (Frugier et al. 2008, Miri et al. 2016, Gamas et al. 2017). CKs have been produced from adenine (isoprenylated) (Sakakibara 2006) and regulate plant development (Werner and Schmulling 2009, Kieber and Schaller 2014, Cortleven et al. 2019). Side-chain cleavage or N-glucosylation permanently ruins cytokinins (iP=develops roots, and tZ= controls LOGs) (Kurakawa et al. 2007). The pathway (histidine/aspartate phosphorylation cascade) of CKs synthesis resembles mechanisms (membrane detection with cellular reactions) observed in bacteria (Werner and Schmulling 2009, Kieber and Schaller 2014). Genes work under CK action due to phosphorylation signaling (from HPs to Tye-B response regulator transcription elements), where histidine kinase receptors (the major element) phosphorylate their remaining (proteins) (Hutchison et al. 2006). The feedback process occurs due to RRAs downstream indication (Sakai et al. 2001).

Cytokinin in Nodule Organogenesis

BAP develops nodulation in crops such as *Medicago sativa* and *Sesbania rostrata* (Lin et al. 2020). LjLHK1 forms nodules without rhizobia infection under cytokinin (Gonzalez-Rizzo et al. 2006, Murray et al. 2007, Tirichine et al. 2007). CK controls nodulation and infection (Miri et al. 2016). Nod element and rhizobia entry produce iP and tZ in *L. japonicus* and *M. truncatula* (van Zeijl et al. 2015, Reid et al. 2016) and numerous IPT decoding genes (Chen et al. 2014, van Zeijl et al. 2015, Reid et al. 2017, Mens et al. 2018). Removing Ipt3 led to reduced nodules in *L. japonicus* (Chen et al. 2014, Reid et al. 2017). No difference in nodule development (between ipt3 and ipt4 plants and the wild) could be observed (Reid et al. 2017). Fast CKs and LOG gene production has been observed during nodulation in response to infection (Mortier et al. 2014, van Zeijl et al. 2015, Reid et al. 2017). Lower nodules formed in *Medicago trucatula* rather than in the wild when MtLog1 was removed or enhanced (Mortier et al. 2014). There are fewer details about LOG genes, but their overexpression helps in nodulation. iP (synthesized by IPT3) in shoots negatively controls nodule development (Sasaki et al. 2014) in *Glycine max* GmIpt5 (Mens et al. 2018).

Partly expressed LjLHK1 and triple *Lotus japonicus* receptors led to nodule formation (Held et al. 2014). Abnormal nodulation also occurs (in CK presence) in highly infected roots (Murray et al. 2007). Lower MtCre1 affects nodulation but not the infection process (Gonzalez-Rizzo et al. 2006, Plet et al. 2011), although MtCHK2–4 partly influences it (Boivin et al. 2016). LjLhk1 (mostly dominant over TCS/TCSn) (Held et al. 2014) shows CK (its usage and pathway overexpression) actively performs nodule development (Murray et al. 2007) in the absence of contagion (Tirichine et al. 2007, Liu et al. 2018b, Heckmann et al. 2011, Reid et al. 2017). MtNIN actively synthesize cytokinin (Vernié et al. 2015).

Gibberellin

Gibberellin biosynthesis and signaling

GA has a role in nodulation (Hayashi et al. 2014). Succeeding GGDP (geranylgeranyl diphosphate) reactions produce gibberellins (GAs) through GA oxidases (Mitchum et al. 2006, Hu et al. 2008, Sun 2008), which actively facilitate various (leaves development and seed germination) plant developmental processes (Sun, 2008, Yamaguchi, 2008). Ways of signal initiation from the nucleus start through GAs (PIFs, GID1 receptors, RGA or RGL, SLY1, and SNE) (Sun 2008, Schwechheimer 2012, Davie`re and Achard 2013).

G1D1 (GIBBERELLIN-INSENSITIVE DWARF1 receptor) doesn't work under GA unavailability (Ueguchi-Tanaka et al. 2005, Griffiths et al. 2006), which directs proteins (DELLA) for gibberellic acid indication, stopping by inactivating phytochrome interacting elements (PIFs) (Sun 2008, Schwechheimer 2012, Davière and Achard 2013). Degradation of DELLA proteins occurs due to GA sensitivity by G1D1 (composite structure assembly of SLY and SNE) (McGinnis et al. 2003, Dill et al. 2004, Ariizumi et al. 2011). Transcripting resultant genes get controlled by its transcriptional elements.

Gibberellin signaling in nodulation

GAs both facilitate (low GA) or inhibit (high GA) nodulation (in *Pisum sativum*, *Medicago truncatula*, and *Lotus japonicus*) (Maekawa et al. 2009) and infectious filament development (Ferguson et al. 2005, Fonouni-Farde et al. 2016a, Jin et al. 2016). Low paclobutrazol (competitor of gibberellic acid) causes more nodules in certain crops (Fonouni-Farde et al. 2016a), while lower nodules are found in legumes (McAdam et al. 2018). GA should be applied at the proper amount because a higher amount results in other structures (protoplast lumps based on IAA or NAA) rather than nodules (Yamagami et al. 2004). The absence of GA synthesizing variants (ls, na, lh, and sln, PsCPS, PsKAO1, PsKO1, and PsGA2OX1) (Yaxley et al. 2001, Davidson et al. 2003, 2004) led to pleiotropic influence (root and shoot) and lower nodules (Ferguson et al. 2005). Nodulation in lh/ko1 could be recovered by GA application (Ferguson et al. 2005). Progeny with extra nodules appear as a result of variants (nod3 and sym 28, 29) crossing with na-1 (Ferguson et al. 2011). Low nodules form in *Medicago trunculata* ga2ox10 plants (*Lotus japonicus* unidentified variants) (Kim et al. 2019). 4–12 dpi (LjGa2ox) have been observed in *Lotus japonicus* (Kouchi, 2004), while 1–12 dpi (MtKao2, MTga3ox1, MtKo1, MtGaox6, MtGa2ox6, MtCps1, and MtGa2ox10) have been observed in *Medicago truncatula* upon rhizobia inoculation (Breakspear et al. 2014, Kim et al. 2019).

DELLA proteins (MtDELLA1, 2, and 3 in M. truncatula) and the SLY gene work in nodulation (under GA) (Maekawa et al. 2009, Fonouni-Farde et al. 2016b, Jin et al. 2016), while their absence (LA and CRY) retards nodulation (Ferguson et al. 2011). LjNin and LjNsp1/2 get inactivated under GA usage (Maekawa et al. 2009). These proteins become active in root meristematic points after 1 day of rhizobium spread (Fonouni-Farde et al. 2016a, 2016b). The appearance of LjGa20ox and nodulation reduction occur due to SLY1A-d activity and LjSly1a overexpression (Maekawa et al. 2009). MtDELLAs (1, 2, and 3) affect nodulation, while association with nodulative genes (MtRip, MtErnl, and MtNin) recovers nodule development (Fonouni-Farde et al. 2016a, 2016b, Jin et al. 2016).

Ethylene

Ethylene biosynthesis and signaling

Ethylene synthesis in gaseous form occurs in various steps as ACC synthase and ACC oxidase (Yamagami et al. 2003, Linkies et al. 2009) that facilitate numerous developmental events such as ageing, ripening, upper and lower plant canopy, and plantlet development in plants (Schaller and Kieber 2002, Schaller 2012). Ser/Thr kinase (CTR1) and ETR/ERS/EEIN4 under ethylene unavailability work together (negatively control ethylene signaling) (Lin et al. 2020). EIN2 (at the C end) becomes disabled as a result of phosphorylation by CTR1 (Ju et al. 2012) or humiliates EIN2 in the F-box (via ETPs action) (Qiao et al. 2009). Genes that are ethylene-dependent in transcription become inactive due to EBFs (that humiliate EIL) and nonfunctional EIN2 (Guo and Ecker 2003, Potuschak et al. 2003). The gene's functionality (Guo and Ecker 2003) could be recovered by EIN3/EIL proteins (via cleavage and nuclear shifting) (An et al. 2010, Wen et al. 2012). EIN2 humiliation could be controlled by phosphorylation (EIN2 with ETR, ERS, and EIN4) and ethylene enabling (which disables ETR, ERS, EIN4, and CTR1) (Bisson and Groth 2010).

Role of ethylene in nodule organogenesis

Inhibition of nodulation in legumes and gene expression occur due to ethylene (ACC), which could be recovered in crops (*Vicia sativa, Medicago trancutala, Pisum sativum*, and *Lotus japonicus*) using AVG or Ag⁺ (Lorteau et al. 2001, Heckmann et al. 2011). Ca^{2+} spiking is also affected (during nitrogen fixation) by ACC in root hairs (Oldroyd et al. 2001). Ethylene affects transcription (unresponsive variants of ethylene) during symbiosis (Larrainzar et al. 2015). MtAcs1 and MtAcs2 activate upon nitrogen fixation (van Zeijl et al. 2015), while rhizobia injection activates ethylene (in 6 hpi) (Reid et al. 2018). Fresh and developed nodules have high levels of MtAcs4, MtAcs7, and MtAcs8 (Larrainzar et al. 2015).

Ethylene also plays a role in nodule placement in *Vicia sativa* at specific points and inhibits at certain points such as Aco buildup (at phloem sides and along cortex), and the SKL variant has nodules on opposite corners of vascular tissues (wild type has nodules on xylem ends) (Penmetsa et al. 2003, 2008). Hypernodulation before rhizobia injection is protected in pericycle cells through ethylene signaling. More nodules formed, akin to the skl/ein2 variants (LjEin2a and LjEin2b, encoding genes in *Lotus japonicus*) (Reid et al. 2018).

Nodulation, ethylene, and agriculture

To overcome water pollution, minimize chemical fertilizers usage, and feed the world's populace, biological methods are being introduced, such as iron supplementation for NF and more yield under L-methionine usage (Arora et al. 2010, Aziz et al. 2015). A reduced effective ethylene role in agriculture has been recorded. ACC activity could be minimized through ACC deaminase, which works after rhizobia infection leads to more nodule development (Gamalero and Glick 2015). While D-serine and ACC work together in nodule formation (in soyabean), and they also drive anoxic metabolism, ACC's beneficial role in agriculture has not been scientifically cleared (Murset et al. 2012). Nitogenase becomes less active (up to 75%) under saline situations, and plants survive through polyamine buildup (cleaning free radicals, activating antioxidant synthesis, and maintaining membrane assembly and cells), while SA usage (enhances lipid peroxidation enzymes) will recover nitrogenase activity by inhibiting polyamine gathering, leading to more ethylene production (Palma et al. 2013). Fertilizers usage influences microbes (at the epidermal-cortical level) but enhances rhizobia activity (in the Pssym15 variant) at the root apex (Jones et al. 2015). Controlling ethylene in agricultural land (disturbing cytokinin) will negatively affect beneficial microbes, which should be kept in mind while altering ecosystems.

Abcisic Acid

ABA (abscisic acid) was found higher in roots adjacent to areas that inhibit nodule development (Ferguson and Mathesius 2014), which could be recovered by abamine (inhibitor of 9-cisepoxycarotenoid dioxygenase) usage to form nodules (Tominaga et al. 2009). ABA enhances or inhibits nodulation and rhizobia infection as its quantity increases under stress. ABA inhibitory roles in various species (*Medicago truncatula, Trifolium repens, Phaseolus vulgaris*, and *Lotus japonicus*) have been recorded (Ding et al. 2008, Suzuki et al. 2004, Khadri et al. 2006, Biswas et al. 2009). ABA minimized in coiled and puffed root hairs (Suzuki et al. 2004). Rhizobia spread stops under ABA activity, leading to gene initiation (ENOD11 and RIP1) responsible for nodules (Ding et al. 2008). ABA limits nodules (in split roots) through a stabilized autoregulation pathway (Biswas et al. 2009). Antioxidant enzyme activity under salinity regulates NF due to ABA preapplication (Palma et al. 2014). ABA (also synthesized in rhizobia) also helps in plants survival under abiotic stress, such as high yield (*Rhizobium leguminosarum*) under water acarcity due to minimized ABA synthesis (higher auxin and GA) (Bano et al. 2010). abi1-1 (which encodes mutated protein phosphatase) affects signaling response of ABA (Gampala et al. 2001, Wu et al. 2003) along with hypernodule progenies (Ding et al. 2008).

Brassinosteroids

Varietal response under BR usage depends on its concentration and plant species. BRs (brassinosteroids) help in nodulation, as observed in the *Pisum sativum* variant (fewer nodules than wild pea) (Ferguson et al. 2005). BRs do not downregulate genes (lk, Psrdn1, Psclv2, or Psnark) or the AON synthetic pathway (pea variants having lower BRs) that cause supernodulative progenies (Foo et al. 2014). High activity of nitrate reductase along nodules and short roots occurs when using multiple amounts of 28-homobrassinolide (soaking Lens culinaris seeds in it) (Shahid et al. 2011). Use of BRs (epibrassinolide or homobrassinolide foliar application) led to more nodules (peanut and French bean) with enhanced nitrate reductase activity (Shahid et al. 2011, Upreti and Murti 2004), while showing the opposite response on *Glycine max* roots (Hunter 2001). Roots and nodule inhibition (in supernodulating *Glycine max*) occur due to brassinolide usage (root injection) (Terakado et al. 2005). Brassinazole (foliar application and addition to media) causes more nodules in wild *Glycine max*. No scientific proof has been presented up to now about BRs role in nodulation.

Salicylic Acid and Jasmonic Acid

There is no clear verification of SA (salicyclic acid) role in nodulation, which depends on multiple factors (such as host species, quantity, growing media, and usage procedure) but has a key role in plants shielding ability against pathogens. SA application has no effect on determinate nodulation in *Glycine max*, *Lotus japonicus*, and *Proteus vulgaris* except in determinate *Glycine max* (Lian et al. 2000, Sato et al. 2002), but minimize indeterminate nodules in other crops (*Pisum sativum, Securigera varia, Medicago sativa,* and *Trifolium refens*) (van Spronsen et al. 2003). Determinate nodule inhibition has been greatly observed in *Glycine max* (super nodulating variant than wild) (Sato et al. 2002). NahG (bacterial SA hydroxylase gene) and lower SA led to more nodules (determinate=*Lotus japonicus* and indeterminate = *Medicago truncatula*) and rhizobia spread, while no nodules and rhizobia spread were observed under SA (Stacey et al. 2006). SA buildup in roots due to inappropriate and nod element (lack with rhizobia strain) usage (*Medicago sativa)* and also in Pssym30 (non-nodulation *Pisum sativum* variant) in the case of rhizobia strain presence rather than wild kind (Brett and Mathesisus 2014). SA causing nodule inhibition could be prevented using an appropriate rhizobia entry procedure for nodule initiation and stopping SA defensive role.

JA has a resistive and curing role in plants (beside inhibiting and initiating nodulation), depending on its amount, host species, etc. JA (jasmonic acid) inhibits nodules (suppressing ENOD11 and RIP1 genes), which could be recovered by AVG (ethylene biosynthesis inhibitor) (Sun et al. 2006), as both hormones (JA and ethylene) suppress gene action (ERF1 and PDFs) (Lorenzo et al. 2003). Nodules (highly NF) develop using abi1-1 in *Medicago sativa* roots (irresponsive to ABA), while nodules (due to ABA) recover through sickle mutation. Nodule inhibition (ENOD40 and NIN gene suppression) occurs due to ABA through various pathways. Shielding and stress reactions (under rhizobia spread) result in nodule quantity and species selection (Santos et al. 2001). Nodule limits (in Ljhar1 and wild *Lotus japonicus)* are due to methyl jasmonate, which prevents LjNIN and contagion strain development (Nakagawa and Kawaguchi 2006), while there are more nodules (in the case of jasmonic acid and blue light) (Suzuki et al. 2011). Nodulation could be resumed (in the phyB variant of *L. japonicus*) using JA (Suzuki et al. 2011). Nodulation under JA (pathway and genes in *Glycine max* leaves) could be recovered via the GmNARK mechanism (Kinkema and Gresshoff 2008, Seo et al. 2006). Nodule inhibition in the Gmnark variant (Seo et al. 2006) could be induced using JA synthetic inhibitors and enhanced JA in leaves (Kinkema and Gresshoff 2008).

Jasmonates induce nodulation (*Rhizobium leguminosarum* and *Bradyrhizobium japonicum*) through activating nod genes (Mabood et al. 2006). Enhanced NF and nodulation are caused by *Bradyrhizobium japonicum* and *Rhizobium leguminosarum* (pre-injection) along with jasmonate (Mabood and Smith 2005, Poustini et al. 2005). JA work in a three-way association (legumes, rhizobia, and herbivores), such as JA resisting herbivores through changing the arrangement of volatile organic substances (Ballhorn et al. 2013).

Management to Develop Nodules in Non-Leguminous Crops

Controlling the hazardous effects of fertilizer, NF is beneficial for the environment and agricultural land too (Rockstrom et al. 2009, Springmann et al. 2018). Auxin (via NIN that points to LjAsl18a/MtLbd16) (Schiessl et al. 2019, Soyano et al. 2019, Xiao et al. 2014, 2019, Bensmihen 2015, Schiessl et al. 2019, Soyano et al. 2019) and actinorhizal bacteria are helpful in nodules and adjacent roots induction via progressive and transcriptional events (Xiao et al. 2014, 2019, Bensmihen 2015, Schiessl et al. 2019, Soyano et al. 2019). Furthermore, roots and nodule development are caused by LjNF-YA1 (that initiates SHI and STY in *Arabidopsis thaliana*) (Hossain et al. 2016). Excessive auxin application led to another assembly (similar to a nodule) (Hiltenbrand et al. 2016, Thomas et al. 2018, 2020), while similar structure formation occurred in legumes too (using auxin inhibitors) (Rightmyer and Long 2011, Li et al. 2014, Ng et al. 2015). Excessive auxin concentration should be avoided to perform roots properly during NF. Meristematic cells (in indeterminate nodules) develop through root cell (in pericycle and then cortex) multiplication (Xiao et al. 2014, Bensmihen 2015, Kohlen et al. 2018), while insistent meristem cells (in determinate nodules) develop due to cell multiplication in only cortex (Xiao et al. 2019).

It shows that the acquisition of the nodule initiating element matters more than its position, as features of the nodule and adjacent root affect the absence of Noot/Cochleata (Ferguson and Reid, 2005, Couzigou et al. 2012, Magne et al. 2018, Shen et al. 2020). For nodule recognition, legume controllers (NIN, NOOT) should be modified (Griesmann et al. 2018, van Velzen et al. 2018). Cortex cell multiplication (legumes) under cytokinin is a great variation, as it also plays a role in nodulation and its recognition (Gauthier-Coles et al. 2019). Furthermore, assembly similar to nodules could be obtained under cytokinin (CCamK or Lhk1) activity (Gleason et al. 2006, Tirichine et al. 2006, 2007). Irrespective of nodule features convening, all other hormones (auxin, GA, Nin, and Nf-ya1) regulate nodulation (Soyano et al. 2013, Fonouni-Farde et al. 2017, Schiessl et al. 2019).

Conclusions and Future Prospects

All phytohormones have a particular and important role in nodulation, its quantity, and other plant developmental processes. Several hormones (auxin, CK, GA, and BRs) have nodulation ability (at specific concentrations) (Ferguson et al. 2005, Oldroyd and Downie 2008). Appropriate hormones working along their specific points of action and their synthetic paths should be studied at the cellular level (using proteins, metabolites, and transcript details). Studying these aspects will open up the proper connection between roots and nodulation, as various species undergo different ways of nodulation. After these systemic observations, there will be accurate legume farming and nodulation (in non-leguminous species) to enhance NF on agricultural lands. NF (the host plant) does not use photosynthetic substances for nodule development. Phytohormones regulate nitrogen and carbon usage during nodulation, similar to how plants alter their substances in response to variable environmental conditions. There is no authentic proof of nod elements and hormone detecting mechanisms, but leguminous crops form nodules as a result of their association with their compatible partner (Den Herder and Parniske 2009).

Overexpression of algae, soil toxicity, and water pollution has been caused by nitrogenous fertilisers, which could be resolved using biotechnology tools. Due to the diversity of knowledge available (research done on different crops under different circumstances), proper hormone interactions could not be defined. For such studies, model crops (legumes) should be selected to find out the proper hormone concentration (exogenous) and its particular point of working, its interaction with peptides, and rhizobia variants responsible for nodule and nodule-like structure formation (spatio-temporal). Hormones secreted by rhizobia are also fruitful (their influence on nodule quantity and functionality is uncertain) but not necessary to drive symbiotic association. There is a need to find out the role of hormones (either unique or typical) in nodulation, the role of genes causing nodulation at early stages, and their connection to the balance of the outer (nod substances, Rhizobium) and inner (hormones, unique elements) signalling pathways.

References

An, F., Zhao, Q., Ji, Y., Li, W., Jiang, Z., Yu, X. et al. 2010. Ethylene-induced stabilization of ETHYLENE INSENSITIVE3 and EIN3-LIKE1 is mediated by proteasomal degradation of EIN_3 binding F-box 1 and 2 that requires EIN_2 in Arabidopsis. Plant Cell., 22: 2384–2401.

Ariizumi, T., Lawrence, P.K. and Steber, C.M. 2011. The role of two Fbox proteins, SLEEPY1 and SNEEZY, in Arabidopsis gibberellin signaling. Plant Physiol., 155: 765–775.

Arora, N.K., Khare, E., Singh, S. and Maheshwari, D.K. 2010. Effect of Al and heavy metals on enzymes of nitrogen metabolism of fast and slowgrowing rhizobia under explanta conditions. World J. Microbiol. Microtechnol., 26: 811–816. doi:10.1007/s11274-009-0237-6

Aziz, M.Z., Saleem, M., Ahmad, Z., Yaseen, M., Naveed, M., Khan, M.Y. et al. 2015. Effect of ethylene on growth, nodulation, early flower induction and yield in mungbean. Int. J. Sci. Eng. Res., 6: 1210–1224.

Ballhorn, D.J., Kautz, S. and Schadler, M. 2013. Induced plant defense via volatile production is dependent on rhizobial symbiosis. Oecologia, 172: 833–846.

Bano, A., Balool, R. and Dazzo, F. 2010. Adaptation of chickpea to desiccation stress is enhanced by symbiotic rhizobia. Symbiosis. 50: 129–133.

Benezech, C., Berrabah, F., Jardinaud, M.-F., Le Scornet, A., Milhes, M., Jiang, G. et al. 2020. Medicago-Sinorhizobium-Ralstonia co-infection reveals legume nodules as pathogen confined infection sites developing weak defenses. Curr. Biol., 30: 351–358.e4.

Bensmihen, S. 2015. Hormonal control of lateral root and nodule development in legumes. Plants., 4: 523–547.

Bisson, M.M.A. and Groth, G. 2010. New insight in ethylene signaling: autokinase activity of ETR1 modulates the interaction of receptors and EIN2. Mol. Plant, 3: 882–889.

Biswas, B., Chan, P.K. and Gresshoff, P.M. 2009. A novel ABA insensitive mutant of Lotus japonicus with a wilty phenotype displays unaltered nodulation regulation. Mol. Plant., 2: 487–499.

Boiero, L., Perrig, D,, MAsciarelli, O., Penna, C., Cassan, F. and Luna, V. 2007. Phytohormone production by three strains of Bradyrhizobium japonicum and possible physiological and technological implications. Appl Microbiol. Biotechnol., 74: 874–880.

Boivin, S., Kazmierczak, T., Brault, M., Wen, J., Gamas, P., Mysore, K.S. et al. 2016. Different cytokinin histidine kinase receptors regulate nodule initiation as well as later nodule developmental stages in Medicago truncatula. Plant Cell Environ., 39: 2198–2209.

Bonfante, P. and Genre, A. 2008. Plants and arbuscular mycorrhizal fungi: an evolutionarydevelopmental perspective. Trends in Plant Science, 13: 492–498.

Brandstatter, I. and Kieber, J.J. 1998. Two genes with similarity to bacterial response regulators are rapidly and specifically induced by cytokinin in Arabidopsis. Plant Cell., 10: 1009–1019.

Breakspear, A., Liu, C., Roy, S., Stacey, N., Rogers, C., Trick, M. et al. 2014. The root hair "infectome" of Medicago truncatula uncovers changes in cell cycle genes and reveals a requirement for auxin signaling in rhizobial infection. Plant Cell., 26: 4680–4701.

Brett, J.F. and Mathesius, U. 2014. Phytohormone Regulation of Legume-Rhizobia Interactions. J. Chem. Ecol., 40: 770–790. DOI 10.1007/s10886-014-0472-7

Campanella, J.J., Smith, S.M., Leibu, D., Wexler, S. and Ludwig-M€uller, J. 2008. The auxin conjugate hydrolase family of Medicago truncatula and their expression during the interaction with two symbionts. J. Plant Growth Regul., 27: 26–38.

Chandler, Lynette, Chandler, L.K. and Dahlquist, C.M. 2015. Functional assessment: Strategies to prevent and remediate challenging behavior in school settings.

Chen, Y., Chen, W., Li, X., Jiang, H., Wu, P., Xia, K. et al. 2014. Knockdown of LjIPT3 influences nodule development in Lotus japonicus. Plant Cell Physiol., 55: 183–193.

Clarke, V.C., Loughlin, P.C., Day, D.A. and Smith, P.M.C. 2014. Transport processes of the legume symbiosome membrane. Front. Plant Sci., 5: 699.

Cooper, J.B. and Long, S.R. 1994. Morphogenetic rescue of Rhizobium meliloti nodulation mutants by trans-zeatin secretion. Plant Cell., 6: 215–225.

Cortleven, A., Leuendorf, J.E., Frank, M., Pezzetta, D., Bolt, S. and Schmeulling, T. 2019. Cytokinin action in response to abiotic and biotic stresses in plants. Plant Cell Environ., 42: 998–1018.

Couzigou, J.-M., Zhukov, V., Mondy, S., Abu el Heba, G., Cosson, V., Ellis, T.H.N. et al. 2012. NODULE ROOT and COCHLEATA maintain nodule development and are legume orthologs of Arabidopsis BLADE-ONPETIOLE genes. Plant Cell. 24: 4498–4510.

Davidson, S.E., Elliott, R.C., Helliwell, C.A., Poole, A.T. and Reid, J.B. 2003. The pea gene NA encodesent-kaurenoic acid oxidase. Plant Physiol., 131: 335–344.

Davidson, S.E., Smith, J.J., Helliwell, C.A., Poole, A.T. and Reid, J.B. 2004. The pea gene LH encodes ent-kaurene oxidase. Plant Physiol., 134: 1123–1134.

Davière, J.-M. and Achard, P. 2013. Gibberellin signaling in plants. Development, 140: 1147–1151.

de Billy, F., Grosjean, C., May, S., Bennett, M. and Cullimore, J.V. 2001. Expression studies on AUX1-like genes in Medicago truncatula suggest that auxin is required at two steps in early nodule development. Mol. Plant Microbe Interact., 14: 267–277.

Dehio, C. and de Bruijn, F.J. 1992. The early nodulin gene SrEnod2 from Sesbania rostrata is inducible by cytokinin. Plant J., 2: 117–128.

Den Herder, G. and Parniske, M. 2009. The unbearable naivety of legumes in symbiosis. Curr. Opin. Plant Biol., 12: 491–499.

Desbrosses, G.J. and Stougaard, J. 2011. Root nodulation: a paradigm for how plant-microbe symbiosis influences host developmental pathways. Cell Host Microbe., 10: 348–358.

Dharmasiri, N., Dharmasiri, S. and Estelle, M. 2005a. The F-box protein TIR1 is an auxin receptor. Nature, 435: 441–445.

Dharmasiri, N., Dharmasiri, S., Weijers, D., Lechner, E., Yamada, M., Hobbie, L. et al. 2005b. Plant development is regulated by a family of auxin receptor F box proteins. Dev. Cell, 9: 109–119.

Dill, A., Thomas, S.G., Hu, J., Steber, C.M. and Sun, T.P. 2004. The Arabidopsis F-box protein SLEEPY1 targets gibberellin signaling repressors for gibberellin-induced degradation. Plant Cell, 16: 1392–1405.

Ding, Y., Kalo, P., Yendrek, C., Sun, J., Liang, Y., Marsh, J.F. et al. 2008. Abscisic acid coordinates Nod Factor and cytokinin signaling during the regulation of nodulation in Medicago truncatula. Plant Cell., 20: 2681–95.

Ding, Y. and Oldroyd, G.E.D. 2009. Positioning the nodule, the hormone dictum. Plant Signal Behav., 4: 89–93.

Dobrev, P.I., Havlicek, L., Vagner, M., Malbeck, J. and Kaminek, M. 2005. Purification and determination of plant hormones auxin and abscisic acid using solid phase extraction and two-dimensional high performance liquid chromatography. J. Chromatograghy, 1075: 159–166.

Drissner, D., Kunze, G., Callewaert, N., Gehrig, P., Tamasloukht, M., Boller, T. et al. 2007. Lyso-Phosphatidylcholine is a signal in the arbuscular mycorrhizal symbiosis. Science, 318: 265–268.

Ferguson B.J., Ross J.J. and Reid J.B. 2005. Nodulation phenotypes of gibberellin and brassinosteroid mutants of pea. Plant Physiol., 138: 2396–2405.

Ferguson, B.J. and Mathesius, U. 2014. Phytohormone regulation of legumerhizobia interactions. J. Chem. Ecol., 40: 770–790. doi: 10.1007/s10886-014- 0472-7

Ferguson, B.J. and Reid, J.B. 2005. Cochleata: getting to the root of legume nodules. Plant Cell Physiol., 46: 1583–1589.

Ferguson, B.J., Foo, E., Ross, J.J. and Reid, J.B. 2011. Relationship between gibberellin, ethylene and nodulation in Pisum sativum. New Phytol., 189: 829–842.

Ferguson, B.J., Indrasumunar, A., Hayashi, S., Lin, M.H., Lin, Y.H., Reid, D.E. et al. 2010. Molecular analysis of legume nodule development and autoregulation. J. Integr. Plant Biol., 52: 61–76.

Ferreira, F.J. and Kieber, J.J. 2005. Cytokinin signaling. Curr. Opin. Plant Biol. 8:518-525.

Fonouni-Farde, C., Diet, A., and Frugier, F. 2016b. Root development and endosymbioses: DELLAs lead the orchestra. Trends Plant Sci., 21: 898–900.

Fonouni-Farde, C., Tan, S., Baudin, M., Brault, M., Wen, J., Mysore, K.S. et al. 2016a. DELLA-mediated gibberellin signalling regulates Nod factor signalling and rhizobial infection. Nat. Commun., 7: 12–636.

Fonouni-Farde, Camille & Kisiala, Anna & Brault, Mathias & Emery, R J & Diet, Anouck & Frugier, Florian. 2017. DELLA1-gibberellin signaling regulates cytokinin-dependent symbiotic nodulation. Plant Physiology, 175: 00919.

Foo, E., Ferguson, B.J. and Reid, J.B. 2014. The potential roles of strigolactones and brassinosteroids in the autoregulation of nodulation pathway. Ann. Bot.

Franssen, H.J., Xiao, T.T., Kulikova, O., Wan, X., Bisseling, T., Scheres, B., and Heidstra, R. 2015. Root developmental programs shape the Medicago truncatula nodule meristem. Development. 142:2941–2950.

Frugier, F., Kosuta, S., Murray, J.D., Crespi, M. and Szczyglowski, K. 2008. Cytokinin: secret agent of symbiosis. Trends Plant Sci., 13: 115–120.

Gage, D.J. and Margolin, W. 2000. Hanging by a thread: invasion of legume plants by rhizobia. Curr. Opin. Microbiol., 3: 613–617.

Gamalero, E. and Glick, B.R. 2015. Bacterial modulation of plant ethylene levels. Plant Physiol. 169: 1–10. doi:10.1104/pp.15.00284

Gamas, P., Brault, M., Jardinaud, M.F. and Frugier, F. 2017. Cytokinins in symbiotic nodulation: when, where, what for? Trends Plant Sci., 22: 792–802.

Gampala, S.S., Hagenbeek, D. and Rock, C.D. 2001. Functional interactions of lanthanum and phospholipase D with the abscisic acid signaling effectors VP1 and ABI1-1 in rice protoplasts. J. Biol. Chem., 276: 9855–60.

Gauthier-Coles, C., White, R.G. and Mathesius, U. 2019. Nodulating legumes are distinguished by a sensitivity to cytokinin in the root cortex leading to pseudonodule development. Front. Plant Sci., 9: 1901.

Gherbi, H., Markmann, K., Svistoonoff, S., Estevan, J., Autran, D., Giczey, G. et al. 2008. SymRK defines a common genetic basis for plant root endosymbioses with arbuscular mycorrhiza fungi, rhizobia, and Frankiabacteria. PNAS, 105(12): 4928–4932.

Giraud, E., Moulin, L., Vellenet, D. et al 2007. Legumes symbioses: absence of Nod genes in photosynthetic bradyrhizobia. Science, 316: 1307–1312.

Gleason, C., Chaudhuri, S., Yang, T., Muñoz, A., Poovaiah, B.W. and Oldroyd, G.E.D. 2006. Nodulation independent of rhizobia induced by a calcium-activated kinase lacking autoinhibition. Nature, 441: 1149–1152.

Gomez-Roldan, V., Roux, C., Girard, D., Bécard, G. and Puech, V. 2007. Strigolactones: Promising plant signals. Plant Signaling and Behavior, 2: 163–164.

Gonzalez-Rizzo, S., Crespi, M. and Frugier, F. 2006. The Medicago truncatula CRE1 cytokinin receptor regulates lateral root development and early symbiotic interaction with Sinorhizobium meliloti. Plant Cell., 18: 2680–2693.

Govindarajulu, M., Pfeffer, P.E., Jin, H.R., Abubaker, J., Douds, D.D., Allen, J.W. et al. 2005. Nitrogen transfer in the arbuscular mycorrhizal symbiosis. Nature, 435: 819–823.

Griesmann, M., Chang, Y., Liu, X., Song, Y., Haberer, G., Crook, M.B. et al. 2018. Phylogenomics reveals multiple losses of nitrogen-fixing root nodule symbiosis. Science, 361: 1743.

Griffiths, J., Murase, K., Rieu, I., Zentella, R., Zhang, Z.-L., Powers, S.J. et al. 2006. Genetic characterization and functional analysis of the GID1 gibberellin receptors in Arabidopsis. Plant Cell., 18: 3399–3414.

Guan, D., Stacey, N., Liu, C., Wen, J., Mysore, K.S., Torres-Jerez, I. et al. 2013. Rhizobial infection is associated with the development of peripheral vasculature in nodules of Medicago truncatula. Plant Physiol., 162: 107–115.

Guether, M., Neuhäuser, B., Balestrini, R., Dynowski, M., Ludewig, U. and Bonfante, P. 2009. A mycorrhizal-specific ammonium transporter from *Lotus japonicus* acquires nitrogen released by arbuscular mycorrhizal fungi," Plant Physiology, 150: 73–83.

Guo, H. and Ecker, J.R. 2003. Plant responses to ethylene gas are mediated by SCFEBF1/EBF2- dependent proteolysis of EIN3 transcription factor. Cell., 115: 667–677.

Harrison, M.J. 2005. Signaling in the arbuscular mycorrhizal symbiosis. Annual Review of Microbiology, 59: 19–42.

Hause, B., Maier, W., Miersch, O., Kramell, R. and Strack, D. 2002. Induction of jasmonate biosynthesis in arbuscular mycorrhizal barley roots. Plant Physiology, 130: 1213–1220.

Hawkins, E.J., Johansen, A. and George, E. 2000. Uptake and transport of organic and inorganic nitrogen by arbuscular mycorrhizal fungi. Plant and Soil, 226: 275–285.

Hayashi, S., Gresshoff, P.M. and Ferguson, B.J. 2014. Mechanistic action of gibberellins in legume nodulation. J. Integr. Plant Biol., 56: 971–978.

Heckmann, A.B., Sandal, N., Bek, A.S., Madsen, L.H., Jurkiewicz, A., Nielsen, M.W. et al. 2011. Cytokinin induction of root nodule primordia in Lotus japonicus is regulated by a mechanism operating in the root cortex. Mol. Plant Microbe Interact, 24: 1385–1395.

Held, M., Hou, H., Miri, M., Huynh, C., Ross, L., Hossain, M.S. et al. 2014. Lotus japonicus cytokinin receptors work partially redundantly to mediate nodule formation. Plant Cell. 26: 678–694.

Herrera-Medina, M.J., Steinkellner, S., Vierheilig, H., Bote, J.A.O. and Garrido, J.M.G. 2007. Abscisic acid determines arbuscule development and functionality in the tomato arbuscular mycorrhiza. New Phytologist, 175: 554–56.

Herridge, D.F., Peoples, M.B. and Boddy, R.M. 2008. Global inputs of biological nitrogen fixation in agricultural systems. Plant Soil., 311: 1–18. doi:10.1007/s11104-008-9668-3

Hiltenbrand, R., Thomas, J., McCarthy, H., Dykema, K.J., Spurr, A., Newhart, H. et al. 2016. A developmental and molecular view of formation of auxin-induced nodule-like structures in land plants. Front. Plant Sci., 7: 1692.

Hodge, A., Campbell, C.D. and Fitter, A.H. 2001. An arbuscular mycorrhizal fungus accelerates decomposition and acquires nitrogen directly from organic material. Nature, 413: 297–299.

Hossain, M.S., Shrestha, A., Zhong, S., Miri, M., Austin, R.S., Sato, S. et al. 2016. Lotus japonicus NF-YA1 plays an essential role during nodule differentiation and targets members of the SHI/STY gene family. Mol. Plant Microbe Interact, 29: 950–964.

Hu, J., Mitchum, M.G., Barnaby, N., Ayele, B.T., Ogawa, M., Nam, E. et al. 2008. Potential sites of bioactive gibberellin production during reproductive growth in Arabidopsis. Plant Cell., 20: 320–336.

Hunter, W.J. 2001. Influence of root-applied epibrassinolide and carbenoxolone on the nodulation and growth of soybean (Glycine max L.) seedlings. J. Agron. Crop Sci., 186: 217–221.

Ioio, R.D., Linhares, F.S., Scacchi, E., Casamitjana Martinez, E., Heidstra, R., Costantino, P. et al. 2007. Cytokinins determine Arabidopsis rootmeristem size by controlling cell differentiation. Curr. Biol. 17: 678–682.

Isayenkov, S., Mrosk, C., Stenzel, I., Strack, D. and Hause, B. 2005. Suppression of allene oxide cyclase in hairy roots of Medicago truncatula reduces jasmonate levels and the degree of mycorrhization with Glomus intraradices. Plant Physiology, 139: 1401–1410.

Javid, M.G., Sorooshzadeh, A., Moradi, F., Sanavy, S.A. and Sanavy, I. 2011. The role of phytohormones in alleviating salt stress in crop plants. Aust. Jour. of Crop Sci., 5(6): 726–734.

Jentschel, K., Thiel, D., Rehn, F. and Ludwig-Müller, J. 2007. Arbuscular mycorrhiza enhances auxin levels and alters auxin biosynthesis in Tropaeolum majus during early stages of colonization. Physiologia Plantarum., 129: 320–333.

Jin, Y., Liu, H., Luo, D., Yu, N., Dong, W., Wang, C. et al. 2016. DELLA proteins are common components of symbiotic rhizobial and mycorrhizal signaling pathways. Nat. Commun., 7: 12433.

Jones, J.M.C., Clairmont, L., Macdonald, E.M., Weiner, C.A., Emery, R.J.N. and Guinel, F.C. 2015. E151 (sym15), a pleiotropic mutant of pea (*Pisum sativum* L.), displays low nodule number, enhanced mycorrhizae, delayed lateral root emergence, and high root cytokinin levels. J. Exp. Bot. Adv. Access, 66: 4047. doi:10.1093/jxb/erv201

Ju, C., Yoon, G.M., Shemansky, J.M., Lin, D.Y., Ying, Z.I., Chang, J. et al. 2012. CTR1 phosphorylates the central regulator EIN2 to control ethylene hormone signaling from the ER membrane to the nucleus in Arabidopsis. Proc. Natl. Acad. Sci. U.S.A., 109: 19486–19491.

Khadri, M., Tejera, N.A. and Lluch, C. 2006. Alleviation of salt stress in common bean (Phaseolus vulgaris) by exogenous abscisic acid supply. J. Plant Growth Regul., 25: 110–119.

Kieber, J.J. and Schaller, G. 2014. Cytokinins. Arabidopsis Book, 12: e0168.

Kim, G.B., Son, S.U., Yu, H.J. and Mun, J.H. 2019. MtGA2ox10 encoding C20-GA2-oxidase regulates rhizobial infection and nodule development in Medicago truncatula. Sci. Rep., 9: 5952.

Kim, S.K., Chang, S.C., Lee, E.J., Chung, W.S., Kim, Y.S., Hwang, S. et al. 2000. Involvement of brassinosteroids in the gravitropic response of primary root of maize. Plant Physiol., 123: 997–1004.

Kinkema, M. and Gresshoff, P.M. 2008. Investigation of downstream signals of the soybean autoregulation of nodulation receptor kinase GmNARK. Mol Plant Microbe Interact., 21: 1337–1348.

Kiseleva, A.A., Tarachovskaya, E.R. and Shishova, M.F. 2012. Biosynthesis of phytohormones in alga. Russ. J. Plant Physiol., 59: 595–610.

Kohlen, W., Ng, J.L.P., Deinum, E.E. and Mathesius, U. 2018. Auxin transport, metabolism, and signalling during nodule initiation: indeterminate and determinate nodules. J. Exp. Bot., 69: 229–244.

Korasick, D.A., Enders, T.A. and Strader, L.C. 2013. Auxin biosynthesis and storage forms. J. Exp. Bot., 64: 2541–2555.

Kouchi, H. 2004. Large-scale analysis of gene expression profiles during early stages of root nodule formation in a model legume, Lotus japonicus. DNA Res., 11: 263–274.

Kurakawa, T., Ueda, N., Maekawa, M., Kobayashi, K., Kojima, M., Nagato, Y. et al. 2007. Direct control of shoot meristem activity by a cytokinin-activating enzyme. Nature., 445: 652–655.

Kurepa, J. and Smalle, J.A. 2022. Auxin/Cytokinin Antagonistic Control of the Shoot/Root Growth Ratio and Its Relevance for Adaptation to Drought and Nutrient Deficiency Stresses. Int. J. Mol. Sci., 23(4): 1933. doi: 10.3390/ijms23041933. PMID: 35216049, PMCID: PMC8879491

Larrainzar, E., Riely, B.K., Kim, S.C., Carrasquilla-Garcia, N., Yu, H.J., Hwang, H.J. et al. 2015. Deep sequencing of the Medicago truncatula root transcriptome reveals a massive and early interaction between nodulation factor and ethylene signals. Plant Physiol., 169: 233–265.

Li, X., Lei, M., Yan, Z., Wang, Q., Chen, A., Sun, J. et al. 2014. The REL3-mediated TAS3 ta-siRNA pathway integrates auxin and ethylene signaling to regulate nodulation in Lotus japonicus. New Phytol., 201: 531–544.

Lian, B., Zhou, X., Miransari, M. and Smith, D.L. 2000. Effects of salicylic acid on the development and root nodulation of soybean seedlings. J. Agron. Crop Sci., 185: 187–192.

Liang, Y. and Harris, J.M. 2005. Response of root branching to abscisic acid is correlated with nodule formation both in legumes and non-legumes. Am. J. Bot., 92: 1675–1683.

Lin, J., Frank, M. and Reid, D. 2020. No Home without Hormones: How Plant Hormones Control Legume Nodule Organogenesis. Plant Comm. 1: 100104.

Linkies, A., Muller, K., Morris, K., Tureckova, V., Wenk, M., Cadman, C.S.C. et al. 2009. Ethylene interacts with abscisic acid to regulate endosperm rupture during germination: a comparative approach using Lepidium sativum and Arabidopsis thaliana. Plant Cell., 21: 3803–3822.

Liu, C.W., Breakspear, A., Roy S. and Murray, J.D. 2015. Cytokinin responses counterpoint auxin signaling during rhizobial infection. Plant Signal. Behavior., 10: e1019982.

Liu, H., Sandal, N., Andersen, K.R., James, E.K., Stougaard, J., Kelly, S. et al. 2018b. A genetic screen for plant mutants with altered nodulation phenotypes in response to rhizobial glycan mutants. New Phytol., 220: 526–538.

Ljung, K., Bhalerao, R.P. and Sandberg, G. 2001. Sites and homeostatic control of auxin biosynthesis in Arabidopsis during vegetative growth. Plant J., 21: 465–474.

Lohar, D.P., Schaff, J.E., Laskey, J.G., Kieber, J.J., Bilyeu, K.D. and Bird, D.M. 2004. Cytokinins play opposite roles in lateral root formation, and nematode and rhizobial symbioses. Plant J., 38: 203–214.

Lomax, T.L., Muday, G.K. and Rubery, P.H. 2001. Auxin transport. In: Plant hormones. Dordrecht: Kluwer., 509–530.

Lorenzo, O., Piqueras, R., Sanchez-Serrano, J.J. and Solano, R. 2003. ETHYLENE RESPONSE FACTOR1 integrates signals from ethylene and jasmonate pathways in plant defense. Plant Cell., 15: 165–178.

Lorteau, M.A., Ferguson, B.J. and Guinel, F.C. 2001. Effects of cytokinin on ethylene production and nodulation in pea (Pisum sativum) cv. Sparkle. Physiol. Plant., 112: 421–428.

Lucas-Garcia, J.A., Probanza, A., Ramos, B., Colón-Flores, J.J. and Gutierrez-Mañero, F.J. 2004. Effects of plant growth promoting rhizobateria (PGPRs) on the biological nitrogen fixation, nodulation and growth of Lupinus albus I. cv. Multolupa. Engineering Life Sciences, 4: 71–77.

Mabood, F. and Smith, D.L. 2005. Pre-inoculation of Bradyrhizobium japonicum with jasmonates accelerates nodulation and nitrogen fixation in soybean (Glycine max) at optimal and suboptimal root zone temperatures. Physiol Plant., 125: 311–323.

Mabood, F., Souleimanov, A., Khan, W. and Smith, D.L. 2006. Jasmonates induce Nod factor production by Bradyrhizobium japonicum. Plant Physiol Biochem. 44: 759–765.

Maekawa, T., Maekawa-Yoshikawa, M., Takeda, N., Imaizumi-Anraku, H., Murooka, Y. and Hayashi, M. 2009. Gibberellin controls the nodulation signaling pathway in Lotus japonicus. Plant J., 58: 183–194.

Magne, K., Couzigou, J.M., Schiessl, K., Liu, S., George, J., Zhukov, V. et al. 2018. MtNODULE ROOT1 and MtNODULE ROOT2 are essential for indeterminate nodule identity. Plant Physiol., 178: 295–316.

Mathesius, U. 2008. Auxin: at the root of nodule development? Funct. Plant Biol., 35: 651–668.

Mathesius, U., Charon, C., Rolfe, B.G., Kondorosi, A. and Crespi, M. 2000. Temporal and spatial order of events during the induction of cortical cell divisions in white clover by Rhizobium leguminosarum bv. trifolii inoculation or localized cytokinin addition. Mol. Plant-Microbe Interac., 13: 617–628.

McAdam, E.L., Reid, J.B. and Foo, E. 2018. Gibberellins promote nodule organogenesis but inhibit the infection stages of nodulation. J. Exp. Bot., 69: 2117–2130.

McGinnis, K.M., Thomas, S.G., Soule, J.D., Strader, L.C., Zale, J.M., Sun, T.-P. et al. 2003. The Arabidopsis SLEEPY1 gene encodes a putative F-box subunit of an SCF E3 ubiquitin ligase. Plant Cell., 15: 1120–1130.

Meakin, G.E., Jepson, B.J.N., Richardson, D.J., Bedmar, E.J. and Delgado, M.J. 2006. The role of Bradyrhizobium japonicum nitric oxide reductase in nitric oxide detoxification in soybean root nodules. Biochemical Society Transactions, 34: 195–196.

Mens, C., Li, D., Haaima, L.E., Gresshoff, P.M. and Ferguson, B.J. 2018. Local and systemic effect of cytokinins on soybean nodulation and regulation of their isopentenyl transferase (IPT) biosynthesis genes following rhizobia inoculation. Front. Plant Sci., 9: 1150.

Miri, M., Janakirama, P., Held, M., Ross, L. and Szczyglowski, K. 2016. Into the root: how cytokinin controls rhizobial infection. Trends Plant Sci., 21: 178–186.

Mitchum, M.G., Yamaguchi, S., Hanada, A., Kuwahara, A., Yoshioka, Y., Kato, T. et al. 2006. Distinct and overlapping roles of two gibberellin 3-oxidases in Arabidopsis development. Plant J., 45: 804–818.

Mortier, V., Wasson, A., Jaworek, P., De Keyser, A., Decroos, M., Holsters, M. et al. 2014. Role of LONELY GUY genes in indeterminate nodulation on Medicago truncatula. New Phytol., 202: 582–593. doi:10.1111/nph.12681

Moubayidin L., Mambro R. and Sabatini S. 2009. Cytokininauxin crosstalk. Trends Plant Sci., 14: 557–562.

Mukherjee, A. and Ané, J.M. 2011. Plant hormones and initiation of legume nodulation and arbuscular mycorrhization. In: Polacco, J.C. and Todd, C.D. (Eds.). Ecological Aspects of Nitrogen Metabolism in Plants, 1st Edn. (Hoboken, N.J., John, W. & Sons, Inc) 354–396.

Murray, J.D. 2011. Invasion by invitation: rhizobial infection in legumes. Mol. Plant-Microbe Interac., 24: 631–639.

Murray, J.D., Karas, B.J., Sato, S., Tabata, S., Amyot, L. and Sczczyglowski, K. 2007. A cytokinin perception mutant colonized by Rhizobium in the absence of nodule organogenesis. Science, 315: 101–104.

Murset, V., Hennecke, H. and Pessi, G. 2012. Disparate role of rhizobial ACC deaminase in root-nodule symbioses. Symbiosis, 57: 43–50. doi:10.1007/s13199-012-0177-z

Nadzieja, M., Kelly, S., Stougaard, J. and Reid, D. 2018. Epidermal auxin biosynthesis facilitates rhizobial infection in Lotus japonicus. Plant J., 95: 101–111.

Nadzieja, M., Stougaard, J. and Reid, D. 2019. A toolkit for high resolution imaging of cell division and phytohormone signaling in legume roots and root nodules. Front. Plant Sci., 10: 1000.

Nakagawa, T. and Kawaguchi, M. 2006. Shoot-appliedMeJA suppresses root nodulation in Lotus japonicus. Plant Cell Physiol., 47: 176–180.

Naz, I., Bano, A. and Hussan, T. 2009. Isolation of phytohormones producing plant Growth promoting rhizobacteria from weeds growing in Khewra salt range, Pakistan and their implication in providing salt tolerance to Glycine max L. Africn Jour of Biotech., 8(21): 5762–5766.

Ng, J.L.P., Hassan, S., Truong, T.T., Hocart, C.H., Laffont, C., Frugier, F. and Mathesius, U. 2015. Flavonoids and auxin transport inhibitors rescue symbiotic nodulation in the Medicago truncatula cytokinin perception mutant cre1. Plant Cell., 27: 2210–2226.

Ng, J.L.P., Welvaert, A., Wen, J., Chen, R. and Mathesius, U. 2020. The Medicago truncatula PIN2 auxin transporter mediates basipetal auxin transport but is not necessary for nodulation. J. Exp. Bot., 71: 1562–1573.

Nizampatnam, N.R., Schreier, S.J., Damodaran, S., Adhikari, S. and Subramanian, S. 2015. microRNA160 dictates stage-specific auxin and cytokinin sensitivities and directs soybean nodule development. Plant J., 84: 140–153.

Nordström, A., Tarkowski, P., Tarkowska, D., Norbaek, R., Astot, C., Dolezal, K. and Sanberg, G. 2004. Auxin regulation of cytokinin biosynthesis in Arabidopsis thaliana: a factor of potential importance for auxincytokinin – regulated development. Proc. Natl. Acad. Sci. USA., 101: 8039–8044.

Oldroyd, G.E.D., Murray, J.D., Poole, P.S. and Downie, J.A. 2011. The rules of engagement in the legume-rhizobial symbiosis. Annu. Rev. Genet., 45: 119–44. doi:10.1146/annurev-genet-110410-132549

Oldroyd, G.E., Engstrom, E.M. and Long, S.R. 2001. Ethylene inhibits the Nod factor signal transduction pathway of Medicago truncatula. Plant Cell., 13: 1835–1849.

Oldroyd, G.E.D. 2013. Speak, friend, and enter: Signalling systems that promote beneficial symbiotic associations in plants. Nat. Rev. Microbiol., 11: 252–263.

Pacios-Bras, C., Schlaman, H.R.M., Boot, K., Admiraal, P., Mateos Langerak, J., Stougaard, J. et al. 2003. Auxin distribution in Lotus japonicus during root nodule development. Plant Mol. Biol., 52: 1169–1180.

Palma, F,. López-Gómez, M., Tejera, N.A. and Lluch, C. 2014. Involvement of abscisic acid in the response of Medicago sativa plants in symbiosis with Sinorhizobium meliloti to salinity. Plant Sci., 223: 16–24.

Palma, F., López-Gómez, M., Tejera, N.A. and Lluch, C. 2013. Salicylic acid improves the salinity tolerance of Medicago sativa in symbiosis with Sinorhizobium meliloti by preventing nitrogen fixation inhibition. Plant Sci., 208: 75–82. doi:10.1016/j.plantsci.2013.03.015

Parry, G., Calderon-Villalobos, L.I., Prigge, M., Peret, B., Dharmasiri, S., Itoh, H. et al. 2009. Complex regulation of the TIR1/AFB family of auxin receptors. Proc. Natl. Acad. Sci. USA., 106: 22540–22545.

Patra, D. and Mandal, S. 2022. Nod–factors are dispensable for nodulation: A twist in bradyrhizobia-legume symbiosis. Symbiosis., 86: 1–15. https://doi.org/10.1007/s13199-021-00826-9

Penmetsa, R.V., Frugoli, J.A., Smith, L.S., Long, S.R. and Cook, D.R. 2003. Dual genetic pathways controlling nodule number in Medicago truncatula. Plant Physiol., 131: 998–1008.

Penmetsa, V.P., Uribe, P., Anderson, J., Lichtenzveig, J., Gish, J.C., Nam, Y.W. et al. 2008. The Medicago truncatula ortholog of Arabidopsis EIN2, sickle, is a negative regulator of symbiotic and pathogenic microbial associations. Plant J., 55: 580–595. doi:10.1111/j.1365-313X.2008.03531.x

Pii, Y., Crimi, M., Cremonese, G., Spena, A. and Pandolfini, T. 2007. Auxin and nitric oxide control indeterminate nodule formation. BMC Plant Biol., 7: 21.

Plet, J., Wasson, A., Ariel, F., Le Signor, C., Baker, D., Mathesius, U., Crespi, M., and Frugier, F. 2011. MtCRE1-dependent cytokinin signaling integrates bacterial and plant cues to coordinate symbiotic nodule organogenesis in Medicago truncatula. Plant J., 65: 622–633.

Potuschak, T., Lechner, E., Parmentier, Y., Yanagisawa, S., Grava, S., Koncz, C. et al. 2003. EIN3-Dependent regulation of plant ethylene hormone signaling by two Arabidopsis F box proteins: EBF1 and EBF2. Cell., 115: 679–689.

Poustini, K., Mabood, F. and Smith, D.L. 2005. Low root zone temperature effects on bean (Phaseolus vulgaris L.) plants inoculated with Rhizobium leguminosarum bv. phaseoli pre-incubated with methyl jasmonate and/or genistein. Acta Agric Scand Sect B Soil Plant Sci., 55: 293–298.

Qiao, H., Chang, K.N., Yazaki, J. and Ecker, J.R. 2009. Interplay between ethylene, ETP1/ETP2 F-box proteins, and degradation of EIN2 triggers ethylene responses in Arabidopsis. Genes Dev., 23: 512–521.

Reid, D., Nadzieja, M., Novák, O., Heckmann, A.B., Sandal, N. and Stougaard, J. 2017. Cytokinin biosynthesis promotes cortical cell responses during nodule development. Plant Physiol., 175: 361–375.

Reid, D.E., Heckmann, A.B., Novák, O., Kelly, S. and Stougaard, J. 2016. CYTOKININ OXIDASE/DEHYDROGENASE3 maintains cytokinin homeostasis during root and nodule development in Lotus japonicus. Plant Physiol., 170: 1060–1074.

Reid, D., Liu, H., Kelly, S., Kawaharada, Y., Mun, T., Andersen, S.U. et al. 2018. Dynamics of Lotus japonicus ethylene production in response to compatible Nod factor. Plant Physiol., 176: 1764–1772.

Rightmyer, A.P. and Long, S.R. 2011. Pseudonodule formation by wild-type and symbiotic mutant Medicago truncatula in response to auxin transport inhibitors. Mol. Plant Microbe Interact., 24: 1372–1384.

Rockström, J., Steffen, W., Noone, K. and Persson, A. 2009. A safe operating space for humanity. Nature., 461: e475.

Ross, J.J., O'Neill, D.P., Smith, J.J., Kerckhoffs, L.H.J. and Elliott, R.C. 2000. Evidence that auxin promotes gibberellin A1 biosynthesis in pea. Plant J., 21: 547–552.

Roy, S., Robson, F., Lilley, J., Liu, C.-W., Cheng, X., Wen, J. et al. 2017. MtLAX2, a functional homologue of the Arabidopsis auxin influx transporter AUX1, is required for nodule organogenesis. Plant Physiol., 174: 326–338.

Sakai, H., Honma, T., Aoyama, T., Sato, S., Kato, T., Tabata, S. et al. 2001. ARR1, a transcription factor for genes immediately responsive to cytokinins. Science., 294: 1519–1521.

Sakakibara, H. 2006. Cytokinins: activity, biosynthesis, and translocation. Annu. Rev. Plant Biol., 57: 431–449.

Santos, R., Herouart, D., Sigaud, S., Touati, D. and Puppo, A. 2001. Oxidative burst in alfalfa-Sinorhizobium meliloti symbiotic interaction. Mol. Plant Microbe Int., 14: 86–89.

Sasaki, T., Suzaki, T., Soyano, T., Kojima, M., Sakakibara, H. and Kawaguchi, M. 2014. Shoot-derived cytokinins systemically regulate root nodulation. Nat. Commun., 5: 4983.

Sato, T., Fujikake, H., Ohtake, N., Sueyoshi, K., Takahashi, T., Sato, A. and Ohyama, T. 2002. Effect of exogenous salicylic acid supply on nodule formation of hypernodulating mutant and wild type of soybean. Soil Sci Plant Nutr., 48: 413–420.

Schaller, G.E. 2012. Ethylene and the regulation of plant development. BMC Biol., 10: 9.

Schaller, G.E. and Kieber, J.J. 2002. Ethylene. Arabidopsis Book, 1: e0071.

Schiessl, K., Lilley, L.S., Lee, T., Tamvakis, I., Kohlen, W., Bailey, P.C. et al. 2019. NODULE INCEPTION recruits the lateral root developmental program for symbiotic nodule organogenesis in Medicago truncatula. Curr. Biol., 29: 3657–3668.e5.

Schwechheimer, C. 2012. Gibberellin signaling in plants—the extended version. Front. Plant Sci., 2: 107.

Seo, H.S., Li, J., Lee, S.Y., Yu, J.W., Kim, K.H., Lee, S.H. et al. 2006. The hypernodulating nts mutation induces jasmonate synthetic pathway in soybean leaves. Mol. Cells., 24: 185–193.

Serova, T.A., Tsyganova, A.V., Tikhonovich, I.A. and Tsyganov, V.E. 2019. Gibberellins Inhibit Nodule Senescence and Stimulate Nodule Meristem Bifurcation in Pea (Pisum sativum L.). Frontiers in Plant Science, 10. ISSN=1664-462X. DOI=10.3389/fpls.2019.00285.

Shahid, M.A., Pervez, M.A., Balal, R.M., Mattson, N.S., Rashid, A., Ahmad, R. et al. 2011. Brassinosteroid (24-epibrassinolide) enhances growth and alleviates the deleterious effects induced by salt stress in pea (Pisum sativum L.). Aust. J. Crop. Sci., 5: 500–510.

Shen, D., Xiao, T.T., van Velzen, R., Kulikova, O., Gong, X., Geurts, R. et al. 2020. A homeotic mutation changes legume nodule ontogeny into actinorhizal-type ontogeny. Plant Cell., 32: 1868–1885.

Singh, V.K., Jain, M. and Garg, R. 2014a. Genome-wide analysis and expression profiling suggest diverse roles of GH3 genes during development and abiotic stress responses in legumes. Front. Plant Sci., 5:789.

Soyano, T., Kouchi, H., Hirota, A. and Hayashi, M. 2013. Nodule inception directly targets NF-Y subunit genes to regulate essential processes of root nodule development in Lotus japonicus. PLoS Genet., 9(3): e1003352.

Soyano, T., Liu, M., Kawaguchi, M. and Hayashi, M. 2021. Leguminous nodule symbiosis involves recruitment of factors contributing to lateral root development. Current Opinion in Plant Biology, 59: 102000.

Soyano, T., Shimoda, Y., Kawaguchi, M. and Hayashi, M. 2019. A shared gene drives lateral root development and root nodule symbiosis pathways in Lotus. Science, 366: 1021–1023.

Spaepen, S., Vanderleyden, J. and Remans, R. 2007. Indole-3-acetic acid in microbial and microorganism-plant signaling. FEMS Microbiol Rev., 31: 425–448.

Sprent, J.I. 2007. Evolving ideas of legume evolution and diversity: a taxonomic perspective on the occurrence of nodulation. New Phytol., 174: 11–25.

Sprent, J.I. 2008. 60 Ma of legume nodulation: what's new? what's changing? J. Exp. Bot., 59: 1081–1084.

Springmann, M., Clark, M., Mason-D'Croz, D., Wiebe, K., Bodirsky, B.L., Lassaletta, L. et al. 2018. Options for keeping the food system within environmental limits. Nature., 562: 519–525.

Stacey, G., McAlvin, C.B., Kim, S.Y., Olivares, J. and Soto, M.J. 2006. Effects of endogenous salicylic acid on nodulation in the model legumes Lotus japonicus and Medicago truncatula. Plant Physiol., 141: 1473–1481.

Stepanova, A.N., Robertson-Hoyt, J., Yun, J., Benavente, L.M., Xie, D.Y., Dole_zal, K. et al. 2008. TAA1-Mediated auxin biosynthesis is essential for hormone crosstalk and plant development. Cell., 133: 177–191.

Sun, J., Cardoza, V., Mitchell, D.M., Bright, L., Oldroyd, G. and Harris, J.M. 2006. Crosstalk between jasmonic acid, ethylene and Nod factor signaling allows integration of diverse inputs for regulation of nodulation. Plant J., 46: 961–970.

Sun, T.P. 2008. Gibberellin metabolism, perception and signaling pathways in Arabidopsis. Arabidopsis Book, 6: e0103.

Suzaki, T., Yano, K., Ito, M., Umehara, Y., Suganuma, N. and Kawaguchi, M. 2012. Positive and negative regulation of cortical cell division during root nodule development in Lotus japonicus is accompanied by auxin response. Development, 139: 3997–4006.

Suzuki, A., Akune, M., Kogiso, M., Imagama, Y., Osuki, K., Uchiumi, T. et al. 2004. Control of nodule number by the phytohormone abscisic acid in the roots of two leguminous species. Plant Cell Physiol., 45: 914–922.

Suzuki, A., Suriyagoda, L., Shigeyama, T., Tominaga, A., Sasaki, M., Hiratsuka, Y. et al. 2011. Lotus japonicus nodulation is photomorphogenetically controlled by sensing the red/far red (R/FR) ratio through jasmonic acid (JA) signaling. Proc. Natl. Acad. Sci. USA., 108: 16837–16842.

Takanashi, K., Sugiyama, A. and Yazaki, K. 2011. Involvement of auxin distribution in root nodule development of Lotus japonicus. Planta., 234: 73–81.

Teale, W.D., Paponov, I.A. and Palme, K. 2006. Auxin in action: signalling, transport and the control of plant growth and development. Nat. Rev. Mol. Cell Biol., 7: 847–859.

Terakado, J., Fujihara, S., Goto, S., Kuratani, R., Suzuki, Y., Yoshida, S. et al. 2005. Systemic effect of a brassinosteroid on root nodule formation in soybean as revealed by the application of brassinolide and brassinazole. Soil Sci. Plant Nutr., 51: 389–395.

Thomas, J., Bowman, M.J., Vega, A., Kim, H.R. and Mukherjee, A. 2018. Comparative transcriptome analysis provides key insights into gene expression pattern during the formation of nodule-like structures in Brachypodium. Funct. Integr. Genomics, 18: 315–326.

Thomas, J., Hiltenbrand, R., Bowman, M.J., Kim, H.R., Winn, M.E. and Mukherjee, A. 2020. Time-course RNA-seq analysis provides an improved understanding of gene regulation during the formation of nodule-like structures in rice. Plant Mol. Biol., 103: 113–128.

Tirichine, L., Imaizumi-Anraku, H., Yoshida, S., Murakami, Y., Madsen, L.H., Miwa, H. et al. 2006. Deregulation of a Ca2+/calmodulin-dependent kinase leads to spontaneous nodule development. Nature., 441: 1153–1156.

Tirichine, L., Sandal, N., Madsen, L.H., Radutoiu, S. and Albrektsen, A.S. 2007. A gain-of-function mutation in a cytokinin receptor triggers spontaneous root nodule organogenesis. Science, 315: 104–107.

Tirichine, L., Sandal, N., Madsen, L.H., Radutoiu, S., Albrektsen, A.S., Sato, S. et al. 2007. A gainof-function mutation in a cytokinin receptor triggers spontaneous root nodule organogenesis. Science, 315: 104–107.

Tominaga, A., Nagata, M., Futsuki, K., Abe, H., Uchiumi, T., Abe, M. et al. 2009. Enhanced nodulation and nitrogen fixation in the abscisic acid lowsensitive mutant enhanced nitrogen fixation1 of Lotus japonicus. Plant Physiol., 151: 1965–1976.

Turner, M., Nizampatnam, N.R., Baron, M., Coppin, S., Damodaran, S., Adhikari, S. et al. 2013. Ectopic expression of miR160 results in auxin hypersensitivity, cytokinin hyposensitivity, and inhibition of symbiotic nodule development in soybean. Plant Physiol., 162: 2042–2055.

Udvardi, M. and Poole, P.S. 2013. Transport and metabolism in legume-rhizobia symbioses. Annu. Rev. Plant Biol., 64: 781–805. doi:10.1146/annurev-arplant050312-120235

Ueguchi-Tanaka, M., Ashikari, M., Nakajima, M., Itoh, H., Katoh, E., Kobayashi, M. et al. 2005. GIBBERELLIN INSENSITIVE DWARF1 encodes a soluble receptor for gibberellin. Nature., 437: 693–698.

Umehara, M., Hanada, A., Yoshida, S., Akiyama, K., Arite, T., Takeda-Kamiya, N. et al. 2008. Inhibition of shoot branching by new terpenoid plant hormones. Nature., 455(7210): 195–200. doi: 10.1038/nature07272. PMID: 18690207

Upreti, K.K. and Murti, G.S.R. 2004. Effects of brassinosteroids on growth, nodulation, phytohormone content and nitrogenase activity in French bean under water stress. Biol Plant., 48: 407–411.

van Noorden, G.E., Ross, J.J., Reid, J.B., Rolfe, B.G. and Mathesius, U. 2006. Defective long-distance auxin transport regulation in the Medicago truncatula super numeric nodules mutant. Plant Physiol., 140: 1494–1506.

van Velzen, R., Holmer, R., Bu, F., Rutten, L., van Zeijl, A., Liu, W. et al. 2018. Comparative genomics of the nonlegume Parasponia reveals insights into evolution of nitrogen-fixing Rhizobium symbioses. Proc. Natl. Acad. Sci. USA., 115: E4700–E4709.

van Zeijl, A., Op den Camp, R.H.M., Deinum, E.E., Charnikhova, T., Franssen, H., Op den Camp, H.J.M., et al. 2015. Rhizobium lipochitooligosaccharide signaling triggers accumulation of cytokinins in Medicago truncatula roots. Mol. Plant., 8: 1213–1226.

Vance, C.P. 2001. Symbiotic nitrogen fixation and phosphorus acquisition. Plant nutrition in a world of declining renewable resources. Plant Physiol., 127: 390–397. doi:10.1104/pp.010331

van-Noorden, G.E., Kerim, T., Goffard, N., Wiblin, R., Pellerone, F.I., Rolfe, B.G. et al. 2007. Overlap of proteome changes in Medicago truncatula in response to auxin and Sinorhizobium meliloti. Plant Physiol., 144: 1115–1131.

Van-Spronsen, P.C., Tak, T., Rood, A.M., van-Brussel, A.A., Kijne, J.W. and Kj, B. 2003. Salicylic acid inhibits indeterminate-type nodulation but not determinate-type nodulation. Mol. Plant Microbe. Interact., 16: 83–91.

Vernie, T., Kim, J., Frances, L., Ding, Y., Sun, J., Guan, D. et al. 2015. The NIN Transcription Factor Coordinates Diverse Nodulation Programs in Different Tissues of the Medicago truncatula Root. Plant Cell., 3410-24.

Vernié, T., Moreau, S.F., Plet, J., Combier, J.P., Rogers, C., Oldroyd, G. et al. 2008. EFD is an ERF transcription factor involved in the control of nodule number and differentiation in Medicago truncatula. Plant Cell., 20(10): 2696–2713.

Vierheilig, H. 2004. Regulatory mechanisms during the plant-arbuscular mycorrhizal fungus interaction. Canadian Journal of Botany, 82: 1166–1176.

Wan, X., Hontelez, J., Lillo, A., Guarnerio, C., van de Peut, D., Fedorova, E. et al. 2007. Medicago truncatula ENOD40-1 and ENOD40-2 are both involved in nodule initiation and bacteroid development. Journal of Experimental Botany, 58: 2033–2041.

Wang, B. and Qiu, Y.L. 2006. Phylogenetic distribution and evolution of mycorrhizas in land plants. Mycorrhiza., 16: 299–363.

Wen, X., Zhang, C., Ji, Y., Zhao, Q., He, W., An, F. et al. 2012. Activation of ethylene signaling is mediated by nuclear translocation of the cleaved EIN2 carboxyl terminus. Cell Res., 22: 1613–1616.

Werner, T., Motyka, V., Laucou, V., Smets, R., Van Onckelen, H. and Schmuelling, T. 2003. Cytokinindeficient transgenic Arabidopsis plants show multiple developmental alterations indicating opposite functions of cytokinins in the regulation of shoot and root meristem activity. Plant Cell., 15: 2532–2550.

Werner, T., and Schm€ulling, T. 2009. Cytokinin action in plant development. Curr. Opin. Plant Biol., 12: 527–538.

Wu, Y., Liu, H., Wang, Q. and Zhang, G. 2021. Roles of cytokinins in root growth and abiotic stress response of Arabidopsis thaliana. Plant Growth Regul., 94: 151–160. https://doi.org/10.1007/s10725-021-00711-x.

Wu, Y., Sanchez, J.P., Lopez-Molina, L., Himmelbach, A., Grill, E. and Chua, N.H. 2003. The abi1-1 mutation blocks ABA signaling downstream of cADPR action. Plant J., 34: 307–15.

Xiao, T.T., Schilderink, S., Moling, S., Deinum, E.E., Kondorosi, E., Franssen, H. et al. 2014. Fate map of Medicago truncatula root nodules. Development., 141: 3517–3528.

Xiao, T.T., van Velzen, R., Kulikova, O., Franken, C. and Bisseling, T. 2019. Lateral root formation involving cell division in both pericycle, cortex and endodermis is a common and ancestral trait in seed plants. Development., 146: 182–592.

Yamagami, M., Haga, K., Napier, R.M. and Iino, M. 2004. Two distinct signaling pathways participate in auxin-induced swelling of pea epidermal protoplasts. Plant Physiol., 134: 735–747.

Yamagami, T., Tsuchisaka, A., Yamada, K., Haddon, W.F., Harden, L.A. and Theologis, A. 2003. Biochemical diversity among the 1-amino-cyclopropane-1-carboxylate synthase isozymes encoded by the Arabidopsis gene family. J. Biol. Chem., 278: 49102–49112.

Yamaguchi, S. 2008. Gibberellin metabolism and its regulation. Annu. Rev. Plant Biol., 59: 225–251.

Yaxley, J.R., Ross, J.J., Sherriff, L.J. and Reid, J.B. 2001. Gibberellin biosynthesis mutations and root development in pea. Plant Physiol., 125: 627–633.

Zahra, N., Hafeez, M.B., Ahmad, M., Iqbal, S., Shaukat, K. et al. 2020. Nitrogen fixation of legumes: biology and physiology. In: Hasanuzzaman, M., Araújo, S. and Gill, S. (eds.). The Plant Family Fabaceae. Springer, Singapore. https://doi.org/10.1007/978-981-15-4752-2_3.

Zhao, Y. 2010. Auxin biosynthesis and its role in plant development. Annu. Rev. Plant Biol., 61: 49–64.

Zhao, Y. 2012. Auxin biosynthesis: a simple two-step pathway converts tryptophan to indole-3-acetic acid in plants. Mol. Plant., 5: 334–338.

Zhao, Y., Christensen, S.K., Fankhauser, C., Cashman, J.R., Cohen, J.D., Weigel, D. and Chory, J. 2001. A role for flavin monooxygenase-like enzymes in auxin biosynthesis. Science, 291: 306–309.

Chandler, J.W. and Werr, W. 2015. Cytokinin-auxin crosstalk in cell type specification. Trends Plant Sci., 20: 292–300.

15

Cytokinin Signaling in Plant Response to Abiotic Stresses

Izhar Ullah,[1] *Muhammad Danish Toor,*[2] *Abdul Basit*[3]
and *Heba I. Mohamed*[4,]*

Introduction

Cytokinins, a type of plant hormone, play a vital and diverse role in regulating various aspects of plant growth and development. They are crucial for fundamental plant processes such as cell division, shoot initiation, leaf aging, and nutrient distribution (Mok and Mok 2001, Kieber and Schaller 2014). Typically cytokinins are synthesized in the roots, and then transported throughout the plant, exerting their effects through intricate signal transduction pathways (Hwang et al. 2012). One significant effect of cytokinins is their ability to stimulate cell division, resulting in an increase in cell numbers and subsequent growth of plant organs. Moreover, cytokinins play a crucial role in shoot initiation, ensuring the development of new shoots from dormant buds. Additionally, these hormones regulate leaf aging by extending the lifespan of leaves and sustaining photosynthetic activity for an extended period. Furthermore, cytokinins facilitate the mobilization of nutrients, aiding in the efficient uptake and transportation of essential elements required for plant growth and metabolism. Extensive research has provided substantial evidence for the remarkable impact of

[1] Department of Horticulture, Faculty of Agriculture, Ondokuz Mayis University, 55139 Atakum, Samsun-Turkey.
[2] Department of Soil Science and Plant Nutrition, Ondokuz Mayis University, 55139 Atakum, Samsun-Turkey.
[3] Department of Horticultural Science, Kyungpook National University, 41566 Daegu, South Korea.
[4] Department of Biological and Geological Sciences, Faculty of Education, Ain Shams, University, Cairo, 11341, Egypt.
* Corresponding author: hebaibrahim79@gmail.com

cytokinins on plant growth. Their application has consistently shown improvements in shoot and root development, promotion of enhanced flowering, and overall increased yield (Werner et al. 2001, Sakakibara 2006). By influencing cellular processes, cytokinins actively contribute to the formation of new tissues and the activation of growth-promoting genes.

In recent studies, researchers have explored the function of cytokinin signaling in plant reactions to abiotic stresses, especially drought. Li et al. (2021) investigated the consequences of overexpressing the cytokinin biosynthesis gene IPT8 and discovered that increased cytokinin stages had more desirable drought tolerance in *Arabidopsis*. Similarly, Mandal et al. (2022) confirmed that overexpression of the cytokinin receptor gene CKI1 advanced drought tolerance by using regulating stomatal closure and improving antioxidant defense. Dong et al. (2021) targeted creeping bentgrass and determined that cytokinin delays leaf senescence and complements drought tolerance by enhancing antioxidant capability and osmotic adjustment. Hu et al. (2021) conducted research on maize and determined that the cytokinin N6-isopentenyl adenine improved drought tolerance through improving photosynthesis and antioxidant protection. Qiu et al. (2022) explored the effects of overexpressing the cytokinin biosynthesis gene OsIPT7 in rice and observed more desirable drought tolerance via advanced photosynthetic capacity and water balance. Chen et al. (2022) investigated tobacco plant life and located that overexpression of the cytokinin biosynthesis gene IPT7 improved drought tolerance by using selling antioxidant capability, decreasing oxidative harm, and facilitating osmotic adjustment. Lastly, Lakhwani et al. (2022) discussed cytokinins as key regulators of drought stress response in flowers, emphasizing their importance in enhancing plant resilience to drought. Abiotic stresses, which include drought, salinity, temperature extremes, and nutrient deficiency, significantly affect plant growth, development, and productivity (Chaves et al. 2003, Mittler 2006).

These environmental pressures disrupt various physiological and biochemical processes in plants, leading to decreased photosynthesis and cell death (Mittler 2006, Suzuki et al. 2014). Abiotic stresses have become a significant concern in agriculture due to their detrimental effects on crop yield and quality, prompting the exploration of methods to enhance plant stress tolerance. Cytokinin signaling pathways have emerged as crucial factors in plant adaptation to abiotic stresses. Several studies have emphasized the role of cytokinins in modulating stress responses and improving stress tolerance in plants (Nishiyama et al. 2011, Tran et al. 2017).

The chapter aims at understanding and utilizing the potential of cytokinin signaling pathways to increase plant stress tolerance, which holds promise for crop improvement and resilience to unfavorable environmental conditions.

Mechanisms of Cytokinin Signaling in Plant Response to Abiotic Stresses

Cytokinin receptors and signal transduction pathways

Cytokinin signaling in plants, especially in response to abiotic stress, involves specific receptors and complex signaling pathways. The various types of cytokinin receptors found in plants include histidine kinase receptors, hybrid-histidine kinase receptors, and cytokinin-binding proteins (Hwang and Sheen 2001). Histidine kinase receptors, the main cytokinin receptors, have an extracellular receptor domain cytokinin binding, which binds the receptor to the cell membrane (meaning the intracellular transmembrane domain). This receptor, which contains histidine and kinase domains, initiates a phosphorylation cascade upon cytokinin binding, resulting in cytokinin signaling (Schaller et al. 2015). Hybrid histidine kinase receptors have additional domains that facilitate cross-talk to different signaling pathways, facilitating different environmental stimuli. Cytokinin binding proteins, lacking the histidine kinase domain, can still interact with cytokinin signaling on other parts of the pathway (Hwang and Sheen 2001).

Cytokinin signal transduction involves multiple steps and interactions with other hormone-signaling pathways. The process begins with cytokinin sensing and binding to receptors such as histidine kinase receptors and cytokinin-binding proteins (Argueso et al. 2010). This interaction results in the autophosphorylation of the histidine kinase receptor, transferring the phosphate groups to specific histidine residues in the receptor and then transferring the phosphate groups to phosphotransferase proteins, which subsequently lead to response regulatory proteins (Argueso et al. 2010, Miyawaki et al. 2006). Phosphorylated response regulatory proteins act as transcription factors, modulating the expression of cytokinin-responsive genes involved in cell division, growth, and stress response (Argueso et al. 2010). In addition, cytokinin-signaling pathways often interact with other hormonal signaling pathways such as auxin and abscisic acid pathways, allowing plants to integrate multiple signals to finely adjust their responses to abiotic stresses (Miyawaki et al. 2006).

Modulation of Cytokinin Signaling under Abiotic Stresses

Altered cytokinin biosynthesis and metabolism

Due to abiotic stresses such as drought, salinity, and intense temperatures, plants undergo various physiological and biochemical changes to adapt and survive. One vital element of those variations is the modulation of cytokinin biosynthesis and metabolism. Cytokinins are phytohormones that play an essential role in plant increase, improvement, and strain responses. Several researches have mentioned changes in cytokinin biosynthesis and metabolism underneath abiotic stress. For example, Tran et al. (2010) investigated the effects of drought on cytokinin metabolism in *Arabidopsis thaliana*. They found that drought leads to a lower degree of active cytokinins, such as trans-zeatin, at the same time as the stages of inactive cytokinins, which include cis-zeatin, expanded. These adjustments in cytokinin metabolism make a contribution to the regulation of plant responses to drought pressure. Similarly, Rivero et al. (2014) targeted salinity and its impact on cytokinin biosynthesis in tomato vegetation. They discovered that salinity stress caused a decrease in the expression of genes concerned with cytokinin biosynthesis, leading to reduced degrees of cytokinins. Additionally, they found that exogenous application of cytokinins progressed the tolerance of tomato flowers to salinity strain.

Changes in cytokinin receptor sensitivity

Abiotic stress also affects the sensitivity of cytokinin receptors, which are important components of cytokinin signaling pathways. Cytokinin receptors recognize the presence of cytokinin and initiate downstream signaling, resulting in physiological responses. A study by Nishiyama et al. (2012) investigated the effect of cold stress on cytokinin receptor sensitivity in *Arabidopsis thaliana*. Cold stress was found to upregulate certain cytokinin receptors, particularly the Arabidopsis histidine kinase 4 (AHK4) receptor. This upregulation increased the sensitivity of the receptor to cytokinin binding, resulting in an enhanced cytokinin response. Furthermore, drought stress has been shown to affect cytokinin receptor sensitivity in maize plants. A study by Pandey et al. (2016) reported that drought stress increased the expression of cytokinin receptors in maize roots. This upregulation enhanced cytokinin storage and subsequent activation of downstream signaling pathways involved in stress response (Fig. 1).

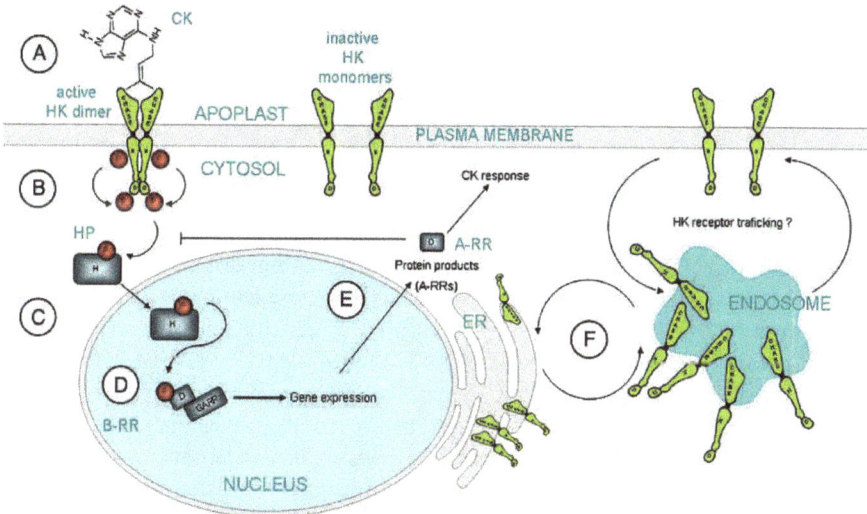

Fig. 1. Illustrates the cytokinin signaling pathway.

Cross-talk with other signaling pathways

Under abiotic stresses, plants integrate multiple signaling pathways to coordinate their responses and maximize survival. Cytokinin signaling pathways often exhibit cross-talk with other signaling pathways, enabling the modulation of stress responses. One example of cross-talk involves the interaction between cytokinin and abscisic acid (ABA) signaling pathways. A study by Nishiyama et al. (2011) revealed that cytokinin and ABA signaling pathways interact antagonistically in response to osmotic stress. They found that cytokinin signaling inhibits ABA-mediated responses, allowing plants to prioritize growth over stress adaptation under favorable conditions. Additionally, a study by Ramegowda et al. (2013) investigated the interaction between cytokinin and ethylene signaling pathways under drought stress in rice plants. They demonstrated that cytokinins antagonistically regulate the ethylene response factor (ERF) transcription factors, which are involved in ethylene signaling. This interaction influences the expression of stress-responsive genes and contributes to the overall stress tolerance of the plants.

Regulation of Gene Expression by Cytokinin Signaling under Abiotic Stresses

Cytokinin-responsive genes involved in abiotic stress tolerance

Cytokinins regulate gene expression to modulate plant responses to abiotic stresses. Numerous studies have identified several cytokinin-responsive genes associated with abiotic stress tolerance. Xie et al. (2016) investigated the role of cytokinin in drought stress response in *Arabidopsis thaliana*. They identified a set of cytokinin-responsive genes involved in stress tolerance, such as DREB transcription factors and LEA (late embryogenesis abundant) proteins. Another study by Zhang et al. (2022) focused on cytokinin signaling in salt stress tolerance in rice. They discovered several cytokinin-responsive genes, including OsAP2-39, OsNHX1, and OsCIPK15, which are involved in ion homeostasis and stress response.

Transcription factors mediating cytokinin-regulated stress responses

Transcription factors are essential for managing cytokinin-driven stress reactions. These factors attach to specific DNA sequences and regulate the activity of genes downstream. For instance, in a study by Tran et al. (2010), scientists explored the role of cytokinin response factors (CRFs) in controlling stress-responsive genes in *Arabidopsis*. They determined that CRFs, like CRF1 and CRF2, positively stimulate stress tolerance by triggering stress-related genes. Another investigation by Li et al. (2017) probed the function of cytokinin-responsive NAC transcription factors in responding to drought stress in rice. They found various NAC transcription factors, including SNAC1 and OsNAC10, that contribute to drought resilience. Transcription factors (TFs) that regulate subsequent defense-related genes can be impacted by mitogen-activated protein kinases (MAPKs). The interplay between MAPKs and TFs has been extensively reviewed, particularly in relation to TFs featuring DNA-binding structures such as the MADS-box, ETS domain, zinc-finger, bZIP, and HMG box domains, which modulate gene expression tied to MAPKs (Treisman, 1996).

Prior research has associated multiple TFs, like PCR1, SRRA/SKN7/PRRr1, HSF1, and MYB, with reactive oxygen species (ROS) signaling, instigating reactions against oxidative stress (Hong et al. 2013). Additionally, the effectiveness of salicylic acid (SA) signaling relies heavily on NPR1 and MPK4, as the malfunction of these genes leads to diminished expression of pathogenesis-related (PR) genes (Loake and Chco, 2007). The pivotal regulator MYC2, a TF containing a bHLH domain, profoundly influences the control of jasmonic acid (JA) signalling (Lorenzo et al.2004). Also, distinct TFs such as ORA59 (from the AP2/ERF family) and two NAC proteins, ANAC055, and ANAC019, stimulate the transcription of various JA and ethylene-responsive genes, thus engaging in the signaling of both plant hormones (Pre et al. 2008). Conversely, WRKY70 downregulates JA-responsive genes (Li et al. 2004). Importantly, the encouragement or inhibition of defence pathways relies on intricate interactions between transcriptional proteins and other proteins (Frerigmann 2016).

Overall, defensive signal transmission coordinates the organized control of downstream stress-responsive genes. Beyond recognizing pathogens and the ensuing signaling that stimulates defense-responsive genes, the exact mechanisms that directly halt pathogenic attack in plants are not completely comprehended. Nevertheless, multiple biological procedures have been identified as participants in ending infections and safeguarding plants. These mechanisms involve reinforcing the cell wall, producing antimicrobial peptides, and generating low molecular weight secondary metabolites (Voigt 2014). Plant secondary metabolites, crucial compounds in plant immunity, assume a notable role in this context (Hartmann 2008). Among these phytochemicals are phytoanticipins and stress-induced phytoalexins, which directly counteract pathogenic intrusion in plants (Ahuja et al. 2012). Additionally, these metabolites also aid in enhancing plant endurance against abiotic stresses like drought, salinity, UV radiation, intense light, and reactive oxygen species (ROS) (Akuja and Ravishankar 2011).

Cytokinin-Induced Epigenetic Modifications in Stress-Responsive Genes

Cytokinins can also induce epigenetic modifications in stress-responsive genes, leading to altered gene expression patterns (Fig. 2). Epigenetic modifications, such as DNA methylation and histone modifications play crucial roles in regulating gene expression. A study by Jeong et al. (2014) explored the epigenetic regulation of cytokinin-responsive genes in response to drought stress in maize. The researchers observed changes in DNA methylation patterns and histone modifications in stress-responsive genes upon cytokinin treatment. Additionally, Zhao et al. (2019) investigated the role of cytokinin in DNA demethylation and subsequent activation of stress-responsive genes in *Arabidopsis*. Their findings revealed that cytokinin treatment leads to active DNA demethylation and enhanced expression of stress-related genes.

Plants face various abiotic stresses together with drought, salinity, high temperatures, and nutrient deficiencies that could adversely affect their flower development and overall productiveness. In order to continue to exist and adapt to hard situations, plant life has advanced mechanisms that contain the signaling and movement of plant hormones, along with cytokinins. Cytokinins, a category of plant hormones, play a vital position in regulating the physiological responses and adaptive mechanisms of vegetation to abiotic stresses. One of the key areas wherein cytokinins are concerned is the regulation of plant water members of the family. Water is a crucial and useful resource for plant increase, and its availability is often constrained beneath abiotic stress situations.

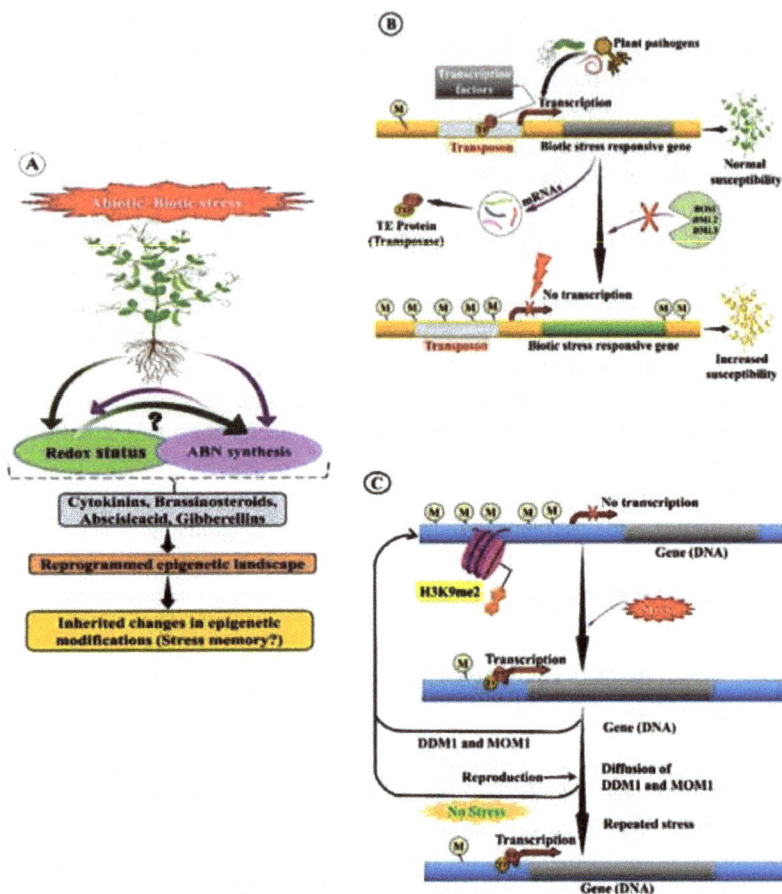

Fig. 2. This figure illustrates how both biotic and abiotic stresses can induce or change DNA methylation and other epigenetic changes in plants (Le et al. 2014, Iwasaki et al. 2014, Sundaram and Wysocka 2020, Ramakrishnan et al.2021). Physiological Responses and Adaptation of Plants to Abiotic Stresses via Cytokinin Signaling.

Cytokinin-mediated regulation of plant water relations includes the modulation of stomatal conductance and the synthesis of osmoprotectants. Stomata, small openings on the leaf floor, manage the alternating of gases and the loss of water through transpiration. Studies carried out by Liu et al. (2019) have shown that cytokinins regulate stomatal conductance, as a result controlling the water loss and assisting flora preserve water stability beneath strain conditions. Additionally, cytokinins enhance the synthesis of osmoprotectants such as proline and sugars, which act as molecular protectants and help preserve cell water balance below stress. Another important role of cytokinins in plant adaptation to abiotic stresses is the induction of antioxidant systems and the scavenging of reactive oxygen species (ROS).

Abiotic stresses often lead to the production of ROS, resulting in oxidative damage in plants. Zhang et al. (2018) and Li et al. (2017) have confirmed that cytokinins promote the production of antioxidant enzymes which include superoxide dismutase (SOD), catalase (CAT), and peroxidase (POD), which detoxify ROS and prevent oxidative stress. Cytokinins also enhance the synthesis of antioxidant compounds, such as ascorbic acid (nutrition C) and glutathione, which are similarly useful resource in ROS scavenging and protection against oxidative damage. Furthermore, cytokinins play a substantial function in enhancing nutrient uptake and usage performance under stress conditions. Abiotic stresses frequently disrupt nutrient availability and uptake, mainly leading to nutrient deficiencies in flora. Cytokinin signaling additionally regulates the expression of nutrient transporters, consisting of nitrate and phosphate transporters, facilitating improved nutrient acquisition by means of roots.

Moreover, cytokinins modulate the allocation of nutrients inside plant life, ensuring efficient utilization and allocation of restricted nutrients throughout stress conditions. In addition to water and nutrient regulation, cytokinins are involved in growth and developmental modifications that permit vegetation to conform to abiotic stresses. Wachsman et al. (2015) and Yamauchi et al. (2010) have highlighted the function of cytokinins in lateral root formation, shoot increase, and the balance between vegetative and reproductive growth. Cytokinins promote lateral root growth, which will increase the basis surface vicinity for progressed water and nutrient uptake. They also affect shoot growth and architecture, permitting plants to allocate resources correctly and optimize pressure responses. Moreover, cytokinins regulate the balance between vegetative growth and reproductive growth, ensuring survival and replica below abiotic strain situations (Table 1).

Applications of Cytokinin Signaling in Improving Plant Abiotic Stress Tolerance

Cytokinin signaling has emerged as a promising avenue for improving plant abiotic stress tolerance. Abiotic stresses such as drought, salinity, heat, and cold can significantly affect plant growth and productivity. Manipulating cytokinin signaling pathways offers potential strategies to enhance stress tolerance and improve plant resilience (Table 2). Here are detailed explanations of the applications of cytokinin signaling in improving plant abiotic stress tolerance (Fig. 3).

Drought Tolerance

Drought, an abiotic stress factor, poses a significant ecological concern with profound impacts on plant photosynthesis, growth, and development (Rizwan et al. 2015, Fahad et al. 2017). Optimal drainage in the soil is essential for the healthy growth, development, and maximum productivity of perennial fruit trees and crops. Even a short period of inadequate drainage can have a long-lasting detrimental effect on the productivity of perennial fruit trees (Fahad et al. 2017, Fuller and Stevens 2019) (Fig. 4).

Table 1. An overview of the significant impact of cytokinins on plant's abiotic stress tolerance.

Mechanism	Findings	References
Investigated the role of cytokinins in drought stress response in *Arabidopsis*.	Cytokinins regulate stomatal closure and enhance root growth under drought conditions.	Nishiyama et al. (2011)
Explored the involvement of cytokinins in salinity stress tolerance in tomato plants.	Cytokinins regulate ion homeostasis and promote the synthesis of compatible solutes to enhance salt tolerance.	Rivero et al. (2007)
Investigated the role of cytokinins in heat stress response in *Arabidopsis*.	Cytokinins regulate the expression of heat shock proteins and antioxidant enzymes to enhance thermotolerance.	Werner et al. (2005)
Explored the involvement of cytokinins in cold stress response in rice.	Cytokinins regulate the expression of cold-responsive genes and enhance cold tolerance in rice plants.	Zhang et al. (2010)
Investigated the role of cytokinins in nutrient deficiency response in *Arabidopsis*.	Cytokinins regulate nutrient uptake, allocation, and remobilization to enhance plant adaptation to nutrient-limited conditions.	Tran et al. (2007)
Explored the cross-talk between cytokinins and abscisic acid (ABA) in drought stress response in maize.	Cytokinins interact with the ABA signaling pathway to regulate stomatal closure and improve drought tolerance.	Pandey et al. (2016)
Investigated the role of cytokinins in oxidative stress response in *Arabidopsis*.	Cytokinins modulate reactive oxygen species (ROS) levels and antioxidant defense systems to enhance oxidative stress tolerance.	Kumar et al. (2012)
Explored the involvement of cytokinins in flooding stress response in soybean.	Cytokinins regulate adventitious root formation and enhance flooding tolerance in soybean plants.	Jeon et al. (2010)
Investigated the role of cytokinins in nutrient signaling and stress response in *Arabidopsis*.	Cytokinins interact with nutrient signaling pathways, such as nitrate and phosphate, to regulate plant growth and stress responses.	Hassler et al. (2018)
Explored the involvement of cytokinins in high-temperature stress response in wheat.	Cytokinins regulate heat stress-responsive genes and enhance thermotolerance in wheat plants.	Hu et al. (2017)
Investigated the interaction between cytokinin and ethylene signaling pathways under drought stress in rice plants.	Cytokinins antagonistically regulate the ethylene response factor (ERF) transcription factors	Ramegowda et al. (2013)

Moreover, drought stress disturbs the carbon metabolism balance, which is the primary source of carbohydrates. This disruption leads to a decrease in the availability of carbon dioxide and the partial closure of stomata at carboxylation sites (Hu et al. 2019). Consequently, there is an increased reliance on shoot respiration to sustain metabolic activity, resulting in reduced carbohydrate reserves in the storage organs of citrus plants (Fahad et al. 2017).

Cytokinins (CKs) play a crucial role in regulating plant responses to drought stress by influencing various physiological processes. They contribute to stomatal

Table 2. Role of cytokinins under abiotic stress.

Plant species	Stress	Functions	References
Hordeum vulgare	Drought stress	Transgenic barley with a bigger root system is created by encouraging the breakdown of CKs in the roots, increasing the crop's tolerance to drought.	Ramireddy et al. (2018)
Agrostis stolonifera	Drought stress	To lessen the impact of drought stress on root growth, overexpressing IPT encourages the production of CKs and activates the ROS scavenging system.	Xu et al. (2016)
Zea mays	Drought stress	The concentration of CKs generated at deep roots during the drought rewatering process affects the compensatory growth of maize.	Wang et al. (2018a, 2018b, 2016)
Populus alba	Drought stress	Poplar local roots have ABA/CK and aquaporins, which control the response to mild drought stress without relying on the ectomycorrhizal fungus *Laccaria bicolor*.	Calvo-Polanco et al. (2019)
Triticum aestivum	Drought stress	CRISPR/Cas9-based gene editing and constitutive overexpression	Wang et al. (2022)
Arabidopsis thaliana	Osmotic stress	In response to osmotic stress, CKs controls *Arabidopsis* root growth via a hormone network that interacts with ABA, ethylene, and IAA.	Gujjar and Supaibulwatana (2019)
Arabidopsis thaliana	Cold stress	The CKs' response variables *Arabidopsis* LR initiation is significantly regulated by the transcription factors CRF2 and CRF3, which are encoded by the gene for APETALA2.	Jeon et al. (2016)
Centaurium erythraea Rafn	Salt stress	The AtCKX transgenic line improves salt tolerance while lowering the amount of CK in the root system and affecting root development.	Trifunović-Momčilov et al. (2020)
Oryza sativa	Salt stress	Application of GA3 or CKs biosynthesis inhibitors can restore CYP71D8L-OE and CYP71D8L root dysplasia phenotypes and demonstrate increased salt tolerance.	Zhou et al. (2020)
Arabidopsis thaliana	High boron stress	By controlling IAA and CKs' response to excessive boron stress, 26S proteasome preserves the viability of the root tip meristem.	Sakamoto et al. (2019)
Zea mays	Low nitrogen stress	Under mild nitrogen stress, BR and IAA work together to control root elongation while CK, ethylene, and ABA have opposing effects.	Lv et al. (2020)

closure, control water transport, maintain cellular balance, and facilitate root development. Manipulating cytokinin signaling provides an opportunity to improve drought tolerance by modulating these processes. For instance, increasing the expression of cytokinin biosynthesis genes or cytokinin receptors can enhance water-

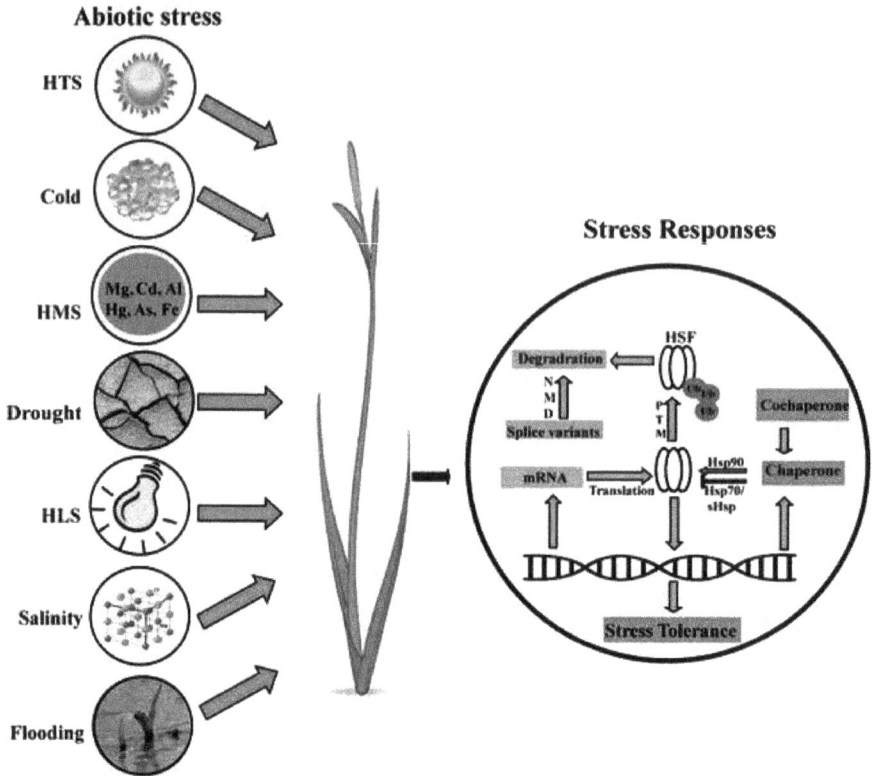

Fig. 3. Responses and molecular adaptation to abiotic stresses (Mishra et al.2018).

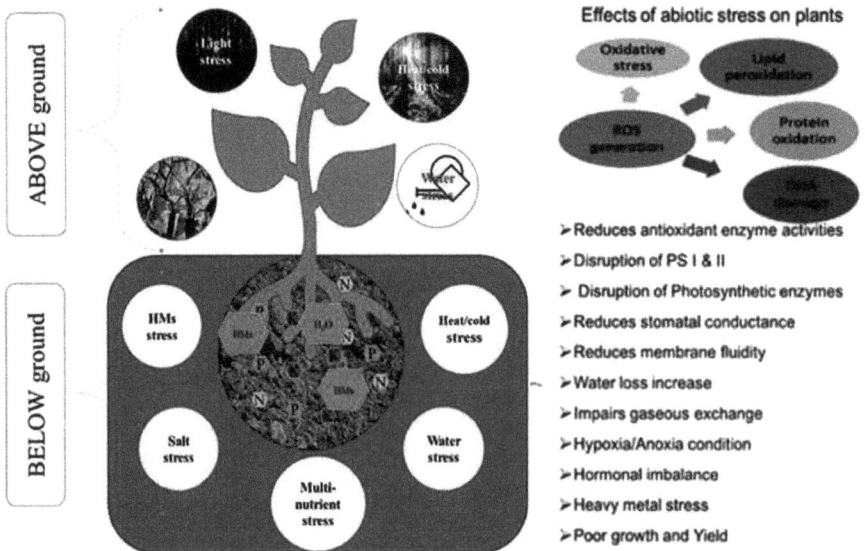

Fig. 4. Adverse effect of abiotic stresses on plants (Sabagh et al. 2021; Kumari et al. 2022).

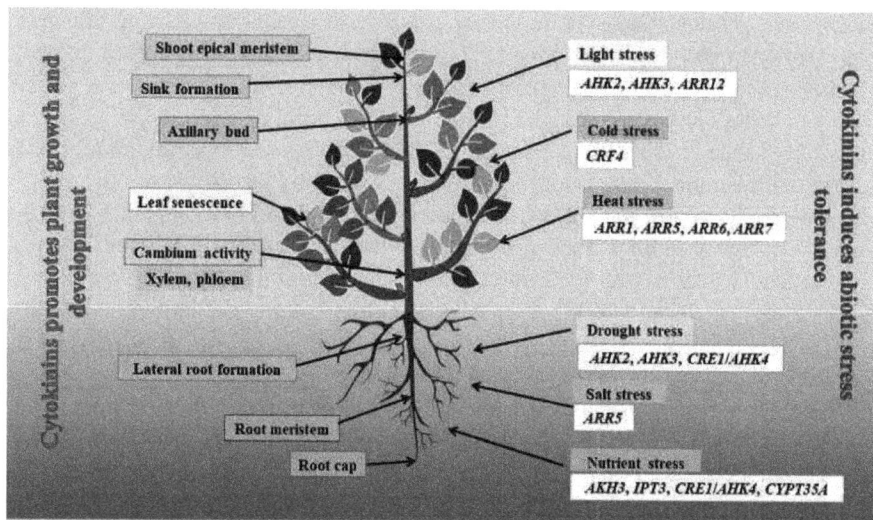

Fig. 5. An overview of the positive role of cytokinins application on the growth and development of plants (Mandal et al. 2022).

use efficiency and stimulate root growth, thereby increasing drought tolerance in crops (Rivero et al. 2022).

Moreover, cytokinins have been found to promote the expression of specific transgenes, including the isopentenyl transferase gene, in transgenic plants. This heightened expression results in significant drought tolerance by delaying senescence and preventing leaf senescence induced by drought. While the accumulation of cytokinins has positive effects on drought tolerance, adverse effects have also been observed.

In *Arabidopsis*, the enzyme CK oxidase/dehydrogenase (CKX) is responsible for cytokinin breakdown. Overexpressing CKX1, CKX2, CKX3, and CKX4 independently in *Arabidopsis* leads to reduced cytokinin levels and consequently enhances drought tolerance in the transgenic lines (Werner et al. 2010, Hamayun et al. 2021b) (Fig. 5).

Salt stress

Cytokinins have proven their ability to counteract the harmful effects of salt stress on plant growth and development. They play a vital role in regulating the movement of ions, maintaining osmotic balance, and supporting antioxidant defense mechanisms. By adjusting how cytokinins communicate within the plant, we can enhance its ability to tolerate salt stress. This involves improving the balance of ions, reducing sodium intake, and encouraging the uptake of potassium. Scientists achieve this by influencing how genes related to cytokinin production or communications are expressed (Wang et al. 2020).

When it comes to managing plant responses to salt stress, various plant hormones are involved. One significant protector in this regard is abscisic acid (ABA) (Zhu 2002). In contrast, cytokinins (CK) typically play a role as inhibitors in response to salt stress. Experiments on *Arabidopsis* plants subjected to salt stress have demonstrated that controlling IPT genes led to a decrease in cytokinin levels (Nishiyama et al. 2011). Plants with lower cytokinin levels demonstrated higher tolerance to salt stress compared to normal plants (Nishiyama et al. 2011, Tran et al. 2007). These resilient plants displayed modified expression of numerous genes related to stress and ABA response (Nishiyama et al. 2012).

One of the genes under cytokinin's regulation in Arabidopsis is HKT1;1. This gene codes for a sodium transporter that prevents the excessive accumulation of sodium ions. Mutants with reduced cytokinin levels exhibited greater expression of HKT1;1 and enhanced resistance to salt stress, highlighting the importance of this regulatory mechanism (Mason et al. 2010, Nishiyama et al. 2011). On the other hand, triggering the overproduction of cytokinins in *Arabidopsis* through the activation of the IPT8 gene led to decreased adaptation to salinity stress. This was associated with higher production of reactive oxygen species (ROS) and reduced expression of enzymes that combat ROS (Wang et al. 2015).

Other studies have found cytokinins to have a positive effect under salinity stress, in contrast to the negative role they were found to play in salt stress resistance in genetic studies. To protect the photosynthetic system and increase flower production in tomato plants under salinity stress, INCYDE, an inhibitor of CKX enzymes, was applied (Aremu et al. 2014). A salt-sensitive rice cultivar showed evidence of cytokinin protection when OsCKX2 was suppressed, which increased cytokinin levels and increased seed yield and decreased yield loss under salinity stress (Ashikari et al. 2005, Joshi et al. 2018).

Heat Stress

High temperatures can harm living organisms by generating reactive oxygen species (ROS) and causing proteins to lose their structure (protein denaturation). This can damage essential parts of cells. Also, plants suffer from reduced photosynthesis due to high heat, leading to an imbalance in their metabolism. To cope with this, plants use various tactics. For instance, they accumulate heat shock proteins (HSP) that act like helpers to prevent protein damage. These strategies are important to safeguard plants from heat stress (Mittler et al. 2011).

Cytokinins are plant chemicals that play a vital role in managing stress from both high and low temperatures. They do this by controlling the production of heat shock proteins, antioxidants, and other genes that manage stress. By influencing these processes, we can improve how plants handle both hot and cold conditions. Elevating cytokinin levels or their sensitivity can be especially helpful. This strengthens heat shock proteins and the plant's defenses against harmful molecules (antioxidant defense mechanisms). Notably, studies with rice, maize, and passion fruit have shown that treating plants with cytokinins boosts their ability to cope with heat and improve their yield (Wu et al. 2017).

Heat stress strongly affects plants' ability to photosynthesize by disturbing their chlorophyll and the efficiency of leaf photosynthesis. It also prompts the creation of harmful ROS and triggers enzymes that degrade proteins, which can ultimately lead to leaf aging. When plants deal with heat, the amount of cytokinins inside their leaves goes up, especially in *Arabidopsis thaliana*. This indicates that more cytokinins are needed for better heat tolerance.

Increased levels of cytokinins can potentially activate the plant's antioxidant system, which fights against harmful reactive oxygen species (ROS) generated during periods of heat stress. Hormone, proteome, and transcriptome investigations have consistently highlighted the significance of cytokinins for plants exposed to heat stress. Elevated cytokinin concentrations have been linked to higher amounts of heat shock proteins (Skalák et al. 2016).

The introduction of a specific gene called isopentenyl transferase (IPT) into *Arabidopsis thaliana* plants has been found to elevate cytokinin levels, contributing to the plant's improved ability to withstand high temperatures (Skalak et al. 2016). The duration for which these increased cytokinin levels are sustained plays a crucial role in determining the plant's resilience to heat stress, as demonstrated by Skalak et al. (2016). Furthermore, enhancing cytokinin levels through the SAG12:ipt gene in *A. stolonifera* plants can enhance heat tolerance by supporting root growth, maintaining chlorophyll content, and delaying leaf aging, all of which collectively contribute to better heat resilience (Xu et al. 2009).

Additional research by Xu et al. (2010) has shown that overexpressing IPT using specific promoters (HSP18:ipt and SAG12:ipt) can lead to an increase in the presence of heat stress proteins, thereby enhancing the plant's capacity to endure high temperatures (Xu et al. 2010). Application of the dexamethasone (DEX) promoter-driven IPT gene overexpression in *Arabidopsis thaliana* plants has been found to stimulate stomata opening and leaf transpiration. These responses are crucial during the initial stages of the plant's reaction to heat stress. However, solely increasing transpiration can offer short-term benefits and may not adequately address the long-term effects of heat stress due to limited water availability within the plant (Wu et al. 2017).

Treating plants with external cytokinins has effects similar to those of naturally occurring ones. Treating plants with exogenous cytokinin zeatin ribose (ZR) has been shown to enhance heat tolerance by promoting root health, increasing antioxidant levels, maintaining higher chlorophyll levels, and bolstering the presence of heat shock proteins (Hu et al. 2020). This treatment also enhances the plants' resistance to heat stress in their reproductive organs.

Using cytokinins (CK) can make plants better equipped for hot temperatures. Varieties with more CK in their leaves can produce flowers even in hot summers, showcasing the protective role of CK. This has potential implications for crop breeding to improve production. CK applications have also been effective in reducing heat stress damage in creeping bentgrass (*Agrostis stolonifera* L.) (Wang et al. 2012). CK improves the plant's antioxidant system by activating enzymes that fight ROS. In rice, heat stress reduces CK in panicles, affecting spikelet formation. Applying CK can counteract this damage and help panicle development (Wu et al. 2017). Inhibitors of CK oxidase/dehydrogenase have shown positive results in *Arabidopsis* (Prerostova et al. 2020). Additionally, introducing the CK-producing gene (isopentenyl transferase) from *Agrobacterium tumefaciens* increases CK levels, making plants better at handling heat stress (Skalák et al. 2016).

Even though forests face threats from climate change, we know little about how heat stress impacts trees. For *Pinus radiata* trees, a conifer, understanding the plant's physiological and hormonal responses to heat stress has revealed insights. Cytokinins help with long-term acclimation to heat, aiding in the recovery of chloroplast function and photosynthesis (Escandón et al. 2016).

Cold stress

The toughening of cell membranes in cold temperatures makes cells more vulnerable to harm (Liu et al. 2016). This results in several changes in the body during exposure to cold stress. These changes include damage to the membrane's integrity, disturbance of water and nutrient equilibrium, and increased outflow of ions. Cold stress triggers processes that regulate gene activity and the way information is used to build proteins, which can either depend on or not rely on the hormone ABA (Prerostova et al. 2021). The drop in temperature leads to the gathering of reactive oxygen species (ROS) due to the reduced performance of protective enzymes, which affects the proper functioning of the ROS cleaning mechanism. The excessive buildup of ROS is detrimental to the cell's membrane, causing interruptions in the cell's chemical processes and the leakage of ions (Sui 2015).

Low temperatures can have a detrimental effect on the reproductive development of plants. For instance, when cold stress occurs during the flowering phase of rice plants (Oryza sativa), it can lead to sterility and reduced yield (Wang et al. 2017). Furthermore, temperatures dropping below the freezing point ($0°C$) cause freezing stress, which can physically damage plants and disrupt their metabolism due to the formation of ice crystals (Cheng et al. 2014).

Cold stress significantly influences energy production and the biochemical needs of plants (Koc et al. 2018). As an example, Zoysia grass in colder regions exhibits better resistance to frost compared to those in warmer areas. This could be attributed to increased carbohydrate storage and the involvement of plant hormones in aiding adaptation to cold conditions (Li et al. 2018).

When exposed to cold stress, *Carpobrotus edulis*, a type of plant, produces more cytokinin. An *Arabidopsis* mutant (amp1) with elevated cytokinin levels also contains this growth-regulating hormone, and this mutant exhibits improved growth and yield when compared to the wild-type (Khan et al. 2017, Fenollosa et al. 2018). Additionally, improving cold resistance in sugarcane through the overexpression of AtCOR15a:ipt delays leaf aging and prevents membrane damage, which helps sugarcane avoid significant losses from freezing damage (Belintani et al. 2012). Recent research conducted by Zwack et al. (2016) suggests that CRF4 has a positive impact on enhancing cold tolerance in *Arabidopsis thaliana*, particularly in plants with a CRF4 mutation when exposed to cold stress. A response mechanism triggered by cold stress may involve increased expression of CRF2 and CRF3, which promotes the growth of lateral roots and counteracts the inhibition of root growth caused by cold, thus bolstering the plant's ability to withstand low temperatures (Jeon et al. 2016).

By controlling the activity of genes like C-repeat binding factors (CBFs) and A-type response regulators (ARRs), ETH appears to attenuate the cold signal. A-type ARRs also appear to be essential in combining cytokinin and ethylene signals to control plant responses to cold stress. However, by affecting the expression of CBFs and A-type ARRs genes, CBFs appear to have a negative impact on the cold signal (Shi et al. 2012). According to these studies, a plant's reaction to cold stress may involve a pathway involving cytokinin signaling and associated transcription factors.

Light stress

In addition to its role in photosynthesis, light also acts as a source of information for plants regarding the time of day and season, as well as influencing their growth direction. However, an excess or deficiency of light can place stress on plants. Recent research has shown that alterations in the day and night cycle can also induce stress in plants (Bhaskar et al. 2021, Li et al. 2021). Cytokinin plays multiple roles in how plants respond to these light-related pressures, which I will briefly explain in the following sections.

While light is essential for photosynthesis, excessive light can cause damage to the photosynthetic system and other cellular components. To counteract the negative effects of excessive light (referred to as high light stress), plants have developed protective mechanisms. These include the movement of chloroplasts as well as leaves in response to light (Takahashi and Badger 2011), as well as cyclic electron transport, which converts extra light energy into heat. Even so, excessive light exposure can lead to an over-reduction in the photosynthetic electron transport chain, which can cause photo-inhibition and reduced photosynthetic efficiency (Yamamoto 2016). One protein particularly affected by excessive light and UV radiation is protein D1, which is a component of photosystem II's reaction center (Edelman and Mattoo 2008).

Low levels of cytokinin in plants have been associated with reduced photo-protection and increased photo-inhibition, mainly due to a significant decrease in the D1 protein level (Cortleven et al. 2014). Further evidence supports cytokinin's protective role in the photosynthetic machinery under intense light conditions (Cortleven and Schmülling 2015). For instance, cytokinin enhances antioxidant-based protection in chloroplasts, thereby prolonging the lifespan of these structures (Procházková et al. 2008). Conversely, inhibiting cytokinin signaling in *Arabidopsis thaliana* by mutating the AHK2 and AHK3 receptor genes has been demonstrated to enhance tolerance to photo-oxidative stress under water deficiency conditions (Danilova et al. 2014).

Light also functions as an information source for the plant's internal clock and governs its activities. Recently, prolonged exposure to light during the night, referred to as photoperiod stress, and has been shown to induce stress symptoms in *Arabidopsis thaliana* plants (Nitschke et al. 2016, Nitschke et al. 2017). Plants lacking in cytokinin were among the first to display this effect, which happened in a predictable order. Long-term exposure to light activated stress-related genes like BAP1 and ZAT12. Jasmonic acid (JA) levels also increased noticeably a few hours into the night, and the following day, because of decreased PSII maximum quantum efficiency (Fv/Fm), there was noticeable leaf lesions. In contrast to cytokinin-deficient plants, wild-type plants showed a milder stress response, highlighting the protective function of cytokinin. The main pathways used by cytokinin are AHK3, ARR2, ARR10, ARR12, and RRBs. Clock mutants, like cca1 lhy, showed a significant stress reaction to photoperiod stress. LHY and CCA1, two important regulators of the circadian clock, showed decreased expression or malfunction in both stress-sensitive clock mutants as well as cytokinin-deficient plants. This highlights the importance of a functional clock in managing photoperiod stress. While this particular type of environmental stress is not widespread in nature, it provides valuable insights into the interplay between cytokinin and stress pathways, topics that have garnered limited attention within the scientific community.

Biotechnological Approaches for Manipulating Cytokinin Signaling in Crops

Manipulating cytokinin signaling in crops using biotechnological methods is an active area of investigation. Cytokinins, which are plant hormones, play a vital role in controlling different growth processes like cell division, shoot creation, and nutrient absorption. By modifying genes responsible for cytokinin receptors, it's possible to adjust how sensitive plants are to cytokinins, leading to desired changes in growth and development. Increasing cytokinin receptors' activity can boost the response to cytokinins, whereas reducing or altering receptors can decrease the response. This approach has been used to enhance agricultural characteristics in various crop types.

Agricultural productivity often faces challenges from factors like extreme temperatures, lack of nutrients, water scarcity, high salinity, and excessive light. Cytokinin-mediated responses in plants to stress largely depend on the amount of this plant hormone (O'Brien and Benkova 2013). Both maintaining lower and higher cytokinin levels contribute to plants adapting to stress. Therefore, precise control of cytokinin concentrations can lead to desired results (Rivero et al. 2007, Werner et al. 2010, Nishiyama et al. 2011, Ha et al. 2012).

The model plant *Arabidopsis thaliana* has been extensively examined to understand how it responds to stress through cytokinins (Ha et al. 2012). The common approach to reduce cytokinin levels is by modifying CKX or IPT gene activity. Increasing CKX genes or disrupting IPT genes leads to lower cytokinin levels. Plants with fewer cytokinins often exhibit denser root growth, weaker main tips, and stunted shoot development. Nishiyama et al. (2011) explained that Arabidopsis plants overexpressing cytokinins (35S:CKX1-35S: CKX4) and ipt1, 3, 5, 7 quadruple mutants were more resistant to salt and drought compared to regular plants. These plants with reduced cytokinins showed stronger primary root growth, leading to better salt and drought resistance (Nishiyama et al. 2011).

Enhancing salt tolerance in transgenic *Gossypium hirsutum* was achieved by overexpressing the *Agrobacterium tumefac*iens IPT gene through the Ghcysp promoter (Liu et al. 2012). This approach also improved plant performance during drought stress in *G. hirsutum* early stage, although it didn't boost yields when drought occurred after flowering (Zhu et al. 2018). In *Oryza sativa*, introducing PSAG39:IPT resulted in early flowering and delayed heading, helping plants deal with moderate water limitations due to drought and increasing grain yields (Zou et al. 2007, Liu et al. 2010). Increased IPT expression also enhanced plant survival during waterlogging and submergence recovery stages (Huynh et al. 2005).

Employing cytokinin-related genes in genetic engineering holds potential for improving stress resilience in crops, supporting sustainable agriculture. Nevertheless, most research has focused on overexpression studies, indicating the need for further investigation into genetic engineering and breeding methods to enhance resistance to abiotic stress by manipulating genes. While these molecular components have been well-studied in non-crop plants like *N. tabacum* and *A. thaliana*, their investigation in economically significant crops is limited. Consequently, researching the mechanisms governing cytokinin signaling in vital crops holds potential for improving abiotic stress resilience. Recent advancements in genome editing, especially with CRISPR, offer new avenues for precisely modifying specific genes to develop stress-tolerant crops. Although only a few studies have described the use of CRISPR and cytokinin-related genes to strengthen stress resilience in commercially valuable crops, genome editing is a potent method for accelerating the advancement of abiotic stress tolerance (Ogata et al. 2020, Wang et al. 2022).

Another approach involves engineering cytokinin biosynthesis and degradation pathways to regulate the internal cytokinin levels in plants. This technique allows precise management of cytokinin levels and can be employed to amplify desired traits such as yield, stress tolerance, and root growth.

Conclusion and Future Prospects in Utilizing Cytokinin Signaling for Crop Improvement

The importance of cytokinin in plant responses to stress is becoming more and more clear. Pathways for cytokinin signaling related to stress and genes that code for enzymes involved in metabolism of cytokinin have been identified as being functionally important, although their downstream components are still largely unknown. Interestingly, many of the parts that are involved in how cytokinin affects development also seem to be connected to stress responses. This suggests that how plants respond to stress and how they develop are closely linked, rather than separate evolutionary parts. To better understand how cytokinins work together with known stress response pathways, we need to learn more about how this hormone manages the balance between growth and defense. To achieve this, we need to use advanced genetic methods and organized analyses to figure out the next steps in the cytokinin signaling pathway and how they relate to the more traditional stress response pathways.

Many of the genes that cytokinin regulates are part of systems that control how plants grow and develop, including their signaling, metabolism, and transportation. Modern technologies have given us new insights into how genes respond to cytokinins, like the genes for cytokinin response factors (CRFs), ARR1, CKXs, ARR12, and also ARR10 (Shi et al. 2013, Abdelrahman et al. 2021). When we study the entire set of genes that are active in a cell at a specific time (transcriptome) and the entire set of proteins (proteome), we can understand how proteins interact and affect different parts of a plant's life. This kind of understanding helps us with various processes in crops. It also helps us see how plants deal with stress that comes from non-living things, like extreme temperatures or a lack of water. Knowing how genes work on a global level and understanding how they're connected to how plants handle tough conditions gives us important targets to think about when we're trying to make crops better.

One thing to keep in mind is that a lot of the research on cytokinins and stress has been done in controlled situations. This means that cytokinins seem to help plants with stress by making them grow differently. However, we still don't know a lot about how cytokinins affect different kinds of plants in their natural environments, especially as the climate changes. We need to do more research to really understand how cytokinins help plants deal with stress in the real world. This kind of understanding will help us take care of the places where we grow crops, and make sure that we can keep making food for a long time.

References

Ahuja, I., Kissen, R. and Bones, A.M. 2012. Phytoalexins in defense against pathogens. Trends Plant Sci., 17: 73–90. doi:10.1016/j.tplants.2011.11.002.

Akula, R. and Ravishankar, G.A. 2011. Influence of abiotic stress signals on secondary metabolites in plants. Plant Signal. Behav., 6: 1720–1731. doi: 10.4161/psb.6.11.17613.

Aremu, A.O., Masondo, N.A., Sunmonu, T.O., Kulkarni, M.G., Zatloukal, M., Spichal, L. et al. 2014. A novel inhibitor of cytokinin degradation (INCYDE) influences the biochemical parameters and photosynthetic apparatus in NaCl-stressed tomato plants. Planta, 240: 877–889.

Argueso, C.T., Raines, T. and Kieber, J.J. 2010. Cytokinin signaling and transcriptional networks. Current Opinion in Plant Biology, 13(5): 533–539.

Ashikari, M., Sakakibara, H., Lin, S., Yamamoto, T., Takashi, T., Nishimura, A. et al. 2005. Cytokinin oxidase regulates rice grain pro-duction. Science, 309: 741–745.

Belintani, N., Guerzoni, J., Moreira, R. and Vieira, L. 2012. Improving low-temperature tolerance in sugarcane by expressing the ipt gene under a cold inducible promoter. Biol. Plant., 56: 71–77. doi:10.1007/s10535-012-0018-1.

Bhaskar, A., Paul, L.K., Sharma, E., Jha, S., Jain, M., Khurana, J.P. et al. 2021. *OsRR6*, a type A response regulator in rice, mediates cytokinin, light and stress responses when over expressed in Arabidopsis. Plant Physiol. Biochem., 161: 98–112. doi:10.1016/j.plaphy.2021.01.047.

Calvo-Polanco, M., Armada, E., Zamarreno, A.M., Garcia-Mina, J.M. and Aroca, R. 2019. Local root aba/cytokinin status and aquaporins regulate poplar responses to mild drought stress independently of the ectomycorrhizal fungus Laccaria bicolor. J. Exp. Bot., 70: 6437–6446.

Chaves, M.M., Flexas, J. and Pinheiro, C. 2009. Photosynthesis under drought and salt stress: regulation mechanisms from whole plant to cell. Annals of Botany, 103(4): 551–560.

Chen, H., Lü, J. and Li, T. 2022. Overexpression of the cytokinin biosynthesis gene IPT7 enhances drought tolerance in tobacco. Plant Physiology and Biochemistry, 170: 358–368.

Cheng, S., Yang, Z., Wang, M., Song, J., Sui, N., Fan, H. et al. 2014. Salinity improves chilling resistance in *Suaeda salsa*. Acta Physiol. Plant., 36: 1823–1830. doi:10.1007/s11738-0141555-3.

Cortleven, A. and Schmülling, T. 2015. Regulation of chloroplast development and function by cytokinin. J. Exp. Bot., 66: 4999–5013. doi:10.1093/jxb/erv132.

Cortleven, A., Nitschke, S., Klaumünzer, M., Abdelgawad, H., Asard, H., Grimm, B. et al. 2014. A novel protective function for cytokinin in the light stress response is mediated by the Arabidopsis histidine kinase2 and Arabidopsis histidine kinase3 receptors. Plant Physiol., 164(3): 1470–1483. doi:10.1104/pp.113.224667.

Danilova, M.N., Kudryakova, N.V., Voronin, P.Y., Oelmüller, R., Kusnetsov, V.V., Kulaeva, O.N. et al. (2014). Membrane receptors of cytokinin and their regulatory role in *Arabidopsis thaliana* plant response to photooxidative stress under conditions of water deficit. Russ. J. Plant Physiol., 61: 434–442. doi:10.1134/s1021443714040062.

Dong, H., Tang, W., Li, Z. and Zhang, D. 2021. Cytokinin delays leaf senescence and enhances drought tolerance in creeping bentgrass (Agrostis stolonifera) by regulating antioxidant capacity and osmotic adjustment. Plant Physiology and Biochemistry, 159: 1–10.

Edelman, M. and Mattoo, A.K. 2008. D1-protein dynamics in photosystem II: The lingering enigma. Photosynth. Res., 98: 609–620. doi:10.1007/s11120-008-9342-x.

Escandón, M., Cañal, M.J., Pascual, J., Pinto, G., Correia, B., Amaral, J. et al. 2016. Integrated physiological and hormonal profile of heat-induced thermotolerance in *Pinus radiata*. Tree Physiol., 36: 63–77. 10.1093/treephys/tpv127.

Fahad, S., Bajwa, A.A., Nazir, U., Anjum, S.A., Farooq, A., Zohaib, A. et al. 2017. Crop production under drought and heat stress: Plant responses and Management Options. Front. Plant Sci., 8: 1147. doi: 10.3389/fpls.2017.01147.

Fenollosa, E., Gamez, A. and Munne-Bosch, S. 2018. Plasticity in the hormonal response to cold stress in the invasive plant *Carpobrotus edulis*. J. Plant Physiol., 231: 202–209. doi:10.1016/j.jplph.2018.09.009.

Frerigmann, H. Advances in Botanical Research. Vol. 80. Elsevier; London, UK. 2016. Glucosinolate regulation in a complex relationship–MYC and MYB–no one can act without each other, pp. 57–97.

Fuller, D.Q. and Stevens, C.J. 2019. Between domestication and civilization: the role of agriculture and arboriculture in the emergence of the first urban societies. Veg. Hist. Archaeobot., 28: 263–282. doi: 10.1007/s00334-019-00727-4.

Gujjar, R.S. and Supaibulwatana, K. 2019. The mode of cytokinin functions assisting plant adaptations to osmotic stresses. Plants-Basel, 8: 542.

Ha, S., Vankova, R., Yamaguchi-Shinozaki, K., Shinozaki, K. and Tran, L.S. 2012. Cytokinins: metabolism and function in plant adaptation to environmental stresses. Trends in Plant Science, 17(3): 172–179.

Ha, S., Vankova, R., Yamaguchi-Shinozaki, K., Shinozaki, K. and Tran, L.S. 2012. Cytokinins: Metabolism and function in plant adaptation to environmental stresses. Trends Plant Sci., 17: 172–179. doi:10.1016/j.tplants.2011.12.005.

Hamayun, M., Hussain, A., Iqbal, A., Khan, S.A., Gul, S., Khan, H. et al. 2021b. *Penicillium Glabrum* Acted as a Heat Stress Relieving Endophyte in Soybean and Sunflower. Polish J. Env. Stud., 2021: 30.

Hartmann T. 2008. The lost origin of chemical ecology in the late 19th century. Proc. Natl. Acad. Sci. USA, 105: 4541–4546. doi:10.1073/pnas.0709231105.

Hassler, S., Lemke, L., Jung, B., Möhlmann, T., Krüger, F. and Schumacher, K. 2018. Functional analysis of the two isopentenyltransferase genes in Arabidopsis revealed that they are differentially involved in cytokinin biosynthesis. Journal of Experimental Botany, 69(2): 327–340.

Higuchi, M., Pischke, M.S., Mahonen, A.P., Miyawaki, K., Hashimoto, Y., Seki, M. et al. 2004. In planta functions of the Arabidopsis cytokinin receptor family. Proceedings of the National Academy of Sciences, 101(23): 8821–8826.

Hong, S.-Y., Roze, L.V. and Linz, J.E. 2013. Oxidative stress-related transcription factors in the regulation of secondary metabolism. Toxins., 5: 683–702. doi:10.3390/toxins5040683.

Hu, Y., Liu, B., Ren, H., Chen, L., Watkins, C.B. and Gan, S. (2021). The leaf senescence-promoting transcription factor AtNAP activates its direct target gene CYTOKININ OXIDASE 3 to facilitate senescence processes by degrading cytokinins. Molecular Horticulture, 1. https://doi.org/10.1186/s43897-021-00017-6.

Hu, L., Hu, T., Zhang, X., Zhang, P. and Zhao, Z. 2021. The cytokinin N6-isopentenyladenine improves drought tolerance in maize (Zea mays L.) by regulating photosynthesis and antioxidant defense. Journal of Agronomy and Crop Science, 207(1): 78–89.

Hu, S., Ding, Y. and Zhu, C. 2020. Sensitivity and responses of chloroplasts to heat stress in plants. Front. Plant Sci., 11: 375. doi:10.3389/fpls.2020.00375.

Hu, W., Ren, T., Meng, F., Cong, R., Li, X., White, P.J. et al. 2019. Leaf photosynthetic capacity is regulated by the interaction of nitrogen and potassium through coordination of CO(2) diffusion and carboxylation. Physiol. Plant, 167: 418–432. doi:10.1111/ppl.12919.

Hu, Y., Jiang, L., Wang, F. and Yu, D. 2017. Jasmonate regulates the inducer of CBF expression–c repeat binding factor/DRE binding factor1 cascade and freezing tolerance in Arabidopsis. Plant Cell, 19(9): 1–16.

Huynh, L.N., Vantoai, T., Streeter, J. and Banowetz, G. 2005. Regulation of flooding tolerance of *SAG12: Ipt* Arabidopsis plants by cytokinin. J. Exp. Bot., 56: 1397–1407. doi:10.1093/jxb/eri141

Hwang, I. and Sheen, J. 2001. Two-component circuitry in Arabidopsis cytokinin signal transduction. Nature, 413(6854): 383–389.

Iwasaki, M. and Paszkowski, J. 2014. Identification of genes preventing transgenerational transmission of stress-induced epigenetic states. Proc. Natl. Acad. Sci. USA, 111: 8547–8552.

Jeon, J., Cho, C., Lee, M.R., Binh, N.V. and Kim, J. 2016. *Cytokinin response factor2* (*CRF2*) and *CRF3* regulate lateral root development in response to cold stress in Arabidopsis. Plant Cell., 28: 1828–1843. doi:10.1105/tpc.15.00909.

Jeon, J., Kim, N.Y., Kim, S., Kang, N.Y., Novák, O. and Ku, S.J. 2010. A subset of cytokinin two-component signaling system plays a role in cold temperature stress response in Arabidopsis. Journal of Biological Chemistry, 285(32): 23371–23386.

Jeong, H., Yang, H. and Yan, S. 2014. Comparative Transcriptome Analysis of Maize Induced by Rhodococcus fascians Infection Reveals a Common Adaptive Response. Journal of Integrative Plant Biology, 56(12): 1072–1086.

Jeon, J., Cho, C., Lee, M.R., Binh, N.V. and Kim, J. 2016. Cytokinin response factor2 (crf2) and crf3 regulate lateral root development in response to cold stress in Arabidopsis. Plant Cell, 28: 1828–1843.

Joshi, R., Sahoo, K.K., Tripathi, A.K., Kumar, R., Gupta, B.K., Pareek, A. et al. 2018. Knockdown of an inflorescence meristem-specific cytokinin oxidase-OsCKX2 in rice reduces yield penalty undersalinity stress condition.Plant, Cell and Environment, 41: 936–946,

Khan, T.A., Fariduddin, Q. and Yusuf, M. 2017. Low-temperature stress: Is phytohormones application a remedy? Environ. Sci. Pollut. Res. Int., 24: 21574–21590. doi:10.1007/s11356-017-9948-7.

Koc, I., Yuksel, I. and Caetano-Anolles, G. 2018. Metabolite-centric reporter pathway and tripartite network analysis of Arabidopsis under cold stress. Front. Bioeng. Biotechnol., 6: 121. doi:10.3389/fbioe.2018.00121.

Kumar, M., Choi, J. and An, G. 2012. Ethylene responsive factor 2 (ERF2) plays a critical role in drought tolerance in rice. Journal of Biological Chemistry, 287(17): 16956–16966.

Kumari, V.V., Banerjee, P., Verma, V.C., Sukumaran, S., Chandran, M.A.S., Gopinath, K.A. et al. 2022. Plant nutrition: An effective way to alleviate abiotic stress in agricultural crops. International Journal of Molecular Sciences, 23(15): 8519.

Lakhwani, D., Pandey, A., Dhar, Y.V., Bag, S.K., Trivedi, P.K. and Asif, M.H. 2022. Cytokinins: A key regulator in drought stress response network in plants. Plant Cell Reports, 41(2): 219–234.

Le, T.-N., Schumann, U., Smith, N.A., Tiwari, S., Au, P.C.K., Zhu, Q.-H. et al. 2014. DNA demethylases target promoter transposable elements to positively regulate stress responsive genes in Arabidopsis. Genome Biol., 15: 458. 316.

Li, J., Brader, G. and Palva, E.T. 2004. The WRKY70 transcription factor: A node of convergence for jasmonate-mediated and salicylate-mediated signals in plant defense. Plant Cell., 16: 319–331. doi: 10.1105/tpc.016980.

Li, K., Pang, C.H., Ding, F., Sui, N., Feng, Z.T., Wang, B.S. et al. 2012. Overexpression of *Suaeda salsa* stroma ascorbate peroxidase in Arabidopsis chloroplasts enhances salt tolerance of plants. S. Afr. J. Bot., 78: 235–245. doi:10.1016/j.sajb.2011.09.006.

Li, S.M., Zheng, H.X., Zhang, X.S. and Sui, N. 2021. Cytokinins as central regulators during plant growth and stress response. Plant Cell. Rep., 40(2): 271–282. doi:10.1007/s00299 020-02612-1.

Li, S., Xie, Z., Hu, C., Zhang, Z. and Luo, Z. 2021. Overexpression of the cytokinin biosynthesis gene IPT8 enhances drought tolerance in Arabidopsis. Environmental and Experimental Botany, 192: 104665.

Li, S., Yang, Y., Zhang, Q., Liu, N., Xu, Q., Hu, L. et al. 2018. Differential physiological and metabolic response to low temperature in two zoysiagrass genotypes native to high and low latitude. PLoS ONE, 13: e0198885. doi:10.1371/journal.pone.0198885.

Li, W., Zhang, J., Zhang, X., Li, X. and Liu, T. 2017. Genome-Wide Identification and Analysis of Cytokinin Oxidase/Dehydrogenase (CKX) Genes Reveal Likely Roles in Pod Development and Stress Responses in Oilseed Rape (Brassica napus L.). International Journal of Molecular Sciences, 18(11): 2451.

Li, L., He, Y., Zhang, Z., Shi, Y., Zhang, X., Xu, X. et al. 2021. OsNAC109 regulates senescence, growth and development by altering the expression of senescence-and phytohormone-associated genes in rice. Plant Molecular Biology, 105: 637–54.

Liu, L., Zhou, Y., Szczerba, M.W., Li, X. and Lin, Y. 2010. Identification and application of a rice senescence-associated promoter. Plant Physiol., 153: 1239–1249. doi:10.1104/pp.110.157123.

Liu, Y., Zhang, M., Meng, Z., Wang, B. and Chen, M. 2019. Research progress on the roles of cytokinin in plant response to stress. International Journal of Molecular Sciences, 21(18): 6574.

Liu, X.X., Fu, C., Yang, W.W., Zhang, Q., Fan, H., Liu, J. et al. 2016. The involvement of TsFtsH8 in *Thellungiella salsuginea* tolerance to cold and high light stresses. Acta Physiol. Plant., 38: 62. doi:10.1007/s11738-016-2080-3.

Loake G. and Grant, M. 2007. Salicylic acid in plant defence-the players and protagonists. Curr. Opin. Plant Biol., 10: 466–472. doi:10.1016/j.pbi.2007.08.008.

Lorenzo, O., Chico, J.M., Sánchez-Serrano, J.J. and Solano, R. 2004. JASMONATE-INSENSITIVE1 encodes a MYC transcription factor essential to discriminate between different jasmonate regulated defense responses in Arabidopsis. Plant Cell., 16: 1938–1950. doi: 10.1105/tpc.022319.

Lv, X.M., Zhang, Y.X., Hu, L., Zhang, Y., Zhang, B., Xia, H.Y. 2020. Low-nitrogen stress stimulates lateral root initiation and nitrogen assimilation in wheat: Roles of phytohormone signaling. J. Plant Growth Regul., 40(1): 436–450.

Mandal, S., Ghorai, M., Anand, U., Samanta, D., Kant, N. and Mishra, T. 2022. Cytokinin and abiotic stress tolerance-What has been accomplished and the way forward?. Frontiers in Genetics, 13: 943025.

Mason, M.G., Jha, D., Salt, D.E., Tester, M., Hill, K., Kieber, J.J. et al. 2010. Type B response regulators ARR1 and ARR12 regulateexpression ofAtHKT1;1and accumulation of sodium in Arabidopsisshoots. Plant Journal, 64: 753–763.

Mishra, D., Shekhar, S., Singh, D., Chakraborty, S. and Chakraborty, N. 2018. Heat shock proteins and abiotic stress tolerance in plants. Regulation of Heat Shock Protein Responses, 41–69.

Mittler, R. 2006. Abiotic stress, the field environment and stress combination. Trends in Plant Science, 11(1): 15–19.

Mittler, R., Finka, A. and Goloubinoff, P. 2011. How do plants feel the heat? Trends Biochem. Sci., 37: 118–125. doi:10.1016/j.tibs.2011.11.007.

Miyawaki, K., Tarkowski, P., Matsumoto-Kitano, M., Kato, T., Sato, S., Tarkowska, D. et al. 2006. Roles of Arabidopsis ATP/ADP isopentenyltransferases and tRNA isopentenyltransferases in cytokinin biosynthesis. Proceedings of the National Academy of Sciences, 103(44): 16598–16603.

Nishiyama, R., Le, D.T., Watanabe, Y., Matsui, A., Tanaka, M., Seki, M. et al. 2012. Transcriptome analyses of a salt-tolerant cytokinin-deficient mutant reveal differential regulation of salt stress responseby cytokinin deficiency.PLoS One, 7: e32124.

Nishiyama, R., Watanabe, Y., Fujita, Y. and Le, D.T. 2011. Analysis of cytokinin mutants and regulation of cytokinin metabolic genes reveals important regulatory roles of cytokinins in drought, salt and abscisic acid responses, and abscisic acid biosynthesis. The Plant Journal, 68(1): 131–146.

Nishiyama, R., Watanabe, Y., Fujita, Y., Le, D.T., Kojima, M., Werner, T. et al. 2011. Analysis of cytokinin mutants and regulation of cytokinin metabolic genes reveals important regulatory roles of cytokinins in drought, salt and abscisic acid responses, and abscisic acid biosynthesis. Plant Cell., 23: 2169–2183. doi:10.1105/tpc.111.087395.

Nishiyama, R., Watanabe, Y., Fujita, Y., Le, D. T., Kojima, M., Werner, T. et al. 2011. Analysis of cytokinin mutants and regulation of cytokinin metabolic genes reveals important regulatory roles of cytokinins in drought, salt and abscisic acid responses, and abscisic acid biosynthesis. The Plant Cell, 23(6): 2169–2183.

Nishiyama, R., Watanabe, Y., Fujita, Y., Le, D.T., Kojima, M., Werner, T. et al. 2012. Analysis of cytokinin mutants and regulation of cytokinin metabolic genes reveals important regulatory roles of cytokinins in drought, salt, and abscisic acid responses, and abscisic acid biosynthesis. Plant Cell, 24(10): 4850–4863.

Nishiyama, R., Watanabe, Y., Fujita, Y., Le, D.T., Kojima, M., Werner, T. et al. 2011. Analysis of cytokinin mutants and regulation of cyto-kinin metabolic genes reveals important regulatory roles of cytokininsin drought, salt and abscisic acid responses, and abscisic acid biosyn-thesis. Plant Cell, 23: 2169–2183.

Nishiyama, R., Watanabe, Y., Leyva-González, M.A., Van Ha, C., Fujita, Y., Tanaka, M. et al. 2011. Analysis of cytokinin mutants and regulation of cytokinin metabolic genes reveals important regulatory roles of cytokinins in drought, salt, and abscisic acid responses, and abscisic acid biosynthesis. Plant Cell, 23(6): 2169–2183.

Nitschke, S., Cortleven, A. and Schmülling, T. 2017. Novel stress in plants by altering the photoperiod. Trends Plant Sci., 11: 913–916. doi:10.1016/j.tplants.2017.09.005.

Nitschke, S., Cortleven, A., Iven, T., Feussner, I., Havaux, M., Riefler, M. et al. 2016. Circadian stress regimes affect the circadian clock and cause jasmonic acid dependent cell death in cytokinin-deficient Arabidopsis plants. Plant Cell., 28: 1616–1639. doi:10.1105/tpc.16.00016.

O'brien, J.A. and Benkova, E. 2013. Cytokinin cross-talking during biotic and abiotic stress responses. Front. Plant Sci., 4: 451. doi:10.3389/fpls.2013.00451

Ogata, T., Ishizaki, T., Fujita, M. and Fujita, Y. 2020. CRISPR/Cas9-targeted mutagenesis of OsERA1 confers enhanced responses to abscisic acid and drought stress and increased primary root growth under nonstressed conditions in rice. PloS One, 15(12): e0243376. doi:10.1371/journal.pone.0243376.

Pandey, B.K., Huang, G., Bhosale, R., Hartman, S., Sturrock, C.J., Jose, L. et al. 2016. Drought induces site-specific phosphorylation of cytokinin signaling proteins. Plant Physiology, 172(3): 1848–1862.

Pandey, S., Bhandari, H.S., Shankar, A., Srivastava, A.K., Singh, A. and Shanker, K. 2016. Interaction of abscisic acid and cytokinins modulates drought stress response in maize (Zea mays L.). Advances in Plants and Agriculture Research, 5(1): 00143.

Pré, M., Atallah, M., Champion, A., De Vos, M., Pieterse C.M. and Memelink J. 2008. The AP2/ERF domain transcription factor ORA59 integrates jasmonic acid and ethylene signals in plant defense. Plant Physiol., 147: 1347–1357. doi: 10.1104/pp.108.117523.

Prerostova, S., Dobrev, P.I., Kramna, B., Gaudinova, A., Knirsch, V., Spichal, L. et al. 2020. Heat acclimation and inhibition of cytokinin degradation positively affect heat stress tolerance of *Arabidopsis*. Front. Plant Sci., 11: 87 10.3389/fpls.2020.00087.

Prerostova, S., Černý, M., Dobrev, P.I., Motyka, V., Hluskova, L., Zupkova, B. et al. 2021. Light regulates the cytokinin-dependent cold stress responses in *Arabidopsis*. Front. Plant Sci., 4: 11608711. doi:10.3389/fpls.2020.608711.

Procházková, D., Haisel, D. and Wilhelmová, N. 2008. Antioxidant protection during ageing and senescence in chloroplasts of tobacco with modulated life span. Cell. Biochem. Funct., 26: 582–590. doi:10.1002/cbf.1481.

Qiu, H., Song, J., Liu, Z., Yan, J., Zhang, L. and Hu, Y. 2022. Overexpression of the cytokinin biosynthesis gene OsIPT7 enhances drought tolerance in rice. Plant Growth Regulation, 96(2): 369–382.

Ramakrishnan, M., Satish, L., Kalendar, R., Narayanan, M., Kandasamy, S., Sharma, A. et al. 2021. The dynamism of transposon methylation for plant development and stress adaptation. International Journal of Molecular Sciences, 22(21): 11387.

Ramegowda, V., Senthil-Kumar, M., Nataraja, K.N., Reddy, M.K. and Mysore, K.S. 2013. Drought stress acclimation imparts tolerance to subsequent diverse stresses in a freshwater floating rice. Plant Physiology and Biochemistry, 63: 74–86.

Ramireddy, E., Hosseini, S.A., Eggert, K., Gillandt, S., Gnad, H. and von Wiren, N. 2018. Root engineering in barley: Increasing cytokinin degradation produces a larger root system, mineral enrichment in the shoot and improved drought tolerance. Plant Physiol. 177: 1078–1095.

Rivero, R.M., Mittler, R., Blumwald, E. and Zandalinas, S.I. 2022. Developing climate-resilient crops: improving plant tolerance to stress combination. The Plant Journal. 2022 Jan; 109(2): 373–89.

Rivero, R.M., Kojima, M., Gepstein, A. et al. 2017. Delayed Leaf Senescence Induces Extreme Drought Tolerance in a Flowering Plant. Proceedings of the National Academy of Sciences, 114(50): 13076–13081.

Rivero, R.M., Gimeno, J., Van Deynze, A., Walia, H. and Blumwald, E. 2010. Enhanced cytokinin synthesis in tobacco plants expressing PSARK::IPT prevents the degradation of photosynthetic protein complexes during drought. Plant Cell Physiology, 51(11): 1929–1941.

Rivero, R.M., Gimeno, J., Van Deynze, A., Walia, H. and Blumwald, E. 2010. Enhanced cytokinin synthesis in tobacco plants expressing PSARK::IPT prevents the degradation of photosynthetic protein complexes during drought. Plant Cell Physiology, 51(11): 1929–1941.

Rivero, R.M., Kojima, M., Gepstein, A., Sakakibara, H., Mittler, R., Gepstein, S. et al. 2007. Delayed leaf senescence induces extreme drought tolerance in a flowering plant. Proc. Natl. Acad. Sci. USA, 104: 19631–19636. doi:10.1073/pnas.0709453104.

Rivero, R.M., Kojima, M., Gepstein, A., Sakakibara, H., Mittler, R., Gepstein, S. et al. 2014. Cytokinin-dependent improvement in transgenic salt tolerance of sweet potato plants. Journal of Experimental Botany, 65(15): 4405–4414.

Rizwan, M., Ali, S., Ibrahim, M., Farid, M., Adrees, M., Bharwana, S.A. et al. 2015. Mechanisms of silicon-mediated alleviation of drought and salt stress in plants: a review. Environ. Sci. Pollut. Res. Int., 22: 15416–15431. doi: 10.1007/s11356-015-5305-x.

Sabagh, A.E., Mbarki, S., Hossain, A., Iqbal, M.A., Islam, M.S., Raza, A. et al. 2021. Potential role of plant growth regulators in administering crucial processes against abiotic stresses. Frontiers in Agronomy, 3: 648694.

Sakakibara, H. 2006. Cytokinins: activity, biosynthesis, and translocation. Annual Review of Plant Biology, 57: 431–449.

Sakamoto, T., Sotta, N., Suzuki, T., Fujiwara, T. and Matsunaga, S. 2019. The 26s proteasome is required for the maintenance of root apical meristem by modulating auxin and cytokinin responses under high-boron stress. Front Plant Sci., 10: 590.

Schaller, G.E., Bishopp, A. and Kieber, J.J. 2015. The Yin-Yang of hormones: cytokinin and auxin interactions in plant development. Plant Cell, 27(1): 44–63.

Shi, Y., Tian, S., Hou, L., Huang, X., Zhang, X., Guo, H. et al. 2012. Ethylene signaling negatively regulates freezing tolerance by repressing expression of CBF and type-A ARR genes in Arabidopsis. Plant Cell., 24: 2578–2595. doi:10.1105/tpc.112.098640.

Skalák, J., Černý, M., Jedelský, P., Dobrá, J., Ge, E., Novák, J. et al. 2016. Stimulation of ipt overexpression as a tool to elucidate the role of cytokinins in high temperature responses of *Arabidopsis thaliana*. J. Exp. Bot., 67: 2861–2873. 10.1093/jxb/erw129

Sobol, S., Chayut, N., Nave, N., Kafle, D., Hegele, M., Kaminetsky, R. et al. 2014. Genetic variation in yield under hot ambient temperatures spotlights a role for cytokinin in protection of developing floral primordia. Plant Cell Environ., 37: 643–657. 10.1111/pce.12184

Sui, N. 2015. Photoinhibition of Suaeda salsa to chilling stress is related to energy dissipation and water-water cycle. Photosynthetica, 53: 207–212. doi:10.1007/s11099-015-0080-y

Sundaram, V. and Wysocka, J. 2020. Transposable elements as a potent source of diverse cis-regulatory sequences in mammalian genomes.Philos.Trans. R. Soc. Lond. B Biol. Sci., 375: 2019034.

Suzuki, N., Rivero, R.M., Shulaev, V., Blumwald, E. and Mittler, R. 2014. Abiotic and biotic stress combinations. New Phytologist, 203(1): 32–43.

Takahashi, S. and Badger, M.R. 2011. Photoprotection in plants: A new light on photosystem II damage. Trends Plant Sci., 16: 53–60. doi:10.1016/j.tplants.2010.10.001.

Tran, L.S.P., Nakashima, K., Sakuma, Y., Osakabe, Y., Qin, F., Simpson, S.D. et al. 2010. Transcriptomic analysis of rice (Oryza sativa L.) in response to drought stress. Journal of Experimental Botany, 61(15): 3451–3464.

Tran, L.S., Nakashima, K., Sakuma, Y., Simpson, S.D., Fujita, Y., Maruyama, K. et al. 2010. Isolation and functional analysis of Arabidopsis stress inducible NAC transcription factors that bind to a drought-responsive cis-element in the early responsive to dehydration stress 1 promoter. The Plant Cell, 22(9): 2480–2498.

Tran, L.S., Nakashima, K., Sakuma, Y., Simpson, S.D., Fujita, Y., Maruyama, K. et al. 2010. Isolation and functional analysis of Arabidopsis stress-inducible NAC transcription factors that bind to a drought-responsive cis-element in the early responsive to dehydration stress 1 promoter. The Plant Cell, 19(10): 3266–3292.

Tran, L.S., Urao, T., Qin, F., Maruyama, K., Kakimoto, T., Shinozaki, K. et al. 2007. Functional analysis of AHK1/ATHK1 and cytokinin receptor histidine kinases in response to abscisic acid, drought, and salt stress in Arabidopsis. Proceedings of the National Academy of Sciences, 104(51): 20623–20628.

Treisman, R. 1996. Regulation of transcription by MAP kinase cascades. Curr. Opin. Cell Biol., 8: 205–215. doi:10.1016/S0955-0674(96)80067-6.

Trifunović-Momčilov, M., Paunović, D., Milošević, S,, Marković, M., Jevremovćc, S., Dragićević, I.Č.et al. 2020. Salinity stress response of non-transformed and AtCKX transgenic centaury (Centaurium erythraea Rafn.) shoots and roots grown *in vitro*. Ann. Appl. Biol., 177: 74–89.

Voigt, C.A. 2014. Callose-mediated resistance to pathogenic intruders in plant defense-related papillae. Front. Plant Sci., 5: 168. doi:10.3389/fpls.2014.00168.

Wachsman, G., Sparks, E.E. and Benfey, P.N. 2015. Genes and networks regulating root anatomy and architecture. New Phytologist, 208(1): 26–38.

Wang, K., Zhang, X. and Ervin, E. 2012. Antioxidative responses in roots and shoots of creeping bentgrass under high temperature: effects of nitrogen and cytokinin. J. Plant Physiol., 169: 492–500. 10.1016/j. jplph.2011.12.007.

Mandal, S., Ghorai, M., Anand, U., Samanta, D., Kant, N., Mishra, T. et al. (2022). Cytokinin and abiotic stress tolerance -What has been accomplished and the way forward? Frontiers in Genetics, 13.

Wang, J., Zhang, Q., Cui, F., Hou, L., Zhao, S., Xia, H. et al. 2017. Genome-wide analysis of gene expression provides new insights into cold responses in *Thellungiella salsuginea*. Front. Plant Sci., 8: 713. doi:10.3389/fpls.2017.00713.

Wang, N., Chen, J., Gao, Y., Zhou, Y., Chen, M., Xu, Z. et al. 2022. Genomic analysis of isopentenyltransferase genes and functional characterization of *TaIPT8* indicates positive effects of cytokinins on drought tolerance in wheat. Crop J., 2214–5141. ISSN. doi:10.1016/j.cj.2022.04.010.

Wang, Y., Shen, W., Chan, Z. and Wu, Y. 2015. Endogenous cytokinin over-production modulates ROS homeostasis and decreases salt stressresistance inArabidopsis thaliana. Frontiers in Plant Science, 6: 1004.

Wang, Y., Tang, J., Ling, Y., Wang, H.Y. and Zhu, Y. 2020. Overexpression of the cytokinin receptor gene CKI1 enhances drought tolerance in Arabidopsis. International Journal of Molecular Sciences, 21(6): 2205.

Wang, X.L., Wang, J.J., Sun, R.H., Hou, X.G., Zhao, W., Shi, J. et al. 2016. Correlation of the corn compensatory growth mechanism after post-drought rewatering with cytokinin induced by root nitrate absorption. Agr. Water Manage, 166: 77–85.

Wang, X.L., Qin, R.R., Sun, R.H., Hou, X.G., Qi, L. and Shi, J. 2018a. Efects of plant population density and root-induced cytokinin on the corn compensatory growth during post-drought rewatering. PLoS ONE, 13: e0198878.

Wang, X.L., Qin, R.R., Sun, R.H., Wang, J.J., Hou, X.G., Qi, L. et al. 2018b. No postdrought compensatory growth of corns with root cutting based on cytokinin induced by roots. Agr Water Manage, 205: 9–20.

Werner, T., Motyka, V., Laucou, V., Smets, R., Van Onckelen, H. and Schmulling, T. 2001. Cytokinin-deficient transgenic Arabidopsis plants show multiple developmental alterations indicating opposite functions of cytokinins in the regulation of shoot and root meristem activity. The Plant Cell, 13(11): 2619–2630.

Werner, T., Nehnevajova, E., Köllmer, I., Novák, O., Strnad, M., Krämer, U. et al. 2010. Root specific reduction of cytokinin causes enhanced root growth, drought tolerance, and leaf mineral enrichment in Arabidopsis and tobacco. Plant Cell., 22: 3905–3920. doi:10.1105/tpc.109.072694.

Werner, T., Nehnevajova, E., Köllmer, I., Novák, O., Strnad, M., Krämer, U. et al. 2010. Root specific reduction of cytokinin causes enhanced root growth, drought tolerance, and leaf mineral enrichment in Arabidopsis and tobacco. Plant Cell, 22: 3905–3920. doi: 10.1105/tpc.109.072694.

Werner, T., Nehnevajova, E., Köllmer, I., Novák, O., Strnad, M., Krämer, U. et al. 2010. Root-specific reduction of cytokinin causes enhanced root growth, drought tolerance, and leaf mineral enrichment in Arabidopsis and tobacco. Plant Cell, 22(12): 3905–3920.

Wu, C., Cui, K., Wang, W., Li, Q., Fahad, S., Hu, Q. et al. 2017. Heat-induced cytokinin transportation and degradation are associated with reduced panicle cytokinin expression and fewer spikelets per panicle in rice. Front. Plant Sci., 8: 371. 10.3389/fpls.2017.00371.

Chen, L., Zhao, J., Song, J. and Jameson, P. E. 2020. Cytokinin dehydrogenase: A genetic target for yield improvement in wheat. Plant Biotechnology Journal, 18(3): 614–630.

Wu, C., Cui, K., Wang, W., Li, Q., Fahad, S., Hu, Q. et al. 2017. Heat-induced cytokinin transportation and degradation are associated with reduced panicle cytokinin expression and fewer spikelets per panicle in rice. Front. Plant Sci., 8: 371. doi:10.3389/fpls.2017.00371.

Wu, C., Cui, K., Wang, W., Li, Q., Fahad, S., Hu, Q. et al. 2017. Heat-induced cytokinin transportation and degradation are associated with reduced panicle cytokinin expression and fewer spikelets per panicle in rice. Front. Plant Sci., 8: 371. doi:10.3389/fpls.2017.00371.

Xie, Z., Nolan, T.M., Jiang, H. and Yin, Y. 2016. AP2/ERF Transcription Factor Regulatory Networks in Hormone and Abiotic Stress Responses in Arabidopsis. Frontiers in Plant Science, 7: 532.

Xu, Y., Gianfagna, T. and Huang, B. 2010. Proteomic changes associated with expression of a gene (ipt) controlling cytokinin synthesis for improving heat tolerance in a perennial grass species. J. Exp. Bot., 61: 3273–3289. 10.1093/jxb/erq149.

Xu, Y., Burgess, P., Zhang, X. and Huang, B. 2016. Enhancing cytokinin synthesis by overexpressing ipt alleviated drought inhibition of root growth through activating ros-scavenging systems in agrostis stolonifera. J. Exp. Bot., 67: 1979–1992.

Xu, Y., Tian, J., Gianfagna, T. and Huang, B. 2009. Effects of *SAG12-ipt* expression on cytokinin production, growth and senescence of creeping bentgrass (*Agrostis stolonifera* L.) under heat stress. Plant Growth Regul., 57: 281–291. doi:10.1007/s10725008-9346-8.

Yamauchi, Y., Ogawa, M., Kuwahara, A., Hanada, A., Kamiya, Y. and Yamaguchi, S. 2004. Activation of gibberellin biosynthesis and response pathways by low temperature during imbibition of Arabidopsis thaliana seeds. Plant Cell., 16(2): 367–78. doi: 10.1105/tpc.018143.

Yamamoto, Y. 2016. Quality control of photosystem II: The mechanisms for avoidance and tolerance of light and heat stresses are closely linked to membrane fluidity of the thylakoids. Front. Plant Sci., 2: 1136. doi:10.3389/fpls.2016.01136.

Zhang, J., Pai, Q., Yue, L., Wu, X., Liu, H. and Wang, W. 2022. Cytokinin regulates female gametophyte development by cell cycle modulation in Arabidopsis thaliana. Plant Science, 324: 111419.

Zhang, H., Li, W., Mao, X. and Jing, R. 2018. A Genome-Wide Survey of Cytokinin Response Factors in Different Crops Identifies OsCRFs as Key Regulators in Rice Salt and Drought Tolerance. Frontiers in Plant Science, 9: 526.

Zhang, Z., Li, Q., Li, Z., Staswick, P.E., Wang, M., Zhu, Y. and He, Z. 2010. Dual regulation role of GH3.5 in salicylic acid and auxin signalling during Arabidopsis-Pseudomonas syringae interaction. Plant Physiology, 152(2): 963–975.

Zhao, X., Chai, Y., Liu, B., Liu, L., Ge, C. and Li, X. 2019. Cytokinin-Stimulated DNA Demethylation Activates Genes Associated with Stress Response in Arabidopsis thaliana. Frontiers in Plant Science, 10: 654.

Zhou, Y.Y., Zhang, Y., Wang, X.W., Han, X., An, Y., Lin, S.W. et al. 2020. Root-specifc nf-y family transcription factor, pdnf-yb21, positively regulates root growth and drought resistance by abscisic acid-mediated indoylacetic acid transport in Populus. New Phytol., 227: 407–426.

Zhu, G., Li, W., Zhang, F. and Guo, W. 2018. RNA-seq analysis reveals alternative splicing under salt stress in cotton, *Gossypium davidsonii*. BMC Genomics, 19: 73. doi:10.1186/s12864-018-4449-8.

Zhu, J.K. 2002. Salt and drought stress signal transduction in plants. Annual Review of Plant Biology, 53: 247–273.

Zou, G.H., Liu, H.Y., Mei, H.W., Liu, G.L., Yu, X.Q., Li, M.S. et al. 2007. Screening for drought resistance of rice recombinant inbred populations in the field. J. Integr. Plant Biol., 49: 1508–1516. doi:10.1111/j.1672-9072.2007.00560.x.

Zwack, P.J., Compton, M.A., Adams, C.I. and Rashotte, A.M. 2016. *Cytokinin response factor 4 (CRF4)* is induced by cold and involved in freezing tolerance. Plant Cell. Rep., 35: 573–584. doi:10.1007/s00299-015-1904-8.

16

Plant Nitric Oxide Signaling under Abiotic Stresses

Alia Telli,[1,*] *Ouiza Djerroudi,*[2] *Salim Azib,*[2] *Ahmed Chaabna*[2] and *Halima Khaled*[3]

Introduction

Due to human activity, the amount of carbon dioxide (CO_2) in the environment has increased. This has an impact on the seasonality, length, and intensity of rainfall patterns as well as increased drought spells, waterlogging, and evapotranspiration. This has a negative impact on the growth, production, and quality of the plants growing in this environment (Onyekachi et al. 2019). Different abiotic stresses such water availability, extreme temperatures, nutrient deficiencies, salinity, heavy metals toxicity, pesticides and UV radiation all have an adverse effects on the development and growth of plants as well as the yield of basic food crops up to 70% (Parihar et al. 2014, Kaur et al. 2023). Abiotic stresses may occur individually, sequentially, or concurrently. However, plants may suffer greater detrimental effects when they are exposed to multiple abiotic stresses, such as cold and drought, heat, and heavy metal, as well as drought and heat, than when they are exposed to each individual stress (Lei et al. 2021).

Abiotic stressors typically increase the generation of reactive oxygen species (ROS), impair the antioxidant defense system, and change the plant cells' redox

[1] Laboratory of preservation of ecosystems in arid and semi-aride areas, Faculty of Nature and Life Sciences, University of KASDI Merbah-Ouargla, PB 511 Ghardaïa road, Ouargla 30000, Algeria.

[2] Laboratory of saharan bio-resources : preservation and valorization, Faculty of Nature and Life Sciences, University of KASDI Merbah-Ouargla, PB 511 Ghardaïa road, Ouargla 30000, Algeria.

[3] Technical institute for the development of Saharan agriculture, PB 27 RP Ain Bennoui Biskra, Algeria.

* Corresponding author: alia.telli@gmail.com

homeostasis. Furthermore, in reaction to environmental challenges, the synthesis and the concentration of methylglyoxal (MG) are raised in plant cells (Bhyan et al. 2020).

Owing to climate change and abiotic constraints, the yield of main crops around the world might be severely influenced (Hassan et al. 2022a). Tolerance of plants to abiotic stress may be increased by choosing abiotic constraints-tolerant cultivars, genetic engineering, and external administration of osmolytes, microorganisms, mineral nutrients, phytohormones, soils nutrients, and best farming methods. Plants release signaling molecules as a form of stress defense that set off a chain reaction of stress-adaptation processes that can cause programmed cell death or plant acclimation. A small but crucial redox signaling molecule called nitric oxide (NO) is involved in a variety of physiological functions in plants, such as germination, growth, flowering, senescence, and environmental constraints (Nabi et al. 2019). This calls for signaling both within and between the affected cells, which, in the near term, modifies critical enzymes and, in the long term, alters the network of genes that regulates the associated metabolic. Together with NO, other tiny molecules including the plant hormones auxin and ethylene also play a role in this signal transduction (Gupta et al. 2022). There is cross-talk between NO and other molecules that regulate vital physiological processes (Hassanuzzaman et al. 2018).

This review highlights the synthesis pathways of NO in plant cells and NO roles in normal and in stressing conditions.

Abiotic Stress

Hans Selye first introduced the notion of stress which was proven and refined in animal models in 1936 and after observing many of his sick patients (Szabo et al. 2017). The stress concept is defined as harmful ecological restrictions for plants. In terms of physics, stress is the force per unit area that results in variation on the object to which it is applied. Stress restricts an individual's ability to operate normally, which in turn diminishes their genetic capacity for growth, development, and reproduction. For plants, Grime (1977) defined stress as the external constraints which limit the rate of dry-matter of all or part of the vegetation. Stress in the context of agriculture is a factor that lowers crop yield and damages biomass through interference with plant homeostasis affecting key metabolic and physiological processes, limiting energy production, and endangering cellular integrity (Jena et al. 2020, Rivero et al. 2021).

The most frequent unfavorable environmental variables that affect and restrict agricultural productivity globally include CO_2 concentration, drought, high soil salinity, heat, cold, oxidative stress, and heavy metal toxicity (Fig. 1).

CO_2 Concentration

The level of atmospheric CO_2 is now 418.82 ppm (https://gml.noaa.gov/ccgg/trends/global.html), due to the burning of fossil energy, deforestation, rapid urbanization and industrialization, and by the end of this century, it will roughly double from current levels and is expected to exceed 800 ppm (Ahammed et al. 2020, Dong et al. 2021, IPCC 2023). The anticipated increases in $[CO_2]$ could lead to a rise in global surface temperature and drought (IPCC 2023). According to studies, increased CO_2

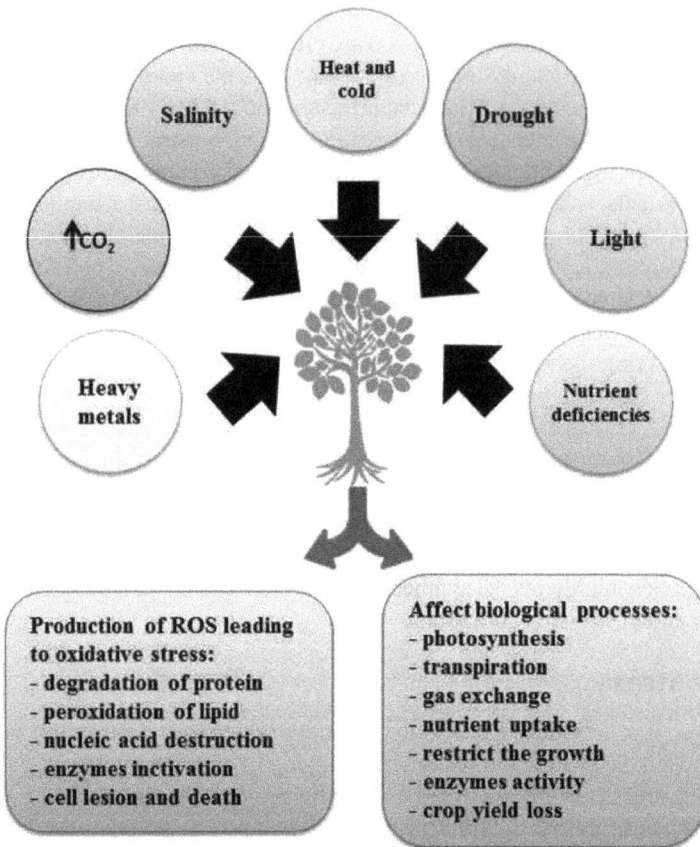

Fig. 1. Abiotic stress affecting plant growth and yield and quality of crop.

concentrations have a direct impact on plant development activities as photosynthesis and gas exchange. The high concentration of CO_2 increases photosynthesis rate and instantaneous water use efficiency, but decreases the transpiration rate and stomatal conductance (Zhang et al. 2021). Elevated CO_2 raises the total carbon amount while lowering the total nitrogen content, which results in a higher carbon to nitrogen ratio and consequently, a reduction in protein and vitamin levels (Ahammed et al. 2020, Singer et al. 2020). It also influences the quality of crops by variably affecting the quantities and biosynthetic gene expression of polyphenols, free amino acids, and macro and micro-elements; elevated CO_2 modifies the nutritional quality of agricultural plants and seed quality (Ahammed et al. 2020, Roy and Mathur 2021). In addition to changing root exudates and the functioning of soil enzymes and microbes in the rhizosphere surroundings, higher levels of carbon dioxide and temperature also have a direct impact on soil chemical and biochemical mechanisms, which affect the absorption and utilization of phosphorus (Guo et al. 2021). Additionally, Rahaman et al. (2019) demonstrated that the increase in greenhouse gases did not promote radial development in tropical trees. The stimulation kinetics of the defense hormones salicylic acid and jasmonic acid, as well as the C3 plant's tolerance to insect herbivory, are all affected by high CO_2 (Gog et al. 2019). Studies have additionally indicated impacts of carbon dioxide high levels simultaneously on the cellular and molecular level modifying the activity of several genes that regulate various metabolic procedures and stress signalling networks (Roy and Mathur 2021). Mndela et al. (2022) found that woody plants will profit from elevated CO_2 via photosynthetic level, productivity, and increased hydration status, but the responses will differ depending on the woody plant characterisitics and the period of exposure to elevated carbon dioxide. Elevated CO_2 alters soil fungal community structure via affecting leaf N and C/N (Chen et al. 2022). The high concentration of CO_2 improved the leaves biochemical constituents, viz. chlorophyll, protein, total sugars, and carbon content whereas it inversely diminished the ascorbic acid content in leaf. Furthermore, elevated CO_2 significantly changed protein, sugars, carbon, and nutrients distribution in leaf, stem, and root tissues (Sharma and Singh 2022). The microbial community of soil is also affected by high concentration of atmospheric CO_2. The work of Norgbey et al. (2022) proved that the gram-positive bacteria and fungi number fell under high N input and elevated CO_2, but gram-negative bacteria rose, indicating that N inputs and elevated CO_2 affected the microbial activity and was associated with nitrogen decrease in the soil.

Drought

One of the most important abiotic stresses is a lack of water. A season without much rain is what is often referred to as a 'drought' in meteorology. Drought stress often happens once the amount of water in the soil is decreased and atmospheric conditions result in a constant loss of water through transpiration or evaporation (Jaleel et al. 2009). The abrupt change in environmental conditions severely limits the water supply and produces drought stress in major agrosystems across the world, particularly in rainfed ecosystems. Agriculture is the most vulnerable to water constraint and is experiencing a large loss in production potential (40 to 60%) in rainfed regions (Saha et al. 2022). It has been demonstrated that drought stress during cultivation affects plants on the molecular, biochemical, physiological, morphological, and ecological level (Bijalwan et al. 2022). Drought stress conditions interrupt photosynthesis, growth, and other critical physiological and biochemical activities (Wahab et al. 2022). The primary consequence of drought stress is a reduction in the water potential and turgor pressure in the developing cells, which results in the absence of turgor pressure required for their development. The growth of root, stem, leaf, and fruit diminishes under drought constraint (Hemati et al. 2022).

Salinity

Among the main environmental dangers that severely restricts growth and agricultural output is salinity stress (Astaneh et al. 2022). Approximately 20% of agricultural areas are salty, and more and more arable land is turning salty as a result of the serious problems with global warming (Miransari and Smith 2019). With more evapotranspiration than rainfall, arid and semi-arid areas have comparatively high soil salinization year-round (Gul et al. 2022). In general, a saline soil is one that has an exchangeable salt content of 15% and an electrical conductivity (EC) of the saturation extract (EC_e) in the root zone that is greater than 4 dS.m^{-1} (or around 40 mM NaCl) at 25°C (Shrivastava and Kumar 2015). Salinity causes Na$^+$ toxicity and ionic imbalance and interferes with crucial metabolic procedures in plant cells, such as protein production, enzymatic reactions, and ribosome activities (Angon et al. 2022). Na$^+$, Ca^{2+}, and Mg^{2+} are the most frequent cations linked to salinity, whereas Cl$^-$,, and are the most prevalent anions. It is thought that Na$^+$ and Cl$^-$ ions are the most significant and both are poisonous to plants, but Na$^+$ particularly damages the soil's physical structure (Safdar et al. 2019). Excessive osmotic stress, ion toxicities, and nutritional disorders are common in plants cultivated in saline soil, and these factors contribute to the deteriorating physical state of the soil and lower plant yield (Kibria and Hoque 2019).

Temperature

One of the significant environmental constraints that limits the general growth and development of plants is the extremes of temperature, either low or high (Saleem et al. 2021). The global surface temperature was 1.09°C [0.95°C–1.20°C] higher in 2011–2020 than 1850–1900, with larger increases over land (1.59°C [1.34°C–1.83°C]) than over the ocean (0.88°C [0.68°C–1.01°C]) (IPCC 2023). Since worldwide climatic projections indicate an increase in mean temperature of 2–4°C during the next half of this century, global climatic variability is anticipated to worsen even further (Ali et al. 2020a). The main stresses that negatively affect plant growth and development and lower yield output in temperate and a few subtropical climates are cold stress (0–15°C) and freezing stress (< 0°C) (Ritonga et al. 2021, Manasa et al. 2022).

Numerous works obviously demonstrate that temperatures exceeding the limits of adaptation significantly affect the metabolism, viability, physiology, and yield of various plants. Plants exposed to extreme temperatures frequently exhibit a common reaction by way of oxidative damage. Nevertheless, the magnitude of harm produced by extreme temperatures relies highly on the period of the unfavorable temperature, the genotypes of exposed plants, and their phase of development (Bhattacharya 2022a, Jaiyeola et al. 2021, Manasa et al. 2022). Cold temperatures often severely restrict the growth (reduce leaf size, stem extension and root proliferation), disturb plant water relationship, imped nutrient uptake, distribution, and productivity of plants (Bhattacharya 2022b and 2022c). Under cold condition, plants undergo a series of changes including the increase of cell membrane permeability, inhibition of protoplast flow, decrease in the photosynthetic rate, and other metabolic changes (Wang et al. 2022, Raza et al. 2023).

According to Jaiyeola et al. (2021), the heat stress may cause plants to experience the consequences that include: raised transpiration, higher water loss, elevated leaf water deficit or stress, decreased leaf water potential, diminished stomatal opening or stomatal conductance, and the accumulation of CO_2 concentration on the leaf surface and in the intercellular (internal) system of the leaf. This further causes stomatal closure or lowered stomatal conductance and, as a result, minimizes stomatal conductance and, ultimately, decreases transpiration and photosynthesis with reduction in Rubisco activity, ultimately hindering plant growth, development, and crop yield.

Light and UV Radiations

Light is regarded as the primary source of energy needed for photosynthesis and numerous additional biological functions, making it a critical environmental factor for plant development (Ahmed et al. 2020). The light use efficiency is known as the ratio of chemical energy accumulated in dry mass of plant shoots to accumulated light energy absorbed by the plant canopy (Kozai 2013). Light stress, which occurs when plants are frequently exposed to excessive or insufficient light levels, affects the agronomic traits of plants by impairing their physiological metabolic processes, such as photosynthesis, antioxidant machinery, and their capacity to fix atmospheric carbon and nitrogen (Yang et al. 2019).

Ultraviolet (UV) radiation consists of three groups: UV-A (315–400 nm), UV-B (280–315nm) and UV-C (100–280 nm). The UV-B region (280–315 nm) is largely absorbed by ozone, and about 20% reaches the earth's surface. Although UV-B is a small portion of sunlight, because of its high energy, it has the ability to harm biological systems. Due to the depletion of the stratospheric ozone layer due to the atmospheric pollution during the past few decades, UV-B in the biosphere has increased (Singh et al. 2017). UV-B radiation makes up only a minor fraction of sunlight, yet it imparts many positive and negative effects on plant growth (Yadav et al. 2020). As a positive effect, UV-B regulates photomorphogenesis including hypocotyl elongation inhibition, cotyledon expansion, and flavonoid accumulation (Shi and Liu 2021). Hence, low intensity of UV-B radiations may be utilized to produce plants, loaded with secondary metabolites, having better reproductive potential, early ripening, and tolerance against fungi, bacteria, and herbivores (Yavaş et al. 2020, Apoorva et al. 2021). However, high intensity UV-B can also harm plants by damaging DNA, triggering the accumulation of reactive oxygen species, and impairing photosynthesis (Shi and Liu 2021, Mareri et al. 2022).

Heavy Metal Toxicity

Heavy metals (HMs) are not destroyable or degradable and persist in the environment for a long duration (Gulzar and Mazumder 2022). Due to both natural and human-caused factors HM exposure in plants has become one of the most prevalent abiotic stresses, causing substantial decreases in plant yield and dangerous health issues. Volcanoes and dust carried by the breeze are two significant natural sources of heavy metal emissions. The main contributors to the contamination of farming soils with heavy metals are the use of pesticides and inorganic fertilizers (Sharma et al. 2020a). HMs including chromium (Cr), cadmium (Cd), mercury (Hg), aluminum (Al), lead (Pb), and arsenic (As), have appeared as the most significant air, water, and soil pollutants, which adversely affect the quantity, quality, and security of plant-based food all over the world (Rahman et al. 2022). Under HM contaminated conditions, plants suffer from various complications, such as nutrient and mineral deficiencies, alteration of various physiological and biological activities, which decreases seed germination, diminishes the plant's growth rate lower leaf area index, increased leaf chlorosis, cell wall injury, etc. (Gulzar and Mazumder 2022, Rahman et al. 2022). HM ions can cause alterations at cellular and molecular levels by replacing essential elements from biomolecules, blocking the functional groups and active sites of enzymes, damaging the membranes and alter transcription patterns of plants (Sharma et al. 2020a). HMs toxicity leads to inductive oxidation damage, retarded growth, development, and oscillation in catalytic proficiency of enzymes, constrained growth of root and shoot in plants and also results in chlorosis (Sharma et al. 2022).

Nutrient Deficiencies

For the growth and evolution of plants, inorganic mineral nutrients must be assimilated. The availability of nutrients directly controls all of the steps that contribute to the development of agricultural yields, including biomass accumulation and partitioning. Basically, plants need 17 elements to produce their best biomass and yield, and some extra elements are helpful for ensuring their survival in adverse conditions and/or enhancing the quality of their commercially valuable produce (Pandey et al. 2021).

Brown et al. (2022) have proposed a new definition of essential or beneficial element in mineral nutrition of plants. A mineral plant nutrient is an element which is essential or beneficial for plant growth and development of the quality attributes of the plant or harvested product, of a given plant species, grown in its natural or cultivated environment. A plant nutrient may be considered essential if the life cycle of a diversity of plant species cannot be completed in the absence of the element. A plant nutrient may be considered beneficial if it does not meet the criteria of essentiality, but can be shown to benefit plant growth and development or the quality attributes of a plant or its harvested product (Brown et al. 2022).

Macronutrients and micronutrients are two categories that are frequently used to categorize essential elements. Macronutrients are composed of carbon, oxygen, hydrogen, nitrogen, phosphorus, potassium, calcium, sulphur, and magnesium, while micronutrients are composed of iron, boron, chlorine, manganese, zinc, copper, molybdenum, and nickel (Ali et al. 2020b, Mesurani and Ram 2020, Shrestha et al. 2020, Xue et al. 2022). The availability of macronutrients like nitrogen, phosphorus, and potassium in the soil has a significant impact on plant development, yield, and crop quality (Ma et al. 2020). Depending on the specific nutrient, a lack can result in chlorosis (yellowing of the leaves), stunted development, or slow growth. High deficiency levels could lead to plant death (Mesurani and Ram 2020). Low levels of micronutrients in plants are linked to their decreasing concentration in soils and/or low bioavailability and presence of abiotic stresses which disturb the proper growth and development of plants (Szerement et al. 2022).

The nutrient deficiencies may occur when one nutrient's availability influences how well other nutrients are absorbed and used, there is nutrient interaction in crop plants. When one nutrient is present in excess concentration in the growth medium, this form of interaction occurs most frequently. The interactions are due either to the formation of precipitates or complexes or to the competition between two elements quite similar pertaining to the location of adsorption, absorption, transport, and function on the surfaces of plant roots or inside plant cells (like Na and K) (Bhattacharya 2021).

The unbalanced mineral nutrition of plants is the result of water deficit and low soil moisture. Indeed, the natural cycles of potassium, phosphorus, and nitrogen can be significantly impacted by greater aridity. Reduced mineralization rates and/or reduced soil nutrient diffusion to the root surface caused by decreased soil moisture with less precipitation may lower plant nutrient absorption (Bhattacharya 2021).

Impact of Abiotic Stress on Plant

All kinds of stressful conditions are translated into oxidative stress into the plant. These abiotic stresses increase the production of reactive oxygen species (ROS) and nitrogen species (RNS), which lead to membrane alteration, protein degradation, lipid per-oxidation, nucleic acid damage, enzymes inactivation, and interrupting various metabolic pathways until cell death occurs (Chaki et al. 2020, Chaudhry and Sidhu 2021, Qamer et al. 2021, Hassan et al. 2022b). The main sites for ROS generation in plant cells are mitochondria, chloroplast, and peroxisomes. In addition, ROS are a by-product of metabolic processes. Under regular circumstances, there is a steadiness between generation and elimination of ROS, but this balance is hampered by different biotic and abiotic stress factors (García-Capparós et al. 2020). However, based on recent studies on agricultural plants, modest levels of ROS signal the induction of tolerance to environmental extremes by changing the expression of protective genes (Qamer et al. 2021).

Among different growth phases of plants, germination is the foremost stage that is affected by abiotic stress (Chaudhary and Sidhu 2021, El Mokhtari et al. 2022). Abiotic stresses either delay or entirely prevent seed germination through osmotic stress and/or ionic toxicity. Some abiotic stresses also increase reactive oxygen species (ROS) in seeds that may be produced during desiccation, germination, and aging (El Mokhtari et al. 2022). The chloroplast is one of the compartments that is most impacted by abiotic stress and senescence. Stressful circumstances result in below-optimal CO_2 assimilation rates, and as a result, light absorption may exceed the requirement for photosynthesis (Pintó-Marijuan and Munné-Bosch 2014).

According to earlier research, the combined effects of environmental factors on photosynthesis, transpiration, and stomatal conductance involve substantial interactions. Numerous physiological and non-physiological activities involved in plant development are significantly impacted by such interactions (Ahmed et al. 2020).

In general, roots experience greater environmental stress than stems. As a result, they may be impacted by such stressors just as much as or even more than the plant's above-ground components. Owing to the limited number of opportunities for root observations, the impact of abiotic stresses on root structure and development has, however, received considerably less research attention than that of the aboveground parts of plants. Roots serve a variety of purposes, including establishing a connection between the plant and its surroundings, absorbing nutrients and water and transporting them to the plant's aboveground organs, secreting certain hormones, and assuring the efficacy of nutrients in the nutrient solution. In order to prevent the plant from being harmed and to make sure the aboveground portion of the plant takes the appropriate measures to adjust to these unfavorable conditions, roots also send some hormonal signals to the body when stressed conditions such as drought, nutrient deficiencies, and salinity exist (Kul et al. 2021).

Both the vegetative and reproductive stages are impacted by cold stress. It causes less germination, slowed development, smaller leaves, and chlorosis at the seedling stage. The main negative impact of cold is shown in cell membrane damage brought on by cold stress-induced dehydration as a result of ice formation (Manasa et al. 2022).

Pollen, which is not well protected, is vulnerable to high UV-B dosages. Flavonoids, which are UV-B absorbing substances, play a complex role in plant reproduction since they are necessary for UV-B protection and appropriate pollen function, as well as for controlling flower and fruit color, which in turn affects pollinator and frugivore visits. Pollinators' responses to UV-B-mediated changes in plants can have direct or indirect effects on pollination, depending on how UV-B radiation affects pollinators (Llorens et al. 2015).

When exposed to high amounts of HMs, plants' significant response is overproducing reactive oxygen species (ROS), which harms the plant's growth and survival (Rahman et al. 2023). Continual exposure of plants to very hazardous quantities of HMs causes a reduction in the amount of water, nutrients, and photosynthesis that plants can perform. Furthermore, chlorosis, growth restriction, and eventually mortality occur in plants that are grown in soil that has high levels of HMs (Yaashika et al. 2022).

Adaptation and Tolerance Mechanisms of Plants

Plants have a sophisticated tolerance mechanism that involves systemic communication, stomatal pore management, preservation of redox equilibrium, and other components to deal with abiotic challenges. Plants generate a variety of tolerance responses that comprises multiple signalling cascades. Additionally, there are receptors and proteins that are necessary for minimizing different types of stressors. Moreover, plants may integrate two or more signalling channels simultaneously or in other plant sections (Singh et al. 2023). Plants combat the oxidative stress via enzymatic and non-enzymatic machinery.

For the purpose to regulate the complex plant metabolic system under both normal and stressed situations, plants have a wide variety of metabolites. These metabolites are further classified to primary metabolites, which are in charge of the principal metabolic procedures that are essential for a plant's survival, and secondary metabolites, which play a supporting role in the development of a plant's ability to interact with its hostile environment but are not required for the primary metabolic reactions for growth and development (Jha and Mohamed 2022). A plant's capacity for adaptability determines its ability to withstand these pressures, and tolerant plants may use a variety of coping mechanisms to mitigate abiotic stress effects (Francini and Sebastiani 2019). In addition, biosynthesis of phytohormones, such as cytokinins, abscisic acid, auxin, jasmonic acid, gibberellin, and ethylene play important role in amelioration of abiotic stress in plants by altering biochemical and physiological process plant tissues (Mushtaq et al. 2020). Crosstalk between different phytohormones is necessary to make strategies for the tolerance of biotic and abiotic stressors. Furthermore, plant growth hormones such as auxin, cytokinin, and gibberellin are also regulated differently under various conditions (Gul et al. 2023).

Plants react to salt stress signals by regulating ion homeostasis, activating the osmotic stress pathway, modulating plant hormone signaling, and altering cytoskeleton dynamics and cell wall composition (Gul et al. 2022). By preserving the osmoregulation of the cell, osmolytes, as well known as osmoprotectants, are mostly located in the cytoplasm and prevent cellular degradation. Osmolytes do not affect other physiological and biochemical processes because they are non-toxic and very soluble (Wahab et al. 2022). These osmolytes include polyamines (putrescine, spermidine, spermine) amino acids (proline), betaine (glycine betaine), sugar (trehalose, fructan), and sugar alcohol (inositol, sorbitol, mannitol, etc.). Due to the fact that the buildup of these substances is proportional to the external osmolarity, they safeguard cellular structures by preserving internal osmotic balance through constant water influx (Kibria and Hoque 2019, Kido et al. 2019). Trehalose (Tre) is one such non-reducing sugar found in bacteria and yeasts, where it serves as source of carbon, and in higher plants and animals, where it acts as osmo-protectant (Hassan et al. 2022b). Additionally, polyamines have a supporting role in salt tolerance mechanisms and adaptation strategies, primarily by stabilizing membranes, neutralizing acids, and reducing the production of ROS (Choudhary et al. 2023). Nevertheless, all plant species can't accumulate or fabricate significant amounts of these compounds under stress which is why genetically engineered plants, with transgenic genes responsible for osmolyte synthesis, could be an alternative for improved crop production (Hasanuzzamn et al. 2019). Enzymes, such as betaine aldehyde dehydrogenase, pyrroline-5-carboxylate reductase, and ornithine-d-aminotransferase play important role in osmotic adjustment in plants under salt stress (Mushtaq et al. 2020).

In response to drought and heat, plant use special mechanisms like leaf rolling, change in leaf orientation to reduce the light absorption, reflecting solar radiation, and regulating stomatal behavior as well as a promoted cuticle and wax layer on the surface of leaves to enhance stomatal resistance (Mafakheri et al. 2021).

Plants have developed a variety of defense mechanisms against cold stress, including reprogramming of genes and transcription factors. As transcription factors that control CBFs (C-repeat binding factors), ICEs (Inducer of CBF Expression) are crucial in maintaining a balance between plant development and stress tolerance (Wang et al. 2022). AP2/ERF (APETALA2/Ethylene responsive factor) transcription factor family is one of the most important cold stress-related TF families that along with other TF families, such as WRKY, bHLH, bZIP, MYB, NAC, and C_2H_2 interrelate to enhance cold stress tolerance (Ritonga et al. 2021).

Three possible pathways for the response to UV-B are shown: the first pathway is mediated by the special UV-B receptor UVR8, the second UVR8-independent pathway is involved in adaptation, and the third is activated by oxidative damage (Golovatskaya and Laptev 2023). Plants have evolved defense strategies under UV-B stress to counteract these detrimental impacts. The most common protective response is an accumulation of secondary metabolites. These provide photoprotection by acting as UV-B absorbing compounds through quenching ROS and RNS (Singh et al. 2023). Plants have developed 'sunscreen' flavonoids that build up when exposed to UV-B stress in order to avoid or reduce destruction (Shi and Liu 2021).

Plants have evolved advanced defense mechanisms to prevent or tolerate the toxic effects of HMs. These defense mechanisms include absorbing and accumulating HMs in cell organelles, immobilizing them by forming complexes with organic chelates, and extracting them by making use of a variety of transporters, ion channels, signaling cascades, and transcription factors, among others (Rahman et al. 2022).

Plant growth regulators (phytohormones) are one of the most important chemical messengers that mediate plant growth and development during stressful conditions (Khan et al. 2021, Zahid et al. 2023). Recently, exogenous application of phytohormones improves the tolerance or resistance of plants against extreme conditions. As regards to various studies, the amount of phytohormones rises considerably under stressful condition, which regulate various physiological processes including plant growth and development, production of osmolytes and secondary metabolites as well as antioxidant molecules (Wan et al. 2016, Hafeez et al. 2021, Li et al. 2021, Rachappanavar et al. 2022). When applied in an exogenous way, including by chemical priming, genetically engineering crops, and using root-associated microbes, they have shown to be effective in helping plants to deal with biotic and abiotic stresses, counteracting their effects and being able to survive and maintain growth even in challenging circumstances (Policarpo Tonelli et al. 2023). Furthermore, exogenous applications of phytohormones are shown to enhance endogenous phytohormones for providing a protective shield against abiotic constraints in plants (Chhaya et al. 2021, Saini et al. 2021). The study performed by Kocaman (2023) indicated that the application of exogenous foliage abscisic acid (ABA) limited the translocation of Cd, Cr, Hg, and Sn into strawberry leaves. Moreover, leaf antioxidant enzyme content and total chlorophyll content indicated a reduction in the negative effects of heavy metal stress on chlorophyll content and yield of strawberry leaves (Kocaman 2023). For a long time, the idea of using organic molecules to increase resistance to abiotic stressors has been discussed.

One strategy to reduce the negative effects of abiotic stressors on plants is the administration of putative exogenous osmotic protecting chemicals like proline (Hosseinifard et al. 2022). Proline assists the preservation of subcellular structures (membranes and proteins), the scavenging of free radicals, and the neutralizing of cellular redox potential during stress in addition to acting as an osmolyte for osmotic adjustment (Zouari et al. 2019). Exogenous treatment using glycine betaine (GB) can boost subsequent growth and yield while also increasing plant species' tolerance to diverse abiotic stressors. Only a tiny portion of the GB supplied to roots is typically transported to the chloroplasts, the majority is often absorbed and stored in the cytosol. GB is transferred from leaves to meristematic tissues, particularly flower buds and shoot apices, and then to actively developing and expanding tissues (Zhang and Yang 2019).

The exogenous application of other organic compounds in order to alleviate abiotic stressors is well studied and known. Melatonin is applied topically and has the following benefits: it shields the photosynthetic machinery, boosts antioxidant defenses, osmoprotectants, and soluble sugar levels, prevents tissue damage and lowers electrolyte leakage, improves ROS scavenging, increases biomass, upholds redox and ion homeostasis, and enhances gaseous exchange. Glutathione spray enhances defense mechanisms, tissue repairs, nitrogen fixation, and upregulates phytochelatins in addition to upregulating the glyoxalase system, decreasing methylglyoxal (MG) toxicity and oxidative stress, and lowering hydrogen peroxide and malondialdehyde buildup (Khalid et al. 2022).

Some beneficial elements exhibited a positive effect on plants. Aluminum (Al), cerium (Ce), cobalt (Co), iodine (I), lanthanum (La), sodium (Na), selenium (Se), silicon (Si), titanium (Ti), and vanadium (V) are emerging as novel biostimulants that can improve crop productivity and nutritional quality while ameliorating responses to environmental stimuli and stressors in some plant species (Gómez-Merino and Trejo-Téllez 2018). Silicon plays important roles in plant productivity by improving mineral nutrient deficiencies (Ali et al. 2020b). When silicon added to plants exposed to UV-B radiation stress, the amount of chlorophyll, soluble sugars, anthocyanins, flavonoids, and UV-absorbing and antioxidant chemicals are increased. Proline concentration, metal toxicity, photosynthesis rate, and lipid peroxidation are additional effects of silicon. During water scarcity, silicon raises ascorbate peroxidase activity, total soluble sugar concentration, relative water content, and photosynthetic rate. Moreover, silicon lowers the levels of malondialdehyde and the activity of peroxidase, catalase, and superoxide dismutase (Mavrič Čermmelj et al. 2022).

The chemical elicitors may be synthetic or natural substances that have been shown to work at very low concentrations as messenger molecules that activate the signaling pathway by binding to receptor molecules (Manikanta et al. 2023).

Nitric Oxide

Nitrogen monoxide (NO) is highly free radical molecule that has long been regarded as a toxic, polluting, and dangerous gas (Weisslocker-Schaetzel et al. 2017). Nitric oxide (NO) is a free gaseous radical with a wide variety of physiological and pathological implications in animal and plant cells (Zhang et al. 2012). It is very reactive due to the unpaired electron present in its molecular orbital; it can directly react with metal complexes, radicals, DNA, proteins, lipids, and other biomolecules (Misra et al. 2011). It is earlier regarded as toxic byproducts of metabolism dependent on oxygen-based respiration. In animal cells, when it was revealed that NO was the signal generated by endothelial cells in response to vasodilators like acetylcholine or bradykinin, the concept of free radicals underwent a paradigm change in the late 1980s. Nitric oxide is a lipophilic molecule that diffuses through membranes (Crawford 2006). In plant cells, NO is an endogenous signal that mediates responses to several stimuli. Stomatal closure, initiated by abscisic acid (ABA), is effected through a complex symphony of intracellular signalling in which NO appears to be one component (Neill et al. 2008, Sun et al. 2019). Recent study revealed that NO represses flowering in *Arabidopsis* (Simpson, 2005). Previous studies have demonstrated the differential modulation of nitric oxide (NO) and hydrogen sulfide (H_2S) content during sweet pepper (*Capsicum annuum* L.) fruit ripening, both of which regulate NADP-isocitrate dehydrogenase activity (Muñoz-Vergas et al. 2020).

Studies have verified that nitric oxide plays a part in controlling the ethylene response during the pre- and post-climacteric phases of fruit maturation. It's noteworthy that nitric oxide links to the ACC oxidase enzyme to produce the persistent ternary complex 'ACC-ACC oxidase-NO.' A decrease in ethylene synthesis in tissues is the result of this signaling process (Mukherjee 2019). It has shown that NO is an important signal involved in plant development, germination, photosynthesis, leaf senescence, pollen growth, and reorientation. NO has a dose-dependent effect on plants, acting both favorably and negatively. Exogenous administration of NO stimulates seed germination, hypocotyl elongation, pollen formation, flowering, and delays senescence at lower concentrations, but at larger concentrations, it damages plants by nitrosating them (Sami et al. 2018).

NO acts as a rapid signaling molecule during a diversity of stress responses and elicits the expression of various redox-regulated genes (Choudhary et al. 2023).

It is reported that phytohormones promote an increase in NO levels by involving both nitrate/nitrite reductase pathway and NOS-like activities (Piacentini et al. 2023).

NO Synthesis

Although both oxidative and reductive processes, which may be either enzymatic or non-enzymatic, are used in the production of nitric oxide in plants, the full mechanism is still unknown (Bhuyan et al. 2020, Gupta et al. 2022).

According to studies done over more than a decade, NO is endogenously produced in plants. It was originally noted that NO could be made without the aid of enzymes. Non-enzymatic NO fabrication could be the result of chemical reactions between N oxides and plant metabolites, of nitrous oxide decomposition, or of chemical reduction of nitrite (NO_2^-)) at acidic pH in apoplast (Wendehenne et al. 2001, León and Costa-Broseta 2020). The cytoplasm, mitochondria, nucleus, and in matrix of the peroxisomes and chloroplasts are where the majority of NO generation occurs in plant cells. While both vascular and nonvascular plants undergo nitrate-dependent NO production in the cytoplasm, which is catalyzed by cytosolic nitrate-reductase (NRs), peroxisomes have been shown to produce arginine-dependent NO catalysed by NOS-like enzymes. It has also been recognized as a source of NO generation from chloroplastic materials. Despite the fact that NOA1 was once thought to be a mitochondrial protein (León and Costa-Broseta 2020).

The synthesis of NO may occur via enzymatic pathways. Oxidation of L-arginine via an NADPH-dependent pathways leads to the biosynthesis of NO. Several works have confirmed nitric oxide synthase (NOS) activity in plants, but the ability to synthesis NO does not depend on NOS. NO seems to be byproduct of denitrification, nitrogen fixation, or respiration (Huang et al. 2002). In the reductive pathway, nitrate reductase (NR) activity converts nitrite to nitrate, which is then reduced to NO by NR. The electron transport process in the mitochondria can change nitrite into NO. According to reports, xanthine oxidoreductase (XOR) uses nitrite to create NO. Salicylhydroxamate (SHAM) or hydroxylamine (HA) can be used to make NO (Fig. 2). The presence of catalase (CAT) or anoxia both reduced the production of NO, but the presence of H_2O_2 or ROS-inducing circumstances increased it (Hassanuzzaman et al. 2018).

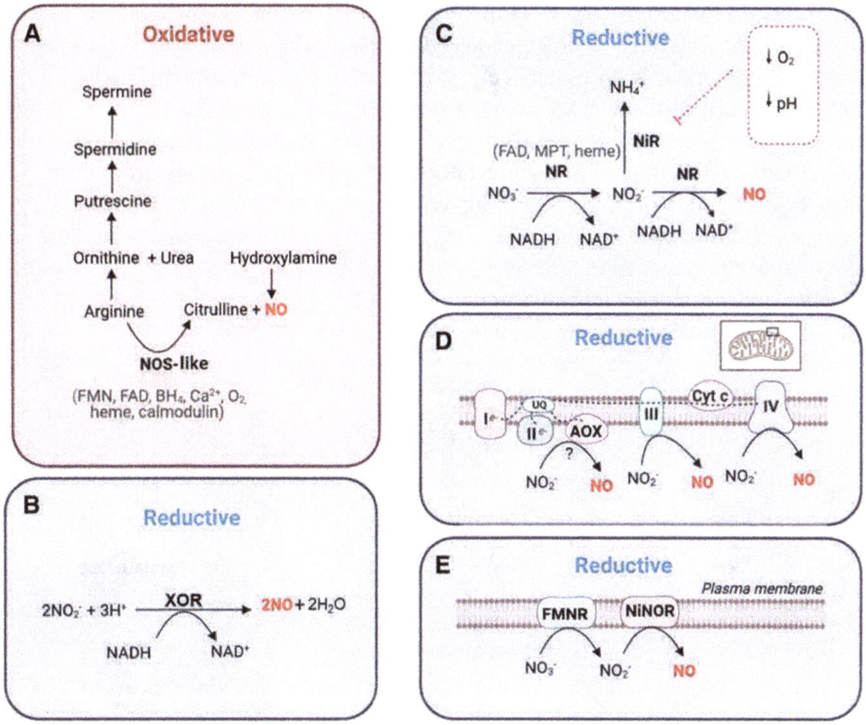

Fig. 2. NO biosynthetic pathways in plants. **(A)** NO is produced by oxidative pathways and reductive pathways. The former include a NOS-like enzyme, a polyamine-mediated pathway and a hydroxylamine pathway, while the latter include NR and xanthine oxidoreductase (XOR). A NOS-like enzyme may use L-arginine as substrate and produce L-citrulline and NO. This activity requires several cofactors such as BH_4, CaM (calmodulin), flavin adenine dinucleotide (FAD), flavin mononucleotide (FMN), , and oxygen. **(B)** XOR catalyzes the reduction of nitrite to NO using NADH or xanthine as reducing substrate. **(C)** NR catalyzes reduction of nitrite (NO_2^-) to NO. The activity requires cofactors such as FAD, hemeiron, and molybdenum-molybdopterin (MPT). Under aerobic conditions, the cytoplasmic nitrate () regulates NR activity, because nitrate competitively inhibits nitrite reduction. Thus, a lower nitrite concentration does not favor its reduction due to an increased Km requirement. Under conditions such as hypoxia, and low pH, the NiR is inhibited, leading to an increased nitrite concentration and its concomitant removal. **(D)** The nitrite generated under hypoxia is transported to mitochondria via a putative nitrite transporter. Under hypoxic conditions, nitrite reduction to NO takes place at complexes III and IV, and possibly AOX. cytochrome c oxidase (COX), inner mitochondrial membrane (IMM). **(E)** NO is also generated by the combined action of a plasma membrane-bound nitrite-NO reductase (PM-NiNOR) and a plasma membrane-bound NR (PM-NR).

NO can diffuse within a cell from the place of production to other areas of the cell where it might trigger an effect by interacting with particular target proteins. As it is lipophilic, NO may accumulate or traverse membranes, or can be stored as transportable compounds like SGNO (S-nitrosoglutathione) (Misra et al. 2011).

NO Donors

Typically, NO donors are employed as exogenous sources rather than in the gas form. This is because direct exposure to NO is technically challenging (Tan et al.

2013). NO is released by active compounds, such as synthetic NO donors. In order to investigate the functions of endogenous NO in plant adaptation to abiotic and biotic stressors, they have been extensively exploited as experimental tools (Lau et al. 2021). Frquently utilized NO donors are 3-morpholinosydnonimine (SIN-1), Sodium nitroprusside (SNP), S-nitrosoglutathione (GSNO), S-nitrosoN-acetylpenicillamine (SNAP), Roussin's Black Salts, and NOR-3($\pm$)-(E)-4-ethyl-2-[(E)-hydroxyimino]-5-nitro-3-hexenamide (Fig. 3) (Floryszak-Wieczorek et al. 2006, Tan et al. 2013). At a comparable donor concentration (500μM) and under light conditions the highest rate of NO generation was found for SNAP, followed by GSNO and SNP. The measured half-life of the donor in the solution was 3 h for SNAP, 7 h for GSNO, and 12 h for SNP (Floryszak-Wieczorek et al. 2006).

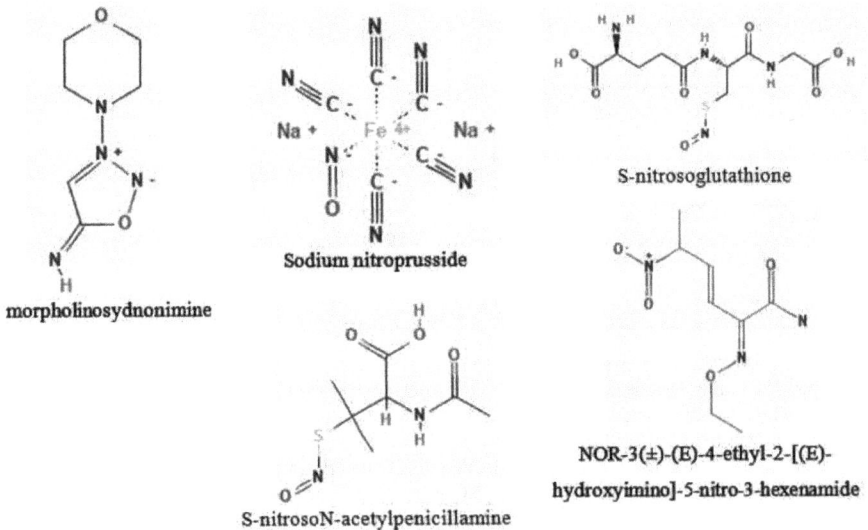

Fig. 3. Structure of some NO donors.

Role of NO in Plant Growth under Optimal Conditions

As one of the smallest diatomic molecules, NO has relatively long half-life (approximately 5 s) in comaparison with other radicals and exhibits a good diffusion rate through hydrophobic and hydrophilic compartments. It can react with thiols, tyrosine residues, metal centers, and ROS (Salgado et al. 2013). It can cause both favorable and unfavorable effects, depending on the NO level and the location of synthesis. NO functions as a regulator of numerous plant hormones, as well as in the synthesis of antioxidant enzymes, stomatal movement, and mineral nutrient flux. These NO-mediated enhancements promote DNA change, transcriptional modifications, and post-translational protein modification directly or indirectly, thereby altering plant requirements (Shah et al. 2023). Since nitration and S-nitrosylation are the latter two alterations that have been further explored in plant species, NO mostly modifies protein activity through posttranslational changes, such as the ligation of NO to transition metals in proteins. But in the intricate signaling pathways including this free radical, changes of other compounds like fatty acids, nucleic acids, cyclic GMP, and phytohormones (for example, cytokinins) also appear to be involved (Asgher et al. 2017).

NO at low concentrations promotes germination, hypocotyl growth, flower development, and delays senescence (Bakshi et al. 2022). It plays also a key role in plant-environment interactions (Zhou et al. 2021). The impacts of NO on plant growth and development were found to be depend to concentration; high doses (40–80 pphm) reduced tomato growth, whereas low doses (0–20 pphm) boosted it, similar results were observed in lettuce and pea plants (Palaval-Unsal and Arisan 2009) (Fig. 4).

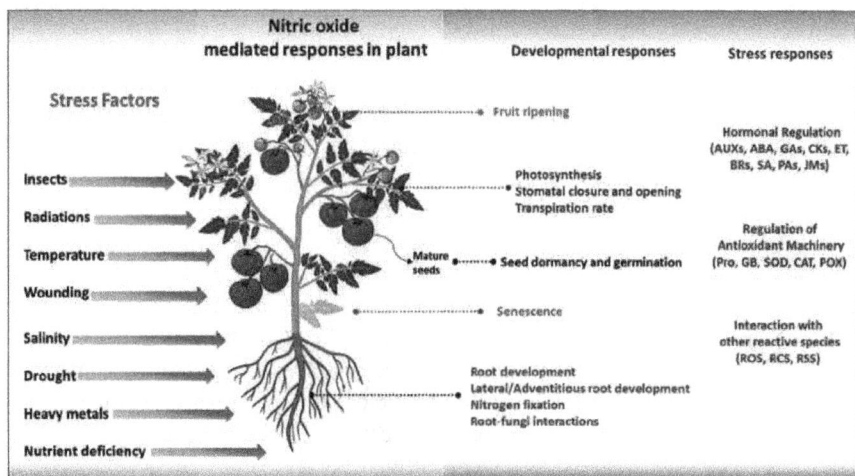

Fig. 4. Nitric oxide (NO)-mediated responses during plant development and plant-environment interactions. (AUX auxins, ABA abscisic acid, GAs gibberellins, CKs cytokinins, ET ethylene, BRs brassinosteroids, SA salicylic acid, PAs polyamines, JMs jasmonates, Pro proline GB glycine betaine, SOD superoxide dismutase, CAT catalase, POX peroxidase, ROS reactive oxygen species, RCS reactive carbonyl species, RSS reactive sulfur species).

Role of NO in Stress Responses of Plants

Little is known about this essential chemical modification in the control of abiotic stress signaling, with the exception of a few studies that support the significance of NO in stress responses. By stimulating their respective scavengers, RNS and NO especially enhance antioxidant reactions, which are beneficial for reducing the risk of oxidative damage that can happen under nutrient duress (Nieves-Cordones et al. 2018). NO essential target is the cytochrome c-dependent respiration and stimulates alternative-pathway respiration with alternative oxidase (AOX) as a terminal electron acceptor. NO participates also in the plant resistance responses by inducing the expression of numerous genes associated to defense (Huang et al. 2002).

In recent studies, the mostly beneficial role of NO against abiotic challenges have been elucidated by observing physiological/biochemical parameters but relatively inadequate research done at the transcripts level or gene regulation subsequently researchers should include it in future (Praveen 2022).

In situation where ROS production is increased, endogenous and/or exogenous NO plays a significant role in ROS detoxification and stimulates the antioxidant response by altering gene transcription and protein function (Sharma et al. 2020b).

Numerous plant species, including grains, legumes, fruit trees, medicinal plants, and vegetables, have been seen to benefit from NO's ability to reduce the negative consequences of drought stress. Although NO's ability to lower oxidative stress and stomatal opening is mostly responsible for its better drought tolerance, it can also have a significant impact on other physiological processes like photosynthesis, proline buildup, and seed germination when there is a water shortage (Santisree et al. 2015). Mata and Lamattina (2001) studied the response of wheat (*Triticum aestivum*) to water stress conditions after treatment with two NO donor (SNP and SNAP). The results of this study demonstrated that the rate of transpiration was decreased with 20% in SNP-treated detached wheat leaves. NO was also able to induce a 35%, 30%, and 65% of stomatal closure in three different species, *Tradescantia* sp. (monocotyledonous) and two dicotyledonous, *Salpichroa organifolia* and fava bean (*Vicia faba*), respectively. The stomatal closure was associated with a 10% rise in RWC in Tradescantia sp. leaves that had been treated with SNP. After the recovery time, ion leakage, a measure of cell injury, was 25% lower in wheat leaves treated with SNP than in control ones (Mata and Lamattina 2001). The work performed by Montilla-Bascón et al. (2017) suggested that there is a potential interplay between NO and polyamine biosynthesis in transgenic barely plants during drought response. This study suggested also NO-ethylene influenced regulatory node in polyamine biosynthesis linked to drought tolerance/susceptibility in barley. Silveira et al. (2017) investigated the generation and accumulation of NO in sugarcane genotypes (drought-sensitive and tolerant genotypes) under water deficit. The sugarcane genotypes IACSP95-5000 (drought tolerant) exhibited a higher root extracellular and leaf intercellular NO content during a water deficit in comparison to the drought-sensitive (IACSP97-7065) genotype of sugarcane. Those differences in intracellular and extracellular NO contents and enzymatic activities were associated with higher leaf hydration in the drought-tolerant genotype as compared to the sensitive one during a water deficit (Silveira et al. 2017).

NO triggers cell signaling by activating a cascade of biochemical events that result in plant tolerance to environmental stresses (Goyal et al. 2021). The high amount of endogenous NO was proved in several species under salt constraint such as *Olea europaea*, *Helianthus annuus*, and *Arabidopsis thaliana* (Fatima et al. 2021). Dinler et al. (2014) proved that the exogenous SNP application protected soybean leaves from salt stress (200 mM NaCl) by increasing relative water content, chlorophyll and abscisic acid content, as well as by lowering stomatal conductance in order to maintain water balance. Wang et al. (2012) demonstrated that the exogenous application of H_2S promoted the germination of alfalfa (*Medicago sativa*) and alleviated salinity damages involving NO pathway. The exogenous application of

NO on a exposed tomato plant to 120 mM of NaCl improved the negative impacts provoked by salinity through increasing significantly the antioxidative enzymes (superoxide dismutase, ascorbate peroxidase, glutathione reductase, and raising some enzymes involved in nitrogen metabolism (nitrate reductase and nitrite reductase)) activities. NO-treated plants showed a higher content in both proline and ascorbate but lower content of H_2O_2 (Manai et al. 2014).

NO production following cold exposure has been reported in numerous plant species, and a series of proteins targeted by NO-based post-translational changes have been discovered. Moreover, important cold-regulated genes have been characterized as NO-dependent, suggesting the crucial importance of NO signalling for cold-responsive gene expression (Puyaubert and Baudouin 2014). Several lines of evidence indicate NO as a key signaling molecule in mediating various plant responses such as photosynthesis, oxidative defense, osmolyte accumulation, gene expression, and protein modifications under heat stress. Additionally, there has been an increase in recent years in the interactions of NO with other signaling molecules and phytohormones to achieve heat tolerance (Parankusam et al. 2017).

In their study, Azizi et al. (2022) demonstrated that the foliar application of NO, especially at high concentration (200 µM) reduced the accumulation of Cd in shoot and roots of plants. Foliar treatment of 200 µM NO ameliorated savory (*Satureja hortensis* L.) plant growth under Cd stress. Proline, chlorophyll, carbohydrates, and peroxidase activity were increased in comparison with control (Azizi et al. 2022). Nitric oxide and melatonin together boost a plant's resistance to metal toxicity. The application of combined MT/NO in *Glycine max* L. under Cd and Pb toxicity increased plant biomass of soybean and modified the stress-resistance system by controlling the activation of antioxidant and molecular transcription factors. MT/NO is reduced metal toxicity through raising the Ca^+ and K^+ and exuding greater organic acids into the rhizosphere. The mRNA expression of the *WARKY27* gene, *MTF-1*, gmNR, and gmGSNOR are all upregulated by MT/NO. To encourage plant cell viability and Ca^{2+} signaling, MT/NO controlled the MAPKs and CDPKs cascades (Imran et al. 2022).

Conclusion

Plants accumulate osmolytes and proteins involved in stress tolerance as a response to stress at both the cellular and molecular levels. Under stress conditions, a variety of genes with various roles are either activated or repressed. Some signalling molecules may play an important role in plant response to stress. Nitric oxide (NO) is one among the signalling compounds which has a profound physiological, biochemical, and molecular effects for plants under both normal and stressed situations. NO can activate the antioxidant system to reduce oxidative harm when there are abiotic stressors. Additionally, NO can control stomatal closure, enhance water absorption, and stimulate root growth—all of which are essential coping mechanisms for stressed-out plants. Understanding the processes that control NO regulation in plants might help create approaches for increasing crop output and strengthening plant defenses against biotic and abiotic stresses. Various NO-donors could be used for plant growth and development even in challenging circumstances, according to earlier studies. Because NO plays such a crucial role in crop yield, researchers were driven to find new NO-donors that are appropriate for large-scale agricultural activities.

References

Ahammed, G.J., Li, X., Liu, A. and Chen, S. 2020. Physiological and Defense Responses of Tea Plants to Elevated CO_2: A Review. Front. Plant Sci., 11: 305. doi:10.3389/fpls.2020.00305.

Ahmed, H.A., Tong, Y.-X. and Yang, Q.-C. 2020. Optimal control of environmental conditions affecting lettuce plant growth in a controlled environment with artificial lighting: A review. S. Afr. J. Bot., 130: 75–89. https://doi.org/10.1016/j.sajb.2019.12.018.

Ali, N., Réthoré, E., Yvin J.-C. and Hosseini S.A. 2020b. The regulatory role of silicon in mitigating plant nutritional stresses. Plants, 9: 1779, doi:103390/plants9121779.

Ali, S., Rizwan, M., Arif, M.S., Ahmed, R., Hasanuzzaman, M., Ali, B. and Hussain, A. 2020a. Approaches in enhancing thermotolerance in plants: An update review. J. Plant Growth Regul., 39: 456–480. https://doi.org/10.1007/s00344-019-09994-x.

Angon, P.B., Tahjib-Ul-Arif, M., Samin, S.I., Habiba, U., Hossain, M.A. and Brestic, M. 2022. How Do Plants Respond to Combined Drought and Salinity Stress?—A Systematic Review. Plants, 11: 2884. https://doi.org/10.3390/plants11212884.

Apoorva, Jaiswal, D., Pandey-Rai, S. and Agarwal, S.B. 2021. Untangling the UV-B radiation-nduced transcriptional network regulating plant morphogenesis and secondary metabolite production. Environ. Exp. Bot., 192: 104655. https://doi.org/10.1016/j.enexpbot.2021.104655.

Asgher, M., Per, T.S., Massod, A., Fatma, M., Freschi, L., Corpas, F.J. et al. 2017. Nitric oxide signalling and its crosstalk with other plant growth regulators in plant responses to abiotic stress. Environ. Sci. Pollut. Res., 24: 2273–2285. DOI 10.1007/s11356-016-7947-8.

Azizi, I., Esmailpour, B. and Fatemi, H. 2022. Exogenous nitric oxide on morphological, biochemical and antioxidant enzyme activity on savory (*Satureja hortensis* L.) plants under cadmium stress. J. Saudi Soc. Agric. Sci., 20: 417–423. https://doi.org/10.1016/j.jssas.2021.05.003.

Bakshi, P., Kaur Kohli, S., Bali, S., Kaur, P., Kumar, V., Sharma, P. et al. 2022. NO and phytohormones cross-talk in plant defense against abiotic stress. pp. 573–596. In: Singh, V.P. et al. (Eds.). Nitric oxide in plant biology: An ancient molecule with emerging roles. Academic Press, Elsevier, UK. https://doi.org/10.1016/B978-0-12-818797-5.00028-5.

Bhattacharya, A. 2021. Soil water deficit and physiological issues in plants. Springer Nature, Singapore. https://doi.org/10.1007/978-981-33-6276-5.

Bhattacharya, A. 2022a. Effect of low temperature on dry matter, partitioning, and seed yield: A review. pp. 629–734. In: Bhattacharya, A. (Ed.). Physiological processes in plants under low temperature stress. Springer Nature, Singapore. https://doi.org/10.1007/978-981-16-9037-2_7.

Bhattacharya, A. 2022b. Lipid metabolism in plants under low-temperature stress: A review. pp. 409–516. In: Bhattacharya A. (Ed.). Physiological processes in plants under low temperature stress. Springer Nature, Singapore. https://doi.org/10.1007/978-981-16-9037-2_5.

Bhattacharya, A. 2022c. Plant growth hormones in plants under low-temperature stress: A review. pp. 517–627. In: Bhattacharya A. (Ed.). Physiological processes in plants under low temperature stress. Springer Nature, Singapore. https://doi.org/10.1007/978-981-16-9037-2_6.

Bhuyan, M.H.M.B., Hasanuzzaman, M., Parvin, K., Mohsin, S.M., Al Mahmud, J., Nahar, K. et al. 2020. Nitric oxide and hydrogen sulphide: two intimate collaborators regulating plant defense against abiotic stress. J. Plant Growth Regul., 90: 409–424. https://doi.org/10.1007/s10725-020-00594-4.

Bijalwan, P., Sharma, M. and Kaushik, P. 2022. Review of the effects of drought stress on plants: A systematic approach. doi:10.20944/preprints202202.0014.v1.

Brown, P.H., Zhao, F.-J. and Dobermann, A. 2022. What is a plant nutrient? Changing definitions to advance science and innovation in plant nutrition. Plant Soil, 476: 11–23. https://doi.org/10.1007/s11104-021-05171-w.

Chaki, M., Begara-Morales, J.C. and Barroso J.B. 2020. Oxidative stress in plants. Antioxidants, 9: 841. Doi:103390/antiox9060481.

Chaudhry, S. and Sidhu, G.P.S. 2021. Climate change regulated abiotic stress mechanisms in plants: A comprehensive review. Pant Cell Rep., https://doi.org/10.1007/s00299-021-02759-5.

Chen, Z., Maltz, M.R., Russell, R., Ye, S., Cao, J. and Shang, H. 2022. Highly elevated CO2 and fertilization with nitrogen stimulates significant *Schima superba* growth and mediates soil microbial community composition along an oligotroph-copiotroph spectrum. J. Soils Sediments, 22: 1555–1571. https://doi.org/10.1007/s11368-022-03167-2.

Chhaya, Yadav, B., Jogawat, A., Gnanasekaran, P., Kumari, P., Lakra, N., Lal, S.K. et al. 2021. An overview of recent advancement in phytohormones-mediated stress management and drought tolerance in crop plants. Plant Gene, 25: 100264. https://doi.org/10.1016/j.plgene.2020.100264.

Choudhary, S., Wani, K.I., Khan, M.M.A. and Aftab, T. 2023. Cellular responses, osmotic adjustments, and role of osmolytes in providing salt stress resilience in higher plants: Polyamines and nitiric oxide crosstalk. J. Plant Growth Regul., 42: 539–553. https://doi.org/10.1007/s00344-022-10584-7.

Crawford, N.M. 2006. Mechanisms for nitric oxide synthesis in plants. J. Exp. Bot., 57(3): 471–478. doi:10.1093/jxb/erj050.

Dinler, B.S., Antoniou, C. and Fotopoulos, V. 2014. Interplay between GST and nitric oxide in the early response of soybean (*Glycine max* L.) plants to salinity stress. J. Plant Physiol., 171: 1740–1747. doi:10.1016/j.jplph.2014.07.026.

Dong, J., Hunt, J., Delhaize, E., Zheng, S.J., Jin, C.W. and Tang, C. 2021. Impact of elevated CO_2 on plant resistance to nutrient deficiency and toxic ions via root exudates: A review. Sci. Tot. Env., 754(1): 142434. https://doi.org/10.1016/j.scitotenv.2020.142434 .

El Moukhtari, A., Ksiaa, M., Zorrig, W., Cabassa, C., Abdelly, C. et al. 2022. How silicon alleviates the effect of abiotic stresses during seed germination: A review. J. Plant Growth Regul., hal-03765797. https://hal.sorbonne-universite.fr/hal-03765797.

Fatima, A., Husain, T., Suhel, M., Prasad, S.M. and Singh, V.P. 2021. Implications of nitric oxide under salinity stress: The possible interaction with other signalling molecules. J. Plant Growth Reg., https://doi.org/10.1007/s00344-020-10255-5.

Floryszak-Wieczorek, J., Milczarek, G., Arasimowicz, M. and Ciszewski, A. 2006. Do nitric oxide donors mimic endogenous NO-related response in plants? Planta, 224(6): 1363–1372. doi:10.1007/s00425-006-0321-1.

Francini, A. and Sebastiani, L. 2019. Abiotic stress effects on performance of horticultural crops. Horticulturae, 5: 67. doi:10.3390/horticulturae5040067.

García-Capparós, P., De filippis, L., Gul, A., Hasanuzzaman, M., Ozturk, M,, Altay, V. et al. 2020. Oxidative stress and antioxidant metabolism under adverse environmental conditions: A review. Bot. Rev., https://doi.org/10.1007/s12229-020-09231-1.

Gideon Onyekachi, O., Ogbonnaya Boniface, O., Felix Gemlack, N. and Nicholas, N. 2019. The Effect of Climate Change on Abiotic Plant Stress: A Review. In: De Oliveira, A.B. (Ed.). Abiotic and Biotic Stress in Plants. doi: 10.5772/intechopen.82681.

Gog, L., Berenbaum, M.R. and Delucia E.H. 2019. Mediation of impacts of elevated CO2 and light environment on *Arabidopsis thaliana* (L.) chemical defense against insect herbivory via photosynthesis. J. Chem. Ecol., 45: 61–73. https://doi.org/10.1007/s10886-018-1035-0.

Golovatskaya, I.F. and Laptev, N.I. 2023. Effect of UV-B radiation on plants growth, active constituents, and productivity. pp. 25–60. In: Husen, A. (Ed.). Plants and their Interaction to Environmental Pollution. Elsevier, https://doi.org/10.1016/B987-0-323-99978-6.00024-8.

Gómez-Merino, F.C. and Trejo-Téllez, L.I. 2018. The role of beneficial elements in triggering adaptive responses to environmental stressors and improving plant performance. In: Vats, S. (Ed.). biotic and abiotic stress tolerance in plants. Springer Nature, Singapore. https://doi.org/10.1007/978-981-10-9029-5_6.

Goyal,, V., Jhanghel, D. and Mehrotra, S. 2021. Emerging warriors against salinity in plants: Nitric oxide and hydrogen sulphide. Physiologia Plantarum, 171(4): 896–908. https://doi.org/10.1111/ppl.13380.

Grime, J.P. 1977. Evidence for the existence of three primary strategies in plants and its relevance to ecological and evolutionary theory. Am. Nat., 111(982): 1169–1194. http://www.cef-cfr.ca/uploads/MEmbres/1977_Grime_AmNat.pdf.

Gul, A., Noor-ul-Huda and Nawaz, S. 2023. Role of phytohormones in biotic vs abiotic stresses with respect to PGRR and autophagy. pp. 41–62. In: Ozturk, M. et al. (Eds.). phytohomones and stress responsive secondary metabolites. Academic Press, Elsevier, London, UK. https://doi.org/10.1016/B978-0-323-91883-1.00016-4.

Gul, Z., Tang, Z.-H., Arif, M. and Ye, Z. 2022. An Insight into Abiotic Stress and Influx Tolerance Mechanisms in Plants to Cope in Saline Environments. Biology, 11: 597. https://doi.org/10.3390/biology11040597.

Gulzar, A.B. Md. and Mazumder, P.B., 2022. Helping plants to deal with heavy metal stress: the role of nanotechnology and plant growth promoting rhizobacteria in the process of phytoremediation. Environ. Sci. Pollut. Res., 29: 40319–40341. https://doi.org/10.1007/s11356-022-19756-0.

Guo, L., Li, Y., Yu, Z., Wu, J., Jin, J. and Liu, X. 2021. Interactive influences of elevated atmospheric CO_2 and temperature on phosphorus acquisition of crops and its availability in soil: A review. Int. J. Plant Prod., 15: 173–187. https://doi.org/10.1007/s42106-021-00138-4.

Gupta, K.J., Kaladhar, V.C., Fitzpatrick, T.B., Fernie, A.R., Møller, I.M. and Loake, G.J. 2022. Nitric oxide regulation of plant metabolism. Mol. Plant., 15: 228–242. https://doi.org/10.1016/j.molp.2021.12.012.

Hafeez, M.B., Zahra, N., Zahra, K., Raza, A., Khan, A., Shaukat, K. et al. 2021. Brassinosteroids: Molecular and physiological responses in plant growth and abiotic stresses. Plant Stress, 2: 100029. https://doi.org/10.1016/j.stress.2021.100029.

Hasanuzzaman, M., Anee, T.I., Bhuiyan, T.F., Nahar, K. and Fujita, M. 2019. Emerging role of osmolytes in enehancing abiotic stress tolerance in rice. pp. 677–708. In: Hasanzzumana, M. et al. (Eds.). Advances in rice research for abiotic stress tolerance. Woodhead Publishing, Elsevier, UK. https://doi.org/10.1016/B978-0-12-814332-2.00033-2.

Hassan, M.U., Nawaz, M., Shah, A.N., Raza, A., Barbant, L., Skalicky, M. et al. 2022b. Trehalose: A key player in plant growth regulation and tolerance to abiotic stresses. J. Plant Growth Regul., https://doi.org/10.1007/s00344-022-10851-7.

Hassan, M.U.I., Rasool, T., Iqbal, C., Arshad, A., Abrar, M., Abrar, M.M. et al. 2022a. Linking plants functioning to adaptive responses under heat stress conditions: A mechanistic review. J. Plant Growth Regul., 41: 2596–2613. https://doi.org/10.1007/s00344-021-10493-1.

Hassanuzzaman, M., Oku, H., Nahar, K., Bhuyan, M.H.M.B., Al Mahmud, J,, Baluska, F. et al. 2018. Nitric oxide-induced salt stress tolerance in plants: ROS metabolism, signalling, and molecular interactions. Plant Biotech. Rep., 12: 77–92. https://doi.org/10.1007/s11816-018-0480-0.

Hemati, A., Moghiseh, E., Amirifar, A., Mofidi-Chelan, M. and Lajayer, B.A. 2022. Physiological effects of drought stress in plants. pp. 113–124. In: Vaishnav A. et al. (Eds.). Plant stress mitigators. Springer Nature, Singapore. https://doi.org/10.1007/978-981-16-7759-5_6.

Hosseinifard, M., Stefaniak, S., Ghorbani Javid, M., Soltani, E., Wojtyla, Ł. and Garnczarska, M. 2022. Contribution of Exogenous Proline to Abiotic Stresses Tolerance in Plants: A Review. Int. J. Mol. Sci., 23: 5186. https://doi.org/10.3390 /ijms23095186.

Huang, X., von Rad, U. and Druner, J. 2002. Nitric oxide induces transcriptional activation of nitric oxide-tolerant alternative oxidase in *Arabidopsis* suspension cells. Planta, 215: 914–923. DOI 10.1007/s00425-002-0828-z.

Imran, M., Khan, A.L., Mun, B.-G., Bilal, S., Shaffique, S., Kwon, E.-H. et al. 2022. Melatonin and nitric oxide: Dual players inhibiting hazardous metal toxicity in soybean plants via molecular and antioxidant signaling cascades. Chemosphere, 308: 136575. https://doi.org/10.1016/j.chemosphere.2022.136575.

IPCC, 2023. Climate change, 2023: Synthesis report of IPCC sixth assessment report (AR6). https://report.ipcc.ch/ar6syr/pdf/IPCC_AR6_SYR_SPM.pdf.

Jaiyeola, P.O., Oluwafemi, F.A., Ayegba, A. and Benibo, I.E. 2021. Review of negative impact of extreme temperature elevations on plant growth, development and crop yield using mathematical equations. IJEAS, 8(4): 42–52. www.ijeas.org.

Jaleel, C.A., Manivannan, P., Wahid, A., Farooq, M., Somasundaram, R. and Panneerselvam, R. 2009. Drought stress in plants: a review on morphological characteristics and pigments composition. Int. J. Agric. Biol., 11: 100–105. http://www.fspublishers.org.

Jena, J., Kumar Sahon, S. and Dash, G.K. 2020. An introduction to abiotic stress in plants. pp. 163–186. In: Rawat, A.K. (Ed.). Advances in Agronomy. AkiNik Publication, New Delhi, India. DOI: https://doi.org/10.22271/ed.book.725.

Jha, Y. and Mohamed, H.I. 2022. Plant secondary metabolites as a tool to investigate biotic stress tolerance in plants: A review. Gesunde Pflanzen, 74: 771–790. https://doi.org/10.1007/s10343-022-00669-4.

Kaur, S., Tiwari, V., Kumari, A., Chaudhary, E., Sharma, A., Ali, U. et al. 2023. Protective and defensive role of anthocaynins under plant abiotic and biotic stresses: An emerging application in sustainable agriculture. J. Biotechnol., 361: 12–29. https://doi.org/10.1016/j.jbiotec.2022.11.009.

Khademi Astaneh, B., Bolandnazar, S. and Zaare Nahanndi, F. 2022. Exogenous nitric oxide protect garlic plants against oxidative stress induced by salt stress. Plant Stress, 5: 100101. https://doi.org/10.1016/j.stress.2022.100101.

Khalid, M., Rehman, H.M., Ahmed, N., Nawaz, S., Saleem, F., Ahmed, S. et al., 2022. Using Exogenous Melatonin, Glutathione, Proline, and Glycine Betaine Treatments to Combat Abiotic Stresses in Crops. Int. J. Mol. Sci., 23: 12913. https://doi.org/10.3390/ijms232112913.

Khan, M.I.R., Ashfaque, F., Chhillar, H., Irfan, M. and Khan, N.A. 2021. The intricacy of silicon, plant growth regulators and other signaling molecules for abiotic stress tolerance: An entrancing crosstalk between stress alleviators. Plant Physiol. Biochem., 162: 36–47. https://doi.org/10.1016/j.plaphy.2021.02.024.

Kibria, M.G. and Hoque, Md.A. 2019. A Review on Plant Responses to Soil Salinity and Amelioration Strategies. Open J. Soil Sci., 9: 219–231. https://doi.org/10.4236/ojss.2019.911013.

Kido, É.A., Ferreira-Neto, J.R.C., da Silva, M.D., Pereira Santos, V.E., da Silva Filho, J.L.B., Benko-Iseppon, M., 2019. Osmoprotectant-related genes in plants under abiotic stress: Expression dynamics, *in silico* genome mapping and biotechnology. pp. 1–40. In: Hossain M.A. et al. (Eds.). Osmoprotectant-Mediated Abiotic Stress Tolerance in Plants. Springer Nature, Switzerland AG. https://doi.org/10.1007/978-3-030-27423-8_1.

Kocaman, A. 2023. Effect of foliar application of abscisic acid on antioxidant content, phytohormones in strawberry shoots, and translocation of various heavy metals. Sci. Horticult., 314: 111943. https://doi.org/10.1016/j.scienta.2023.111943.

Kozai, T., 2013. Resource use efficiency of closed plant production system with artificial light: concept, estimation and application to plant factory. Proceedings of the Japan Academy, Series B, 89(10): 447–461.

Kul, R., Ekinci, M., Turan, M., Ors, S. and Yildirim, E. 2021. How Abiotic Stress Conditions Affects Plant Roots. IntechOpen. doi:10.5772/intechopen.95286.

Lau, S.-E., Hamdan, M.F., Pua, T.-L., Saidi, N.B. and Tan, B.C. 2021. Plant Nitric Oxide Signaling under Drought Stress. Plants, 10(2): 360. https://doi.org/10.3390/plants10020360.

Lei, C., Bagavathiannan, M,, Wang, H., Sharpe, S.M., Meng, W. and Yu, J. 2021. Osmopriming with polyethylene glycol (PEG) for abiotic tolerance in germinating crop seeds: A review. Agronomy, 11(11): 2194. https://doi.org/10.3390/agronomy11112194.

León, J. and Costa-Broseta, Á. 2020. Present knowledge and controversies, deficiencies, and misconceptions on nitric oxide synthesis, sensing, and signalling in plants. Plant Cell Environ., 43: 1–15. https://doi.org/10.1111/pce.13617.

Li, Z., Zhou, J., Dong, T., Xu, Y. and Shang, Y. 2021. Application of electrochemical methods for the detection of abiotic stress biomarkers in plants. Biosens. Bioelectron. J., 182: 113105. https://doi.org/10.1016/j.bios.2021.113105.

Llorens, L., Badenes-Pérez, F.R., Julkunen-Tiitto, R., Zidorn, C., Fereres, A. and Jansen, M.A.K. 2015. The role of UV-B radiation in plant sexual reproduction. Perspect. Plant Ecol. Evol. Syst., 17: 243–254. https://doi.org/10.1016/j.ppees.2015.03.001.

Ma, N., Dong, L., Lü, W., Lü, J., Meng, Q. and Liu, P. 2020. Transcriptome analysis of maize seedling roots in response to nitrogen-, phosphorus-, and postassium deficiency. Plant Soil, 447: 637–658. https://doi.org/10.1007/s11104-019-04385-3.

Mafakheri, M., Kordrostami, M. and Al-Khayri, J.M. 2021. Plant abiotic stress tolerance mechanisms. pp. 29–59. In: Al-Khayri, J.M. et al. (Eds.). Nanobiotechnology. Springer Nature, Switzerland. https://doi.org/10.1007/978-3-030-73606-4_2.

Manai, J., Kalai, T., Gouia, H. and Corpas, F.J. 2014. Exogenous nitric oxide (NO) ameliorates salinity-induced oxidative stress tomato (*Solanum lycopersicum*) plants. J. Soil Sci. & Plant Nutr., 14(2): 433–446. http://dx.doi.org/10.4067/S0718-95162014005000034.

Manasa, L.S., Panigraphy, M., Panigraphi, K.C.S. and Rout, G.R. 2022. Overview of cold stress regulation in plants. Bot. Rev., 88: 359–387. https://doi.org/10.1007/s12229-021-09267-x.

Manikanta, Ch.L.N., Ratnakumar, P., Manasa, R., Pandey, B.B., Vaikuntapu, P.R., Guru, A. et al. 2023. Chemical elicitors-A mitigation strategy for maximize crop yields under abiotic stress. pp. 271–291 In: Ghorbanpour, M. and Shahid, M.A. (Eds.). Plant Stress Mitigators: Types, Techniques and Functions. Academic Press, Elsevier, UK. https://doi.org/10.1016/B978-0-323-89871-3.00013-6.

Mareri, L., Parrotta, L. and Cai, G. 2022. Environmental Stress and Plants. *Int. J. Mol. Sci.*, 23, 5416. https://doi.org/10.3390/ijms23105416.

Marvič Čermmelj, A., Golob, A., Vogel-Mikuš, K. and Germ, M. 2022. Silicon mitigates negative impacts of drought and UV-B radiation in plants. Plants, 11(1): 91. https://doi.org/10.3390/plants11010091

Mata, C.G. and Lamattina, L. 2001. Nitric oxide induces stomacal closure and enhances the adaptive plant response against drought stress. Plant Physiol., 126(3): 1196–1204. doi: 10.1104/pp.126.3.1196

Mesurani, P. and Ram, V.R. 2020. Plant nutrition and its role in plant growth: A review. *JIJRMEET*, 8(8): 1–7. www.raijmr.com.

Miransari, M. and Smith, D. 2019. Sustainable wheat (*Triticum aestivum* L.) production in saline fields: a review. Crit. Rev. Biotechnol., DOI:10.1080/07388551.2019.1654973.

Misra, A.N., Misra, M. and Singh, R., 2011. Nitric oxide ameliorates stress responses in plants. Plant Soil Environ., 57(3): 95–100. https://www.agriculturejournals.cz/pdfs/pse/2011/03/01.pdf.

Mndela, M., Tjelele, J.T., Madakadze I.C., Mangwane, M., Samuels, I.M., Muller, F. et al. 2022. A global meta-analysis of woody plant responses to elevated CO_2: Implications on biomass, growth, leaf N content, photosynthesis and water relations. Ecol. Process., 11: 52. https://doi.org/10.1186/s13717-022-00397-7.

Montilla-Bascón, G., Rubiales, D., Hebelstrup, K.H., Mandon, J., Harren, F.J.M., Cristscu, S.M. et al. 2017. Reduced nitric oxide levels during drought stress promote drought tolerance in barley and is associated with elevated polyamine biosynthesis. Sci. Rep., 7: 13311. https://doi.org/10.1038/s41598-017-13458-1.

Mukherjee, S. 2019. Recent advancements in the mechanism of nitric oxide signaling associated with hydrogen sulfide and melatonin crosstalk during ethylene-induced fruit ripening in plants. Nitric. Oxide, 82: 25–34. https://doi.org/10.1016/j.niox.2018.11.003.

Muñoz-Vergas, M.A., Gonzàlez-Gordo, S., Palma, J.M. and Corpas, F.J. 2020. Inhibition of NADP-malic enzyme activity by H_2S and NO in sweet pepper (*Capsicum annuum* L.) fruits. *Physiol. Plant.*, 168, 278-288. doi:10.1111/ppl.13000.

Mushtaq, Z., Faizan, S. and Gulzar B. 2020. Salt stress, its impacts on plants and the strategies plants are employing against it: A review. J. Appl. Biol. Biotechnol., 8(3): 81–91. DOI:10.7324/JABB.2020.80315.

Nabi, R.B.S., Tayade, R., Hussain, A., Kulkarni, K.P., Imran, Q.M., Mun, B.-G. and Yun, B.-W. 2019. Nitric oxide regulates plant responses to drought, salinity, and heavy metal stress. Environ. Exp. Bot., 161: 120–133. https://doi.org/10.1016/j.envexpbot.2019.02.003.

Neill, S., Barros, R., Bright, J., Desikan, R., Hancock, J., Harrison, J. et al. 2008. Nitric oxide, stomatal closure, and abiotic stress. J. Exp. Bot., 59(2): 165–176. doi:10.1093/jxb/erm293.

Nieves-Cordones, M., López-Delacalle, M., Ródenas, R., Martínez, V., Rubio. F. and Rivero, R.M. 2018. Critical responses to nutrient deprivation: a comprehensive review on the role of ROS and RNS, Environ. Exp. Bot. https://doi.org/10.1016/j.envexpbot.2018.10.039.

Norgbey, E., Murava, R.T., Rajasekar, A., Huang, Q., Zhou, J. and Robinson, S. 2022. Effects of anthropogenic nitrogen additions and elevated CO_2 on microbial community, carbon and nitrogen content in a replication wetland. Environ. Monit. Assess., 194: 575. https://doi.org/10.1007/s10661-022-10229-y.

Pandey, R., Vegavasi, K. and Hawkesford, M.J. 2021. Plant adaptationto nutrient stress. Plant Physiol. Rep., 26(4): 583–586. https://doi.org/10.1007/s40502-021-00636-7.

Panaval-Unsal, N. and Arisan, D. 2009. Nitric oxide signalling in plants. Bot. Rev., 75: 203–229. DOI 10.1007/s12229-009-9031-2.

Parankusam, S., Adimulam, S.S., Bhatnagar-Mathur, P. and Sharma, K.K. 2017. Nitric Oxide (NO) in Plant Heat Stress Tolerance: Current Knowledge and Perspectives. Front. Plant Sci., 8: 1582. doi:10.3389/fpls.2017.01582.

Parihar, P., Singh, S., Singh, R., Singh, V.P. and Prasad S.M. 2014. Effect of salinity stress on plants and its tolerance strategies: a review. Environ. Sci. Pollut. Res., 22: 3739. DOI:10.1007/s11356-014-3739-1.

Piacentini, D., Della Rovere, F., Lanni, F., Cittadini, M., Palombi, M., Fattorini, L. et al. 2023. Brassinosteroids interact with nitric oxide in the response of rice root systems to arsenic stress. Environ. Exp. Bot., 209: 105287. https://doi.org/10.1016/j.envexpbot.2023.105287.

Pintó-Marijuan, M. and Munné-Bosch, S. 2014. Photo-oxidative stress markers as a measure of abiotic stress-induced leaf senescence: advantages and limitations. J. Exp. Bot., 65(14): 3845–3857. Doi:10.1093/jxb/eru086.

Policarpo Tonelli, F.M., Policarpo Tonelli, F.C. and Severino Lemos, M. 2023. Exogenous application of phytohormones to increase plant performance under stress. pp. 275–285. In: Ozturk, M. et al. (Eds.). Phytohormones and stress responsive secondary metabolites. Academic Press, Elsevier, UK. https://doi.org/10.1016/B978-0-323-91883-1.00004-8.

Praveen, A., Nitric oxide mediated alleviation of abiotic challenges in plants. Nitric Oxide, 128: 37–49. https://doi.org/10.1016/j.niox.2022.08.005.

Puyaubert, J. and Baudouin, E. 2014. New clues for a cold case: nitric oxide response to low temperature. Plant, Cell and Environ., 37: 2623–2630. doi: 10.1111/pce.12329.

Qamer, Z., Chaudhary, M.T., Du, X., Hinze, L. and Azhar M.T. 2021. Review of oxidative stress and antioxidative defense mechanisms in *Gossypium hirsutum* L. response to extreme abiotic. J. Cotton. *Res.*, 4, 9. https://doi.org/10.1186/s42397-021-00086-4.

Rachappanavar, V., Padiyal, A., Sharma, J.K. and Gupta, S.K. 2022. Plant hormone-mediated stress regulation responses in fruit crops—A review. Sci. Hortic., 304: 111302. https://doi.org/10.1016/j.scienta.2022.111302.

Rahaman, M., Islam, M., Gebrekirstos, A. and Bräuning, A. 2019. Trends in tree growth and intrinsic water-use efficiency in tropics under elevated CO_2 and climate change. Trends, 33: 623–640. https://doi.org/10.1007/s00468-019-01836-3.

Rahman, S.U., Li, Y., Hussain, S., Hussain, B., Kha, W.-ud-D., Riaz, L. et al. 2023. Role of phytohormones in heavy metal tolerance in plants: A review. Ecol. Ind., 146: 109844. https://doi.org/10.1016/j.ecolind.2022.109844.

Rahman, S.U., Nawaz, M.F., Gul, S., Yasin, G., Hussain, B., Li, Y. et al. 2022. State-of-the-art MOICS strategies against toxic effects of heavy metals in plants: A review. Ecotoxicol. Environ. Saf., 242: 113952. https://doi.org/10.1016/j.ecoenv.2022.113952.

Raza, A., Charag, S., Najafi-Kakavand, S., Abbas, S., Shoaib, Y., Anwar, S. et al. 2023. Role of phytohormones in regulating cold stress tolerance: Physiological and molecular approaches for developing cold-smart crop plants. Plant Stress, 8: 100152. https://doi.org/10.1016/j.stress.2023.100152.

Ritonga, F.N., Ngatia, J.N., Wang, Y., Khoso, M.A., Farooq, U. and Chen, S. 2021. AP2/ERF, an important cold stress-related transcription factor family in plants: A review. Physiol. Mol. Biol. Plants, 27(9): 1953–1968. https://doi.org/10.1007/s12298-021-01061-8.

Rivero, R.M., Mittler, R., Blumwald, E. and Zandalinas, S.I. 2021. Developing climate-resilient crops: improving plant tolerance to stress combination. Plant J., 109(2): 373–389 doi:10.1111/tpj.15483.

Roy, S. and Mathur, P. 2021. Delineating the mechanisms of elevated CO_2 mediated growth, stress tolerance and phytohormonal regulation in plants. Plant Cell. Reports, 40: 1345–1365. https://doi.org/10.1007/s00299-021-02738-w.

Safdar, H., Amin, A., Shfiq, Y., Ali, A., Yasin, R., Shoukat, A. et al. 2019. A review: Impact of salinity on plant growth. Nat. & Sci., 17(1): 34–40. http://www.sciencepub.net/nature.

Saha, D., Choyal, P., Mishr, U.N., Dey, P., Bose, B., Prothihba, M.D. 2022 Drought stress responses and inducing tolerance by seed priming approach in plants. Plant Stress, 4, 100066. https://doi.org/10.1016/j.stress.2022.100066.

Saini, S., Kaur, N. and Pati, P.K. 2021. Phytohormones: key players in the modulation of heavy metal stress tolerance in plants. Ecotoxicol. Environ. Saf., 223: 112578. https://doi.org/10.1016/j.ecoenv.2021.112578.

Saleem, M., Fariduddin, Q. and Janda, T. 2021. Multifaceted role of salicylic acid combating cold stress in plants: A review. J. Plant Growth Regul., 40: 464–485. https://doi.org/10.1007/s00344-020-10152-x.

Salgado, I., Martínez, M.C., Oliveira, H.C. and Frungillo, L. 2013. Nitric oxide signaling and homeostasis in plants: a focus on nitrate reductase and S-nitroglutathione reductase in stress-related. Braz. J. Bot., 36(2): 89–98. DOI 10.1007/s40415-013-0013-6.

Sami, F., Faizan, M., Faraz, A., Siddiqui, H. and Yusuf, M. 2018. Nitric oxide-mediated integrative alterations in plant metabolism to confer abiotic stress tolerance, NO crosstalk with phytohormones and NO-mediated post translational modifications in modulating diverse plant stress. Nitric Oxide, 73: 22–38. https://doi.org/10.1016/j.niox.2017.12.005.

Sanstisree, P., Bhatnagar-Mathur, P. and Sharma, K.K. 2015. NO to drought-multifunctional role of nitric oxide in plant drought: Do we have all the ansewers? Plant Sci., 239: 44–55. http://dx.doi.org/10.1016/plantsci.2015.07.012.

Sharma, A., Kumar, V., Shahzad, B., Ramakrishnan, M., Singh Sidhu, G.P., Bali, A.S. et al. 2020a. Photosynthesis response of plants under different abiotic stresses: A review. J. Plant Growth Regul., https://doi.org/10.1007/s00344-019-10018-x.

Sharma, A., Soares, C., Sousa, B., Martins, M., Kumar, V., Shahzad, B. et al. 2020b. Nitric oxide-mediated regulation of oxidative stress in plants under metal stress: A review on molecular and biochemical aspects. Physiol. Plant., 168: 318–344. doi:10.1111/ppl.13004.

Sharma, P., Dutta, D., Udayan, A., Nadda, A.K., Lam, S.S. and Kumar, S. 2022. Role of microbes in bioaccumulation of heavy metals in municipal solid waste: Impacts on plant and human being. Environ. Pollut., 305: 119248. https://doi.org/10.1016/j.envpol.2022.119248.

Sharma, R. and Singh, H. 2022. Alteration in biochemical constituents and nutrients partitioning of *Asparagus racemosus* in response to elevated atmospheric CO_2 concentration. Environ. Sc. Pollut. Res., 29: 6812–6921. https://doi.org/10.1007/s11356-021-16050-3.

Shi, C. and Liu, H. 2021. How plants protect themselves from ultraviolet-B radiation stress. Plant Physiol., 187: 1096–1103. Doi:10.1093/plphys/kiab245.

Shrestha, J., Kandel, M., Subedi, S. and Shah, K.K. 2020. Role of nutrients in rice (*Oryza sativa* L.): A review. Agrica, 9: 53–62. DOI:10.5958/2394-448X.2020.00008.5.

Shrivastava, P. and Kumar, R. 2015. Soil salinity: A serious environmental issue and plant growth promoting bacteria as one of the tools for its alleviation. Saudi J. Biol. Sci., 22: 123–131. http://dx.doi.org/10.1016/j.sjbs.2014.12.001.

Silveira, N.M., Hancock, J.T., Frungillo, L., Siasou, E., Marcos, F.C.C., Salgado, I. et al. 2017. Evidence towards the involvement of nitric oxide in drought tolerance of sugarcane. Plant Physiology and Biochemistry, 115: 354–359. https://doi.org/10.1016/j.plaphy.2017.04.011.

Simpson, G.G. 2005. NO flowering. BioEssays, 27: 239–241. DOI:10.1002/bies.20201.

Singer, S.D., Soolanayakanahally, R.Y., Foroud, N.A. and Kroebel, R. 2020. Biotechnological strategies for improved photosynthesis in a future of elevated atmospheric CO_2. Planta, 251: 24. https://doi.org/10.1007/s00425-019-03301-4.

Singh, P., Singh, A. and Choudhary, K.K. 2023. Revisiting the role of phenylpropanoids in plant defense against UV-B stress. Plant Stress, 7: 100143. https://doi.org/10.1016/j.stress.2023.100143.

Singh, S., Uddin, M., Khan, M.M.A., Chishti, A.S., Singh, S. and Bhat, U.H. 2023. The role of plant-derived smoke and kirrikinolide in abiotic stress mitigation: An omic approach. Plant Stress, 7: 100147. https://doi.org/10.1016/j.stress.2023.100147.

Singh, V.P., Chatterjee, S., Kataria, S., Joshi, J., Datta, S., Vairale, M.G. et al. 2017. A review on responses of plants to UV-B radiation related stress. pp. 75–92. In: Singh, V.P. et al. (Eds.). UV-B radiation: from environmental stressor to regulator of plant growth. John Wiley & Son Ltd, doi:10.1002/9781119143611.ch5.

Sun, L.R., Yue, C.M. and Hao, F.S. 2019. Update on roles of nitric oxide in regulating stomatal closure, plant signalling & behaviour, 14: 10, e1649569, DOI:10.1080/15592324.2019.1649569.

Szabo, S., Yoshida M., Filakovszky, J. and Juhasz, G. 2017. "Stress" is 80 years old: from Hans Selye original paper in 1936 to recent advances in GI ulceration. Curr. Pharm. Des., 23(27): 4029–4041. Doi: 10.2174/1381612823666170622110046.

Szerement, J., Szatanik-Kloc, A., Mokrzycki, J. and Mierzwa-Hersztek, M. 2022. Agronomic biofortification with Se, Zn, and Fe: An effective strategy to enhance crop nutritional quality and stress defense—A review. J. Soil Sci. & Plant Nutr., 22: 1129–1159. https://doi.org/10.1007/s42729-021-00719-2.

Wahab, A., Abdi, G., Saleem, M.H., Ali, B., Ullah, S., Shah, W. et al. 2022. Plants' Physio-Biochemical and Phyto-Hormonal Responses to Alleviate the Adverse Effects of Drought Stress: A Comprehensive Review. Plants, 11: 1620. https://doi.org/10.3390/plants11131620.

Wang, X., Song, Q., Liu, Y., Brestic, M. and Yang, X. 2022. The network centered on ICEs play roles in plant cold tolerance, growth and development. Planta, 255: 81. https://doi.org/10.1007/s00425-022-03858-7.

Wang, X., Li, L., Cui, W., Xu, S., Shen, W. and Wang, R. 2012. Hydrogen sulphide enhances alfalfa (*Medicago sativa*) tolerance against salinity during seed germination by nitric oxide pathway. Plant and Soil, 351: 107–119. https://doi.org/10.1007/s11104-011-0936-2.

Wani, S.H., Kumar, V., Shiram, V. and Sah, S.K. 2016. Phytohormones and their metabolic engineering for abiotic stress tolerance in crop plants. Crop J., 4: 162–176. https://doi.org/10.1016/j.cj.2016.01.010.

Weisslocker-Schaetzel, M., André, F., Touazi, N., Foresi, N., Lembrouk, M., Dorlet, P. et al. 2017. The NOS-like protein from the microalgae *Ostreococcus tauri* is a genuine and ultrafast NO-producing enzyme, Plant Sci., 265: 100–111. https://doi.org/10.1016/j.plantsci.2017.09.019.

Wendehenne, D., Pugin, A., Klessig, D.F. and Durner, J. 2001. Nitric oxide: comparative synthesis and signaling in animal and plant cells. Trends in Plant Science, 6(4): 177–183. doi:10.1016/s1360-1385(01)01893-3.

Xue, Y., Zhu, S., Schultze-Kraft, R., Liu, G., Chen, Z., 2022. Dissection of Crop Metabolome Responses to Nitrogen, Phosphorus, Potassium, and Other Nutrient Deficiencies. Int. J. Mol. Sci., 23: 9079. https://doi.org/10.3390/ijms23169079.

Yaashika, P.R., Senthil Kumar, P., Jeevanantham, S. and Saravanan, R. 2022. A review on bioremediation approach for heavy metal detoxification and accumulation in plants. Environ. Pollut., 301: 119035. https://doi.org/10.1016/j.envpol.2022.119035.

Yadav, A., Singh, D., Lingwan, M., Yadukrishnan, P., Masakapalli, S.K. and Datta, S. 2020. Light signaling and UV-B-mediated plant growth regulation. JIPB, 62(9): 1270–1292. doi:10.1111/jipb.12932.

Yang, B., Tang, J., Yu, Z., Khare, T., Srivastav, A., Datir, S. and Kumar, V. 2019. Light stress responses and prospects for engineering light stress tolerance in crop plants. J. Plant Growth Regul., 38: 1489–1506. https://doi.org/10.1007/s00344-019-09951-8.

Yavaş, I., Ünay, A., Ali, S. and Abbas, Z. 2020. UV-B radiation and secondary metabolites. TURJAF, 8(1): 147–157. DOI: https://doi.org/10.24925/turjaf.v8i1.147-157.2878.

Zahid, G., Iftikhar, S., Shimira, F., Ahmed, H.M. and Kaçar, Y.A. 2023. An overview and recent progress of plant growth regulators (PGRs) in the mitigation of abiotic stresses in fruits: A review. Sci. Horticul., 309: 111621. https://doi.org/10.1016/j.scienta.2022.111621.

Zhang, B., Zheng L.P. and Wang J.W. 2012. Nitric oxide elicitation for secondary metabolite production in cultured plant cells. Appl. Microbiol. Biotechnol., 93: 455–466. DOI 10.1007/s00253-011-3658-8.

Zhang, J., Deng, L., Jiang, H., Peng, C., Huang, C., Zhang, M. et al. 2021. The effects of elevated CO_2, elevated O_3, elevated temperature and drought on plant leaf gas exchanges: A global meta-analysis of experimental studies. Environ. Sci. Pollut. Res., 28: 15274–15289. https://doi.or/10.1007/s11356-020-11728-6.

Zhang, T. and Yang, X. 2019. Exogenous glycine betaine-mediated modulation of abiotic stress tolerance in plants: Possible mechanisms. pp. 141–152. *In*: Hossain, M.A. et al. (Eds.). Osmoprotectant-Mediated abiotic stress tolerance in plants. Springer Nature, Switzerland AG. https://doi.org/10.1007/978-3-030-27423-8_6.

Zhou, X., Joshi, S., Khare, T., Patil, S., Shang, J. and Kumar, V. 2021. Nitric oxide, crosstalk with stress regulators and plant abiotic stress tolerance. Plant Cell Rep., 40: 1395–1414. https://doi.org/10.1007/s00299-021-02705-5.

Zouari, M., Ben Hassena, A., Trabelsi, L., Ben Rouina, B., Decou, R. and Labrousse, P. 2019. Exogenous proline-mediated abiotic stress tolerance in plants: Possible mechanisms. pp. 99–121. In: Hossain, M.A. et al. (Eds.). Osmoprotectant-mediated abiotic stress tolerance in plants. Springer Nature, Switzerland AG. https://doi.org/10.1007/978-3-030-27423-8_4.

17

UV Light Stress Induces Phenolic Compounds in Plants

Darem Sabrina,[1] *Telli Alia,*[2,]* *Benhamouda Hicham*[3] and
Benslama Mohamed[1]

Introduction

Phenolic compounds are secondary metabolites with an aromatic ring bearing a hydroxyl substituent. The primary monomers in polyphenols are phenolic rings, which are generally classified as phenolic acids and phenolic alcohols; most are originated from plants. Polyphenols are a broad heterogeneous group from a chemical point of view: some are soluble only in organic solvents, and some are water, carboxylic acids, and glycosides soluble. Other phenolics are insoluble polymers (Sambangi 2022).

In keeping with their chemical diversity, phenolics play various essential roles in the plant. Most phenolic substances significantly affect various physiological activities in plants during growth and development. Also, they are known to have an essential role in plant defense mechanisms, such as defensive role against abiotic and biotic stressful conditions. Abiotic stress encloses all stress from environmental changes such as high or low light and temperature, ultraviolet (UV) radiation, nutrient deficiencies, and drought or flood-like conditions. Biotic stress includes infection from a microbial pathogen, attack by herbivorous organisms, and increased

[1] Laboratory of Soil and Sustainable Development, Department of Biology, Badji Mokhtar, University of Annaba, P.O. Box 12, 23000 Annaba, Algeria.
[2] Laboratory of Protection of Ecosystem in Arid and Semi-arid Area, University of KASDI Merbah, P.O. Box 511, Ghardaia Road, Ouargla, Algeria.
[3] Faculty of life and earth sciences, University of Ghardaia P.O. Box 455, 47000 Ghardaia, Algeria.
* Corresponding author: alia.telli@gmail.com

production of oxidative species and free radicals within cells (Audil et al. 2019, Mei et al. 2018).

In plants, phenolic compounds are synthesized in response to specific environmental alerts or stresses. One of the largest groups of phenolics is flavonoids; they are considered exceptionally responsive to ultraviolet (UV) radiation (Koski and Ashman 2014). Plants sense radiation through several photoreceptors, absorbing blue light (λ = 455–492 nm) and UV-A radiation (λ = 315–400 nm) mainly (Hanns and Dorothee 2003, Nigle and Dyla 2003, Verdaghuer et al. 2016, Mathew Robson et al. 2014, Llorens et al. 2015). In collaboration with the receptors in the visible light range, a number of UV-B (λ = 280–315 nm) receptors mediate various reactions (Balakumar et al. 1993). In addition to their regulatory functions, high-irradiance visible light and UV light can harm plants. This is especially true when plants are unexpectedly exposed to an environment with bright light to which they are not adapted (Anna et al. 2019, Pan and Guo 2016, Magness 1920). Several works of literature reported that UV-B radiation induces the buildup of polyphenols in leaf epidermal cells' vacuoles and cell walls (Jansen 2002). The greatest classes of phenolic molecules found to absorb UV light while allowing through the radiation required for photosynthesis are called flavonoids.

The damage caused by UV radiation or high-intensity light exposure, either alone or in combination, leads plants to develop various mechanisms to avoid the harmful effects, foremost by filtering UV wavelengths, repairing UV-induced damage and scavenging ROS (Pristov et al. 2013, Ken et al. 2016, Jainendra et al. 2019). The primary goal of this overview is to clarify the impact of visible, UV-A and UV-B, as well as UV-C wavebands of solar radiation on the accumulation of phenolic composition in plants. This review purpose was to illustrate the involvement of defensive polyphenols in plant UV radiation responses.

Definition

In 1980, the term polyphenols was introduced, replacing the older term 'vegetable tannin' (Haslam 1998); which in the scientific literature is used primarily to refer to the application of plant extracts in the manufacture of leather, making it an important etymological term. However, Bate-Smith and Swain's idea is probably the most suitable (Haslam and Cai 1994). They supported White's earlier hypotheses and classified vegetable tannins as water-soluble phenolic compounds having molecular masses between 500–3,000 Da, besides giving the usual phenolic reactions, they have special properties such as the ability to precipitate alkaloids, gelatin and other proteins (Haslam and Cai 1994).

Phenolic compounds or polyphenols are secondary plant metabolites, characterized by the presence of at least one benzene ring (aromatic rings) linked directly by a free hydroxyl group (Fig. 1) or in a function: ether, ester, heteroside. They can combine with proteins to form complexes. The word 'polyphenol' should be used to define plant secondary metabolites that are derived only from the shikimate-derived phenylpropanoid and/or polyketide pathway(s), have more than one phenolic ring, and lack any nitrogen-based functional group in their most basic structural expression" (Belscak-Cvitanovic et al. 2018).

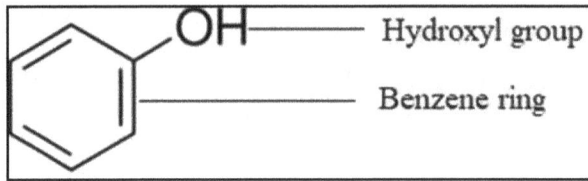

Fig. 1. Structure of the phenol ring.

Physicochemical Properties of Polyphenols

The physico-chemical properties of polyphenols are mainly related to their chemical structure. They can act as antioxidants, structural polymers (lignin), defense against UV radiation, or aggression by pathogens and pests. They also play an important function as communication molecules between plants and other organisms (signal molecules (salicylic acid and flavonoids) and defense response chemicals (tannins and phytoalexins)), both above and below ground. All phenolic compounds absorb strongly in both UV (ultraviolet) portion of the spectrum, and the visible range. Each phenolic compound class has specific absorption properties. For example, phenols and phenolic acids show spectral absorption maxima in between 250 and 290 nm; flavones and flavonols show absorption maxima between 250 and 350 nm. Normally, polar organic solvents can dissolve plant phenolics. While the comparable aglycones are frequently not, the majority of phenolic glycosides are water soluble. In proportion to the number of hydroxyl groups present, water solubility rises, with a few exceptions (Cheynier 2005, Belscak-Cvitanovic et al. 2018, Dunja et al. 2021).

Biosynthesis of Polyphenol

The biosynthetic pathways of phenolic compounds are widely known. The shikimic acid pathway is responsible for the formation of phenylalanine and tyrosine and the deamination of these amino acids leads to hydroxycinnamic acids, whose CoA esters are in turn responsible for most classes of phenolic compounds (Fig. 2).

The shikimate pathway has seven steps to produce the final product chorismate. Phosphoenolpyruvate and erythrose-4-phosphate are joined in the first step to produce 3-deoxy-o-arabino-heptulosonate 7-phosphate (DAHP), a 2-deoxy-D-glucose-6-phosphate derivative. In the next stage, the highly substituted cyclohexane derivative 3-dehydroquinate is produced by swapping DAHP with 3-dehydroquinate synthase. The shikimate pathway's final stages introduce side chains and two of the three double bonds necessary to convert this cyclohexane into a benzene ring, which is a characteristic of aromatic compounds. The third and fourth steps in the shikimate pathway are dehydroquinate to 3-dehydroshikimate by 3-dehydroquinate dehydratase (DHD) and the reversible reduction of 3-dehydroshikimate to shikimate via NADPH via shikimate dehydrogenase (SDH) (SDH). These reactions are catalyzed by DHD and SDH, which are likewise parts of the AROM complex in fungi, catalyze these processes. Furthermore, in E. coli the combination DHD-SDH enzyme has a single

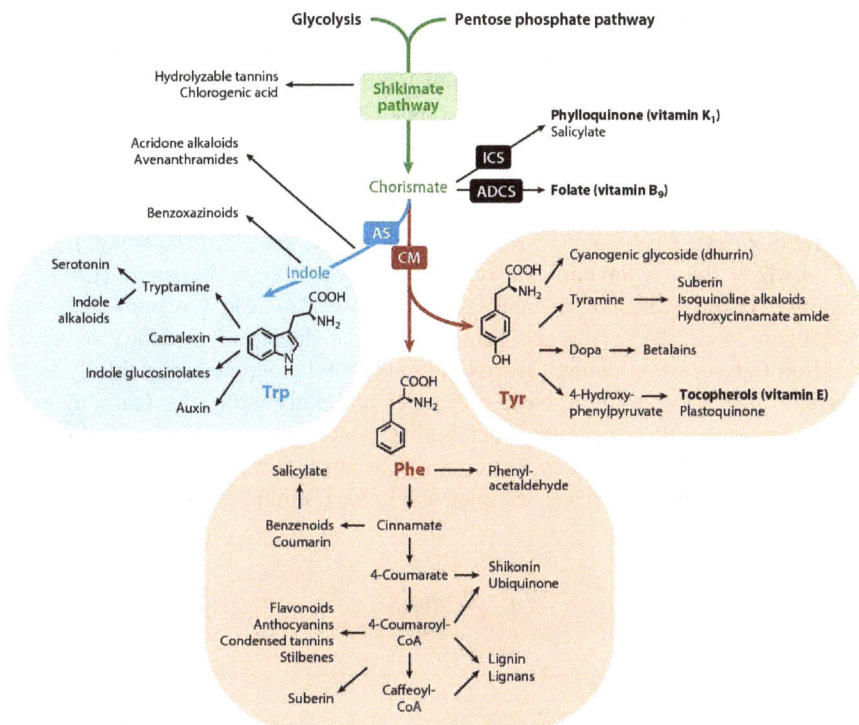

Fig. 2. The aromatic amino acid pathways promote the synthesis of many natural products in plants. The shikimate pathway (shown in *green*) produces chorismate, a common precursor for the tryptophan (Trp) route (*blue*), the phenylalanine/tyrosine (Phe/Tyr) pathways (*red*), and the pathways leading to folate, phylloquinone, and salicylate. Trp, Phe, and Tyr are additionally converted to a diverse array of plant natural products that play critical roles in plant physiology, some of which are important nutrients in human diets (*bold*). Other abbreviations: ADCS, aminodeoxychorismate synthase; AS, anthranilate synthase; CM, chorismate mutase; CoA, coenzyme A; ICS, isochorismate synthase (Maeda and Dudareva 2012).

function. Additionally, in plants, this pair has two distinct functions. The fifth step in the shikimate pathway is the synthesis of shikimate 3-phosphate. Shikimate kinase catalyzes the phosphorylation of shikimate at the C3 hydroxyl group, with ATP serving as a substrate. 5-endolpyruvylshikimate 3-phosphate synthase (EPSP) and chorismate synthase catalyze the sixth and seventh steps, respectively. By converting the enopyruvyl molecule of phosphoenolpyruvate into shikimate 3-phosphate, EPSP (also known as 3-phodphoshikimate 1-carboxyvinyltransferase) controls the last stage of the shikimate pathway. Chorismate synthase is the last enzyme in the shikimate pathway. It catalyzes the conversion of EPSP to chorismate, a precursor of SMs. In this final phase, the 1,4-anti-elimination of the 3-phosphate and C6-pro-R hydrogen from EPSP forms the second double bond in the ring to generate chorismate. In higher plants, chorismate serves as a precursor for tryptophan, tyrosine, phenylalanine, salicylate, phylloquinone, and folate; it is regulated by enzymes such as chorismate mutase, iso-chorismate synthase, anthranilate synthase, and amino-deoxychorismate synthase (Santos-Sánchez et al. 2019).

Classification of Polyphenols

Natural phenolic compounds can be grouped into several classes which are differentiated firstly by the complexity of the basic carbon skeleton (ranging from a simple C6 to highly polymerised forms); C6 (simple phenol, benzoquinones), C6—C1 (phenolic acid), C6—C2 (acetophenone, phenylacetic acid), C6—C3 (hydroxycinnamic acid, coumarin, phenylpropanes, chromones), C6—C4 (naphthoquinones), C6—C1—C6 (xanthones), C6—C2—C6 (stilbenes, anthraquinones), C6—C3—C6 (flavonoids, isoflavonoids, neoflavonoids), (C6—C3—C6)2,3 (bi-, triflavonoids), (C6—C3)2 (lignans, neolignans), (C6—C3)n (lignins), (C6)n (catechol melanins), and (C6—C3—C6)n (condensed tannins) (Lattanzio et al. 2013) (Fig. 3). Secondly by the degree of modification of this skeleton (degree of oxidation, hydroxylation, methylation, etc.), and thirdly by the possible links between these basic molecules and other molecules (carbohydrates, lipids, proteins, other secondary metabolites which may or may not be phenolic compounds). Hybrid phenolics, which are substances that combine phenolics with other natural chemicals such as terpenes and lipids (Dimitrios and Vassiliki 2019).

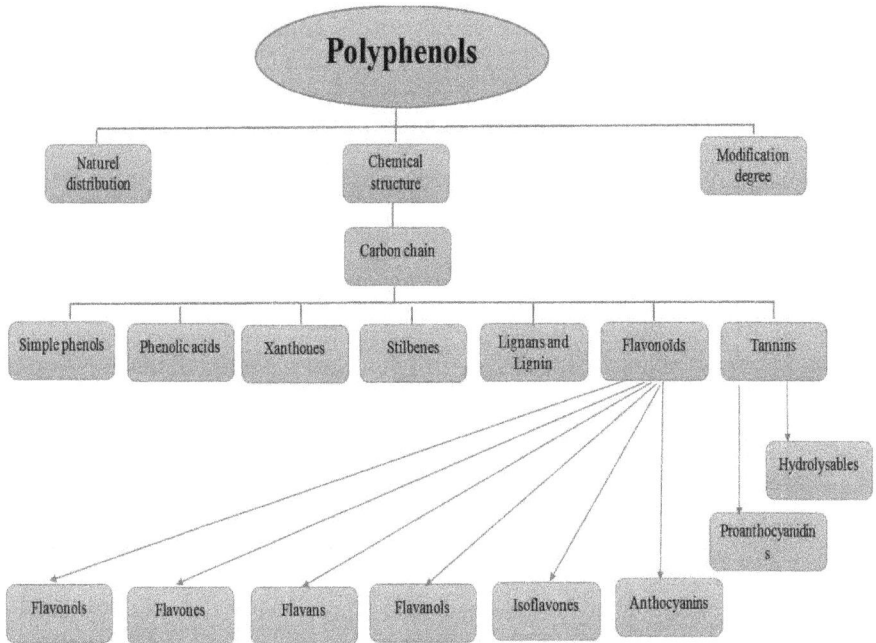

Fig. 3. Polyphenolic classes based on carbon chain.

Simple phenols (C6)

Simple phenols (catechol, guaiacol, phloroglucinol) are rather rare in nature with the exception of hydroquinone which exists in several families (Ericaceae, Rosaceae), catechol has been discovered in the leaves of Gaultheria species. Pop et al. found Bergenia crassifolia L. to contain between 15% and 23% of arbutin by dry weight of the plant, making it the richest source of the substance. In the peel of numerous citrus fruits, a glucoside known as phloroglucinol has been found (Pop et al. 2009).

Phenolic compounds (C6—Cn)

Depending on the length of the chain containing the carboxyl group, phenolic acids can be divided into hydroxybenzoic acids , hydroxycinnamic acids, and other hydroxyphenyl acids (acetic, propanoic, and pentaenoic).

The hydroxybenzoic acids (p-hydroxybenzoic, protocatechic, vanillic, gallic, syringic, salicylic, and gentisic) are derived from benzoic acid and have a basic formula of C6-C1. They are particularly well represented in Gymnosperms and Angiosperms from which they are often released after alkaline hydrolysis of plant material, in particular lignin and certain tannins. Hydroxybenzoic acids frequently exist as esters or glucosides, such as salicylic acid, whose glucoside or methyl or glucosyl esters are likely to represent forms of storage or circulation in the plant (Herrmann and Nagel 1989, Macheix et al. 2005).

One of the less frequent C6—C2 compounds, phenolic ketones have been discovered as plant ingredients on occasion. Picein, the primary constituent of all examined spruce needles *Picea abies* L., is also found in *Larix decidua* Mill., *Populus balsamifera*, and *Salix* spp. Hydroxyphenylacetic acid is found in bamboo shoots both free and as a glucoside. In *Xanthoxylum* spp., a phloroacetophenone derivative called xanthoxylin has been discovered (Münzenberger et al. 1995, Nyman et al. 2000, Tomas et al. 2000, Potter et al. 2002).

The hydroxycinnamic acids are a significant class whose fundamental structure (C6-C3) is derived from cinnamic acid. The degree of hydroxylation of the benzene ring and its potential alteration by secondary reactions (by methylation in ferulic or sinapic acids) is an essential element of these compounds' chemical reactivity. Furthermore, the presence of a double bond in the side chain results in two isomeric series (cis or Z trans or E) whose biological properties may be different. However, the trans forms are naturally predominant and it is possible that the cis forms are only extraction artefacts (Herrmann and Nagel 1989).

Coumarins are oxygenated heterocycles with the basic structure C6—C3 of benzo-2-pyrone. They were first isolated by Vogel in 1820 from *Coumarouna odorata*. Nowadays, more than 800 different plant and microbial species have yielded about 1000 different coumarin molecules. In plants, they are identified in the Apiaceae, Asteraceae, Fabaceae, Rosaceae, Rubiaceae, Rutaceae, and Solanaceae. From the structural point of view, they are categorized into simple coumarins with substituents on the benzene ring, simple hydroxycoumarins, furanocoumarins, and pyranocoumarins which are similar to furanocoumarins but have a six-membered pyran ring (AFC 2008).

The basic molecules of the hydroxycinnamic series are p-coumaric acid (and its isomers, o- and m-coumaric acids), caffeic acid and ferulic acid and its 5-hydroxy derivative, and finally sinapic acid. The group is often referred to under the common name of 'Phenylpropanoids' (Herrmann and Nagel 1989).

Naphthoquinone (C6—C4) is a kind of quinone pigment found in nature that is generated from naphthalene. There are several isomeric naphthoquinones known, the most notable of which is 1,2-naphthoquinone. 1,4-Naphthoquinone has a pungent odor comparable to benzoquinone and produces volatile yellow triclinic crystals. Avicenniaceae, Bignoniaceae, Boraginaceae, Droseraceae, Ebenaceae, Juglandaceae, Nepenthaceae, and Plumbaginaceae are the most important higher plant families producing naphthoquinones. They are produced by a number of biosynthetic routes. It is almost insoluble in cold water, mildly soluble in petroleum ether, and more so in polar organic solvents. It becomes reddish-brown in alkaline solutions. Vitamin K is a 1,4-naphthoquinone derivative. It is a planar molecule with a single aromatic ring attached to a quinone component (Gaultier and Hauw 1965).

Phenolic compounds C6—Cn—C6

Xanthones

It belongs to a family of polyphenolic chemicals (C6-C1-C6) that are often extracted from higher plants and microbes. The monomeric xanthones are classified into four principal categories: simple xanthones, xanthone glycosides, prenylated xanthones, and miscellaneous xanthones (Sultanbawa 1979, Gales and Damas 2005, Vieira and Kijjoan 2005).

The biosynthetic pathways to xanthones have recently been addressed (Fig. 4). When the xanthones from the Guttiferae are studied, it is discovered that the shikimic

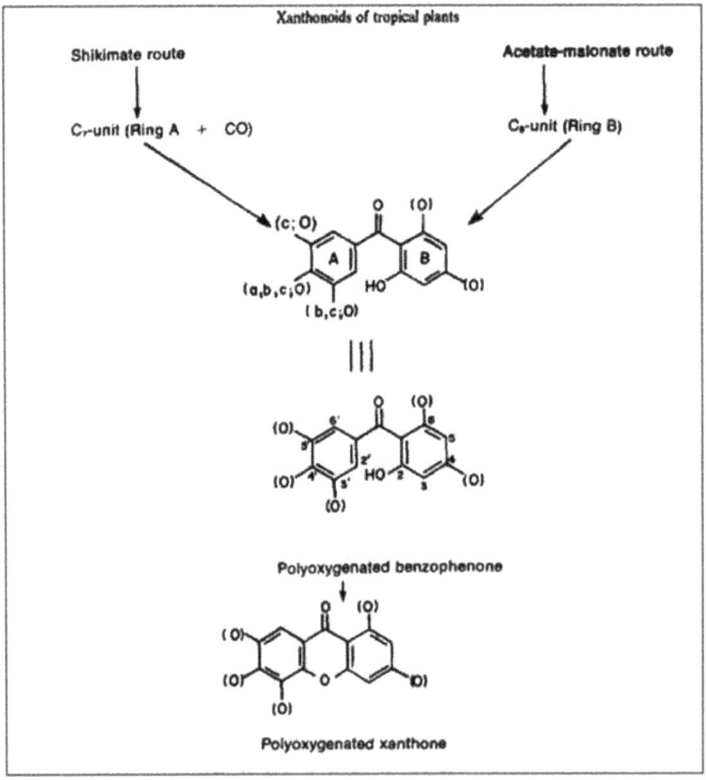

Fig. 4. Biosynthesis of xanthones in higher plants (Sultanbawa 1979).

acid pathway provides ring A and the connected CO group (C7 unit), whilst the acetate-malonate polyketide pathway provides ring B (C6 unit). Polyhydroxy benzophenones or their biogentically counterparts might thus act as intermediates in the synthesis of xanthones. These notions are encapsulated in Fig. 3, which depicts the most often seen mono ((a), di-0a,b), and tri-(c,c,c) locations for the shikimate-derived ring A (Gales and Damas 2005).

Stilbenes

The stilbene family members have the C6—C2—C6 structure. The most widely known stilbenoid is resveratrol, a constituent of red wine, because of its cardioprotective and anticancer effects (cis and trans-resveratrol). However, this chemical is also found in a variety of other plants, including peanuts and the berries of some shrubs (Izhaki 2002, Chen et al. 2014, Qign Liu et al. 2018) (Fig. 5).

Stilbenes are produced by the phenylpropanoid-polymalonate route. The initial stage of the process is shared by stilbenoids and flavonoids, with the two biosynthetic pathways separating at the cyclization of a styryl-3,5,7-triketoheptanoic acid. A chalcone is produced by C-acylation, and additional modifications lead to flavonoids, but an aldol condensation of the same intermediate polyketide yields a stilbene-2-carboxylic acid, which is an unstable intermediate in paths to a variety of stilbenoids (Evans et al. 1979, Faisal et al. 2017) (Fig. 6).

Stilbenes primary physiological activities include phytoalexins and growth regulators. Preexisting stilbenes may aid in the defense of plant tissues against fungus, insects, and other creatures. Furthermore, infections by suitable organisms or a variety of abiotic stimuli such as UV radiation can promote the production of a

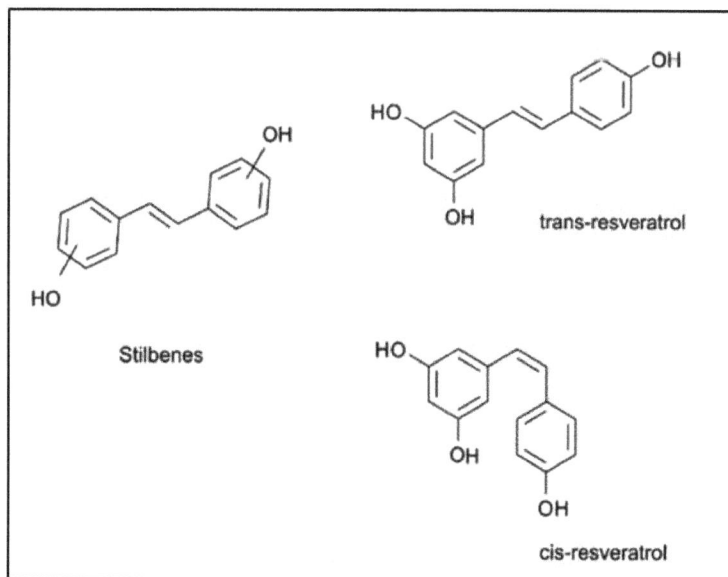

Fig. 5. Basic structure of stilbenes (1,2-diarylethenes; left) and *cis-* and *trans*-resveratrol (right) (Singla et al. 2019).

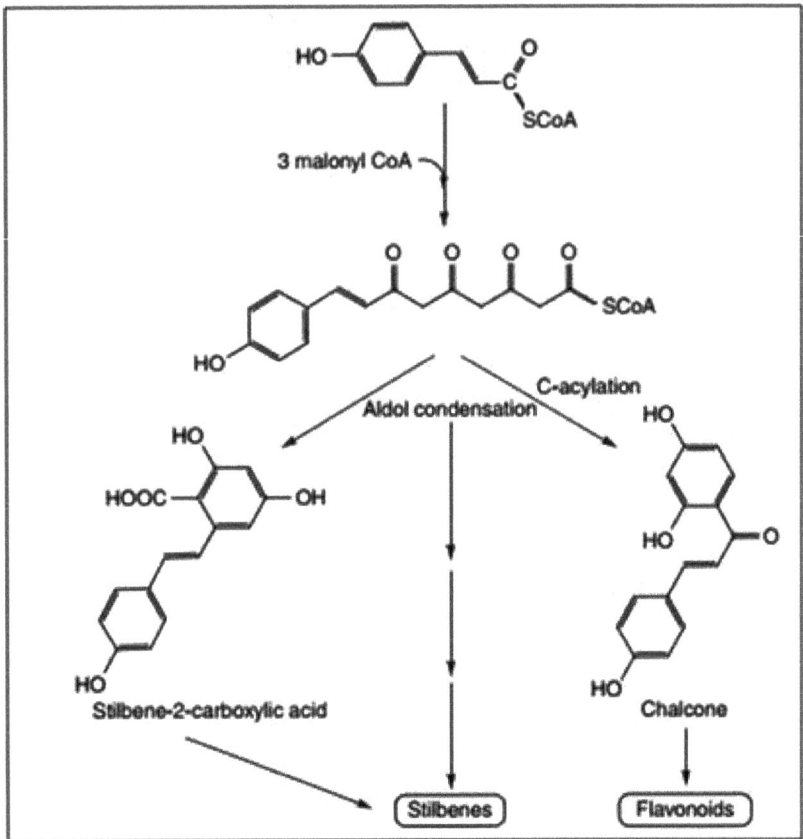

Fig. 6. Schematic of stilbene biosynthesis (Singla et al. 2019).

number of antifungal stilbenes. These stilbene phytoalexins include resveratrol and its derivatives found in Trifolium, Arachis, and Vitaceae members (Faisal et al. 2017)

Anthraquinones have a chemical structure that contains an anthracene ring (tricyclic aromatic) with two carbonyls in positions C9 and C10. Rubiaceae, Rhamnaceae, Fabaceae, Polygonaceae, Bignoniaceae, Verbenaceae, Scrophulariaceae, and Liliaceae are plant families that contain substantial amounts of anthraquinones (Izhaki 2002). The most well-known member of the class is aloe   emodin   [1,8-dihydroxy-3-(hydroxymethyl)anthraquinone,   which   shows antineoplastic action on a variety of malignant cells. Anthrones are reduced versions of anthraquinones that lack one carbonyloxygen, and both types frequently coexist in plants (Evans et al. 1979).

Lignans and lignins

Monolignols are cinnamic acid derivatives that act as precursors for phenylpropanoid compounds such as lignans and lignin (Fleuriet et al. 2005). Lignans have a structural representation of (C6-C3)2; the (C6-C3) unit is called a propylbenzene. They are produced by plants by the oxidative dimerisation of two coniferous alcohol molecules.

The resultant metabolites are known as lignans when this dimerisation includes oxidative bonding across the C-8 of the propenyl side chains of two connected coniferous alcohol units, forming the (C8-C8) bond. The term neolignan is employed to characterize all other types of linkage. When there is no direct (C-C) bond between the (C6-C3) units but linked by an ether oxygen atom, the compound is called oxineolignan. Other forms of lignans are sesquineolignans (three (C6-C3) units) and dineolignans (four (C6-C3) units) (Stalikas 2007). Lignans are essentially identified in oil seeds (Fleuriet et al. 2005). The lignans matairesinol (17), secoisolariciresinol, and others have been found in red wine made from Vitaceae family grapes (Nurmi et al. 2003), while neolignam biphenyls are isolated from *Magnolia officinalis* (Seo et al. 2015) and oxineolignans from *Bursera tonkinensis* of the Burceraceae family (Bertella and Luterbacher 2020).

Lignin (from the Latin *lignum* meaning wood) is a polyphenolic macromolecule, which is one of the main components of wood along with cellulose and hemicelluloses (Jawaid and Abdul Khalil 2011). This biopolymer is present mainly in vascular plants and in some coralligenous red algae, suggesting an evolutionary convergence of lignin biosynthesis between these algae and tracheophytes (Martone et al. 2009). Its main functions are to confer rigidity and mechanical strength to cell walls, as well as to provide water impermeability and resistance to decomposition (Fernandes Pereira 2015) (Fig. 7).

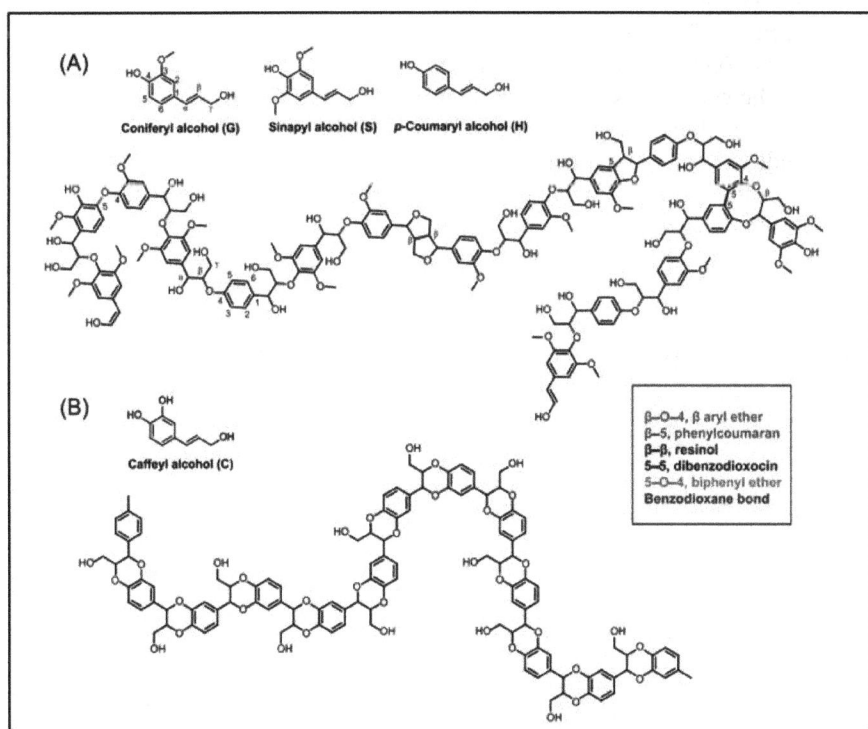

Fig. 7. Model chemical structures of lignin (Bertella and Luterbacher 2020).

Lignins result from the three-dimensional polymerisation of three basic phenolic monolignol molecules which are coumarylic, coniferyl, and sinapyl alcohols (Fig. 6), corresponding to p-coumaric, ferulic, and sinapic acids respectively. In the polymer itself, the incorporated monolignols are then referred to as H (derived from p-coumaryl alcohol), G (derived from coniferyl alcohol) and S (derived from sinapyl alcohol) units, which can be measured after lignin degradation. The relative simplicity of the structure and number of basic monomers contrasts strongly with the complexity of the final polymer (Pandey 1999, Brunow and Lundquist 2010). Indeed, despite an overall homogeneity that gives all lignins characteristic properties, there is also a high variability, which mainly concerns three points:

- Frequency of the three monomers in the final polymer.
- The heterogeneity of the bonds between the different monomers.
- Heterogeneity due to heteropolymerisation, that is, the incorporation during or after oxidative polymerisation of the three basic monolignols (Lucas et al. 2000).

Flavonoids

Flavonoids are a class of secondary metabolites that are widely distributed in the plant kingdom. They are almost universal pigments in plants and are partly responsible for the colouring of flowers, fruits and sometimes leaves. They are found dissolved in the vacuole of cells as heterosides or as constituents of particular plastids, the chromoplasts (Ghedira 2005). The term flavonoid covers a very wide range of natural polyphenolic compounds. There are about 6500 flavonoids divided into 12 classes (Stöckigt et al. 2002) and their number is increasing. By definition, flavonoids are compounds that share the structure of diphenylpropane (C6-C3-C6) (De Rijke et al. 2006); the three carbons serving as a junction between the two benzene rings noted A and B generally form an oxygenated heterocycle C (Panche et al. 2016).

De Rijke et al. (2006), classified flavonoids into 6 families involving flavonols, flavones, flavans, isoflavones, anthocyanins and flavanols. Within these six families, two types of structures have been identified, that of flavonoids in the strict sense, whose structure has the aromatic ring B in position 3 on the C3 chain, and that of isoflavonoids, whose aromatic ring B is in position 2 on the C3 chain. There are different structures of flavonoids, including: flavones, flavonols, flavanones, flavanonols, flavans, flavan-3-ols, flavylium, chalcones, aurones, isoflavones, isoflavonols, isoflavans, pterocarpans, 3-arylcoumarins, coumestanes, rotenoids (Bruce et al. 2000, Mamyrbékova-Békro et al. 2008, Bradley et al. 2012).

Tannins

The term tannin was introduced at the end of the eighteenth century to define the organic substances present in the aqueous extracts of leaves, flowers, stems and fruits. The molecular weight of tannins varies between 500 and 2000 Dalton and 3000 for the most complex structures. Tannins are very abundant phenolic compounds in angiosperms, gymnosperms and dicotyledons (König et al. 1994; Arapitsas 2012). They are characterized by an astringent taste. Tannins are divided into two groups:

- Condensed tannins, formed by pro anthocyanidins (in the form of oligomers).
- Hydrolysable tannins, esters of phenolic acids and glucose

Hydrolysable tannins are esters of carbohydrates or acid phenols, or acid phenol derivatives, the carbohydrate molecule is usually glucose, but in some cases other carbohydrate molecules may be present. Phenols, the carbohydrate molecule is usually glucose, but in some cases polysaccharides. This group of tannins is characteristic of dicotyledons, and is found in all organs: roots, stems, leaves or fruits before ripening. Due to their numerous (-OH) groups, these tannins dissolve to a greater or lesser extent (in the form of dissolve to a greater or lesser extent (depending on their molecular weight) in water, forming colloidal colloidal solutions. These tannins consist of a central core of glucose and a side chain (in position 1, 2, 3, 4 or 6 on the glucose) comprising one to (n) monomers (S) of phenolic acid. Carbon-to-carbon bonds between the rings (bi-phenyl bonds made by oxidative coupling), lead to more rigid branched molecules of decreased solubility called ellagic tannins (Khanbabaee and van Ree 2001; Crozier et al. 2006; Wang et al. 2012; Fraga-Corral et al. 2020 ; Virtanen and Karonen 2020) (Fig. 8).

Condensed tannins, also known as polyphenols or proanthocyanidins, are widely used in human food (fruit, vegetables, tea). They are oligomers or polymers of flavan-3-ols which have the property of releasing anthocyanins in an acidic medium at high temperature by breaking the inter-monomeric bond. They do not hydrolyse under the action of mineral acids but form insoluble compounds called phlobaphenes or tannin red when boiled. The complex structure of condensed tannins consists of monomeric repeating units (flavan-3-ols) that vary in their asymmetric centre and degree of oxidation. The natural forms of the monomeric carbons of the flavan-3-ols differ in the stereochemistry of the The natural forms of the monomeric carbons of flavan-3-ols are differentiated by the stereochemistry of the asymmetric carbons C2 and C3 and by the degree of hydroxylation of the B ring. A distinction is thus made between catechins (di-hydroxylated) and gallo-catechins (tri-hydroxylated) (Hernes and Hedges 2004, Amaral-Labat et al. 2013, Kardel et al. 2013, Pedro et al. 2019) (Fig. 9).

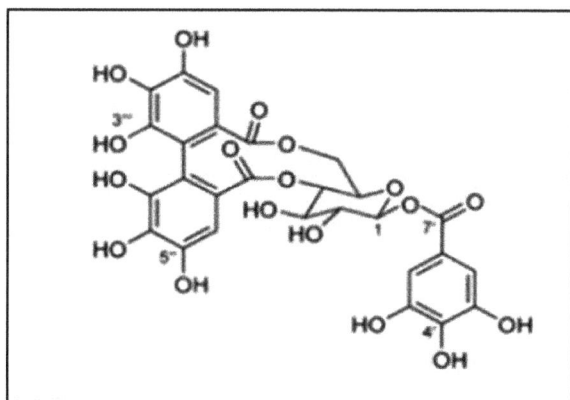

Fig. 8. Structures of 1-*O*-galloyl-4,6-(−)-hexahydroxydiphenoyl-β-d-glucose (strictinin) (R = H) and 1-*O*-digalloyl-4,6-(−)-hexahydroxydiphenoyl-β-d-glucose (R = gallate) (Engelhardt 2013).

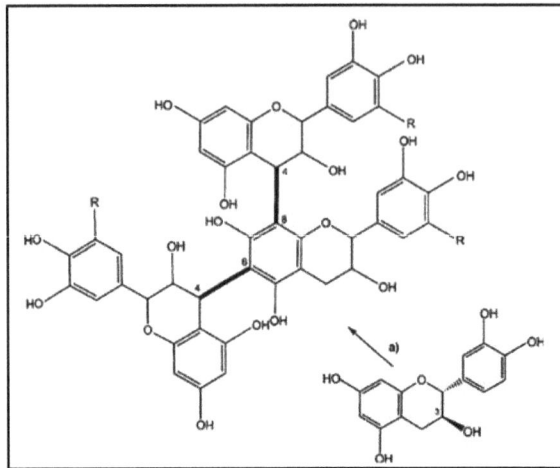

Fig. 9. General structure of condensed tannins: (R=H or OH): a) a flavan-3-ol monomer, catechin (R=H) (Engelhardt 2013, Falcão and Machado Araújo 2011).

Hybrid phenolic

The class of hybrid phenolics might be divided into phenolic terpenes and phenolic lipids. They are all made up of very uncommon yet highly active biological and chemical substances.

The phenolic terpenes are classified into three families: phenolic monoterpenes, diterpenes, and triterpenes. Carvacrol and thymol are typical phenolic monoterpenes of the Lamiaceae family. The essential oils of all of the main Lamiaceae genera include carvacrol, and the essential oils of certain species contain up to 95% of the chemical. Carvacrol has been identified as a bioactive molecule that has been researched as an antibacterial and antiviral agent (Baser 2008, Ramawat and Mérillon 2013, Gilling et al. 2014).

These polyphenols provide a strong chemical arsenal against chronic pathologies by influencing multiple physiological processes such as cellular redox potential, enzyme activity, cell proliferation, and signaling transduction pathways. The table below illustrates the biological activities effect of polyphenols belonging to different classes (Table 1).

UV light effect on phenolic compounds

Light is a fundamental abiotic component required by plants for photosynthesis and hence the production of carbon skeletons and energy for secondary metabolite synthesis. When the light intensity exceeds the chloroplast's carbon fixation capacity, free radicals are produced, a condition known as photo-oxidative stress. If the antioxidant system is unable to remove free radicals, this is known as photo-oxidative stress. Get rid of them. Among the secondary metabolites, a few are directly implicated in the processes of defense against photooxidative stress. These include

Table 1. Biological properties of some polyphenols (Hideo et al. 1993, Sohn et al. 2004, Sammomiya et al. 2005 , Tripoli et al. 2007, Win et al. 2008, Hirata et al. 2009, Ilhan et al. 2009, Kim et al. 2009, Sperry and Smith 2010, Siyu et al. 2020, Bokelmann 2022).

Carbon skeleton	Classes	Example	Chemical structure	Origin	biological activity
C6	Simple phenols	catechol (A), Hydroquinone (B)	A, B	Bearberry Ericales, Lamiales, and Asterales	Antibacterial, antioxidant
C6-C1	Hydroxybenzoic acids	Vanillic (A), Gallic (B), Salicylic (C) Acids	A, B, C	Strawberry-spice	Antibacterial, anti-ulcer, antiparasitic, antifungal, and antioxidant
C6–C2	Hydroxyphenylacetic acids	Acetophenones (A), Phenylacetic (B) acids	A, B	Bamboo shoots	

Table 1. contd. ...

... Table 1. contd.

Carbon skeleton	Classes	Example	Chemical structure	Origin	biological activity
C6–C3	Hydroxycinnamic acids	Coumarins (A) -caffeic (B) and ferulic acids (C)		Citrus, olives, coffee beans, fruits, potatoes, carrots, and propolis	Vascular protectors, anti-inflammatory, anti-parasitic analgesic, and anti-edematous
C6–C4	Naphthoquinone	Juglone (A), lawsone (B)		Nuts, henna plant (*Lawsonia inermis*)	Antibacterial, immunosuppressive, contraceptive, antimicrobial, antiulcer, anticancer, antimalarial, anti-inflammatory, trypanocidal, leishmanicidal, molluscicidal, and antifungal activities

Table 1. contd. ...

C6-C1–C6	Xanthones	Norathyriol (A), gambogic acid (B),		higher plants, fungi, ferns, and lichens.	antitumor activities, antibacterial, antithrombotic, vasodilatory, and anti-inflammatory
C6-C2–C6	Stilbenes, anthraquinones	phytoalexins resveratrol (A), chrysophanol (B)		Itadori tea, grapes, wine, peanuts, and soy; Aloe vera	Antitumor, antioxidant, antiviral, and phytoestrogenic, and anti-depressive
C6-C3–C6	Flavonoids, isoflavonoids	Aurantinidine, flavone		Red peppers, chamomile, celery, ginkgo biloba, fruits, vegetables, and flowers	Antitumor, antiparasitic, vasodilatory, antibacterial, anticarcinogenic, anti-inflammatory, analgesic, hypotensive, antiviral, diuretic, osteogenic, antioxidant, antiatherogenic, antithrombotic, anti-allergic

Table 1. contd. ....

Carbon skeleton	Classes	Example	Chemical structure	Origin	biological activity
(C6–C3)2	Lignans, neolignans	Genistein, Burseneolignan		leguminous plants, pine	Anti-inflammatories, analgesics
(C6–C3)n	Lignins	coniferyl alcohol, sinapyl alcohol		Wood, fruit stones	antioxidant, antitumor,
(C15)n	Tannins	Catechin (A), Procyanidin (B)		Bark of trees, wood, leaves, buds, stems, fruits, seeds, roots, and plant galls, green tea	Stabilising effects on collagen, antioxidant, anti-tumor, antifungal and anti-inflammatory effects, and antioxidant

carotenoids, anthocyanins, and many flavonoids. flavonoids. Phenolic molecules have a direct antioxidant action as well, and their importance in preventing oxidative stress is well established.

UV light is part of the sun spectrum and is classified into three wavebands: UV-C (200–280 nm), UV-B (280–315 nm), and UV-A (315–400 nm). The development of plants is possible only in the presence of the ozone layer which plays the role of a screen vis-à-vis UV radiation in the stratosphere. This layer, by playing the role of a UV filter, absorbs UV-C solar rays and part of UV-B radiation; Photoreceptors perceive different sections of the solar light spectrum (Paik and Huq 2019). Red/far-red light is received by phytochromes (phyA–phyE), blue light by cryptochromes (CRY1, CRY2, and CRY3), phototropins (PHOT1, PHOT2), and F-box containing flavin-binding proteins (ZEITLUPE, FKF1/LKP2), and UV-B light by the UVR8 receptor (Baudry et al. 2010, Roeber et al. 2020).

UV-B radiation is biologically active, but it is also the most aggressive. This radiation can interfere with growth, development, photosynthesis, flowering, pollination, and transpiration; they induce morphological changes at the leaf level or/and the entire plant. For example, leaf curling, to reduce the leaf area affected by this radiation, is a photo-morphogenic response that is observed at low doses of UV-B. Leaf thickening is another protective strategy that may be accompanied by a redistribution of chlorophyll from the adaxial surface (Demkura et al. 2010, Hoffmann et al. 2015, Escobar-Bravo et al. 2017).

The quality of light can also affect plant bioactive constituent and secondary metabolites (SM) synthesis. The three basic components of light are photoperiod (length), power (sum), and quality (repeat/recurrence) (Carvalho et al. 2010, Karppinen et al. 2014). Different plant species react differently to light intensity and quantity (Morales et al. 2010). Concerns have been raised about the effects of UV radiation (UV-B: 280–320 nm), which has been shown to have a considerable impact on the content and concentration of plant SMs such as alkaloids (Carvalho et al. 2010), terpenoids, flavonoids (Karppinen et al. 2014), cyanogenic glycosides, tannins, and anthocyanins (Morales et al. 2010, Gouvea et al. 2012). Plants may adapt to changes in light radiation by accumulating and releasing a wide range of SMs, including phenolic compounds, triterpenoids, and flavonoids, many of which have high economic and utility value due to their antioxidant properties (Koes et al. 2005, Jaakola and Hohtola 2010).

Plants store phenolic acids and flavonoids in mesophyll and epidermal cell vacuoles during light stress via photosynthetic machinery and metabolism (Tattini et al. 2004, Conéjéro et al. 2014, Szymańska et al. 2017), which regulates the control of phenylpropane phenolic levels in Xanthium species. When maize plants are exposed to UV-B radiation, the expression of the P1, B, and PL1 genes increases, resulting in the biosynthesis of the transcriptional regulators anthocyanin and 3-deoxy-flavanoid, which in turn regulate the activity of the ZmFLS1 protein, which converts dihydroflavonols, dihydroquercetin, and dihydrokaempferol, respectively, into flavonols, quercetin, and kaempferol. UV-B exposure enhanced the quercetin-to-kaempferol ratio in numerous petunia (Koes et al. 2005), birch (Taylor 1965), and Crocus taxa cultivars. The amount of hydroxylation on the B-ring differs between the flavonols quercetin and kaempferol, with quercetin being dihydroxylated and kaempferol being monohydroxylated. UV-B absorbance frequently decreased as the amount of hydroxylation increased (Ghasemzadeh et al. 2010, Jang et al. 2017). This is consistent with the fact that light-responsive dihydroxy flavonoids have a significantly greater ability than monohydroxy flavonoids to reduce reactive oxygen species ROS generation and quench ROS after they have been created (Carvalho et al. 2010).

Increased UV-B levels led to an increase in the ratio of quercetin and kaempferol in different cultivars of petunia (Koes et al. 2005), birch (Taylor 1965), and Crocus taxa []. The flavonols quercetin and kaempferol differ from each other only in the degree of hydroxylation on the B-ring, with quercetin being dihydroxylated and kaempferol monohydroxylated. Generally, as the level of hydroxylation increased, the absorption of UV-B decreased (Raúl et al. 2010). This is consistent with the fact that light-responsive dihydroxy flavonoids are far more effective at suppressing the production of ROS than monohydroxy flavonoids are at doing so, as well as quenching ROS that have already generated (Giraud et al. 2012).

Identically, in *Pinus contorta*, anthocyanin amounts were observed to be lower when the plant was grown under short sunlight conditions compared to long sunlight conditions, but proanthocyanin and flavan-3-ol rates were only marginally impacted by modification in sunlight period (Isah 2019). *Ipomoea batatas* showed an important increase in phenolic acids such as hydroxybenzoic acids and hydroxycinnamic and flavonoids like flavonols, anthocyanins, and catechins content after prolonged light exposure (León-Chan et al. 2017).

According to León-Chan et al. (2017), low temperature and UV-B radiation enhance chlorophyll breakdown and accumulation of carotenoids, chlorogenic acid, flavonoids apigenin-7-O-glucoside, and luteolin-7-O-glucoside in bell pepper plant leaves. They observed that UV-B radiation increases the content of flavonoids in leaves, while low temperature combined with UV-B radiation improve the amount of chlorogenic acid in leaves. They also observed that luteolin-7-O-glucoside is involved in the quenching of ROS generated as a result of UV-B radiation exposure at low temperatures. However, Peng et al. (2017) show that, flavone O-glycosides are regulated by flavone 7-Oglucosyltransferase and flavone 5-O-glucosyltransferase under light stress. They argued that allelic variety confers UV-B tolerance in plants in nature.

Bridgen (2016) found that UV-C irradiation has a consistent and distinct influence on flowering and growth performance. Our results show that brief periods of UV-C exposure to young plants in a greenhouse are helpful in regulating plant development. We concluded that when administering UV-C light to greenhouse-grown plants, the dose rates are crucial to the plants' response. The adequate weekly dosage, as little as 15 minutes each week, will govern the growth response of a plant. Furthermore, too much UV-C irradiation will burn plants, while too little will have no impact. Furthermore, at suitable dose rates, UV-C radiation has been shown in this study to reduce total plant height. Several species grew shorter in response to UV-C light than control plants that received standard greenhouse illumination. UV-C light can also help plants branch out. UVC light, at suitable dose rates, increases branching and the number of flowers produced in some species. This reduces the need to squeeze plants and use plant growth regulators. However, in other circumstances, greater branching is coupled by delayed flowering. Plants' flowering times can be influenced by UV-C light. Depending on the plant type and dose rate, UV-C irradiation can either delay or trigger quicker flowering.

Conclusion

Because they are immobile, plants must withstand abiotic conditions like drought, salt, extremely high temperatures, and intense radiation. The distribution of plants is severely constrained by these stresses, which affects how they grow and develop. Being sessile, plants create a variety of natural defenses to deal with particular environmental challenges. Producing and storing phenolic chemicals, which help in the defense against visible UV-A, UV-B, and UV-C solar radiations, is one of these adaption mechanisms.

The multilayered nature of plant responses to abiotic stresses is highlighted by recent advances in our understanding of the molecular mechanisms underpinning these responses. A number of processes, including sensing, signaling, transcription, translation, and post-translational protein modifications, are involved. Plants produce phenolics, which are flavonoids that function as photoprotectors, to filter damaging radiation. The crucial function of phenolic chemicals in plants' defense against harmful UV radiation is illustrated in a variety of literary genres. It has long been assumed that the accumulation of these chemicals results from UV-B light.

Anthocyanins are a subclass of flavonoids that are mostly absorbed in the green spectrum (500–565 nm), which lowers the total amount of PAR (400–700 nm) that hits the chloroplasts and speeds up photosynthetic recovery after light stress. Anthocyanins can also absorb UV rays when they are acylated, and they are typically the main phenolic component that is in charge of scavenging ROS. Additionally, anthocyanin production is known to be induced by light stress. The latter could increase one's tolerance for UV and visible light exposure. Despite the fact that the plant may incur costs during the production of these phenolic compounds, flavonoids play a critical function in photoprotection.

References

Amaral-Labat, G., Grishechko, L.I., Fierro, V., Kuznetsov, B.N., Pizzi, A. and Celzard, A. 2013. Tannin-based xerogels with distinctive porous structures, Biomass and Bioenergy, V., 56: 437–445. doi: https://doi.org/10.1016/j.biombioe.2013.06.001.

Anna, B.B., Grzegorz, B., Marek, K., Piotr, G. and Marcin, F. 2019. Exposure to High-Intensity Light Systemically Induces Micro-Transcriptomic Changes in Arabidopsis thaliana Roots. Int. J. Mol. Sci. V.20(20): 5131. doi:10.3390/ijms20205131.

Arapitsas, P. 2012. Hydrolyzable tannin analysis in food, Food Chemistry, V. 135(3): 1708–1717. doi:https://doi.org/10.1016/j.foodchem.2012.05.096

Athanasiadou, S., Kyriazakis, I., Jackson, F. and Coop, R.L. 2001. Direct anthelmintic effects of condensed tannins towards different gastrointestinal nematodes of sheep: *in vitro* and *in vivo* studies. Vet Parasitol. V.99(3): 205–19. doi:10.1016/s0304-4017(01)00467-8.

Audil, G., Ajaz, A.L. and Noor, I.W. 2019. Biotic and Abiotic stresses in plants. Book. doi:10.5772/intechopen.85832.

B. Sperry J. and B. Smith A. 2010. Chapter 160 - Chemical Synthesis of Diverse Phenolic Compounds Isolated from Olive Oils, Ed. Victor R. Preedy, Ronald Ross Watson, Olives and Olive Oil in Health and Disease Prevention, Academic Press, pp. 1439–1464. doi:https://doi.org/10.1016/B978-0-12-374420-3.00160-1.

Balakumar, T., Hani Babu Vincent, V. and Paliwal, K. 1993. On the interaction of UV-B radiation (280–315 nm) with water stress in crop plants, V.87(2): 217–222. doi:10.1111/j.1399-3054.1993.tb00145.x.

Baser, K.H.C. 2008. Biological and pharmacological activities of carvacrol and carvacrol bearing essential oils. Curr. Pharm. Des., V.14: 3106–19. doi:https://doi.org/10.2174/138161208786404227.

Baudry, A., Ito, S., Song, Y.H., Strait, A.A., Kiba, T., Lu, S. et al. 2010. F-box proteins FKF1 and LKP2 act in concert with ZEITLUPE to control Arabidopsis clock progression. Plant Cell., V.22(3): 606–22. doi:10.1105/tpc.109.072843.

Belscak-Cvitanovic, A., Durgo, K., Huđek, A., Bačun-Družina, V. and Komes, D. 2018. Overview of polyhenols and their properties. Polyphenols: Properties, Recovery, and Applications, 3–44. doi:https://doi.org/10.1016/B978-0-12-813572-3.00001-4.

Bokelmann, J.M. 2022. 67 - St. John's Wort (Hypericum perforatum): Flowering Buds and Tops, Ed., Medicinal Herbs in Primary Care, Elsevier, pp. 569–577. doi:https://doi.org/10.1016/B978-0-323-84676-9.00067-2.

Bradley J. Miller, Tanya Pieterse, Charlene Marais Barend and Bezuidenhoudt, C.B. 2012. Ring-closing metathesis as a new methodology for the synthesis of monomeric flavonoids and neoflavonoids Tetrahedron Letters, 53(35): 4708–4710, doi:10.1016/j.tetlet.2012.06.094.

Bridgen, M.P. 2016. Using ultraviolet-C (UV-C) irradiation on greenhouse ornamental plants for growth regulation, Acta Hortic. 1134. ISHS, pp. 48–56. doi:10.17660/ActaHortic.2016.1134.7.

Bruce, W., Fokerts, O., Garnaat, C., Crasta, O., Roth, B. and Bowen, B. 2000. Expression profiling of the maize flavondoid pathway genes controlled by estradiol-inducible transcripton factors CRC and P, Plant Cell V.12(1): 65–80. doi:10.1105/tpc.12.1.65.

Carvalho Isabel S., Cavaco T., Carvalho Lara M. and Duque, P. 2010. Effect of photoperiod on flavonoid pathway activity in sweet potato (Ipomoea batatas (L.) Lam.) leaves, V.118(2): 384–390. doi:10.1016/j.foodchem.2009.05.005.

Carvalho, R., Takaki, M. and Azevedo, R.A. 2011. Plant pigments: the many faces of light perception. V.33(2): 241–248. doi:10.1007/s11738-010-0533-7.

Catalgol, B., Batirel, S., Taga, Y. and Ozer, N.K. 2012. Resveratrol: French paradox revisited. Front Pharmacol., 1–18. doi: https://doi.org/10.3389/ fphar.2012.00141.

Cheynier, V. 2005. Polyphenols in foods are more complex than often thought. The American Journal of Clinical Nutrition, 81(1): 223–229. doi: 10.1093/ajcn/81.1.223S.

Conéjéro, G., Noirot, M., Talamond, P. and Verdeil, J.L. 2014. Spectral analysis combined with advanced linear unmixing allows for histolocalization of phenolics in leaves of coffee trees. Frontiers in Plant Science, 5: 39. doi:10.3389/fpls.2014.00039.

Coumarin in flavourings and other food ingredients with flavouring properties - Scientific Opinion of the Panel on Food Additives, Flavourings, Processing Aids and Materials in Contact with Food (AFC). doi:10.2903/j.efsa.2008.793.

Crozier, A., Jaganath, I.B. and Clifford, M.N. 2006. Phenols, Polyphenols and Tannins: An Overview. Chap. In: Plant Secondary Metabolites: Occurrence, Structure and Role in the Human Diet, Ed. Blackwell Publishing Ltd., Hoboken, NJ, pp. 1–24. doi:10.1002/9780470988558.ch1.

De Rijke, E., Out, P., Niessen, W.M.A., Ariese, F., Gooijer, C. and Brinkman, U.A.T. 2006. Analytical separation and detection methods for flavonoids. Journal of Chromatography A, 1112: 31–63. doi: 10.1016/j.chroma.2006.01.019.

Demkura, P.V., Abdala, G., Baldwin, I.T. and Ballaré, C.L. 2010. Jasmonate-dependent and-independent pathways mediate specific effects of solar ultraviolet B radiation on leaf phenolics and antiherbivore defense. Plant Physiol., 152: 1084–1095. doi: 10.1104/pp.109.148999.

Dimitrios, T. and Vassiliki, O. 2019. Classification of Phenolic Compounds in Plants, chap. In: Polyphenols in plants. Ed. Elsevier, pp. 263–284. https://doi.org/10.1016/B978-0-12-813768-0.00026-8.

Dunja, Š., Erna, K., Ivana, Š., Valerija, V.B. 3 and Branka, S. 2021. The Role of Polyphenols in Abiotic Stress Response: The Influence of Molecular Structure. Review. Journal, Ed. MDPI, I.118 doi:https://doi.org/10.3390/plants10010118.

Edwin Dorrestijn Lucas, J.J. Laarhoven, Isabel W.C.E. Arends and Peter Mulder. 2000. The occurrence and reactivity of phenoxyl linkages in lignin and low rank coal. Journal of Analytical and Applied Pyrolysis, 54(1): 153–192. doi:10.1016/s0165-2370(99)00082-0.

Engelhardt, U.H. 2013. Chemistry of Tea. Reference Module in Chemistry, Molecular Sciences and Chemical Engineering, Elsevier, doi:https://doi.org/10.1016/B978-0-12-409547-2.02784-0.

Escobar-Bravo, R., Klinkhamer, P.G.L. and Leiss, K.A. 2017. Interactive Effects of UV-B Light with Abiotic Factors on Plant Growth and Chemistry, and Their Consequences for Defense against Arthropod Herbivores. Front. Plant Sci., 8: 278. doi:10.3389/fpls.2017.00278.

Faisal, M., Saeed, A., Shahzad, D., Fattah, T.A., Lal, B., Channar, P.A. 2017. Enzyme inhibitory activities an insight into the structure–activity relationship of biscoumarin derivatives. Eur. J. Med. Chem. 141: 386–403. doi: https://doi.org/10.1016/j.ejmech.2017.10.009.

Falcão, L. and Machado Araújo, M.E. 2011. Tannins Characterisation in New and Historic Vegetable Tanned Leather Fibres by Spot Tests. Journal of Cultural Heritage, V.12(2): 149–156. doi:10.1016/j.culher.2010.10.005.

Fleuriet, A., Jay-Allemand, C. and Macheix, J.J. 2005. Composés phénoliques des végétaux un exemple des métabolites secondaires d'importance économique. Presses Polytechniques et Universitaires Romandes, pp. 121–216.

Fraga-Corral, M., García-Oliveira, P., Pereira, A.G., Lourenço-Lopes, C., Jimenez-Lopez, C., Prieto M.A. et al. 2020. Technological Application of Tannin-Based Extracts. Molecules, V.25(3): 614. doi:10.3390/molecules25030614.

Gales, L., Damas, A.M. 2005. Xanthones—a structural perspective. Curr. Med. Chem., 12: 2499–515. . doi:10.2174/092986705774370727.

Gaultier, J. and Hauw, C. 1965. Structure de l'α-Naphtoquinone. Acta Crystallographica, 18(2): 179–183. doi:10.1107/S0365110X65000439.

Ghasemzadeh, A., Jaafar, H.Z. and Rahmat, A. 2010. Antioxidant activities, total phenolics and flavonoids content in two varieties of Malaysia young ginger (Zingiber officinale Roscoe). Molecules, 15(6): 4324–4333. doi:https://doi.org/10.3390/molecules15064324.

Ghedira, K. 2005. Les flavonoïdes: structure, propriétés biologiques, rôle prophylactique et emplois en thérapeutique. Phytotherapie, 3(4): 162. doi:10.1007/s10298-005-0096-8.

Gilling, D.H., Kitajima, M., Torrey, J.R. and Bright, K.R. 2014. Antiviral efficacy and mechanisms of action of oregano essential oil and its primary component carvacrol against murine norovirus. J Appl Microbiol, V.116: 1149–63. doi:https://doi.org/10.1111/jam.12453.

Giraud, E., Ivanova, A., Gordon, C.S., Whelan, J. and Considine, M.J. 2012. Sulphur dioxide evokes a large scale reprogramming of the grape berry transcriptome associated with oxidative signalling and biotic defence responses. Plantcell Environ., V.35: 405–417. doi:10.1111/j.1365-3040.2011.02379.x

GOSTA Brunow and Knut Lundquist. 2010. Functional Groups and Bonding Patterns in Lignin (Including the Lignin-Carbohydrate Complexes),CRC Press. doi:10.1201/ebk1574444865-c7.

Gouvea, D.R., Gobbo-Neto, L. and Lopes, N.P. 2012. The influence of biotic and abiotic factors on the production of secondary metabolites in medicinal plants. Plant Bioact. Drug Discov. Princ. Pract. Perspect., 17: 419–452.

Hanns, F. and Dorothee, S. 2003. Ultraviolet-B Radiation-Mediated Responses in Plants. Balancing Damage and Protection. Plant Physiology, V.133(4): 1420–1428, doi:https://doi.org/10.1104/pp.103.030049.

Haslam, E. and Cai, Y. 1994. Plant polyphenols (vegetable tannins): gallic acid metabolism. Natural Product Reports, 11: 41–66. doi:https://doi.org/10.1039/NP9941100041.

Haslam, E. 1998. Practical Polyphenolics—From Structure to Molecular Recognition and Physiological Action, Cambridge University Press, Cambridge.

Hernes, J.P. and Hedges, J.I. 2004. Tannin signatures of barks, needles, leaves, cones, and wood at the molecular level11Associate editor: C. Arnosti, Geochimica et Cosmochimica Acta,V.68(6): 1293–1307. doi:https://doi.org/10.1016/j.gca.2003.09.015.

Herrmann, Karl and Nagel, Charles W. 1989. Occurrence and content of hydroxycinnamic and hydroxybenzoic acid compounds in foods, V.28(4): 315–347. doi:10.1080/10408398909527504.

Hideo O., Akio M., Yasuko Y., Mitsuru N. and Yoshimasa T. 1993. Antioxidant activity of tannins and flavonoids in Eucalyptus rostrata, Phytochemistry, 33(3): 557–561. doi: https://doi.org/10.1016/0031-9422(93)85448-Z.

Hirata, T., Fujii M, Akita K, Yanaka N, Ogawa K, Kuroyanagi M. et al. 2009. Identification and physiological evaluation of the components from citrus fruits as potential drugs for anti-corpulence and anticancer. Bioorg. Med. Chem., Vol. 1; 17(1): 25–8. doi: 10.1016/j.bmc.2008.11.039.

Hoffmann, A.M., Noga, G. and Hunsche, M. 2015. High blue light improves acclimation and photosynthetic recovery of pepper plants exposed to UV stress. Environ. Exp. Bot., 109: 254–263. doi:10.1016/j. envexpbot.2014.06.017.

Ilhan, G., Erdem, Y. and Shigeru, I. 2009. An anti-ulcerogenic flavonol diglucoside from Equisetum palustre L., Journal of Ethnopharmacology, 121(3): 360–365. doi:https://doi.org/10.1016/j. jep.2008.11.016.

In, S.-J., Seo, K.H., Song, N.-Y., Lee, D.-S., Kim, Y.-C. and Baek, N.-I. 2015. Lignans and neolignans from the stems of Vibrunum erosum and their neuroprotective and anti-inflammatory activity. Archives of Pharmacal Research. Jan, 38(1): 26–34. doi: 10.1007/s12272-014-0358-9.

Isah, T. 2019. Stress and defense responses in plant secondary metabolites production. Biol. Res., 52: 39. doi:http://orcid.org/0000-0002-2112-530X.

Izhaki, I. 2002. Emodin–a secondary metabolite with multiple ecological functions in higher plants. New Phytol. 155: 205–17. doi: https://doi.org/ 10.1046/j.1469-8137.2002.00459.x.

Chen, R., Zhang, J., Hu, Y., Wang, S., Chen, M. and Wang, Y. 2014. Potential antineoplastic effects of aloe-emodin: a comprehensive review. Am. J. Chin. Med., 42: 275–88. doi:https://doi.org/10.1142/ S0192415X14500189.

Qing Liu, Y., Song Meng, P., Chao Zhang, H., Liu, X., Xi Wang, M., Wu Cao, W. et al. 2018. Inhibitory effect of aloe emodin mediated photodynamic therapy on human oral mucosa carcinoma *in vitro* and in vivo. Biomed Pharmacother, 97: 697–707. doi:https://doi.org/10.1016/j. biopha.2017.10.080.

Evans, F.J., Lee, M.G. and Games, D.E. 1979. Electron impact, chemical ionization and field desorption mass spectra of some Anthraquinone and Anthrone derivatives of plant origin. Biol. Mass Spectrom 6: 374–80. doi:https://doi.org/10.1002/bms.1200060903.

Jaakola, L. and Hohtola, A. 2010. Effect of latitude on flavonoid biosynthesis in plants. Plant cell Environ., 33: 1239–1247. doi:https://doi.org/10.1111/j.1365-3040.2010.02154.x.

Jainendra, P., Rajneesh, H., Ahmed, I., Singhl Deepak, K., Prashant R., Vinod K. et al. 2019. Oxidative Stress and Antioxidant Defense in Plants Exposed to Ultraviolet Radiation. Chap. doi:https://doi. org/10.1002/9781119468677.ch16.

Jang, H.J., Lee, S.J., Kim, C.Y., Hwang, J.T., Choi, J.H., Park, J.H. et al. 2017. Effect of sunlight radiation on the growth and chemical constituents of Salvia plebeia R. Br. Molecules, 22(8): 1279. doi:10.3390/molecules22081279.

Karamali Khanbabaee and Teunis van Ree. 2001. Tannins: classification and definition, Nat. Prod. Rep., 18: 641–649 doi::10.1039/b1010611.

Kardel, M., Taube, F., Schulz, H., Schütze, W. and Gierus, M. 2013. Different approaches to evaluate tannin content and structure of selected plant extracts-review and new aspects. J. Appl. Bot. Food Qual., 86(1): 154–166. doi:10.5073/JABFQ.2013.086.021.

Ken, Y., Tomoko, K. and Frantisek, B. 2016. UV-B Induced Generation of Reactive Oxygen Species Promotes Formtion of BFA-Induced Compartments in Cells of Arabidopsis Root Apices. Front. Plant. Sec. Plant Physiology. doi:https://doi.org/10.3389/fpls.2015.01162.

Kim, J.Y., Lim, H.J., Lee da, Y., Kim, J.S., Kim, D.H., Lee, H.J. et al. 2009. *In vitro* anti-inflammatory activity of lignans isolated from Magnolia fargesii. Bioorg. Med. Chem. Lett., Vol.1; 19(3): 937–40. doi:10.1016/j.bmcl.2008.11.103.

Koes, R., Verweij, W. and Quattrocchio, F. 2005. Flavonoids: a colorful model for the regulation and evolution Karppinen of biochemical pathways, Trends in Plant Science, 10(5): 236–242. doi:https:// doi.org/10.1016/j.tplants.2005.03.002.

König, S., Schellenberger, A., Neef, H. and Schneider, G. 1994. Specificity of coenzyme binding in thiamin diphosphate-dependent enzymes. Crystal structures of yeast transketolase in complex with analogs of thiamin diphosphate. J. Biol. Chem., 269(14): 10879–82. PMID: 8144674.

Koski, M. and Ashman, T. 2014. Dissecting pollinator responses to abiquotus unltraviolet floral pattern in the wild. Func. Ecol., 28: 868–877. doi:https://doi.org/10.1111/1365-2435.12242.

Karppinen, L.K., Luengo Escobar, A., Häggman, H., H. and Jaakola. 2014. Light-controlled flavonoid biosynthesis in fruits. Front. Plant Sci., 5: 534. doi:https://doi.org/10.3389/fpls.2014.00534

Lattanzio, Ramawat, Kishan Gopal and Mérillon, Jean-Michel. 2013. Natural Products Phenolic Compounds: Introduction. doi:10.1007/978-3-642-22144-6_57.

León-Chan, R.G., López-Meyer, M., Osuna-Enciso, T., Sañudo-Barajas, J.A., Heredia, J.B., León-Félix, J. 2017 .Low temperature and ultraviolet-B radiation affect chlorophyll content and induce the accumulation of UV-B-absorbing and antioxidant compounds in bell pepper (Capsicum annuum) plants. Environmental and Experimental Botany, 139: 143–151. doi:10.1016/j. envexpbot.2017.05.006.

Llorens, L., Badenes-Pérez, F.R., Julkunen-Tiitto, R., Zidorn, C., Fereres, A. and Jansen Marcel, A.K. 2015. The role of UV-B radiation in plant sexual reproduction, Perspectives in Plant Ecology, Evolution and Systematics, 17(3): 243–254. doi: https://doi.org/10.1016/j.ppees.2015.03.001.

Jawaid, M. and Abdul Khalil, H.P.S. 2011. Cellulosic/synthetic fibre reinforced polymer hybrid composites: A review, Carbohydrate Polymers, 86(1)1: 1–18. doi: 10.1016/j.carbpol.2011.04.043.

Macheix, D., Annie, F. and Christian Jay-Al. 2005. Les composés phénoliques des végétaux: un exemple de métabolites secondaires d'importance économique, pp. 31–176.

Maeda, H. and Dudareva, N. 2012. The shikimate pathway and aromatic amino acid biosynthesis in plants. Annu. Rev. Plant Biol., 63: 73–105. doi:10.1146/annurev-arplant-042811-105439

Magness J.R. 1920. Effect of Light Exposure on Plant Growth, Botanical Gazette, 70(3): 246–248. doi:https://www.jstor.org/stable/2470260.

Mamyrbékova-Békro J.A., Konan, K.M., Békro, Y.A., Djié Bi, M.G., Zomi Bi, T.J., Mambo V. et al. 2008. Phytocompounds of the Extracts of Four Medicinal Plants of Côte d'´Ivoire and Assessment of their Potential Antioxidant by Thin Layer Chromatography. European Journal of Scientific Research, 24(2): 219–228.

Marcel, A.K. Jansen. 2002. Ultraviolet-B radiation effects on plants: induction of morphogenic responses, 116(3): 423–429. doi:10.1034/j.1399-3054.

Martone, P.T., Estevez, J.M., Lu, E., Ruel, Denny, K.M.W., Somerville, C. and Ralph, J. 2009. Discovery of Lignin in Seaweed Reveals Convergent Evolution of CellWall Architecture. Current Biology, 19(2): 169–175. doi:10.1016/j.cub.2008.12.031.

Matthew Robson, T., Karel, K., Otmar, U. and Marcel, A.K.J. 2014. Re-interpreting plant morphological responses to UV-B radiation. Wiley Journal. doi:https://doi.org/10.1111/pce.12374.

Mei, H., Cheng-Q.H. and Nai-Zheng, D. 2018. Abiotic stresses: General Defenses of Land Plants and Chances for Engineering Multistress Tolerance. Rev. Journal of Frontier. Plant. doi:https://doi. org/10.3389/fpls.2018.01771.

Morales, L.O., Tegelberg, R., Brosché, M., Keinänen, M., Lindfors, A. and Aphalo, P.J. 2010. Effects of solar UV-A and UV-B radiation on gene expression and phenolic accumulation in *Betula pendula* leaves. Tree Physiol., 30: 923–934. doi:https://doi.org/10.1093/treephys/tpq051.

Münzenberger, B., Kottke, I. and Oberwinkler, F. 1995.Reduction of phenolics in mycorrhizas of Larix decidua Mill. Tree Physiol., 15(3): 191–6. doi:10.1093/treephys/15.3.191.

Nigel, D. and Dylan Gwynn-J. 2003. Ecological roles of solar UV radiation: towards an integrated approach, Trends in Ecology & Evolution, 18(1): 48–55. doi:https://doi.org/10.1016/S0169-5347(02)00014-9.

Nurmi, T., Voutilainen, S., Nyyssönen, K., Adlercreutz, H. and Salonen, J.T. 2003. Liquid chromatography method for plant and mammalian lignans in human urine. Journal of Chromatography, 798: 101–110. doi:10.1016/j.jchromb.2003.09.018.

Nyman, T., Julkunen-Tiitto, R. 2000. Manipulation of the phenolic chemistry of willows by gall inducing sawflies. Proceedings of the National Academy of Sciences of the United States of America. Nov;97(24): 13184–13187. doi:10.1073/pnas.230294097. PMID: 11078506.

Nyman, T. and Julkunen-Tiitto, R. 2000. Manipulation of the phenolic chemistry of willows by gall-inducing sawflies. Proceedings of the National Academy of Sciences of the United States of America. Nov;97(24): 13184–13187. DOI:10.1073/pnas.230294097.

Paik, I. and Huq, E. 2019. Plant photoreceptors: Multi-functional sensory proteins and their signaling networks. Semin Cell Dev. Biol., 92: 114–121. doi:10.1016/j.semcdb.2019.03.007.

Pan, J. and Guo, B. 2016. Effects of Light Intensity on the Growth, Photosynthetic Characteristics, and Flavonoid Content of Epimedium pseudowushanense B.L.Guo. Molecules, 4; 21(11): 1475. doi:10.3390/molecules21111475.

Panche, A.N., Diwan, A.D. and Chandra, S.R. 2016. Flavonoids: an overview. J. Nutr. Sci., 5: e 47. doi:10.1017/jns.2016.41.

Pandey, K.K. 1999. A study of chemical structure of soft and hardwood and wood polymers by FTIR spectroscopy. Journal of Applied Polymer Science, 71(12): 1969–1975. doi:10.1002/(sici)1097-4628(19990321)71:12<1969::aid-app6>3.0.co;2-d.

Paulo Henrique Fernandes Pereira, Morsyleide de Freitas Rosa, Maria Odila Hilário Cioffi et Kelly Cristina Coelho de Carvalho Benini, Vegetal fibers in polymeric composites: a review. Polímeros, 25: 2015-jan-feb, pp. 9–22 doi:10.1590/0104-1428.1722.

Pedro L. de Hoyos-Martínez, Juliette M., Jalel Labidi J. and El Bouhtoury, F.C. 2019. Tannins extraction: A key point for their valorization and cleaner production. Journal of Cleaner Production, 206: 1138–1155. doi:https://doi.org/10.1016/j.jclepro.2018.09.243.

Peng, M., Shahzad, R., Gul, A., Subthain, H., Shen, S. Lei, L. et al. 2017. Differentially evolved glucosyltransferases determine natural variation of rice flavone accumulation and UV-tolerance. Nature Communications, 8(1): 1–2. doi:10.1038/s41467-017-02168-x.

Pop, C., Vlase, L. and Tamas, M. 2009. Natural resources containing arbutin. Determination of arbutin in the leaves of *Bergenia crassifolia* (L.). Fritsch. acclimated in Romania. Not. Bot. Horti. Agrobot. Cluj-Napoca, 37: 129–32. doi:https://doi.org/10.15835/nbha3713108.

Potter, G.A., Patterson, L.H.,Wanogho, E., Perry, P.J., Butler, P.C., Ijaz, T. et al. 2002. The cancer preventative agent resveratrol is converted to the anticancer agent piceatannol by the cytochrome P450 enzyme CYP1B1. Br. J. Cancer, 86: 774–778. doi:10.1038/sj.bjc.6600197.

Pristov, J.B., Jovanović, S.V. and Mitrović, A. 2013. UV-irradiation provokes generation of superoxide on cell wall polygalacturonic acid. Physiologia Plantarum, Wiley Online Library, 148: 574–581. doi:https://doi.org/10.1111/ppl.12001.

Ramawat, Kishan Gopal and Mérillon, Jean-Michel. 2013. Natural Products ǁ Phenolic Compounds: Introduction. 10.1007/978-3-642-22144-6(Chapter 57), 1543–1580. doi:10.1007/978-3-642-22144-6_57.

Raúl F. Guerrero, Belén Puertas, María I. Fernández, Miguel Palma, Emma Cantos-Villar 2010. Induction of stilbenes in grapes by UV-C: Comparison of different subspecies of Vitis. , 11(1), 0–238. doi:10.1016/j.ifset.2009.10.005.

Roeber, V.M., Bajaj, I., Rohde, M., Schmülling, T. and Cortleven, A. 2020. Light acts as a stressor and influences abiotic and biotic stress responses in plants. Plant, Cell & Environment published by John Wiley & Sons Ltd., pp. 645–664. doi:10.1111/pce.13948.

Sambangi, P. 2022. Phenolic Compounds in the Plant Development and Defense: An Overview. Chp. In Plant stress Physiology. doi:10.5772/intechopen.102873.

Sanders, T.H., McMichael, R.W. and Hendrix. K.W. 2000. Occurrence of resveratrol in edible peanuts. J Agric. Food Chem, 48: 1243–6. doi:https://doi.org/10.1021/jf990737b.

Santos-Sánchez, F.N., Salas-Coronado, R., Hernández-Carlos, B. and Villanueva-Cañongo, C. 2019. Shikimic Acid Pathway in Biosynthesis of Phenolic Compounds. IntechOpen. doi:1.0.5772/intechopen.83815.

Sannomiya, M., Fonseca, V.B., da Silva, M.A., Rocha, L.R., Dos Santos, L.C., Hiruma-Lima, C.A. et al. 2005. Flavonoids and antiulcerogenic activity from Byrsonima crassa leaves extracts. J Ethnopharmacol., 10; 97(1): 1–6. doi:10.1016/j.jep.2004.09.053.

Shizuo, T. 2011. Polyphenol Content and Antioxidant Effects in Herb Teas, Chinese [22] Medicine, Vol.2 No. 1 doi:10.4236/cm.2011.21005.

Singla, K., Dubey K., Garg A., Sharma K. and Fiorino, M. 2019. Natural Polyphenols: Chemical Classification, Definition of Classes, Subcategories, and Structures. Journal of AOAC International 102(5): 1397–1400. doi:https://doi.org/10.5740/jaoacint.19-0133.

Siyu, S., Jiasi, W., Yue, G., Yu, L., Dong, Y. and Ping, W. 2020. The pharmacological properties of chrysophanol, the recent advances. Biomedicine & Pharmacotherapy, Vol. 125. doi:https://doi.org/10.1016/j.biopha.2020.110002.

Sohn, H.Y., Son, K.H., Kwon, C.S., Kwon, G.S. and Kang, S.S. 2004. Antimicrobial and cytotoxic activity of 18 prenylated flavonoids isolated from medicinal plants: Morus alba L., Morus mongolica Schneider, Broussnetia papyrifera (L.) Vent, Sophora flavescens Ait and Echinosophora koreensis Nakai. Phytomedicine. 11(7–8): 666–72. doi::10.1016/j.phymed.2003.09.005. PMID: 15636183.

Stalikas, C.D. 2007. Extraction, separation and detection methods for phenolic acids and flavonoids. Journal of Separation Science, 30: 3268–3295. doi:10.1002/jssc.200700261.

Stefania Bertella and Jeremy S. Luterbacher. 2020. Lignin Functionalization for the Production of Novel Materials, review. Trends in Chemistry-Cell press, 2(5). Elsevier. doi:https://doi.org/10.1016/j.trechm.2020.03.001.

Stöckigt, J., Sheludko, Y., Unger, M., Gerasimenko, I., Warzecha, H. and Stöckigt, D. 2002. High-performance liquid chromatographic, capillary electrophoretic and capillary electrophoretic-electrospray ionization mass spectrometric analysis of selected alkaloid groups. Journal of Chromatography A, 967: 85–113. doi:10.1016/s0021-9673(02)00037-7.

Sultanbawa, M.U.S. 1979. Xanthonoids of tropical plants. Tetrahedron report number 84, vol. 36. doi:10.1016/s0040-4020(01)83114-8.

Vieira, L.M.M. and Kijjoa, A. 2005. Naturally-occurring xanthones: recent developments. Curr. Med. Chem. 12: 241346. doi:10.2174/092986705774370682.

Szymańska, R., Ślesak, I., Orzechowska, A. and Kruk, J. 2017. Physiological and biochemical responses to high light and temperature stress in plants. Environmental and Experimental Botany, 139: 165–177. doi:10.1016/j.envexpbot.2017.05.002.

Tattini, M., Galardi, C., Pinelli, P., Massai, R., Remorini, D., Agati, G. 2004. Differential accumulation of flavonoids and hydroxycinnamates in leaves of Ligustrum vulgare under excess light and drought stress. New Phytologist, 163(3): 547–561. doi:10.1111/j.1469-8137.2004.01126.x.

Taylor, A.O. 1965. Some Effects of Photoperiod on the Biosynthesis of Phenylpropane Derivatives in Xanthium. Plant Physiol., 40(2): 273–80. doi:10.1104/pp.40.2.273.

Tomas, B., Francisco, A. and Clifford, M.N. 2000. Dietary hydroxybenzoic acid derivatives—nature, occurrence and dietary burden. Journal of the Science of Food and Agriculture, 80(7): 1024–1032. doi:10.1002/(sici)1097-0010(20000515)80:7<1024::aid-jsfa567>3.0.co;2-s.

Tripoli, E., La Guardia, M., Giammanco, S., Di Majo, D. and Giammanco, M. 2007. Citrus flavonoids: Molecular structure, biological activity and nutritional properties: A review, Food Chemistry, 104(2): 466–479. doi:https://doi.org/10.1016/j.foodchem.2006.11.054.

Verdaguer, D., Jansen Marcel, A.K., Liorens, L., Morales Luis, O. and Neugart, S. 2016. UV-A radiation effects on higher plants: exploring the known unknown. Plant Science, doi:10.1016/j.plantsci.2016.11.014.

Virtanen, V. and Karonen, M. 2020. Partition Coefficients (logP) of Hydrolysable Tannins. Molecules, V. 25(16): 3691. doi:10.3390/molecules25163691.

Wang, D., Kasuga, J. and Kuwabara, C. 2012. Presence of supercooling-facilitating (anti-ice nucleation) hydrolyzable tannins in deep supercooling xylem parenchyma cells in Cercidiphyllum japonicum. Planta, pp. 747–759. doi:10.1007/s00425-011-1536-3.

Win, N.N., Awale, S., Esumi, H., Tezuka, Y. and Kadota, S. 2008. Novel anticancer agents, kayeassamins C-I from the flower of Kayea assamica of Myanmar. Bioorg. Med. Chem., 15; 16(18): 8653–60. doi:10.1016/j.bmc.2008.07.091.

18

Triacontanol
Role in Abiotic Stress Resistance and Tolerance

Moh Sajid Ansari,[1,]* Adnan Khan,[1] Abrar A. Khan[1] and Heba I. Mohamed[2]

Introduction

The global agricultural landscape is currently confronted with a formidable challenge in crop production (López-Marqués et al. 2020). As projected by the Food and Agriculture Organization (FAO 2017), the world's population is expected to reach 10 billion, and the demand for cereals and livestock production is anticipated to surge by over 60% (Springmann et al. 2018). Achieving agricultural growth hinges significantly on enhancing productivity, primarily through increasing crop yields. However, it's noteworthy that the period of the Green Revolution marked a notable phase when substantial yield increments were realized (López-Marqués et al. 2020). Subsequent to that era, the rate of yield enhancement has experienced a gradual decline.

In the realm of plant biology, various environmental and abiotic stressors, including heavy metal exposure, extreme temperatures (both cold and heat stress), drought, flooding, and salinity stress, have emerged as pivotal constraints in the global agricultural system. These stressors impose significant limitations on the economic yields, as well as the quality and quantity, of diverse crop species. Over time, extensive research efforts have been undertaken to unravel the intricate mechanisms governing stress tolerance in plants (Roychoudhury and Chakraborty 2013). Given the multifaceted nature of abiotic stress tolerance mechanisms in plants, which are

[1] Department of Botany, Faculty of Life Sciences, Aligarh Muslim University, Aligarh, India.
[2] Biological and Geological Science Department, Faculty of Education, Ain Shams University, Cairo 11341, Egypt.
* Corresponding author: sajidamu12@gmail.com

governed by a complex interplay of multiple genes, substantial ongoing research endeavors are focused on developing plant varieties endowed with heightened resistance to such stressors. Notably, stress resistance can be orchestrated through the induction of antioxidant reservoirs, the accumulation of osmo-protectants, and the production of defensive proteins (Roychoudhury et al. 2009, Basu et al. 2010).

Triacontanol (TC) is a natural wax component found in the epicuticular layer of the epidermal cell wall, possessing plant growth-regulating properties. Its beneficial effects encompass the promotion of plant growth, stress tolerance, and yield enhancement (Tomba et al. 2020, Ahmad et al. 2021, Islam et al. 2021). A plethora of studies have been conducted to investigate the impact of abiotic stresses on crops. These stressors have the potential to cause a number of problems, including the production of reactive oxygen species (ROS), membrane breakdown, reduction in the efficiency of photosynthesis, and altered nutrient intake. By interfering with physiological, biochemical, and molecular mechanisms, these disturbances can have a severe negative impact on crop growth and development, ultimately resulting in a sharp decline in production (Dresselhaus et al. 2018, dos et al. 2020).

The process of seed priming induces a physiological state that promotes germination and enhances uniform seedling emergence. This priming effect is achieved through the modulation of hormone levels, metabolic activities, dormancy mechanisms, and membrane permeability (Bose et al. 2018, Jisha et al. 2013, Aboutalebian et al. 2017, Aymen et al. 2018). Plants can increase their biotic and abiotic resilience by exogenously applying a variety of bioactive chemicals to their leaves or during seed priming. This improvement is due to enhanced antioxidant defense system activity, the Osmo protective processes, and stress-related protein regulation. Additionally, bioactive compounds support enhanced photosynthesis, improved nutrient uptake, changes in hormonal profiles, and enhancements in crop quality, productivity, and stand (Waqas et al. 2019, Zulfiqar et al. 2021).

Furthermore, Triacontanol (TC) exhibits the capability to modulate physiological and biochemical processes in plants across a spectrum of environmental conditions. Specifically, when applied in the presence of abiotic stress factors, TC contributes to the suppression of oxidative stress and lipid peroxidation, regulation of programmed cell death, enhancement of the antioxidant system, improved nutrient uptake, amino acid accumulation, chlorophyll synthesis, soluble sugar content, relative water content, and photosynthetic activity (Ali et al. 2020). TC effectively mitigates the detrimental impacts of abiotic stress by orchestrating the expression of various stress-responsive genes (Sarwar et al. 2021). Notably, TC has demonstrated its regulatory prowess across a diverse range of physiochemical processes, both under normal growth conditions and in the presence of stressors, in various crop species. This versatility has been observed in crops such as *Glycine max* (soybean) (Krishnan and Kumari 2008), *Helianthus annuus* L. (sunflower) (Aziz et al. 2013), and wheat (Perveen et al. 2014).

Chemical Properties of Triacontanol (TC)

Triacontanol is a saturated long-chain primary fatty alcohol consisting of 30 carbon atoms. Specifically, it is a derivative of triacontane, wherein one of the terminal

methyl hydrogen atoms is substituted with a hydroxy group. Its chemical formula is
$C30H61OH$, and its molecular mass is calculated to be 438.81 g/mol. Triacontanol
possesses a density of 0.777 g/ml at 95°C and a melting point of 87°C (189°F;
360 K). This compound was initially extracted from *Medicago sativa* L. (alfalfa)
shoots (Ries et al. 1977). It is commonly referred to as triacontanol (TC) but may
also be known by alternative names such as benzyl alcohol, melissyl alcohol, or
myricyl alcohol.

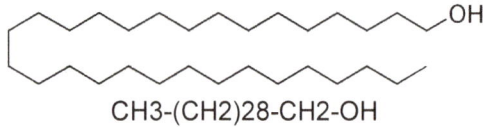

CH3-(CH2)28-CH2-OH

Fig. 1. The structural formula of Triacontanol (TC) is (Source: https://en.wikipedia.org/wiki/1-
Triacontanolce).

Roles of Triacontanol under Various Abiotic Stresses

Triacontanol represents a relatively recent addition to the realm of plant hormones,
yet its significance in plant metabolism is extensive. It has been established that the
augmentation of stress tolerance in numerous plant species under diverse abiotic
stressors can be attributed to the endogenous elevation of cellular triacontanol levels.
This enhanced stress resilience is underpinned by improvements observed across
several biochemical and physiological parameters, as summarized in Tables 1-5 and
Fig. 2. For instance, the improvement in growth was accompanied by elevated levels
proline and amino acid, improved leaf relative water content (LRWC), and improved
ion accumulation at a concentration of 200 ppm of TC in *Juniperus procera* that had
been subjected to water stress conditions (Kibatu et al. 2014).

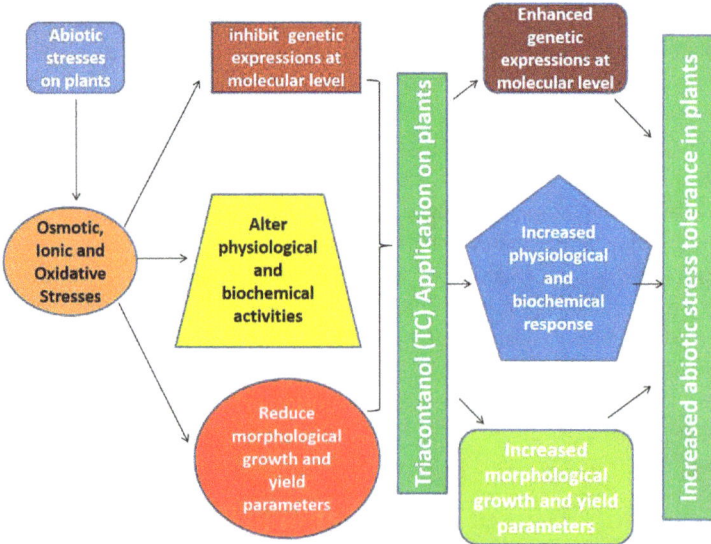

Fig. 2. Illustrative depiction: role of Triacontanol (TC) under various abiotic stresses on plants and its
mitigation.

Similarly, when grown in salt-stressed soils, triacontanol at a concentration of 10 mmol proved effective in preserving superior growth, bolstering physio-biochemical characteristics, augmenting the accumulation of leaf photosynthetic pigments, and increasing the build up of total soluble proteins in *Glycine max* L. (Soybean) (Krishnan and Kumari 2008). Exogenous administration of TC, in addition to its endogenous effects, has been shown to improve resistance to abiotic stress in many plant species. This can be accomplished via seed priming or foliar spraying (Ali et al. 2017, Aziz and Shahbaz 2015, Khanam and Mohammad 2018).

According to Perveen et al. (2012), priming seeds with TC at concentrations of 0, 10, and 20 mol increased growth, Leaf water relative content (LRWC), membrane permeability, and buildup of phenolic contents glycine betaine (GB), free proline, and free amino acids, in wheat plants. All of these concentrations of exogenously administered TC were helpful in promoting salt tolerance, with 10 M being the most potent amount. The effects of TC exogenous treatment on sunflower plants growing under saline stress were also studied by Aziz and Shahbaz (2015). By increasing both enzymatic and nonenzymatic antioxidant defense systems, they observed that foliar treatment at all doses of TC was efficient in reducing the negative effects of salt stress.

Additionally, Li et al. (2016 a,b) observed that foliar spraying of TC improved the contents of leaf sucrose and chlorophyll as well as the activities of antioxidant enzymes (like glutathione and ascorbate) levels in plants of the *Oryza sativa* species that were under water stress. A variety of plant species, *Erythroxylum coca* (coca) (Sitinjak and Pandiangan 2015), *Zea mays* L. (maize) (Perveen et al. 2016), and such as *Helianthus annuus* L. (sunflower) (Aziz et al. 2013), have shown that TC foliar spray is effective in enhancing various development and physio-biochemical procedures. This has helped to significantly reduce the deleterious effects of different biotic and abiotic stresses.

Triacontanol (TC), a substance employed during this study as a seed primer, is a saturated primary alcohol referred to as a plant growth regulator (PGR), and it has been shown to activate a variety of physiological and biochemical mechanism in crop plants (Karam et al. 2017, Singh et al. 2012). Numerous crops, including *Oryza sativa* L. (rice), and *Zea mays* L. (maize) have been found to benefit from the growth-promoting effects of triacontanol, even at relatively low concentrations (Naeem et al. 2012, Ahmad et al. 2013). TC is now being used to increase plant tolerance to a variety of abiotic stressors, including salt, heavy metal, and chilling conditions (Zaid et al. 2020, Islam et al. 2021, Naeem et al. 2012).

Notably, under abiotic stress conditions, the exogenous application of TC induces growth stimulation, augments the content of photosynthetic pigments, and enhances the accumulation of compatible osmolytes (Borowski et al. 2009, Perveen et al. 2013). Moreover, TC bolsters antioxidant defense mechanisms that are both enzymatic and non-enzymatic (Zaid et al. 2020, Perveen et al. 2013, Maresca et al 2017). It's worth mentioning that TC can also alleviate stress-related hazards by regulating the expression of specific genes (Islam et al. 2021). The primary objective of this study is to assess the efficacy of TC in enhancing plant tolerance.

Mechanisms of Exogenous Triacontanol (TC) Application on Plant Growth

Triacontanol (TC) has been thoroughly studied for its beneficial effects on a broad range of metabolic activities, including enzyme activity, germination of seeds, seedling development, and photosynthesis. By controlling gene expression, TC also plays a crucial part in promoting and enhancing resistance to diverse abiotic stresses. It is noteworthy that the discovery of a secondary messenger of TC, L (+)-adenosine, specifically as 9-b-L (+)-adenosine or 9H-purine-6-amine, 9-bL-ribofuranosyl, has offered crucial insights into the first stages of TC's activity in plants (Ries et al. 1990). Understanding the mechanism underlying how TC affects plants was greatly advanced by the identification of TC-mediated L (+)-adenosine formation/ release, also known as TRIM. According to research done by Ries et al. (1990), the L (+)-adenosine obtained from TC-treated plants is identical to that obtained from untreated control plants and has the same impact on the metabolic processes of plants. According to Ries and colleagues (1990, 1991), TC has the power to increase the proportion of L (+)- to D (–)-adenosine in the tonoplast. According to Olsson and Pearson (1990), the principal source of L (+)-adenosine synthesis in plants is thought to be adenosine monophosphate. In untreated plants, L (+)-adenosine (in a non-racemic form) may coexist alongside D (–)-adenosine in an inactive racemic combination, according to studies utilizing adenosine deaminase. Non-racemic adenosine appears to be released in TC-treated plants, impacting numerous plant metabolic processes, but racemic adenosine D (–)-adenosine) fails to activate such activities (Ries et al. 1990, Ries 1991).

The effects of exogenously administered L (+)-adenosine on plant physio-biochemical processes have also been studied by Ries and Wert (1992), Ries et al. (1993). According to several studies (Aftab et al. 2010, Keramat et al. 2017, Islam et al. 2020), externally applied TC quickly absorbs plant cells through the epidermal cell membrane and causes the production of L (+)-adenosine. It indicates that L (+)-adenosine causes an increase in calcium ion concentration, most likely inside the tonoplast. Calmodulin proteins may then attach to elevated calcium ions, activating them as a result (Fig. 3). The actions of kinases and phosphatases can then be affected by activated calmodulin proteins, which can also directly impact transcription factors including MYB2, CAMTA3, and GTL, among others. This cascade ultimately leads to the regulation of gene expression, encompassing genes related to photosynthesis, stress mitigation, antioxidant defense systems, and osmolyte accumulation. These coordinated reactions help explain the improved yield and growth seen in TC-mediated plants in both typical and stressful circumstances (Fig. 3). This explanation generally agrees with Chen et al.'s (2002, 2003) and Islam et al.'s (2020) results.

Impact of Triacontanol on Alleviating Abiotic Stresses in Plants

Abiotic restrictions, such as exposure to salt stress, water stress, heavy metals, temperature variations (chilling and heat stress), and transplant shock have a

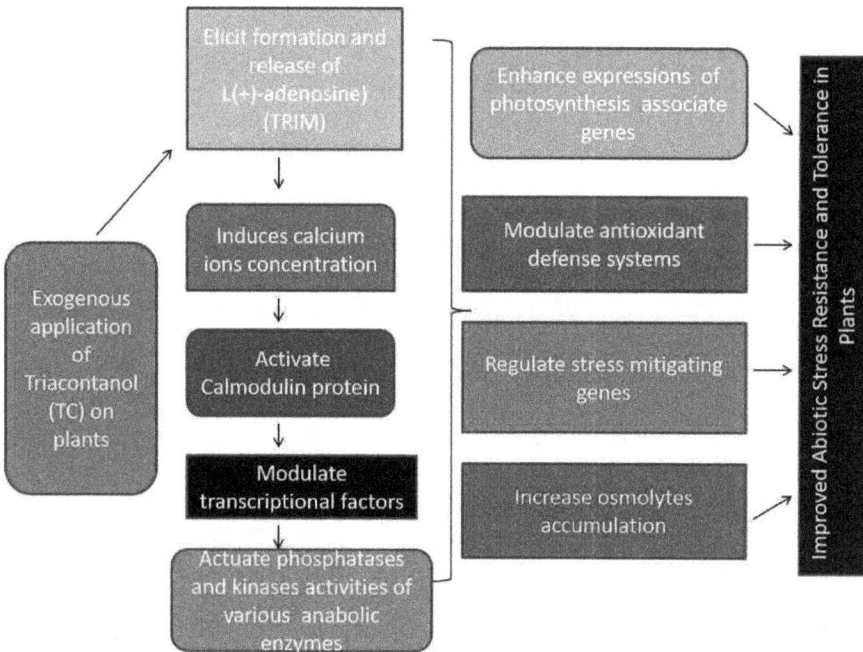

Fig. 3. Illustrating the mechanisms of exogenous Triacontanol (TC) application on plant growth.

substantial impact on the physiological, morphological, and biochemical aspects of a plant's overall development and output (Egamberdieva et al. 2017, Ahanger et al. 2017, 2018). Triacontanol (TC) is a signaling chemical that plants use to elicit defense mechanisms against a variety of abiotic stresses, according to a large body of research (Waqas et al. 2016, Naeem et al. 2012) (Fig. 1, Tables 1–5). The type of stress, the particular plant species involved, and the concentration of TC used are all important considerations when analyzing the effects of TC under such challenging circumstances. The ensuing section provides a succinct summary of the outcomes documented by researchers in this regard.

Salt stress

Salt stress stands as a paramount environmental factor, wielding substantial influence over crop growth and yield on a global scale (Negi et al. 2020, Zörb et al. 2019). Its pervasive impact extends to approximately 20% of agricultural lands, encompassing a striking 33% of irrigated agricultural territories (FAO 2017). Salt stress is primarily induced by the accumulation of neutral salts like sodium chloride (NaCl) and sodium sulfate (Na_2SO_4) in the soil matrix (Van Zelm et al. 2020). Elevated levels of NaCl in the soil can lead to a diminishment of soil moisture, imparting detrimental effects upon plant life. These repercussions manifest as toxicity stemming from the presence of sodium and chloride ions within the plant's physiology (Ahmad et al. 2022b). In response to the adversities imposed by salt stress, plants invoke a repertoire of adaptive mechanisms, including the modulation of gene expression and the activation of hormonal pathways (Raza et al. 2022, Feng et al. 2023).

In general, salt stress causes osmotic stress, metabolic disturbances, ionic imbalance, ionic toxicity, and the reactive oxygen species (ROS) generation which have a negative impact on crop growth development, and productivity (Islam et al. 2020, Per et al. 2017, 2018,). The increase of ROS in plant tissues is a typical side effect of salt stress, and this buildup causes oxidative stress-related damage, such as lipid peroxidation, denaturation of proteins, nucleic acid impairment, enzyme inactivation, and which disrupts cellular integrity (Islam et al. 2020, Per et al. 2017).

Extensive examination of the available literature underscores the constructive regulatory role played by Triacontanol (TC) in bolstering plant salt tolerance. For instance, Krishnan and Kumari (2008), for instance, documented that the treating of *Glycine max* L. with 10 mM n-TC exerted a significant restorative effect on metabolic processes. This treatment notably augmented parameters including relative water content (RWC), photosynthetic pigment, proteins, soluble sugars, and nucleic acid levels. Similarly, in the study conducted by Perveen et al. (2013), it was seen that TC effectively alleviated the deleterious consequences of 150 mM salinity stress in *Triticum aestivum* L. This amelioration manifested through improvements in chlorophyll content, net photosynthesis (PN), transpiration rate (E), stomatal conductance (gs), electron transport rate, alongside a concurrent reduction in relative membrane permeability.

Furthermore, an analogous investigation exhibited that foliar application of TC significantly bolstered salt tolerance in wheat cultivars (Perveen et al. 2014). This intervention led to a substantial increase in plant biomass and peroxidase activity, while concurrently reducing the levels of malondialdehyde (MDA) and hydrogen peroxide (H2O2). In addition, Aziz et al. (2013) reported that the application of TC at concentrations of 50 and 100 µM ameliorated the adverse impacts induced by 150 mM salinity stress on various gas exchange attributes, growth parameters, and the efficiency of PS-II in sunflower. Aziz and Shahbaz (2015) similarly observed that TC effectively mitigated salt-induced toxicity in sunflower by enhancing the superoxide dismutase (SOD), peroxidase (POD), and glutathione reductase (GR) activity are important antioxidant enzymes. This mitigation was accompanied by an increased buildup of proline and leaf glycine betaine (GB) level of concentration.

Additionally, Perveen et al. (2017) conducted a study with maize hybrids, applying 2 and 5 µM TC to the foliage. Their investigation revealed that the concentration of 5 µM notably enhanced salinity tolerance. This enhancement was manifested through significant improvements in growth parameters, leaf nitrate reductase (NR) activity, shoot potassium content, and soluble protein levels. Concurrently, it led to a reduction in levels of relative membrane permeability, hydrogen peroxide (H2O2), malondialdehyde (MDA), and sodium ions (Na+) uptake. In a separate study, Khanam and Mohammad (2018) implemented two spray treatments of 1 µM TC on the foliage of Mentha piperita L. The results demonstrated that TC effectively mitigated salt-induced damage by enhancing various growth-related attributes. This included increased chlorophyll content, improved gas exchange parameters (photosynthesis rate (PN), transpiration rate (E), stomatal conductance (gs), internal CO2 concentration, elevated leaf proline levels, as well as enhanced nitrogen (N),

Table 1. Roles of triacontanols under salt stress.

Plant species	Abiotic stress	Triacontanol application	Plant response beneficial effect	References
Cucumis sativus L. (Cucumber)	Salt stress	0.80 mg L^{-1}	Triacontanol functions as an effective scavenger of reactive oxygen species (ROS) by enhancing the activity of antioxidant enzymes such as superoxide dismutase (SOD), peroxidase (POD), and catalase (CAT), in addition to influencing the levels of compatible solutes including proline, glycine, betaine, and phenolic compounds.	Sarwar et al. (2021)
Raphanus sativus L. (Radish)	Salt stress	10 μM	Application of triacontanol had a stimulatory effect on all the aforementioned parameters.	Cavusoglu et al. (2008)
Triticum aestivum L. (Wheat)	Salt stress	0, 10, and 20 μM	Enhancements were observed in growth parameters, yield, activities of antioxidant enzymes, chlorophyll contents, and leaf water relations.	Perveen et al. (2014)
Mentha arvensis L. (Corn mint)	Salt stress	10^{-6} M	Enhancements were noted in growth, photosynthesis, carbonic anhydrase activity, NPK content, peltate glandular trichome density, essential oil production, menthol content, and overall yield.	Khannam and Mohammad (2018)
Helianthus annuus L. (Sunflower)	Salt stress	0,50, and 100 μM	Triacontanol enhanced shoot and root fresh weights and length, transpiration rate, water use efficiency, stomatal conductance, and assimilation rate.	Aziz et al. (2013)
Glycine max L. (Soybean)	Salt stress	10 μM	Enhanced growth was accompanied by improvements in various physio-biochemical parameters, including the content of chlorophyll pigments, nucleic acids, total soluble sugars, and proteins.	Krishnan and Kumari (2008)
Momordica charantia L. (Bitter gourd)	Salt stress	5 and 10 ppm	Enhancements were observed in the studied yield attributes.	Sureshkumar et al. (2016)
Helianthus annus L. (Sunflower)	Salt stress	10 mmol	Enhancement activities of POD, SOD, and GR enzymes, as well as the levels of free proline and GB, and total soluble protein content.	Aziz and Shahbaz (2015)

Table 1 contd. ...

... Table 1 contd.

Plant species	Abiotic stress	Triacontanol application	Plant response beneficial effect	References
Triticum aestivum L. (Wheat)	Salt stress	0, 10, and 20 μmol	Enhancement growth, leaf relative water content, membrane permeability, total free amino acids, free proline, GB, and total phenolic contents.	Perveen et al. (2012)
Oryza sativa L. (Rice)	Salt stress	10 μg/L	Plant biomass, leaf photosynthetic rate, chlorophyll content, protein content, and ion accumulation demonstrated improvement.	Chen et al. (2002)
Zea mays L. (Maize)	Salt stress	0, 2, and 5 μM	Enhanced growth, increased NR activity, elevated proline, higher levels of total phenolics and soluble proteins, elevated shoot potassium content, and reduced relative membrane permeability, as well as decreased levels of H_2O_2, MDA, and shoot sodium ion content.	Perveen et al. (2017)
Zea mays L. (Maize)	Salt stress	11.2 μM	Triacontanol mitigated the reduction in growth and increased nitrogen assimilation, as well as the activities of antioxidant enzymes, flavonoids, phenols, and proline.	Ertani et al. (2013)
Brassica juncea L. (Indian mustard)	Salt stress	150 μM	Triacontanol (150 M) and H2S (25 M), both used alone and in combination, increased plant growth and yield parameters under salt stress.	Verma et al. 2023
Capsicum annuum L. (Hot pepper)	Salt stress	50 μM and 75 μM	Triacontanol significantly enhanced plant growth, including attributes such as plant height, shoot length, leaf area, and both fresh and dry biomasses, by modulating the aforementioned physio-biochemical parameters.	Sarwar et al. (2022)
Spinacia oleracea L. (spinach)	Salt stress	25 nM and 1 μM	Triacontanol enhanced the germination energy and capacity, as well as the shoot and root biomass of young plants.	Tompa et al. (2022)

phosphorus (P), and potassium (K) contents. Moreover, the application of TC led to heightened antioxidant enzyme activities and positively influenced both yield and the quality attributes of the crop.

The mechanism underlying TC-mediated salinity tolerance in plants primarily involves the modulation of antioxidant enzyme activity. This modulation establishes a balance between the accumulation of reactive oxygen species (ROS) and their scavenging, ultimately safeguarding cell membranes from cellular injury caused by salt (Karam and Keramat 2017). To sum up, it is evidently discernible that the exogenous application of TC plays a pivotal role in promoting plant growth and ameliorating salt-induced oxidative damage. This is achieved through mechanisms such as the augmentation of compatible solute accumulation, the enhancement of antioxidant enzyme activity, ROS limitation, improved mineral nutrient acquisition, and the mitigation of lipid peroxidation.

Heavy Metal Stress

Heavy metals pose a significant environmental threat as a result of their widespread contamination of agricultural land and the environment, primarily due to their excessive release through industrialization and anthropogenic activities (Ghazaryan 2020). Metal and metalloid stress has emerged as a serious environmental concern, reducing agricultural output and posing a number of health risks to living beings (Wani et al. 2018, Ullah et al. 2018). The continual introduction of industrial waste water contaminated with metals, waste disposal, agricultural residues, and sewage contaminants has transformed once-fertile agricultural soils into barren landscapes, significantly impacting crop growth, development, and overall yield. In response to metal and metalloid-induced damage, plants possess inherent mechanisms of tolerance that enable them to withstand such stressors up to a certain threshold (Ahanger et al. 2018, Anjum et al. 2014).

Recent research has delved into the promising potential of Triacontanol (TC) in ameliorating the adverse impacts of heavy metal stress, yielding consistently favorable outcomes. For instance, Muthuchelian et al. (2001) conducted a study where foliar spray treatments of TC, at a concentration of 1 mg kg^{-1}, effectively curtailed cadmium (Cd)-induced damage in *Erythrina variegata* L. These treatments led to significant improvements in multiple parameters, including fresh weight and dry weight, photosynthetic pigments, CO_2 fixation, as well as the activities of Rubisco, Photosystem (PS-I & PS II), and nitrate reductase (NR).

Similarly, the pre-soaking of Maize grains with TC, at a concentration of 35 ppm, exhibited remarkable mitigation of Cd-induced toxicity, resulting in enhancements across various growth-related attributes (Ahmad et al. 2013). In a separate exploration, the pretreatment of coriander seedlings with TC concentrations of 5, 10, or 20 μmol L^{-1} effectively alleviated arsenic toxicity by elevating the activities of antioxidant enzymes (Karam et al. 2016). Furthermore, another investigation found that the pretreatment of coriander seedlings with 10 μM TC mitigated oxidative damage induced by arsenic, achieved by modulating the activities of oxidative markers and non-enzymatic antioxidants (Karam et al. 2017). These consistent findings substantiate the efficacy of TC in counteracting the deleterious effects of heavy metal stress, underscoring its potential as a valuable asset in strategies aimed at mitigating heavy metal stress in plants as shown in Table 2.

Furthermore, Maresca et al. (2017) have observed those foliar spray treatments with concentrations of 10 and 20 μM TC result in a substantial enhancement in Cd tolerance by efficiently controlling the actions of antioxidant enzymes, both enzymatic and non-enzymatic. Furthermore, Keramat et al. (2017) reported that

Table 2. Roles of triacontanols under heavy metal stress.

Plant species	Abiotic stress	Triacontanol application	Plant response beneficial effect	References
Brassica napus L. (Oilseed rape)	Heavy metal stress	0, 10, and 20 μM	Triacontanol resulted in enhanced elevated total chlorophyll content, shoot fresh weight, increased enzymatic and non-enzymatic antioxidant activities, and a reduction in H_2O_2 and MDA levels.	Maresca et al. (2017)
Mentha arvensis L. (Corn mint)	Heavy metal stress	1 μM	Enhanced tolerance was observed through the modulation of photosynthetic pigments, leaf carbonic anhydrase, plant biomass, osmolyte accumulation, contents, hydrogen peroxide contents, and antioxidant activities.	Zaid et al. (2020)
Triticum aestivum L. (Wheat)	Heavy metal stress (Aresnic)	1 μM	Enhanced the growth and yield, photosynthetic pigments, flavonoid content, anthocyanin content, relative water conductivity, stomatal conductance, and internal CO_2 concentration.	Ali and Perveen (2020)
Coriandrum sativum L. (Coriander)	Heavy metal stress (Arsenic)	10 and 100 μM	Triacontanol enhanced nonenzymatic antioxidant potentials, ROS markers, including EL, MDA, and H_2O_2 contents.	Asadi karam et al. (2017a)
Brassica napus L. (Oilseed rape)	Heavy metal stress (Cadmium)	10 and 20 μM	Triacontanol promoted growth, enhanced the uptake of mineral nutrients, and improved yield and quality attributes. Enhanced activities of antioxidant enzymes like ascorbate, glutathione, and phyto-chelatin.	Asadi karam et al. (2017a)
Phaseolus vulgaris L. (common bean)	Heavy metal stress (Lead)	20 μmol L^{-1}	Triacontanol increased, photosynthetic rate, transpiration rate, stomatal conductance, and mineral contents, while concurrently decreasing Pb accumulation in seedlings.	Ahmad et al. (2023)

foliar application spray with 10 M TC generates increased arsenic tolerance in Corundum. This enhancement can be attributed to TC's capability to regulate the redox conditions of plant mechanisms, primarily via their antioxidant defense route involving ascorbate and glutathione. Furthermore, Zaid et al. (2020) have also observed that foliar applications of 1 μM TC to *Mentha arvensis* L. (two cultivars of Kushal and Kosi) significantly enhance tolerance to heavy metal stress. This improvement is manifested through increased plant biomass, optimization of photosynthesis-related parameters, improved mineral metabolism, and accumulation of osmolytes. Simultaneously, the treatment leads to a reduction in oxidative stress, as evidenced by decreased levels of TBARS (Thiobarbituric acid reactive substances), while promoting the induction of antioxidant defense systems.

Water stress

Water scarcity is anticipated to become more prevalent and severe in numerous regions worldwide due to climatic shifts (Singh et al. 2014). This phenomenon will result in reduced irrigation water availability for agricultural purposes. Intercropping stands out as an effective agricultural management strategy to conserve irrigation water while minimizing adverse environmental impacts as shown in Table 3. It represents a sustainable agricultural approach geared towards enhancing resource utilization efficiency (Saad et al. 2022). Water stress is a formidable environmental constraint, impacting various growth, stomatal conductance, photosynthesis content, and the long-term viability of cellular elements are all examples of physiological and biochemical mechanisms. This is mostly due to the formation of reactive oxygen species (ROS) and phytocompounds (Hasanuzzaman et al. 2017, Farooq et al. 2009, Osakabe et al. 2014).

A comprehensive analysis of the existing research underscores the pivotal role of Triacontanol (TC) in conferring tolerance to drought, floods, and moisture stress on agricultural plants. For example, when exposed to PEG-induced drought stress, seedlings of *Triticum aestivum* L. from the WL 2265 and Sonalika cultivars, treated with a solution containing 30% TC, an aliphatic alcohol, exhibited regulated physiological changes. These changes were manifested as enhanced seedling growth, increased seed germination, soluble sugars, PEP-carboxylase activity, and accumulation of free amino acids (Thind 1991). Additionally, Muthuchelian et al. (1997) explored the growth and photosynthetic processes of water-stressed plants in different experiment Ery*thrina variegata L.* seedlings after applying 1 gm^{-3} TC to their leaves. Their experiments unveiled that TC fortified stress tolerance by shoot and root growth, leaf area index, photosynthetic components, stomatal conductance carbon dioxide fixation, and Rubisco activity. Additionally, when olive varieties were subjected to a foliar spray of 40 ppm TC, they effectively mitigated injuries induced by water stress, significantly elevating water potential and osmolyte contents (Thakur et al. 1998).

It was found that the use of a 10 g L^{-1} TC spray dramatically reduced membrane damage in the context of jack pine seedlings, thereby decreasing the negative impacts of drought stress. When TC was applied to water-stressed Cowpea seedlings, Raghava and Raghava (2010) undertook a thorough examination of the germination abilities of the seeds. Their findings revealed that even at low concentrations (0.4–0.6 mL/L), TC exhibited a significant positive impact on various germination parameters. These parameters included relative seed germination, speed, ability, length of the radical, and overall plant growth. Consequently, these improvements contributed to enhanced water stress tolerance in cowpea plants. Furthermore, research by Sanadhya et al.

Table 3. Roles of triacontanols under water stress.

Plant species	Abiotic stress	Triacontanol application	Plant response beneficial effect	References
Pyrus malus (Apple)	Water stress	10 ppm	Enhanced ion accumulation, nutrient uptake, photosynthetic efficiency, and fruit yield.	Mohd. Zubair et al. (2018)
Zea mays L. (Maize)	Water stress	0.1 mg/L	Enhanced antioxidant enzyme activity, ion buildup, and characteristics of seed quality.	Ali et al. (2017)
Oryza sativa L. (Rice)	Water stress	0, 1, 5, and 10 µmol	Enhanced activities of antioxidant enzyme, ascorbate, glutathione, and chlorophyll and sucrose contents.	Li et al. (2016a,b)
Zea mays L. (Maize)	Water stress	5 µmol	Enhanced activities of antioxidant like CAT and POD, along with higher levels of proline, total soluble proteins, and total phenolics.	Perveen et al. (2016)
Theobroma cacao L. (Cacao tree)	Water stress	0, 0.1, 0.5, 1, and 2 mL/L	Enhanced growth, yield, and physiological characteristics.	Sitinjak and Pandiangan (2015)
Juniperus procera L. (African pencil-cedar)	Water stress	200 ppm	Enhancement in growth, yield, and physiological and biochemical characteristics.	Kibatu et al. (2014)
Lupinus luteus L. (Yellow lupin)	Water stress	0.5 and 1 mg/L	Enhanced rise in the photosynthetic rate, chlorophyll content, leaf relative water content, and lateral shoot.	Borowska and Prusiński (2012)
Sesamum indicum L. (Sesame), *Arachis hypogaea* L.(peanut)	Water stress	1.0 ml/L	Triacontanol and potassium silicate have been reported to play significant roles in mitigating water stress in comparison to the control group. This impact is observed in terms of yield, water usage efficiency, land equivalent ratio, and the overall benefits.	El-Mehy et al. 2023

(2012) demonstrated that priming Mung bean seeds with TC resulted in an increased tolerance to PEG-induced drought stress. This treatment notably elevated enhanced shoot and root lengths, fresh and dry seedling weight, and germination rates.

Similarly, in a study by Suman et al. (2013), the treatments of 5 and 10 µg TC were found to be effective in mitigating PEG-induced drought toxicity in *Oryza sativa* L. (rice) seedlings. This treatment led to enhancements in seeding length and germination, as well as fresh and dry weights. Furthermore, it positively influenced the mechanisms of key enzymes (SOD, CAT, and POD) in the seedlings. Additionally, Perveen et al. (2016) conducted an investigation into the growth of maize cultivars exposed to drought stress and treated with leaf-applied TC at concentrations of 2 and 5 µM. Their results highlighted the adverse impact of drought stress on growth attributes, as well as alterations in proline, nitrate reductase (NR) activity, total phenolics, and levels of soluble proteins. However, the application of 5 µM TC effectively modulated these parameters, ultimately enhancing drought tolerance in wheat cultivars.

Drought Stress

Drought stress stands as a prominent environmental constraint, exerting significant adverse effects on the growth and yield, quality and quantity, and nutritional attributes of a wide array of oilseed crops as shown in Table 4. These oilseed crops encompass sesame (Ozkan and Kulak 2013), peppermint (Khorasaninejad et al. 2011), canola (Raza et al. 2015, Akram et al. 2017), groundnut (Karimian et al. 2015), sunflower (Eslami et al. 2015), maize (Ali et al. 2013), chamomile (Farhoudi et al. 2014), oilseed rape (Raza et al. 2017), carthamus (Nazari et al. 2017), and soybean (Shukla et al. 2017).

Drought stress exerts a disruptive influence on many physiological functions, involving membrane integrity disruption (Zang et al. 2019) and enzyme activity mechanisms (Kong et al. 2021), ultimately resulting in the inhibition of plant growth. It is noteworthy that while drought experienced during the vegetative stage of growth can reduce plant biomass, water deficiency during the flowering stage has a considerably more profound impact on yield (Sadiq et al. 2017). The buildup of osmolytes that are compatible in response to abiotic stressors represents a common adaptive strategy employed by plants (Elewa et al. 2017). Additionally, the presence of a robust antioxidant defense system is indicative of a plant's tolerance to abiotic stresses (Khan et al. 2017).

Drought stress induces physio-chemical and morphological alterations in plants, detrimentally affecting agricultural production in arid and semiarid regions. In light of the escalating global water scarcity threat and rapid population growth, it becomes imperative to devise strategies that can meet the burgeoning worldwide need for food (Ijaz et al. 2019). One of the pivotal responses triggered by drought stress, chiefly mediated by phytohormones and their intricate signaling networks, is the enhanced production of phytocompound (Chen et al. 2021). The judicious application of plant growth regulators, such as triacontanol (TC), in minute quantities has emerged as an effective approach for augmenting plant growth and crop yields (Pang et al. 2020). TC, extensively documented for its pivotal role in plant responses to abiotic stresses like drought (Alharbi et al. 2021, El-Beltagi et al. 2022), effectively mitigates the stress-induced effects in plants. This mitigation is achieved through several mechanisms, including the enhancement of plant biomass, the accumulation of chlorophyll pigments, improved acquisition of mineral nutrients, augmented accumulation of compatible solutes, and the fortification of both enzymatical and non-enzymatical antioxidant defense systems (Perveen et al. 2016). Notably, the foliar spray of TC has been demonstrated to elicit substantial enhancements in plant height, leaf area, and biomass in hot pepper plants (Sarwar et al. 2022). Moreover, research findings consistently report that foliar application of TC ameliorates the detrimental impacts of abiotic stresses on wheat (Perveen et al. 2014), *Zea mays* L. (maize) (Perveen et al. 2016), and *Oryza sativa* L. (rice) (Alharbi et al. 2021).

Chilling and Heat Stress

Climate change brings about fluctuations in temperature patterns, including episodes of chilling, freezing, and heat stress as shown in Table 5. These temperature variations

Table 4. Roles of triacontanols under drought stress.

Plant species	Abiotic stress	Triacontanol application	Plant response beneficial effect	References
Zea mays L. (Maize)	Drought stress	0, 2, and 5 µM	Triacontanol was applied topically to increase chlorophyll content, phenolic content, proline, antioxidant activity like CAT, and while decreasing oxidative stress indicators including hydrogen peroxide and malondialdehyde content.	Perveen et al. (2016)
Linum usitatissimum L. (Flaxseed)	Drought stress	1.0 µM	Triacontanol enhanced in both growth and yield, chlorophyll a and b contents, along with enhanced activities of key antioxidant enzymes such as SOD, and CAT. Moreover, total phenolics, total soluble proteins, anthocyanin contents, were all significantly elevated.	Perveen et al. (2022)
Fragaria x *ananassa* (Strawberry)	Drought stress	1 ppm	Triacontanol led to a substantial enhanced in plant growth and yield, antioxidant activities like, CAT, POX, and SOD. Significantly elevated chlorophyll contents, improved gas exchange, and enhanced water use efficiency.	El-Beltagi et al. (2022)
Brassica juncea L. (Indian Mustard)	Drought stress	20 µM	Triacontanol application resulted in enhanced levels of total soluble sugars, and reduced glutathione, and proline content in stressed plants. Furthermore, the activities of key antioxidant enzymes, including APX, GR, SOD, CAT, and PAL.	Ahmad et al. (2021)
Oryza sativa L. (Rice)	Drought stress	35 ppm	Triacontanol exhibited the capacity to enhance both plant growth and the plant's ability to tolerate drought stress. This enhancement was attributed to Triacontanol's role in regulating stomatal conductance, thereby influencing stomatal closure.	Alharbi et al. (2021)

are well-documented for their capacity to disrupt crucial physiological and metabolic processes in plants. These processes encompass a wide range of physiological and metabolic responses within plants, including seed germination, transpiration, protein functionality, photosynthesis, seedling growth, enzyme inactivation, the accumulation of reactive oxygen species (ROS), alterations in membrane integrity, and tissue damage. Ultimately, these adverse effects can lead to plant mortality (Nahar et al. 2015, Ahanger et al. 2018, Hasanuzzaman et al. 2017). In response to such challenging environmental circumstances, triacontanol (TC) emerges as a stress shields, playing a significant role in enhancing plant resilience (Naeem et al. 2012).

For instance, study by Cavusoglu and Kabar (2007) conducted research demonstrating that the exogenous application of 10 µM TC effectively mitigated the detrimental effects of high-temperature stress on both radish germination and the fresh mass of barley. Furthermore, Borowski and Blamowski (2009) expanded on this understanding by investigating the impact of TC on plants subjected to chilling

Table 5. Roles of triacontanols under different abiotic stresses (heat, chilling, and transplant shock).

Plant species	Abiotic stress	Triacontanol application	Plant response beneficial effect	References
Vigna radiata L. (Mung bean)	Heat stress	11 µM	Advantageous effects of triacontanol, both in the presence and absence of heat stress, and regulation of defense hormone levels on mung-bean plants.	Waqas et al. (2016)
Cucumis sativus L. (Cucumber)	Chilling stress	0, 0.01, and 0.10 mg dm^{-3}	Triacontanol effectively modulated, photosynthetic pigments (Chl a and b) and stomatal conductance, transpiration rate, antioxidants activities like POD, and CAT, while reducing EL and proline content.	Borowski (2009)
Oryza sativa L. (Rice)	Transplant shock	10 µM	Triacontanol enhanced photosynthetic pigments chlorophyll and sucrose content. Furthermore, it significantly enhanced the antioxidants activities like CAT, POD emphasizing the significant role of POD in effectively scavenging H$_2$O$_2$ during the recovery time.	Li et al. (2016)
Brassica oleracea L. (Cabbage)	Transplant shock	10 µM	Enhance morphological characteristics, uptake of mineral content along with biomass production. Triacontanol enhanced plant growth and yield, morphological properties, mineral content, and absorption.	Ahmad et al. (2023)

stress, specifically Tusli. Their findings highlighted TC's pronounced ameliorating effect, evident through improvements in plant fresh and dry weight photosynthetic pigments, photosynthetic rate (PN), transpiration rate (E), quantum efficiency, and stomatal conductance (gs). Building upon this, Borowski (2009) extended the research to cucumber seedlings exposed to chilling stress, revealing TC's regulatory influence on leaf chlorophyll content, PN, gs, E rate, leaf proline content, catalase (CAT), peroxidase (POD), and electrolyte leakage (EL) activities as a result of instant chilling stress condition in cucumber.

Furthermore, Waqas et al. (2016) added to our knowledge of TC's significance by revealing that foliar-applied TC improved heat stress response. This improvement was demonstrated by its ability to boost growth, as well as changes in the endogenous levels of ABA and JA, amino acids, and nutritional content in Mung bean. When considering the collective body of research in this area, it becomes evident that TC serves as a valuable tool in mitigating temperature-induced stress in plants. It achieves this by orchestrating various physiochemical processes and influencing the concentrations of antioxidants, osmo-protectants, and phytohormones.

Transplant Shock

Transplant shock is a common occurrence in agricultural practices and is often characterized by a temporary stagnation in plant growth and developmental processes. However, the application of triacontanol (TC), known for its stress-modulating properties, has shown promise in alleviating the negative consequences associated with transplant shock in plants. For instance, Li et al. (2016a) conducted research demonstrating that a foliar spray treatment with 10 µM TC effectively mitigated transplant shock inhibits development and causes oxidative damage (ROS) in rice seedlings. This ameliorative effect was attributed to the significant enhancement of sucrose content, catalase (CAT) activities, and the optimization of the redox state of peroxidase (POD), ascorbic acid (ASA), and reduced glutathione (GSH). Transplant shock, characterized by disruptions in multiple metabolic processes within transplanted seedlings stemming from water content changes and root trimming (Sasaki and Gotoh 1999), leads to a transient hindrance in seedling growth and development, widely recognized as a consequence of this phenomenon. Recognizing the multifaceted roles of TC in plants and the detrimental impact of transplant shock on seedlings, it has been postulated that the foliar application of TC before mechanical transplanting could attenuate the adverse effects associated with this phenomenon. This study aims to investigate the role of TC and its potential effects on enhancing the yield of machine-transplanted *Oryza sativa* (rice).

Recognizing the multifaceted functions of TC in plants as well as the detrimental effects of transplant shock on seedlings, it has been postulated that foliar delivery of TC prior to mechanically transplanting might mitigate the negative consequences associated with this phenomenon. The purpose of this study is to look at the significance of TC and its possible impacts on improving the yield of machine-transplanted *Oryza sativa* (rice).

Conclusion

The application of exogenous Triacontanol (TC) holds significant promise in augmenting plant stress resilience, growth rates, and overall productivity. TC acts as a catalyst, prompting the release of non-racemic L (+)-adenosine, also known as TRIM, a secondary messenger predominantly found in the tonoplast. This, in turn, initiates the influx of vital ions, including calcium (Ca^{++}), magnesium (Mg^{++}), and potassium (K^+), by facilitating the opening of channels within the plasma membrane. Elevated levels of calcium engage with Calmodulin proteins, activating key transcription factors such as CAMTA3, MYB2, and GTL. These activated factors subsequently amplify the activities of genes and enzymes involved in anabolism, leading to marked improvements in photosynthesis and overall productivity. Furthermore, the application of exogenous TC has been observed to significantly bolster a plant's ability to withstand abiotic stresses. This is achieved through various mechanisms, including the upregulation of antioxidant enzyme activity, the reduction of lipid peroxidation, and the mitigation of reactive oxygen species (ROS) production. TC application also fosters the accumulation of osmolytes within plant cells, reduces electrolytic leakage, and helps maintain optimal osmotic balance.

Additionally, TC has demonstrated its effectiveness in enhancing yield attributes, elevating yield quality, and promoting biofuel synthesis, both under standard growth conditions and during exposure to stressful environments. While these findings are promising, further research endeavors are imperative to unveil the intricate aspects of TC biosynthesis, delineate the receptors responsible for TC signaling, elucidate TC and TRIM's roles in regulating plant metabolism, identify the rate-determining steps within these processes, and explore the involvement of additional transcription factors.

References

Aboutalebian, M.A. and Nazari, S. 2017. Seedling emergence and activity of some antioxidant enzymes of canola (Brassica napus L.) can be increased by seed priming. J. Agri. Sci., 155: 1541–1552.

Aftab, T., Khan, M.M.A., Idrees, M., Naeem, M., Singh, M. and Ram, M. 2010. Stimulation of crop productivity, photosynthesis, and artemisinin production in *Artemisia annua* L. by triacontanol and gibberellic acid application. J. Plant Interac., 5: 273–281.

Ahanger, M.A., Ashraf, M., Bajguz, A. and Ahmad, P. 2018. Brassinosteroids regulate growth in plants under stressful environments and crosstalk with other potential phytohormones. J. Plant Grow. Reg., 37: 1007–1024.

Ahanger, M.A., Tomar, N.S., Tittal, M., Argal, S. and Agarwal, R.M. 2017. Plant growth under water/salt stress: ROS production; antioxidants and significance of added potassium under such conditions. Physiol. Mole. Bio. Plants., 23: 731–744.

Ahmad, H.F.S., Hassan, H.M. and El-Shafey, A.S. 2013. Effect of cadmium on growth, flowering, and fruiting of Maize (Zea mays L.) and possible roles of triacontanol in alleviating cadmium toxicity. Egy. J. Bot., 53: 23–44.

Ahmad, I., Zhu, G., Zhou, G., Song, X., Hussein Ibrahim, M.E. and Ibrahim Salih, E.G. 2022. Effect of N on growth, antioxidant capacity, and chlorophyll content of sorghum. Agron. 12: 501.

Ahmad, J., Ali, A.A., Al-Huqail, A.A. and Qureshi, M.I. 2021. Triacontanol attenuates drought-induced oxidative stress in Brassica juncea L. by regulating lignification genes, calcium metabolism, and the antioxidant system. Plant Physiol. Bio., 166: 985–998.

Ahmed, S., Ahmad, M., Sardar, R. and Ismail, M.A. 2023. Triacontanol priming as a smart strategy to attenuate lead toxicity in Brassica oleracea L. Intern. J. Phyto., 25(9): 1173–1188.

Ahmed, S., Amjad, M., Sardar, R., Siddiqui, M. H., and Irfan, M. 2023. Seed priming with triacontanol alleviates lead stress in Phaseolus vulgaris L. (Common Bean) through improving nutritional orchestration and morpho-physiological characteristics. Plants., 12(8): 1672.

Akram, N. A., Iqbal, M., Muhammad, A., Ashraf, F., Al Qurainy, F. and Shafiq, S. 2017. Aminolevulinic acid and nitric oxide regulate oxidative defense and secondary metabolisms in canola (*Brassica napus* L.) under drought stress. Protopl., 12(1–2): 1–2.

Alharbi, B.M., Abdulmajeed, A.M. and Hassan, H. 2021. Biochemical and molecular effects induced by triacontanol in acquired tolerance of rice to drought stress. Genes., 12(8): 1119.

Ali, H.M.M., Khan, Z.H. and Afzal, I. 2017. Exogenous application of growth-promoting substances improves growth, yield, and quality of spring maize (*Zea mays* L.) hybrids under late-sown conditions. Bullet. Bio. Allie. Sci. Res., 2(3): 1–9.

Ali, H.M.M. and Perveen, S. 2020. Effect of foliar-applied triacontanol on wheat (*Triticum aestivum* L.) under arsenic stress: A study of changes in growth, yield, and photosynthetic characteristics. Physiol. Mole. Bio. Plants., 26: 1215–1224.

Ali, Q., Anwar, F., Ashraf, M., Saari, N. and Perveen, R. 2013. Ameliorating effects of exogenously applied proline on seed composition, seed oil quality, and oil antioxidant activity of maize (Zea mays L.) under drought stress. Int. J Mole. Sci., 14: 818–835.

Anjum, N.A., Gill, S.S., Gill, R., Hasanuzzaman, M., Duarte, A.C., Pereira, E. et al. 2014. Metal/metalloid stress tolerance in plants: Role of ascorbate, its redox couple, and associated enzymes. Protopl., 251: 1265–1283.

Asadi Karam, E., Keramat, B., Asrar, Z. and Mozafari, H. 2017a. Study of the interaction effect between triacontanol and nitric oxide on alleviating oxidative stress arsenic toxicity in coriander seedlings. J. Plant Inter., 12: 14–20.

Asadi Karam, E., Maresca, V., Sorbo, S., Keramat, B. and Basile, A. 2017b. Effects of triacontanol on the ascorbate- glutathione cycle in Brassica napus L. exposed to cadmium-induced oxidative stress. Ecotox. Environ. Safety., 144: 268–274.

Aymen, E.M. 2018. Seed priming with plant growth regulators to improve crop abiotic stress tolerance. pp. 95–106. In Advan. in Seed Priming. Springer.

Aziz, R. and Shahbaz, M. 2015. Triacontanol-induced regulation in the key osmo-protectants and oxidative defense system of sunflower plants at various growth stages under salt stress. Int. J. Agri. Biol., 17(5): 881–890.

Aziz, R., Shahbaz, M., and Ashraf, M. 2013. Influence of foliar application of triacontanol on growth attributes, gas exchange, and chlorophyll fluorescence in sunflower (*Helianthus annuus* L.) under saline stress. Paki. J. Bot., 45(6): 1913–1918.

Basu, S., Roychoudhury, A., Saha, P.P. and Sengupta, D.N. 2010. Differential antioxidative responses of indica rice cultivars to drought stress. Plant Grow. Reg., 60(1): 51–59.

Borowska, M. and Prusiński, J. 2012. Effect of triacontanol on the productivity of yellow lupin (*Lupinus luteus* L.) plants. J. Cent. Euro. Agri., 12(4): 673–683.

Borowski, E. 2009. Response to chilling in cucumber (*Cucumis sativus* L.) plants treated with triacontanol and Asahi SL. Acta Agrobot., 62: 165–172.

Borowski, E. and Blamowski, Z.K. 2009. The effects of triacontanol 'TRIA' and Asahi SL on the development and metabolic activity of sweet basil (Ocimum basilicum L.) plants treated with chilling. Folia Horti. 21: 39–48.

Bose, B., Kumar, M., Singhal, R.K. and Mondal, S. 2018. Impact of seed priming on the modulation of physico-chemical and molecular processes during germination, growth, and development of crops. pp. 23–40. In Advances in Seed Priming. Springer.

Cavusoglu, K. and Kabar, K. 2007. Comparative effects of some plant growth regulators on the germination of barley and radish seeds under high-temperature stress. Eura. J Bio., 1: 1–10.

Çavuşoğlu, K., Kiliç, S. and Kabar, K. 2008. Effects of some plant growth regulators on leaf anatomy of radish seedlings grown under saline conditions. J. App. Biol. Sci., 2(2): 47–50.

Chen, X., Yuan, H., Chen, R., Zhu, L., Du, B., Weng, Q. et al. 2002. Isolation and characterization of triacontanol-regulated genes in rice (Oryza sativa L.): Possible role of triacontanol as a plant growth stimulator. Plant Cell Physiol., 43: 869–876.

Chen, X., Yuan, H., Chen, R., Zhu, L. and He, G. 2003. Biochemical and photochemical changes in response to triacontanol in rice (Oryza sativa L.). Plant Grow. Reg., 40: 249–256.

Chen, Y., Chen, Y., Guo, Q., Zhu, G., Wang, C. and Liu, Z. 2021. Effects of drought stress on the growth, physiology, and secondary metabolite production in *Pinellia ternata* Thunb. Pak. J. Bot., 53(3): 833-840.

dos Santos, T.B., Ribas, A.F., de Souza, S.G.H., Budzinski, I.G.F. and Domingues, D.S. 2022. Physiological responses to drought, salinity, and heat stress in plants: A review. Stresses, 2: 113–135.

Dresselhaus, T. and Hückelhoven, R. 2018. Biotic and abiotic stress responses in crop plants. Agron., 8: 267.

Egamberdieva, D., Wirth, S.J., Alqarawi, A.A., Abd_Allah, E.F. and Hashem, A. 2017. Phytohormones and beneficial microbes: Essential components for plants to balance stress and fitness. Fronti. Micro. 8: 2104.

El-Beltagi, H.S., Ismail, S.A., Ibrahim, N.M., Shehata, W.F., Alkhateeb, A.A., Ghazzawy, H.S. et al. 2022. Unraveling the effect of triacontanol in combating drought stress by improving growth, productivity, and physiological performance in strawberry plants. Plants., 11(15): 1913.

Elewa, T.A., Sadak, M.S. and Saad, A.M. 2017. Proline treatment improves physiological responses in quinoa plants under drought stress. Bio. Res., 14: 21–33.

El-Mehy, A.A., Abd-Allah, A.A., Kasem, E.E. and Mohamed, M.S. 2023. Mitigation of the impact of water deficiency on intercropped sesame and peanut systems through foliar application of potassium-silicate and triacontanol. Egy. J. Agri. Res., 101(2): 292–303.

Ertani, A., Schiavon, M., Muscolo, A. and Nardi, S. 2013. Alfalfa plant-derived bio-stimulant stimulates short-term growth of salt-stressed Maize (Zea mays L.) plants. Plant Soil., 364(1–2): 145–158.

Eslami, M. 2015. The effect of drought stress on oil percent and yield and the type of sunflower (*Helianthus annuus* L.) fatty acids. Agri. Sci. Develop., 4: 1–4.

FAO. 2017. Agriculture Organization of the United Nations. The future of food and agriculture: Trends and challenges. Rome: FAO. doi:10.1093/jxb/erx011.

Farhoudi, R., Lee, D.J. and Hussain, M. 2014. Mild drought improves growth and flower oil productivity of German chamomile (*Matricaria recutita* L.). J Ess. Oil-Bear. Plants. 17(1).

Farooq, M., Wahid, A., Lee, D.J., Ito, O. and Siddique, K.H. 2009. Advances in drought resistance of rice. Criti. Revi. Plant Sci., 28: 199–217.

Feng, D., Gao, Q., Liu, J., Tang, J., Hua, Z. and Sun, X. 2023. Categories of exogenous substances and their effect on alleviation of plant salt stress. Euro. J Agro., 142: 126656. doi:10.1016/j.eja.2022.126656.

Ghazaryan, K., Movsesyan, H., Gevorgyan, A., Minkina, T., Sushkova, S., Rajput, V. et al. 2020. Comparative hydrochemical assessment of groundwater quality from different aquifers for irrigation purposes using IWQI: A case-study from Masis province in Armenia. Groundwater Sust. Devel. 11, 100459. doi:10.1016/j.gsd.2020.100459.

Hasanuzzaman, M., Nahar, K., Bhuiyan, T.F., Anee, T.I., Inafuku, M., Oku, H. et al. 2017. Salicylic acid: An all-rounder in regulating abiotic stress responses in plants. pp. 31–75. In: El-Esawi, M.A. (Ed.). Phytohormones-signaling mechanisms and crosstalk in plant development and stress responses. InTech, Croatia. https://doi.org/10.5772/intechopen.68213.

Ijaz, F., Riaz, U., Iqbal, S., Zaman, Q.U., Ijaz, M.F., Javed, H. et al. 2019. Potential of rhizobium and PGPR to enhance growth and fodder yield of berseem (*Trifolium alexandrinum* L.) in the presence and absence of tryptamine. Pak. J Agri. Res., 32(2): 398–406.

Islam, S., Zaid, A. and Mohammad, F. 2020. Role of triacontanol in counteracting the ill effects of salinity in plants: A review. J. Plant Grow. Reg. https://doi.org/10.1007/s00344-020-10064-w

Islam, S., Zaid, A. and Mohammad, F. 2021. Role of Triacontanol in Counteracting the Ill Effects of Salinity in Plants: A Review. J. Plant Grow. Reg., 40: 1–10.

Jisha, K.C., Vijayakumari, K. and Puthur, J.T. 2013. Seed priming for abiotic stress tolerance: An overview. Acta Physiol. Plant., 35: 1381–1396.

Karam, E.A. and Keramat, B. 2017. Foliar spray of triacontanol improves growth by alleviating oxidative damage in coriander under salinity. Indi. J. Plant Physiol., 22: 120–124.

Karam, E.A., Keramat, B., Asrar, Z. and Mozafari, H. 2016. Triacontanol induced changes in growth, oxidative defense system in Coriander (Coriandrum sativum) under arsenic toxicity. Ind. J. Plant Physiol., 21: 137–142.

Karam, E.A., Keramat, B., Asrar, Z. and Mozafari, H. 2017. Study of interaction effect between triacontanol and nitric oxide on alleviating of oxidative stress arsenic toxicity in coriander seedlings. J Plant Interact., 12: 14–20.

Karimian, M.A., Dahmardeh, M., Bidarnamani, F. and Forouzandeh, M. 2015. Assessment of quantitative and qualitative factors of peanut (*Arachis hypogaea* L.) under drought stress and salicylic acid treatments. Biological Forum—An Internat. J., 7(1): 871–878.

Keramat, B., Sorbo, S., Maresca, V., Asrar, Z., Mozafari, H. and Basile, A. 2017. Interaction of triacontanol and arsenic on the ascorbate-glutathione cycle and their effects on the ultrastructure in *Coriandrum sativum* L. Env. Exp. Bot., 141: 161–169.

Khan, A., Anwar, Y., Hasan, M.M., Iqbal, A., Ali, M., Alharby, H.F. et al. 2017. Attenuation of drought stress in Brassica seedlings with exogenous application of Ca^{2+} and H_2O_2. Plants., 13: 20.

Khanam, D. and Mohammad, F. 2018. Plant growth regulators ameliorate the ill effect of salt stress through improved growth, photosynthesis, antioxidant system, yield, and quality attributes in *Mentha piperita* L. Acta Physiol. Planta., 40(11): 1–13.

Khorasaninejad, S., Mousavi, A., Soltanloo, H., Khodayar, H.K. and Khalighi, A. 2011. The effect of drought stress on growth parameters, essential oil yield, and constituent of peppermint (*Mentha piperita* L.). J. Medi. Plants Res., 5(22): 5360–5365.

Kibatu, T., Mamo, D. and Getachew, G. 2014. Effect of Alfalfa (*Medicago sativa*) organic extract on the growth of pencil Cedar (*Juniperus procera*) seedlings. Asian J. Plant Sci. Res., 4(6): 47–51.

Kong, H., Zhang, Z., Qin, J., and Akram, N. A. 2021. Interactive effects of abscisic acid (ABA) and drought stress on the physiological responses of winter wheat (*Triticum aestivum* L.). Pak. J Bot. 53(5): 1545–1551.

Krishnan, R. R., and Kumari, B. D. 2008. Effect of N-triacontanol on the growth of salt-stressed soybean plants. J Biosci. 19(2): 53–62.

Li, X., Zhong, Q., Li, Y., Li, G., Ding, Y., Wang, S., and Chen, L. 2016a. Triacontanol reduces transplanting shock in machine-transplanted rice by improving the growth and antioxidant systems. Front. Plant Sci. 7: 872.

López-Marqués, R. L., Nørrevang, A. F., Ache, P., Moog, M., Visintainer, D., Wendt, T., et al. 2020. Prospects for the accelerated improvement of the resilient crop quinoa. J Exp. Bot. 71: 5333–5347. doi: 10.1093/jxb/eraa285.

Maresca, V., Sorbo, S., Keramat, B., and Basile, A. 2017. Effects of triacontanol on ascorbate- glutathione cycle in Brassica napus L. exposed to cadmium-induced oxidative stress. Eco. Env. Saf. 144, 268–274.

Mittler, R. 2022. ROS and Redox Signaling in Cell-to-Cell and Systemic Responses of Plants. Free Radi. Bio. Medi. 189, 1.

Mohd. Zubair, Hussain, S. S., Munib-U-Rehman, and Baba, J. Ah. 2018. Influence of solubor, biozyme, and triacontanol on leaf and fruit nutrient content of apple cv. Red delicious. Inter. J Agri. Sci. 14(1), 85–91.

Muthuchelian, K., Bertamini, M., and Nedunchezhian, N. 2001. Triacontanol can protect *Erythrina variegata* from cadmium toxicity. J Plant Physiol. 158, 1487–1490.

Muthuchelian, K., Murugan, C., Nedunchezhian, N., and Kulandaivelu, G. 1997. Photosynthesis and growth of *Erythrina variegata* as affected by water stress and triacontanol. Photosynthetica. 33, 241–248.

Naeem, M., Khan, M. M. A., and Moinuddin. 2012. Triacontanol: A potent plant growth regulator in agriculture. J Plant Inter. 7: 129–142.

Nahar, K., Hasanuzzaman, M., Ahamed, K. U., Hakeem, K. R., Ozturk, M., and Fujita, M. 2015. Plant responses and tolerance to high-temperature stress: Role of exogenous phyto-protectants. In K. R. Hakeem (Ed.), Crop Production and Global Environmental Issues (pp. 385–435). Springer.

Nazari, M., Mirlohi, A., and Majidi, M. M. 2017. Effects of drought stress on oil characteristics of *Carthamus* species. J Ameri. Oil Chem. Soc. 94: 247–256.

Negi, P., Pandey, M., Dorn, K. M., Nikam, A. A., Devarumath, R. M., Srivastava, A. K., et al. 2020. Transcriptional reprogramming and enhanced photosynthesis drive inducible salt tolerance in sugarcane mutant line M4209. J Exp. Bot. 71, 6159–6173. doi: 10.1093/jxb/eraa339.

Olsson, R. A., and Pearson, J. D. 1990. Cardiovascular purinoceptors. Physiol. Reviews. 70: 761–845.

Ozkan, A., and Kulak, M. 2013. Effects of water stress on growth, oil yield, fatty acid composition, and mineral content of *Sesamum indicum*. J Ani. Plant Sci. 23(6): 1686–1690.

Pang, Q., Chen, X., Lv, J., Li, T., Fang, J., and Jia, H. 2020. Triacontanol promotes the fruit development and retards fruit senescence in strawberry: A Transcriptome Analysis. Plants. 9: 488.

Per, T. S., Khan, M. I. R., Anjum, N. A., Masood, A., Hussain, S. J., and Khan, N. A. 2018. Jasmonates in plants under abiotic stresses: Crosstalk with other phytohormones matters. Env. Exp. Bot. 145: 104–120.

Perveen, S., Iqbal, M., Nawaz, A., Parveen, A., and Mahmood, S. 2016. Induction of drought tolerance in *Zea mays* L. by foliar application of triacontanol. Pak. J Bot. 48(3): 907–915.

Perveen, S., Iqbal, M., Parveen, A., Akram, M. S., Shahbaz, M., Akber, S., and Mehboob, A. 2017. Exogenous triacontanol-mediated increase in phenolics, proline, activity of nitrate reductase, and shoot K+ confers salt tolerance in maize (*Zea mays* L.). Braz. J Bot. 40: 1–11.

Perveen, S., Shahbaz, M., and Ashraf, M. 2012. Is pre-sowing seed treatment with triacontanol effective in improving some physiological and biochemical attributes of wheat (*Triticum aestivum* L.) under salt stress? J App. Bot. Food Qual. 85(1): 41–48.

Perveen, S., Shahbaz, M., and Ashraf, M. 2013. Influence of foliar-applied triacontanol on growth, gas exchange characteristics, and chlorophyll fluorescence at different growth stages in wheat under saline conditions. Photosynt. 51: 541–551.

Perveen, S., Parvaiz, M., Shahbaz, M., Saeed, M. and Zafar, S. 2022. Triacontanol positively influences growth, yield, biochemical attributes, and antioxidant enzymes of two linseed (*Linum usitatissimum* L.) accessions differing in drought tolerance. Pak. J. Bot.. 54(3): 843–853.

Perveen, S., Shahbaz, M. and Ashraf, M. 2014. Triacontanol-induced changes in growth, yield, leaf water relations, oxidative defense system, minerals, and some key osmo-protectants in *Triticum aestivum* under saline conditions. Turk. J. Bot., 38(5): 896–913.

Perveen, S., Shahbaz, M. and Ashraf, M. 2013. Influence of foliar-applied triacontanol on growth, gas exchange characteristics, and chlorophyll fluorescence at different growth stages in wheat under saline conditions. Photosynt., 51: 541–551.

Raghava, N. and Raghava, R.P. 2010. Effect of miraculan on seed germination parameters in cowpea under water stress. Biosci. Biotech. Res., 7: 353–358.

Rajasekaran, L.R. and Blake, T.J. 1999. New plant growth regulators protect photosynthesis and enhance growth under drought of jack pine seedlings. J Plant Grow. Reg., 18: 175–181.

Rajput, V.D., Minkina, T., Kumari, A., Harish, Singh, V.K., Verma, K.K. et al. 2021. Coping with the Challenges of Abiotic Stress in Plants: New Dimensions in the Field Application of Nanoparticles. Plants., 10: 1221.

Raza, A., Tabassum, J., Fakhar, A. Z., Sharif, R., Chen, H., Zhang, C. et al. 2022. Smart reprogramming of plants against salinity stress using modern biotechnological tools. Critical Reviews in Biotech. doi:10.1080/07388551.2022.2093695

Raza, M.A., Shahid, A.M., Saleem, M.F., Khan, I.H., Ahmad, S., Ali, M. et al. 2017. Effects and management strategies to mitigate drought stress in oilseed rape (*Brassica napus* L.): A review. Zemdirbyste, 104(1): 85–94.

Raza, M.A.S., Shahid, A.M., Ijaz, M., Khan, I.H., Saleem, M.F. and Ahmad, S. 2015. Studies on canola (*Brassica napus* L.) and camelina (*Camelina sativa* L.) under different irrigation levels. J. Agri. Bio. Sci., 10(4): 130–138.

Ries, S. 1991. Triacontanol and its second messenger 9-b-L (+)-adenosine as plant growth substances. Plant Physiol., 95: 986–989.

Ries, S., Savithiry, S., Wert, V. and Widders, I. 1993. Rapid induction of ion pulses in tomato, cucumber, and maize plants following a foliar application of L (+)-adenosine. Plant Physiol., 101: 49–55.

Ries, S. and Wert, V. 1992. Response of maize and rice to 9-beta- (+) adenosine applied under different environmental conditions. Plant Grow. Reg., 11: 69–74.

Ries, S., Wert, V., O'Leary, N.F.D. and Nair, M. 1990. 9-b-L (+) Adenosine: a new naturally occurring plant growth substance elicited by triacontanol in rice. Plant Grow. Reg., 9: 263–273.

Ries, S.K., Wert, V., Sweeley, C.C. and Leavitt, R.A. 1977. Triacontanol: a new naturally occurring plant growth regulator. Science, 195(4284): 1339–1341.

Roychoudhury, A. and Chakraborty, M. 2013. Biochemical and molecular basis of varietal difference in plant salt tolerance. Ann. Res. and Revi. Bio., 422–454.

Roychoudhury, A., Basu, S. and Sengupta, D.N. 2009. Effects of exogenous abscisic acid on some physiological responses in a popular aromatic indica rice compared with those from two traditional nonaromatic indica rice cultivars. Acta Physiol. Planta., 31(5): 915–926.

Saad, A.H., El Naim, A.M., Ahmed, A.A., Ibrahim, K.A., Islam, M.S., Al-Qthanin, R.N. et al. 2022. Response of sesame to intercropping with groundnut and cowpea. Comm. Soil Sci. Plant Anal., 1–12.

Sadiq, M., Akram, N. and Ashraf, M. 2017. Foliar applications of alpha-tocopherol improve the composition of fresh pods of *Vigna radiata* subjected to water deficiency. Turk. J Bot., 41: 244–252.

Sanadhya, D., Kathuria, E. and Malik, C.P. 2012. Effect of drought stress and its interaction with two phytohormones on *Vigna radiata* seed germination and seedling growth. Inter. J Life Sci., 1: 201–2017.

Sarwar, M., Anjum, S., Alam, M.W., Ali, Q., Ayyub, C.M., Haider, M.S. et al. 2022. Triacontanol regulates morphological traits and enzymatic activities of salinity-affected hot pepper plants. Sci. Rep., 12(1): 1–8.

Sarwar, M., Anjum, S., Ali, Q., Alam, M.W., Haider, M.S. and Mehboob, W. 2021. Triacontanol modulates salt stress tolerance in cucumbers by altering the physiological and biochemical status of plant cells. Sci. Rep., 11(1): 1–10.

Sasaki, R. and Gotoh, K. 1999. Characteristics of rooting and early growth of transplanted rice nursling seedlings of different ages with different leaf numbers. Japa. J. Crop Sci., 68: 194–198. doi:10.1626/jcs.68.194.

Shukla, P.S., Shotton, K., Norman, E., Neily, W., Critchley, A.T. and Prithiviraj, B. 2017. Seaweed extracts improve the drought tolerance of soybeans by regulating stress-response genes. AoB Plants., 10(1).

Singh, M., Khan, M.M. and Moinuddin, N.M. 2012. Augmentation of nutraceuticals, productivity, and quality of ginger (Zingiber officinale Rosc.) through triacontanol application. Plant Biosystems— An Intern. J Deal. Asp. Plant Bio., 146: 106–113.

Singh, V.P., Mishra, A.K., Chowdhary, H. and Khedun, C.P. 2014. Climate change and its impact on water resources. pp. 525–569. In Modern Water Resources Engineering. Humana Press, Totowa, NJ.

Sitinjak, R.R. and Pandiangan, D. 2015. The effect of plant growth regulator triacontanol on the growth of cacao seedlings (*Theobroma cacao* L.). AGRIVITA J Agri. Sci., 36(3): 260–267.

Springmann, M., Clark, M., Mason-D'croz, D., Wiebe, K., Bodirsky, B. L., Lassaletta, L., et al. 2018. Options for keeping the food system within environmental limits. Nature. 562, 519–525. doi: 10.1038/s41586-018-0594-0.

Suman, K., Kondamudi, R., Rao, Y.V., Kiran, T.V., Swamy, K.N., Rao, P.R. et al. 2013. Effect of triacontanol on seed germination, seedling growth, and antioxidant enzyme in rice under polyethylene glycol-induced drought stress. Andh. Agri. J., 60: 132–137.

Sureshkumar, R., Karuppaiah, P., Rajkumar, M. and Sendhilnathan, R. 2016. Influence of plant growth regulators on certain yield and quality attributes of bitter gourd (*Momordica charantia* L.) in the rice fallow of Cauvery delta region. Inter. J Cur. Res., 8(5): 30293–30295.

Thakur, A., Thakur, P.S. and Singh, R.P. 1998. Influence of paclobutrazol and triacontanol on growth and water relations in olive varieties under water stress. Indian Journal of Plant Physiology, 3: 116–120.

Thind, S.K. 1991. Effects of a long-chain aliphatic alcohol mixture on growth and solute accumulation in water-stressed wheat seedlings under laboratory conditions. Plant Grow. Reg., 10: 223–234.

Tompa, B., Balint, J. and Fodorpataki, L. 2022. Enhancement of biomass production, salinity tolerance, and nutraceutical content of spinach (*Spinacia oleracea* L.) with the cuticular wax constituent triacontanol. J App. Bot. Food Qual., 95: 121–128.

Ullah, A., Manghwar, H., Shaban, M., Khan, A.H., Akbar, A., Ali, U. et al. 2018. Phytohormones enhanced drought tolerance in plants: a coping strategy. Env. Sci. Poll. Res., 25: 33103–33118.

Van Zelm, E., Zhang, Y. and Testerink, C. 2020. Salt tolerance mechanisms of plants. Ann. Rev. Plant Bio. 71: 403–433. doi:10.1146/annurev-arplant-050718-100005

Verma, T., Bhardwaj, S., Raza, A., Djalovic, I., Prasad, P.V. and Kapoor, D. 2023. Mitigation of salt stress in Indian mustard (*Brassica juncea* L.) by the application of triacontanol and hydrogen sulfide. Plant Sig. Beh., 18(1): 2189371.

Wani, W., Masoodi, K.Z., Zaid, A., Wani, S.H., Shah, F., Meena, V.S. et al. 2018. Engineering plants for heavy metal stress tolerance. Rendi. Lince. Sci. Fisi. Nat., 29: 709–723.

Waqas, M., Shahzad, R., Khan, A.L., Asaf, S., Kim, Y.H., Kang, S.M. and Lee, I.J. 2016. Salvaging effect of triacontanol on plant growth, thermotolerance, macronutrient content, amino acid concentration and modulation of defense hormonal levels under heat stress. Plant Physiol. Bio., 99: 118–125.

Waqas, M., Korres, N.E., Khan, M.D., Nizami, A.-S., Deeba, F., Ali, I. and Hussain, H. 2019. Advances in the Concept and Methods of Seed Priming. pp. 11–41. In Priming and Pretreatment of Seeds and Seedlings: Implication in Plant Stress Tolerance and Enhancing Productivity in Crop Plants. Springer.

Zaid, A., Mohammad, F. and Fariduddin, Q. 2020. Plant growth regulators improve growth, photosynthesis, mineral nutrient, and antioxidant system under cadmium stress in menthol mint (*Mentha arvensis* L.). Physiol. Mole. Bio. Plants., 26: 25–39.

Zhang, C., Shi, S., Liu, Z., Yang, F. and Yin, G. 2019. Drought tolerance in alfalfa (*Medicago sativa* L.) varieties is associated with enhanced antioxidative protection and declined lipid peroxidation. J Plant Physiol., 232: 226–240.

Zörb, C., Geilfus, C.M. and Dietz, K.J. 2019. Salinity and crop yield. Plant Bio., 21: 31–38. doi:10.1111/plb.12884

Zulfiqar, F. 2021. Effect of Seed Priming on Horticultural Crops. Sci. Horti., 286: 110197.

Index

For Product Safety Concerns and Information please contact our EU
representative GPSR@taylorandfrancis.com
Taylor & Francis Verlag GmbH, Kaufingerstraße 24, 80331 München, Germany

www.ingramcontent.com/pod-product-compliance
Lightning Source LLC
Chambersburg PA
CBHW060745220326
41598CB00022B/2330

* 9 7 8 1 0 3 2 4 8 5 3 1 7 *